AF538313

Societal Impact of Spaceflight

SOCIETAL IMPACT *of* SPACEFLIGHT

Steven J. Dick and Roger D. Launius
Editors

National Aeronautics and Space Administration
Office of External Relations
History Division
Washington, DC

2007

NASA SP-2007-4801

Library of Congress Cataloging-in-Publication Data

Societal impact of spaceflight / Steven J. Dick and Roger D. Launius, editors.
p. cm. — (The NASA history series, NASA SP-2007-4801)
Includes bibliographical references and index.
1. Astronautics and civilization. 2. Astronautics—United States—History. I. Dick, Steven J. II. Launius, Roger D.
CB440.S63 2007
303.48'3--dc22

2007042444

For sale by the Superintendent of Documents, U.S. Government Printing Office
Internet: bookstore.gpo.gov Phone: toll free (866) 512-1800; DC area (202) 512-1800
Fax: (202) 512-2104 Mail: Stop IDCC, Washington, DC 20402-0001

ISBN 978-0-16-080190-7

Contents

SECTION IV APPLICATIONS SATELLITES, THE ENVIRONMENT, AND NATIONAL SECURITY

SECTION V SOCIAL IMPACT

SECTION VI SPACEFLIGHT, CULTURE AND IDEOLOGY

Section VII Afterword

Introduction

Fifty years after humanity first broke the gravitational bonds of Earth, the societal impact of spaceflight is a compelling subject whose time has come. It was recognized early in the Space Age that spaceflight would affect society. NASA's founding document, the National Aeronautics and Space Act of 1958, specifically charged the new Agency with eight objectives, including "the establishment of long-range studies of the potential benefits to be gained from, the opportunities for, and the problems involved in the utilization of aeronautical and space activities for peaceful and scientific purposes." Although the Space Act has been often amended, this provision has never changed, and still remains one of the main objectives of NASA.[1] Despite a few early studies, the mandate to study societal impact went unfulfilled as NASA concentrated on the many opportunities and technical problems of spaceflight itself.

It is time to take up the challenge once again. Multidecade programs to explore the planets, build and operate large space telescopes and space stations, or take humans to the Moon and Mars, require that the public have a vested interest. The same is true of the space activities now spread around the world. But whether or not the ambitious space visions of the United States and other countries are fulfilled, the question of societal impact over the past 50 years remains urgent and may in fact help fulfill current visions or at least raise the level of debate.

The subject of the societal impact of spaceflight, however, is not as simple as it may seem. Questions abound. Has the Space Age in fact had an impact on society? If so, what are those influences? What do we mean by "impact," "society," and "spaceflight"? And, realizing that society is not monolithic, what parts of society might have been affected? Conversely, has society had an effect on spaceflight? To put it another way, in the currently popular mode of counterfactual history, What would be different had there been no Space Age?[2]

It is with such questions in mind that the NASA History Division and the National Air and Space Museum's Division of Space History jointly organized a conference on the subject in Washington, DC from September 19–21, 2006. Because the scope of the societal impact of spaceflight is enormous, the planners had their work cut out for them in trying to establish some thematic coherence rather than merely presenting a hodgepodge of papers. The themes that emerged, all infused with the underlying questions above, form the sections contained in this book.

1. The National Aeronautics and Space Act and its complete legislative history may be found at *http://www.hq.nasa.gov/office/pao/History/spaceact-legishistory.pdf.* The passage quoted here is on page 6.

2. In the Prologue to his book *The Spaceflight Revolution* (NASA History Series SP-4308, Washington, DC, 1995), James Hansen (one of the authors in this volume) discusses at some length the importance and uses of counterfactual "what if" history in the context of spaceflight.

First, it would seem obvious that certain turning points in the history of spaceflight must have had an impact: Sputnik, the Moon landing, and the Space Shuttle disasters are etched in memory for better or worse. But unpacking the nature and extent of that impact is no simple task. Secondly, a commercial and economic component to spaceflight is undeniable. It ranges from a far-reaching aerospace industry at one end of the spectrum to the famous (and sometimes literally legendary) "spinoffs" at the other end; it is a part of national and international political economy; and it has sometimes measurable but often elusive effects on daily life and commerce. Economic impact is closely related to a third area: applications satellites, which are in turn often inseparable from environmental issues and national security. Imaging the Earth from space and global space surveillance have played an arguably central role in the increasingly heated debate over climate change, and changed the manner in which national security issues are understood and interpreted. Just how central is a matter that only historical analysis can reveal. In a fourth domain, that of social impact, space activities have affected science, math, and engineering education; embodied questions of status, civil rights, and gender among other social issues; and led to the creation of "space states" such as California, Florida, and Texas. Finally, spaceflight has affected culture in multiple ways, ranging from worldviews altered or completely transformed by the images of Earth from space and the spectacular views of space from Earth-orbiting spacecraft, to our place in the universe made possible by studies of cosmic evolution and the search for extraterrestrial life and the embodiment of these and other themes in literature and the arts. Several essays in this volume also address issues of spaceflight, ideology, and culture, in particular the space movement and its links to ideas of progress and utopia.

These overarching themes in turn raise further questions. What is the difference between social impact and cultural impact? What is the interplay between spaceflight and those enduring American values of pioneering, progress, enterprise, and rugged individualism? How does this interplay differ from experiences in the Soviet/Russian, European, or Chinese milieu? How has spaceflight affected conceptions of self and others, as well as our understanding of purpose in the universe? In the end, all the themes in this volume form overlapping domains, and the attentive reader will find a synergy between the thematic sections in the book.

Although we believe we have captured many of the overarching themes, gaps undoubtedly remain, and at a lower level there is certainly no claim to be comprehensive, only an offer of representative exemplars from the major themes. In the area of commercial impact, for example, aside from applications satellites only one paper (Jennifer Ross-Nazzal) explicitly addresses a commercial spinoff of the space program—the area most people think of immediately when and if they think at all about spaceflight and society. History, rather than public affairs, has an important role to play here in analyzing commercial impacts. An entire volume could be devoted to this subject alone, and further volumes in the NASA History Series will do so.

The themes of this volume also tie into deeper threads of contemporary intellectual argument. One has to do with the meaning of culture. More than 50 years ago two anthropologists collapsed 164 distinct definitions of culture into one: "[C]ulture is a product; is historical; includes ideas, pattern, and values; is selective; is learned; is based upon symbols; and is an abstraction from behavior and the products of behavior."[3] More recently Clifford Geertz defined culture as "an historically transmitted pattern of meanings embedded in symbolic forms by means of which men [people] communicate, perpetuate and develop their knowledge about and attitudes toward life."[4] According to Harvard biologist E. O. Wilson—famed for his work on sociobiology—each society creates culture and is created by it.[5] In short, culture and society are moving targets, evolving with time and in space (perhaps literally in outer space); not only does Chinese culture differ from Western culture, both were different 50 years ago than they are now.

Another broadly related intellectual theme is postmodernism, the construction of our worldview. In the context of this volume one might well ask about the societal and cultural impact *on* spaceflight rather than *of* spaceflight. Glen Asner points out in his paper that little attention is given to the possibility of reverse effect in this volume, despite explicit requests in the call for papers (John Logsdon, with his examination of the impact of the post-Cold War environment, is one exception). As Asner puts it "The concept of societal impact is problematic to the extent that it is based on an assumption that the influence of spaceflight on society is more worthy of analysis than other conceptualizations of the relationship, such as the influence of society on spaceflight or the mutual shaping of spaceflight and society."[6] He suggests possibilities for examining the history of spaceflight by focusing on status, race, and gender in the context of work, the local community, and education. This means recognizing as viable subjects for historical analysis all individuals and social groups involved in space endeavors, regardless of their social standing. Martin Collins makes a similar point in the final paper in this volume, where he notes that Sputnik, and by extension other events in the history of spaceflight, was "a manifestation and symbol of deeper structures of economic and cultural order." We would do well to ponder his call for "clarifying explanatory aims and tools—of placing spaceflight *in* history."[7]

---

3. Alfred L. Kroeber and Clyde K. M. Kluckhohn, *Culture: A Critical Review of Concepts and Definitions*, Papers of the Peabody Museum, Harvard University, v. 47, no. 12, pp. 643-4, 656 (Cambridge MA: The Peabody Museum, 1952).

4. Clifford Geertz, *The Interpretation of Cultures* (New York: Basic Books, 1973), p. 289. For more on the debate over the nature and meaning of culture see Adam Kuper, *Culture: The Anthropologists' Account* (Harvard University Press: Cambridge, MA, 1999). For debated differences between the concepts of culture and society a good starting point is Nigel Rapport and Joanna Overing, *Social and Cultural Anthropology: The Key Concepts* (Routledge: London and New York, 2000), entries on "culture" and "society".

5. E. O. Wilson, *Consilience: The Unity of Knowledge* (Knopf: New York, 1998)

6. Glen Asner, this volume.

Despite the importance of the subject, very few systematic studies of the societal impact of space exploration have been undertaken over the last 50 years. One exception that stands out from four decades ago is *The Railroad and the Space Program: An Exploration of Historical Analogy*. Funded by NASA through the American Academy of Arts and Sciences, *The Railroad and the Space Program* focused on the uses of historical analogy to illuminate the problem of societal impact. Confident in the use of historical analogy as suggestive, but not predictive, of the future, the authors of the volume elaborated on two technological events whose beginnings were separated in time by 150 years. The railroad was, they said, an engine of social revolution that had its greatest impact only 50 years after the start of the railways in America. As a transportation system, the railway had to be competitive with canals and turnpikes and, 20 years after the start of railways in America, more miles of canals were being built than railroads. It was not at all clear that railroads could be economically feasible. However, though many technological, economic and managerial hurdles needed to be overcome, railroads are still with us. In the course of the nineteenth century they represented human conquest of natural obstacles, with consequences for humans' view of nature and our place in it. Moreover, secondary consequences often turned out to have greater societal impact than the supposed primary purposes for which they were built.

The space program has had, and still has, it technological challenges, and the economic benefits may be even longer-term than those of the railroad. But by conquering the third dimension of space, as aviation did to a very small extent in the thin skin of Earth's atmosphere and as the railroad did in two geographical dimensions, in the long run the space program may have an impact that exceeds that of the railroad. Although originally suspicious of parallels with the past, present, and future, the authors in the end saw "the possibility of moving up onto a level of abstraction where the terrain of the past is suggestive of the topography of the present and its future projection."[8] They cautioned that in taking such an approach, as much empirical detail should be used as possible and analogies drawn from vague generalities should be avoided. Four decades later, *The Railroad and the Space Program* still makes for relevant reading.

In addition to that early study, there have been sporadic forays. On the occasion of the 60th anniversary of the British Interplanetary Society, NASA was heavily involved in a special issue of its journal devoted to "the impact of space on culture."[9] There NASA scientists Charles Elachi (now Director of the Jet Propulsion Laboratory) and W. I. McLaughlin, as well as historian Sylvia Kraemer, among others, discussed the impact of space endeavors on space science, politics, the fine

7. Martin Collins, this volume.

8. Bruce Mazlish (ed). *The Railroad and the Space Program: An Exploration in Historical Analogy*. (Cambridge, MA: MIT Press, 1965).

arts, and education. In 1994 the Mission from Planet Earth program in the Office of Space Science at NASA sponsored a symposium entitled "What is the Value of Space Exploration?" A variety of speakers ranging from Carl Sagan to Stephen Jay Gould discussed the scientific, economic, cultural, and educational impact of space exploration.[10]

More recently, in 2005 the International Academy of Astronautics (IAA), which has a commission devoted to space and society, sponsored the first international conference on space and society in Budapest, Hungary.[11] The IAA and the European Space Agency (ESA) jointly sponsored a study published as *The Impact of Space Activities upon Society*,[12] in which well-known players on the world scene briefly discussed their ideas of societal impact, ranging from the practical to the inspirational.

In addition to these activities, the authors of more general studies of spaceflight have on occasion tackled the subject of societal impact. In her book *Rocket Dreams: How the Space Age Shaped Our Vision of a World Beyond*, Marina Benjamin argues that space exploration has shaped our worldviews in more ways than one. "The impact of seeing the Earth from space focused our energies on the home planet in unprecedented ways, dramatically affecting our relationship to the natural world and our appreciation of the greater community of mankind, and prompting a revolution in our understanding of the Earth as a living system," she wrote. Benjamin thinks it no coincidence that the first Earth Day on 20 April 1970 occurred in the midst of the Apollo program; or that one of the astronauts developed a new school of spiritualism while others have also been profoundly affected spiritually; or that people "should be drawn to an innovative model for the domestic economy sprung free from the American space program by NASA administrator James Webb." Space exploration shapes world views and changes cultures in unexpected ways; by corollary, so does lack of exploration.[13]

Others have demonstrated the complex relation of space goals to social, racial, and political themes (see Kim McQuaid in this volume). One such study is De Witt Kilgore's *Astrofuturism: Science, Race and Visions of Utopia in Space*, where the author examines the work of Wernher von Braun, Willy Ley, Robert Heinlein, Arthur C. Clarke, Gentry Lee, Gerard O'Neill, and Ben Bova, among others, in what he calls the tradition of American astrofuturism.[14]

---

9. British Interplanetary Society, "The Impact of Space on Culture," *Journal of the British Interplanetary Society*, 1993; 46(11).

10. NASA. What is the value of space exploration? July 18–19, 1994, NASA History Reference Collection.

11. IAA, 2005. Meeting agenda at *http://www.iaaweb.org/iaa/Publications/budapest2005fp.pdf*

12. European Space Agency, *The Impact of Space Activities upon Society*, ESA BR-237, 2005.

Finally, we fully recognize that this volume is centered on Western culture and especially the United States. And although Western space programs may have had worldwide effects by their very scope and nature, we consider this analysis only a beginning and hope it will generate more robust discussion and comparison with the impact of space programs in other parts of the world. It also needs to be said that this conference and this volume were decidedly not designed as commercials for NASA or spaceflight in general. As scholars, our goal is not propaganda, but to use rigorous scholarly methods to examine societal impact. Only then can we begin to hope to measure the real impact of spaceflight.

In closing, we wish to thank our organizing committee, which included the staff of the NASA History Division (Glen Asner, Nadine Andreassen, Colin Fries, Stephen Garber, John Hargenrader, and Jane Odom), Roger Launius and his staff at the Smithsonian National Air and Space Museum (NASM), Linda Billings, Giny Cheong, John Cloud (National Oceanic and Atmospheric Administration [NOAA]) and a variety of others from whom we sought advice. We thank Scott Pace, NASA Associate Administrator for Program Analysis and Evaluation; Donald Lopez, NASM Deputy Director; and Ted Maxwell, NASM Associate Director for Collections and Research, all of whom gave opening remarks at the meeting. Our thanks also to our session chairs: William Becker (George Washington University), Dwayne Day (National Research Council), Cathy Lewis (NASM), Michael Ciancone (NASA Johnson Space Center), and William Sims Bainbridge (National Science Foundation). Our thanks to NASA's Printing and Design office for seeing this volume through the press. And finally, our thanks to the Smithsonian Institution's Hirshhorn Museum, which provided an appropriately artistic and congenial venue for the meeting.

Steven J. Dick, NASA Chief Historian
Roger D. Launius, National Air and Space Museum

Washington, DC December 2007

13. Marina Benjamin, *Rocket Dreams: How the Space Age Shaped our Vision of a World Beyond* (Free Press: New York, 2003).

14. De Witt Douglas Kilgore, *Astrofuturism: Science, Race and Visions of Utopia in Space* (University of Pennsylvania Press: Philadelphia, 2003).

Section I

Societal Impact of Spaceflight in Context

CHAPTER 1

HAS SPACEFLIGHT HAD AN IMPACT ON SOCIETY? AN INTERPRETATIVE FRAMEWORK

Howard E. McCurdy

As a person who works with political scientists, I must confess that the effort to assess the societal impact of spaceflight reminds me a bit of the story about the mayor who reduced crime. You may recall that Rudolph Giuliani, the get-tough-on-crime U.S. attorney for southern New York State, narrowly defeated incumbent David Dinkins for the New York City mayoralty post in 1993. At the time, crime in New York City seemed to be out of control. Giuliani embraced the "broken window" theory of crime prevention, drawn from a 1982 article by James Q. Wilson and George L. Kelling and promoted by William J. Bratton, Giuliani's head of police. In essence, the theory suggests that tolerance of low-level vandalism (broken windows) encourages additional petty crime and eventually more serious offenses. Giuliani and Bratton adopted a "zero-tolerance" policy toward petty crimes such as graffiti marking, subway turnstile jumping, and "squeegee men" who demanded payment for cleaning the windshields of automobile drivers stuck in traffic. Upon implementation of the policy—a turning point in the history of New York City—crimes rates dropped suddenly and dramatically and continued to fall thereafter.[1]

The story set off a frenzy of methodological investigation among social scientists interested in the societal impact of Rudolph Giuliani's policy toward crime. From the scientific point of view, Giuliani had proposed a theory. As good social scientists, analysts used the tools of statistical analysis and econometrics to compare the explanatory power of Giuliani's theory relative to other theories of crime. The findings become elaborate at this point, but in general were not kind to the idea that Giuliani's zero-tolerance policy affected crime. For example, economist Steven D. Levitt with co-author Stephen J. Dubner suggest that the drop in crime could more easily be explained by demographic factors such as a decline in the number of angry young males.[2]

---

1. James Q. Wilson and George L. Kelling, "Broken Windows," *Atlantic Monthly* (March 1982), pp. 29–38.
2. Steven D. Leavitt and Stephen J. Dubner, *Freakonomics: A Rogue Economist Explores the Hidden Side of Everything* (New York: William Morrow, 2005); see also George L. Kelling and Catherine Coles, *Fixing Broken Windows: Restoring Order and Reducing Crime in Our Communities* (New York: Martin Kessler, 1996).

I was reminded of the Giuliani story while scanning back issues of NASA's *Spinoff* publication. For the past 30 years, NASA's Commercial Technology Program has produced a book-sized publication that annually lists 40 to 50 space technologies adopted by the commercial sector. The effects imputed to the Apollo flights to the moon alone are impressive. According to the publication, Project Apollo contributed to the development of computed axial tomography (CAT) scan machines, kidney dialysis, cordless power tools, athletic shoe designs, freeze-dried foods, and the cool suits worn by National Association for Stock Car Auto Racing (NASCAR) race drivers. In total, NASA officials have identified 1,400 space technologies that have "benefited U.S. industry, improved the quality of life and created jobs for Americans." I must admit that I approach such claims of societal impact with the same degree of skepticism that social scientists direct at Giuliani's theory of crime prevention. Perhaps the NASA Space Flight Program gave us freeze-dried foods and other such benefits; perhaps it did not. Without extensive investigation of a scientific sort, it is difficult to tell.[3]

In some ways, claims of societal impact tell us more about ourselves than they do about the societal changes we think we observe. Giuliani's theory, embraced by the few Republicans remaining in a hugely Democratic city, might say more about the social preference of upper-middle-class Americans for neatness and order than the prevention of crime. Giuliani's zero-tolerance policy may or may not have affected a drop in crime, but it did make New York City a friendlier place for middle- and upper-middle-class families. In a similar way, images of the space program reveal much about the fabric of American society. The images tell us a great deal about who we think we are and where we believe we might like to go.

In preparing *Space and the American Imagination*, I concentrated on the latter.[4] I tried to place visions about space travel, which are plentiful, into the broader context of the social movements upon which they draw. Thus, efforts to view space as the "final frontier" could be viewed as an attempt to revitalize the values thought to flow from the experience of westward migration in North America. The fact that so many advocates of spaceflight emphasize the frontier analogy says something about the impact they would like to have upon American society. The next step, obviously, requires an examination of the impact that the experience actually produces relative to the expectations proffered.

Both subjects—the study of impacts and the examination of expectations—present methodological challenges. In this chapter, I will comment upon those challenges and the manner in which they affect our effort to understand the societal impact of spaceflight. The chapter deals with the methodological challenges presented by efforts to understand the material consequences of spaceflight, its cultural effects, and its unanticipated consequences.

3. NASA Scientific and Technical Information (STI), "Apollo's Contributions to America," 21 October 2005, *http://www.sti.nasa.gov/tto/apollo.htm* (accessed 27 August 2006).

4. Howard E. McCurdy, *Space and the American Imagination* (Washington: Smithsonian Institution Press, 1997).

Assessing Impacts

Among the 1,400 spinoffs ascribed to the U.S. civil space program, one of the most interesting involves the relationship between spaceflight and the computing industry. The relationship illustrates the difficulties of assessing impact. NASA scientists and engineers installed integrated circuits (ICs) in their lunar and planetary spacecraft prior to the widespread use of these devices. The people designing the Apollo flight computer, for example, incorporated ICs—an achievement driven by their realization that clunky, universal automatic computer (UNIVAC)-type computing machines would be too large for a spacecraft with severe mass constraints. The consequent utilization of ICs for a wide range of Earthly applications has been called "one of the most significant occurrences in the history of mankind."[5] It is tempting to see a relationship between spaceflight and the IC/personal computer (PC) revolution, and one can find occasional references to PCs as a spinoff of the space program, along with Teflon® and Velcro® straps.[6]

As might be anticipated, the actual relationship between spaceflight and computing is more complex. No simple cause-and-effect relationship can be shown. As the author of one NASA History Office publication concludes:

> Since NASA is well known as an extensive user of computers—mainly because spaceflight would not be possible without them—there is a common sense that at least part of the reason for the rapid growth and innovation in the computer industry is that NASA has served as a main driver due to its requirements. Actually, the situation is not so straightforward. In most cases, because of the need for reliability and safety, NASA deliberately sought to use proven equipment and techniques . . . [G]eneralizations cannot be made, other than that there was no conscious attempt on the part of NASA in its flight programs to improve the technology of computing.[7]

Social scientists view statements about impacts arising from historical events with a great deal of suspicion. Methodologically, such statements take the form of interrupted time-series analysis. This is one of the weakest forms of policy analysis and one that social scientists often urge investigators to avoid. When done in a retrospective fashion, the technique can be quite misleading. Knowing that a change in society followed a

5. Wikipedia, "Integrated Circuits," *http://en.wikipedia.org/wiki/Integrated_circuit*, 25 August 2006 (accessed 27 August 2006).

6. See John Savard, "Microcomputers as a Space Spinoff," *http://www.gatago.com/sci/space/policy/21491345.html*, 27 June 2006; and Eleanor A. O'Rangers, "NASA Spin-offs: Bringing Space Down to Earth," http://www.space.com/adastra/adastra_spinoffs_050127.html, 26 January 2005 (both accessed 27 August 2006).

7. James E. Tomayko, *Computers in Spaceflight: The NASA Experience*, NASA History Office, Contractor Report 182505, March 1988, p. 2.

turning point in no way verifies that the event caused the change. The change could be due to other factors or it could have occurred in the absence of the intervening event. Absent statistical controls or experimental methods, it is quite impossible to know.

Policy analysts utilize a number of techniques that compensate for the limitations of interrupted time-series analysis. They engage in comparative studies; insist that statements about cause and effect be grounded in theory; and require analysts to derive predictions from the hypotheses proposed.

Comparative analysis helps to remove many of the shortcomings associated with the study of societal impacts. Conclusions drawn from a single set of events in a single society are the equivalent of hypotheses tested using a sample of one. Such tests have a wide margin of error. The incorporation of information from other settings can broaden the analysis and enhance its reliability. What may appear to be anomalies in one setting may seem common when viewed comparatively. The unexpected difficulties of producing a large, cheap, reliable space shuttle in the United States were repeated with the Soviet Buran spacecraft, which turned out to be so expensive that Soviet officials abandoned the program. Conversely, what appears to be common may turn out to be unique. One of the universal benefits of human spaceflight, for example, is thought to be national prestige. This has both external (impressing other nations) and internal (building national confidence) dimensions.[8] Assessing whether such activity actually produces such effects can be enhanced by examining the process in many nations, including the reactions of those that do not engage in human space travel. What seems to be generally believed (that spaceflight confers prestige) can be tested for its effects. Comparative work of this sort is underway by Asif Siddiqi, James T. Andrews, James Hansen, and Margaret Weitekamp, and much of it appears in this publication.

Grounding statements in theory and making predictions based on those theories also helps. The history of spaceflight suffers from no lack of predictions; notoriety often flows to those persons whose predictions turn out to be true. One of the most notable is Arthur C. Clarke's anticipation of communication satellites, famously presented in a 1945 edition of *Wireless World*. In that publication, Clarke pointed out that a communication station placed in geostationary orbit "could act as a repeater to relay transmissions between any two points on the hemisphere beneath" and that three such stations would provide "complete coverage of the globe." Clarke did not actually predict the use of such stations—"[S]uch an undertaking may seem

8. See John M. Logsdon, *The Decision to Go to the Moon: Project Apollo and the National Interest* (Chicago: University of Chicago Press, 1976) and Roger D. Launius, "Compelling Rationales for Spaceflight: History and the Search for Relevance," in *Critical Issues in the History of Spaceflight*, Steven J. Dick and Roger D. Launius, ed. (Washington, DC: NASA SP-2006-4702, 2006), pp. 37–70.

fantastic," he said—but, rather, pointed out its feasibility and advantages relative to ground-based transmitters. Nonetheless, the article is generally credited as having anticipated the use of communication satellites and is often presented as part of narratives assessing the impact of such.[9]

Regrettably, such statements are not very helpful in assessing societal impact—even when the statements are true. A correct prediction offered in the absence of a supporting theory is as unreliable as an ex post facto statement about cause and effect. Such a prediction is subject to a number of methodological pitfalls, the most striking being what is known as the "Jeane Dixon effect." Dixon was a psychic who famously predicted the election and assassination of President John F. Kennedy. John Allen Paulos, a Temple University professor and commentator on the general public's misunderstanding of mathematic principles, noted how the science of probabilities ensures that someone like Dixon will make a fair number of correct predictions if that person makes a sufficiently large number of forecasts. Adding to the accumulated tally of her errors, Dixon predicted that World War III would begin in 1958; that labor leader Walter Reuther would run for president in 1964; and that the Soviet Union would win the race to the Moon.[10] The Dixon effect refers to the tendency of observers viewing events with the advantage of hindsight to overlook false forecasts while applauding the ones that did come true.

By grounding a prediction in theory, the suggestion of cause and effect can be assessed twice. The effect can be checked on the basis of whether or not it occurred and the theory can be checked for its underlying logic. A correct prediction, such as those that Dixon did make, cannot be judged to reveal an effect if the underlying theory is flawed. Dixon derived her predictions from the practice of astrology, a clearly misdirected theory. Clarke offered his speculations regarding communication satellites without regard to any theory at all. Like numerous other pieces anticipating some development in spaceflight, Clarke's article speculates neither on the likelihood that his proposal might be adopted nor on the possible impact of worldwide communication. He merely comments on the technical feasibility of orbiting communication stations and predicts that the coverage they would provide would be cost-effective relative to the ground-based systems the stations would replace. It should be noted, in this respect, that Clarke also predicted that his communication platforms would take the form of space stations with people on-board and that their development would be expedited by the use of nuclear-powered rockets by 1965.[11]

9. Arthur C. Clarke, "Extra Terrestrial Relays: Can Rocket Stations Give World Wide Radio Coverage?" *Wireless World* (October 1945), p. 306. See also Irving Fang, *A History of Mass Communication: Six Information Revolutions* (Boston: Focal Press, 1997), p. 210, and Donald H. Martin, *Communication Satellites*, 4th ed. (El Segundo, CA: Aerospace Press, 2000).

10. John Allen Paulos, *Beyond Numeracy: Ruminations of a Numbers Man* (New York: Alfred A. Knopf, 1991), p. 40.

11. Clarke, "Extra Terrestrial Relays," pp. 306, 308.

A sound assessment of societal impact is enhanced to the degree that the analysis is rooted in sound theory, derives testable predictions from the theory, and utilizes more than one case as a basis for testing the statements proffered. Such standards are hard to apply to the material effects of spaceflight, such as commercial products or applications in the realm of national defense. Assessing cultural impacts is even more challenging.

CULTURAL EFFECTS

Events in spaceflight have social, cultural, and ideological effects. In many ways, these are more interesting than the material spinoffs from the space program, since they involve both imagination and effect. In an odd sort of way, the effects of spaceflight influence their causes. Put more simply, what people imagine might happen in space serves as a basis for making it occur. The anticipation of cultural impacts thus provides the motivation to undertake the activities necessary to produce the change. Many chapters in this book are concerned with the social, cultural, and ideological effects of events in space.

The cultural effects of spaceflight (a term meant to also include social and ideological effects) bounce between the relativism inherent in postmodern analysis and the reality of space physics. Postmodern analysis postulates the notion that people ultimately determine the types of worlds in which they live through the thoughts they have; physics presents principles that are hard to violate. One is relative, the other deterministic.

By imagining space or, more specifically, anticipating the events that will occur there, people may shape their future. The direction of that shaping can be conservative or radical. I would like to suggest that the dominant forms of spaceflight anticipation, especially in the United States, are conservative. In America, expectations about space have been offered as a means of reinforcing the dominant values in society, including many that existed before space travel began. This may help to explain why modern conservatives are more supportive of space exploration than are American liberals.

Expectations regarding the cultural effects of spaceflight are often expressed metaphorically. Metaphors are figures of speech that contain an implied comparison, easing the challenge of explaining strange and often unfathomable phenomena to an often inattentive public. The comparison of spaceflight to terrestrial expeditions of discovery, for example, casts the complexity of interplanetary travel in terms the general public can more readily understand. In the United States, spaceflight has been described using metaphors that characterize the most salient features of American life. The metaphors are many. The exploration of space, we are told, will be like frontier life—resurrecting the experience of westward migration in an extraterrestrial realm. The exploration of space will provide sources of business opportunity in the same way that industrial and postindustrial developments gave the United States the most prosperous economy in the history of the world. Space will be the new military "high ground," similar to the Roman roadways and the aviation hardware that conferred national power upon the nations that pursued the supporting technologies. Spaceflight—or at least the investigative part of it—will help to maintain the scientific revolution that made empiricism the primary means for studying natural phenomena. Spaceflight will continue to serve as a demonstration of national prowess, in the same manner that expositions and world fairs have provided national demonstrations of technology. Spaceflight will allow a "revenge of the nerds," elevating the status of people who did not have much social standing during their adolescent years. These metaphors confer expectations regarding the impact of spaceflight, especially in America.

Although the use of metaphors eases the task of explaining prospective impacts of spaceflight, it also gives those expectations a distinctly conservative flavor. If spaceflight continues over many centuries, it might produce transformations as radical as those that the Renaissance imposed on the medieval world. Spaceflight might lead to fundamental alterations in the human species, or to scary new discoveries that result in a total reorganization of society. It might be like nothing we have ever experienced before. Science fiction writers such as H. G. Wells, Isaac Asimov, and Arthur C. Clarke have explored some of these possibilities.[12] The dominant metaphors (at least those presented in the United States) do not anticipate radical change. Instead, the American vision of spaceflight promises to conserve the values associated with the continental frontier, our business civilization, the scientific revolution, national security, overall progress based on technology, and the tendency to elevate scientists, engineers, and other experts to positions of power in society. Collectively, these are distinctly American values.

The rhetoric of spaceflight demonstrates the presence of these expectations, at least in the United States. America is thought to be a frontier nation, with many of its characteristics shaped by the presence of open land and the absence of established institutions such as those found in feudal Europe. The innovative spirit, the preference for democracy, and the absence of social barriers that would otherwise impede cooperation and perpetuate inequality are all thought to flow from the American frontier. At least, that is how it has appeared to many of the people whose European ancestors arrived in America after 1600.[13] Space travel is commonly presented as a means of extending

12. See the concepts of evolutionary biology in H. G. Wells, *The Time Machine and War of the Worlds*, Frank D. McConnell, ed. (New York: Oxford University Press, 1977); global transformation in Arthur C. Clarke, *Childhood's End* (New York: Harcourt Brace & World, 1953); and psychohistory in Isaac Asimov, *Foundation* (New York: Gnome Press, 1951). For an analogous view, also see Ray Kurzweil, *The Singularity Is Near: When Humans Transcend Biology* (New York: Viking, 2005).

13. Frederick Jackson Turner, "The Significance of the Frontier in American History," in John M. Farager, *Rereading Frederick Jackson Turner* (New York: Henry Holt, 1994).

these traditions. "Without a frontier from which to breathe new life," Robert Zubrin argues, "the spirit that gave rise to the progressive humanistic culture that America has represented for the past two centuries is fading." Zubrin advocates the settlement of Mars as a means of perpetuating the values associated with the American frontier.[14]

According to political scientist Dwight Waldo, America is a distinctly business-oriented civilization. This is a central feature of American life. Wealth and power flow from the strength of business enterprise and the corporations around which the economy is structured. Not surprisingly, Americans advocate corporate methods as the best means for organizing the global economy and the government bureaucracies that regulate it. Americans also anticipate the extension of business activities into space. In 2001, journalist Lou Dobbs announced that space will provide the next great business frontier; it will create "entirely new forms of technology, new forms of manufacturing, new forms of recreation, and even new materials," he said. According to this prophesy, space commerce will provide business opportunities as large as those emerging from the Internet revolution, and will cease to be the province of government agencies interested primarily in science and exploration. In this sense, spaceflight serves to extend the values associated with corporate capitalism.[15]

America's status as a world superpower is largely based on a military apparatus that relies upon technology to project force and reduce risk. In this context, the cosmos is consistently represented as the new military "high ground." Senate Majority Leader and later President Lyndon B. Johnson embraced this point of view when he helped launch America's entry into space by declaring that "control of space means control of the world." Advocates of both robotic and human spaceflight continue to use national security arguments as a justification for U.S. supremacy in this realm.[16]

Americans are quintessential progressives, generally accepting the Promethean notion that progress as a whole is good for humankind and that such progress typically occurs through advances in science and technology. Historically, not all cultures have embraced the doctrine of progress through technology. Some groups elevate the attainment of spirituality through religious faith and salvation, a perspective that exhibits mosques and cathedrals rather than rocket ships as symbols of perfection. In the eighteenth century, the doctrines of natural rights and reason formed the basis for the concept of human perfection. Space travel and its various

14. Robert Zubrin, *The Case for Mars: The Plan to Settle the Red Planet and Why We Must* (New York: Free Press, 1996), p. 297. See also National Commission on Space (Thomas O. Paine, chair), *Pioneering the Space Frontier* (New York: Bantam Books, 1986).

15. Dwight Waldo, *The Administrative State: A Study of the Political Theory of American Public Administration* (New York: Ronald Press, 1948), p. 5; Lou Dobbs with H. P. Newquist, *Space: The Next Business Frontier* (New York: Pocket Books, 2001), p. 2.

16. Statement of Democratic Leader Lyndon B. Johnson to the Meeting of the Democratic Conference on 7 January 1958, Statements of LBJ Collection, Box 23, Lyndon Baines Johnson Library, Austin, Texas; see also Office of Science and Technology Policy, Executive Office of the President, The White House, "U.S. National Space Policy," 31 August 2006.

spinoffs emphasize a view of progress rooted in the Age of Enlightenment and the scientific revolution that accompanied it. These are cultural choices, unattached to any absolute requirement that human civilizations advance in one particular way.

The cultural manifestations of spaceflight also help to answer the classic social question: Who should rule? Throughout history, this question has been answered in different ways. In some societies, priests rule; in others, hereditary monarchs. Plato favored Guardians, who ruled on the basis of their innate understanding of the Good. The doctrine of technological progress favors rule by experts, in which scientists, engineers, and other experts employ objective methods to discover the "one best way" of organizing social affairs, typically accompanied by an emphasis on the need for efficiency in a machine-based civilization. The concept that experts should build and operate the machinery of a technological society seems intuitively obvious, but it is not a choice that has been pursued by all civilizations at all times.[17]

Two concepts help frame the use of metaphors as a means of explaining both past and anticipated impacts of spaceflight. One is the doctrine of American exceptionalism; the other is the vocabulary of postmodern analysis. The doctrine of American exceptionalism, rooted in works such as those by Alexis de Tocqueville, Frederick Jackson Turner, Louis Hartz, and Aaron Wildavsky, traces the power of the American experience to relatively unique material and social conditions. These include the absence of feudal institutions and the existence of an open frontier—conditions thought to encourage equality, cooperation, creativity, democracy, and a liberal tradition as the term is used in its classical, Lockean sense.[18] Spaceflight, in this regard, is presumed to provide an analogous force, encouraging the perpetuation of traditions thought to have made America unique. Without such a continuing force, advocates of the doctrine suggest, America will become more like the rest of the world. The doctrine of American exceptionalism is speculative and controversial, but serves as a larger framework through which the presumed impact of spaceflight can be examined.

Postmodern analysis provides a number of concepts useful for examining the manner in which imagination shapes future events. Proponents of this perspective emphasize the roles that the broadcast media and similar technologies play in decentralizing power and framing ideas within the public at large. Under these conditions, ideas are thought to take shape in the minds of the beholders and lack an

17. See Sylvia D. Fries, *NASA Engineers and the Age of Apollo* (Washington: NASA SP-4104, 1992); Homer H. Hickam, *Rocket Boys: A Memoir* (New York: Delacorte Press, 1998); Jeff Kanew, *Revenge of the Nerds* [film] (20th Century Fox, 1984), and Waldo, *The Administrative State*.

18. Alexis de Tocqueville, *Democracy in America* (New York: Vintage Books, 1954); Turner, "The Significance of the Frontier in American History"; Louis Hartz, *The Liberal Tradition in America: An Interpretation of American Political Thought Since the Revolution* (New York: Harcourt, Brace & World, 1955); Aaron Wildavsky, *The Rise of Radical Egalitarianism* (Washington: American University Press, 1991). See also Seymour Martin Lipset, *American Exceptionalism: A Double-Edged Sword* (New York: W. W. Norton, 1996).

objective reality. This is not a totally new concept and may be associated with periods other than the postmodern one. In separate works, Joseph Corn and Roderick Nash demonstrate the manner in which mental images have shaped the development of aviation and the American environmental conservation movement. Nash, in particular, shows how modern conservation required for its birth and sustenance a reformulation of the popular conception of wilderness. By reframing "wilderness" from a condition of savage peril to a citadel of spiritual renewal, writers and artists made new government policies possible. A similar process guided the history of aviation. People imagined effects from aviation that far exceeded the material benefits of this new technology and which, in turn, helped to elicit government support.[19]

Expectations regarding spaceflight are expressed through a number of forms familiar to people engaged in postmodern analysis. One is hyperreality, or the reappropriation of familiar cultural symbols through the mass media. Thus, Gene Roddenberry presented the original Star Trek television series not so much as a work of science fiction but as a reinterpretation of the Hollywood Western in an extraterrestrial setting. As his director's notes reveal, this was a deliberate decision. The Hollywood Western was a proven product; resetting it in space helped to ensure an audience for what might have otherwise been a quickly forgotten series.

The concept of simulacrum also guides postmodern analysis. This concept characterizes the process of making imperfect copies of original forms, as a paint-by-numbers kit might reproduce a work of art by Vincent Van Gogh. Visions of spaceflight abound with simulacrum, from winged spaceships that resemble jet fighters to robots that often resemble human beings.

Postmodern analysis provides a framework through which visions of spaceflight may be examined in a context that is larger than the subject itself. The postmodern concept of cultural relativism rejects the traditional notion that societies progress in predictable ways, as from agrarian to industrial, in favor of the more existential belief that people become what they choose to be. This directly contradicts the dominant interpretation in which spaceflight is seen as moving along a forward line of progress that nature provides.[20] The postmodern framework accepts aspirations for space travel as social creations that vary according to the predispositions of the beings that create them. These contrasting points of view add conceptual depth to what might otherwise remain a relatively narrow assessment of impacts in a single field.

---

19. Joseph J. Corn, *The Winged Gospel: America's Romance with Aviation, 1900–1950* (New York: Oxford University Press, 1983); Roderick F. Nash, *Wilderness and the American Mind*, 4th ed. (New Haven: Yale University Press, 2001).

20. See Arthur C. Clarke, *2001: A Space Odyssey* (New York: New American Library, 1968).

Implemented visions of spaceflight eventually confront physical conditions; the laws of physics provide the ultimate methodological check on anticipated effects. Some of the more interesting checks occur in the social realm. Take, for example, the widespread belief that space represents some sort of "final frontier." This line of reasoning draws heavily on the American mythology of frontier life. Yet many other societies have confronted physical frontiers – and not always with the same results. An obscure but interesting article in the *Journal of the British Interplanetary Society*, using a comparative perspective, suggests that the conditions present in extraterrestrial colonies may lead to social and political effects quite different than those remembered from the American frontier. In America, frontiers are thought to have promoted equality, cooperation, and rural independence. Conditions in space, however, may lead to the creation of societies that are autocratic, corporate, and feudal in nature. This is certainly the history of civilizations, such as ancient Egypt, that employed hydraulic technologies to open barren lands. In this respect, any extraterrestrial colonies that actually arise may less resemble the mythical conditions thought to exist on the American frontier than the Egyptian-like civilization presented in Roland Emmerich's classic science fiction film *Stargate*.[21]

In presenting the ultimate justification for spaceflight, advocates such as Carl Sagan and Robert Goddard argued that it would be necessary for the survival of humankind. Carl Sagan insisted that no technological civilization could expect to live long without moving onto other planets, whereas Robert Goddard observed that humans would eventually need to disperse Earthly life forms before the Sun grew cold. Asked to address the British Interplanetary Society, philosopher and science fiction writer Olaf Stapledon posed a critical challenge in this regard. "If one undertakes to discuss what man ought to do with the planets," Stapledon said, "one must first say what one thinks man ought to do with himself."[22]

Put another way, exactly what aspects of human society do the advocates of spaceflight propose to preserve? The answer, taken generally from the words of spaceflight advocates, is that they plan to conserve the values associated with American exceptionalism and capitalist democracy. These are the frames through which spaceflight is most commonly viewed in America and they tend to create the principal expectations regarding the societal impacts that spaceflight is presumed to have.

---

21. David Sivier, "The Development of Politics in Extraterrestrial Colonies," *Journal of the British Interplanetary Society* (September/October 2000); see also Karl A. Wittfogel, *Oriental Despotism* (New Haven: Yale University Press, 1957) and Roland Emmerich, *Stargate* [film] (Metro-Goldwyn-Mayer, 1994).

22. Carl Sagan, *Pale Blue Dot: A Vision of the Human Future in Space* (New York: Random House, 1994); Esther C. Goddard and G. Edward Pendray, ed., *The Papers of Robert H. Goddard* (New York: McGraw-Hill, 1970) Vol. 3, p. 1612; Olaf Stapledon, "Interplanetary Man?" in Robert Crossley, *An Olaf Stapledon Reader* (Syracuse, NY: Syracuse University Press, 1997), pp. 232–233.

UNANTICIPATED CONSEQUENCES

Finally, how does one deal with unanticipated impacts? Some effects appear outside the cultural anticipation imposed on spaceflight and do not receive a decent share of predictions in advance. Commenting on the nature of the universe in general, British geneticist, biometrician, and physiologist J. B. S. Haldane suggested that in some respects it "is not only queerer than we suppose, but queerer than we *can* suppose."[23] For some of its ultimate effects, the impact of spaceflight may turn out to be stranger than people are able to imagine in advance.

Two recent developments help to illustrate this situation. The first is the so-far disappointing pursuit of extraterrestrial life. The widespread expectation that spaceflight will result in the discovery of extraterrestrial life permeates the early literature on spaceflight, from the contemplation of environmental conditions on Mars to the presentation of alien forms in science fiction.[24] In a fashion similar to other metaphors imposed on space travel, the vision of a universe teeming with life derives much of its force from the widespread expectation that expeditions in the extraterrestrial realm will be similar to earlier ventures in the terrestrial one. Terrestrial explorers returned with tales of exotic species and strange cultures, fueling expectations that extraterrestrial journeys would reveal the same.

Throughout the first 50 years of spaceflight, at least, this expectation has not been fulfilled. Confounding widespread expectations, robotic spacecraft have revealed the surface of Mars to be essentially sterile, not the "abode of life" that writers such as Percival Lowell and Willy Ley portrayed. Inspection of Venus, which was often portrayed in pre-Space Age writings as a Paleozoic planet, has exposed a hellish place much too warm to permit the development of complex life.[25]

Just as the anticipation of observed impacts can be checked with reference to their underlying theories, so the significance of unexpected effects can be gauged by the emergence of new hypotheses. Grand experiments, including those taking the form of government policy, often produce results unanticipated by previous theory or experience. Such results, where significant, commonly prompt the presentation of new theories which, in turn, can be tested in conventional ways. The appearance of a new theory serves as an important marker for the presence of a significant unanticipated result or event.

---

23. J. B. S. Haldane, *Possible Worlds and Other Papers* (New York: Harper & Row, 1927), p. 298.

24. See Steven J. Dick, *Life on Other Worlds: The 20th-Century Extraterrestrial Life Debate* (New York: Cambridge University Press, 1998).

25. See Percival Lowell, *Mars as the Abode of Life* (New York: Macmillan, 1908); Isabel M. Lewis, "Life on Venus and Mars?" *Nature Magazine* (September 1934), p. 134.

During the early stages of space exploration, statements anticipating the ubiquity of extraterrestrial life forms were common. Defending the search for extraterrestrial life in a 1975 issue of *Scientific American*, Carl Sagan and Frank Drake announced that "Our best guess is that there are a million civilizations in our galaxy at or beyond the earth's present level of technological development." By 1990, this expectation had been sufficiently degraded to allow Peter Ward and Donald Brownlee to issue a contrary hypothesis. Life on Earth, they suggested, might be a result of a combination of events with a very low probability of occurrence. Simple life forms might be widespread in the universe, they allowed, but "[C]*omplex* life—animals and higher plants—is likely to be far more rare than is commonly assumed."[26]

The emergence of another new theory accompanied a different impact that was largely unanticipated in early writings about spaceflight. Prior to the Space Age, few people wrote extensively about the effect that viewing the whole Earth from a distance would have on human conceptions of their home planet, in spite of the obvious analogy provided by the intellectual shift accompanying the movement from the Aristotelian to the Copernican vision of the cosmos. With the advent of spaceflight, new images of Earth appeared. Apollo astronauts provided the most dramatic ones, from the 1968 photograph of Earth rising over the Moon to the iconic 1972 whole Earth image that decorates the Earth Day flag.

These images coincided with the emergence of the Gaia hypothesis—the strange new suggestion that the whole Earth and its biota might have the capacity to regulate conditions in such a manner as to produce conditions favorable to the maintenance of life. James Lovelock formulated this hypothesis in the early 1960s partly as a response to requests from NASA to develop instruments that could detect signatures of life in planetary atmospheres.[27] The Gaia hypothesis did not receive much attention until images of the whole Earth began to appear. Imagining Earth as a single, self-regulating system is much easier when one sees the whole planet as it appears from afar. In addition to sponsoring the research that spawned this theory, spaceflight might have created a perspective that hastened its acceptance. Again, this particular effect had not been much anticipated.

---

26. Carl Sagan and Frank Drake, "The Search for Extraterrestrial Intelligence," *Scientific American* (May 1975), pp. 80–89; Peter D. Ward and Donald Brownlee, *Rare Earth: Why Complex Life Is Uncommon in the Universe* (New York: Copernicus, 2000), p. xiv.

27. See James Lovelock, "Gaia As Seen Through the Atmosphere," *Journal of Atmospheric Environment* 6 (1972), pp. 579–580; Lynn Margulis, *Symbiotic Planet: A New Look at Evolution* (New York: Basic Books, 1998).

SUMMARY

The assessment of societal impacts arising from the interjection of any new set of events can be quite challenging, no matter where it is conducted. The temptation to draw connections where none exist, or to ignore the implications of unanticipated effects, is strong, outweighed (one hopes) by the desire of analysts to tell the story as truthfully as possible.

This chapter offers a number of methods for deepening the study of societal impacts insofar as they arise from the spaceflight venture and improving the reliability of the conclusions drawn. The use of interrupted time-series analysis—commonly characterized as turning points—contains weaknesses that can be partly overcome through comparative analysis. The examination of predictions can be enhanced by insisting that they be examined in the context of supporting theories. In a similar manner, the significance of unanticipated societal impacts can be measured through the acceptance rate of new theories suggested by the precipitating events. When assessing the cultural effects of spaceflight, findings can be strengthened by observing the material and ideological characteristics of the societies in which those effects are presumed to occur. As noted in this chapter, many of the cultural effects ascribed to spaceflight in the United States have the quality of conserving values thought by Americans to distinguish their nation. It would be interesting to know whether the promotion of spaceflight in other nations has reinforced social values different than those found in the United States.

Section II

Turning Point Impacts

Chapter 2

What Are Turning Points in History, and What Were They for the Space Age?[1]

Roger D. Launius

Debates over "turning points" in history have sometimes become quite difficult and controversial among observers of the past. At sum they signify, represent, and define lasting changes in the climate of the times. The definition of turning points is exceptionally idiosyncratic, and their delineation also shifts over time as perspectives change and events become more distant. For most people who look back on the twentieth century, 1929 and 1941 demonstrated turning points as the nation changed in fundamental ways in response to the beginning of the Great Depression and as the United States entered World War II. On the other hand, 1963 and 1987 were probably not turning points despite the Kennedy assassination and the stock market crash, respectively. Therefore, to a very real extent turning points reflect the sea change that follows an event rather than the event itself. Additionally, not all turning points need be marked by a dramatic event. For instance, no one event marked the shift from the conformist 1950s to the radical 1960s and 1970s, although many observers agree that these decades were indeed turning points.

In the context of spaceflight, what are the turning points? Most would probably agree that the launch of Sputnik in 1957 represented a turning point, although later in this essay I will make a case in opposition to this belief. But what about the Kennedy decision to go to the Moon, the Moon landings themselves, the first flight of the Space Shuttle, the losses of *Challenger* and *Columbia*, and the rise of China as a player in human spaceflight? This list might be expanded indefinitely. This essay explores what constitutes a turning point in history and examines some turning points in the history of the Space Age.

---

1. The author thanks the following scholars for offering helpful suggestions about this essay: David C. Arnold, William E. Burrows, Erik Conway, Jonathan Coopersmith, Deborah G. Douglas, Donald C. Elder, Mark A. Erickson, James Rodger Fleming, Amy Foster, Anne Collins Goodyear, Adam L. Gruen, Richard P. Hallion, Peter L. Hays, J. D. Hunley, Stephen B. Johnson, Katy Kudela, Laura Lovett, Dick Myers, Anna K. Nelson, Randy Papadopolous, Erik P. Rau, Philip Scranton, James Spiller, James A. Vedda, and David Ward.

DEFINING A TURNING POINT

In a recent search of Amazon.com for the words "turning point" in the titles of books, I found 1,134 relevant titles. These ranged from *The Turning Point: Jefferson's Battle for the Presidency* by Frank van der Linden to *The Higher Freedom: A New Turning Point in Jewish History* by David Polish, to *The Right Moment: Ronald Reagan's First Victory and the Decisive Turning Point in American Politics* by Matthew Dallek.[2] And the term is hardly new. Hoffman Nickerson used it in 1928 to describe the battle of Saratoga during the American Revolution.[3] Postmodern scholars such as Fritjof Capra have employed it as well.[4] It appears in historical work of all types and varieties, schools and subjects, and grade levels and sophistication. Indeed, the concept of a turning point is ubiquitous in the literature of history. And not just in the written word—professors, pundits, politicians, and plebeians all use it in all manner of settings and circumstances. Many course offerings at the nation's colleges and universities include "turning point(s)" in their titles.

At a core level, a turning point may be defined as an event or set of events that, had it not happened as it did, would have prompted a different course in history. Dictionaries define it as "a point at which a significant change occurs."[5] The classic youngster's encyclopedia, *World Book*, defines it as "a point at which a notable or decisive change takes place; critical point; crisis: The Battle of Gettysburg was a turning point in the Civil War."[6] The use of the term comes up in the most interesting places. *Encyclopædia Britannica* incorporates 560 entries in which the term is used. Not so unusual is the statement that the Battle of Midway "marked the turning point of the military struggle between" the United States and Japan in 1942, and "the year 1206 was a turning point in the history of the Mongols and in world history: the moment when the Mongols were first ready to move out beyond the steppe."[7] More obtuse are such interpretations as the death of Antiochus in 129 BC marking "a turning point in the history of the eastern Mediterranean: Greco-

2. Frank van der Linden, *The Turning Point: Jefferson's Battle for the Presidency* (Golden, CO: Fulcrum Publishers, 2000); David Polish, *The Higher Freedom: A New Turning Point in Jewish History* (Chicago: Quadrangle Books, 1965); Matthew Dallek, *The Right Moment: Ronald Reagan's First Victory and the Decisive Turning Point in American Politics* (New York: Free Press, 2000).

3. Hoffman Nickerson, *The Turning Point of the Revolution; or, Burgoyne in America* (New York: Houghton Mifflin, 1928).

4. Fritjof Capra, *The Turning Point: Science, Society, and the Rising Culture* (New York: Simon and Schuster, 1982).

5. "Turning Point," Merriam-Webster Online Dictionary, *http://www.m-w.com/dictionary/turning%20point* (accessed 21 August 2006).

6. World Book Encyclopedia and Learning Sources, *http://www.worldbook.com/wb/dict?lu=turning%20point* (accessed 21 August 2006).

7. "Midway, Battle of," Encyclopædia Britannica online, *http://search.eb.com/eb/article-9052586?query=turning%20point&ct=eb* (accessed 21 August 2006); "Genghis Khan," Encyclopædia Britannica online, *http://search.eb.com/eb/article-41207?query=turning%20point&ct=eb* (accessed 21 August 2006).

Macedonian domination received a decisive blow; it would survive for only 46 more years."[8] Tying the demise of Greek domination nearly two generations later to the death of Antiochus seems tenuous at best.

From a sociological perspective, a turning point represents a lasting shift in the *zeitgeist* or "spirit of the age." Several ingredients must be present. The shock to the system of civilization is profound and it may be measured in several ways. According to sociologist Ted Goertzel, one of the most reliable indicators is the response of the financiers. "Financial markets are one of the quickest and most sensitive indicators of a country's mood," he noted. "Panic can move quickly after a shock . . . and markets can spiral out of control." Public opinion polls may also take the temperature of the society and its reaction to some major event,[9] but those will work only for recent events where the data and structures that Goertzel understands are available. Clearly, there is no manner in which the Mongol invasions of Genghis Kahn, the death of Antiochus, or even the Battle of Midway can be assessed using financial data and public opinion polls.

Political scientists would employ analytical models such as Frank Baumgartner's and Bryan Jones's punctuated equilibrium analysis, which suggests that the policy process is comprised of long periods of stability that are then interrupted by predictable periods of instability which lead to major policy changes. Baumgartner and Jones describe "a political system that displays considerable stability with regard to the manner in which it processes issues, but the stability is punctuated with periods of volatile change." In times of stability the public is limited in its ability to effect change to the overall system, and most people are not even focused on making changes because they are relatively content with the current situation. Only in times of unique crisis and instability do enough members of society rise up to undertake fundamental change, often from a perceived threat or dramatic event.[10] A turning point, therefore, results from a punctuation in the equilibrium of everyday life. This theory—clinical and sterile as it might actually be—has been applied to all manner of decisive events in history and is consistently reaffirmed in the discipline of political science.

Other social science disciplines approach the issue of marked change in different ways and with differing analytic tools, but all, it seems, recognize a turning point in the stream of time as little more than an artificial construct that facilitates interpreting dramatic changes in society. Indeed, it seems as subjective a term as "scientific revolution" was for Thomas Kuhn, who defined it as a "noncumulative developmental episode in which an older paradigm is replaced in whole or in part

8. "Iran, Ancient," Encyclopædia Britannica online, *http://search.eb.com/eb/article-32135?query=turning%20point&ct=eb* (accessed 21 August 2006).

9. Ted Goertzel, "September 11, 2001: A Turning Point for America's Future?" undated paper in possession of author.

10. Frank R. Baumgartner and Bryan D. Jones, *Agendas and Instability in American Politics* (Chicago: University of Chicago Press, 1993), pp. 3–24; Frank R. Baumgartner and Bryan D. Jones, "Agenda Dynamics and Policy Subsystems," The Journal of Politics 53 (November 1991): pp. 1044–1074.

by an incompatible new one."[11] As with "scientific revolution," assigning turning point status to an event is very much up to the individuals analyzing it and its effects. Indeed, people at the time may well not recognize a turning point as such. As historian Erik Rau remarked:

> [H]istorians today think of the Battle of Saratoga as a turning point in the history of the American Revolution, but many at the time would have had no reason to believe this. This makes the turning point of Saratoga no less real to us in understanding Saratoga, but it may not have influenced very many people's behavior on the ground at the time. You can't see Saratoga as a turning point until after the war is over and you take stock of what happened. A turning point is ultimately a construct of historical reflection, and a historical unit of analysis, rather than an event that reveals itself to the people living through it at the time.[12]

Another analogous term that has gained credence in recent years is the singularity-rooted balance of equations, which is now applied far beyond its original application and is a statement of the power of nomenclature in modern society. Again, there is no firm definition acceptable to all.[13]

Of course, when considering turning points in history we are treading a path well-worn by earlier historians, some of whom were illustrious in their own time and still evoke hushed tones of reverence in seminars on historiography. At sum, the issue of a turning point in history is really about assigning significance to historical events, and many in this profession have pondered this problem. Carl L. Becker, for one, explored this in his seminal paper, "What Are Historical Facts?" first presented at the Research Club of Cornell University on 14 April 1926. Using as an example Julius Caesar's crossing of the Rubicon in 49 BCE, Becker argues that we have chosen to single this out and assign it significance, indeed marking it as a turning point in Roman history. Why? Many others had crossed the Rubicon at many other times, yet they are unremembered. Why is Caesar's crossing in the year 49 BCE significant? Only considered in the context of what were the significant results of his entry into Rome may we begin to explore this event. And considered in relation to the web of interconnection, it is actually a symbol standing for the historical record—a convenient shorthand—that allows us to explain significance. Becker reasoned that any "historical fact is not the past event, but a symbol which enables us to recreate it imaginatively."[14]

---

11. Thomas H. Kuhn, *The Structure of Scientific Revolutions* (Chicago: University of Chicago Press, 1962), p. 92.
12. Erik P. Rau e-mail to author, "Turning Points in History," 17 August 2006, copy in possession of author.
13. A noncosmological use of this term may be found in Ray Kurzweil, *The Singularity Is Near: When Humans Transcend Biology* (New York: Viking, 2005).
14. Carl L. Becker, "What Are Historical Facts?" in *Detachment and the Writing of History: Essays and Letters of Carl L. Becker*, ed. Phil L. Snyder (Ithaca, NY: Cornell University Press, 1958), pp. 41–64; quotes from pp. 45–46.

Becker traveled into similar territory in his presidential address before the American Historical Association in 1931, where he declared "Everyman his own Historian." He asserted that history is an artificial extension of memory and "in this sense is story, in aim always a true story; a story that employs all the devices of literary art (statement and generalization, narration and description, comparison and comment and analogy) to present the succession of events in the life of man, and from the succession of events thus presented to derive a satisfactory meaning." He added that "in every age history is taken to be a story of actual events from which a significant meaning may be derived."[15] Turning points in history are all about assigning significance to events of the past, and they are exceptionally slippery and idiosyncratic to the individuals assigning that significance. At the same time, some historians handle this issue with style and grace and aplomb.

One example of the difficult task of assigning significance to events will suffice, and the process will conjure an image of a turning point. At the five-year anniversary of the 11 September 2001 attacks on the World Trade Center and the Pentagon, most people would probably consider this instance as a clear point of demarcation in which the trajectory of the world as we understood it shifted appreciably. In the aftermath of 9/11, feelings of insecurity at home and hysteria in Washington abounded. Major changes in governmental policies and partisan politics resulted. A sense that the nation as a superpower might be at risk abounded and the response needed to be swift and decisive. Military action resulted, some of it taking a course unanticipated by those planning it. There were hearings and finger-pointing, and floodgates of government funding opened for all manner of presumed security-enhancing programs and intelligence specialists. Additionally, President George W. Bush was criticized for the 9/11 attacks and his failure to prepare for such an eventuality.[16]

But is it appropriate to view 9/11 as a turning point?" At one level, perhaps, but some have argued that this event was simply one chapter of a much longer story. As Cambridge University historian Brendan Simms recently commented:

> Without the attacks on the World Trade Center and the Pentagon, we may say with a reasonable degree of confidence that airline travel would have been easier. But beyond that, it becomes difficult to speculate. Some sort of attempt to topple Hussein was brewing in any case. Oil prices would still have risen given the increase in global, particularly Chinese and Indian, demand. The Iranian nuclear issue would be equally acute. And needless to say, the issue of Palestine would still be with us.

---

15. Carl L. Becker, "Everyman His Own Historian," *American Historical Review*, 37 (January 1932): pp. 221–236, quote from 231–232.

16. This includes everything from such polemics as Gore Vidal, "The Enemy Within," *The Observer* (London), 27 October 2002, to more the reasoned analysis of The National Commission on Terrorist Attacks Upon the United States, *The 9/11 Commission Report* (Washington, DC: U.S. Government Printing Office, 22 July 2004).

Simms agrees with former Chinese Premier Zhou En-lai's quip about the significance of the French Revolution: it is too early to tell.[17]

Likewise, Rutgers University sociologist Ted Goertzel questions 9/11 as a turning point in history. He cites polls suggesting that U.S. attitudes were mostly unchanged by the attacks and that efforts to return to normalcy motivated many people affected. He found that "the stock market recovered quickly from the shock of 9/11" and that the "domestic political climate does not seem to have shifted." In only one major area did Goertzel find a significant shift in national perspective, noting that the "country's foreign policy mood has shifted from introverted to extroverted." Indeed, he found that the following major elements remained firm both before and after 9/11:

- American military hegemony is strong;
- The stock market recovered from its initial shock;
- America is firmly in an extroverted foreign policy mood; and
- Western "sensate" popular culture seems irresistible.

For Goertzel, 9/11 as a turning point is more nuanced and not nearly as straightforward as many have suggested.[18]

With the foregoing discussion, it appears that turning points in history resemble so many other constructs in history, such as periodization, dialectic, causation, and significance, in their lack of firm definition. Undoubtedly, however, they are part of the toolbox used by historians and they appear throughout the master narrative of human history. Since turning points in history seem remarkably similar to beauty (that is, they exist in the eye of the beholder, thereby demonstrating the need for sagacious historians), do they still offer useful frames of reference for historical study? I asked several friends, colleagues, fellow travelers, and critics to offer their thoughts on turning points in space history, and what I received was a remarkable set of broad observations. Many of the ideas presented proved remarkably reflective and some were profound. As Dick Myers observed, "Like so many things in our existence, the definition depends upon the context . . . I think that one defines it in the concrete, not the abstract." In considering the histories of the space age, historians working in this arena have the power to define turning points however they wish. They will "be unique to that topic . . . [and] are defined by the context in which they occur or are said to occur—the context in which historians, etc. are explaining and analyzing and trying to understand."[19]

17. Brendan Simms, "9/11: Historic Turning Point, or Bump in the Road?" *Los Angeles Times*, 10 September 2006.

18. Adam Clymer, "U.S. Attitudes Altered Little by Sept. 11, Pollsters Say," *New York Times*, 20 May 2002; Ted Goertzel, "9/11 As a Turning Point in History," undated PowerPoint presentation in possession of author.

19. Dick Myers e-mail to author, "Turning Points in History," 16 August 2006, copy in possession of author.

Historian Philip Scranton carried this line of thinking a bit further. He suggested that there might be multiple framings of historical turning points:

> [F]irst from the perspective of contemporary actors (then refracted through the longer term assessments of historians—hence Sputnik was a major break for those working in the world of 1957–58, but not so big a deal fifty years on) and alternatively, the turning points historians construct in their narratives and periodizations, years or events which may not have seemed such a big deal to the folks at the time. Once in a while (I'd try 1968) both actors and historians agree that there's a major shift that's been launched. That frames a third, probably pretty small, category.[20]

Art historian David Ward offered an additional thought on this subject. He noted that the concept of turning points had value for political, diplomatic, military, and economic history, but was much less useful in social and other types of history. As Ward commented, it would be "rather hard to pin down the moment when modernism [in art] arrived."[21]

Deborah G. Douglas criticized the concept of turning points in history and suggested that they represent

> [. . .] the spaces/places in time that the historical community feels it has some fundamental understanding of and can therefore use in analysis and, more importantly, in our narratives. Depending on your disciplinary point of view, you may find yourself attracted or repelled by the particular term turning point but I suspect that has more to do with the time scale of your study and your literary tastes.[22]

She allowed that "the concept is popular but it is also formulaic and didactic—too amateurish, really—for good writers and readers."

Turning points are also representative of the dominant culture in which they are situated. For example, how would noted Marxist historian Howard Zinn interpret the turning points usually associated with the twentieth century? His warning is apropos in this context: "There is an underside to every Age about which history does not often speak, because history is written from records left by the privileged. We learn . . . about the thinking of an age from its intellectual elite."[23] Moreover, how would a Vietnamese scholar approach a history of the

20. Philip Scranton e-mail to author, "Turning Points in History," 17 August 2006, copy in possession of author.

21. David Ward e-mail to author, "Turning Points in History," 21 August 2006, copy in possession of author.

22. Deborah G. Douglas e-mail to author, "Turning Points in History," 16 August 2006, copy in possession of author.

23. Howard Zinn, *The Politics of History* (Boston: Beacon Press, 1970), p. 102.

1960s or, more to the point, a how would a Chinese scholar focusing on aerospace history? The reality is that turning points lack clear cohesion across a broad spectrum. Graphically demonstrating the lack of clear definition and meaning of a turning point, Douglas suggested a game:

> Assemble on cards a large number of events that might be considered turning points in space history. Shuffle the deck, pick 10 at random, and spend 5 minutes making up a story. Do it again a couple of times and compare your stories. Are you fitting your 'turning points' into your preconceived narrative or do you have vastly different stories?[24]

This approach might yield really interesting results and is grist at least for a session at one of the major historical conferences.

Despite the ease with which we might appropriately dispense with turning points as a useful analytical tool in history, they are everywhere in the national discourse. In twentieth-century America, events commonly assigned turning point status include the following, ranked by opinion leaders in a poll conducted in 1999.

TOP 25 NEWS STORIES OF THE TWENTIETH CENTURY[25]		
1	U.S. drops atomic bomb	1945
2	Men first walk on the Moon	1969
3	Japan bombs Pearl Harbor	1941
4	Wrights fly first airplane	1903
5	U.S. women win the right to vote	1920
6	JFK assassinated in Dallas	1963
7	Nazi Holocaust exposed	1945
8	World War I begins	1914
9	Court ends "separate but equal"	1954
10	U.S. stock market crashes	1929
11	Antibiotic penicillin discovered	1928
12	DNA's structure discovered	1953
13	Soviet Union dissolves	1991
14	President Nixon resigns	1974
15	Germany invades Poland	1939
16	Communists take over Russia	1917
17	Ford creates assembly line	1913
18	Soviets launch first satellite	1957
19	Einstein conceives relativity	1905
20	Birth control pill OK'd by FDA	1960
21	New polio vaccine works	1953
22	Hitler named chancellor	1933
23	M. L. King, Jr., assassinated	1968
24	Allies invade France on D-Day	1944
25	Deadly AIDS disease identified	1981

24. Douglas e-mail to author, "Turning Points in History," 16 August 2006.

25. "Stories of the Century, 1900-2000," Newseum, *http://www.newseum.org/century/finalresults.htm* (accessed 13 September 2006).

For the period since 2000, almost certainly the 2001 terrorist attacks on the World Trade Center and the Pentagon and the 2003 invasion of Iraq would be assigned important status as turning points in history.

The reality is that accepting all of these events, as significant as many are, as turning points demonstrates the less than useful nature of the term. Certain events are immediately considered turning points, such as Pearl Harbor and the atomic bomb, whereas others are assigned this status only in retrospect, such as the stock market crash and the oil embargo. Turning points of national significance probably take place less often than this list suggests, and the probability that any individual would witness more than a handful of them during his or her lifetime is small. Instead, the 25 events listed here are within the memory of many people still alive, and even those of us a little younger can remember more than dozen of them.

Constructing Turning Points in Space History

Rather than playing the game as outlined by Debbie Douglas, let me suggest some turning points in the history of spaceflight. I will then analyze three of them, "turning" the concept on its "pointed" head: 1) a recognized turning point which I will argue might not be one after all; 2) an event not usually thought of as a turning point but which I will assert is appropriately considered one; and 3) an event that was immediately labeled a turning point at the time it took place but, as time passes, appears much less so than previously thought.

Based on inputs from several close observers of the history of spaceflight, major turning points in the field may include the following:

1. Robert Goddard's first liquid-fueled rocket (1926).
2. Development of ballistic missiles (1944).
3. Launch of Sputnik (1957).
4. Flight of Yuri Gagarin (1961).
5. JFK's announcement of Apollo landing decision (1961).
6. Launch of the first operational applications satellites (1962).
7. Apollo 11 lunar landing (1969).
8. Nixon's Space Shuttle decision (1972).
9. First flight of the Space Shuttle (1981).
10. *Challenger* accident (1986).
11. Demise of the Soviet Union as competitor in space (1991).
12. Decision to turn the Space Station into a multinational program involving Russia (1992).
13. *Columbia* accident (2003).
14. Bush's announcement of the Vision for Space Exploration (2004).
15. Flight of SpaceShipOne (2004).

What is most interesting about this list, compiled from inputs from many sources, is the lack any mention of planetary exploration or Earth science, and only a passing reference to applications satellites. Most are also political turning points, a few are technological, and none is social or scientific in focus. What is included (and especially what is excluded) in this list represents a fascinating avenue for further exploration, but I must leave that for another time and place.

Sputnik

Virtually everyone would agree that the launch of Sputnik 1 on 4 October 1957, represented an undisputed turning point in space history. Most observers chart the beginning of the Space Age from that date. Indulge me while I argue an alternative position—that it did not represent a turning point at all but, rather, a continuation of the events that had been moving along the same path from at least World War II. In the summer of 1957, six months into Dwight D. Eisenhower's second term and before the Sputnik turning point in history, the president asked the National Security Council (NSC) to review the U.S. space program to ensure that the level of investment and progress being made was adequate. He intended to field the first intercontinental ballistic missiles (ICBMs) and reconnaissance satellites by the time he left office. These capabilities in the new high ground of space would

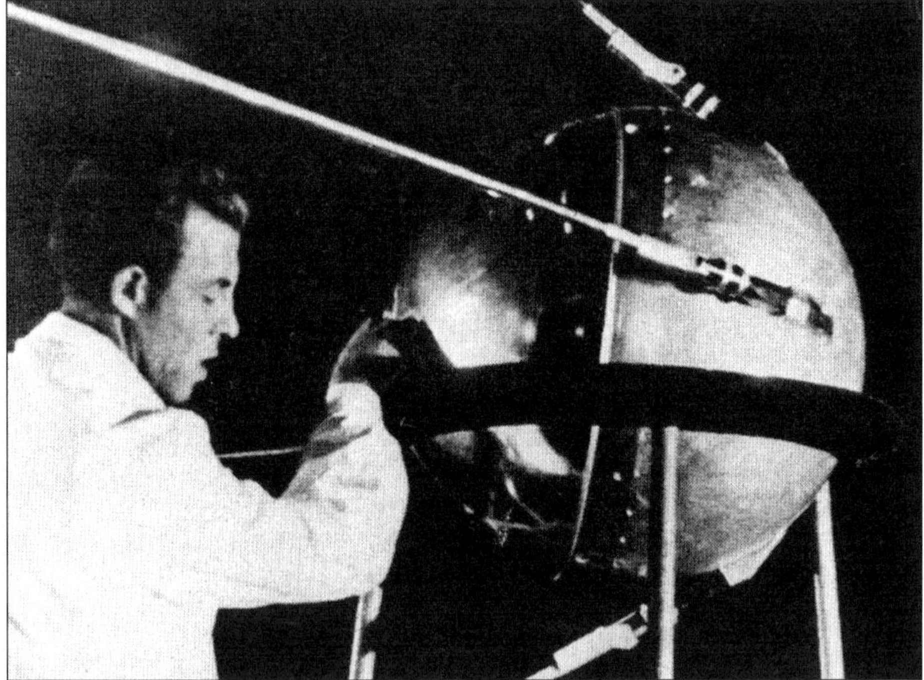

Figure 2.1— The launch of Sputnik 1 is usually viewed as the beginning of the space age and a critical turning point in history. Is it conceivable that it was less pivotal than usually thought? (NASA photo no. GPN-2002-000166).

ensure that the United States could compete effectively with the Soviet Union in the cold war rivalry that gripped the world. Eisenhower learned that between 1953 and 1957 the nation had spent $11.8 billion on military space activities, mostly on ballistic missile and reconnaissance satellite development. "The cost of continuing these programs from FY 1957 through FY 1963," the NCS reported, "would amount to approximately $36.1 billion, for a grand total of $47 billion."[26]

By any measure, this should be considered a significant investment on the part of the Eisenhower administration, and it suggests that Eisenhower had developed a strategy for ensuring U.S. technological comparability, and eventual superiority, in the global game of one-upmanship and rivalry that was the cold war. When adjusted for inflation, only Presidents Ronald Reagan and Bill Clinton, surprisingly, made similar investments in space technology.[27] Those assets also found use on both the military and civilian sides of the space program during subsequent years.[28] In an irony of proportions too great to ignore, in 1 October 1957, after the launch of Sputnik 1, Eisenhower found himself branded by the Democrats as an incompetent for allowing the Soviet Union to beat the U.S. into orbit by launching the first satellite. For example, Eisenhower argued that "I am always a little bit amazed about this business of catching up. What you want is enough, a thing that is adequate. A deterrent has no added power, once it has become completely adequate, for compelling the respect of any potential opponent for your deterrent and, therefore, to make him act prudently."[29]

Moreover, Eisenhower had long followed a path toward the development of launch vehicles for use in the ICBM program; satellite technology for reconnaissance and communications; infrastructure required to support these activities such as tracking and launch facilities; and utilitarian science that either directly supported those missions or was a natural byproduct of them. An example of such a byproduct was when, early in the military rocket research program, scientists won the opportunity to place on some of the test vehicles instruments that provided data about the upper atmosphere, solar and stellar ultraviolet radiation, and the aurora. This became a very successful scientific program that was carried out with limited

26. S. Everett Gleason, "Discussion at the 329th Meeting of the National Security Council, Wednesday, July 3, 1957," 5 July 1957, NSC Records, Dwight D. Eisenhower Presidential Papers, Eisenhower Library, Abilene, KS, p. 2.

27. Reagan spent $233.02 billion on space issues in his eight years in office. Clinton spent $230.14 billion during his eight years in office. In contrast, Eisenhower's spending was $183.69 billion. All of these are in inflation-adjusted dollars. Calculated using data in Appendix E-1A, "Space Activities of the U.S. Government," *Aeronautics and Space Report of the President, Fiscal Year 2003 Activities* (Washington, DC: NASA NP-2004-17-389-HQ, 2004), p. 140.

28. Much has been made of dual-use technology over the years, and space access has been an especially important part of this capability. On space access, see *To Reach the High Frontier: A History of U.S. Launch Vehicles*, ed. Roger D. Launius and Dennis R. Jenkins (Lexington: University Press of Kentucky, 2002).

29. "The President's News Conference of 3 February 1960," *Public Papers of the Presidents of the United States: Dwight D. Eisenhower*, 1960 61 (Washington, DC: U.S. Government Printing Office, 1964), p. 24.

fanfare and funding. As a result, scientists taking part in this program used all the military's captured V 2s, persuaded the Department of Defense (DOD) to develop new sounding rockets to replace them, and continued to use the nation's rocket development program for scientific research throughout the 1950s.[30]

Eisenhower's space program also placed considerable emphasis on satellite technology. During the mid-1950s, the president was preoccupied with the need to conduct surveillance of Soviet Union activities and its growing nuclear capability. This led to the development of both surveillance aircraft and satellites on an aggressive basis in the 1950s. As the 1960 downing of the U-2 reconnaissance airplane revealed, however, aircraft overflights had severe shortcomings. A spacecraft was much less vulnerable. Eisenhower authorized the Vanguard satellite program in part because he wanted to establish the principle of overflight (namely, that a satellite did not intrude upon a nation's airspace when crossing its territory and was not subject to interception), and an internationally supported scientific satellite served this purpose better than any military launch.[31]

Nothing summarizes this balanced, measured approach toward space activities better than a statement Eisenhower made in 1959 at a meeting with top advisors. He outlined three major goals that had to be accomplished:

> The first is that we must get what Defense really needs in space; this is mandatory. The second is that we should make a real advance in space so that the United States does not have to be ashamed no matter what other countries do; this is where the super booster is needed. The third is that we should have an orderly, progressive scientific program, well balanced with other scientific endeavors.[32]

Within the context of this philosophy, Eisenhower was willing to expend resources sufficient to meet major objectives, but not to open the floodgates of government expenditures for activities that he believed did not have a viable component.

30. The military created the V 2 Upper Atmosphere Panel in 1946 to oversee this activity. In 1948 it became the Upper Atmosphere Rocket Research Panel, and in 1957 the Rocket and Satellite Research Panel. See Lyman Spitser Jr., "Astronomical Advantages of an Extra-Terrestrial Observatory," *The Astronomy Quarterly* 7 (September 1946): pp. 19–20; James A. Van Allen, *Origins of Magnetospheric* Physics (Washington, DC: Smithsonian Institution Press, 1983); Homer E. Newell, *Beyond the Atmosphere: Early Years of Space Science* (Washington, DC: NASA SP 4211, 1980); George K. Megerian, "Minutes of V-2 Upper Atmosphere Research Panel Meeting," V-2 Report No. 13, 29 December 1947; George K. Megerian, "Minutes of Meeting of Upper Atmosphere Rocket Research Panel," Panel Report No. 35, 29 April 1953, both in NASA Historical Reference Collection.

31. R. Cargill Hall, "The Origins of U.S. Space Policy: Eisenhower, Open Skies, and Freedom of Space," *Colloquy*, 14, no. 3 (December 1993); R. Cargill Hall, "Origins of U.S. Space Policy: Eisenhower, Open Skies, and Freedom of Space," in *Exploring the Unknown: Selected Documents in the History of the U.S. Civil Space Program*, Vol. I, gen. ed. John M. Logsdon (Washington, DC: NASA SP-4407, 1995), chapter 2.

32. Brig. Gen. A. J. Goodpaster, "Memorandum of Conference with the President, October 12, 1959," 23 October 1959, Records of the White House Office of Science and Technology, Box 12, Eisenhower Library, Abilene, KS.

Eisenhower was also not unreceptive to increases in funding for space activities purely to further scientific understanding. The experience of approval of the International Geophysical Year (IGY) satellite effort is instructive on this score. As early as 1950, a small group of scientists in the United States began discussing among themselves the possibility of using Earth-circling satellites to obtain scientific information about the planet.[33] In 1952, urged on by these same American scientists, the International Council of Scientific Unions (ICSU) proposed the IGY, a cooperative scientific endeavor to study solar–terrestrial relations during a period of maximum solar activity. Some 67 nations agreed to conduct cooperative experiments to study solar–terrestrial relations during a period of maximum solar activity in 1957–1958.

In October 1954, at the behest of essentially this same group of U.S. scientists, the ICSU challenged nations to use their missiles being developed for war to launch scientific satellites to support the IGY research program. In July 1955, largely the same enclave of American scientists convinced Eisenhower that the United States should respond to the ICSU call for participation in the IGY by launching a scientific satellite. Eisenhower's decision called for existing organizations within the DOD to develop and launch a small scientific satellite, "under international auspices, such as the International Geophysical Year, in order to emphasize its peaceful purposes[;] . . . considerable prestige and psychological benefits will accrue to the nation which first is successful in launching a satellite . . . especially if the USSR were to be the first to establish a satellite." The result was Project Vanguard, carried out under the supervision of the Naval Research Laboratory. Eisenhower also approved a budget of $23.5 million, modest but considered adequate for the effort by scientific and technical personnel consulted by the administration.[34]

Although some have asserted that Sputnik represented the "shock of the century," there did not seem to be much shock immediately after the launch of Sputnik 1. Most recognized that it did not pose a threat to the United States and

33. This group included Lloyd Berkner, Joseph Kaplan, Fred Singer, James Van Allen, and Homer Newell. The fingerprints of these core leaders are all over every decision relative to the IGY satellite program and the U.S. decision by Eisenhower to sponsor a satellite. See the discussion of this effort in Constance McLaughlin Green and Milton Lomask, *Vanguard: A History* (Washington, DC: Smithsonian Institution Press, 1971), pp. 6–39; Rip Bulkeley, *The Sputniks Crisis and Early United States Space Policy: A Critique of the Historiography of Space* (Bloomington, IN: Indiana University Press, 1991), pp. 89–122; R. Cargill Hall, "Origins and Early Development of the Vanguard and Explorer Satellite Programs," *Airpower Historian* 9 (October 1964): pp. 102–108.

34. National Security Council, NSC 5520 "Draft Statement of Policy on U.S. Scientific Satellite Program," 20 May 1955; United States National Committee for the International Geophysical Year 1957-1958, "Minutes of the First Meeting, Technical Panel on Earth Satellite Program, 20 October 1955," both in NASA Historical Reference Collection; Don Irwin to Mr. Rockefeller and General Parker, "Pentagon Briefing on Earth Satellite Program," 12 October 1955; Richard Hirsch to Elmer B. Staats, "Pentagon Meeting on Earth Satellite Program," 13 October 1955, both in White House Office of Special Assistant for National Security Affairs, NSC, OCB Central Files, Box 11, "OCB 000.9 (National & Physical Sciences)," Eisenhower Library. Abilene, KS.

thus no one took immediate action to respond to it. Instead, congratulations ensued and people were excited by the Soviet success. At the same time, Eisenhower acknowledged the need to "take all feasible measures to accelerate missile and satellite programs."[35] He also moved to assure the American public that all was well, accepting the findings of representatives of the International Affairs Seminars of Washington who reported on 15–16 October 1957:

> If there was any trauma following the Russian sputnik [sic], it occurred in Washington and not among the general public. Washington, for its part, took its cue from the newspapers and other issue makers. The misevaluation by leadership of the extent of public interest, as measured by the amount of news, coverage and the words of the issue makers, led to words and actions which further confused the issue. This situation points up the general problem for a democracy of: who is the 'public' to which leadership attends and who in fact do the issue makers represent?[36]

As it turned out, failure to appreciate the ability of Eisenhower's political enemies to use Sputnik as a wedge issue in the 1958 campaign hurt his administration.

In his first press conference after the launch of Sputnik 1, on October 9 Eisenhower calmed speculation and said it did not raise his apprehension "[. . .] one iota. I see nothing at this moment, at this stage of development, that is significant in that development as far as security is concerned."[37] Others in the administration did the same.[38] *The New York Times* disparaged the Soviet "attempt to persuade people, especially in Asia and Africa, that Moscow has taken over world leadership in science."

Life magazine was no less derogatory, warning that, at best, the "Sputniks give this old Communist swindle a new lease of plausibility."[39] What concerns that the public might have had about Sputnik 1 died down in the latter part of October 1957. For instance, there was little discussion of the satellite issue in the popular press during the latter part of the month and it did not come up in the president's press conference of 30 October 1957.[40]

While advocates of more aggressive space activities and political opponents of the White House still criticized, public confidence in the nation's leadership did not

35. Dwight D. Eisenhower, *The White House Years: Waging Peace* (Garden City, NY: Doubleday, 1965), p. 211.

36. International Affairs Seminars of Washington, "American Reactions to Crisis," 15–16 October 1958, NASA Historical Reference Collection.

37. *Facts on File*, XVII, no. 884, p. 330.

38. Ibid., p. 331; Richard M. Nixon, *The Memoirs of Richard Nixon* (New York: Grossett & Dunlap, 1978), p. 111.

39. "Soviet Claiming Lead in Science," *New York Times*, 5 October 1957: p. A2; "A Proposal for a 'Giant Leap,'" *Life*, 16 November 1957: p. 53.

40. *Public Papers of the Presidents of the United States: Dwight D. Eisenhower, 1957* (Washington, DC: U.S. Government Printing Office, 1958), pp. 774–787; NASA clippings file, "October 1957," NASA Historical Reference Collection.

seem to suffer appreciably until Sputnik 2 was launched on 3 November 1957. This time the Soviet Union counted coup on the United States with an impressive 1,121 pound spacecraft that included a dog named Laika. If anything, the turning point in history came following the 6 December 1957 failure of the Vanguard launch. After the two successful Soviet Sputniks, and this rather spectacular failure on national television, dramatic actions appeared necessary. Accordingly, it seems that Sputnik may not have been such a significant turning point in history as many have thought. It represented one stage of a succession of activities in the history that we all understand, nothing more.

What would have been different had there not been a Sputnik? The U.S. rocketry programs were well in hand in 1957 and there is every reason to believe they would have continued on as they did.[41] The same is true of the satellite reconnaissance effort.[42] Space science was being pursued expeditiously through a variety of avenues; even with efforts to send probes to the Moon, and except for an acceleration of effort probably would have been continued along pretty much the path that came with this turning point.[43] Communications satellites were being pursued by AT&T and might have even achieved success earlier had there been less government involvement.[44] In all, Sputnik has been assigned significance far beyond what it truly deserves.[45]

---

41. See Eugene M. Emme, ed., *The History of Rocket Technology: Essays on Research, Development, and Utility* (Detroit, MI: Wayne State University Press, 1964); Richard P. Hallion, "The Development of American Launch Vehicles Since 1945," in *Space Science Comes of Age: Perspectives in the History of the Space Sciences*, Paul A. Hanle and Von Del Chamberlain, ed. (Washington, DC: Smithsonian Institution Press, 1981), pp. 115–134; Roger D. Launius, "Between a Rocket and a Hard Place: The Challenge of Space Access," in *Space Policy in the 21st Century*, W. Henry Lambright, ed. (Baltimore, MD: Johns Hopkins University Press, 2002), pp. 15–54.

42. Three important books on the early satellite reconnaissance program have been published: Dwayne A. Day, John M. Logsdon, and Brian Latell, ed., *Eye in the Sky: The Story of the Corona Spy Satellite* (Washington, DC: Smithsonian Institution Press, 1998); Robert A. McDonald, *Corona Between the Sun and the Earth: The First NRO Reconnaissance Eye in Space* (Bethesda, MD: ASPRS Publications, 1997); Curtis Peebles, *The Corona Project: America's First Spy Satellites* (Annapolis, MD: Naval Institute Press, 1997). See also William E. Burrows, *Deep Black: Space Espionage and National Security* (New York: Random House, 1987); Jeffrey T. Richelson, *America's Secret Eyes in Space: The U.S. Keyhole Spy Satellite Program* (New York: Harper and Row, 1990).

43. The early history of this effort is well told in Homer E. Newell, *High Altitude Rocket Research* (New York: Academic Press, 1953); R. Cargill Hall, "Early U.S. Satellite Proposals," *Technology and Culture* 4 (Fall 1961): pp. 410–434; R. Cargill Hall, "Origins and Development of the Vanguard and Explorer Satellite Programs," *Airpower Historian* 11 (October 1964): pp. 101–112; R. Cargill Hall, *Lunar Impact: A History of Project Ranger* (Washington, DC: NASA SP 4210, 1977); David H. DeVorkin, *Science with a Vengeance: How the Military Created the US Space Sciences After World War II* (New York: Springer-Verlag, 1992).

44. This is the thesis of David J. Whalen, *The Origins of Satellite Communications, 1945–1965* (Washington, DC: Smithsonian Institution Press, 2002).

45. One could make the case that considerable resources were spent on useful activities such as science and technology, education, and retraining of workforces. Even so, some scholars minimize its long-term effect. See Herbert Kliebard, *The Struggle for the American Curriculum*, 2nd ed. (New York: Routledge, 1995); Andrew Fraknoi, "Space Science Education in the U.S.: The Good, the Bad, and the Ugly," contained in this collection; "The Nationalization of U.S. Science," *Fortune* (September 1976): p. 158.

Kennedy's Role

What about an event that is not considered a turning point in space history, but perhaps should be? The assassination of John F. Kennedy looms large in the history of the United States during the middle part of the twentieth century, no doubt, but what role did it play in the unfolding of the history of spaceflight? If Kennedy had not been assassinated, would anything relative to Apollo have changed? Few refer to this event as something of significance in the history of Apollo, but it may well be that Kennedy's death solidified support for the Moon landings. Despite public support for Apollo, we know that Kennedy had expressed concerns about the program and the funds that it sucked out of the treasury. In late May 1961, his budget director had warned JFK of the large price tag of Apollo and, when he met Nikita Khrushchev in Vienna the following month, Kennedy suggested that the United States and the Soviet Union explore the Moon as a joint project. The Soviet leader reportedly first said "No," then replied, "Why not?" and then changed his mind again, saying that disarmament was a prerequisite for U.S.–USSR cooperation in space.[46] In the fall of 1963, in what might be considered an American version of glasnost more than 20 years before the term became famous, JFK aggressively pursued a venture to turn the Apollo program into a joint effort. He privately met with NASA Administrator James Webb on 18 September and told him to prepare for a joint program. As Webb recalled, "He didn't ask me if he should do it; he told me he thought he should do it and wanted to do it and that he wanted some assurance from me as to whether he would be undercut at NASA." On 20 September 1963, Kennedy made a well-known speech before the United Nations, in which he again proposed a joint human mission to the Moon. He closed by urging, "Let us do the big things together." Publicly, the Soviet Union was noncommittal. *Pravda*, for example, dismissed the 1963 proposal as premature. Some have suggested that Khrushchev viewed the American offer as a ploy to open up Soviet society and compromise Soviet technology. Whatever the case Kennedy was assassinated in November, 1963 and Khrushchev was deposed the next year, and nothing came of the offer.[47] Thereafter Lyndon B. Johnson and NASA Administrator James E. Webb constantly defended the Apollo program as the dying wish of this slain president.

---

46. Dodd L. Harvey and Linda C. Ciccoritti, *U.S.-Soviet Cooperation in Space* (Miami, FL: University of Miami Center for Advanced International Studies, 1974), pp. 78–79. A State Department memo covering the two leaders' discussion in Vienna does not mention Khrushchev's fleeting acquiescence, instead focusing on Khrushchev's desire to have progress in disarmament before consenting to a joint lunar landing program. See 6/4/61 Memcon between JFK and Khrushchev, 6/4/61, Luncheon, Soviet Embassy, Vienna in the Kennedy Presidential Library, Box 126, NASA Historical Reference Collection.

47. *Public Papers of the Presidents of the United States, John F. Kennedy, 1963*, p. 695, cited in Harvey and Ciccoritti, *U.S.-Soviet Cooperation in Space*, p. 123; "Text of President Kennedy's Address on Peace Issues a U.N. General Assembly," *New York Times*, 21 September 1963: C6; Yuri Karash, "The Price of Rivalry in Space," *Baltimore Sun*, 19 July 1994: p. 11A; Walter A. McDougall, *. . . the Heavens and the Earth: A Political History of the Space Age* (New York: Basic Books, 1985), p. 395.

That was a very powerful argument to be made in the political arena and they achieved success in protecting the program, even as everything else at NASA began to suffer budget cuts from the mid-1960s onward.

Had Kennedy served two full terms, it is quite easy to envision a point in the mid-1960s, probably near the time that Project Gemini was successfully underway, in which he might have decided that the international situation that sparked announcement of a lunar landing by the "end of the decade" had passed and he could have safely turned off the landing clock. Had he done so, Apollo might have stretched out for many more years, and perhaps have ultimately been successful; but, just as likely, it could have become something akin to the current, open-ended Space Station program without clear objectives and no time frame for completion. JFK's assassination, therefore, could be interpreted as a turning point in the history of spaceflight although it is not usually accepted as such.

Figure 2.2—The decision of John F. Kennedy to land Americans on the Moon by the end of the decade is viewed as a pivotal event in the history of the Space Age. But evidence suggests that he was reconsidering this decision at the time of his assassination in November 1963. Had his death not occurred and he been allowed to serve a full term or perhaps two terms in the White House, how might the Moon landing program have unfolded? Accordingly, the Kennedy assassination may be an unrecognized turning point in the history of the space program. Here Kennedy is depicted in a motorcade with Mercury astronaut John Glenn, the first American to orbit Earth in February 1962. (NASA photo no. GPN-2002-000050).

Apollo

Finally, there are events that were hailed at the time as turning points in history and accepted as such by virtually everyone, but which now invite reconsideration. The most obvious that I would point to is the Apollo 11 landing on the Moon on 20 July 1969. Immediately thereafter, President Richard Nixon told an assembled audience that the dates encompassing the flight of Apollo 11 represented the most significant week in the history of Earth since the creation.[48] Perhaps he was caught up in the moment, but at least at that time the president expressed the view that this was both a path-breaking and permanent endeavor, a legacy of accomplishment that future generations would reflect upon as they plied intergalactic space and colonized planets throughout the galaxy. Undoubtedly, he believed it a turning point in history. Others did as well. Apollo suggested that America had both the capability and the wherewithal to accomplish truly astounding peaceful goals. All it needed was the will.[49]

Now, more than a generation removed from the last of the Apollo missions to visit the Moon in December 1972, that turning point appears far less significant than it did during the time of Apollo. Advocates of human exploration have tended to view the astronauts who landed on the Moon as people akin to fifteenth-century voyagers of discovery such as Christopher Columbus—the vanguards of sustained human exploration and migration. But as time progresses, those first space ventures may well prove to be more like Leif Erickson's voyages from Scandinavia several hundred years earlier—an exploratory dead end.

MAXIMS OF TURNING POINTS IN SPACE HISTORY

I would close this essay by offering 10 maxims for anyone considering the place of turning points in the history of spaceflight.[50]

1. Turning points signify a critical juncture in the coalescence of a set of events that signal a shift in the stream of history.
2. Turning points most often represent attempts by observers to assign significance to events, either at the time or thereafter. Depending on perspective, countervailing issues, and subsequent developments, they may shift or become meaningless or meaningful. They are also subject to "political spin" and the vicissitudes of the "master narrative."

48. *CBS Evening News Transcript, 10:56:20 PM EDT, 7/20/69* (New York: CBS News, 1969), p. 159.

49. *Congressional Quarterly* (25 July 1969): p. 1311; *The Futurist* (October 1969): p. 123. On the possibilities raised by Apollo, see Roger D. Launius, "Perfect Worlds, Perfect Societies: The Persistent Goal of Utopia in Human Spaceflight," *Journal of the British Interplanetary Society* 56 (September/October 2003): pp. 338–349.

50. The following is based on the comments of Richard P. Hallion, "Turning Points in Aerospace History: Some Thoughts," 16 August 2006, copy in possession of author.

Figure 2.3—Astronaut Buzz Aldrin, lunar module pilot of the first lunar landing mission, poses for a photograph beside the deployed United States flag during an Apollo 11 extravehicular activity (EVA) on the lunar surface. The first Moon landing is universally viewed as a pivotal event in history. With the passage of time, and the failure to continue lunar exploration beyond the Apollo program, it seems less a turning point and more a "blip" in the flow of history. How should it be interpreted in the future? (NASA photo no. GPN-2001-000012).

3. Turning points provide a short-hand of analysis that may be used effectively, but too often they mash hackneyed and amateurish analysis. They are like George W. Pierson's characterization of the "Frontier Thesis" of Frederick Jackson Turner: "too optimistic, too romantic, too provincial, and too nationalistic" to be of great utility for the historian's task.[51]
4. Turning points adhere to the standard of definition employed by Justice Potter Stewart when confronted with defining pornography: "I shall not today attempt further to define the kinds of material . . . but I know it when I see it."[52] They defy definition, and one person's turning point might conceivably be another person's stasis. It might be appropriate to apply something like the Saffir-Simpson Hurricane Scale to identify importance and severity of turning points on a 1 to 5 scale.

51. Quoted in Richard Hofstadter, *The Progressive Historians: Turner, Beard, Parrington* (New York: Alfred A. Knopf, 1969), p. 149.

52. *Jacobellis v. Ohio*, 378 U.S. 184 (1964).

5. Turning points are often vague and imprecise. Their very elasticity offers a clue to their attraction as well as a core reason for using them with care. They may be invoked to argue for or against virtually anything, and accordingly they represent a form of historical mirage and incoherence.
6. Turning points of the most useful variety are those used in the most simple, concrete situations.
7. Turning points do not necessarily provide useful indicators of subsequent success or failure for the actors involved in their story.
8. Turning points most often signify a linear conception of history that rarely represents the reality of a complex, parallel, multicausational evolution of history.
9. Turning points are often psychological in focus, firing those experiencing them to undertake a different approach to what had previously been the norm. These are appropriately considered the fault lines in the stream of history. It might goad them to action, or it might lull them to complacency. It never fosters the status quo.
10. Turning points too often lead those invoking them into accepting a progressive interpretation of the past in which ideas of exceptionalism and advancement reign.[53]

This last maxim is especially significant. American history has been dominated by a vision of progress, of moving from nothing to something—essentially the opposite of the law of entropy in physics. As historian Richard P. Hallion recently remarked, repeated acceptance of the turning point concept

> [. . .] implies a teleological, linear, sequential 'achievement of events' leading inexorably in a certain direction, usually defined as 'progress.' In fact, this ignores the inherently disordered nature of the historical progress, which reflects chance, national circumstance, individual action (and we must remember that, at heart, all history is the working of people through time), and which results in a typically simultaneous and parallel pattern of development, one in which exploitation and innovation is at least equally as important is invention.[54]

With the overburdening dominance of American exceptionalism as a guiding principle of American historiography, it behooves historians to weigh carefully the usefulness—versus the possible confusions—of the use of the term turning point in historical explanation. This is especially the case for space historians as we enter a season of significant anniversaries of major events in the history of the space age, such as the 2007 50th-anniversary of the launch of Sputnik 1.

53. On American exceptionalism, see Kerwin Lee Klein, *Frontiers of Historical Imagination: Narrating the European Conquest of Native America, 1890–1990* (Berkeley, CA: University of California Press, 1997); Richard T. Hughes, *Myths America Lives By* (Urbana, IL: University of Illinois Press, 2003).

54. Hallion, "Turning Points in Aerospace History: Some Thoughts," p. 4.

Or perhaps I am obsessing over this issue; Debbie Douglas has suggested to me that this might be the case. In a note, she suggested that we might consider several questions when thinking about turning points as we travel along the river that we call the history of the Space Age. I close with these questions about the study of space history, admitting that I have few answers and those that I do possess may be satisfactory only to myself. When considering the Space Age:

1. Is space a "major river" (e.g., the Mississippi, the Yangtze, or the Amazon) or something a little less (e.g., the Columbia or the Rhine)?
2. What are the things I need to "know before I go" and what are the "must sees" once I arrive?
3. How will I see and understand the world differently because of this "experience?"[55]

This last question stands at the center of the historical discipline. Our answers could have profound implications for those studying this subject.[55]

55. Douglas e-mail to author, "Turning Points in History," 16 August 2006.

CHAPTER 3

IN SEARCH OF A RED COSMOS: SPACE EXPLORATION, PUBLIC CULTURE, AND SOVIET SOCIETY

James T. Andrews

In the Soviet 1920s, a proliferation of popular books, newspaper articles, and pamphlets on air and spaceflight filled the popular press and Soviet readers became part of a cosmopolitan readership throughout Europe engaged in news on exploration of the cosmos. Indeed, as I have argued in my book *Science for the Masses*, this only continued a pre-Revolutionary fascination with the stars, heavens, and the universe beyond. Astronomy and amateur space societies proliferated in Soviet Russia until the Stalinist 1930s and genuinely were generated from below, independently from the state.[1]

However, to a certain degree, two catalytic time periods changed that public response—both 1935, in Stalin's times, and 1957, in Khrushchev's. In 1935, Stalin and the Central Committee sanctioned Konstantin Tsiolkovskii to give a taped speech on May Day from Red Square, which would be broadcast all over the former Soviet Union. Tsiolkovskii's speech would be used by the regime to boast the preeminence of early Soviet rocket theorists over Western thinkers. Along with Stalin's Soviet nationalist cultural campaigns, it would begin a contest with the West of technological superiority that wrenched the early popular enthusiasm for space flight into a politicized and ultimately nationalized context. By 1957, with the launching of Sputnik 1, the Khrushchev regime and its successors would continue that program, only this time directing memorial celebrations to earlier rocket theorists; launching popular campaigns from above in the press and journals; mythologizing cosmonauts and physicists alike; and urging Soviet citizens to engage in the contest with the West, while focusing on its "national" resonance.

This article will begin by analyzing in more detail how the early, more cosmopolitan fascination with spaceflight in Russia shifted to become directed from above in the shaping of popular consciousness of spaceflight after both 1935

1. See James T. Andrews, *Science for the Masses: The Bolshevik State, Public Science, and the Popular Imagination in Soviet Russia, 1917–1934* (College Station, TX: Texas A&M University Press, Russian and East European Studies Series, 2003).

and 1957. It will also attempt to theorize how one can deconstruct that campaign in a censored state, and whether there still remained the genuine, popularly driven response and enthusiasm to space exploration during the Stalin and Khrushchev eras. Indeed, some ordinary Russians as well as well-known cultural critics criticized the campaigns to place space exploration on the national cultural agenda. Furthermore, this paper will explore how the popularization of space exploration in Soviet Russia may have also had a genuine inspirational effect on future physicists regardless of the political context within which these texts and campaigns were created from above. Yet, ultimately this was a dialogical tension between state and society, and although the public attempted to respond in independent ways, the monumental shifts from 1935 through 1957 nevertheless served to constrain the Soviet public's enthusiasm while it directed it into "proper channels."

AIR- AND SPACEFLIGHT, THE COSMOS, AND THE POPULAR IMAGINATION FROM TSARIST TO STALIN'S SOVIET TIMES

On a cold, wintry day during Lenin's regime in 1921, a long line of people waited, freezing in the Moscow snow to hear another lecture in a series on the planet Mars; it would be presented at the famed Moscow Polytechnic Museum by the astronomer A. A. Mikhailov.[2] Soviet citizens in the 1920s had flocked to hear talks on astronomy, air flight, and popular rocketry, and frequented museums in both capital and provincial cities to expand their knowledge on these topics. These densely populated lectures and long lines in the 1920s were not anomalous because, since as far back as the late nineteenth century, Russians had been fascinated by popular scientific themes. In the late nineteenth and early twentieth centuries, Tsarist Russia witnessed an explosion of scientific and amateur societies that helped sponsor lectures and events on popular topics such as air flight, astronomy, and the cosmos beyond. These societies proliferated before the onslaught of World War I and the Russian revolution, while their membership grew as well. By the 1920s, after the Bolshevik Revolution of 1917 and the Russian Civil War (1918–1920), a period called the New Economic Policy (1921–1927) allowed for a mixed economy to flourish and thus books, pamphlets, and even some newspapers could be published independently of the state.[3] Within this economic and political context of the Soviet 1920s, air- and spaceflight, along with astronomy, became not only popular themes in the mass media—they literally became crazes.

2. "Otchet M.O.L.A. na pervoe polugodie 1921 goda," Gosudarstvennyi Arkhiv Rossiskoi Federatsii—State Archive of the Russian Federation (hereafter cited as GARF), f. 2307, op. 2, d. 371, l.69.

3. For a critical overview of the transitional qualities of the period of the New Economic Policy (NEP), see William G. Rosenberg, "Introduction: NEP Russia as a 'Transitional' Society," in *Russia in the Era of NEP, Explorations in Soviet Society and Culture,* ed. Sheila Fitzpatrick et al. (Bloomington, IN: Indiana University Press, 1991), pp. 1–12.

In Soviet Russia during the 1920s, professors such as N.A. Rynin in Leningrad became almost full-time popularizers of spaceflight, in particular, while the public eagerly consumed journal and newspaper articles devoted to this topic.[4] Rynin, a prolific writer on Russian rocketry and astronautics, was also interested in organizing public astronautical societies in the 1920s. In the late 1920s he began to write and publish a multivolume encyclopedia on cosmonautics that placed him at the forefront of the popularization of rocketry in Russia.[5]

This Soviet aeronautical craze was certainly part of a pan-European phenomenon, as the reporting of aeronautical feats in Europe were popular news items and were anticipated well ahead of time. This fixation with air flight in both the European and Russian public media of the 1920s was similar to the way that U.S. and Soviet rocket flights were both elaborately portrayed by television reporters and eagerly anticipated by a viewing audience in the 1960s and 1970s. Western technological developments were practically revered in the Soviet newspapers of the 1920s, and thus readers were exposed to news on global developments in aeronautics and rocketry. America itself was portrayed as a symbol and emblem of how technology was transforming modern culture, and Soviet readers believed they were part of a cosmopolitan readership that could synthesize European, American, and Russian developments in rocketry and aeronautics in general.[6]

Though interest in spaceflight had predated the 1917 Russian revolution, certain groups in the Soviet 1920s (such as the Biocosmists) believed in the importance of spreading ideas on interplanetary travel for public consumption. The Biocosmists were interested in space travel as a means to achieve immortality, and they included amongst their group the renowned geochemist and science popularizer V. I. Vernadskii. This group also included, amongst their diverse members, the space visionary K. E Tsiolkovskii, a mathematics teacher from Kaluga, Russia. Besides Tsiokovskii, other followers of this group included influential Bolsheviks such as Leonid Krasin (the designer of the Lenin Mausoleum) and Valerian Muraviev (editor at the Central Institute of Labor in Moscow and a devout follower of Frederick Taylor). The Biocosmists could, to some extent, aptly be described as millenarians

4. N. A. Rynin, *Mechty, legendy, i pervye fantasii* (Leningrad, 1930).

5. N. A. Rynin, *Interplanetary Flight and Communication (A Multi-Volume Encyclopedia)* (Israeli Program of Scientific Translation, published for NASA, Jerusalem, 1970). For an overview of the life of N. A. Rynin, See Frank H. Winter, "Nikolai Alexeyevich Rynin (1877–1942), Soviet Astronautical Pioneer: An American Appreciation," in *Earth-Oriented Applied Space Technology*, 2, no. 1 (1982): pp. 69–80.

6. For a look at how America was portrayed in the Russian press and journals, see Jeffrey Brooks, "The Press and Its Message: Images of America in the 1920s and 1930s," in *Russia in the Era of NEP, Explorations in Soviet Society and Culture*, ed. Sheila Fitzpatrick et al. (Bloomington, IN: Indiana University Press, 1991); also Hans Rogger, "America in the Russian Mind," *Pacific Historical Review* 47 (February 1979): pp. 27–51.

and utopians, as they had a belief in the unbound ability of man to transform nature as well as to explore and colonize the cosmos.[7]

The Biocosmists were heavily influenced by the ideas and writings of the Russian pre-Revolutionary philosopher Nikolai Fedorov. Fedorov had worried that Earth was overcrowded and believed that humans could overcome this Malthusian pressure by exploring and colonizing space. Fedorov's vague notions of space travel as a way to achieve immortality for the human race was at the crux of his mystical utopian ideas and were very popular among Russian intellectuals.[8] One of Fedorov's most avid disciples was the space visionary Konstantin Tsiolkovskii. According to the science journalist Victor Shkolovskii, Fedorov had hoped Tsiolkovskii would popularize notions of space flight and rocketry amongst the Russian reading public.[9] In the Soviet period, the Biocosmists became devout followers of Fedorov, and they spread his (and Konstantin Tsiolkovskii's) ideas in the popular media for an eager readership willingly consuming articles on space travel.[10]

However, during the Soviet 1920s, professional science educators also served as popularizers of space flight and rocketry. Those Russian intellectuals, such as the Leningrad physics professor Ia. I. Perel'man, had more didactic purposes in mind. Perel'man, for instance, published many articles on rocket science and space travel in the several widely distributed popular journals he edited, such as *In Nature's Workshop*. These articles had an educational focus, attempting to explain the basics of gravitational forces and rudimentary rocketry to a popular audience.[11] Perel'man was particularly interested in spreading the ideas of the space visionary Konstantin Tsiolkovskii, and popularized Tsiolkovskii's theories on space flight in his widely read book entitled *Mezhplanetnoe puteshestvie* (*Interplanetary Travel*). Perel'man adamantly defended the notion of space flight against skeptics, explaining to readers how rockets could potentially overcome gravitational forces by projectiles traveling

7. For a good overview of the participants and focus of the Biocosmists and other utopian groups, see Richard Stites, *Revolutionary Dreams: Utopian Vision and Experimental Life in the Russian Revolution* (New York: Oxford University Press, 1989), pp. 168–170.

8. For an analysis of Fedorov and his school of mysticism, see Peter Wiles, "On Physical Immortality," *Survey*, nos. 56/57 (1965): pp. 132–134.

9. See Victor Shklovskii, "Kosmonavtika ot A do Ia," in *Literaturnaia gazeta* (7 April 1971): p. 13.

10. For an analysis of the philosophical roots of Russian cosmism, see Michael Hagemeister, "Russian Cosmism in the 1920s and Today," in *The Occult in Russian and Soviet Culture*, Bernice Glatzer Rosenthal, ed. (Ithaca, NY: Cornell University Press, 1997). Hagemeister argues that the city of Kaluga, Russia, where Tsiolkovskii lived most of his life, was a center for cosmism whose followers professed a belief in the omnipotence of science and technology. According to some Biocosmists, such as Tsiolkovskii, by traveling to outer space the human race could lose its corporeality and gain a type of immortality in infinite space and time. See A. L. Chizhevskii, "Stranitsy vospominanii o K. E. Tsiolkovskom," *Khimiia i zhizn'*, no. 1 (1977): pp. 23–32.

11. For an example of these types of articles, particularly those explaining the basis of rocketry and overcoming the Earth's gravitational forces, see Ia. I. Perel'man, "Za predely atmosfery," in *V masterskoi prirody,* nos. 5-6 (1919): pp. 32–33.

at high speeds with the use of liquid fuels (something Tsiolkovskii had dreamed of earlier).[12] Perel'man was also editor of the popular-science journal *Priroda i liudi* (*Nature and People*), which also carried articles on science and the cosmos. During the 1920s, Perel'man had served in the Soviet Commissariat of Enlightenment (Ministry of Education), where he worked on school curricula reform. There he made great strides in introducing the basics of physics, mathematics, and astronomy into secondary school curricula—a crucial building block for young students in understanding rocketry and space discovery.[13]

Though Perel'man fought hard to substantiate the importance of rocketry in the public mind, on some level the fascination with air flight had already forged an interested and impressionable public. State and privately commissioned (by each journal or newspaper) reader surveys in the 1920s offer historians detailed responses to reader interests. This survey data showed there was a genuine fascination with rocketry and that air flight and space exploration were extremely popular topics amongst readers. Interestingly enough, the surveys pointedly show how readers were actively exposed to news and information on air- and spaceflight from Western European and American sources. However, during the Stalinist 1930s and 1940s, this would soon change.[14]

By the mid-1930s, a cultural shift had occurred in Russia under Stalin, dubbed by the 1940s historian Nicholas Timasheff as "The Great Retreat." Timasheff, and some current cultural historians, have argued that during high Stalinism Russia embodied a retreat away from socialist cultural norms back toward greater Russian, more nationalistic themes.[15] It is within this context that the Soviet aeronautical feats during the 1930s were glorified and popularized through propagandistic means by the Soviet press.[16] During the earlier 1920s, international aeronautical feats (especially those in the West) were covered with the same frequency as equivalent Russian achievements. However, during the Stalinist 1930s and 1940s, prior to the Sputnik era, Russians began to witness a departure toward an increasingly nationalistic, triumphal manner.

12. See Ia. I. Perel'man, *Mezhplanetnoe puteshestvie* (Leningrad, 1923).

13. See editor's biographical entry in *V masterskoi prirody*, nos. 5–6 (1919).

14. For an overview of these sociological reader surveys, particularly focusing on reader questionnaires, see M. Rappeport, "Chto dala nasha anketa?" *Nauka i teknika*, 13 January 1926. For a look at the specific reader surveys of one popular scientific journal in the 1920s, see "Nasha anketa," *Iskra*, no. 6 (June 1927): pp. 38–39.

15. See Nicholas Timasheff, *The Great Retreat: The Growth and Decline of Communism in Russia* (New York: E. P. Dutton & Co., 1946). For a more current analysis of cultural practices during the Stalinist 1930s, see Sheila Fitzpatrick, *Everyday Stalinism: Ordinary Life in Extraordinary Times, Soviet Russia in the 1930s* (New York: Oxford University Press, 1999).

16. In the popular journals, the 1930s were characterized as years of "Stakhanovite Socialist Aviation." In the summer of 1936, Chkalov, Baidukov, and Beliakov made their historic, nonstop flight in a Soviet ANT-35. In 1936, Levanovskii and Levchenko flew from Los Angeles to Moscow, and Molokov flew along the arctic seaboard of the USSR. See L. Khvat, *Besprimernyi perelet* (Moscow, 1936). Also see "Po stalinskomu marshrutu," *Chto chitat'*, no. 2 (1936): pp. 45–47.

It is during this era that the visionary rocket and space theorist K. E. Tsiolkovskii was asked to give his catalytic speech on the future of human space travel on May Day, 1935, from Red Square. Though catalytic moments are, individually, critical junctures in history, Tsiolkovskii's speech must be contextualized within the greater Russian cultural nationalism propagated at the time by the Stalinist regime. Nonetheless, this was no ordinary speech; its repercussion was extraordinary amongst the public, politicians, and physicists alike. His taped speech was also broadcast by radio throughout the former Soviet Union, across 11 time zones, with an enormous social impact. Both Stalin, and later Khrushchev, would use the figure of Tsiolkovskii to focus on the superiority of Soviet technology over Western capitalism and its scientific system. However, both during this speech and at times prior to this event, Tsiolkovskii used these Soviet public venues to promote his own ideas about the future possibility of space flight. This speech was given while impressive Soviet airplanes flew above Red Square, and Tsiolkovskii described them as "steel dragonflies" which were only a tip of a more profound iceberg.[17]

Though events like this were certainly propagandistic public spectacles (see figure 3.1), scientists and future physicists alike were still very impressed with the secondary, depoliticized vision (or meaning) that Tsiolkovskii's ideas embodied. In his memoirs, the nuclear physicist and science advisor to Gorbachev, Roald Z. Sagdeev, acknowledged the duality embedded in these Soviet public spectacles. On one hand, he believed Stalin used Tsiolkovskii's 1935 broadcast from Red Square to further build the notion of the superiority of Soviet technology in the ensuing arms and space race. On the other hand, Tsiolkovskii's work became better known in the 1930s and 1940s, and many future space scientists read his popular work voraciously. Sagdeev argues that on 1 May 1935, enthusiastic Soviet citizens (including his parents, educated scientific academics) were enthralled by the speech.[18]

In a recollection related to Sagdeev's above, Valentin Glushko, designer of Energiya and many rocket engines that operated on Tsiolkovskii's dream of using liquid propellants, to some extent corroborates Sagdeev's perspective in his own memoirs. Glushko corresponded with Tsiolkovskii as a teenager and was inspired by his popular books in the 1920s and 1930s. Glushko believed that, mixed in with the Soviet propaganda and nationalist fervor propagated from above, was sheer enthusiasm and pride on the part of future scientists (and young space enthusiasts)

---

17. K. E. Tsiolkovskii, "Osyshchestvliaetsia mechta chelovechestva, Pervomaiskoe prevetstvie K. E. Tsiolkovskogo na plenke," a speech taped in his office/laboratory Kaluga, Russia in the last week of April 1935. The speech is transcribed in *K. E. Tsiolkovskii, Sbornik posviashchennyi pamiati znamenitogo deiatelia nauki* (Kaluga, 1935).

18. See Roald Z. Sagdeev, *The Making of a Soviet Scientist: My Adventures in Nuclear Fusion and Space from Stalin to Star Wars* (New York: John Wiley & Sons, Inc., 1994), pp. 4–6, 181–182.

from below.[19] This reflects somewhat on the popular surge of both interest in spaceflight (which continued in Stalin's time) and the symbiosis that coalesced this public interest with the nationalist drive from above. Many physicists as well as ordinary citizens made pilgrimages to Kaluga, Russia to see Tsiokovskii before his death in September of 1935. Tsiolkovskii's funeral in provincial Russia was almost a type of national, cathartic dirge and thus a reflection of the spontaneous interest in local space heroes.

This genuinely popular adulation for space heroes continued into the Khrushchev era as well. The eminent historian of Russian science, Loren R. Graham, reported in his recent memoirs that he had a similar impression on 12 April 1961, when he marched through Red Square at the celebration for the cosmonaut Yuri Gagarin sponsored by the Soviet leadership. Graham found this to be a mix of propagandistic spectacle from above and sincere, heartfelt public outpouring of support from below. As Graham looked back at that day and canonization, he also ruminated on the views of Soviet citizens and their pride in Gagarin:

> In later years when the Soviet Union became [a] decrepit and failing society, I often recall that day as the apogee in Soviet citizens' belief that they held the key to the future of civilization. The celebrations on the street were genuine and heartfelt. Soviet science was, they were sure, the best in the world, and Soviet rockets succeeded where American ones failed.[20]

Space Pervades the Soviet Consciousness: Sputnik, the Khrushchev Era, and the Public Sphere

During the era of the Second World War, and during Soviet reconstruction in the late 1940s and 1950s, Soviet aeronautical and cosmonautic feats were, to some extent, relegated to the periphery of the public landscape while the country was rehabilitated physically, politically and psychologically. But with the Khrushchev era and the dawn of Sputnik in 1957, the country witnessed a return to the nationalistic fervor of Soviet aeronautical and space development; again, as momentous as 1957 was, it built on the Stalin years but this time the regime orchestrated the public and social response more elaborately.

With the launching of Sputnik 1 in 1957, as part of the myriad of celebratory events, a host of journals filled pages with laudatory articles on Soviet rocketry, the history of spaceflight, and the life of the new cosmonaut. They included eclectic

19. See Valentin Glushko's reminiscences in his grandiose history of the Soviet space program, *The Soviet Encyclopedia of the Cosmos* (Moscow: Nauka, 1974).

20. See Loren R. Graham, *Moscow Stories* (Bloomington, IN: Indiana University Press, 2006), pp. 18–19.

journals such as *Ogonek* (*Little Flame*), literary journals such as *Literaturnaia gazeta* (*Literary Journal*), and more politicized, official ones such as *Kommunist* (*The Communist*) and *Partiinaia zhizn'* (*Party Life*). While most writers (and journalists) glorified Soviet achievements in space, there were the occasional letters to editors (which were actually published in newspapers such as *Komsomol'skaia Pravda*) that questioned the public support of the space effort, but they were generally anomalous to the norm.[21]

All the same, public debate on the efficacy of the space program did exist in the popular press under Khrushchev. Sometimes ordinary, concerned citizens wrote letters to editors of newspapers, such as *Komsomol'skaia Pravda*, that questioned why so much funding was shunted to the space program at a time when salaries for workers in factories were woefully low and consumer items were so scarce.[22] Other letters were queries regarding whether automatons could accomplish similar feats conducted by human cosmonauts in outer space. Many of these types of letters, in general, also questioned the safety of space travel in rockets for Soviet cosmonauts.[23]

With the above exceptions aside, however, public discourse on the space program was mostly constrained, and even limited to voices with large public reputations (such as major writers of literary significance). Some literary figures, such as Il'ia Ehrenburg, were concerned about how technology and the space race obscured the importance of other aspects of Soviet life on Earth, such as the development of literature and the arts, and questioned the substantial funds and government subsidies put into these technical arenas.[24] These critiques by literary figures as well as citizens may have been a repercussion or reflection of the Khrushchev "thaw"—the limited loosening of controls on artistic and public expression in the Soviet Union from 1953 until approximately 1962.[25] Furthermore, they may have reflected the need for a more outspoken segment of the cultural intelligentsia to remind the public of Russia's great artistic tradition (which should not be masked by its recent technological feats). All the same, these critiques, as well as ordinary citizens' letters mentioned above, were never outright diatribes against the regime's achievements in spaceflight, and much of the public discourse still remained, in a censored state, oriented toward glorifications of those achievements.

21. See Paul R. Josephson, "Rockets, Reactors, and Soviet Culture," in *Science and the Soviet Social Order*, ed. Loren R. Graham (Cambridge, MA: Harvard University Press, 1990), pp. 180–185.

22. For an example of this, see a worker's letter to *Komsomol'skaia Pravda* published under the name Aleksei N., "Ne rano li zaigryvat's lunoi," *Komsomol'skaia Pravda*, 11 June 1960: p. 1.

23. For an example of articles (as well as letters to editors) in the popular Soviet press and journals on the controversy of humans vs. automatons being sent into space, see B. Danilin, "Kto poletit v kosmos—chelovek ili avtomat?" *Molodaia gvardiia* 1 (1961): pp. 204–208.

24. See Il'ia Ehrenburg, "O lune, o zemle, o serdtse," *Literaturnaia gazeta* 1 (January 1960): pp. 3–4.

25. For an overview of the cultural and public/civic thaw under Nikita Khrushchev, see Priscilla Johnson, *Khrushchev and the Arts: The Politics of Soviet Culture* (Cambridge, MA: The MIT Press, 1964).

The historian Paul Josephson, in his analysis of the public ramifications of nuclear, atomic, and space science, argues that celebrations and mass rallies (particularly in Moscow) became an important site for the Soviet "masses" to become involved in the spectacle of display for Soviet "big science."[26] Planetariums hosted lectures on outer space; writers produced short stories with exaggerated platitudes for adults and children; and Soviet composers created popular songs (especially short *chastushki*) celebrating Sputnik to be sung to children at schools.[27] However, official institutions such as the Academy of Sciences became the greatest proponents and conduits for disseminating more detailed public lectures on the significance of these achievements. It was S. P. Korolev, the director of the post-WWII Soviet rocket program and, in actuality, the real father of the Russian space program, who was asked to direct these celebrations at the Academy; he was also asked to give the 1957 keynote commemorative speech for the capstone series of events planned in the era of Sputnik which honored Soviet space legends such as Konstantin Tsiolkovskii (the grandfather of the Russian space program—*Ded cosmosa*).[28]

What is interesting about the various speeches given by academics such as Korolev, however, is that although they were prescribed to mythologize great feats in Soviet rocketry (and help build a pantheon of iconic figures in Soviet space history), the actual speeches focused as much on small (yet significant) scientific contributions these figures made. For instance, Korolev's 1957 speech glorifying Tsiolkovskii certainly painted him within the Soviet paradigm of one of the "first" to conceive of rockets with liquid fuel. However, Korolev also spent as much time in his speech, if not more, discussing the more pertinent contribution of Tsiolkovskii's mathematical equations on the velocity of rockets leaving Earth's atmosphere.[29]

---

26. See Josephson's analysis of public display of big science in the former Soviet Union in his excellent chapter entitled "Rockets, Reactors, and Soviet Culture," in *Science and the Soviet Social Order,* Graham, ed.

27. See S. Ostrovskii, "Pesenka o sputnike," *Kul'turno-prosvetitel'naia rabota* 1 (1958): pp. 30–33. These children's *chastushki* were two- or four-line folk verses to be sung in an upbeat tempo with fervor. See G. Liando, "Nebesnye chastushki," *Kul'turno-prosvetitel'naia rabota* 1 (1958): p. 34.

28. In the 1940s during the war, but primarily after the war and into the 1950s, the Soviets make unsubstantiated claims of national priority in scientific discoveries. These claims ranged from the ludicrous assertion of the invention of the electric light, radio, and telegraph, to more specific scientific assertions of Soviets discoveries in a variety of disciplines, such as structural chemistry. Loren Graham believes most of these claims were abandoned later in the Brezhnev era in the 1960s and 1970s. However, he rightfully asserts that a few of those disciplinary claims (particularly revolving around certain scientific figures) should be investigated more seriously and need to be further analyzed in isolation of the general nationalistic assertions. See Loren R. Graham, *Science in Russia and the Soviet Union: A Short History* (New York: Cambridge University Press, 1993), pp. 142–143. These assertions are relevant to this public debate since the Soviets glorified their early theorists of cosmonautics, such as Tsiokovskii, claiming at times that they were the first to conceive of rocket flight.

29. See S. P. Korolyev, "On the Practical Significance of the Scientific and Engineering Propositions of Tsiolkovskii in Rocketry." Lecture given on 17 September 1957, based on the centennial celebrations of the birth of Tsiolkovskii held in Moscow, in *K. E. Tsiolkovskii, Izbrannyie trudy* (Moscow, Academy of Sciences Publishers, 1963), pp. 16–18.

Though nationalistic in orientation, these public speeches at the Academy sought the small kernel of scientific truth, so to speak, while downplaying the greater Soviet myth. Academician Boris Chertok, an engineer and the deputy director under Korolev, later described Korolev's speech on Tsiolkovskii as critical to the rocket community, if not overlooked at the time. Chertok, in his recently published memoirs, admitted that he and Korolev agreed that it was Tsiolkovskii's velocity equation that was his real, lasting legacy of scientific contribution to the future of rocketry.[30] Chertok admitted that the regime exaggerated these iconic figures and, at times, those such as Korolev tried to focus on real scientific contributions generally overshadowed by the regime. Ironically, it was Chertok himself who believed that myth and reality are nebulous concepts and those lines were sometimes blurred historically. In his memoirs, Chertok recanted a story about how mythic Tsiolkovskii actually was, despite his real scientific contributions to rocketry:

> Of the first missile decade, the last three years were certainly the most interesting in terms of science and engineering. The people who joined the missile programs during 1954–56 to a great extent determined the subsequent development of our cosmonautics program. While these people were still relatively young, someone's quip caught their fancy. According to our personal history forms, our personnel fall into one of two categories: they are either Tsiolkovskii's best students or individuals whose youth isn't their main shortcoming.[31]

Epilogue: The Mythology of Soviet Cosmonautics and its Social and Cultural Impacts and Ramifications

By the height of the Khrushchev era in the early 1960s, and after Yuri Gagarin's historic piloted circling of outer space, the Soviet paradigm as propagated in public went beyond national enthusiasm toward emphasizing how the regime made quantum leaps to outpace the West. In April of 1961, just after Yuri Gagarin became the first human being to rocket into space orbit, the Soviets held a gala diplomatic banquet in the Kremlin in his honor. At the event, the beaming Soviet premier, Nikita Khrushchev, embraced Gagarin and then made a toast. He said, "We used to go barefoot and without clothes and arrogant Western theoreticians predicted that bast-shoed Russians would never become a great power."[32] Furthermore, he said, "once-illiterate Russia, which many regarded as a barbaric country, had now

30. Boris Chertok, *Rockets and People*, vol. 1, NASA History Series, NASA SP-2005-4110 (Washington, DC, 2005), p. 3.

31. Boris Chertok, *Rockets and People, Creating a Rocket Industry*, vol. 2, NASA History Series, NASA SP-2006-4110 (Washington, DC, 2006), p. 168.

32. *Pravda*, 15 April 1961: p. 2.

pioneered the path into space."[33] This speech, published the next day in *Pravda* for all Soviet citizens to read, propagated a notion that the Soviets overcame great adversity to show the West how they could lead in the space race. Although this speech maintained the triumphal tone of the Stalin era, it went beyond that to emphasize the "Promethean" nature and "quantum leaps" of Russia's advances

Ironically, Khrushchev's boastful speech disregarded the real legacies his regime inherited. That is, Russia had a long history of not only rocket design and invention stretching back to the Tsarist era, but also an enthusiastic, engaged public that was fascinated with global discoveries in aeronautics and rocketry going back to the Tsarist era and the cosmopolitan 1920s. In fact, Russia had a tradition in the Tsarist era of public display of rocketry going back to the eighteenth century. The Romanov dynasty was especially well known for being fond of fireworks displays at public festivities in St. Petersburg, which may have been a catalyst for public interest in rocketry. This interest, as mentioned above, was fostered in the late Tsarist and early Soviet press and popular journals of the time, and well into the late 1920s.

Though this cosmopolitanism and enthusiasm may have changed in the Stalinist 1930s and 1940s as the regime propagated more nationalistic myths of Soviet scientific triumphs, a fascination with and national pride in space discovery and rocketry was still maintained in the minds of average Soviet citizens and physicists alike. By the time of the Khrushchev thaw, with its limited public debate, citizens may have questioned both the efficacy of the Soviet space program and its propagandistic celebrations, but they (like most citizens globally) maintained that fascination which stretched back to the eras of the Tsars and Lenin. Though we may not be able to document that "fascination from below" with the same set of sociological reader surveys and social-scientific data available to historians for the pre-WWII era, and though that imagination clearly was highly constrained and orchestrated in a censored state "from above," it existed (legacy intact) in memories and oral testimony all the same.

33. Ibid.

Figure 3.1—Photo of a 1933 public demonstration of the Stalin-era technical society, Aviation and Chemistry. Konstantin Tsiolkovskii (space visionary), second from the left, was more frequently asked to take part in these Stalin-era festivities glorifying the regime as the 1930s unfolded. These festivities were part of a larger movement to expand public spectacles, while focusing on the achievements of Soviet science and technology. Photo courtesy of Russian Academy of Sciences Archive in Moscow. (From ARAN, fond 555, op. 2, d. 149, l.3.)

Chapter 4

Live from the Moon: The Societal Impact of Apollo

Andrew Chaikin

In October 2006, newspapers and Web sites around the world carried the news that a 37-year-old mystery had apparently been solved: Computer analysis seemed to reveal that the word "a," long thought to be missing from Apollo 11 commander Neil Armstrong's first words after setting foot on the Moon, had been spoken after all.[1] Now there was no need for editors to insert parentheses in Armstrong's famous quote, "That's one small step for a man, one giant leap for mankind." Aside from the relief it gave grammarians—as Armstrong himself acknowledged, the sentence would not have made sense without the missing article—the story illustrated the enduring cultural impact of Project Apollo.

Our society's experiences of the first human voyages to another world have continued to evolve, even though the last of these pioneering journeys took place more than three decades ago, in December 1972. That is understandable, considering the fact that Apollo was among the most memorable events of the twentieth century. Many observers have identified the Moon missions as one of civilization's crowning achievements; Apollo has also been called the greatest technological feat of the last millennium. As one would expect for an event of such magnitude, the cultural impact of Apollo has been multifaceted. It was an event of international importance and yet it touched countless lives on an intensely personal level. Apollo was set in motion by geopolitical, cold war concerns that had little to do with exploration: President John F. Kennedy saw the lunar landing challenge as a way to best the Soviet Union and show the world the strength of a free society. But like all explorations, Apollo had some consequences that were largely unanticipated, including the profound effects of seeing Earth from lunar distance.

1. For example, "Armstrong 'got moon quote right'," BBC News, 2 October 2006, *http://news.bbc.co.uk/2/hi/americas/5398560.stm* (accessed 13 August 2007).

A LEAP IN PERSPECTIVE

When Americans awoke on Christmas morning of 1968, they were greeted by more than presents under the tree. Newspaper, radio, and television reports were filled with the momentous news that three astronauts—the crew of NASA's Apollo 8 mission—were on their way back to Earth after becoming the first human beings to orbit the Moon. For the U.S. space program, Apollo 8 represented a major step toward achieving the national goal, set by President Kennedy in 1961, of landing men on the Moon before the end of the 1960s. The news that three astronauts had flown around our nearest celestial neighbor sparked feelings of national pride and accomplishment, but it was also clear that other impacts were being felt from this extraordinary feat.

In particular, there was a new awareness of Earth and its place in universe. Through their spacecraft windows, the crew of the Moonward-bound Apollo 8 had seen their home planet shrink until it was small enough to cover with an outstretched thumb. But it wasn't just Earth's small size that impressed the astronauts; it was the fact that our planet was so clearly alive and, in that way, apparently alone in the universe. In contrast to the bleak and lifeless Moonscape, Earth represented, in the words of Apollo 8 command module pilot Jim Lovell, "a grand oasis in the big vastness of space."

The fact that millions of people heard Lovell's words live, during a television broadcast by the astronauts from lunar orbit, was one way in which Apollo was unlike any previous exploration in human history. No longer did the populace have to wait for news reports to trickle in from the frontier; now they were eyewitnesses to the event and its impact unfolded in real time. Only the astronauts could actually see what Earth looked like from 230,000 miles away, but anyone following the mission could share, in some measure, that unprecedented leap in perspective. Poet Archibald MacLeish expressed this in a reflection entitled "Riders on Earth Together, Brothers in Eternal Cold" that was printed on the front page of *The New York Times* on Christmas Day:

> *To see the earth as it truly is, small blue and beautiful in that eternal silence where it floats, is to see ourselves as riders on the earth together, brothers on that bright loveliness in the eternal cold—brothers who know now they are truly brothers.*[2]

This transcendent idea stood in stark contrast to the previous events of 1968. The nation was becoming increasingly divided over such issues as the escalation of the Vietnam War, racial tensions, and the troubles of the inner cities. In a year that saw more than its share of horrors, including the assassinations of Martin Luther King,

2. Archibald MacLeish, "A Reflection: Riders on Earth Together, Brothers in Eternal Cold," *The New York Times,* 25 December 1968: p. 1.

Jr. and Robert Kennedy, Apollo 8 provided an uplifting end. One of the countless telegrams received by the astronauts after their return said, "You saved 1968."[3]

Still, not all reactions to the Moon mission were positive. Atheist Madeline Murray O'Hare protested the fact that during their second lunar-orbit telecast on Christmas Eve the astronauts had read from Genesis; she later sued, unsuccessfully, to block any form of public prayer by astronauts during space missions. But for most Earthlings the impact of Apollo 8's new perspective on their home was lasting and positive. It was captured in stunning clarity in the astronauts' photographs, one of which—an image of Earthrise taken by Bill Anders—was later made into a U.S. postage stamp. Many observers have noted the timing of Apollo 8 with respect to the increase in environmental activism; futurist Stewart Brand maintains it is no coincidence that the first Earth Day, a nationwide observance of environmental issues, took place in April 1970, some 16 months after Americans first saw how their world looked from the Moon. It is worth noting that for the astronauts themselves the arresting sight of their world as a lovely and seemingly fragile "Christmas ornament" rising above the lunar horizon was the greatest surprise of the mission. Anders would later recall thinking, "We came all this way to explore the Moon, and the most important thing is that we discovered the Earth."[4]

SHIFTING PRIORITIES

As momentous as Apollo 8 was, its historical impact was equaled, even surpassed, by that of Apollo 11, the first landing of humans on another world. When Neil Armstrong and Buzz Aldrin took history's first Moonwalk on 20 July 1969, an estimated 600 million people—one-fifth of the world's population—witnessed it on live television and radio. It was difficult not to feel the enormity of the event, and some observers viewed it as a turning point in the course of civilization—especially science fiction writers, many of whom had envisioned the event in the decades before it happened. One was Robert Heinlein, who had penned the story for the 1950 film *Destination Moon*; on the day of the Moonwalk he appeared as a guest on CBS News' television coverage of the mission. "This is the greatest event in all the history of the human race up to this time," Heinlein said. "Today is New Year's Day of the Year One. If we don't change the calendar, historians will do so."[5]

And yet no one could ignore the fact that the first Moon landing, taking place at a time of continuing turmoil in the United States, was also evoking dissent. On

3. Frank Borman has mentioned this telegram in a number of interviews. See, for example, *American Experience*, PBS, *http://www.pbs.org/wgbh/amex/moon/peopleevents/e_1968.html* (accessed 13 August 2007).

4. Bill Anders interview with the author, July 1987.

5. *Man on the Moon* [DVD] (Apollo 11 chapter), Marathon Music and Video, distributed by EDI, Eugene, OR, 2003.

the day before the Apollo 11 launch, Ralph Abernathy, chairman of the Southern Christian Leadership Council, came to the Kennedy Space Center with a small group of protesters to draw attention to the plight of the nation's poor. And in New York City on the day of the landing, a member of Harlem's black community voiced the same concern to a network TV reporter:

> *The cash they wasted, as far as I'm concerned, in getting to the moon, could have been used to feed poor black people in Harlem, and all over this country. So, you know, never mind the moon; let's get some of that cash in Harlem.*[6]

A defense of the Apollo expenditures (the estimated total was $24 billion) came from Arthur C. Clarke, the writer and futurist who had collaborated with director Stanley Kubrick to create the screenplay for Kubrick's 1968 epic science fiction film *2001: A Space Odyssey*. In his comments, Clarke looked to Apollo's long-term benefits:

> *I think in the long run the money that's been put into the space program is one of the best investments this country has ever made . . . This is a downpayment on the future of mankind. It's as simple as that.*[7]

Another celebrated science fiction visionary, Ray Bradbury, was even more forceful. He encountered negative sentiments while appearing on a televised panel discussion on the Moonwalk in London. There he found himself confronted by criticism of Apollo from his fellow guests, who included Irish political activist Bernadette Devlin. Bradbury responded with his own big-picture perspective, as he later described it:

> *This is the result of six billion years of evolution. Tonight, we have given the lie to gravity. We have reached for the stars . . . And you refuse celebrate? To hell with you!*[8]

In general, reactions to Apollo 11 were divided between those who felt that reaching for the Moon was out of step with urgent needs on Earth, and those who insisted that Apollo's cost was outweighed by its long-term significance. How people gauged the importance of the first human footsteps on another world depended very much on who was being asked.

Still, it was clear that Apollo 11's cultural imprint was indelible. The quote that accompanied Neil Armstrong's first lunar footstep became instantly immortal and became the source of countless spinoffs. (Less than a month after the Moonwalk, one appeared at New York's Shea Stadium, where a teenaged fan cheered the Mets baseball team and their manager, Gil Hodges, on the way to their first winning season

---

6. Ibid.

7. *CBS News Transcript, 10:56:20 PM, EDT, 7/20/69: The Historic Conquest of the Moon as Reported to the American People by CBS News over the CBS Television Network* (New York: CBS News, 1970), p. 60.

8. Ray Bradbury speech, San Francisco Palace of Fine Arts, 10 July 1972 (recording provided to the author by space artist Don Davis).

with a homemade sign: "One Small Step for Hodges, One Giant Leap for Met-kind.")[9] Another way in which the success of Apollo 11 entered the culture was in a new phrase that entered the language: "If they can put a man on the Moon, why can't they [fill in the blank]?" The blank ranged from curing cancer to solving the problems of the inner cities.[10] This questioner was usually unaware of the fact that, unlike many of the difficult problems of the day, the Moon landing was a feat of engineering for which the enabling scientific discoveries had already been made. But it underscored the way in which the success of Apollo 11 had permanently altered the public's sense of what a group of human beings dedicated to a single goal was capable of accomplishing.

Acknowledgement of the Moon landing also showed up in popular songs written shortly afterward. Singer-songwriter Joni Mitchell wrote in her song, "Willy,"

He stood looking thru the lace at the face on the conquered Moon.[11]

Less enduring was a composition entitled "American Moon," which proclaimed,

Apollo Eleven delivered our heavenly right to say,
"The man in the moon is a citizen of the U.S.A."
Stand up and brag for your grand old flag
Waving on the moon tonight, oh yes,
Waving on the moon tonight.[12]

"American Moon" reflected the way in which Apollo 11 could be viewed through the lens of nationalism: By winning the space race with the Soviet Union, Apollo had given a boost to the nation's prestige in the world and, for many Americans, a heightened a sense of national pride. But seen through another lens, particularly that of the nation's disadvantaged, the view was starkly different. To black poet Gil Scott-Heron, Apollo was emblematic of the nation's racial inequalities. He expressed this in "Whitey on the Moon," which begins,

A rat done bit my sister Nell.
(with Whitey on the moon)
Her face and arms began to swell.
(and Whitey's on the moon)
I can't pay no doctor bill (but Whitey's on the moon)
Ten years from now I'll be payin' still.
(while Whitey's on the moon)[13]

9. Joseph Durso, "Mets Complete a Four-Game Sweep of Padres with Pair of 3-to-2 Victories," *The New York Times,* 18 August 1969: p. 41.

10. See, for example, "The Moon and Middle America," *Time,* 1 August 1969; Evan Jenkins, "For Ph.D.'s, No End to Lean Years," *The New York Times,* 8 January 1973: p. 72.

11. Joni Mitchell, "Willy," *Ladies of the Canyon,* 1970 (CD 1990, Reprise, Burbank, CA, 6376-2).

12. Lyrics are printed on the Web site of the Smithsonian's National Museum of American History at *http://americanhistory.si.edu/ssb/6_thestory/6d_views/main6d1b_ll.html* (accessed 13 August 2007). The author saw the song performed by Judy Carne on *The Ed Sullivan Show* on 26 October 1969.

13. *Now and Then: The Poems of Gil Scott-Heron* (Edinburgh, Scotland: Payback Press, 2000), p. 21.

APOLLO FADES FROM VIEW

For many people, the Moon program began and ended with Apollo 11; the night they saw two Americans leave their footprints on another world would prove to be their only vivid memory of Apollo. But in truth, Apollo was far from over. Six more landing attempts followed, all but one of which were successful, while the magnitude of Apollo's lunar explorations quickened at a truly extraordinary pace. In the summer of 1971, just two years after Armstrong and Aldrin explored a bland acre of Moonscape for two and a half hours, Apollo 15 astronauts Dave Scott and Jim Irwin were living on the Moon for three full days. Their Moonwalks lasted up to seven hours—a full working day on the surface of the Moon—and they drove a battery-powered rover for miles across the surface and even up the side of a lunar mountain, where they picked up rocks almost as old as the solar system itself. By that time, however, Apollo had largely faded from the nation's consciousness.

Public interest in the Moon program had begun to fall off after Apollo 11. In November 1969, Apollo 12 astronauts Pete Conrad and Alan Bean achieved history's second lunar landing and made two Moonwalks. Once again there were live pictures from the surface of the Moon, this time in color. But in news coverage there were expressions of apathy; one Tennessee resident was quoted as saying, "It's old hat, it's not like the first time."[14] In a sense, that reaction was predictable, given the fact that Apollo's stated objective of achieving a lunar landing before the end of the 1960s had already been accomplished. In a culture attuned to "firsts," even the second occurrence of something as extraordinary as landing men on the Moon could not generate the same level of excitement.

But there were other factors that exacerbated the decline in interest. One was the sheer strangeness of the events. Unlike science fiction writers (and their readers), most Americans had little familiarity with space technology and although TV commentators struggled mightily to convey the nuts and bolts of the Apollo program, arcane concepts like space rendezvous were, literally and figuratively, over viewers' heads. In addition, NASA (and for that matter, the astronauts themselves) tended to emphasize the technical elements of the program rather than the human experiences that would have been easier for the public to relate to. Then there was lunar science, which increasingly became the focus of both the astronauts and mission planners as the landings progressed. Talk of breccias and vesicles, of coarse-grained basalt and plagioclase feldspar was not easy for nonscientists to follow. The cultural divide between scientists and the rest of the populace was nothing new—it had been described a decade earlier by C. P. Snow in an essay entitled *The Two Cultures*—but Apollo seemed to throw that gap into vivid relief.[15]

---

14. Douglas Robinson, "Second Moon Visit Stirs Less Public Excitement," *The New York Times*, 20 November 1969: p. 30.

15. This effect of Apollo is described by Kerry Joels in "Apollo and the Two Cultures," in *Apollo: Ten Years Since Tranquillity Base* (Washington, DC: National Air and Space Museum, 1979).

Still another factor was the increasing dissent over the war in Vietnam. Among young people especially, one of the effects of the war was a mistrust of the government's use of technology; there was little distinction between the technology of warfare and the technology of space exploration. And though Apollo's successes had briefly raised the national mood, they could not compete with Americans' mounting preoccupation with the war. In April 1970, when an oxygen tank aboard Apollo 13 exploded 200,000 miles from Earth, the crisis sparked a resurgence of interest in the Moon program, as NASA struggled to save Jim Lovell and his crew from a lonely death in space. But even then it was clear that the war was still on many people's minds. During the crisis a student at Duke University was asked by a professor, "Do you think they'll get them back?" The student responded by talking about the American troops in Vietnam.[16] A couple of weeks after Apollo 13's safe return, four students were killed by National Guardsmen during an antiwar demonstration at Kent State University, intensifying the country's conflict over the war. Interest in Apollo never recovered to earlier levels.

This was despite the fact that TV pictures of the astronauts' activities on the Moon had greatly improved; the last three Apollo landings carried a higher-quality color TV camera which could be controlled from Earth. And the scenery visited by these teams was some of the most impressive landscapes on the Moon, with towering mountains, a winding canyon, and giant boulders. But even this could not reverse the dwindling tide of public attention. By the time of the final Apollo missions, television networks no longer covered the Moonwalks in their entirety.

THE ASTRONAUTS REVEALED

Apollo ended at a time when many Americans were turning inward. The 1970s saw increasing numbers of people engaged in a search for self-awareness and the realization of one's own potential. Although the quest for enlightenment often spilled over into self-indulgence, prompting writer Tom Wolfe to christen the 1970s "the Me decade," there was more interest than ever before in personal experience as a gateway to understanding. So in the mid-1970s, when several Apollo astronauts began to describe their experiences before, during, and after their lunar missions, they found an extremely receptive audience. People wanted to know what the astronauts thought and felt as they left Earth far behind, orbited the Moon, and walked on its alien surface. They wanted to hear how they had been affected by their incredible voyages. Underneath their curiosity, many harbored a hope that somehow these men had been transformed by their journeys.

And in a couple of instances, the experiences of a lunar astronaut fulfilled that wish. Apollo 15's Jim Irwin returned from the Moon and revealed that he had felt

16. This incident was described to the author by science journalist Mark Washburn in 1988.

the presence of God there. He soon left NASA to pursue a new life as a Baptist minister, working to share his faith. But no Apollo veteran's testimony matched the mood of the period better than Apollo 14's Ed Mitchell. Long interested in psychic phenomena, Mitchell had conducted an experiment in extrasensory perception during the trip to and from the Moon with a handful of volunteers on Earth. Made public only after he and his crewmates returned, Mitchell's experiment spawned a flurry of media reports. In the years that followed, Mitchell said he had experienced a shift in consciousness during the trip home from the Moon, giving him a profound awareness of the universe as a conscious, evolving entity. In 1972 Mitchell left NASA to pursue the scientific study of consciousness and psychic phenomena.[17] One of his first subjects was Uri Geller, whose claimed feats of telekinesis (such as spoon bending) made him an international celebrity in the early 1970s.

Needless to say, these kinds of experiences and activities did not fit with the public's image of the astronauts. During the Apollo program the media, especially *Life* magazine, had portrayed the astronauts as all-American heroes with rock-solid temperaments, heartland opinions, and unwavering morals. Their wives, meanwhile, were expected to maintain composure at all times, despite the stresses imposed on them by their husbands' dangerous profession and by the demands of sudden celebrity. As the 1960s progressed, however, the *Life* magazine image of the astronauts and their wives began to seem incongruous with emerging trends in popular culture. The Apollo missions took place at a time when the antihero was on the rise, as exemplified by such films as *Cool Hand Luke* (1967), *Bullitt* (1968), and *Midnight Cowboy* (1969). Against that backdrop, some saw the astronauts as hopelessly square, in a camp with what newly elected president Richard Nixon had called "the silent majority." Certainly, that did not make it any easier for those segments of the public to relate to the Apollo missions.

So it came as a bit of a shock—and, some observers said, a relief—when astronaut memoirs revealed very human traits.[18] For Apollo 11's Buzz Aldrin, it wasn't the Moon that changed him most dramatically, but what awaited him back on his home planet. In his 1974 autobiography *Return to Earth*, Aldrin candidly related his struggle with alcoholism and manic depression, which came in the wake of the intense public attention he received after Apollo 11.[19] If that was not in line with the public's perceptions of astronauts, then neither was the Apollo 15 astronauts' involvement in a plan to sell first-day covers carried on their Moon mission to a

17. Mitchell has described his experiences in his own book, *The Way of the Explorer* (New York: G. P. Putnam's Sons, 1996).

18. Howard Muson, "Comedown from the Moon—What Has Happened to the Astronauts," *The New York Times*, 3 December 1972, Sunday Magazine: p. 37.

19. Edwin E. Aldrin, Jr. with Wayne Warga, *Return to Earth* (New York: Random House, 1973).

German stamp dealer, as described in Jim Irwin's book *To Rule the Night*.[20] In 1977 even more damage was done to the astronauts' image by *The All-American Boys,* the memoir of Walt Cunningham, who had been a crewman on Earth-orbit Apollo 7 mission in 1968; it described, among other things, his colleagues' extramarital affairs.[21] And in his 1979 landmark portrait of the early astronauts, *The Right Stuff,* Tom Wolfe showed the humanity not only of the men but also their wives, who had secretly harbored fears and anxieties they never shared with the public—or each other.

For all the upheavals surrounding their image, the astronauts themselves remained relatively anonymous despite their status as the only humans in history to have visited another world. In 1975 a new television ad campaign by American Express featured celebrities whose faces were largely unknown to the public; one was Apollo 12 commander Pete Conrad. Looking into the camera, Conrad asked, "Do you know me? I walked on the Moon." And Conrad was not the only one who went through his post-Apollo life so unrecognized.

Nevertheless, even as the astronauts themselves receded from the public's consciousness, new reminders of their journeys became firmly embedded in pop culture. When a new network called Music Television (MTV) debuted in the summer of 1981, its trailers featured pictures and film clips of astronauts on the Moon. For MTV's target audience, some of whom had been small children at the time of Apollo 11, these images must have seemed like just another slice of 1960s nostalgia.

History Doesn't Repeat

By the time Moonwalkers were bouncing across MTV trailers, the Moon itself was nowhere to be seen in NASA's activities. In April 1981, after a six-year hiatus in American human spaceflight missions, NASA achieved the first flight of its reusable Space Shuttle. Buoyed by NASA's promise that the Shuttle would make spaceflight routine, many people responded with high enthusiasm. And in the first several years of Shuttle missions there was plenty of action to excite space buffs; they could witness spacewalking Shuttle astronauts repairing satellites and flying through the void with self-propelled maneuvering units. (Moonwalking, on the other hand, was now coming to be known as a dance move performed by Michael Jackson.)

The fact that the Shuttle never ventured beyond low-Earth orbit was lost on some Americans, for whom going to the Moon had become synonymous with the idea of spaceflight. No longer exposed to the intricacies of Apollo, people underestimated the difficulty of lunar voyages and did not realize that no one had been to the Moon since

20. James B. Irwin with William A. Emerson, Jr., *To Rule the Night: The Discovery Voyage of Astronaut Jim Irwin* (Boston: G. K. Hall, 1974; Philadelphia: J. B. Lippincott Co., 1973).

21. Walter Cunningham with Mickey Herskowitz, *The All-American Boys* (New York: Macmillan Co., 1977).

1972. This was evident from a comment by Apollo 17 commander Gene Cernan, who said in late 1986, "People don't have any concept. They think all astronauts do the same thing, go to the Moon. [They say,] 'Oh, doesn't the shuttle go to the Moon?'" [22] The reality was that not only were humans no longer going to the Moon, but NASA had no plans to send them there.

And by late 1986 the dream of using the Shuttle to make spaceflight routine had been shattered by the explosion of the Shuttle *Challenger* in January of that year. In the wake of the disaster, there was renewed focus on NASA's uncertain future; even before the accident, a presidential commission had been convened to study possible long-range plans for the Agency to pursue. Proposed scenarios included a return of humans to the Moon, followed by piloted expeditions to Mars. However much enthusiasm there might have been for this prescription—and there was a great deal among space advocates—there were uncertainties about how to enact it. Was it simply a matter of convincing the president to call for such a program, as John Kennedy had in 1961? Many space advocates thought so, and believed history would repeat itself.

They got a chance to test their belief on 20 July 1989, the 20th anniversary of the Apollo 11 lunar landing, when President George H. W. Bush declared that America was going back to the Moon, "this time to stay," and after that to Mars. The so-called Space Exploration Initiative (SEI) called for a 30-year effort. But because of the plan's projected cost, public reaction was decidedly mixed and there was strong Congressional resistance. Even NASA itself failed to embrace the plan. SEI never got off the ground.

More than a decade later, Bush still remembered in detail the stinging defeat of his space initiative, especially the expectation he'd been given by space advocates that he could launch a major space program just by following Kennedy's example. Summing up the experience, he told NASA Administrator Sean O'Keefe in 2003, "I was set up. We were all set up."[23]

As it turned out, the Apollo model for how to launch an ambitious space program was not valid. And by 1991, with the end of the cold war, the old forces that had given rise to Apollo were no longer in the equation. In the wake of SEI's failure there was a growing awareness of something sharp-eyed observers had noted long before: Apollo had been a historical anomaly.

APOLLO AS A MULTIGENERATIONAL EXPERIENCE

By the 1990s the generation that had been children during the Apollo missions had grown up. It included countless scientists and engineers, and even a number of astronauts and flight controllers, who had been inspired to pursue their careers by the

22. Author interview with Gene Cernan, December 1986.

23. Sean O'Keefe, personal communication, December 2003.

Moon missions. It also included storytellers who were moved to revisit the Apollo saga. One was film director Ron Howard, who brought Apollo's most dramatic mission, 1970's harrowing Apollo 13 flight, to the screen in 1995. With a cast headed by Oscar-winner Tom Hanks as Jim Lovell, the film depicted the struggle, in space and on the ground, to rescue Lovell and his crew. One of the year's top-grossing films, *Apollo 13* was more than a retelling of events; it was a celebration of the courage and ingenuity that characterized the entire Apollo program. The astronauts were not the only heroes featured; the film also spotlighted the mission controllers in Houston in all their engineering-nerd glory. The film even spawned a couple of new catchphrases: "Houston, we have a problem"[24]—the first announcement from Hanks' Jim Lovell that Apollo 13 was in trouble—and the rallying cry spoken by Ed Harris as Flight Director Gene Kranz, "Failure is not an option."

Apollo 13 brought the drama of the Moon program to a new generation of young people and reminded adults of what they had lived through but might not have fully absorbed. Writing about the film in *The New York Times*, science writer John Noble Wilford, who had covered the Apollo missions, saw it as a reminder of a particular spirit of exploration. "One can imagine that the story of Apollo 13, perhaps now or in other retellings by generations to come," wrote Wilford, "will evoke a time when people took risks to reach grand goals, a time when the astronauts were themselves lionized and we still embraced heroes."[25]

The resurgence of interest in Apollo was furthered when Hanks went on to produce a 12-part miniseries for HBO on Apollo, *From the Earth to the Moon* (based in part on this writer's book about the Apollo astronauts and their missions, *A Man on the Moon*). In a sense, the various retellings of the Apollo saga, so many years after the events, filled an important cultural gap. With all the distractions now long gone—the political discord, the distrust of technology, the antihero culture—the public was ready, at last, to celebrate the Moon program with new appreciation and understanding of what it had accomplished.

The Moon Hoax

Even decades after it happened, some were unwilling to celebrate Apollo, especially the people who believed it had never happened. This was not a new phenomenon; even at the time of the first Moon landings there were some who insisted Apollo was a government hoax and that the Moonwalks had been filmed somewhere in

24. The film's dialogue was a slight rewording of the actual transmission from Apollo 13, "Houston, we've had a problem."

25. John Noble Wilford, "When We Were Racing With the Moon," *The New York Times*, 25 June 1995: p. 2.

the Nevada desert.[26] (A group called the International Flat Earth Society, meanwhile, refused to be swayed by the astronauts' own reports and photos showing that the planet is, in fact, a sphere.) In 1978, the film *Capricorn One* portrayed a faked Mars mission; although the director, Peter Hayms, did not believe Apollo had been faked, he was nevertheless fascinated by the notion that such a hoax was possible. Interestingly, he had written the script in 1972 but met strong resistance to the idea in Hollywood. By the late 1970s, when he sold the film, that resistance was gone—in part, because of a new level of distrust of government in the wake of the Watergate scandal that had made the idea of a faked space program more acceptable to studios.[27]

Capricorn One was well received by audiences, but the idea that Apollo itself had been a hoax was never embraced by a large percentage of Americans. A 1999 Gallup poll revealed that "[T]he overwhelming majority of Americans (89%) do not believe the U.S. government staged or faked the Apollo Moon landing. Only 6% of the public believes the landing was faked and another 5% have no opinion."[28] Still, the hoax theory continued to have a presence in the culture, as evidenced in February 2001 when the Fox TV network aired a program called "Conspiracy Theory: Did We Land on the Moon?"[29] The producers used faux-scientific analyses of the astronauts' photographs to "disprove" their validity; in truth, the show's popularity revealed, above all, a lack of scientific literacy among its followers.

Despite its obvious flaws, the hoax theory has persisted largely because the Apollo missions were so difficult for most people to relate to. In 1969 the writer Norman Mailer, commissioned by *Life* magazine to cover Apollo 11, had observed after watching the Moonwalk, "The event was so removed, however, so unreal, that no objective correlative existed to prove it had not conceivably been an event staged in a television studio—the greatest con of the century . . . ." In the same breath, however, Mailer acknowledged the impossibility of carrying out such a hoax. "It would take criminals and confidence men mightier, more trustworthy and more resourceful than anything in this century or the ones before. Merely to conceive of such men was the surest way to know the event was not staged."[30] Years later, Neil Armstrong put it more simply: "It would have been harder to fake it than to do it."[31]

26. John Noble Wilford, "A Moon Landing? What Moon Landing?" *The New York Times*, 18 December 1969: p. 30.

27. Benedict Nightingale, "What If a Mars Landing Were Faked? Asks Peter Hyams," *The New York Times*, 28 May 1978: p. D10.

28. *http://www.galluppoll.com/content/?ci=1993&pg=1* (accessed 13 August 2007).

29. For an in-depth discussion of the Fox broadcast and the Moon hoax theories, see *http://www.badastronomy.com/bad/tv/foxapollo.html* (accessed 13 August 2007).

30. Norman Mailer, *Of a Fire on the Moon* (Boston: Little, Brown, 1970), p. 130.

31. Neil Armstrong, personal communication, 2003.

CONCLUSION: AHEAD OF ITS TIME

Today, more than three decades after the program ended, Apollo remains a unique event in the history of space exploration. There is no shortage of reminders of the Moon voyages, including DVDs of the Moonwalks, memorabilia, and other Apollo-related products. Actual Apollo hardware and lunar samples are on display at museums around the world.[32] But the reality of humans walking on the Moon has receded into our past. There is something strangely out of place about an event so futuristic that happened so long ago. Gene Cernan described this feeling in his 1999 autobiography, *The Last Man on the Moon*, when he wrote, "Sometimes it seems that Apollo came before its time. President Kennedy reached far into the twenty-first century, grabbed a decade of time and slipped it neatly into the 1960s and 1970s."[33]

If the first human voyages to the Moon had taken place the way science fiction writers and space visionaries had predicted—after a step-by-step progression from the first satellites to the first human spaceflights, then the establishment of a space infrastructure including reusable space shuttles and permanent space stations in Earth orbit—they would not have seemed so unreal. The public would have had decades to get used to the reality of spaceflight, and space technology would have become a familiar part of the culture. As it actually happened, however, the populace was relatively unprepared for what took place from December 1968 to December 1972, when humans journeyed from their home planet to another celestial body.

In January 2004, President George W. Bush announced the Vision for Space Exploration, including return to the Moon. Unlike his father's ill-fated Space Exploration Initiative, the new program did not rely on substantial increases in the NASA budget. And it came at a time when public support for the space program was high.[34] If all goes according to plan, astronauts will be back on the Moon no later than 2020 and our culture will once again be faced with absorbing the reality of humans walking on another world.

Will Apollo turn out to be the momentous punctuation mark in human history that Heinlein, Clarke, and Bradbury predicted at the time of Apollo 11? It seems inescapable that it will, because no matter how far humans are able to go in their quest to explore the universe, the Apollo missions will stand as the opening chapter. Future generations will no doubt judge the program's significance not only by what Apollo achieved, but what it led to. If, as space visionaries have long maintained, human

32. For a listing of museum displays of Apollo command modules, see "Location of Apollo Command Modules," National Air and Space Museum Web site, *http://www.nasm.si.edu/collections/imagery/apollo/spacecraftcm.htm* (accessed 13 August 2007).

33. Gene Cernan and Don Davis, *The Last Man on the Moon: Astronaut Eugene Cernan and America's Race in Space* (New York: St. Martin's Press, 1999), p. 344.

34. Leonard David, "NSB Report Finds Steady Public Support For NASA," 25 May 2004, *http://www.space.com/spacenews/archive04/nsbarch_052504.html* (accessed 13 August 2007).

expansion into the solar system to become a multiplanet species is inevitable, then Apollo will surely be seen as the first "giant leap" of that journey.

In our time, however, Apollo's greatest impact has yet to be completely incorporated into the culture. Sending the first explorers to the Moon showed us that humans have the ability to accomplish seemingly impossible things when they work together. And the testimony and photographs of the men who made those voyages revealed Earth as a precious oasis of life in a vast and hostile universe, a world to be cherished and protected. Absorbing these lessons is often at odds with the short-term focus of our day-to-day culture. But Apollo's impact will always be there to be revisited and re-experienced, and to guide us in charting our long-term future.

Author's acknowledgment: I would like to thank my wife, Vicki, fellow writer and Apollo devotee, for providing many helpful comments during the writing of this chapter.

Chapter 5

Framing the Meanings of Spaceflight in the Shuttle Era

Valerie Neal

Among public policy analysts and pundits, it is conventional wisdom (and has been almost since the Space Shuttle appeared) that the United States lacks a unifying societal consensus about the fundamental purpose or goal of contemporary human spaceflight. Thirty years and more than 115 missions after the first Shuttle orbiter *Enterprise* made its debut in 1976, the debate continues along much the same lines as it began: What purpose justifies the cost and risk of placing people in space? In what intellectual framework does this enterprise make civic sense? Both the proponents and opponents of human spaceflight have struggled to express a credible, broadly persuasive rationale that appeals to or reflects supporting societal values.

Advocates in the Mercury-Gemini-Apollo era of the 1960s were not so tasked. The politics of the time and presidential leadership gave rise to two readily intelligible frames of reference for nascent human spaceflight: a competitive space race with the Soviet Union, and a pioneering venture into a new frontier. Both resonated with the American public's hopes, fears, and values. NASA did not need to craft a compelling rationale for sending people into space; politicians and the media purveyed these messages.

Steeped in cold war anxieties about a possibly mortal adversary, citizens could understand the importance of an all-out thrust into space, especially after the Soviets made the first forays there. There was little disputing whether it was worth the cost and risk; the affirmative response accorded with a people accustomed to victory and anxious about the bomb. A *Time* magazine cover in December, 1968, with an astronaut and cosmonaut sprinting toward the Moon, captured the patriotic urgency of this race against time, perceived as a race for survival against communism.

Likewise, the effort to reach the Moon resonated with a widely held view of America as a pioneering nation with a frontier heritage. President Kennedy and his speechwriters masterfully worked with this deeply ingrained sense of national identity as a metaphor for exploring the new ocean of space. Racing and pioneering merged in triumphant images of the U.S. flag and astronauts on the dim landscape of another world.

But what vision came after? Without a race to win or a frontier to conquer, continued human spaceflight demanded a new purpose that made sense as a national endeavor. How would NASA make the case and what role would the media play in defining its purpose? How would society find meaning in continued spaceflight? Could human spaceflight fit into other frames?

Over the past five decades NASA, the media, and interested sectors (aerospace industry, scientific community, political figures, grass-roots groups, and others) plus thoughtful individuals have engaged in an ongoing process of asserting and contesting the value of human spaceflight by advancing a variety of visions or metaphors meant to answer such questions and sway public opinion. The continual effort to define the purpose of human spaceflight and reach a societal consensus on its value can be viewed as an extended exercise in the social construction of meaning. In the Shuttle era, at least five reference frames have been crafted, promoted, critiqued, refined, accepted, rejected, or transformed in the process of shaping and communicating the meaning of human spaceflight. These frames reveal much about what Americans hope for—and doubt—in our national ventures into space.

FRAME ANALYSIS AS AN INTERPRETIVE TOOL

To pursue these questions about the meaning of Shuttle-era human spaceflight, it is helpful to apply some concepts, terms, and techniques from the literature of "frame analysis" that has become prominent in social science disciplines, especially in media studies and the study of social movements.[1] In this context human spaceflight can be considered a social movement that has an action agenda, an imperative to muster resources, and a need to mobilize public support in order to carry out its agenda. NASA is the hub of this social movement, with aerospace companies, space societies, other government entities, and auxiliaries in the advocacy community, including some in the media.

To analyze how social movements motivate public support, some scholars focus on framing processes, and they use the term "framing" for the "construction of meaning." Framing is the packaging of messages that resonate with core values and appeal to supporters. A "collective action frame" is a construct of ideas and

1. Erving Goffman, *Frame Analysis* (New York: Harper & Row, 1974); William A. Gamson and Andre Modigliani, "Media Discourse and Public Opinion on Nuclear Power," *American Journal of Sociology* 95 (1989): pp. 1–37; William A. Gamson, David Croteau, William Hoynes, and Theodore Sasson, "Media Images and the Social Construction of Reality," *Annual Review of Sociology* 18 (1992): pp. 373–393; Zhongdang Pan and Gerald M. Kosicki, "Framing Analysis: An Approach to News Discourse," *Political Communication* 10 (1993): pp. 55–73; Robert D. Benford and David A. Snow, "Framing Processes and Social Movements: An Overview and Assessment," *Annual Review of Sociology* 26 (2000): pp. 611–639.

meanings based on shared beliefs and values that will motivate support.[2] It is the conceptual analogy to a structural framework or a picture frame. The space race and the space frontier are such conceptual frames.

Frames are "the basic frameworks of understanding available in our society for making sense out of events"; they help to render events meaningful, organize experience, guide action, and simplify and condense aspects of the world.[3] They are intended to motivate support and disarm opposition, to inspire adherents, and to legitimize the activities and campaigns of a social movement. Frames provide context for a proposed action or policy. Opponents may contest or challenge them with counter-frames.[4]

The mobilizing potency of a frame lies in its credibility and resonance. It must be consistent with the facts and goals of the movement, and it must resonate with the beliefs, values, and interests of the targeted support community or constituents. Even more broadly, it should have "narrative fidelity" or coherence with cultural assumptions and myths in the public domain. Activists use cultural resources—beliefs, values, myths—as a "tool kit" to make their cause appealing and believable, and audiences also use them to gauge resonance.[5]

Because framing is an intentional process, frames need not be static. They can evolve as circumstances change, either to account for unexpected events or to better appeal to the target community. To mobilize support, a frame may need to be fairly elastic.[6]

Social movement activists are not the only ones developing frames of meaning. Media discourse also participates in the process of constructing meaning. Analysis of media discourse relative to a variety of social movements (e.g., the women's movement, nuclear power, civil rights) reveals sophisticated frames or "interpretive packages" that are promulgated to make sense of issues and events. Like frames, interpretive packages have a central organizing idea, often presented in shorthand through symbols, metaphors, visual images, and icons. The media provide both an accessible forum for public consideration of issues and for suggested interpretations that help to shape the social construction of meaning.[7]

---

2. Benford and Snow, "Framing Processes," cited above, explicate these and other key concepts and vocabulary in frame analysis scholarship.

3. Goffman, *Frame Analysis*, 10; Benford and Snow, "Framing Processes," pp. 613–614.

4. Benford and Snow, "Framing Processes," discuss frame disputes, contested framing processes, and counter-framing, pp. 625–627.

5. Benford and Snow, "Framing Processes," pp. 619–622, 629; Gamson and Modigliani, "Media Discourse," pp. 1–10.

6. Benford and Snow, "Framing Processes," discuss framing processes and dynamics, pp. 622–632; Scott Davies, "The Paradox of Progressive Education: A Frame Analysis," *Sociology of Education* 75, no. 4 (October 2002): pp. 269–286, esp. 270–273, refers to "two faces" (stable and changeable) of frames.

7. Gamson and Modigliani, "Media Discourse"; Gamson et al., "Media Images"; Nadya Terkildsen and Frauke Schnell, "How Media Frames Move Public Opinion: An Analysis of the Women's Movement," *Political Research Quarterly* 50, no. 4 (December 1977): pp. 879–900; Frank D. Durham, "News Frames as Social Narratives: TWA Flight 800," *Journal of Communication* 48, no. 4 (Autumn 1998): pp. 100–117.

This paper applies frame analysis concepts to human spaceflight during the three-plus decades of the Shuttle era. Primary sources for this analysis are selected elements of societal discourse that helped shape or curb public expectations of contemporary spaceflight—in this study, NASA's publicity materials, *The New York Times* (news, editorials, and opinion pieces), and editorial cartoons from a variety of papers. *The New York Times* was selected for its breadth of coverage of Shuttle missions and spaceflight, its often critical editorial stance, and the long tenure of reporter-analyst John Noble Wilford, who often wrestled with the meaning of human spaceflight. Other newspapers, magazines, and electronic media that could be fruitfully explored are not included in this brief study; likewise, speeches, transcripts of Congressional hearings, and other official documents might be examined for a broader study. Among the techniques of frame analysis is close textual study with attention to keywords and themes, a rhetorical approach that is suitable for the sources examined.

FRAMING HUMAN SPACEFLIGHT: A NEW ERA IN SPACE TRANSPORTATION

With the Space Shuttle, NASA introduced a new frame of reference to justify human spaceflight and capture popular interest and political support. It was "A New Era" in space transportation, setting human spaceflight into a long tradition of optimistic, progressive, utopian visions of a brighter future. The cultural context for a new age or new era extends to the origins of America as a new world, a key concept of national identity. The frame of newness also harkened to a history of American innovation in transportation; automobiles and aircraft had already brought about new eras in travel, with widespread social impact. Placing human spaceflight and the Shuttle into this frame—radically different from the pioneering race of the 1960s—gave it a familiar appeal.

NASA promoted this theme through varied channels, including informative, colorful public affairs brochures disseminated to the media and elsewhere. As soon as the decision to develop the Space Shuttle was made in 1972, NASA began to frame the new era for the public. Artist Robert McCall was commissioned to paint scenes of typical Shuttle missions for a brochure that literally framed new ways of doing things in space.[8] A 1977 pamphlet titled *The Shuttle Era* claimed, "Now a new era nears . . . the coming of age in space" when people will be able to do important work there in ways never before possible.[9] At about the same time, the Shuttle contractor Rockwell International began to release public relations materials to promote "A Promising New Era."[10]

8. *Space Shuttle* (Washington, DC: National Aeronautics and Space Administration Educational Publication EP-96, June 1972).

9. *The Shuttle Era*, Space Shuttle Fact Sheet NASA-S-76-815A (Washington, DC: National Aeronautics and Space Administration, March 1977).

10. Rockwell International Space Division, *Space Shuttle Transportation System: A Promising New Era for Earth*, September 1976, and *Space Shuttle: A Promising New Era for Earth*, January 1977.

Crews from several missions in the 1980s relished their role in delivering and repairing satellites, adopting such business-like monikers as Ace Satellite Repair Co. and Ace Moving Co., with "We pick up and deliver" and "The sky's no limit" as mottos.

Routine space transportation was the central tenet of the new era. In this frame, spaceflight would no longer be a pioneering adventure; it would become commonplace and practical, in Earth orbit, not outward-bound. In a burst of metaphors, NASA claimed that people would travel a highway to space in a workhorse shuttle vehicle that would operate like an airliner. That mixed image might have been a clue that the new-era routine transportation frame was strained.

NASA further elaborated the concept of routine access to space with purposes that could appeal to special interests and make sense to the public at large. Commercial enterprise could use the shuttle to cash in on space by launching satellites or developing manufacturing capabilities there. Knowledge would increase as observatories were placed in orbit or scientists conducted laboratory science in space. National security would be enhanced by regular delivery of defense department payloads. All these activities on the Shuttle would lead to practical benefits on Earth. NASA thus plugged into the frame something to appeal to each necessary constituency—business, science, and military—and purposes that moreover would resonate with the public.

With promised economic, scientific, and security benefits, citizens could understand a practical approach to human spaceflight. Add to that the typical American consumer's desire for the latest-model vehicle or the newest technology, as well as Americans' regard for the nation's transportation systems, and the new era of routine space transportation was a potent frame for human spaceflight on the Space Shuttle. In this context, the purpose of human spaceflight was not exploration; it was useful work. The Shuttle served as icon for this whole frame of meaning. To see the stubby-winged shape of the orbiter or the whole launch configuration with boosters and fuel tank was to recognize the new era—and new meaning—of human spaceflight. Humans were curiously absent from these early depictions; the Shuttle vehicle, often called a spaceplane or a space truck, symbolized the practical new purpose of people in space.

The New York Times director of science news John Noble Wilford was among the first journalists to introduce the Shuttle-era frame of reference to the public. His 1977 feature article, "Another Small Step for Man: Shuttling into Space," laid a bridge from the past to the future as the first Shuttle *Enterprise* engaged in atmospheric flight tests. Echoing Neil Armstrong's famous words on the Moon, Wilford placed the Shuttle on the next rung of the ladder to humanity's destiny in space and recognized it as a revolution in space travel. He foresaw that the "era of the spaceplane" meant hauling orbital freight on regular flights and handling satellites by the three Rs—release, retrieve, repair. The Shuttle would not be used for exploration. But, because it would offer the ability to do new things in space, the Shuttle might have a far-reaching impact, as did the automobile and airplane.[11] At the end of the 1970s decade (just a bit prematurely), Wilford announced a variant of the new era concept: the "Commuting Age Dawns in Space."[12]

When the new era truly dawned in 1981 as Space Shuttle *Columbia* roared into orbit, the new frame of reference crafted by NASA and presented in the media was in place. There might have been a different meaning construction—perhaps a mythic journey or another metaphor—but none other was offered. Already there were skeptics and critics, but the news media in unison trumpeted a new era of routine transportation to space.

Framing Human Spaceflight: A Business

A corollary to the new era of routine space transportation also was promoted: spaceflight as a business. NASA claimed that the reusable Shuttle would lower the cost of spaceflight and make transportation to and from Earth orbit economical. The foundation for Shuttle-era spaceflight would be a business model inspired by the commercial airline industry. NASA managers studied airline operations and sought to drum up the customer market, contracted with payload owners for orbital flights, plotted the mission manifests, and calculated the operating margins to turn spaceflight into, if not a flourishing, at least a break-even business. With a sufficient number of vehicles and frequency of flights, the Shuttle might bring down the cost of spaceflight and pay for itself.

This business-model frame served to defend the Shuttle against critics who argued that the program was unnecessary and too expensive, and it dovetailed well with the

11. John Noble Wilford, "Another Small Step for Man: Shuttling into Space," *The New York Times*, 7 August 1977, Sunday Magazine: pp. 7, 28, 54 ff.

12. John Noble Wilford, "Commuting Age Dawns in Space," *The New York Times*, 30 December 1979.

concept of routine transportation for useful work in space. Transportation businesses on Earth—interstate trucking, railroads, shipping, as well as airlines—were familiar analogues to give meaning to a space transportation enterprise. This blend of concepts exemplified frame enhancement or frame elaboration, a strategy for broadening the appeal of a social action agenda, often by appropriating some elements of an adversary's position. Human spaceflight in this frame did not mean adventure and exploration; it meant efficiently running a business for practical benefits if not profits.

The business-model frame proved vulnerable to critique by standard business accounting and auditing principles; it invited measurement of costs and gains. NASA had provided the quantifiable metrics for judging the performance of human spaceflight: flight rates and flight costs. As the Shuttle became operational in the 1980s, it was not difficult for stakeholders in the business to do cost-benefit audits and assess the return on investment in human spaceflight. The value of work performed by the astronauts was more difficult to measure quantitatively, so the cost of operating the Shuttle served as the primary measure for judging the value of human spaceflight. Thus, the business frame that was meant to promote also became a frame for critiquing spaceflight.

REALITY CHECK: THE EARLY SHUTTLE ERA IN PRACTICE

A brief survey of reporting and editorializing about spaceflight during the first five years of the Shuttle era shows how these two theoretical frames of meaning fared in practice. Reactions to the first 23 Shuttle missions (1981–1985) in *The New York Times* served as "reality checks" for assessing actual spaceflight in the new era within the routine transportation and business frames. Greeting the Shuttle as a bold new approach to human spaceflight and the first mission as a triumphant return to space, the paper proclaimed "Columbia . . . Opening a New Era of Space Flight."[13] Yet chief Shuttle observer John Noble Wilford cautioned from the outset that the future was by no means certain; it might prove difficult to fulfill the optimistic predictions of the new era.

A week before *Columbia*'s first launch, Wilford published another long, thoughtful essay, this one on "Space and the American Vision."[14] Four years had elapsed since his "Shuttling into Space" article—years during which the Shuttle had been plagued with technical problems, cost increases, and delays. Wilford again framed the meaning of the new era of human spaceflight, but now the routine transportation scheme did not seem as plausible or resonant as before, and the Shuttle had not even flown yet. There was a note of ambivalence about the Shuttle

13. Articles in *The New York Times* by Wilford and others, April 1981; headline from 15 April 1981: pp. A21.

14. John Noble Wilford, "Space and the American Vision," *The New York Times*, 5 April 1981, Sunday Magazine: pp. 14 ff, 118 ff.

era in his rhetoric as he tried to reconcile America's spacefaring destiny with the spaceplane's mundane mission of hauling orbital freight.

Because 13 of the first 23 missions were indeed freight-hauling flights to deliver satellites for commercial customers, *The New York Times* reporters generally conveyed Shuttle mission news within the routine transportation frame, featuring the three Rs of space trucking (release, retrieve, repair). But they also made room in stories for questions about the cost of Shuttle missions and reported all manner of technical glitches and delays that belied the concept of routine spaceflight. The terms "failure," "delay," and "problem" repeated frequently in news accounts subtly challenged the accepted frames of reference and sowed doubts about the fit between these frames and reality. Yet the Shuttle came to be understood as a space truck delivering large cargos to orbit—an image that some of the astronaut crews happily fostered—and successive satellite deliveries helped to establish a semblance of routine spaceflight.[15]

Attention to five missions in 1984 and 1985 elevated the space truck to new heights of interest by putting humans squarely in the focus. These missions added a Buck Rogers gloss to the notion of routine work in space and made vivid the role of human spaceflight.[16] The common theme of these servicing and salvage missions was satellites gone awry—humans to the rescue. The drama of astronauts flying away from the Shuttle in jet backpacks, grappling errant satellites, wrestling them into the payload bay, and then conducting repairs, put a human face on the new-era frame. The Shuttle image broadened from delivery truck to tow truck to service station, and the astronauts earned credit as orbital repairmen. Extravehicular activity (EVA) figured heavily in these missions and was a visibly effective way to demonstrate human capability in space. The missions showed off new astronaut tools—the piloted maneuvering unit backpack, the remote manipulator system robotic arm, the power hand tools—that gave working in space a vivid dexterity. The message in the media, and from NASA, was that "nothing like this has ever been done before."

By the end of 1985, with 23 Shuttle missions completed, *The New York Times* (and other news media) had validated the new era of routine space transportation concept as the meaning frame for human spaceflight. However, a noticeable current of critique ran through some of the news reports, and more so in editorials and opinions. Alert journalists noted that about two-thirds of the launches had been

15. Typically *The New York Times* ran a news article each day of each mission; several in the days just before launch and after landing; at least one article for every delay or significant problem; and occasional analytical pieces. The mission-related coverage during the 1981–85 period totaled hundreds of articles.

16. The five missions were, in 1984, the 10th (STS 41-B), featuring first flights in the Manned Maneuvering Unit; the 11th (STS 41-C), the Solar Max observatory repair mission; the 14th (STS 51-A), the first satellite retrieval to return the Westar and Palapa communications satellites; and in 1985, the 16th (STS 51-D), another satellite delivery mission, and the 20th (STS 51-I) to deliver three satellites and retrieve/repair another. See *The New York Times* articles by Wilford and others in January–April and November 1984, and April and August–September 1985.

The quintessential frames for the meaning of human spaceflight are images of a single astronaut poised against black space, the vivid Earth, or the landscape of another world. They resonate with adventure, risk, courage, heroism, discovery, and beauty.

delayed by weather or technical problems; several missions had been delayed in returning or brought home early for the same reasons; and five years into the new era the launch schedule was always subject to change. By these measures, "routine" transportation seemed ephemeral. Of the satellites deployed from the Shuttle, enough had failed to reach their intended orbits or operate properly that salvage missions were required, making the satellite deployment role for the Shuttle look less rosy. Worrisome repeated problems such as damaged tiles, fluid leaks, computer malfunctions, locked brakes, and blown tires also clouded the picture of routine transportation. Occasional serious anomalies discovered after landing—evidence of a fire and explosion in the engine compartment, a large hole in a wing with partial melting of the structure—gave pause for observers to wonder how safe the Shuttle really was.[17] Despite the frequency and variety of missions in this new era, evidence mounted that human spaceflight was not yet routine.

Only a few of the early Shuttle missions provoked editorial commentary in *The New York Times*, which began to challenge the concepts of routine space transportation and useful human spaceflight. A skeptical editorial—"Is the Shuttle Worth Rooting For?"—appeared on the eve of the first Shuttle launch. While acknowledging the Shuttle as "an unquestionable technological achievement," the editors noted that it was "a technology in search of a mission" that might become a white elephant. The

17. Ninth mission (STS-9, 1983), aft compartment fire upon landing; 16th mission (STS 51-D, 1985), damaged wing.

reason for their ambivalence: uncertainty that the Shuttle would really cut the cost of operating in space.[18] A few days later, the editors tempered their end-of-mission congratulations with the question, "Now that we own a successful space shuttle, what do we do with it?" Their standard: "What can a reasonable society afford?"[19] The next editorial on the Shuttle suggested limiting the number of spaceplanes to allow for continued planetary exploration.[20]

To mark the third successful Shuttle mission, *The New York Times* acknowledged that *Columbia* "almost succeeded in placing the stamp of routine on shuttling into space," but charged that NASA was not using the magnificent machine with sufficient style. It deserved a purpose greater than trucking freight. In this instance, reality fit within the routine transportation frame but the frame itself was challenged as unimaginative. However, no alternate frame was tendered.[21]

The tension between spacefaring and freight-hauling was a latent stress on the new-era routine transportation and business frames of reference. Wilford's occasional reflections on the Shuttle missions showed the stress fractures in these frames, and revealed how they were becoming dissonant, rather than resonant, with some important societal values. "This is no adventure in exploration; this is a freight run," he wrote upon witnessing the eighth launch. It did not inspire the same thrill as a mission to the Moon. He began to try to reframe human spaceflight by defining for it a purpose worthy of a spacefaring people with a tradition of exploration. With NASA under pressure to make spaceflight an economical business, he argued that the nation should aspire to a new vision of its future in space. Although the Shuttle and a future space station would expand human activities in space, he looked to the robotic voyages of discovery in the solar system as the model for inspiring wonder and rekindling the spirit of the Apollo era.[22]

Before 20 missions had flown, Wilford wrote a piece measuring actual performance against promise, in effect measuring the fit between the routine space transportation frame and reality. Using such metrics as number of missions projected vs. accomplished and number of satellites scheduled vs. orbited, he showed the large gap between hopes and reality. These discrepancies were prompting a reevaluation of the Shuttle program by its customers and critics, and even its proponents. Regardless of spectacular achievements, the frame for human spaceflight in the Shuttle era was getting out of alignment with reality.[23]

18. "Is the Shuttle Worth Rooting For?" *The New York Times*, 9 April 1981: p. A22.

19. "Down to Earth," *The New York Times*, 14 April 1981: p. A30.

20. "What Does the 'S' in NASA Mean?" *The New York Times*, 4 November 1981: p. A30.

21. "Too Fine a Machine," *The New York Times*, 31 March 1982: p. A30.

22. John Noble Wilford, "Big Business in Space," *The New York Times*, 18 September 1983, Sunday Magazine: pp. 46–47 ff., 50, 83.

23. John Noble Wilford, "Gap Between Early Hope and Present Accomplishment Grows Large; Space Shuttle Re-evaluated," *The New York Times*, 14 May 1985: p. C1.

Other observers also subjected Shuttle-era human spaceflight to a cost-benefit analysis and found that the numbers did not add up to economical space transportation. Historian of technology Alex Roland published one of the most strident critiques of this type in the popular magazine *Discover*. In "The Shuttle: Triumph or Turkey?" he appraised its cost, technical failures, maintenance demands, uncertain schedule, deployment mishaps, and other shortcomings against the promises of routine space transportation, and found the sophisticated, versatile Shuttle wanting. "Judged on cost, the shuttle is a turkey .... It costs too much to fly .... And cost is the principal criterion by which it should be judged." In setting a cost-benefit frame over the Shuttle, Roland was not reframing human spaceflight itself; indeed, he did not comment on the value of the missions or crews. Rather, he faulted the vehicle—the icon of human spaceflight—to attack the credibility of NASA's new-era and business frames for the unrealized promise of routine, reliable, economical space transportation.[24]

Editorial cartoons from this period also had perspectives on the new-era frame of reference, as they quite literally distilled an idea or opinion within an inked frame. Editorial cartoonists across the country treated the Shuttle and human spaceflight as subjects.[25] In the early 1980s many of them responded to the concept of routine space transportation with pride or humor. They tended to treat the first Shuttle mission as a patriotic and technical triumph, featuring Uncle Sam and the U.S. flag on track toward America's destiny in space. Some depicted passengers lined up with a Shuttle timetable, waiting for pickup. Others drew the Shuttle as a space truck and astronauts as handymen on the satellite delivery and servicing missions. They depicted the foibles of launch delays and technical problems—a Shuttle on the launch pad covered in cobwebs, suited astronauts growing old while waiting to fly, tiles falling off the Shuttle, a tanker truck of superglue at the pad, a countdown clock with a ridiculously high number.

The editorial cartoonists, inspired by the news and their own idiosyncratic perspective on things, independently endowed the Shuttle and human spaceflight with meaning inside the frames they drew.[26] Their charter for the Shuttle, as for other topics, was to distill the essential meaning of things stripped of hype. Perhaps earlier than others, they began to see (and lampoon) a misalignment of NASA's frame of reference and reality.

24. Alex Roland, "The Shuttle: Triumph or Turkey?" *Discover* (November 1985): pp. 29–49; quotes, 45.

25. The NASA Historical Reference Collection at NASA Headquarters in Washington, DC, contains many cartoon files catalogued by year and topic in the series Cartoons.

26. Various scholars have examined editorial cartoons as effective keys to frames of meaning: William A. Gamson and David Stuart, "Media Discourse as a Symbolic Contest: The Bomb in Political Cartoons," *Sociological Forum* 7, no. 1 (March 1992): pp. 55–86; Edward T. Linenthal, *Symbolic Defense: The Cultural Significance of the Strategic Defense Initiative* (Urbana, IL: University of Illinois Press, 1989); Thomas H. Bivins, The Body Politic: The Changing Shape of Uncle Sam," *Journalism Quarterly* 63 (Spring 1987): pp. 13–20; Roger A. Fischer, "Oddity, Icon, Challenge: The Statue of Liberty in American Cartoon Art, 1879–1986," *Journal of American Culture* 9, no. 4 (Winter 1986): pp. 63–81.

REFRAMING HUMAN SPACEFLIGHT: SCIENTIFIC RESEARCH

Social movement scholars have defined several processes for invigorating or strengthening a contextual frame to make it less vulnerable to criticism and more appealing to supporters. Clarification and expansion of the concept (frame amplification and frame extension) can be effective strategies for protecting a core concept and expanding its appeal to a broader community.[27]

As editorial and opinion writers began to critique the practice and meaning of human spaceflight in the Shuttle era, NASA did what social action movements often do to maintain support. It began to extend the frame, stretching its elastic boundaries to include other appealing elements. As soon as the Shuttle became operational, NASA began to retool for another big engineering project. Presidential approval to begin development of an orbital station complex came in 1984. Human spaceflight now encompassed not only the Shuttle but also a space station, promoted as "the next logical step" to a "permanent presence" in space.[28]

This expanded package of meaning protected the Shuttle as essential to the assembly and routine supply of the space station, and both were deemed essential for the continuation of human spaceflight. However, to avoid a completely circular justification for the Shuttle and station, NASA elaborated the purpose of human spaceflight to include scientific research, a dimension of useful work that would bring benefits through new knowledge. This elaboration evolved in relation to three human spaceflight programs: Spacelab, Space Station *Freedom*, and the International Space Station.

Scientific research was a secondary theme in the early Shuttle era. Just four of the first 25 Shuttle missions had focused on science instead of commercial or national security payloads.[29] In the 1990s science became a major focus on half of the missions, with some 30 flights completely dedicated to research and other flights carrying at least a few experiments. The Spacelab suite of laboratory module and instrument pallets, developed by the European Space Agency and installed in the payload bay, effectively turned the Shuttle into a temporary orbital research station generally staffed by Ph.D.'s. These missions included experiments in various disciplines where flight crews could carry out research with the goal of pushing the frontiers of knowledge.[30]

A primary scientific objective was to study space motion sickness and adaptation

---

27. David A. Snow, E. Burke Rochford, Jr., Steven K. Worden, and Robert D. Benford, "Frame Alignment Processes, Micromobilization, and Movement Participation," *American Sociological Review* 51, no. 4 (August 1986): pp. 464–481, esp. 469–473.

28. *Space Station Freedom Media Handbook* (Washington, DC: NASA, May 1992).

29. Spacelab 1 (STS-9, 1983), Spacelab 3 (STS 51-B, 1985), Spacelab 2 (STS 51-F, 1985), and Spacelab German D-1 (STS 61-A, 1985), the 9th, 17th, 19th, and 22nd shuttle missions. See *The New York Times* articles November–December, 1983; April–May and July–August, 1985.

30. Examples of public affairs material framing human spaceflight as scientific research are the NASA Marshall Space Flight Center pamphlet *Spacelab*, 13-M-883, which describes the facility and its uses, and the NASA Information Summaries PMS-008A (Hqs), "Space Station," August 1988.

to weightlessness—topics that put the spotlight on the role of humans in space. Another was to investigate the properties of materials and processes in microgravity. Investigations in life and materials science included both basic and applied research. These Shuttle missions refined the ability of astronaut crews to collaborate with scientists on the ground while carrying out experiments, thus opening the space environment to hundreds of researchers. Enabling members of the worldwide scientific community to participate directly in space missions broadened the appeal of human spaceflight in those disciplines based on laboratory methods. Astronomers and space physicists generally were not persuaded that human spaceflight was necessary; automated instruments and satellites were more effective and less expensive means for conducting their research.

NASA and the media began to stretch the human spaceflight frame beyond the Shuttle, seeing the Shuttle-borne laboratory as a precursor to a space station. The new-era frame now began to imply a very long-term, perhaps permanent human presence in space. The effort to promote a space station, known first as *Freedom* and then as the International Space Station, relied on the key ideas of orbital research, "cutting-edge science," a "world-class laboratory," "frontiers of knowledge," and other superlatives to bolster the meaning of continued human spaceflight. The purpose of human spaceflight on the space station was to advance science, which would yield discoveries for benefits on Earth and enable future exploration. If the stretch occasionally seemed improbable—that research on the space station might lead to cures for cancer or AIDS or osteoporosis—it also showed that NASA was seeking new constituencies, especially women, to garner public support for an expensive new program.[31]

The New York Times editorial column stridently challenged this framing of human spaceflight on the grounds of cost, size, purpose, utility, scientific potential, necessity, and logic. Especially during the precarious years of the late 1980s and early 1990s when the space station program was in political trouble, *The New York Times* urged its cancellation and a redirection of human spaceflight. Calling the proposed orbital research station an extravagant folly and the arguments for a permanent human presence there specious, the editors found in it no compelling national purpose or social value. *The New York Times* attempted to reframe its meaning as a grandiose fiasco. Only when the station was scaled down in size and purpose did the editors briefly give it credence but never full support.[32]

31. NASA Press Release 92-92, "Goldin Says America Needs Space Station Freedom Now," 24 June 1992; NASA Press Release 92-119, "NASA, NIH Sign Agreement on Joint, Space-Related Research," 21 July 1992; Boeing, "The Space Station Brochure," early 1990s; "Space Station Freedom: Gateway to the Future," NASA publication NP-137, 1992; "The International Space Station: The NASA Research Plan," NASA NP-1998-02-232-HQ, 1998.

32. Examples of strident critiques of the space station basis for human spaceflight that appeared in *The New York Times* include "NASA's Black Hole in Space," 29 March 1990: p. A22; "Space Yes; Space Station No," 6 June 1991: p. A24; "NASA's Untouchable Folly," 14 July 1991: p. E18; "The Wrong Space Station," 29 July 1992: p. A20; "Is NASA Among the Truly Needy?" 6 March 1995: p. A14. Two qualified exceptions were "How to Put Space in Its Place," 12 December 1990: p. A22 and "Space, In Proportion," 6 March 1991: p. A24.

Influential voices outside *The New York Times* also doubted the value of the space station and the meaning of human spaceflight in scientific research. Space scientist James A. Van Allen was one of the earliest and most earnest critics. He made the point, often repeated in *The New York Times*, that "the overwhelming majority of scientific and utilitarian achievements in space have come from unmanned, automated and commandable spacecraft." Robotic satellites and planetary probes had advanced the frontiers of knowledge quite successfully and at far less cost than people could. Van Allen argued that the space station would seriously *diminish*, not expand, opportunities for scientific advances. He found the human spaceflight-for-science frame to be disingenuous and the high value placed on piloted flight to be excessive.[33]

Van Allen suggested that the cultural obsession with human spaceflight defied reason when the motive was science, but he granted the power of popular interest in science fiction and the space program's potential for creating real adventure. Arguments of scientific productivity, however, did not derail the space station and, 20 years after Van Allen wrote, his critique has been partly vindicated. Instead of "the tidal wave of basic science" that NASA had predicted for the space station, a trickle has flowed.[34] Circumstances have required crews to spend more time operating and maintaining the International Space Station than exploiting its capabilities for laboratory science. If there have been discoveries from cutting-edge experiments aboard the station, they have not been well advertised. A reality check of this frame now would likely show it out of alignment with its premises and less resonant with societal values than at its origin.

FRAME SHIFT: HUMAN SPACEFLIGHT AS HEROISM

Scholars of meaning construction in social movements and the media note that occasionally an event creates some perturbation in the prevalent meaning frame of an issue. Such a crisis may provoke reconsideration or even reconstruction of meaning. A crisis becomes a critical discourse moment that can change the basis of meaning, introduce new values, and prompt a shift to a new frame of meaning.[35] Such a critical moment occurred in January, 1986.

The year began with news that the Voyager 2 spacecraft had reached the neighborhood of Uranus, its first planetary encounter since leaving Saturn five years earlier. Images from the spacecraft showed new moons, rings, colors, mountains, craters, and other intriguing features. As NASA and the media hailed this ongoing mission

33. James A. Van Allen, "Space Science, Space Technology and the Space Station," *Scientific American*, 254, no. 1 (January 1986): pp. 32–39.

34. NASA administrator Daniel Goldin quoted in NASA Press Release 92-92, "Goldin Says America Needs Space Station Freedom Now," 24 June 1992.

35. Gamson and Modigliani, "Media Discourse and Public Opinion," and Benford and Snow, "Framing Processes."

of discovery, *The New York Times* published two editorial odes to Voyager as space exploration "at its most intelligent and productive" and "at its best." By comparison, human spaceflight seemed adrift, with NASA flying politicians and a teacher to hold public attention. In a terrible coincidence, the second of these pieces appeared on January 28, the morning of *Challenger's* final launch. Its barbed closing line chided, "If NASA wants lasting public support for a vigorous space program, the wonder of seeing new worlds will do it a lot more good than soap opera elevated to Earth orbit."[36]

What happened that morning, witnessed by millions of television viewers, was nothing as trivial as a soap opera. The catastrophic loss of the Shuttle and death of seven crewmembers barely a minute after liftoff seared the nation, shaking national pride and confidence about the technology and safety of human spaceflight. The dimensions of the tragedy broadened and deepened during the weeks of investigation, with stunning revelations of flawed technology and questionable decision making within NASA.

The *Challenger* accident shattered the new-era frame of routine spaceflight. What some had suspected suddenly became clear—space transportation was not yet routine, measured not by a dry financial cost-benefit analysis but by the cost of human life. The risk of spaceflight had been absent in the new-era frame of reference. That this was a basic freight-hauling mission to deliver a satellite—a task that did not inherently require a human crew—made their deaths even more tragic. Spaceflight deemed as routine as airline flight implied safety. As the pace of Shuttle missions had quickened, the public had understandably become complacent about spaceflight, perhaps the inevitable result of the frame of reference that had given meaning to the Shuttle era.

With the accident and loss of life, the disparity between reality and the conceptual frame of meaning for human spaceflight was too great to hold. It lost credibility and resonance in the shock of tragedy. The astronauts' deaths demanded greater significance than the space truck rationale could provide. Both the Shuttle and human spaceflight would be questioned and revalued, first to make sense of the tragedy and then to reconceive America's future in space.

The public search for meaning immediately defaulted to the 1960s frame of pioneering exploration and heroism on the space frontier. From President Reagan's consoling remarks to media coverage, official tributes, and personal mourning, the theme was courage and sacrifice in the cause of exploration.[37] The very purpose that the Shuttle did *not* actually have—exploration—became the cause for which the *Challenger* crew sacrificed their lives. Invoking the quest of exploration elevated the *Challenger* mission to a noble cause and valued the deaths as heroic. The routine space transportation frame could not bestow that meaning.

---

36. "Adrift in Space," *The New York Times*, 7 January 1986: p. A20 and "On to Neptune," *The New York Times*, 28 January 1986: p. A24.

37. Transcript of President Reagan's statement to the nation, reprinted in *The New York Times*, 29 January 1986: p. A9.

The New York Times reported the details of the accident and subsequent investigation and also immediately began to offer perspective on the news. An analysis piece "Should U.S. Continue to Send People Into Space?" appeared as soon as January 30 under a heading "Issue and Debate." John Noble Wilford's articles included reflections on human vulnerability, trust in technology, and the unrelenting dangers and risks of exploration as germane to a reappraisal of spaceflight. This bleak time in the space program was an opportunity to set new goals and a clearer mission for the Shuttle and beyond. He noted that human spaceflight was bound to continue, because "With the loss of the *Challenger* and its crew of seven, we learned, to our surprise, how much these adventures into space, into the future, mean to us as a people."[38]

Editorial cartoons telegraphed the societal impact of spaceflight as scores of cartoonists responded to the *Challenger* tragedy.[39] The primary themes, as in the president's address, were national sorrow and heroism, variously depicted as Uncle Sam with head bowed, the flag on the Moon at half-mast, or an eagle shedding a tear. Some cartoonists framed the accident in a spiritual dimension, showing the Shuttle as a constellation, the astronauts as new stars, or the Shuttle and crew entering heaven. There were no cartoons featuring a space truck or astro-delivery-nauts, no suggestions of routine spaceflight. A few editorial artists who also wrote about responding to the *Challenger* accident described the meanings they sought to distill within their drawings as the fragility of mankind's wings, shattered faith in space technology, or inexpressible sorrow for a profound loss to the nation.[40]

Framing the *Challenger* accident within the heroic cause of exploration—really a return to the meaning frame of the 1960s—was powerful, perhaps instinctive. It gave meaning to a shocking tragedy and resonated with societal values of patriotism and faith that offered consolation for the present and hope for the future. The exploration frame appealed to public sentiment, which translated into expressions of increased public support for the space program. In the immediate aftermath of the *Challenger* accident, the supportive public and the Shuttle's critics seemed to be oddly in accord in revaluing the meaning of human spaceflight as exploration, not freight-hauling and similar practical work.

38. David Rosenbaum, "Should U.S. Continue to Send People Into Space?" *The New York Times*, 30 January 1986: p. A18; John Noble Wilford, "Faith in Technology Is Jolted, but There Is No Going Back," *The New York Times*, 29 January 1986: p. A7; John Noble Wilford, "The Challenger's Fate, the Shuttles' Future," *The New York Times*, 2 February 1986: p. E1; John Noble Wilford, "America's Future in Space After the Challenger," *The New York Times*, 16 March 1986: p. 85 ff.

39. Files in the History Office at NASA Headquarters in Washington, DC, contain some 150 *Challenger*-related cartoons published in 1986.

40. Examples include Garner's drawing in *The Washington Times*, 19 January 1986: 11A; Swann's drawing in *The Huntsville Times*, 19 January 1986; Marlette's drawing in *The Charlotte Observer*, 28 January 1986 and in Doug Marlette, *Shred This Book!* (Atlanta: Peachtree Publishers, 1988), pp. 86–88; Ohman's drawing in *Newsweek*, 10 February 1986: p. 21, and in Jack Ohman, *Back to the '80s* (New York: Simon & Schuster, 1986), pp. 136–137.

The new-era routine spaceflight frame had originated with NASA and then was promoted to the public. However, the reframing of human spaceflight after the *Challenger* accident seems to have arisen outside the Agency. *The New York Times* became a forum for reappraising the state of human spaceflight by publishing its own perspectives and those of several prominent citizens. Immediately after the accident, a *New York Times* editorial addressed "The Challenge Beyond Challenger" with thoughts for reordering the nation's priorities in space. The coincidence of the Shuttle's destruction and Voyager's success illustrated a need to establish goals in space and use humans only when necessary. As most of the tasks for the Shuttle crews could be performed better by rockets or automation, a better goal for human spaceflight might be a mission to Mars to "satisfy humanity's sense of adventure." This surprising proposal, given that robots could also explore Mars, was a concession that humans might have some role in space more justifiable than then-current roles.[41]

For weeks, *The New York Times's* editorials and op-eds reflected on both the routine spaceflight reference frame and the need to reorient the role of human spaceflight. In their quest to find a justifiable purpose for sending people into space, the only one tentatively suggested was a piloted mission to Mars.[42] As a critical discourse moment, the *Challenger* accident prompted a shift from the routine spaceflight frame to its direct opposite: exploration.

Frame Transformation: Human Spaceflight as Exploration

From 1986 into the 1990s, and then again after the 2003 *Columbia* accident, considerable energy went into transforming the meaning of human spaceflight. Shuttle flights continued to carry out satellite delivery and science missions, and then preparatory and actual space station missions. Human spaceflight continued within the meaning frames of transportation and science, but on another track a new frame—exploration—was taking shape through various task force/advisory committee studies and media discourse. The framers shaped this concept largely in antithesis of the others, a counter-frame based on opposition to the status quo. Their purpose was to transform the meaning of human spaceflight by situating it within a different set of traditions and values.[43]

---

41. *The New York Times*, 31 January 1986: p. A30.

42. "Risk and Routine," *The New York Times*, 7 February 1986: p. A34; Tom Wicker, "Icon and O Rings," *The New York Times*, 18 February 1986: p. A23; "The Seal on NASA's Fate," *The New York Times*, 22 February 1986: p. A22; "The Frailties of Machines and Men," *The New York Times*, 2 March 1986: p. E22; "How to Regain Face in Space," *The New York Times*, 28 May 1986: p. A22.

43. Frame transformation is discussed in Benford and Snow, "Framing Processes" and Snow et al., "Frame Alignment Processes."

Within weeks of the *Challenger* accident, an alternative plan for human spaceflight appeared. The National Commission on Space, created by Congress and appointed by the president, released a report of its year-long project to develop an exciting vision and goals for the twenty-first century. Ambitious and optimistic, it was an antidote to the malaise spawned by the accident. This new vision was crafted in public dialogues around the country as the commissioners sought to hear what citizens expected of their space program. In a word—exploration.

The advisory commission's report, *Pioneering the Space Frontier*, focused on exploration and settlement within the solar system as the extended home of humanity. American leadership could open this new frontier to science, technology, and economic enterprise. The elaborate plan envisioned a massive infrastructure: space station, different types of vehicles and spaceports, a lunar outpost, a Mars base, and related technologies. The Shuttle era was confined to an orbital beltway near Earth, but in the future era humans would move out on a "highway to space" and a "bridge between worlds," to set up residence and do useful work producing propellants and other life- necessary resources. This vision was a hybrid of the familiar frontier and transportation frames for human spaceflight applied to a new setting and purpose. Colorful cover art and illustrations engagingly framed this rather industrialized vision of the space frontier, published as a report dedicated to the *Challenger* crew, who in President Reagan's words were "pulling us into the future."[44]

NASA also engaged in its own reappraisal of the future of human spaceflight. Astronaut Sally Ride chaired an internal agency planning group that prepared a report on *Leadership and America's Future in Space*.[45] It, too, proposed an eventual human mission to Mars, but at a more measured pace and scale than the national commission had proposed. These and other studies were gestures toward a transformational vision of human spaceflight beyond the Shuttle era, but they were not converted to action plans.

Near the one-year anniversary of the *Challenger* accident, an encouraging piece by space scientist Carl Sagan appeared in *The New York Times*. "It's Time to Go to Mars," he wrote, in a systematic program of exploration advancing from robotic rovers to sample-return missions and then to "the first human footfalls on another planet." Unlike the national commission's vision of productive industry on Mars, Sagan's vision focused on the values of adventure, excitement, inspiration, valor, prestige, and purpose in the space frontier. He argued that exploration of Mars for the sake of knowledge could revitalize the moribund space program and make possible a new goal, "establishing humanity as a multiplanet species."[46]

---

44. *Pioneering the Space Frontier: The Report of the National Commission on Space* (New York: Bantam Books, 1986).

45. Sally K. Ride, *Leadership and America's Future in Space* (Washington, DC: NASA, August 1987).

46. Carl Sagan, "It's Time to Go to Mars," *The New York Times*, 23 January 1987: p. A27.

Variants of the exploration of Mars arose as forces inside and outside NASA tried to reframe the purpose of human spaceflight. The Mars goal seemed a worthy commitment for astronauts, and it might align the human and robotic flight programs in a complementary rather than competitive enterprise. It could also reassert American leadership in space in an inspiring, challenging adventure. By spring, 1987, John Noble Wilford could report in *The New York Times* that "momentum is building in the space agency and among . . . leaders to make Mars the next major goal of the American civilian space program."[47] Exploration, specifically the exploration of Mars, had gained credibility and resonance as the future meaning of human spaceflight.

The 20th anniversary of the Apollo 11 landing highlighted the discrepancy between current human spaceflight and aspirations for a new purpose. President George H. W. Bush marked the anniversary in 1989 by endorsing a spacefaring initiative to return to the Moon and move on to Mars. Apparently formalizing the frame shift from Shuttle-era concepts to exploration, the announcement was more rhetoric than mandate, for he set no schedule and made no funding commitment for such an enterprise. It met with skepticism among political leaders and space policy analysts as too costly. *The New York Times* dismissed it as "Mr. Bush's giant step back in space . . . a failure of imagination" because it sounded like Apollo redux without a compelling reason.[48] The president's new frame for the meaning of human spaceflight seemed rickety but it did authorize NASA to chart a path out of Earth orbit through a new space exploration initiative.

Despite the ferment, the transformation process was slow, and in the meantime human spaceflight was still riding the Shuttle and preparing a space station. The New York Times published numerous impatient, frustrated editorials on the theme "stuck in Earth orbit for no good reason." The editorial page framed the Shuttle as fragile, vulnerable, neither fully safe nor fully reliable, with nowhere to go. The planned space station was decried as an extravagant folly, a "black hole," a fiasco, purposeless or a "potpourri of purposes," grandiose, unsuitable for anything except being a place for the Shuttle to go. The space agency was "an aged and faltering institution," ailing and "pinched in scope and vision." The drumbeat message: Cancel the space station and do something more imaginative than carry astronauts and cargo to low Earth orbit.[49] *The New York Times's* editorial position framed human spaceflight as properly

47. J. N. Wilford, "Exploration of Mars Is Advised as Goal for NASA," *The New York Times*, 18 March 1987: p. B6 and "The Allure of Mars Grows as U.S. Searches for New National Goal," *The New York Times*, 24 March 1987: p. C1.

48. "Mr. Bush's Giant Step Back in Space," *The New York Times,* 21 July 1989: p. A28.

49. Typical *The New York Times* editorials in this vein include "NASA's Black Hole in Space," 29 March 1990: p. A22; "Rethink Space," 21 July 1990: p. 20; "Those Hisses on the Launching Pad," 19 September 1990: p. 22; "Space Yes, Space Station No," 6 June 1991: p. A24; "NASA's Untouchable Folly, 14 July 1991: p. E18; "The Wrong Space Station," 29 July 1992: p. A20; "Is NASA Among the Truly Needy?", 6 March 1995: p. A14.

grounded in wonder, imagination, excitement, frontiers, discovery, and a clear goal worthy of risking human life.

As momentum built in the media and advisory and advocacy bodies for missions to the Moon and Mars, NASA began to elaborate and extend the space station concept to more explicitly embrace exploration.[50] In the 1990s the Agency started describing the orbital research center as a stepping-stone or a bridge to future exploration. The language of future spaceflight borrowed from the builder's lexicon, as planners worked on "blueprints" and "architectures" for exploration. Adding an overlay of exploration to the space station partially reframed it to disarm critics and strengthen support.

The New York Times editors disagreed with that gloss, as did another NASA critic, science journalist Timothy Ferris, writing for the op-ed page. *The New York Times* charged that the space station was not designed to be a way station to other worlds, a launching pad for planetary exploration, or a stepping-stone to anywhere. "The shocking surprise is how little the station would contribute to the nation's long-range space goals," really only life science research.[51] The week before the first element of the International Space Station was placed in orbit in 1998, Ferris wrote a critical op-ed piece titled "NASA's Mission to Nowhere." In his view, the station "touted as a giant leap into space and a step toward the stars in truth . . . is little more than a Motel 6 in low [E]arth orbit . . . . [I]t will be of almost no use in getting to Mars, the Moon, or anywhere else—except into debt." Ferris argued that a far better plan would be to abandon the space station and mount "an international effort to put a colony on Mars" to make humanity a two-planet species. It could have great scientific value and also be a grand adventure, a future where we "really get somewhere."[52]

With effort focused on assembling and operating the space station, the space exploration initiative withered until another critical discourse moment forced the issue again. The second Shuttle tragedy, the loss of *Columbia* and crew during reentry in February 2003, again thrust the purpose of human spaceflight into the media spotlight for debate whether this type of orbital mission was worth the risk and cost of human lives. Again the public responded to the tragedy by revering the astronauts as heroic explorers, and editorial cartoonists depicted the apotheosis of the Shuttle and crew as stars in the heavens.[53]

---

50. Other reports from this period included *Report of the Advisory Committee on the Future of the U.S. Space Program* (Augustine Committee) (Washington, DC: U.S. Government Printing Office, 1990) and *America at the Threshold: America's Space Exploration Initiative*, Report of the Synthesis Group (Stafford Committee) (Washington, DC: U.S. Government Printing Office, 1991).

51. Space Yes; Space Station No," *The New York Times,* 6 June 1991: p. A24; "NASA's Untouchable Folly," *The New York Times*, 14 July 1991: p. E18; "The Wrong Space Station," *The New York Times*, 29 July 1992: p. A20.

52. Timothy Ferris, "NASA's Mission to Nowhere," *The New York Times*, 29 November 1998: p. WK9.

53. Cartoon series file 2003 in the NASA Historical Reference Collection at NASA Headquarters.

As the exploring Voyager mission starkly contrasted with the earlier Shuttle tragedy, the call of distant worlds also beckoned after the *Columbia* tragedy. *The New York Times* responded to the tragedy not with a call to halt human spaceflight but to redirect it to exploration. "Curiosity and the quest for knowledge . . . make it inevitable that humans will continue to venture into space . . . to engage in the sheer thrill of exploration and new discoveries."[54] Soon robust robots roaming on Mars captured public attention with the vicarious thrill of exploration, in contrast to the handicapped human spaceflight program. Editorial cartoonists depicted the Shuttle as physically decrepit, geriatric, on life support, with the astronauts idled on Earth while robots explored Mars.[55]

In 2004 President George W. Bush urged a new vision for space exploration for the future beyond the Shuttle and space station. Like the space exploration initiative 15 years earlier, this presidential charter stimulated planning studies inside and outside NASA. But this time NASA took the challenge seriously enough to reorganize for action, aiming for a transformation of both the rhetoric of human spaceflight and the agency itself. Emphasizing sustained and affordable programs to satisfy the spirit of discovery, planners have been careful not to make exaggerated claims about the benefits of exploration. Human spaceflight now is being framed not as a practical or a business enterprise but more lyrically, as exploration resonant with mystery, curiosity, adventure, and reinvigoration after a long stay in Earth orbit.

NASA's slogans for the space exploration vision, "The New Age of Exploration" and "A Renewed Spirit of Discovery," herald a return to a cultural tradition of exploration that expands knowledge and fuels wonder. This framing approach differs rhetorically from the previous initiative; publicity materials depict people on Mars as explorers, not as miners, and prose addresses compelling questions of scientific and societal importance more than technology. It is too early to know if or how that renewal will occur, but the current vision for space exploration seems to be reasonably framed for broad appeal. It takes human and robotic explorers out of competition and elevates scientific discovery as their shared goal. More modest in promises than earlier frames yet potentially more heroic, exploration aims at the worthier purpose that critics and advocates of human spaceflight have long demanded.[56]

54. "The Call of Distant Worlds," *The New York Times*, 9 February 2003: WK14.

55. Cartoon series files 2003–2005 in the NASA Historical Reference Collection at NASA Headquarters.

56. *The Vision for Space Exploration* (Washington, DC: National Aeronautics and Space Administration, 2004).

CONCLUSION

This frame analysis of the Shuttle era has focused on the social construction of meanings for human spaceflight. Five meaning frames have been probed: a new era of routine transportation, business, scientific research, heroism, and exploration. NASA was the primary, but not sole, shaper of these meanings; the media, represented here by *The New York Times*, and the public also exerted a strong influence by critiquing the fit between frames and reality. When a frame became dissonant with societal expectations, either NASA subtly revamped it or the media and public pressured for change.

The varied meanings of human spaceflight in the Shuttle era can be interpreted as arising from processes of frame development, frame extension, frame shifts, and transformations—all strategies used by social action movements to appeal to and sustain their supporters and also used by media to give readers a context for thinking about issues. These frames helped society make sense of the costly, risky endeavor of human spaceflight by anchoring it in traditions and values that matter to citizens. Curiously, the keenest consensus about the meaning of human spaceflight arose not from its successes but from the two Shuttle tragedies. These critical moments forced a societal discourse about the defining purpose of human spaceflight that prompted reframing and transformation. It seems ironic that robotic planetary missions also inspired efforts to reframe the meaning of human spaceflight as exploration.

That in the course of more than 30 years the meaning of human spaceflight has been malleable may attest to societal wisdom and adaptability to changing circumstances, or it may indicate a restless desire to try something new. In any case, human spaceflight remains anchored in American culture and resilient in meaning.

Chapter 6

Space in the Post–Cold War Environment

John M. Logsdon

In a thoughtful report issued in December 1992, one year after the dissolution of the Soviet Union, a blue-ribbon panel of the Vice President's Space Policy Advisory Board observed that "[T]he U.S. civil and national security space programs have evolved within a policy framework that reflected the international tensions, as well as the economic and technological constraints and alliance relationships of the Cold War period." The panel suggested that "[T]he end of the Cold War, advances in technology, and other developments present new opportunities for cooperation and progress in space." Given this reality, the group found it necessary to "transform the U.S. space program to meet the challenges of the new post Cold War era."[1] The report suggested the steps needed for such a transformation; its recommendations have as much relevance now as they did in 1992.

The title of this report, "A Post Cold War Assessment of U.S. Space Policy," carried with it an underlying assumption. That assumption was that the "cold war"—the protracted geopolitical, ideological, and military struggle that emerged after World War II between the United States and its allies and the Soviet Union and its allies, that never erupted into direct military conflict between the United States and the Soviet Union, and that ended with the collapse of the Soviet Union in 1991—

1. Vice President's Space Policy Advisory Board, "A Post Cold War Assessment of U.S. Space Policy: A Task Group Report," December 1992, pp. v, vii. During the administration of President George H. W. Bush, there was in the Executive Office of the President a National Space Council, chaired by Vice President Dan Quayle and supported by a small staff. Supporting the Space Council was a Vice President's Space Policy Advisory Board, which was activated only in mid-1992. The members of the Task Group that prepared this report, in addition to the author of this paper, were all individuals with long and diverse experience in the space sector. They included Laurel Wilkening, Chair; James Abrahamson; Edward "Pete" Aldridge; Joseph Allen; Daniel Fink; John Foster, Jr.; Edward Frieman; Don Fuqua; Donald Kutyna; and Bruce Murray. The report can be found at *http://history.nasa.gov/33080.pt1.pdf* and *http://history.nasa.gov/33080.pt2.pdf* (accessed 9 November 2006).

had had an important and continuing influence on the content and character of the U.S. space program. Given the credentials of the panel, the members of which had long and varied involvement with the space sector, its assessment ought to reflect a reasoned perspective regarding the impact that the end of the cold war should have on the way the United States would henceforth carry out its space efforts.

The panel's report made a key observation that is central to the argument of this paper—that "the quest for leadership has been a fundamental objective of the U.S. space program." Since the end of World War II, the "U.S. ability to influence the shape and flow of events around the world has been a core national interest." Successive presidents "have recognized the contributions that the U.S. space program made to the perception of the United States as a leading nation." With respect to the cold war, "in the 1960s, and for most of the next two decades, space leadership clearly meant besting the USSR in visible, challenging space exploration endeavors." But, the panel observed, it was global space leadership that was the basic goal, with U.S.–Soviet space relations an important, but not the only, venue for achieving that leadership. At the time the panel report was issued in 1992, the then-current national space policy stated that "a fundamental objective guiding United States space activities has been, and continues to be, space leadership."[2]

If this perspective is valid, then the end of the cold war would not have changed the importance of space leadership as the underlying goal of the U.S. space program, although different means for achieving that goal would have had to be pursued. If that is the case, then the end of the cold war would be less of a watershed in U.S. space policy than is usually thought. This paper will provide evidence in support of this conclusion.

The relationship between the cold war and U.S. efforts toward leadership in space is thus far from straightforward. The quest for global leadership, rather than direct U.S.–Soviet competition, has been the primary political influence on the evolution of the U.S. space program. As the 1992 panel observed, during those times in the 1957–1991 period when the Soviet Union loomed as a direct peer competitor of the United States, global space leadership indeed meant leadership in comparison to the Soviet Union. But at times when U.S.–USSR relations were not actively competitive, such as after the Cuban missile crisis, the period of U.S.–Soviet détente during the Nixon administration, and the latter years of the Reagan administration when the spirit of *glasnost* colored relationships between the two countries, the U.S. quest for space leadership continued. That did not change with the disappearance of the Soviet Union in 1991.

2. Ibid., pp. 1-2. National Space Policy NSPD-1, 2 November 1989, *http://www.au.af.mil/au/awc/awcgate/nspd1* (accessed 8 November 2006).

The Cold War and the U.S. Civilian Space Program

The focus of this paper is on the U.S. civilian space program, and particularly that major portion of the civilian program carried out by NASA. Certainly the U.S.–Soviet strategic and military rivalry during the cold war was a major influence on U.S. national security space efforts, but those will not be discussed in what follows. So, rather than repeat the word "civilian" below, the reader should assume that all references are to the civilian sector of U.S. space activities. The issue to be discussed is how the intertwining between the desire for global space leadership and the need to demonstrate space superiority vis-à-vis the Soviet Union shaped U.S. space efforts in the 1957–1991 period.

President Dwight D. Eisenhower, even after assessing the international and domestic political impacts of Soviet space successes in the aftermath of Sputnik, came to the conclusion that space leadership, particularly in highly visible space achievements, was not needed to preserve U.S. global standing overall. His efforts to avoid having a U.S. space program driven primarily by competition with the Soviet Union have been well documented. As two careful analysts commented, "[G]iven the political pressures for an all-out space race with the Soviet Union, the degree to which Eisenhower controlled the space policy agenda in the late 1950s stands as a considerable achievement." Even so, "[I]t would be inaccurate . . . to suggest that he was ever really in command of events." In fact, they conclude, "[E]arly U.S. space policy was indeed heavily determined by what the Soviet Union did."[3] The Eisenhower administration in January 1960 issued a formal statement of national space policy that reflected the tension between trying to develop a U.S. space effort based on its inherent merits and one that was competitive with the USSR. The policy suggested that:

> [T]o minimize the psychological advantages which the USSR has acquired as the result of space accomplishments, select from among those current or projected U.S. space activities of intrinsic military, scientific or technological value, one or more projects which offer promise of obtaining a demonstrably effective advantage over the Soviets and, so far as is consistent with solid achievements in the overall space program, stress these projects in present and future programming.[4]

---

3. David Callahan and Fred I. Greenstein, "The Reluctant Racer: Eisenhower and U.S. Space Policy," in *Spaceflight and the Myth of Presidential Leadership*, Roger D. Launius and Howard E. McCurdy, ed. (Urbana, IL: University of Illinois Press, 1997), pp. 39–40.

4. National Aeronautics and Space Council, "U.S. Policy on Outer Space," 26 January 1960, in *Exploring the Unknown: Selected Documents in the History of the U.S. Civil Space Program*, Volume I, Organizing for Exploration, John M. Logsdon et al., ed. (Washington, DC: National Aeronautics and Space Administration Special Publication-4407, 1995), pp. 367–368.

This statement captures well the ambivalent stance of the Eisenhower administration; while desiring a space program of substantive value, it was virtually impossible to avoid the influence of Soviet achievements in space because of their propaganda impacts on U.S. interests abroad and on national morale at home.

Although President John F. Kennedy is best known with respect to space for challenging the Soviet Union to a race to the Moon, the reality is that he, too, was ambivalent about linking space achievement to cold war competition; he saw space cooperation between the U.S. and the Soviet Union as an alternate path to U.S. leadership. Kennedy's first inclination upon taking office was to use space as an area for tension reduction with the Soviet Union; in his Inaugural Address, the new president addressed the Soviet leadership, saying "Let both sides seek to invoke the wonders of science instead of its terrors. Together let us explore the stars."[5] In the early months of the Kennedy administration, there was a concerted effort to find feasible areas of U.S.–USSR space cooperation. But the 12 April 1961 flight of the first human in orbit, Soviet cosmonaut Yuri Gagarin, and its international and domestic aftermath convinced Kennedy that he had to compete in space with the Soviet Union in order to avoid a significant loss of U.S. national prestige and to demonstrate that the United States, not the Soviet Union, was the superior technological and military power.[6]

Kennedy's advisors were blunt in their linkage between space achievement and cold war competition. In their 8 May 1961 memorandum recommending that Kennedy set a lunar landing as a national goal, NASA Administrator James Webb and Secretary of Defense Robert McNamara argued that "[O]ur attainments [in space] are a major element in the competition between the Soviet system and our own .... In this sense, [they] are part of the battle along the fluid front of the Cold War." As he announced his decision to go to the Moon, Kennedy equated the venture with U.S. leadership, saying it was "time for this nation to take a clearly leading role in space achievement."[7]

The tension between the imperative to beat the Soviet Union to the Moon and the desire for overall space leadership was implicit in the program that the president approved in May 1961, which had as its central focus the lunar landing objective but also called for an across-the-board acceleration of U.S. space efforts. This tension surfaced in an argument between Kennedy and NASA Administrator James Webb as the president met with his advisors on 21 November 1962 to discuss the NASA budget. Kennedy declared "[T]his is important for political reasons, international political reasons. This is, whether we like it or not, in a sense a race ....

5. Public Papers of The Presidents of the United States: John F. Kennedy, 1961 (Washington, DC: U.S. Government Printing Office, 1962), p. 2.

6. See John M. Logsdon, *The Decision to Go to the Moon: Project Apollo and the National Interest* (Cambridge, MA: MIT Press, 1970) for a detailed account of this decision.

7. James E. Webb and Robert McNamara, "Recommendations for Our National Space Program: Changes, Policies, Goals," 8 May 1961, reprinted in Logsdon, *Exploring the Unknown*, p. 444. John F. Kennedy, "Urgent National Needs," Speech to a Joint Session of Congress, 25 May 1961 in ibid., p. 453.

Everything that we do ought to really be tied into getting onto the Moon ahead of the Russians." Webb retorted "Why can't it be tied to preeminence in space?"

As he prepared to leave the meeting, the president asked Webb to prepare a letter stating his position on why space preeminence, and not just being first to the Moon, should be the country's goal: "I think in the letter you ought to mention how the other programs which the Agency is carrying out tie into the lunar program, and what their connection is, and how essential they are to the target dates we're talking about, and if they are only indirectly related, what their contribution is to the general and specific things possibly we're doing in space."[8]

Webb's letter was sent to the president on November 30. In it, Webb said that in his view "[T]he objective of our national space program is to become preeminent in all important aspects of this endeavor and to conduct the program in such a manner that our emerging scientific, technological, and operational competence in space is clearly evident." Webb emphasized that "[T]he manned lunar landing program, although of highest national priority, will not by itself create the preeminent position we seek."[9]

President Kennedy seems to have accepted the basic argument made by James Webb—that preeminence in space should be the guiding objective of the national space program. In a 17 July 1963 press conference, Kennedy responded to a press report that the Soviet Union was not planning to send its cosmonauts to the Moon, saying, "The point of the matter always has been not only of our excitement or interest in being on the moon; but the capacity to dominate space, which would be demonstrated by a moon flight, I believe, is essential to the United States as a leading free world power. That is why I am interested in it and that is why I think we should continue."[10]

John Kennedy never gave up on the hope that the space relationship between the U.S. and the Soviet Union could be changed from competition to cooperation. With the October 1962 successful outcome of the Cuban missile crisis in hand, in 1963 Kennedy sought to engage the Soviet leadership in reducing global tensions through such agreements as the Limited Test Ban Treaty. Space was part of this "peace offensive." On 20 September 1963, Kennedy went before the General Assembly of the United Nations and said "[I]n a field where the United States and the Soviet Union have a special capacity—in the field of space—there is room for new cooperation ... I include among these possibilities a joint expedition to the moon."[11]

---

8. "Transcript of Presidential Meeting in the Cabinet Room of the White House, November 21, 1962." This transcript can be found at *http://history.nasa.gov/JFK-Webbconv/* (accessed 7 November 2006).

9. James E. Webb, Administrator, NASA, Letter to the President, 30 November 1962, in Logsdon, *Exploring the Unknown*, p. 461.

10. "News Conference 58," John F. Kennedy Library and Museum, *http://www.jfklibrary.org/Historical+Resources/Archives/Reference+Desk/Press+Conferences/003POF05Pressconference58_07171963.htm* (accessed 25 August 2006).

11. Public Papers of the Presidents of the United States: John F. Kennedy, 1963, (Washington, DC: U.S. Government Printing Office, 1964), pp. 567–568.

Kennedy's top advisor Theodore Sorenson later explained this apparent switch in policy:

> I think the President had three objectives in space. One was to ensure its demilitarization. The second was to prevent the field to be occupied [by] the Russians to the exclusion of the United States. And the third was to make certain that American scientific prestige and American scientific effort were at the top. Those three goals all would have been assured in a space effort which culminated in our beating the Russians to the moon. All three of them would have been endangered had the Russians continued to outpace us in their space effort and beat us to the moon. But I believe all three of those goals would also have been assured by a joint Soviet-American venture to the moon. The difficulty was that in 1961, although the President favored the joint effort, we had comparatively few chips to offer. Obviously the Russians were well ahead of us at that time . . . . But by 1963 our effort had accelerated considerably. There was a very real chance we were even with the Soviets in this effort. In addition, our relations with the Soviets, following the Cuban missile crisis and the test ban treaty, were much improved—so the President felt that, without harming any of those three goals, we now were in a position to ask the Soviets to join us and make it efficient and economical for both countries.[12]

Like Dwight Eisenhower before him, John F. Kennedy tried to avoid direct competition with the Soviet Union as the defining feature of the U.S. space effort, in his case by trying several times during his brief presidency to turn space into an area for cooperative tension reduction rather than zero-sum competition. Even so, much more than Eisenhower, Kennedy was willing to accept the alternative of U.S.–USSR competition if the cooperative option was not feasible. It is impossible to know what might have happened in this respect if Kennedy had been able to complete two terms as president. But with his assassination, the Apollo program came to be seen as one of his legacies, and there was no possibility of shifting it to a cooperative undertaking. Getting to the Moon before the Soviet Union became the defining goal of the U.S. space effort between 1963 and 1969. When, in 1968, it appeared as if the Soviet Union might send cosmonauts *around* the Moon, without landing, before the United States, the Apollo schedule was modified to insert the Apollo 8 circumlunar mission in December 1968. Although the public record supports the argument that this shift was made for programmatic reasons

12. Theodore Sorenson interview by Carl Kaysen, 26 March 1964, in John F. Kennedy Presidential Library, Boston, MA.

having to do with the fact that the lunar landing module was not ready for a scheduled December 1968 test flight, some (including members of the Apollo 8 crew) have suggested that the threat of being beaten to the Moon by the Soviets was an important factor in the decision to fly Apollo 8 to lunar orbit.[13]

The next major opportunity for determining the character of the U.S. space effort came in 1969, as it became clear that the U.S. would soon achieve Kennedy's goal of a lunar landing "before the decade is out." On February 13 of that year the new president, Richard M. Nixon, asked for a "definitive recommendation on the direction which the U.S. space program should take in the post-Apollo period."The president chartered a Space Task Group chaired by Vice President Spiro Agnew to prepare that recommendation; the group's report was submitted to President Nixon on September 15. It noted that "for the short term, the race with the Soviets has been won" and that "[P]ublic frustration over Soviet accomplishments in space, an important force in support of the Nation's acceptance of the lunar landing in 1961, is not now present. Today, new Soviet achievements are not likely to have the effect of those in the past." Based on this reasoning, the Space Task Group proposed that the political goal of the post-Apollo program should be "to promote a sense of world community" by expanded international participation in U.S. space efforts, rather than to pursue another unilateral demonstration of U.S. strength through space achievements.[14]

The absence of visible Soviet competition in space at the end of the 1960s made such an approach feasible and reduced somewhat the political saliency of the U.S. space effort to overall U.S. foreign policy objectives.[15] As they discussed the significant cuts to the NASA budget that had been made in the immediate aftermath of the Apollo 11 and 12 missions, President Richard Nixon told NASA Administrator Thomas Paine that "[O]ne of our main troubles . . . is that the Soviets have not been flying dramatic missions for a long time" and that "[I]t was an unfortunate truth that new Soviet spectaculars were what the public needed to get interested in U.S. space activities."[16]

Such an approach reflected a more muted view of the impact of the cold war per se on U.S. space efforts. Rather than make bilateral space competition "part of the battle along the fluid front of the Cold War," the United States would use its space

13. See Robert Zimmerman, *Genesis: The Story of Apollo 8* (New York: Dell, 1999) for a discussion of the various factors leading to the decision to fly the mission.

14. Space Task Group Report to the President, "The Post-Apollo Space Program: Directions for the Future," September 1969, Appendix A, pp. 7, 16, 27.

15. Although the U.S. intelligence community was aware of Soviet development of systems capable of sending cosmonauts to the Moon and of the failures of those systems, this information was not publicly available, and the Soviet Union denied that it had a lunar landing program.

16. Thomas O. Paine, "Meeting with the President," 22 January 1970; Memorandum for the Record, 22 January 1970, NASA Collection, University of Houston, Clear Lake Library.

capabilities as part of its strategy of global leadership, potentially in partnership with many other nations. As long as the Soviet Union remained a strong military and political power, there would be a challenge to U.S. leadership, but the events of the 1960s, from the Cuban missile crisis to the Apollo 11 lunar landing, had changed the nature of the cold war threat and its impact on U.S. space activities. The link between space capabilities and the U.S. global image was not lost on Nixon and his closest advisors. One of them, Caspar "Cap" Weinberger, commented as additional cuts to the NASA budget were being contemplated in mid-1971 that such cuts would provide confirmation that "our best years are behind us, that we are turning inward, reducing our defense commitments, and voluntarily starting to give up our super-power status, and our desire to maintain world superiority." Nixon responded, "I agree with Cap."[17]

Weinberger's memorandum came in the midst of the debate over whether to develop the Space Shuttle as the next major U.S. space program. There is little specific mention of U.S.–Soviet competition in the arguments NASA put forth in trying to convince the White House to go forward with the Shuttle, although NASA Administrator James Fletcher did suggest in his "best-case" paper that "Man has learned to fly in space, and man will continue to fly in space. This is a fact. And, given this fact, the United States cannot forgo its responsibility—to itself and to the free world—to have a part in manned space flight." He added, "For the U.S. not to be in space, while others do have men in space, is unthinkable, and a position which America cannot accept."[18]

Rather than continue to compete with the Soviet Union in space during the 1970s, the United States pursued a cooperative strategy. The Space Task Group had suggested that "[I]n the case of the USSR, experience over the past ten years makes clear that the central problem in developing space cooperation is political rather than technical or economic."[19] As part of its strategy of détente with the Soviet Union, the Nixon administration approved the Apollo–Soyuz Test Project; the 1975 "handshake in space" was intended to symbolize a new era in U.S.–Soviet relations, both in space and overall. This initial high-profile cooperative venture was potentially to be followed by a docking between a Space Shuttle and a Soviet *Salyut* space station and then by joint development of a larger space station.[20] However, this cooperation fell victim to increased U.S.–USSR tensions in the wake of the Soviet invasion of Afghanistan, and was never pursued.

17. Caspar Weinberger, Memorandum for the President, "Future of NASA," 12 August 1971 in Logsdon, *Exploring the Unknown*, p. 547.

18. James C. Fletcher, "The Space Shuttle," 22 November 1971 in ibid., p. 556.

19. Space Task Group, p. 17.

20. A. P. Aleksandrov, USSR Academy of Sciences, and A. M. Lovelace, NASA, "Agreement between the USSR Academy of Sciences and the National Aeronautics and Space Administration of the USA on Cooperation in the Area of Manned Space Flight," 11 May 1977, in *Exploring the Unknown: Selected Documents in the History of the U.S. Civil Space Program*, Volume II, External Relationships, John M. Logsdon, Dwayne A. Day, and Roger D. Launius, ed. (Washington, DC: NASA Special Publication 4407, 1996), p. 215.

Like Dwight Eisenhower 20 years earlier, President Jimmy Carter was not convinced that civilian space leadership was an essential element of U.S. global power. In his first space policy statement, issued on 11 May 1978, Carter listed "United States space leadership" as his third priority for civil space efforts. Later that year, another White House policy statement noted that "[I]t is neither feasible nor necessary at this time to commit the US to a high-challenge, highly-visible space engineering initiative comparable to Apollo."[21] In this view, Carter was an exception to the judgment of the four presidents who had preceded him that space leadership was important.

As he entered the White House in 1981, President Ronald Reagan brought with him a strongly anticommunist perspective that colored his stance toward the Soviet Union in the first several years of his presidency. The U.S.–Soviet agreement on space cooperation that had been initiated in 1972 and renewed in 1977 was allowed to lapse in 1982. In the first Reagan administration statement of space policy, issued that same year, "space leadership" was once again identified as one of the "basic goals" of U.S. space activities.[22]

As NASA sought presidential approval for development of a space station in late 1983, Administrator James Beggs told President Reagan that "President Kennedy's decision to go to the Moon chartered a course that resulted in leadership in space for the United States"; that "President Nixon, against the wishes of many, continued America's commitment to leadership in space by approving the Space Shuttle"; and that "this focus on leadership in space was reaffirmed in your Space Policy." Beggs suggested that "[I]n the 1990s, leadership in space will have a new dimension, something perhaps that Presidents Nixon and Kennedy could not foresee when they committed America to leadership in space . . . [T]he new dimension will be the presence of the private sector in space." Beggs referred to the *Salyut* space station as "the centerpiece of the Soviet program" and said, "[W]hat worries me is what the Soviets are up to. What are they planning to fly in the late 1980s and the 1990s? Will they be successful in their plans to dominate space?" Beggs concluded his sales pitch by noting that "[O]ur leadership in space these past 25 years told the world that America was strong and that America accepted the challenge of space, and that she was equal to the responsibilities of leadership." Asking the president to approve space station development, Beggs concluded his presentation by saying, "[T]he stakes are enormous: leadership in space for the next 25 years." The final viewgraph accompanying his presentation to the president showed an artist's conception of the

21. Presidential Directive/NSC-37, "National Space Policy," 11 May 1978 in Logsdon, *Exploring the Unknown*, p. 574; and Zbigniew Brzezinski, Presidential Directive/NSC-42, "Civil and Further National Space Policy," 10 October 1978, in ibid., p. 576.

22. White House, National Security Decision Directive Number 42, "National Space Policy," 4 July 1982 in Logsdon, *Exploring the Unknown*, p. 590.

space station with the highlighted legend "a highly visible symbol of U.S. strength." [23] Once again, the goal of space leadership and cold war competition were intertwined for a sympathetic president.

President Reagan not only approved space station development; as he announced his decision in his 25 January 1984 State of the Union address, he also said, "NASA will invite other countries to participate so we can strengthen peace, build prosperity, and expand freedom for all who share our goals."[24] Although Canada and Europe had made contributions to the development of the Space Shuttle in the 1970s, this announcement escalated international cooperation in the development of the next major U.S. space program to a central feature of U.S. space strategy, marking a definite transition from the unilateral demonstration of national power that had fueled the Apollo program to an approach where the U.S. would demonstrate its leadership as the managing partner in a long-term, highly visible, multilateral undertaking. Still, the invitation to participate was limited to U.S. allies; the existence of the cold war still conditioned the U.S. move toward a cooperative approach.

By the end of Ronald Reagan's second term in office, the end of the cold war was well in sight, as the reforms of Mikhail Gorbachev took hold and the Soviet Union struggled with its internal economic and political problems. Even so, in the aftermath of the Soviet launch of its very large Energia booster in May 1987, following on the launch of the core of the Mir space station a year earlier, Time magazine headlined the cover story of its 5 October 1987 issue: "Surging Ahead: The Soviets Overtake the U.S. as the No. 1 Spacefaring Nation." The article suggested that the Soviet Union "had surged past the U.S. in almost all areas of space exploration" and that "if unchallenged, Moscow is likely to become the world's dominant power in space by the twenty-first century."[25] Twenty-five years earlier, this sort of report might have provoked a debate over how to respond to a new Soviet space challenge, but there was no such reaction in 1987. The U.S. space effort was focused on recovering from the January 1986 Challenger accident and on getting started with the space station, and Soviet space achievements were not perceived as a major threat to U.S. interests.

Rather, the United States revived its space cooperation agreement with the Soviet Union in 1987, and a year later Mikhail Gorbachev suggested to Ronald

---

23. NASA, "Revised Talking Points for the Space Station Presentation to the President and the Cabinet Council," 30 November 1983 in Logsdon, *Exploring the Unknown*, pp. 595–600.

24. President Ronald Reagan, "Speech on the State of the Union," 25 January 1984, *http://www.reagan.utexas.edu/archives/speeches/1984/12584e.htm* (accessed 8 November 2006).

25. Michael D. Lemonick, "Surging Ahead: The Soviets Overtake the U.S. as the No. 1 Spacefaring Nation," in *Time*, 5 October 1987, *http://www.time.com/time/magazine/printout/0,8816,965658,00.html* (accessed 13 September 2006).

Reagan that the two countries cooperate in a human mission to Mars.[26] The final Reagan administration statement of space policy, issued on 11 February 1988, stated that "[A] fundamental objective guiding United States space activities has been, and continues to be, space leadership." The statement went on to say that "[L]eadership in an increasingly competitive international environment does not require United States preeminence in all areas .... It does require United States preeminence in key areas of space activity critical to achieving our national security, scientific, technical, economic, and foreign policy goals."[27]

THE IMPACT OF THE END OF THE COLD WAR: LEADERSHIP THROUGH COOPERATION

This lengthy excursion into the three-way relationship between the cold war, the U.S. quest for space leadership, and the choices that have defined the U.S. civilian space program was intended to demonstrate that even before the end of the cold war the quest for global leadership, rather than direct U.S.–Soviet competition, had been the primary political influence on the evolution of the U.S. space program. Cold war competition was, of course, the single most important contextual factor influencing this quest for leadership in the 1957–1991 period, but it was a secondary, not fundamental, consideration.

If this analysis is accepted, then the end of the cold war should have had a significant, but not decisive, impact on space in the post-cold war era. An obvious impact, of course, was that the Russian Federation, which inherited most Soviet space capabilities, became a politically attractive space partner for the United States rather than a peer competitor. This was especially the case, given the economic problems faced by the new Russian government headed by Boris Yeltsin and the desire of President George H. W. Bush and the successor Clinton administration to support Yeltsin's democratic reforms.

The United States was quick to recognize the changed situation. In 1992, the United States and Russia reached initial agreement to have the U.S. Space Shuttle rendezvous with the Soviet Mir space station; this initiative resurrected a cooperative concept that had been agreed to 15 years earlier. Then, in 1993, the White House embraced a proposal suggested by the NASA Administrator and the Russian space leadership to invite Russia to join the space station program together with the United States "friends and allies" that had been partners in the program since its inception. In

26. Memorandum of Conversation, "The President's First One-on-One Conversation with General Secretary Gorbachev," 29 May 1988, *http://www.margaretthatcher.org/archive/displaydocument.asp?docid=110610* (accessed 8 November 2006).

27. White House, Office of the Press Secretary, "Fact Sheet: Presidential Directive on National Space Policy," 11 February 1988, in Logsdon, *Exploring the Unknown*, p. 602.

essence, the U.S. and the Soviet Union merged their future human spaceflight efforts; the activity which had been the central focus for competition for more than 30 years became a highly visible arena for post-cold war cooperation.[28]

A more fundamental impact of the end of the cold war was the need for a redefinition of the meaning of space leadership, from being superior in space to the Soviet Union in all or most areas, to some other definition. The 1992 report "A Post Cold War Assessment of U.S. Space Policy" recognized this need. It noted that "[W]ith the end of the Cold War . . . the term 'space leadership' takes on new meaning." It suggested that "[T]o remain a leading nation in space continues to be in the U.S. interest." The report also recognized that "[S]pace leadership must be earned. By maintaining unsurpassed technological capabilities in key areas and using those capacities effectively and efficiently, the United States will have the capability to act independently, visibly, and impressively when and where it chooses." The Task Group concluded that

> Future space leadership, then, requires combining challenge, openness, quality of execution, and productive application of results. Proceeding ahead with a well-conceived, successfully executed national space program aimed at concrete objectives that are scientifically, economically, and socially beneficial, and that serve important U.S. interests, is the best way to ensure leadership in space. Leadership, in this sense, becomes both a goal in itself and the result of excellence in formulating goals for space and achieving them as planned.
>
> It is this concept of leadership that should guide future U.S. activities in space.[29]

Although there had been a growing emphasis on U.S. leadership in cooperative space activities beginning with the 1969 Space Task Group report, which increased with the 1984 decision to make the space station a cooperative undertaking, the 1992 report suggested that the U.S. develop a "cooperative strategy" as a "central feature of its future approach to overall space policy." The panel recommended that

> The United States should take the initiative in shaping a common international agenda in selected areas of civilian and national security space activity . . . . Enhanced international cooperation should be sought not only for its programmatic benefits, but also because it is the preferred way for the United States to influence the direction of future space undertakings around

28. For a discussion of the evolution of U.S.–Russian cooperation in human spaceflight, see John M. Logsdon and James Millar, "U.S.–Russian Cooperation in Human Spaceflight: Assessing the Impacts," *Space Policy* (August 2001).

29. Vice President's Space Policy Advisory Board, "A Post Cold War Assessment," pp. 13, 15.

> the world. Broader national security, political, technological, and economic benefits for the United States can flow from a carefully crafted 'cooperative strategy' . . . .[30]

In the 15 years since these words were written, they appear to have been reflected in the U.S. strategy for space. A statement of national space policy issued in 1996 reflected a definition of leadership as both a desirable goal and a product of excellence in formulating and executing the nation's space program. That policy noted that "[F]or three decades, the United States has led the world in the exploration and use of outer space" and that "[W]e will maintain this leadership role by supporting a strong, stable, and balanced national space program." The 1996 policy also recognized the desirability of enhanced international cooperation, saying that "[T]he United States will seek greater levels of partnership and cooperation in national and international space activities."[31]

Whether the U.S. civilian space program has over the past 15 years been implemented at the level of excellence that translates into recognized leadership is, at minimum, questionable, but that is a topic for a different paper. The U.S. approach to fulfilling its leadership role in the partnership now known as the International Space Station has also had its ups and downs. In recent years, the national and NASA leadership appear to have recognized the importance of getting the space program back on a positive track if the United States is to be more than a leader in rhetoric. The most important step in this direction, of course, was the White House decision to make human and robotic exploration of the solar system the overriding goal of the U.S. civilian space program. International participation in space exploration under U.S. leadership was an important element of that decision. The White House also made the tough decision that its leadership role in space required the United States to honor its international commitments with respect to the space station, even as pressures to retire the Space Shuttle and accelerate progress in space exploration argued for a different decision. As the 1992 report suggested, NASA has taken the lead in crafting a "cooperative strategy" with respect to space exploration, and space agencies from around the world are working with NASA to flesh out the substance of that strategy.

Recent events, then, suggest that the 1992 assessment of what actions would best serve U.S. interests with respect to the contributions of its space efforts to broader national goals is now being pursued. The Vice President's Space Policy Advisory Board concluded that the end of the cold war called for a U.S. space program based on excellence in formulation and execution, one which was carried out in concert with other nations. The most recent national space policy, approved on 31 August 2006, states that the first priority goal of the U.S. in space is to "strengthen the nation's

---

30. Ibid., p. 42.

31. The White House, National Science and Technology Council, "Fact Sheet: National Space Policy," 19 September 1996, *http://history.nasa.gov/appf2.pdf* (accessed 9 November 2006).

space leadership." The policy also states that it is in the U.S. interest to "encourage international cooperation with foreign nations and/or consortia on space activities that are of mutual benefit and that further the peaceful exploration and use of space, as well as to advance national security, homeland security, and foreign policy objectives."[32]

Leadership in space has been an important goal for the United States for almost 50 years. The path to that leadership for the first 30-plus years of the Space Age, during the cold war, was primarily by besting the Soviet Union in visible space achievements. Even so, from at least 1969 on, there has been a cooperative aspect to U.S. space strategy. With the end of the cold war, leadership in space cooperation became the primary path to leadership overall. That shift in focus, from competition to cooperation, is the primary space impact of the end of the cold war.

32. The White House, Office of Science and Technology Policy, "Fact Sheet: National Space Policy" 6 October 2006, *http://www.ostp.gov/html/US National Space Policy.pdf* (accessed 13 July 2007).

Chapter 7

The *Taikonaut* as Icon: the Cultural and Political Significance of Yang Liwei, China's First Space Traveler

James R. Hansen

In 2005 the government of the People's Republic of China (PRC) sponsored the development of a new video game featuring heroes from Chinese history. The plan was to wean Chinese young people off their growing addiction to Western video games and replace it with something appropriate to Chinese values. Unlike American video games in which players slay dragons, fight aliens, beat up bad guys (or, more likely, be the bad guys themselves), in the new game "Chinese Heroes" players click on icons of select Chinese heroes to learn about their noble experiences and carry out healthy and constructive tasks like moving bricks and darning socks. An official with China's General Administration of Press and Publication, which sponsored the game's development by a Shanghai gaming company, hoped the game "will teach players about Chinese ethics."[1]

Five heroes are featured in the video game:

- Bao Zheng: an eleventh century statesman renowned for his battle against government corruption, strong sense of fair play, ability to tell truth from falsehood, and determination to mete out justice without fear or favor;
- Yue Fei, a twelfth century general who, with only 800 soldiers, defeated an invading army 500,000-strong. Before he left home to join the army at age 18, his mother allegedly tattooed four characters on his back which meant "Serve the country loyally," a constant reminder to protect China at all costs;
- Zheng He, the eunuch admiral of the Ming dynasty whose "treasure ships" sailed across the Indian Ocean to Africa in the early fifteenth century;

1. Kou Xiaowei, an official with China's General Administration of Press and Publication, quoted in "Game On for Chinese Heroes," 29 August 2006, *http://english.aljazeera.net* (accessed 12 June 2006).

- Zheng Chenggong, a pirate who seized Taiwan from Dutch colonial rule in 1661; and
- Lei Feng, the People's Liberation Army soldier and faithful Party member credited by Chairman Mao for his cheerful selflessness and modesty.

How popular the game "Chinese Heroes" will become for the estimated 20 million Chinese now playing video games daily is questionable. It may not be serious competition for American video games such as "Grand Theft Auto," in which the starring role is played by an "unstoppable bad-ass" who wreaks havoc in a gritty Miami Vice-like environment and where the player can customize his character with every manner of tattoo, "jack" a cop's car, watch a pimp "beat down" a prostitute, and start an epic gun battle using a flamethrower, grenades, sniper rifle, Colt-45 pistol, AK-47, or sawed-off shotgun. The director of the Beijing Internet Addiction Treatment Centre, Tao Ran, has expressed doubt that Chinese Heroes will appeal very much to China's youth, saying, "Teenagers seek adventure and fulfillment in dramatic and skill-demanding games. If hero games do not focus on killing and domination, gamers will definitely not play them."[2]

One very powerful way that China has successfully combined graphic violence with its traditional appreciation for certain select types of heroes is in its martial arts movies. Notable within this extremely popular genre is the 2003 Oscar-nominated film, *Hero*, the most expensive Chinese film ever made. Set during the Warring States Period (shortly before the unification of China in the third century BC), the film tells the story of assassination attempts on the king of Chin by three legendary warriors who seek revenge for Chin's subjugation of their lands. The king justifies his actions as necessary for the peace of China, a justification that the sole surviving assassin (played by actor Jet Li) ultimately understands and accepts. Only a strong leader, the first "emperor," can unite all of China, and only through unification can the Chinese people ever escape civil war and find peace.

Western critics assailed the film's message as "totalitarian" and "pro-Chinese reunification" (vis-à-vis Taiwan), and for promoting a "sinister ethic that blatantly justifies the murder and repression of political opponents." That is why, critics said, the Beijing government so strongly endorsed the film—because the Chin emperor stood for today's rulers.[3] But other observers viewed the film differently, saying that it was a tale of sacrifice and love, one that embraced Confucian values and the ancient Chinese ethic that the very best people in society must care for the people first. As director Zhang Yimou said about his film, "The final assassin understands

2. Ibid. See also "A Clash of Cyber Cultures," *The Standard: China's Business Newspaper*, 24 September 2005, *http://www.thestandard.com.hk* (accessed 12 June 2006).
3. See, for example, Shelly Kraicir's review of the film, "Absence as Spectacle: Zhang Yimou's 'Hero,'" in *Cinema Scope Magazine* 5 (Spring 2003): p. 9.

that if he doesn't kill the Emperor, it's better for the people, because the civil war will end. The number one martial arts fighter decides not to kill the king, for the sake of peace."[4]

Ten months after the film opened in Hong Kong, the Chinese launched another "Hero," this time in the form of 38-year-old Yang Liwei. The rocket carrying the *yǔhángyuán* (or Chinese word for astronaut) was a Long March 2F, one in a series of guided missiles named after the historic journey of 6,000 miles of 1934–1935 in which an army of 80,000 soldiers led by Mao Zedong, surrounded by the Nationalist army of Chiang Kai-shek, fled their bases in south China and escaped to the north, with only some 8,000 surviving the trek. The Long March became the central metaphor of Chinese revolutionary mythology and a source of inspiration for all subsequent Red Guards.[5]

After several years of speculation about the possibility of a piloted spaceflight by the Chinese, the launch finally came on 15 October 2003, 56 years to the month after the launch of the world's first satellite, the Soviet Union's Sputnik (4 October 1957), from Jiuquan Satellite Launch Center on the high Gobi Desert, some 1,600 kilometers northwest of Beijing—about as far off the beaten path as one can find even in a country as large as China. Yang Liwei's spacecraft, the *Shenzhou V* (Divine Vessel V), closely resembled the Russian *Soyuz*, which hardly surprised Western observers given that Chinese engineers had been working closely with the Russians since 1994 and had benefited from access to complete blueprints and the full-scale *Soyuz* spacecraft.[6]

---

4. Zhang Yimou, quoted in Liza Bear, "Fighting for Peace (and Art Films), Zhang Yimou on 'Hero,'" 27 August 2004, *indieWIRE, http://www.indiewire.com* (accessed 12 June 2006).

5. On the Long March of 1934–1935, see Harrison E. Salisbury, T*he Long March: The Untold Story* (New York: Harper & Row, 1980); Jean Fritz, *China's Long March* (New York: Putnam, 1988); and Benjamin Yang, *From Revolution to Politics: Chinese Communists on the Long March* (Boulder, CO: Westview Press, 1990). For a firsthand Chinese account of the Long March, see Changfeng Chen, *On the Long March with Chairman Mao* (Beijing: Foreign Language Press, 1972.) On the Long March family of rockets, see Hormuz P. Mama, "China's Long March Family of Launch Vehicles," *Spaceflight* 37 (September 1995).

6. The best overview of the Chinese space program can be found in Brian Harvey, *China's Space Program: From Conception to Manned Spaceflight* (Chichester, U.K.: Springer, 2004). Harvey also provides a good bibliography. One of the most active publishing analysts of the Chinese missile and space programs is Great Britain's Philip S. Clark. In the U.S., James R. Oberg follows Chinese developments closely; several of his papers have appeared in IEEE *Spectrum*. See also "Testimony of James Oberg: Senate Science, Technology, and Space Hearing: International Space Exploration Program, 27 Apr. 2004," *http://www.spaceref.com*. See also Chen Lan, "Dragon in Space: A History of China's Shenzhou Manned Space Programme," *Spaceflight* 46 (April 2004): pp. 137–144.

Inside the *Shenzhou V* spacecraft, the *taikonaut*, as Western journalists had come to call a Chinese astronaut,[7] spent a little more than 21 hours in space, orbiting Earth 14 times before reentering the atmosphere and parachuting down onto the steppe of central inner Mongolia. The instant Yang Liwei made orbit over the Pacific at 9:10 a.m. Beijing time on October 15, the Chinese knew they had accomplished something remarkable: their countryman, Yang Liwei, had made the history books, joining the elite company of Russia's Yuri Gagarin and America's Alan Shepard as "first men" into space.[8]

Who was Yang Liwei? What was his background and training? Why was he selected for China's pioneering mission? How did Chinese society react to the news of his spaceflight? What sort of icon did Yang become; what sort of hero did he represent? Just as there is no way to fathom what the U.S. space program has meant to America without understanding what Americans wanted from their heroes—what they projected onto their heroes over the past 45 years—there is also no way to understand what the Chinese are after in space, or here on Earth, without understanding the iconography that has developed around their astronauts.

What is known about Yang Liwei in the West is what official Chinese sources have told us.[9] Yang was born on 21 June 1965, in northeast China's Lioaning Province, a major industrial region not far from the Yalu River that forms the Chinese border with North Korea. Yang's mother was a teacher and his father an economist, meaning that his family, by Chinese standards, was by no means humble or poor. According to official biographies, Yang had a "happy and tranquil childhood" and was "very intelligent and a good team leader of his playmates."[10] Excelling in mathematics and math competitions, Yang scored high on entrance exams and went to the best

7. The Western term taikonaut blends the Chinese word for outer space (taikong) with the English word "astronaut." The term is not used by the Chinese themselves. Apparently the term was coined in May 1998 by a Malaysian journalist, Chiew Lee Yih, who first used it in newsgroups. The Chinese word term for space, taikong, literally means "great emptiness." In the Chinese language itself, the term *y'uhángyuán* ("universe navigator") has long been used for "astronaut." Official English texts issued by the Chinese government actually use the term "astronaut."

8. "China Puts Its First Man in Space," BBC News (15 October 2003), *http://www.news.bbc.co.uk*; "World Acclaims Launch," China Daily, 13 October 2003, *http://www.chinaembassy.org*; John Pomfret, "China Launches Its First Manned Space Mission," *The Washington Post*, 15 October 2003: p. A1; Brian Berger, "Shenzhou 5 Launch Introduces New Dynamic in Space," *SPACENEWS*, 20 October 2003: p. 6; Deborah Zabarenko, "United States Shrugs Off China Space Launch," *Reuters English News Service*, 22 January 1999; Philip S. Clark, "Shen Zhou 5—Flying the Red Flag," *Spaceflight* 46 (February 2004): pp. 54–60; Tim Furniss, "China Celebrates First Manned Orbital Mission," *Spaceflight* 45 (December 2003): pp. 488–489.

9. On what is known about the life of Yang Liwei prior to his *Shenzhou V* mission, see Brian Harvey, *China's Space Program*, pp. 2–14. See also "The Making of China's Astronaut," Xinhaunet, 17 October 2003, *http://www.chinaview.com*, and "First Chinese Astronaut Talks about Eight Years in Training," 24 July 2006, *http://www.spacedaily.com* (both accessed 12 June 2006).

10. "The Making of China's Astronaut," 17 October 2003, *http://news.xinhuanet.com* (accessed 12 June 2006).

middle school in his county. Joining the People's Liberation Army at age 18, Yang was recruited by one of the Chinese Air Force's top aviation colleges, where he earned the highest grade in every class he took. Upon graduation, he became a fighter pilot and was rated as an "elite" member of his Air Force division.

If there is a Chinese equivalent to "the Right Stuff," Yang Liwei had it. As an attack aircraft pilot, he demonstrated his crisis management during a very low-flying exercise over northwest China's barren Xinjiang region. Losing one of his jet engines, Yang, "with great calmness," radioed his situation, skillfully climbed to 4,921 feet (1,500 meters), managed to get his plane over snow-covered Mt. Tienshan, and landed "without hesitation"[11] even after his other engine had flamed out. Climbing out of the cockpit, dripping wet with sweat, amid cheers from his colleagues, Yang was greeted by his division commander who awarded him with an on-the-spot promotion.

In all, Lt. Col. Yang accumulated more than 1,350 hours of flying time in the Air Force. In 1996, from a pool of 1,500 candidate pilots, he was chosen for spaceflight training along with 12 others, and went to a special institute in Beijing where he passed a rigorous 30-course curriculum. "To establish myself as a qualified astronaut," Yang was later quoted as saying, "I studied harder than even in my college years and received tougher training that even that which made me a fighter pilot."[12] During the first two years of training, he reportedly never went to bed before midnight and rarely even left the training center. In a bid to improve his English, he often called his wife from his apartment in Space City, asking her to help him practice. So dedicated was he to training that his wife once found him moving rapidly in circles at home on a swivel chair. His training directors described him as "a sober-minded person with a superb capability for self-control." In a critical series of final simulations leading to his selection for *Shenzhou V*, Yang identified and remedied all the "faults" his instructors had set up for him. After each, when the instructor asked him whether he had made any errors, Yang confidently replied, "No errors at all!" A psychologist who asked him how he would feel if he were to fly a real spacecraft got the answer, "I'll be more relaxed than talking to you, so let me go for the flight!"[13]

The Chinese government tried to keep the identity of its first taikonaut a secret, but a few days before the launch Yang's identity was discovered and his picture published in the Hong Kong newspaper *Wen Wei Po*.[14] Originally, Beijing agreed to a live broadcast of the launch, but apparently lost its nerve at the last minute. Thirty minutes after *Shenzhou V* successfully reached orbit, the government's flagship

---

11. Ibid.

12. Ibid.

13. Ibid.

14. Harvey, *China's Space Program*, p. 2.

television station cut into regular programming to make the proud announcement. Televised replays of the launch quickly followed, beginning a day of saturation coverage in the Chinese media. Yang was shown walking out of his quarters in his flight suit and getting on a bus taking him to the launch pad. Waiting for him at the bus was the president of China, Hu Jintao, who made a few remarks about the great significance of the mission for China. In the U.S. space program, remarks made by astronauts at launch and during their missions—such as "Light this candle," "The clock has started," and "Godspeed, John Glenn"—became colloquial. The most widely reported remark made by Yang Liwei came when he met his president at the bus. "I will not disappoint our Motherland," he said. "I will complete each movement with total concentration, and I will gain honor for the People's Liberation Army and for the Chinese nation."[15]

What the Chinese people seem to have appreciated most during the flight of *Shenzhou V* were Yang Liwei's communications with his 8-year-old son, Ningkong.[16] In a Confucian society (which, of course, China has remained despite its communism), the father–son relationship is fundamental not only for the family but for all society and politics.[17] Whereas in the U.S. the most memorable in-flight comments from America's astronauts have rarely had much to do with children or family, in China a great emphasis was placed on Yang talking lovingly while in space to his "dear wife" and "dear son." What Americans seemingly most remember have been the vintage, off-color vernacular of our "space cowboys"—comments like Gus Grissom's "No bucks, no Buck Rogers" or "The issue here is not pussy; the issue is monkey," or Alan Shepard's "Please, dear God, don't let me f*** up." These earthy types of American expressions, it seems clear, will never pass from the lips of any *taikonaut*.

15. Quoted in Pomfret, "China Launches Its First Manned Space Mission," p. A17.

16. On the celebrity of Yang Liwei's son Ningkong see Joe McDonald, "Chinese Astronaut an Instant Hero But Where Is He?" *Associated Press*, 22 October 2003, *http://www.space.com* (accessed 12 June 2006).

17. On Confucian values in modern China, see Thomas A. Wilson, ed., *On Sacred Grounds: Culture, Society, Politics, and the Formation of the Cult of Confucius* (Cambridge, MA: Harvard University Asia Center, 2002); Honghe Liu, *Confucianism in the Eyes of a Confucian Liberal: Hsu Fu-kuan's Critical Examination of the Confucian Political Tradition* (New York: P. Lang, 2001); Howard Giskin and Bettye S. Walsh, eds., *An Introduction to Chinese Culture through the Family* (Albany, NY: State University of New York Press, 2001); and Theodore Huters, R. Bin Wong, and Pauline Yu, eds., *Culture & State in Chinese History: Conventions, Accommodations, and Critiques* (Stanford, CA: Stanford University Press, 1997). On the persistence of Confucianism within Communist China, see Zongli Tang and Bing Zuo, *Maoism and Chinese Culture* (New York: Nova Science Publishers, 1996); Edward Friedman, *National Identity and Democratic Prospects in Socialist China* (Armonk, NY: M. E. Sharpe, 1995); and Deborah Davis-Friedmann, *Long Lives: Chinese Elderly and the Communist Revolution* (Stanford, CA: Stanford University Press, 1991).

In the Chinese press, a great deal was made about Yang as husband and family man. According to his wife, Zhang Yumei,[18] a middle-school teacher, Yang is a caring husband, and to his son, a hero. Following the launch, his wife said she would never forget the expression in her husband's eyes when she was about to be carried into the operating room for a kidney biopsy in 2001: "Just at the moment when I was to enter the operating theater, I saw the expression of extreme care, love, and regret like I've never seen. I felt as if a knife had pierced my heart."[19] During her recovery, Yang sat constantly at her bedside—that is, until time came to leave for more *taikonaut* training when, as a supremely dedicated member of the People's Liberation Army, he declined an offer from his commander to postpone it. Following his return from space, a picture showing Yang and his wife embracing appeared in virtually every Chinese newspaper. Its caption said that his wife asked her husband what wonderful things he saw in space. "I saw our planet," he told her. "It's so beautiful, like you."

From all across China came an outpouring of national pride over the spaceflight of Yang Liwei. *The People's Daily*, the official newspaper of the Chinese Communist Party, ran 100,000 extra copies that were quickly snapped up. In some towns, there were spontaneous parades and demonstrations. Schoolchildren exhibited their pictures of spaceships and astronauts. Hundreds of wall posters appeared, many of them combining themes of twenty-first century technology with more traditional styles of socialist realism. Postage stamps were printed in Yang's honor. *The People's Liberation Army Daily* trumpeted: "For China this is the beginning and there will be no end."[20]

A week after this flight, Yang, accompanied by his son Ningkong, opened an exhibition in Beijing of his *Shenzhou V* capsule, spacesuit, and parachute—an exhibit that subsequently made a road show across China. Yang's immediate political value lay in Hong Kong, where his visit, at the special invitation of the regional government, lasted a full six days.[21] For 155 years Hong Kong had been a British colony, until its sovereignty transferred in 1997 and it became a "special administrative region" of the PRC. Under the terms of the Sino-British Joint Declaration, China promised that Hong Kong with its 6.8 million people would enjoy a relatively high degree of autonomy until at least the year 2047. Under the "One Country, Two Systems"

---

18. It is curious (though not in the least significant) that the name of Yang Liwei's wife, Zhang Yumei, is so close to the name of the director of the film "Hero" (and other celebrated Chinese films), Zhang Yimou.

19. Quoted in Pomfret, "China Launches Its First Manned Space Mission," p. A17.

20. Quoted in John Pomfret, "China's First Space Traveler Returns a Hero," *The Washington Post*, 16 October 2003: p. A22.

21. Andrew Brown, "China Astronaut on Charm Offensive," 1 November 2003, *http://www.cnn.com*; Joe McDonald, "China's First Man in Space Going Public," *Associated Press*, 24 October 2003, *http://www.space.com*; "HK Newspapers Hail Space Hero Yang Liwei," *Xinhaunet*, 1 November 2003, *http://news.xinhaunet.com* (all accessed 12 June 2006).

policy, Hong Kong would retain its own legal system, currency, customs policy, cultural delegation, international sport teams, and immigration laws. Late into 2003, however, the issue of unification remained deeply troubled. Morale in Hong Kong was low; its economy weak due to perceived government meddling; and officials of the new regime were so unpopular that the city had been hit by unprecedented antigovernment protests, not to mention an outbreak of severe acute respiratory syndrome (SARS) that in one stretch had attacked 685 and killed 116 people.

Colonel Yang's visit was widely interpreted—more before it *began* than after it ended—as a cynical bid to promote Chinese nationalism by elevating Hong Kong's confidence and restoring the city's image. Several Hong Kong and Taiwan newspapers criticized the *taikonaut*'s visit as a thinly veiled attempt to boost pro-Beijing political parties in the region's upcoming elections. Correspondents reported that many Hong Kong residents were indifferent to Yang's feat.[22] "It's nothing new—America did it years ago," a local businessman was quoted as saying. "I won't feel anything just because of his visit," admitted a downtown shopkeeper.[23] "It's just a gimmick," declared an accounting clerk. A 21-year-old female university student stated, "I always liked Britain better." As for Yang's spaceflight, she said it was being "blown totally out of proportion."[24]

But even before Yang arrived, there were signs that many people in Hong Kong were not so jaded. A survey by the Hong Kong Federation of Youth Groups showed that 71 percent of Hong Kong young people felt excited and more proud of being Chinese by the news of the *Shenzhou V* flight. Support for his visit came from nearly 50 organizations in Hong Kong, not all of them so pro-Beijing, and his itinerary attracted people of all ages and from walks of life. For the first time in the history of Hong Kong's Science Museum, an exhibit was kept open around the clock, for four straight days, to meet the popular demand. When the *taikonaut* arrived, several thousand people waving Chinese and Hong Kong flags lined the streets. "It's worth the wait," said a 73-year-old man. "I never thought I would live to see the day that China could proudly stand alongside the United States and Russia as nations that sent a man into space!" "I think they should make a cartoon strip of Yang," said an 11-year-old boy.[25] "I just want to shake hands with Uncle Yang," said an elementary school student. A young female student picked by her schoolmates

22. "Hong Kong Ho-Hum over Chinese Astronaut's Visit," *Taipei Times*, 1 November 2003: p. 5; "Kids and Flags but Still No Genuine Cheer for Astronaut," *Taipei Times*, 2 November 2003: p. 5.

23. Quotes in "HK Welcomes China's Space Hero," BBC News, 31 October 2003, *http://news.bbc.co.uk* (accessed 12 June 2006).

24. Quotes in Min Lee, "First Chinese Taikonaut Visits Hong Kong," *Associated Press*, 31 October 2003, *http://www.space.com* (accessed 12 June 2006).

25. Quotes in Alan Low, "Spaceman Yang Gets Star Treatment," *The Age*, 2 November 2003, *http://www.theage.com.au* (accessed 12 June 2006).

to ask a question ofYang explained to a TV reporter, "I know it will be difficult, but I want to be the first woman astronaut of the nation."[26]

Chinese authorities hoped that the taikonaut would be an inspirational force for all Chinese—especially in Hong Kong, where the recent mood had been downbeat. "The moral encouragement that you have brought to the Hong Kong people is enormous," declared Hong Kong's chief executive, Tung Chee-hwa, at the Science Museum's ribbon-cutting. At Hong Kong Stadium, a capacity crowd of 40,000 gave Yang a standing ovation as he entered and was driven around the stadium in a golf cart. Inside was a party of local pop celebrities and movie stars, including Jackie Chan, with whom he sang a song. Jackie Chan asked the children of Hong Kong to take Yang as their model, stressing that success in life depended upon dedication and heartfelt effort.[27] Yang did not make a speech at the stadium, but the crowd was pleased just to see him in the flesh. "I'm just happy he is here," said one young man, "because he was brave enough to fly into space, and that courageous act has brought prestige for Chinese everywhere." [28]

In a country where the concept of celebrity was relatively new but quickly growing—including the likes of pop singer Gao Feng, NBA basketball player Yao Ming, *Crouching Tiger, Hidden Dragon* film star Zhang Ziyi, fashion model Sun Na, Olympic diving star Guo Jinjing, and badminton superstar Lin Dan—it was clear inside the stadium that the people saw in the *taikonaut* not just celebrity but a man with a special aura or mystique. "Uncle Yang looks more handsome in person than on the TV screen," said a primary school student who came to the stadium with her father.[29] Following the event, the *Chinese Language Daily* commented, "Yang is not just a star. The welcome he received from Hong Kong residents exceeded that of any star. He is the superstar supported by Hong Kong residents of different age groups and different walks of life."[30]

Interestingly, the qualities ofYang's character that appealed most to the Chinese were not those connected to the glamour and glitz of the silver screen or pop music, but to characteristics more essentially Chinese. One Hong Kong schoolteacher said,

26. Quoted in "Hong Kong People Greet Hero Yang Liwei," *People's Daily*, 1 November 2003, *http://www.english.people.com.cn* (accessed 12 June 2006).

27. Immediately after Yang Liwei's visit to Hong Kong, Hong Kong's Moral and Civic Education Section of the Education and Manpower Bureau made available on its Web site a battery of teaching resources based on Yang's *Shenzhou V* flight. The title of the Web site material was "Perseverance and Commitment: Space Mission ofYang Liwei." See also "Yang Liwei's Space Mission Online for Moral Education," *The People's Daily*, 8 December 2003, *http://www.china.org.cn* (accessed 12 June 2006).

28. Quote in Low, "Spaceman Yang Gets Star Treatment."

29. Quoted in "Yang Liwei Meets HK Community," *Chinese Language Daily*, 2 November 2003, *http://www.chinadaily.com.cn* (accessed 12 June 2006).

30. See Low, "Spaceman Yang Gets Star Treatment."

"It is really an unforgettable experience to see Yang in Hong Kong, and I admire his modesty and calmness very much." Newspaper article after article described Yang as "looking healthy and respectful and speaking in appropriate terms, with honest and cordial attitude." In return, Yang spoke very gently and respectfully about his hosts. In brief remarks at the museum exhibition, the *taikonaut* said he was "deeply moved" by the warmth of his reception from the "big Chinese family." At another event, he stated that "The Hong Kong compatriots make me feel at home. Their enthusiasm has made my heart beat faster than when I was in the spacecraft."[31] Leaving the city after six days, he paid his final respects, "Hong Kong is much more beautiful than I had imagined. It is like a pearl."

Before he left, Yang visited the Bank of China Tower, in the central city, where the bank president presented a check of a half-million Hong Kong dollars (about $64,000 USD) to the China Space Foundation, in support of China's research and development of space technology. Yang also went on a sightseeing visit to Tsing Ma Bridge, the world's longest road-and-rail suspension bridge, which links the new Hong Kong International Airport to Kowloon and Hong Kong Island.[32] In both places, people saw the human embodiment of the Chinese space program juxtaposed with other vital symbols of China as a progressive force in the world.

Leaving Hong Kong, Yang Liwei next traveled to Macao, China's second Special Administrative Unit, whose sovereignty had transferred from Portugal to Beijing in 1999. Some of Macao's problems in 2003 were different from Hong Kong's—notably involving labor unrest due to economic transformations that had transformed the oldest European colony in China from a tiny fishing village with gambling dens into a well-established tourist spot with huge, modern casinos. But, as with Hong Kong, national unification under Beijing's leadership was the overarching concern. A visit from the new Chinese hero might help.

During a two-day visit, Yang visited several of Macao's historic landmarks and then met for an entire afternoon with 1,000 students and teachers, answering questions. He visited the recently installed Macao Garrison of the People's Liberation Army and attended a luncheon hosted by the garrison. As a result of those two days, an ad hoc consortium in Macao raised more than 14 million *patacas* (about $1.75 million USD), for the China Space Foundation—an organization that promotes China's space industries but which is not to be confused with the China National Space Administration, the counterpart to America's NASA. Most of this money came from "political and business dignitaries" (Macao's chief executive was reported to have made a personal contribution of 300,000 patacas), but some of it (how much was not reported) came from students and common folks. According

31. Quoted in "Yang Liwei Meets HK Community."

32. "China's First Spaceman Tours Hong Kong," *Xinjuanet*, 2 November 2003, *http://www.pladaily.com.cn* (accessed 12 June 2006).

to Beijing, this sizeable donation from the 450,000 people residing in the Macao Special Administrative Region "embodied a strong sense of patriotism of the Macao compatriots." On a per capita basis, the gift represented a donation of 31 patacas (or $4 USD) for every resident. To put that into perspective, if every American today gifted $4 to the U.S. space agency, NASA would cash a check for some $1.2 billion.[33]

Leaving Macao, "Great Hero Yang" next stopped in the northern coastal metropolis of Tianjin, the largest open-water seaport in North China, which, along with Beijing, Shanghai, and Chongqing constitutes one of the PRC's four administrative municipalities. Why Tianjin after Hong Kong and Macao? Perhaps it was because Tianjin had been one of the places slapped with a travel advisory by the World Health Organization for its SARS outbreak a few months earlier. Some of the most violent protests against locating SARS clinics in local communities had taken place in Tianjin, and it may have been that Beijing wanted a visit from the *taikonaut* to rejuvenate the city's spirit. It may also have had something to do with the fact that Tianjin is the center for many of China's pillar industries: machinery, electronics, chemicals, and metallurgy.

The impact of Yang Liwei's historic spaceflight ranged far beyond the first three spots strategically selected for his immediate post-flight tour. It triggered nothing less than a nationwide frenzy—what one Western observer called a "flowering of patriotic kitsch." In Shanghai, an estimated half-million people queued in freezing conditions to see China's first astronaut. At a high profile rally in Beijing, Yang was conferred the title of "Space Hero." A decree issued by the General Political Department of the Chinese People's Liberation Army instituted Yang as a "model" for all Chinese soldiers. The decree ordered all members of the PLA and the Chinese People's Armed Police to learn from Yang and his "heroic achievement." "Military activities in various forms should be conducted to study the spirit of the astronauts," the decree said.[34]

One of the most interesting things that happened after the *Shenzhou V* flight was the immediate commercialization of Yang's name. Everything from rice to milk to action figures quickly bore the taikonaut's image, name, or title.[35] The Chinese government tried to put a stop to this by trademarking and copyrighting Yang's name and likeness, but with limited effect.[36] Even Yang's home county in northeast China got into the act. Trying to cash in on his fame, Suizhong county leaders registered his name as a trademark for local produce, selling "Great Hero Yang" lettuce and cabbage and renaming a special white pear after him.

---

33. "China's Space Hero Visits Hong Kong Garrison," *PLA Daily,* 5 November 2003, *http://www.english.chinamil.com* (accessed 12 June 2006).

34. "China's First Astronaut Crowned 'Space Hero,'" *Xinhaunet*, 7 November 2003, *http://www.chinaview.com* (accessed 12 June 2006).

35. Sangwon Suh, "Hong Kong: Selling Chinese Milk and Patriotism," *International Herald Tribune*, 9 March 2004, *http://www.iht.com* (accessed 12 June 2006).

36. Barry Gottlieb, "Don't Touch That Name," *Alternet*, 28 October 2003, *http://www.alternet.org* (accessed 12 June 2006).

At one point Beijing felt it had to put him on ice for a while, to temper the individual side of the achievement which it at first had been aggressively promoting. Commenting on Yang's public absence, a professor of public policy at Qinghua University in Beijing said, "It is normal that Yang Liwei has been regarded as a national hero and a good example for the young to learn from." However, "[T]he government should make sure there aren't excessive reports about one individual, because behind the success there was a whole project and system supporting the mission." "Yang Liwei's name will long be recalled," said a message posted on a Web site run by the Party newspaper, *People's Daily,* "while nobody will talk about the politicians!" [37]

But in the spring of 2004 the attention on "Great Hero Yang" again heightened when the *taikonaut* toured the U.S. In New York, he met with Secretary-General of the United Nations Kofi Annan and presented two U.N. flags he had carried with him on *Shenzhou V*. In Washington, Yang visited the office of Florida Senator Bill Nelson, the only serving member of Congress to have flown in space; while there, he also met Apollo 11 astronaut Buzz Aldrin. Yang toured the Kennedy Space Center, met Mickey Mouse at Disney World,[38] and got a VIP's view of the Johnson Space Center in Texas. Yang's American tour was widely reported in the Chinese press and was even shown on Chinese television.

What seems clear is that Yang's hero status signified some sort of sea-change in Chinese society and politics, because such publicity for a living person had been almost unknown in China's communist system prior to Yang. China had lauded "national martyrs" such as Wang Wei, the fighter pilot who died in a 2001 collision with a U.S. Navy plane, but when looking for people to serve as communist "models," the party usually picked plumbers and bus drivers for brief fame as "model workers." It has tried hard not to celebrate the cult of any individual other than leaders of the regime.

But there was Yang after his spaceflight, an instant hero, an icon, lionized in the state-run press not only as the country's first person in space but also as an elite pilot, a star student, Communist Party member, devoted family man, and national treasure. Yang's was an image crafted seemingly for a ruling party in need of a high-tech hero to bolster Chinese nationalism and pep up its own reputation—the same party whose very existence depended on the *group* being more important than any *individual*, and whose power often depended on its leaders hogging the spotlight.

Compared to China's bland, group-oriented leadership, Yang apparently had struck an extraordinarily responsive chord. Even his 8-year-old son became a celebrity, showing up over and over again in the Chinese media. On one occasion, party officials visited his school and honored his class with the honorary title "Space Squadron." Standing beside

37. Quoted in McDonald, "Chinese Astronaut an Instant Hero But Where Is He?" *Associated Press*, 22 October 2003.

38. "China's First 'Taikonaut' Rides Mission SPACE at Epcot, Walt Disney World in Florida," Disney Press Release, 28 May 2004, *http://www.hongkongdisneyland.com* (accessed 12 June 2006).

39. Quoted in McDonald, "Chinese Astronaut an Instant Hero," *Associated Press*, 22 October 2003.

a model rocket, young Ningkong gave a speech praising his father's accomplishment. "People asked me if I was afraid about Daddy going into space and I said 'not a bit,' because I knew that China's space technology was very advanced and Daddy was really awesome," he said. "I want to be like Daddy and travel to outer space some day." [39]

Whether or not Yang Liwei's son ever travels to space, it seems more and more clear that other Chinese youth of his generation will be doing exactly that—to orbit, to a Chinese space station, and perhaps to the Moon and to Mars. Though impossible for any Western analyst to predict with confidence what the Chinese will do next in space, or when they will do it, it seems clear from the public reaction to the October 2003 *Shenzhou V* flight that the Chinese people are excited by their prospects in space—excited in some ways that Beijing did not fully anticipate and could not fully control. Interestingly, when the two-man crew of *Shenzhou VI* flew into space in October 2005, the government essentially hid those *taikonauts* from view. There were a few celebratory events in Hong Kong and China, but nothing like Yang Liwei's road show, over which the government felt for a time that it had lost control.[40]

Whether Yang's visits to Hong Kong and Macao effectively served the nationalistic and political purposes Beijing had in mind is not so certain. Chinese officials commented at the time that they hoped his visits would instill pride for the larger "Motherland" among the residents of China's two Special Administrative Regions, but although the events with Yang sold out and were hugely popular, they do not seem to have had exactly the desired effect. Indeed, people living in those recently transferred sovereignties celebrated the *taikonaut* but without associating him all that much with the Beijing government or by warming to the mainland's way of life.

Evidence of this dissonance can be seen in the reaction to the subsequent announcement that the Chinese national anthem and a 45-second video featuring Yang and Olympic diving star Guo Jinjing were to be broadcast on Hong Kong's Chinese-language TV stations every night. Co-produced by Hong Kong's Committee on the Promotion of Civic Education and its Commission on Youth, the video (the soundtrack of which was *The March of the Volunteers*, the Chinese national anthem) aimed at "enhancing the sense of national unity." But critics in Hong Kong asked, "Does the government really believe that civil awareness can be raised by broadcasting the national anthem? Following this logic, will the government increase the number of times the video is shown if they believe people's civil awareness is not high enough?" One Hong Kong political commentator remarked, "People will only treat it as propaganda."[41]

---

40. On *Shenzhou VI*, see Edward Cody, "China Sends 2nd. Manned Spacecraft Into Orbit," *The Washington Post*, 12 October 2005: p. A2; Guy Gugliotta, "China Keeps Reaching for the Stars," *The Washington Post*, 15 October 2005: p. A12; and "Astronauts' Safe Return Sparks Big Celebrations," *China Daily*, 17 October 2005, *http://english.sina.com/china* (accessed 12 June 2006).

41. Quotes in Martin Wong, "Hong Kong: National Anthem to be Broadcast Before News," *South China Morning Post*, 1 October 2004, *http://www.asiamedia.ucla.edu* (accessed 12 June 2006); "Hong Kong: Chinese National Anthem Video Draws Fire," *South China Morning Post*, 7 October 2004: p. 1.

The effects of Yang's visits, then, were not exactly what Beijing was after; rather, the effects are best understood as a cascade of unintended consequences that may, if examined carefully, tell us something very important about how China has been changing, and will continue to change, as the twenty-first century heirs to the "Middle Kingdom" move out slowly but surely into exploring the universe.[42]

It would be prudent, especially for those whose heritage is not Chinese, to be extremely cautious in drawing conclusions about what the PRC might do in the future in terms of space exploration, based on what we think we know about China's past. One of the principal lessons of Chinese history that has been related over and over again in the West concerns the overseas voyages of Admiral Zheng He, one of the heroes of the PRC's new video games. As the lesson goes, a Confucian faction, after gaining control over the Ming court in the early fifteenth century, put an abrupt halt to the grand naval expeditions. The conservatives felt that "barbarian" nations offered little of value to add to the prosperity already present in the Middle Kingdom, and, anyway, it was improper for decent Chinese to go abroad while their parents were still alive. Western historians have speculated on how differently world history might have turned out had the Ming emperors sustained a vigorous colonial policy instead.

Advocates of a vigorous American human spaceflight program have made a similar object lesson of China's turning away from foreign ventures: by withdrawing from exploration, American society, too, will stagnate and open itself to exploitation by others. Space enthusiast Robert Zubrin, in his 1999 book *Entering Space*, declared:

> By accepting the challenge of the outside world, Western civilization blossomed outward to dominate the globe. In contrast, the grand Chinese civilization grew demoralized in its stagnation and implicit acceptance of inferior global status and decayed, ultimately to be completely disrupted and remade by expansive Western influences.
>
> Only twenty-five years ago, the United States, following in the footsteps of the Ming emperors, abandoned its own pioneering program of space exploration. At that time, American leaders could console themselves with the equivalent of the advice of the Ming court bureaucrats—exploration is too expensive, and nothing of value exists beyond what is familiar.

---

42. Another bellwether of the underlying changes taking place in China today may be seen in the newest Chinese history textbooks being introduced into some schools in Shanghai. Instead of straight-jacketing students with texts based in communist ideology and the teachings of Mao, the emphasis of the new textbooks is on producing innovative thinking and preparing students for global discourse. Not that history and politics have been completely disentangled; far from it. But the new textbooks reflect a new emphasis, one that is indicative of what one of its authors, a Shanghai scholar, calls a sea-change in thinking about what students need to know. See Joseph Kahn, "Where's Mao? Chinese Revise History Books," *The New York Times*, 1 September 2006.

"Now we know better," Zubrin has insisted. "The universe has presented us with its challenge. To remain who we are, we must accept. We must enter space."[43]

But Zubrin's analysis and others like it, which suggest there is something *implicit* in the Confucian mindset and within the social order of China's "inner space" that ultimately works against exploration of "outer space," may be fallacious. Rather than any inherent Chinese cultural inertia favoring the familiar and avoiding the unexpected, perhaps the underlying factor forcing China's Ming emperors to withdraw from their foreign ventures was something quite different, and very particular, historically speaking. For example, in his 2005 book, *Why Geography Matters*, Michigan State University geography professor Harm de Blij argued that, just as the Ming were poised to round the Cape of Good Hope and enter the Atlantic, "disaster struck at home." A geological event, a "Little Ice Age" in north China, resulted in major famine and social disorder. The Ming rulers were forced to end their maritime expeditions, ordering the country's shipyards to build only barges that could navigate China's internal waterways with cargoes of rice. If the environmental crisis had not occurred, China might very well have become the world's dominant colonial power.[44]

It thus seems inappropriate, if not foolish, to believe that anything innate in the country's historical character will stop China from becoming one of the world's predominant space powers. If the iconography surrounding *taikonaut* Yang Liwei is any sort of reliable indicator, Chinese society is already well on its way toward successfully mixing its traditional Confucian values, communist ideology, and drive for economic and high-tech industrial competitiveness into an effective recipe for an expansive program of human spaceflight.

43. Robert Zubrin, *Entering Space: Creating a Spacefaring Civilization* (New York: Tarcher/Putnam, 1999), p. xi. For more from Zubrin on the Mings' turning away from overseas exploration, see pp. 18–20.

44. Harm de Blij, *Why Geography Matters: Three Challenges Facing America: Climate Change, the Rise of China, and Global Terrorism* (New York: Oxford University Press, 2005), pp. 78–79.

Section III

Commercial and Economic Impact

Chapter 8

Commercial and Economic Impact of Spaceflight: An Overview

Philip Scranton

"Present efforts to accelerate the transfer of military/space technologies to commercial application appear handicapped by insufficient knowledge of how technology is applied at the level of the firm . . . most contributions [do not] come in the form of direct and readily identifiable results of a particular effort." Denver Research Institute, *The Commercial Application of Missile/Space Technology* (Denver, CO: DRI, September 1963), vi, p. 1.

"Over the past quarter century, two-thirds of our space dollars have been invested in manned spaceflight, with little to show for the investment save circus. By and large it has been wonderful circus, just as the [early aviators'] barnstorming was, but hardly more productive . . . The real payoff in space—the work of the communications satellites, the weather and earth resources satellites, the scientific probes—has been funded by the remaining one-third of the civilian space budget." Alex Roland, "Barnstorming in Space," in *Space Policy Reconsidered*, Radford Byerly, ed. (Boulder, CO: Westview Press, 1989), p. 42.

Apart from wowing the public and energizing thousands of American engineers and scientists, what is spaceflight actually *for*? Apart from creating platforms for military surveillance and intelligence functions, what is spaceflight actually *for*? Apart from sustaining layers of enterprises and competencies in the aerospace, instrumentation, electronics, materials, project management, and consulting segments of the economy (along with a modest cohort of university-based researchers, including some historians), what is NASA's work actually *for*? If not articulated in just this fashion, similar questions about the private sector payoffs of public sector spending on space have persisted from the Agency's earliest years. By 1962, NASA had created a "technology utilization program" in response to congressional sniping, and soon began releasing annual reports of technology spinoffs.[1] Yet recently an economist argued that neither NASA's direct or indirect impact on the economy (in terms of employment, multiplier effects, spinoffs, or statistically-estimated influences on growth) could be established as sufficiently sizable to justify ongoing costs.[2] Even so, from NASA's earliest years, observers have argued that there is no effective way to reach quantitatively sound answers to questions about economic impact and the effective use of public funds for aerospace. Instead, as Roger Launius suggested in a recent historiographical study, settled camps of celebrants and critics seem to commence with their conclusions and search about for evidence to support them.[3] This is not likely to change, but I hope not to reinforce that practice here.

Like many public programs, NASA's work has been expected to be productive not simply in terms of internal criteria (setting and reaching goals, effective management, etc.), but also in terms of wider contributions to society and the nation. One persistent strain in the calls for this broader influence and performance has been the expectation that NASA's work will generate substantial and substantive commercial and economic innovations and opportunities. The folkloristic notion that the main thing spaceflight delivered to American consumers was the powdered breakfast drink Tang[4] evokes the narrow domains in which such expectations have often been defined and the relatively unsystematic way in which actual contributions have been gauged in differently positioned literatures.

This essay, which introduces a group of more tightly focused and empirically rich studies, will undertake to outline a framework that may help situate the economic and technical contributions that NASA's work provided to American business and thereby to the nation's citizens. Contrary to consumer expectations, virtually all these contributions have been indirect, as a Denver Research Institute (DRI) study explained in the early 1960s, and hence imperceptible to most observers. Second, as Alex Roland

1. Alfred Alibrando, "NASA Technology Utilization Scrutiny Due," *Aviation Week and Space Technology* 79 (5 August 1963): pp. 28–29. See also "Aerospace in Perspective: Utilization," *Space/Aeronautics* 43 (January 1965): pp. 80–87, and *NASA Spinoffs: Thirty Year Commemorative Edition*, Sarah Gall and Joseph Pramberger, ed. (Washington, DC: NASA Technology Transfer Division, 1992).

2. Molly K. Macaulay, "The Economics of Space," in *Space Politics and Policy: An Evolutionary Perspective,* Eligar Sadeh, ed. (Dordrecht, The Netherlands: Kluwer, 2002), pp.181-200, see esp. pp. 183–185. Of course, economic justification is constrained by seriously imperfect quantification and measurement issues, and cannot reach economic dimensions of spacefaring that are not quantifiable, such as enhancing the climate for or urgency of innovation, except through dubious proxies (e.g., R&D spending) whose limited utility, though long recognized, has not undercut their wide use.

3. Roger D. Launius, "Historical Dimensions of the Space Age," in ibid., pp. 3–26, esp. 4–11.

4. Indeed, this association is erroneous in the specifics as well, in that Tang was used in but not developed for NASA flights. See *http://en.wikipedia.org/wiki/Tang_(drink)* (accessed 11 July 2006). General Foods created Tang circa 1957, but it became famous when used in the Gemini program in 1965. Similar delusions about the NASA origins of Teflon® and Velcro® abound, but frame the popular expectations for space outcomes—consumer goods as economic spinoffs.

has grumbled, it does seem that our obsession with human flights, along with their vast expense, has overshadowed efforts at communications exostructure building which have had long-term significance—notably, as almost all parties agree, multiple satellite networks (agency, private, and military), the subject of another section in this book.

Two contextual claims lead off my discussion. First, spacefaring innovations were embedded in a more complex, indeed transnational, culture of technological experimentation that intensified in the early years of the cold war. As a result, explained the DRI report, "[N]o clear line can be drawn between space and non-space knowledge [bases] because the two are closely interwoven."[5] Thus, identifying the space program's "firsts" is a rather inconsequential exercise. Instead, it makes more sense to highlight those situations where NASA projects added critical momentum and capability to nascent innovations, providing essential test-beds for them (and the funding for revision and redesign), and to explore projects where the complexity of NASA-posed problems galvanized cross-disciplinary amalgams of technique and materials, with implications for the industrial world outside. For example, the DRI's research identified 185 spaceflight spinoffs by 1962, including plastic coatings, microminiature welding devices, and high-capacity infrared sensors already installed in manufacturing facilities.[6] Thus, one key background argument here is that a sizable share of NASA undertakings can plausibly be described as "experimental development" projects, in which exploratory engineering, risk-laden fabrication and testing, and integrative scientific practices strove to overcome the liabilities inherent in complexity, incomplete information, inhospitable conditions of human work and artifact usage, and the necessity of customized production.

Second, though we still have inadequate information regarding many of its dimensions, the military's role in space projects and technologies enlarged NASA's programs, such that NASA-anchored innovations had an impact on the development of defense and intelligence capabilities (and the hardware and software behind

5. Denver Research Institute, *The Commercial Application of Missile/Space Technology* (Denver, CO: Denver Research Institute, 1963), p. iv. For a contemporary appreciation of the DRI study, see "Scientific Horizons to Watch," *Factory* 122 (February 1964): pp. 88–91. For a later, detailed analysis by DRI staffers, see Robert H. Waterman and Lloyd G. Marts, "Space-Related Technology: Its Commercial Use," in *Advances in Space Science and Technology*, Vol. 6 (New York: Academic Press, 1964), pp. 173–244.

6. "Hitchhiking on Space Technologies," *Steel* 153 (23 December 1963): pp. 26–30. NASA's Louis Fong then estimated that the space effort had yielded some 850 "innovations thought to have industrial potential." Among the 185 that had been utilized, *Steel* reported specifics on 22 items. Sundstrand had found oil field applications for a low-speed pump "originally designed for missiles," Hills-McCanna was selling a commercial version of an aerospace ball valve, and General Dynamics had already sold 5,000 units of a mechanical warning device indicating shock levels on packages, created for missile shipment containers.

them) and, by extension, on the enterprises designing and fabricating such devices.[7] As researchers noted more than 40 years ago: "Because of the interaction among [the military, NASA, other agencies, industry, and universities], attribution of a given technological advance to a particular source is often impossible."[8]

The idea for military uses of space evidently arose in the immediate postwar years, as the Douglas Aircraft Company's Research and Development division (which soon became RAND) informed the Army Air Forces that it might be possible to create and launch a "world-circling spaceship," which could play a role in the technology-intensive wars of the future. The proposal, filed away for more than a decade, resurfaced after Sputnik and, much altered, laid the foundations for the military satellite command and control networks "designed specifically to support the nation's top secret National Reconnaissance Program [NRP]."[9] Operated in parallel with NASA satellite projects, the NRP was another element in the rivalries between the Air Force and the spaceflight agency. The question of space weapons is outside our scope here, but "the non-weapon military uses of space," as Colin Gray termed them, have been diverse and significant. From early, high-resolution film canister devices and the Vanguard/Minitrack system of the 1950s[10] through the fourth-generation, ten-ton Big Bird satellite in the 1970s, surveillance, missile tracking, and data gathering from space were clear military priorities, along with advance warning capabilities of attack, assessment of Soviet bloc activity, secure communications, navigation assistance, weather monitoring, extensive mapping, and continuing "measurement of the [E]arth's gravitational and magnetic fields (essential for the accuracy of inertially-guided missiles)."[11] Both NASA and military agencies engaged much the same set of

7. See Peter L. Hays, "NASA and the Department of Defense: Enduring Themes in Three Key Areas," in *Critical Issues in the History of Spaceflight*, Steven J. Dick and Roger D. Launius, eds. (Washington, DC: NASA SP-2006-4502, 2006), pp. 199–238. In discussion after his presentation at the conference this volume reprises, Hays suggested that the military (nonpublic) segment of NASA budgets was as large or larger than the civilian segment in many years following the end of the Apollo era. For another Hays analysis of these issues, see Peter L. Hays, "Space and the Military," in Sadeh, ed., *Space Politics and Policy*, pp. 335–369.

8. Denver Research Institute, *Commercial Application of Missile/Space Technology*, p. vi.

9. David Christopher Arnold, *Spying from Space* (College Station, TX: Texas A&M University Press, 2005), p. 7. For a contemporary overview of military space activity, see "Aerospace in Perspective: Control," *Space/Aeronautics* 43 (January 1965): pp. 88–101.

10. The first recovery of a photographic capsule was reportedly from Discoverer 13 in August 1960. See Colin Gray, *American Military Space Policy* (Cambridge, MA: Abt Books, 1982), pp. 23–25.

11. Ibid., p. 35.

prime and subcontractors for satellite projects, which created opportunities for, but not guarantees of, technical cross-fertilization between public and classified initiatives.[12]

In consequence, it may be crucial, in thinking about commercial/economic impacts of spaceflight, to posit NASA as the globe's largest dispersed technology-development unit, inextricably blending national security concerns, engineering ambitions, policy negotiations, and scientific aspirations into projects that energetic, at times urgent, collaborators shaped into material forms and functions. From this perspective, in the long run counting spinoff items may be less rewarding than recognizing NASA's historic role as the hub of Big Engineering in America, reshaping the model for high-tech contracting (and subcontracting), and as the nation's public research and development (R&D) division, taking on challenges and complexities that no corporate enterprise would or could have shouldered during the cold war.[13]

In the remainder of this discussion, I will sketch three domains where spaceflight's nonsatellite commercial and economic impacts can be situated, then will focus briefly on three of the third domain's components and add a fourth. In considering areas where spaceflight has economic implications, we may loosely divide this terrain into segments which attend to (1) the impact of *operations in space*; (2) the impact on *enterprises* of producing and managing space projects; and (3) the impact of operations derived from *experimental development for space*.

The first segment represents the realm of "space commerce," dominated by satellite capabilities and the information and revenue streams they generate. In addition, commercial users must cover transportation charges for putting artifacts in

12. The technical press was intensely interested in the components manufacturing potential for satellite operations. Exemplary articles from the early 1960s include "Future Space Satellites and Missions," *Electronic Industries* 21 (June 1962): pp. K12–K16, and "How to Get Your Products into Orbit," *Steel* 153 (29 July 1963): pp. 79–84, with specifications on satellite components at pp. 82–83. See also Octave Romaine, "OAO: NASA's Biggest Satellite Yet," *Space/Aeronautics* 37 (February 1962): pp. 54–58; and Donald Fink, "GE Designs Navigation Satellite Network," *Aviation Week & Space Technology* 80 (30 March 1964): pp. 49–51. The substantial Earth-based tracking and monitoring systems that had to be built in tandem with satellite launches received equal attention, not least because they involved, in some measure, familiar construction tasks. See "Space Tracking Is Big Business," *Electronics* 32 (6 March 1959): pp. 24–25; "Earth Based Electronics," *Electronics* 34 (17 November 1961): pp. 108–118. For satellite manufacturing, see H. L. Wuerfel and R. P. Dunphy [RCA], "The Relay Communications Satellite: A Study in the Achievement of High Reliability," *Industrial Quality Control* 22 (January 1966): pp. 355–368. For a British view stressing the need for European satellites with applications different from those envisioned by the U.S., see Alan H. Stratford, "The Economics of Telecommunications Satellite Systems," *Journal of the Royal Aeronautical Society* 66 (June 1962): pp. 364–370. For a later NASA overview, see Edgar Cortright, "Space Science and Applications: Where We Stand Today," *Astronautics & Aeronautics* 4 (June 1966): pp. 42–48. (By 1966, NASA's Cortright explained, "more than 60 successful satellites and deep-space probes" had been launched [p. 42].)

13. For a sense of the speed and uncertainty of technological development at this time, see Ronald Kohl, "Technology in Turmoil," *Machine Design* 38 (29 September 1966): pp. C1–C20. Arguably, the U.S. Air Force pioneered Big Engineering in and after World War II in aircraft, propulsion, and avionics. Nuclear submarine work brought the Navy alongside before NASA was created.

space, most of which appear to accrue to government units, some in partnerships with private sector organizations. We also have had the recent appearance of individual space travel (ballistic launches only, I believe), though again with the revenues flowing from private to public coffers, usually inadequate to cover costs and not as yet generating anything like nodes of investment opportunity. As political scientist James Vedda has pointed out: "In space commerce, the quest continues for affordable, reliable, and flexible access to space." Here we find "the slowest rate of improvement of all space technologies," even as the sophistication and significance of satellites has soared.[14]

Outside of communications, where waves of innovation have followed like clockwork, notions that commercial materials processing, pharmaceutical development, or crystal growing would become the base for space manufacturing never got past the experimental or prototyping phase. In the early 1980s, Johnson & Johnson partnered with McDonnell Douglas to attempt purifying a drug component in space, while 3M and John Deere arranged materials testing studies and Fairchild laid plans for an industrial space platform. These schemes vaporized even before the *Challenger* accident in 1986, as markets had been substantially overestimated and costs severely underestimated. In 1984 the annual market for space materials testing or production had been gauged at $40 billion or more; a few years later that number shrank to $2 billion, but realistically was closer to zero. Beyond the stunning costs of getting into space with appropriate materials and tools lay the challenges of creating human-free operations (as peopled production in space added orders of magnitude to costs) and of dealing with the narrow range of products that could be sold at prices which could realize profits. Nor were questions of quality and reliability easily solved. In the late 1980s, drug-industry consultant John Naugle judged that only goods priced above $2,000 per pound could be economically manufactured in space, and only if there was a $500 million annual market awaiting them (to help offset operating costs of $100–$200 million). As he summarized:

> A space manufacturer thus must pay for a long, costly R&D process, make a large initial capital investment, and wait a long time for a return on his investment. During the R&D process, his ground-based competitors will be hard at work to produce the product more cheaply . . . These factors combine to suggest why pharmaceutical manufacturers have not leaped to enter the market in space processing.[15]

14. James Vedda, "Space Commerce," in Sadeh, ed., *Space Policies*, pp. 201–227, quote from p. 210. Stephen Johnson orally confirms government losses in providing commercial launch services.

15. John Naugle, "A Manufacturer's View of Commercial Activity in Space," in *Economics and Technology in US Space Policy*, Molly K. Macauley, ed. (Washington, DC: Resources for the Future, 1987), pp. 70, 76. For satellite innovations, helpful sources include Office of Technology Assessment, *Civilian Space Policy and Applications* (Washington, DC: U.S. Congress, Office of Technology Assessment, 1982), especially Chapter 8, and Stephen Johnson, "Space Business," in Sadeh, ed., *Space Politics and Policy*.

Naugle also argued that three situational shortcomings further hampered planning for the industrialization of space: the lack of "assured transportation" up and back; the problem of "launch insurance"; and the absence of a spacelab where firms could "conduct proprietary experiments."[16] Two decades down the road it does not appear that space-based processing is much nearer to being realized.

Second, spaceflight has had durable, creative impacts on corporations that secured prime and subcontracts for vehicles, craft, components, instrumentation, engines, et al. Not only did serving NASA invite key cold war technology-intense firms to diversify beyond weapons systems—creating new divisions and goals, space demand also played a role in building markets for design and program management specialists (e.g., TRW's Systems Technology division and Intellitronics Laboratories)[17] and for technology/innovation consultants. As Stephen Johnson is focusing on business in considerable detail, a quick view of McDonnell in 1959 and 1966 will offer a snapshot of one firm.

In the 1959 fiscal year, McDonnell Aircraft Company (MAC), a St. Louis powerhouse, recorded sales of $436 million, with an orders backlog of $650 million. It had just undertaken its first Mercury spacecraft contract for a modest $20 million, less than 5 percent of the firm's annual turnover. At the time, MAC's space efforts were confined to the Quail decoy missile and the Talos ramjet (subcontracted from Bendix), whereas its core competencies lay in jet aircraft design and assembly—the F4H Phantom II, the RF-101C Voodoo for reconnaissance photography, the F101B interceptor, and the F3H Demon series of all-weather fighters.[18]

Seven years later, MAC was winding up its role in the Gemini program (5 successful flights in 11 months) and had adapted a Gemini capsule for use with the planned Air Force Manned Orbiting Laboratory (which never flew; cancelled in 1969). It also had undertaken extended, self-financed research into "advanced orbital spacecraft, laboratories, [and] military systems in space," investing another $1.3 million in "studies related to Mars exploration." If MAC's aircraft backbone remained the Phantom (seven models in production, 99 units contracted in 1966 for the USAF [$272 million]), the company had branched out into work on Navy surface-to-surface missiles, hypersonic air-breathing planes (with NASA), an aircraft collision-avoidance instrumentation system,[19] and a critical-patient monitoring system for hospitals (derived from piloted flight instrumentation). At least two dozen additional projects were being researched, ranging from special materials to human

16. Naugle, in *Economics and Technology*, p. 78.

17. Aerospace Industries Association of American, *Aerospace Year Book 1960* (Washington, DC: American Aviation Publications, 1960), pp. 191–192.

18. Ibid., pp. 129–131. See also "McDonnell Is Optimistic on Future," *Aviation Week* 72 (9 May 1960): p. 103.

19. The 1992 *NASA Spinoffs*, 61, volume credits collision avoidance instrumentation to an Ames-FAA collaboration in the 1980s and does not mention MAC's earlier effort.

performance simulation and plasma physics.[20] Put simply, spacefaring transformed MAC from a military hardware builder into a multidimensional, edge technology developer. As I have argued elsewhere, to date historians have underinvested in examining the business trajectories of major aerospace manufacturers and the industrial districts they often anchored.[21]

The impacts of *experimental development* for space are more distributed and perhaps diffuse, but they were extensive. This refers to the transformations of materials, processes, and instruments that aerospace design and fabrication triggered for particular purposes, but which were available for adoption and adaptation throughout industry. The 1963 Denver team identified six areas of transfer from space to commercial sectors: stimulation of research; improved processes or techniques; improved products; new products; increased availability of materials and instrumentation; and cost reductions. One of these was speculative and ultimately marginal (cost reductions), and one was obvious (stimulating research), whereas products, new or improved, have often become the focal point of commercialization arguments. In general, criticism focused at the product innovation dimension has paid considerable attention to issues of technical complexity and ostensibly narrow markets, but the sheer volume of the spinoffs that NASA publications have heralded may suggest that the durable impact here was through extensive, cross-sectoral, technical fertilizations, rather than in delivering market blockbusters.[22]

Here we will follow DRI's concern for novel or improved processes, techniques, materials, and instrumentation, innovation efforts that commenced at the outset of

---

20. Aerospace Industries Association of America, *The 1967 Aerospace Year Book* (Washington, DC: Spartan Books, 1967), pp. 131–133. For complementary, contemporary views of this process, see "'Arms Makers" Reshuffled by Shift to Missiles,' *Financial World* 113 (3 February 1960): pp. 3, 27; "New Business from Space Age," *FW* 116 (15 February 1961): pp. 4–5; and "Air Force Contractors: Who's Tending the Store," *Forbes* 85 (1 June 1960): pp. 25–28. For similar portraits of individual firms, see "Profitable Plain Jane" [Northrop], *Forbes* 88 (1 September 1961): pp. 26–27, and "Martin Marietta—Space Leader," *FW* 117 (6 April 1962): pp. 8, 24.

21. For a challenging policy perspective on this issue, see Robert Butterworth, *Growing the Space Industrial Base: Policy Pitfalls and Prospects*, Air War College, Maxwell Paper no. 23, (Maxwell AFB, September 2000). See also Joan Bromberg, *NASA and the Space Industry* (Baltimore:The Johns Hopkins University Press, 1999). A well-known and crucial liability in the economics of space business was the unreliability of the monopsonistic market (there being only one buyer) and persistent uncertainties about whether unplanned costs necessary for redesigns and fixes in experimental development would be covered by NASA. Note that McDonnell (above) spent its own funds on preliminary experiments regarding future aerospace projects in the 1960s; this, of course, affected the bottom line and aerospace stock prices were mercurial throughout this era.

22. The Institute noted that "the more subtle forms of technology transfer have had and will continue to have the greatest impact—not the direct product type of transfer which is most often publicized." (*Commercial Application*, p. 5) See also *NASA Spinoffs*, 1992, and Sean O'Keefe, *Spinoff 2002, 40th Anniversary* (Washington, DC: NASA Technology Utilization Program, 2002). One Web site visited in July 2006 indicates that something on the order of 1,400 spinoffs have been documented in its publications: *http://www.sti.nasa.gov/tto/shuttle.htm* (accessed 16 July 2006).

space initiatives and continue to influence present practice. From day one, operational requirements for spaceflight components were sufficiently demanding that both new materials and revamped processes for fabrication had to be devised. In addition, challenges on the organizational side of project efforts promoted a variety of managerial techniques, some new, some solidified after earlier cold war experimentation, and some whose limits NASA work suggested. These will be briefly highlighted.

On the *manufacturing process* front, we can note innovations such as chemical milling and high-energy forming (cited in the 1963 Denver assessment referenced in the epigraph of this chapter), as well as electron-beam, thermal, numerical control, ultra-cold, and electrical discharge machining; electrolytic grinding; plasma and induced magnetic field welding; plus stretch, magnetic, and shear forming.[23] Chemical milling initially proved valuable in shaping large components (such as missile bulkheads) to close tolerances. Regarded in 1961 as a "recent development," it would shortly become a major element in fabricating computer components, for it involved protecting part of a surface with a masking agent, immersing the object in "an etching bath, which may be acidic or basic, to remove [material] from specific areas to produce the desired configuration," then stripping off the masking agent. This technique was particularly suited to "[d]esigns normally not considered producible by mechanical means," and could serve as an alternative to "mechanical milling [for] complicated shapes."[24] According to J. B. Mohler, the Boeing engineer who outlined the process and offered a dozen pointers for implementation, "all the common metals and most of the less common" had been successfully shaped through chemical milling, though he did not anticipate its use with semiconductors in integrated circuit production, which soon overshadowed its relevance to aerospace components.[25]

More dramatic, though likely having a narrower impact, was high-energy-rate forming (HERF, also known as explosive forming), again initially connected

---

23. W. D. Nelson et al., "Trends in Aerospace Manufacturing," *Metal Progress* 79 (April 1961): pp. 106–110; "Technology Review, State of the Art: Production Engineering," *Space/Aeronautics* [R&D handbook] 38 (1963): pp. I-3 to I-10; Leo Gatzek [North American], "Fabricating Metals for Space Vehicles," *Mechanical Engineering* 87 (October 1965): pp. 46–49; "How Manufacturing Methods Cope with Future Space Projects," *Iron Age* 184 (1 October 1959): pp. 86–88. For recent assessments of several of these innovations, see A. G. Mamalis et al., "Electromagnetic Forming and Powder Processing: Trends and Developments," *Applied Mechanics Reviews* 57 (July 2004): pp. 299–324; K. H. Ho et al., "State of the Art in Wire Electrical Discharge Machining,"_*International Journal of Machine Tools and Manufacture* 44 (2004): pp. 1247–1259; and J. C. Rebelo et al, "An Experimental Study on Electro-Discharge Machining," *Journal of Materials Processing Technology* 103 (2000): pp. 389–397.

24. J. B. Mohler [Boeing], "Introduction to Chemical Milling," *Materials_in Design Engineering* 53 (April 1961): pp. 128–132. Quotes from pp. 128–129.

25. Ibid., 129. Mohler listed "Aluminum, magnesium, titanium, tool steels, stainless, steels, monel, Inconel, and various superalloys" as ready targets, with less experience as yet in chemically milling "molybdenum, tungsten, beryllium, and tantalum." On chemical milling and electronics, see William T. Harris, *Chemical Milling: The Technology of Cutting Materials by Etching* [Oxford Series on Advanced Manufacturing] (Oxford, U.K.: Oxford University Press, 1976).

with missile building. Used early on to create hemispheric domes for Polaris and Minuteman missiles, explosive forming involves fashioning a die that mirrors the shape desired, tightly fixing a component blank to the die, and evacuating air from the space between the two pieces, lowering the apparatus into a large vat of water, then setting off an underwater explosive charge just above the blank's geometric center.The water pressure blast at hundreds of feet-per-second (vs. roughly 5 fps [1.6 m/sec]for conventional forming) shapes the blank instantly to the die form and the product needs little further machining. By 1961 Ryan Aeronautics had made domes 4.5 feet (1.37 m) across with this process, and the company's James Orr speculated that dies 50 to 60 feet (15.2 to 18.3 m) in diameter could be constructed "in ground," like swimming pools, to create very large metallic components. Explosive forming was not a cost-saver, though; rather, it improved quality. Aerojet-General estimated that making 42-inch (1.07 m) missile domes by standard methods (welding smaller pieces) cost $60 each, whereas explosive forming raised the per-item charge to $100. The difference was that diameter, contour, and thickness tolerances were far closer. An Aerojet spokesman added: "[C]onventional methods simply aren't satisfactory for many of today's missile requirements." Initially, the refractory metals (titanium, molybdenum, et al.) could not be explosively formed except at high temperatures, which made the water bath approach impossible; here Ryan substituted sand successfully, and fashioned a prototype titanium Army helmet, using embedded electric rods to heat both the die and blank. HERF-work continues to the present, though more often for complex, customized items than for serial production.[26]

26. Explosive forming was employed, confirming Orr's ambitions, to create the domes for the Saturn rockets' fuel tanks. Photos in Roger Bilstein's *Stages to Saturn* (rep. ed., Gainesville, FL: University Press of Florida, 1996), p. 221, and Mike Gray's account of fabrication, (*Angle of Attack* [New York: W.W. Norton, 1992], pp. 155–156) indicate that HERFed "gores (curved, tapered wedges) were formed in a 60,000-gallon water tank," then welded together. On explosive forming, see "NASA Studies High-Energy Forming," *American Machinist* 105 (2 October 1961): p. 85; "HERF: Metalworking's New Frontier," *American Machinist* 101 (3 September 1962): pp. 93–104; James Orr [Ryan], "Explosive Forming" *SAE Journal* 69 (June 1961): pp. 57–59; Floyd Cox, "Ryan's Experience in Explosive Forming," *Metal Progress* 80 (August 1961): pp. 71–73; "How Explosives Form Space-Age Parts," *Steel* 148 (5 June 1961): pp. 86–88; C. W. Gipe [Ryan], "High Energy Processes Shape Space-Age Parts," *SAE Journal* 69 (March 1961): pp. 44–45; L. C. Stuckenbruck and C. H. Martinez [North American], "Explosive Forming in the Missile Industry," *Machinery* 67 (November 1961): pp. 99–105; and R. Gorcey et al. [Rocketdyne], "Progress Report on Developments in Explosive Forming," *Machine Design* 33(13 April 1961): pp. 188–190. For a critical perspective, see "Exploding the Myths of Explosive Forming: Springback Is Still with Us," *American Machinist* 105 (26 June 1961): p. 79. "Fabricating Fuel Tanks for the Saturn Rocket," *Modern Metals* 18 (June 1962): pp. 46–48 is helpful on work in rocket building. For more on hot-forming titanium, see C. F. Morris, Jr. [Convair], "Titanium Alloys Hot-Formed on Gas-Heated Stretch Press Dies," *Space/Aeronautics* 34 (August 1960): pp. 117–118, 122, 126. Regarding current uses, Google Scholar search using "explosive forming" and "high energy forming" turned up more than 700 books, articles, and reports. For a comprehensive overview, see D. J. Mynors and B. Zhang, "Applications and Capabilities of Explosive Forming," *Journal of Materials Processing* 125/126 (2002): pp. 1–25.

Attending to other manufacturing process novelties must await a different occasion, but it is worth noting that older industrial techniques also got spaceflight boosts and a few twists: welding and metal spinning, for example. Large, cylindrical rocket motor cases could not be formed in one piece, and thus plates had to be welded into rings and the rings had to be welded together (along with interconnecting forgings called Y-rings) to form columns. This was a substantial fabrication challenge, as W. D. Abbott noted in 1966:

> All material and all welds must . . . be tough, ductile and free of linear defects, particularly surface defects. The magnitude of the problem is more apparent when it is recognized that there are more than 3000 sq. ft. of surface and 500 ft. of weld on a 156 in. diameter rocket motor case . . . All the material and every inch of weld must be completely free of linear defects and gas holes or porosities no larger than 0.04 in. in diameter to guarantee reliability.[27]

The Saturn rocket case was 60 percent larger (260 inches [6.6 m] in diameter) and a misery to weld. Its tanks were a special nightmare: "At a time in history when a flawless weld of a few feet was considered miraculous, the S-2 called for a half a mile of flawless welds."[28] Big artifacts, big fabrication problems.

Moreover, what was being welded in many applications were not just standard steels, but particularly challenging specialized alloys such as the "maraging" steel used in rocket cases. The finished welds needed to have "essentially the same cleanliness, mechanical properties and fracture toughness as the base material." Neither gas metal-arc nor submerged-arc processes could handle these tasks. This triggered refinement of plasma arc approaches to welding, which used an inert gas shield, a "nonconsumable tungsten electrode" and a "constricting nozzle . . . to concentrate the available thermal energy on a relatively small area of the workpiece." The result was faster, deeper welds with no contamination and remarkably few flaws.[29] Further

---

27. W. D. Abbott [Excelco Developments, Inc.], "18% Nickel Maraging Steel Fabrication Application to Large Rocket Motor Cases," *Welding Journal* 45 (July 1966): pp. 595–598, quote from p. 596.

28. Gray, *Angle of Attack*, p. 154. On procedures for Saturn welding, see Charles Garland [Sun Shipbuilding], "Fabrication of 260 Inch Diameter Rocket Cases with Maraging Steel," *Welding Journal* 45 (January 1966): pp. 22–29. The alloy's 18 percent nickel gave it "toughness," which meant stability under pressures of 200,000 psi and an ability at that strain to sustain cracks up to 1 inch long without failure (p. 23). See also, for the Atlas, W. P. McGregor [Convair], "Inert-Gas Tungsten-Arc Spot Welding in Missile Production," *Machinery* 67 (December 1960): pp. 119–121.

29. L. J. Privoznik and H. R. Miller [Westinghouse], "Evaluation of Plasma Arc Welding for 120 in. Diameter Rocket Cases," *Welding Journal* 45 (September 1966): pp. 717–725, quotes from p. 718; R. E. Heise, "Advanced Fabrication Methods," *Metal Progress* 80 (July 1961): pp. 105–107. A titanium alloy liquid hydrogen testing tank, built at Beech Aircraft in 1962 (24 ft. long by 8 ft. diameter) involved several of the novel techniques. Its end closures were "stretch formed to contour, chemically milled and welded together" using an argon, plasma welding technique. See "Titanium Liquid Hydrogen Tank," *Light Metal Age* 20 (August 1962): p. 7.

elaboration of precision welding has continued over the last four decades, widening industrial practice and generating a considerable technical literature.[30]

Metal-spinning is an antique skill[31] in which a vertically rotating blank comes in contact with a laterally moving bar, shaping the blank into a cone, a hemisphere, or another three-dimensional curvilinear form. Installed at Boeing to attack the problems of shaping difficult-to-machine alloys and renamed "flowspinning," it was used to create nose cones, bulkheads, and rocket cases. For rocket cases, the spinning lathe created a dome with vertical sides, then the top was cut off to leave a ring that needed no welding and thus was stronger all around. (This worked only up to case diameters of 50 inches [1.27 m].) Here a traditional metalworking capability was stretched in scale and range to new applications, including work with demanding alloys and high-precision components.[32]

New *materials* appropriate to space environments and launch stresses were at the root of many process innovations and reorientations. As Convair's C. F. Morris explained: "By and large, conventional tooling and manufacturing methods are inadequate for the high strength alloys—Rene 41, Vascojet, titanium, various stainless steels, etc.—that have recently come into use for high temperature aerospace

30. For a useful overview, see E. Craig, "The Plasma Arc Welding Process: A Review," *Welding Journal* 67 (February 1988): pp. 19–25. For the varieties of welding innovations, see Patricio Mendez and Thomas Eagar, "Welding Processes for Aeronautics," *Advanced Materials and Processes* 159 (May 2001): pp. 39–43. A Google Scholar search for "plasma arc welding" turned up nearly 1,100 technical papers and reports.

31. Metal spinning was introduced in the United States by the 1840s and was used to make "gold, silver and pewter hollowware and chalices" for generations. Only after World War I were experiments made with "tougher metals." See *http://www.jobshop.com/techinfo/papers/metalspinpaper.shtml* (accessed 18 July 2006). It was used as well on nuclear submarine components (Raymond Spiotta, "Age-Old Art Helps Build Nuclear Submarines," *Machinery* 67 (March 1961): pp. 102–105.

32. "Boeing Flowspins the Superalloys," *Steel* 148 (February 20, 1961): pp. 62–64; "Lockheed Wheezey-Gheezey Sizes Missile Shells," *Steel* 148 (June 12, 1961): p. 145; Lewis Zwissler [Aerojet], "Spinning Makes Stronger Rocket Cases," *Metal Progress* 78 (December 1960): pp. 70–77. For more recent applications, see E. Quigley and J. Monaghan, "Metal Forming: An Analysis of Spinning Processes," *Journal of Materials Processing Technology* 103 (June 2000): pp. 114–119. Another pre-World War II process given fresh impetus by space fabrication needs was powder metallurgy. See C. G. Goetzel and J. B. Rittenhouse [Lockheed], "Powder Metallurgy Applications in Space Vehicle Systems," *Journal of Metals* 57 (August 1965): pp. 876–879 (with 30 citations to related studies). For more information, see Gordon Dowson, *Powder Metallurgy: The Process and Its Products* (New York: Hilger, 1990). A related manufacturing dynamic which cannot be covered here is the creation of "clean rooms" for production and assembly of delicate electronic and instrumentation components. A quick Google and Google Scholar search turned up very little research on the history of the clean room, but Sandia Labs claims it built one in 1960 as a weapons spinoff. A 1959 trade journal article documents a clean room at Kearfott, a subsidiary of General Precision, in Little Falls, NJ ["'Totally Pure Air' Required for Plant," *Air Conditioning, Heating and Refrigeration News* 88 (5 October 1959): pp. 6–7], and a 1963 essay in Machine Design pointed out that beyond dust and particles, organic hazards to aerospace hardware also had to be attacked ["Of Mold and Missiles," *Machine Design* 35 (18 July 1963): pp. 146–150]. I am not aware of an integrated study of the clean room's history, but Marshall University's Dan Holbrook presented a preliminary inquiry into this topic at the 2006 Society for the History of Technology conference.

structures."[33] Yet beyond heat stresses, materials innovations also had to overcome challenges presented by other conditions: very low temperatures, no atmosphere, zero gravity, sudden shocks, cyclical stresses, and vibration in its varied forms. In 1960–1961, Hughes Aircraft created test chambers to explore metals' reactions to some of these conditions, trying to anticipate space-based materials difficulties. At near zero atmospheric pressure, they found "odd pits forming on the [metallic] specimen[s]," and identified a phenomenon called "metal evaporation," which could lead to unintended metal plating as released molecules attached themselves randomly to other surfaces. For Hughes, the radical implication was that "whole sections of a vehicle can disappear during an extended space voyage if the wrong metal is employed." In addition, lubricants vaporized in zero gravity and metals would "cold weld when . . . left in contact for a few days." Several "tough plastics, like vinyl . . . get brittle and crumble in space," Hughes found, but Teflon was not among them, accounting for its utility.[34] Such environmental conditions only added to the need for special materials that could stand heat, cold, vibration, etc. Hence, extensive trials of steel-based and nickel-chromium alloys were fundamental, in tandem with work on refractory metal alloying elements, ceramic-metal amalgams, honeycomb structures, and lightweight metal substitutes including resins, ceramics, rigidified fabrics, fiberglasses, and space-stable polymers.[35]

By the mid-1960s, concerns arose that metals and alloys were reaching their limits of manipulability. In a 1966 *Space/Aeronautics* state-of-the-art review, metallurgists and technology managers noted the following:

- Regarding high-strength steel alloys, "lack of adequate fracture toughness, stress corrosion and . . . stress cracking resistance, weldability [and] nondestructive inspection techniques."

---

33. Morris, "Titanium Alloys," p. 117. Candidate materials were created at what seems a remarkable rate in the early spacefaring years. One annual review in 1961 described 92 new materials, ranging from alloys to lubricants to coatings, offered by 78 different firms, ranging from GE and Crucible Steel to Rollway Bearing and WaiMet Alloys. "Condensed Review of Some Recently Developed Materials," *Machinery* 68 (October 1961): pp. 115–124. See also Jack Hauck and John Vaccari, "Aerospace Materials: Today and Tomorrow," *Materials in Design Engineering* 62 (August 1965): pp. 97–103.

34. "How Metals React in Space," *Steel* 149 (2 October 1961): pp. 58–59. Zinc, cadmium, and magnesium alloys evaporated (reducing light weight magnesium's attractiveness), whereas iron, steel alloys, titanium, platinum and tungsten did not show these patterns (adding to light-weight titanium's value). See also Eric Linden, "Aerospace Electronic Materials," *Electro-Technology* 68 (1961): pp. 125–131, esp. p. 128, and N. H. Langdon, "Polymers Out In Space," [two parts] *Rubber Journal and International Plastics* (4 February 1961): pp. 174–178 and (11 February1961): pp. 210–213.

35. R. M. Treco, "How Space-Age Energy Sources Spark Rise of New Materials," *Iron Age* 187 (2 March 1961): pp. 87–89; A. Hurlich and J. F. Watson [Convair], "Selection of Metals for Use at Cryogenic Temperatures," *Metal Progress* 79 (April 1961): pp. 65–72; "Space: A Galaxy of New Materials Concepts," *Steel* 153 (1 July 1963): pp. 32–39; "*Structures and Materials*," Space/Aeronautics 46 (mid-July 1966): pp. 80–92, 166, 168–169 (three pages of citations). This last article emphasizes the relationship between new materials and structural dilemmas for launch, reentry, and space operations vehicles. See also "Bundled into Orbit," *Plastics World* 20 (March 1962): p. 46 and "Fabricating for Planes and Missiles," *Plastics World* 20 (February 1962): pp. 50–51.

- Regarding superalloys, "further increase in temperature limits . . . appears very doubtful." Attempts to improve strength and stability by reducing chromium content have "increased hot corrosion and oxidation problems."
- Regarding critical light metal alloys, for aluminum, an "increase in strength without decrease in ductility" was needed; for titanium, ductility at extremely low temperatures was a problem; and for beryllium, brittle fracture issues had not been resolved.
- Regarding the refractory metals, problems with "metallurgical stability and preservation of mechanical properties in [a] variety of high-temperature environments."[36]
- In consequence, designers had been placing increased emphasis on coatings, composites, and unconventional materials. For example, by 1963 Aerojet experimented with a "filament-wound Fiberglas" version of its 260-inch-diameter (6.6 m) rocket case, but encountered substantial problems, some of them with materials providers. Charles Walance of Hughes Aircraft's Aerospace Group complained: "We don't want to make materials and components. We're systems people. But we have found, in space-work, that we must do an alarming number of jobs ourselves because industry is unable or unwilling to attempt to meet our needs."[37]

The Air Force appears to have stepped in at about this time, creating its Advanced Filaments and Composites Division in January 1965. General Bernard Schriever argued for simultaneous efforts on multiple composite components and formulations, looking for a "cascade" effect—"the extra performance gain which is achieved when the many individual gains are examined all together." The Air Force estimated it had taken $350–$400 million to bring titanium from experimental work to regular use in aircraft and aerospace devices, and believed "comparable development of composites will cost at least as much."[38] Again, composites (particularly resin-based ones and those using fiberglass) had been floating at the technological

36. "Structures and Materials," *Space/Aeronautics* 91 (1966). On refractories, see "Refractory Metals Emerging as Structural Materials," Steel 148 (8 May 1961): pp. 115–130.

37. "Space: A Galaxy," pp. 32, 35. This doesn't seem to square with the flood of new materials (see note 31), though problems of communications and testing may have been involved. For a later overview of advanced materials and aerospace initiatives, see "Testimony of Gregory Eyring, OTA, July 7, 1988," *National Critical Materials Policy*, Hearing before the Subcommittee on Transportation, Aviation and Materials of the Committee on Science, Space, and Technology, HR, 100th. Congress, Second Session, No. 1241, GPO: Washington, DC, 1988, pp. 49–66.

38. Michael L. Yaffee, "Composite Materials Offer Vast Potential for Structures," *Aviation Week and Space Technology* 82 (3 May 1965): pp. 38–53, quotes from pp. 39, 46. For a brief look at the history of industrial composites, see A. Brent Strong and Michael Miles, "Metals vs. Composites: Breaking down the Elements," available at *http://mlf.byu.edu/pages/papers/files/metalPlastic.php*, and A. Brent Strong, "History of Composite Materials: Opportunities and Necessities," available at *http://mlf.byu.edu/pages/papers/files/history.php* (both accessed 18 July 2006).

edges of engineering practice since just before the World War II, but aerospace demand proved propulsive through creating a renewed, complex environment of experiment and application.[39] Thus, the materials revolution percolated through spacefaring initiatives and was partly propelled by them.

Creativity in *instrumentation* is remarkably well documented as critical to space project capabilities, though in my view the historical issues are relatively under-conceptualized and have been researched in a rather scattered way. Similar to composites, instrumentation advances got a big boost in the 1940s. Two GE engineers reported in 1947 that "the exacting requirements for electronic and other control and measuring devices during the recent war stimulated rapid progress in the improvement of electric instruments and associated components."[40] Yet by the early 1960s, instrument builders were fretting over what they termed "the measurement gap," echoing the "missile gap" rhetoric of the late 1950s. Their concern even had a cold war resonance, when Aerojet's V. R. Boulton quoted a Soviet source approvingly:

> It is known that the uniformity, correctness, and correct employment of measures and measuring devices is a matter of great importance for the national economy, since the use of incorrect measures and measuring devices causes unproductive losses, leads to an increase in production rejects, and to an incorrect assessment of material values.[41]

The U.S. industry had grown by 50 percent in four years since 1958 (estimated sales rising from $4 billion to $6 billion),[42] spurred by the surge of aerospace contracting, but the high-precision requirements of space projects forced a complex problem to the surface. Boulton explained:

---

39. Brent Strong notes that "[I]n the frantic days of the war, among the last parts of an aircraft to be designed were the ducts. Since all the other systems were already fixed, the ducts were required to go around the other systems, often resulting in ducts that were convoluted, twisting, turning, and place[d] in the most difficult to access locations. Metal ducts just couldn't easily be made in these 'horrible' shapes. Composites seemed to be the answer. The composites were hand laid-up on plaster mandrels which were made in the required shape. Then after the resin had cured, the plaster mandrels were broken out of the composite parts. Literally thousands of such ducts were made in numerous manufacturing plants clustered around the aircraft manufacturing/assembly facilities." See "History of Composite Materials," note 38.

40. D. B. Fisk and J. M. Whittenton, "Progress in Instrument Design," *GE Review* 50 (October 1947): pp. 8–11, quote from p. 8.

41. Orval Linebrink [Battelle], "The Measurement Gap," *ISA Journal* 8 (February 1961): pp. 38–41; V. R. Boulton, "Economics of Instrumentation Precision for Space Vehicle Development," *Aerospace Engineering* 20 (March 1961): pp. 30–31, 64–67, quote on p. 30 from Soviet journal Measurement Techniques. See also L. E. Howlett, "The Crisis in Measurement," *The Engineering Journal* 44 (April 1961): pp. 73–76.

42. "Instruments Chart Rapid Growth," *Iron Age* 192 (19 September 1963): pp. 70–71. By 1966, a technical journal estimated instrumentation sales for R&D alone (not counting production and monitoring of installed systems) at $4.5 billion annually. See "Instrumentation and the Management of R&D," *Research/Development* 17 (August 1966): pp. 23–36, data on p. 27.

> Ideally a measuring system should be an order of magnitude more repeatable and accurate than the system under test . . . so that variations in the results from test to test represent primarily the variations in performance of the item being tested. Realistically, the repeatability of most measuring systems is hardly as good as that of the systems they are used to evaluate—at least this is true of present day rocket propulsion systems . . . . The uncertainty in [our] evaluation results . . . is at present much too large.[43]

Measurement was the heart of instrumentation, of course, being concerned with mass, time, temperature, dimensions, force, stress, vibration, speed, altitude/depth, volume, and flow, gauged through devices operated electrically, optically, physically, sonically, etc. Unsurprisingly, the uncertainty that concerned Boulton also surfaced in defining the boundaries of instrumentation, as a function and as an industry. During the 1950s and 1960s, monitoring and control technics, electric/electronic information processing tools (computers), and simulators were all regarded by insiders as elements of the burgeoning instrument family.[44]

NASA's Keith Glennan nicely captured the broader situation as the transition from Mercury to Gemini and Apollo evolved in 1961:

> Proud as we are of our space technology, we must also be sensible to its failings. The demand for precision measurements is outstripping the best that U.S. science and technology can provide. For upcoming space ventures, NASA needs to measure such things as engine parts to one-millionth of an inch tolerance, and rocket engine thrust to 0.5% accuracy at one million pounds.[45]

Devices for measuring and reporting spacecraft location through telemetry, for accurately assessing temperature, pressure, radiation, stresses and strains, and a dozen other dimensions of mission artifact conditions before and after launch had to be created or adapted, then tested for precision, and evaluated to discern whether what was being measured was directly salient to a critical factor. The rise of reliable, precise, and speedy instrumentation as a key dimension of technical practice preceded NASA's inauguration, but its momentum accelerated at a rapid pace once piloted spaceflight became a national priority. In addition, the host of biomedical spinoffs increasingly described in NASA's publications has its origins in space program monitoring and stressing of human subjects, commencing with flight simulators in Mercury's early months and flowing through to biomedical testing

---

43. Boulton, pp. 30–31. Boulton observed that Aerojet's testing expenses (20,000 tests over five years on 20 weapons systems) amounted to as much as 70 percent of project costs—a proportion that could be reduced dramatically through improved instrumentation accuracy.

44. A. G. McNish, "Fundamentals of Measurement," *Electro-Technology* 71 (May 1963): pp. 113–128; "How We'll Measure Tomorrow," Steel 151 (31 December 1962): pp. 20–24.

45. . T. K. Glennan, "Taking the Measure of Space," *ISA Journal* 8 (February 1961): p. 39.

at SpaceLab and during Shuttle flights. In NASA's *30th Spinoffs Review* (1992), instruments comprised a third of the items discussed (28 of 78), which suggests the durable significance of innovation in this domain.[46]

Finally, NASA projects provided test platforms or incubators for a number of managerial techniques as well: project management and team-tasking, high-level quality control, reliability analyses, and handling concurrency/redesign challenges. The project management dimension, building on Air Force and Navy missile development practices, is well known. General Sam Phillips brought the Program Evaluation and Review Technique (PERT) to the latter stages of Project Mercury, and using this integrated means of charting interdependent task trajectories was significant in Gemini's and Apollo's ability to meet deadlines, most notably the end-of-decade Moon landing target. Along the way, NASA and the Department of Defense (DOD) devised a cost-sensitive upgrading of PERT and a series of now widely used protocols: Work Breakdown Structure (1962), the Earned Value System (1963), and the probabilistic Graphical Evaluation and Review Technique (1966).[47]

Perhaps equally important, aerospace initiatives presented unique management problems, which reframed the notion of a project and the position of project manager. Two parameters dominated:

> First, a specific performance requirement must be met at a single point in time under conditions of substantial uncertainty and high risk—from a technical as well as from a time-and-cost point of view. Second, the project is large, urgent and important enough to demand the concentration of company talents to fulfill contractual requirements. The project manager is the keystone of this pyramid of human and material resources.[48]

Alongside formalized techniques, effective project managers rapidly developed ad hoc and temporary, informal lines of communication in order to tackle problem solving, assembled special cross-disciplinary teams to attack bottlenecks, and limited insertion of PERT and related monitoring devices. As Robert Rados, who was involved in the Tiros weather satellite for NASA, explained, the project manager has

46. NASA's *1992 Spinoffs* volume (note 1) addresses 22 biomedical technologies, 14 industrial productivity, and 12 consumer/home/recreation devices, among the 78 spinoffs profiled. Interestingly, the 2002 40th anniversary *Spinoffs* list (note 22) reported just 10 biomedicals (perhaps reflecting the reduction of piloted spaceflight), 6 industrial productivity enhancers, and 5 consumer/home items. Meanwhile, IT and communications goods and capabilities numbered 14, while 7 safety and security innovations made the list (some specifically referencing 9/11). Interestingly, instruments dropped to about one-quarter of the 50 spinoffs detailed in the 2002 volume. On simulators, see Melvin Sadoff and Charles Harper [NASA-Ames], "Piloted Flight-Simulator Research: A Critical Review," Aerospace Engineering 21 (September 1962): pp. 50–63.

47. Grant Cates, "Improving Project Management with Simulation and Completion Distribution Functions," Ph.D. dissertation, Industrial Engineering, University of Central Florida, 2004, Chapter 1.

48. George Steiner, "The Project Manager's Problems," *Astronautics and Aeronautics* 4 (June 1966): pp. 75–76.

to "program for success," which involves securing the "material basis for continued progress" on the work and "creat[ing] an atmosphere that favors ambitious, forward looking decisions rather than restrictive and cautious ones."[49] Actualizing the concept of the innovation manager was crucial, as Lockheed head L. Eugene Root emphasized in his 1962 presidential address to the Institute of Aeronautical Sciences. A key danger in merging the engineer/scientist and the manager lay in the skeptical, analytical, and conservative outlook which is the desired initial result of scientific training—question all propositions, demand proof through more documentation, send back for future study. This is a deadly trap for managers in a frontier technology. Most great decisions have been difficult to support by the facts available at the time. Innovative project managers had to "demonstrate the ability to make decisions" in the absence of "conclusive feasibility studies," in urgent, high-risk situations with multiple uncertainties. This problem set lay well outside the terrain of routinized corporate management.[50]

The importance of quality control and components reliability was evident to all parties in aerospace manufacturing and spaceflight operations, but here NASA practice soon indicated the limits of industrial approaches rather more than their utility. Three problems surfaced to bedevil projects: (1) insufficient numbers of units or iterations in use to make statistically significant reliability or refined quality control procedures effective; (2) complexity of devices which presented an "overwhelming number of different possible modes of failure"; and (3) the persistent pattern of artifact redesign which made it impossible to have a stable object for quality and reliability testing.[51] As engineers from Chrysler's missile division recounted:

> The large number of unavoidable engineering changes superimposed on the 60,000 to 70,000 parts in a specific missile system would make it virtually impossible to cope with the resultant complexities on a practical basis. Experience has shown that engineering changes during the first production year of

---

49. Engelbert Kirschner, "The Project Manager," Space/Aeronautics 43 (February 1965): pp. 56–64, quote from 64. This is a remarkable review of work practice among three PMs: Rados on the Tiros satellite at NASA, Abraham Schnapf at Tiros-RCA, and William Chalmers on TRW's Vela project. On integrative, task-based teams at Rocketdyne, see "Top-Speed Technology Puts Ideas into Orbit," *Factory* 122 (September 1964): pp. 84–89.

50. L. Eugene Root, "Our Expanding Scientific and Technological Challenge," *Aerospace Engineering* 21 (December 1962): pp. 8–9, 68–70 (quotes from p. 68).

51. W. W. Hohenner, "Manned Rocket Flight Requires A Trade-Off Between Performance and Complexity," *Westinghouse Engineer* 20 (March 1960): pp. 54–58 (quote from p. 56); Dorian Shainin, "Managing a Reliability Program," *Aerospace Engineering* 1 (December 1962): pp. 64, 92–93; E. J. Lancaster and R. E. Biedenbinder [USAF Ballistic Systems Division], "A Critique of Quality Assurance Activity in Air Force Ballistic Missile Programs," *Industrial Quality Control* 18 (January, February, and March 1962): pp. 9–13 (Jan.), pp. 5–9 (Feb.), and pp. 5–9 (Mar.). See also Edward Sharp [NASA-Lewis], "Quality to Meet the Demands of Space," *IQC* 16 (July 1959): pp. 9–11 and John Condon [NASA], "NASA's Reliability Requirements," *IQC* 22 (December 1965): pp. 287–289

> a missile weapon system average out at approximately 1,100 changes per week, resulting in as many as 800 modifications in inspection and testing instructions.[52]

At times like these, the commercial impact of space initiatives was to indicate the limits of managerial practice and technique and the need, in complex task environments, to combine personalized and detail-oriented management.

If we return at the close to the positions articulated in this essay's epigraphs, we may well agree with the Denver team that the commercial impact of NASA innovations was indirect and specialized, and that separating NASA contributions from those of military technology projects fragments the web of connections and exchanges that experimental development featured. To be sure, exploring aerospace technology's uses, plus the dynamics and constraints involved in its fabrication, can help alleviate the handicap of "insufficient information . . . at the level of the firm." Yet, while recognizing how thin historical research still is into these applications, we would surely be hard-pressed to concur with Alex Roland's assertion that the nation "has little to show for [its] investment save circus." Beneath the satellites, probes, and human spaceflights, for a generation or more extensive innovations in process, materials, and instrumentation have flowed outward from NASA projects and resonated through the industrial economy. Their scope can more readily realized than their scale can be measured, but their significance is evident.

52. J. F. Patrick and F. G. Brune, "Quality Instructions for Missile Production," *Industrial Quality Control* 19 (September 1962): pp. 19–22 (quote from p. 19). Patrick and Brune outlined a data-processing approach to tracking revisions in "inspection and test instructions" but admitted that "major problems of correlation, up-dating, recording, and distribution of information still remain."

Chapter 9

The Political Economy of Spaceflight

Stephen B. Johnson

Political economy has a long, distinguished history, going at least as far back as the eighteenth century. Adam Smith's *Wealth of Nations* (1776) is the most famous of these early works; it discussed the complex relationships between economic and political activities while at the same time laying the foundation for classical economic theory. At that time, and indeed continuing to this day, the question of appropriate government policies to spur economic activity to national advantage has remained of paramount importance to national leaders. Spaceflight is a prime candidate for political-economic analysis, largely because the government-industry nexus has remained tightly interlocked. Free-market or laissez-faire policies have seldom if ever applied to space activities, and the classical economic theories that assume the existence of a free market have correspondingly little traction in describing the economics of spaceflight. By contrast, political economic approaches are directly applicable, due to the continuous and intimate interactions of government, industry, and academia in space activities. This essay offers a preliminary exploration of the political economy approach to the subject of spaceflight, so as to provide a few paths upon which future researchers may tread.[1]

1. This approach has been tried at least once before. See M. A. Holman, *The Political Economy of the Space Program* (Palo Alto, CA: Pacific Books, 1974). Many things have changed since that time, most prominently the massive growth of telecommunications and navigation, and the end of the cold war.

WHY PAY TO GO INTO SPACE?

To begin to understand the political economy of spaceflight, investigators must first understand the reasons why humans are willing to devote significant resources to going into space. Put another way, what do people "demand" that going into space can "supply"?[2]

Rocketeers achieved the first trips into space with the German V2 ballistic missile during World War II. Given that other weapons existed to deliver explosives at long distances, what characteristic made the development of ballistic missiles appealing? It was not merely the ability to deliver at a long distance, because the German Luftwaffe had bombers for that purpose. Rather, it was the ability to do this automatically (without human pilots) and at such high speeds as to make interception impossible (their relative invulnerability in flight) that made them of such great interest. Many nations realized that once the V2 attained operational status a new and enormously destructive weapon now existed, one possessing great speed and invulnerability.[3]

Immediately, space became the new "high ground," coveted by military and intelligence organizations for the same reasons they have always wanted to control the high ground. From high locations, one can see for very long distances to monitor the activities of adversaries. Vast resources have gone into the development of reconnaissance satellites of various kinds. The very first satellite development program in the U.S. was an Air Force reconnaissance program. Although the military was the first to point telescopes and cameras at Earth, science too can take advantage of the high ground to observe various natural and human-made changes to Earth's land, seas, and atmosphere. Weather prediction is vastly improved by viewing the atmosphere from space.

Finally, the sheer difficulty of going into space has posed a challenge and an adventure, which makes for dramatic entertainment and, if successful, garners respect. The space race between the U.S. and the Soviet Union had many facets, but the human flight program in particular was significantly influenced by the challenge and drama of putting and maintaining humans in space. Launches have been and remain a very dangerous affair; the space race in the 1960s put time pressures on both sides to cut corners and try ever-more complex and risky activities, from

2. This is another way of looking at rationales for spaceflight, as in Roger Launius' recent article on the subject. This section differs from Launius' account in that I search for specific characteristics of the space environment that make it useful, whereas Launius' article stresses the reasons typically given by politicians, leaders, and others. Those reasons and rationales often simply assume, without any discussion or direct statement, the characteristics I discuss here. See Roger D. Launius, "Compelling Rationales for Spaceflight? History and the Search for Relevance," in *Critical Issues in the History of Spaceflight,* Steven J. Dick and Roger D. Launius, ed. (Washington, DC: NASA SP-2006-4702, 2006), pp. 37–70.

3. Benjamin King and Timothy Kutta, *Impact: The History of Germany's V-Weapons in World War II* (Rockville Centre, NY: Sarpedon, 1998); Michael J. Neufeld, *The Rocket and the Reich: Peenemünde and the Coming of the Ballistic Missile Era* (New York: The Free Press, 1995).

extravehicular activities (EVAs) to putting men on the Moon. Though less dramatic, the Soviets set many long-duration space records on the *Salyut* and *Mir* space stations as their response to American lunar success.

The robotic space race also had its sense of adventure and exploration, as the U.S. and Soviet Union sent probes farther and farther out into the solar system, with the U.S. in particular making amazing discoveries that caught the attention of the world. More recently, the ability to create "virtual exploration" through Mars rovers in the 1990s and 2000s brought the space program into the homes of anyone with a computer hooked up to the World Wide Web. This form of exploration became increasingly popular to a generation brought up on computer games and the Internet.

The drama of human flight and the wonder of astronomical discoveries interest young people as well as adults. This fact provides another justification for spaceflight: its ability to lure young people to careers in science and engineering. Developed economies require scientists and engineers to function and to continue to spur economic growth through the development of new technologies, making education in mathematics, science, and engineering a priority. Spaceflight lures students into these fields (the "space and dinosaurs" phenomenon), and hence one of its justifications for government spending is as a mechanism to increase the supply of technically capable citizens.

Technical advances often bring economic opportunities and growth. Without question, spaceflight has created or propagated a number of technical advances, some of which have had significant impacts on Earth. Examples include digital imagery enhancement, which is now often used in medical applications; fireproof space suit materials, which are now used in firemen's gear; and testing of food for astronauts, which contributed to food testing programs worldwide. These so-called spinoffs have frequently been cited as reasons for funding the space program. What has made space a particularly effective generator of spinoffs is the fact that the space environment is extreme or unusual in a variety of ways. These differences force scientists and engineers to think in new ways about how to accomplish tasks in space, which in turn create novel technologies and new ideas. Spinoffs have thus become another classical argument for government funding of space activities.[4]

4. NASA publishes its annual magazine *Spinoff* to advertise this aspect of space. See also David Baker, *Inventions from Outer Space: Everyday Uses for NASA Technology* (New York: Random House, 2000). The counter-argument is that you can spend the money directly on those applications on Earth and get more "bang for the buck." The counter-counter-argument is that just allocating money to those Earth applications does not guarantee that people would have conceived of those solutions. In other words, the uniqueness of the application requires people to think differently than they otherwise would do, regardless of the money spent. My opinion is that spending money in two different places gets you two different sets of technology, both of which may be useful but in different ways. There has also much debate about the exact quantity and quality/nature of space spinoffs. Some argue that space programs are very conservative and hence are not really innovators. Others note that in the 1950s and 1960s they were not so conservative. My view leans more toward the innovative side. Some areas have been consistently at the forefront of technologies, such as telecommunications, autonomy, lightweight materials, and certain human physiology applications.

In the long run, space offers the potential of the resources of the universe, which will likely become increasingly important as Earth's resources dwindle through exploitation and use. To date, the only space resource that has been used has been solar power to power spacecraft, but the potential for mineral resources on asteroids and other celestial bodies exists, should the advance of technologies and the economics on Earth or in space make it viable.

One truly final argument has attracted lower levels of interest and funding, although the argument itself goes back to the beginning of modern space activities. In the late nineteenth century, Lord Kelvin argued that the theories of thermodynamics implied that the Sun was only some tens of millions years old and is continuing to cool, implying that life on Earth had not been around that long and furthermore could not survive much longer.[5] Since then, scientific discoveries and theories have furthered arguments about the potential future of longevity of life on Earth. By the 1950s astrophysicists theorized that the Sun will eventually become a red giant star, burning out any life on Earth, though billions of years in the future.[6] However, by the 1990s, scientists had evidence that a massive asteroid hitting Earth millennia ago had caused the extinction of the dinosaurs and many other species on Earth. A similar strike would likewise almost certainly destroy humanity. Finally, the invention of the atomic bomb and the hydrogen bomb by the early 1950s showed that humans had the potential to destroy all human life with human-made weapons.

For all of these ills, many space enthusiasts argue that humans must leave the planet in order to survive. In this case, space is simply "anywhere but Earth," since in the long run Earth is doomed. A more positive view is that humans can build a new utopia away from the perils and contaminating influences of Earth. In both its negative survival and positive utopian forms, this argument entices some to support space advocacy groups and to become involved in the space program. Although this "ultimate motivation" is quite important for many individuals to become involved in space activities and to get space activities started when little funding is available, in practice this has had little economic impact because funding generally requires an appeal to hard-headed politicians and non-space leaders who require near-term, practical uses. The main discernible results of these concerns are modest increases of scientific funding in the 1990s and early twenty-first century to search for near-Earth asteroids.

In summary, space has several enticing features that lure military, civilian, and commercial organizations and individuals to spend money to take advantage of them. Understanding the political and economic processes and structural features by which this money is spent is the task of political economy. Several approaches to

5. Joe D. Burchfield, *Lord Kelvin and the Age of the Earth* (Chicago: University of Chicago Press, 1975, 1990), p. 43 in the 1990 edition.

6. Jean-Louis Tassoul and Monique Tassoul, *A Concise History of Solar and Stellar Physics* (Princeton, NJ: Princeton University Press, 2004).

space economics and commerce appear in the research and trade literature, each of which sheds light on different aspects of the subject.

The Uses and Abuses of Economic Data

To gauge the economic impact of space activities, it is frequently desirable to provide quantitative measures.[7] As with many other efforts to quantify social activities, this is an activity fraught with methodological dangers and problems. Enumerating a few of the more common complications of providing quantitative measures will provide a counterbalance to the danger of believing too literally some of the more typical quantitative measures provided by various government and nongovernment sources.

In general, we can characterize economic activities as chains of suppliers and consumers. For example, DISH Network provides television service in my home, and I am thus a consumer of a particular satellite service and DISH Network is the supplier. DISH Network in turn purchases satellites from manufacturers such as Boeing or European Aeronautic Defence and Space Company (EADS). In this linkage, DISH Network becomes the consumer and Boeing or EADS the supplier. Similarly, Boeing and EADS purchase thousands of components from various subsystem and component vendors, and so on.

One unfortunately common problem is to count the purchases or revenues at more than one location along the chain, thus doubling or tripling the estimated sector size. There are at least two ways to battle this problem. The simplest is to pick one consistent link in the chain and measure at only that location. Depending on the interest of the analyst, different links will be of more or less interest as the measuring location of choice. This works well to measure the size of a given space sector, such as satellite manufacturing or space telecommunications. Another method is to count only the locations where the money is actually spent on people, which ideally correlates to measuring the number of personnel involved at each link along the chain. This method is helpful in determining structural changes in a sector, such as determining the relative sizes and change in sizes of different companies in a supply chain. For example, analysis shows that in the European aerospace industry, prime contractors have grown in comparison to subsystem contractors in the early years of the twenty-first century.[8]

Another issue is selection of revenues or expenditures to measure money flows. Although it is typical and often required for companies to provide revenue figures, it may be more useful to measure expenditures because expenditures also include

7. The best overview of space economic data sources is Henry R. Hertzfeld, *Space Economic Data* (Washington, DC: U.S. Department of Commerce, Office of Space Commercialization, December, 2002).

8. ASD-Eurospace, *Facts and Figures, The European Space Industry in 2005* (Paris: ASD-Eurospace, June, 2006).

funding acquired through stock sales, bank loans, or other borrowing mechanisms. For example, if one tried to measure the economic impact of the Communications Satellite Corporation in the mid-1960s, measuring revenues would be deceptive because its revenues were very small in comparison to expenditures, as it spent funds raised from stock offerings.[9] A similar problem applies to assessing the impact of the Iridium venture in the 1990s. In this case, the company went bankrupt and thus its money spent to acquire telecommunications satellites ultimately came from outside investors who received little or no return on their investment. In both cases, these expenditures created a temporary spike in satellite communications economic activity.

Government data sources, when available, are often reliable measures but they only cover certain topics of interest to those government institutions. Thus one can get accurate figures on government expenditures, including those to government contractors. Some organizations, such as NASA, gather their statistics into historical data books that are extremely useful for this sort of research. The U.S. Department of Defense's (DOD) space expenditures are well documented and can be extracted for those who have the patience to go through the defense budget line by line, except for "black" programs, which are hidden throughout the defense budget. NASA is required to publish its annual *Aeronautics and Space Report to the President*, which provides a quick and useful overview of government air and space expenditures. The Department of Transportation publishes quarterly reports that provide excellent data on the politics and economics of launchers, whereas the Department of Commerce's Office of Space Commercialization performs a variety of studies and provide occasional reports on major space sectors. The U.S. Census Bureau has its own economic classification system for industries in the U.S., with aerospace products and parts manufacturing (North American Industry Classification System [NAICS] 3364) being the primary category in which space activities are classified. Other nations usually have similar documentation, but access to the data can be difficult for noncitizens and, even when available, the nuances of each system require significant amounts of time and effort to learn. Thus, an international picture of government space expenses is remarkably difficult to acquire. The Space Policy Institute of George Washington University performed an annual survey of civilian space programs and results published annually in *Aerospace America*. Unfortunately, in recent years, this survey has been halted due to increasing national barriers to release of this information.[10]

For corporations, space business is often only one of several product lines, and extracting the proportion that is space-related cannot be easily done unless the companies themselves release data that separates them. *Space News* has annual surveys leading to publication of its "top 50" space manufacturers, and other lists for the top telecommunications operators,

9. David J. Whalen, *The Origins of Satellite Communications 1945-1965* (Washington, DC: Smithsonian Institution Press, 2002).

10. Hertzfeld, *Space Economic Data*, p. 14; e-mail, Hertzfeld to author, 8 April 2007.

direct broadcast companies, and so on. The data are only as reliable as the companies' efforts and willingness to provide accurate data, but are readily available and occasionally useful.

Industry associations also collect a variety of economic data on the aerospace industry. The Aerospace Industries Association (AIA) publishes its annual *Aerospace Facts and Figures* for the United States, while Eurospace provides its *Eurospace Facts and Figures* for the space industry in Europe. The two associations use different methodologies for somewhat different objectives. The AIA data focus on overall aerospace manufacturer revenues and employment, and separates them into air and space domains. Eurospace uses a sophisticated methodology to track expenditures at different industrial levels to do structural analyses, as well as to track the overall revenue and employment figures.

Finally, it is important to use launch statistics, which each nation is required by the Registration Convention to supply to the United Nations. These data, by their nature, are a straightforward "apples-to-apples" comparison of what actually goes into space, without the complications of currency conversions or measurement complexities (though one needs to distinguish between launch attempts and launch successes). Simple comparisons and assessments of launches and the satellites placed in orbit provide an excellent counterbalance to economic data. A couple of examples show the criticality of using these data. One major problem in the assessment of space activities is estimating the economic importance of secret reconnaissance programs, since the economic data remains classified. To get around this problem, by using declassified information about the Corona program and its design we can estimate the rough costs of a Corona satellite.[11] The cost of Thor-Delta launchers of the period can also be estimated from civilian launch data. Combining this with the number of Corona/reconnaissance launches that occurred (which can be extracted from the launch data), one can estimate annual expenditures on these programs.

Another troubling economic problem is estimation of the economic value of Soviet and Russian or Chinese programs. In the case of the former Soviet Union or China, budget figures are not available or, in the few cases where they are, they are unreliable due to the lack of convertibility between capitalist and socialist economic systems. In the case of Russia, the value of the ruble is extremely low compared to the dollar or euro, so reports of Russian government expenditures on space programs lead to a significant underestimate of its economic importance in Russia and worldwide. For all of these types of cases, launch data can provide a means to provide a first order of magnitude estimate.

Overall, it is important to compare the data from these various sources to assess their reasonability, while recognizing the means by which, and the purposes for which, they were gathered. Only in this way can we avoid glaring mistakes and also avoid the all-too-easy belief in the absolute validity of the various enumerations that exist.

11. Despite official classification of Corona satellite costs, this came out in public literature in an interview of a major participant, Frank Buzard, as $5 million per Corona and about $500,000 per camera, for a total of some $5.5 million per Corona satellite. "An Interview with Frank W. Buzard," *Quest* 13/4 (2006): p. 36.

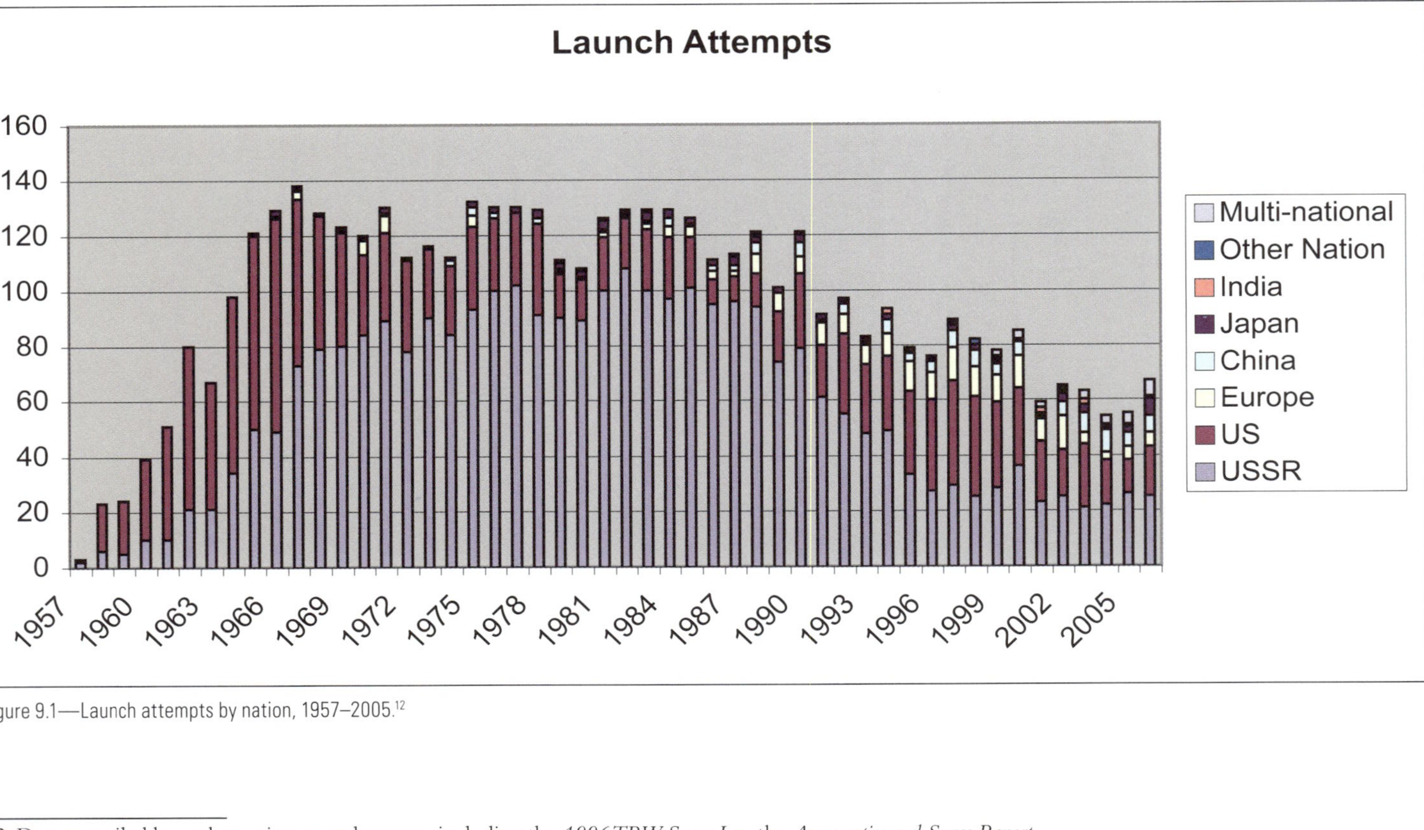

Figure 9.1—Launch attempts by nation, 1957–2005.[12]

12. Data compiled by author, using several sources, including the *1996 TRW Space Log*, the *Aeronautics and Space Report of the President* (annual), the *Quarterly Launch Reports*, Department of Transportation (quarterly), and European Space Vehicle Launcher Development Organisation records from the ESA Historical Archives of the European University Institute. There are several discrepancies between these sources, which I have tried to resolve.

NATIONAL VIEW

One common viewpoint from which to assess space programs is a "national view," in which the space activities of each nation are compared. Thus one can compare national government space funding levels, commercial contracts of corporations categorized by nation, the space policies adopted by various nations, the number of satellites launched by each nation, and so on. The emphasis of this sort of classification is usually to assess the relative effectiveness of a nation's government policies as compared to other nations' policies, typically to improve economic competitiveness vis-à-vis other nations.

To take a simple example, an annual tabulation of all orbital launch attempts—categorized by nation and projected from 1957 to the present—gives a simple indicator of the relative scale of investment by each nation in space activities. Shown in figure 9.1, this way of assessing space activities shows that the U.S. and the Soviet Union and its successors have been by far the dominant space powers. Although this is not particularly surprising, what is surprising is how many more spacecraft the Soviet Union launched during the 1970s and 1980s than did any other nation. This seems to indicate that the Soviet Union was the dominant space power during this time. In addition, any thought of the supposed recent demise of Russia as a space power should be exorcised by the continued frequency of Russian launches, both to launch spacecraft from other nations as well as satellites from Russia and the other former Soviet republics.

When we compare instead national government budgets as shown in figure 9.2, we get an entirely different view of the relative economic significance of the U.S. versus the Soviet Union. Here, the U.S. appears as the overwhelmingly dominant space power. Instead of a time series, I here only show a single year because it is extraordinarily

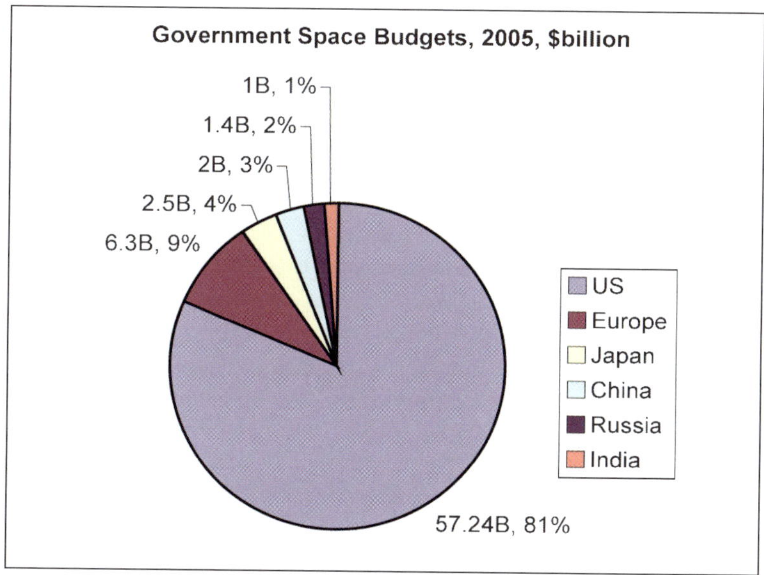

Figure 9.2—Government space spending, 2005.[12]

difficult to compare budget figures for capitalist nations to their communist counterparts. Even when the budget figures can be compared—when both nations are "capitalist" and their exchange rates theoretically reflect the relative value of their currency—these figures remain problematic. Is it really true, for example, that Russia's program is less important at an international level than, say, Japan's? This seems implausible, given the number of launches of each nation.[13]

Even such simple comparisons as government spending have inherent difficulties. To make a fully legitimate comparison, each nation must gather the relevant statistics and those statistics must be comparable from nation to nation. In general, however, the statistics, even when they are gathered, are not strictly comparable. In some cases, the statistics exist but remain classified. Reconnaissance and intelligence budgets are a classic example. Or, in the case of Russia, some of its government space institutions are "commercial" in that they sell their services. For example, is U.S. government funding of Russia to build International Space Station modules to be tracked as a U.S. or Russian figure? Similarly, if South Korea's government purchases a communications satellite from a European vendor, does this reflect South Korean or European space competitiveness? It depends on the question the analyst is trying to answer.

Another way to assess the relative importance of a nation's space program would be to compare the number of citizens employed in that nation's space industry. This is probably a more reliable way to make apples-to-apples comparisons, but the data are difficult to acquire.

Although most national space policies are focused on particular space sectors such as communications or navigation, a few affect many space activities. The most prominent of these in the U.S. are the International Traffic in Arms Regulations (ITAR). Since space technologies are by their nature tied to a host of military concerns, they are subject to ITAR. This has been particularly important in the export of communications satellites. In 1999, Congress, responding to perceived technology transfer of American satellite technologies to China, placed communications satellites on the "munitions list," governed by ITAR. The transfer of export licensing authority from the Department of Commerce to the Department of State significantly hurt American manufacturers, as the regulations have slowed and in some cases prevented export of American satellites, to the benefit of European companies.[14]

The purpose of these national comparisons is often to assess relative national economic competitiveness so as to determine the effectiveness of space policies. Although in a general sense these figures provide a relative scale of activities by different nations, they are largely useless for policy assessment. This is because space policies largely refer to specific sectors of space, such as navigation, space launch, remote sensing, human flight, or science. In other cases, the policies refer to intergovernment

13. ASD-Eurospace, *Facts and Figures, The European Space Industry in 2005* (Paris: ASD-Eurospace, June, 2006), p. 12. European Governments spent €2.662 billion on space products in 2005.

14. James A. Vedda, "Space Commerce," in *Space Politics and Policy: An Evolutionary Perspective,* ed. Eligar Sadeh (Dordrecht, The Netherlands: Kluwer Academic Publishers, 2002), pp. 219–220.

relations, technology transfer, regulations over foreign corporate ownership, and so on. It is very difficult to translate from a policy decision or proposal to an unambiguous impact on the overall space budget or launch figures. These figures are generally too high to be of direct utility. We must use other means to be politically useful.

FUNCTIONAL VIEW

Another typical way to assess space activities is by functional view: military, civilian, and commercial. This tripartite division recognizes that there are three very different motivations, and usually three very different sets of organizations, that are involved in space.

The military, which for the purposes of this paper will include civilian intelligence-gathering organizations, exists to provide national security. For millennia, the military has had its own unique set of institutions, starting with armies and navies, expanding to include air forces in the early twentieth century, and space assets with the advent of the Space Age. Initially controlled by the military, space endeavors were the province of one or more of the armed services.

In the Soviet Union there was no strict separation of military and civilian efforts and, of course, no commercial activities in the socialist state. The Ministry of Armaments controlled the development of ballistic missile and space programs until 1965, when they were transferred to the Ministry of General Machine Building. The Air Force trained cosmonauts, and from 1959 to 1981 the Strategic Missile Forces, a branch of the military, operated all ballistic missile and space systems.[15] Was the launch of Yuri Gagarin in April 1961 a "civilian" activity, even though it was operated by the military? Similarly, the Soviet space station program of the 1970s and 1980s consisted of stations for both military (Almaz) and civilian (DOS—Long Duration Station) purposes, but both programs were operated by the military and all were called *Salyut* and proclaimed to be civilian systems.

Initial space efforts in the U.S. were divided between the three military services: Army, Navy, and Air Force. All three services began developing ballistic missiles and satellites. The Army's primary space organizations, the Army Ballistic Missile Agency and the Jet Propulsion Laboratory, were transferred to NASA by 1960, at which point Army space efforts focused on antiballistic missile systems and satellite communications. While the Navy's Vanguard program and personnel were transferred to NASA, the Navy managed to retain most of its space organizations, which concentrated on space navigation, communications, and with the National Reconnaissance Office, reconnaissance. The Air Force garnered the lion's share of military space programs, with the primary roles for space system development and space launch. It was involved with space reconnaissance, communications, navigation, tracking, and a host of other functions.

15. Asif A. Siddiqi, *Challenge to Apollo: The Soviet Union and the Space Race, 1945–1974* (Washington, DC: NASA SP-2000-4408, 2000), pp. 891–895.

The U.S. civilian space program came into official existence in 1958 when NASA was created to manage the civilian space effort separately from military space activities. Civilian space programs are intended for the "public good," including programs for weather forecasting, technology development, national prestige, education, and science. Separation of civilian from military space efforts in the U.S. is not always obvious. For example, the U.S. Navy ran the Vanguard program, which was intended for "civilian" scientific purposes. Although generally categorized as a military space activity because the Navy ran the program, its function was civilian in intent. Many science and technology programs are similarly difficult to classify. So, for example, the Clementine probe to the Moon was a military program to test technologies but it performed civilian science functions. Many geodetic satellites are run by civilian institutions but were mainly motivated by military purposes, such as ballistic missile targeting. It has been customary to consider all nonmilitary government space expenditures (that is, money spent by any government institution not part of the DOD, with the exception of intelligence) as "civilian."

Other civilian space programs came into being in a number of nations soon thereafter, sometimes as national civilian agencies, such as France's Centre National d'Études Spatiales (CNES), and sometimes divided among a variety of institutions, as occurred in the United Kingdom, West Germany, and Japan.

Commercial space efforts began very early in the U.S., with the launch of American Telephone & Telegraph's (AT&T) privately funded the Telstar satellite in 1962. The International Telecommunications Satellite Organization (Intelsat) came into being in 1964 as a multinational consortium to run the international communications satellite network. Although the U.S. created a private corporation, the Communications Satellite Corporation, to manage its shares in the new organization, other nations usually assigned government organizations to represent their interests in Intelsat. This is because most nations ran their telephone networks as national public services. So, although Intelsat has usually been classified as "commercial" by American analysts, other nations considered it a civilian activity.

Many private corporations depend mainly or even totally on government funding. For example, the private space company United Space Alliance, a joint venture of Boeing and Lockheed Martin that services the Space Shuttle, is totally dependent on the government. Is it "commercial"? A number of space companies that did not depend on government funding, such as PanAmSat and DirecTV came into existence by the 1980s and 1990s.

Other commercial-civilian hybrids abound. Arianespace and SPOT Image were established as commercial companies but were partly owned by the French government, whereas the launchers and satellites they operated were developed by governments (the European Space Agency [ESA] for Arianespace, and France for SPOT Image).[16]

---

16. J. Krige, A. Russo, and L. Sebesta, *A History of the European Space Agency 1958–1987*, Volume II (Noordwijk, The Netherlands: ESA SP-1235, 2000), pp. 472–481.

After the collapse of the Soviet Union, the Russian government institution Khrunichev sold launch services to other nations and to private entities, and owns shares of private companies. In the space launch industry, it is customary to classify any launch that is up for bid by launch providers as commercial. This can be true even if the launch provider (supply) and the satellite owner (demand) are government-owned! Commercial often means "competed," not necessarily "private." Other confusing cases have existed in the past, such as when private companies Martin Marietta and General Dynamics competed with the U.S. government (NASA) and with the European public-private hybrid company Arianespace to launch Intelsat satellites in the early 1980s. Certainly there was competition, but in this competition private entities such as Martin Marietta and General Dynamics stood little chance against the rival governments. Of course, one can also question how "commercial" Martin Marietta and General Dynamics really were, since the vast bulk of their funding (and many upgrades to their launch vehicles) were funded by the U.S. government.[17]

One government's military or civilian expenditure is often a commercial company's profit or another government's revenue, so it is often a matter of choice or convenience to decide whether an expenditure or revenue is military, civilian, or commercial. Even with all of the complications just described, the tripartite division of military, civilian, and commercial does provide more visibility into the relative scale of activities, now divided by major function: national security, public good, or economic profit. This gives a somewhat more informative view of the relative scale of activities for these different national functions.

Figure 9.3 shows the relative size of military, civilian, and commercial activities in the U.S. in 2005, as defined by government expenditures for military and civilian purposes, and by revenues for commercial activities. Strictly speaking, these are not the same things and so the comparison is not as direct as one would like, but the statistics on commercial activities are usually based on revenues, not expenditures. If one could construct similar statistics for other nations, national comparisons could be made regarding the relative priorities of each nation. It is quite clear that after the collapse of the Soviet Union, U.S. military space expenditures far exceed those of any other nation, with Russia probably remaining in second place and China third, with Europe not too far behind China. Figure 9.4 shows the relative proportions of European space activities, which shows the much smaller proportion of expenditures on military space as compared to civilian and commercial activities. Expenditures by all other nations are significantly smaller than these four. Israel's space activities also have a large military slant, and recently Japan has funded military space activities for communications and reconnaissance.

17. Joan Lisa Bromberg, *NASA and the Space Industry* (Baltimore, MD: Johns Hopkins University Press, 1999), pp. 124–132.

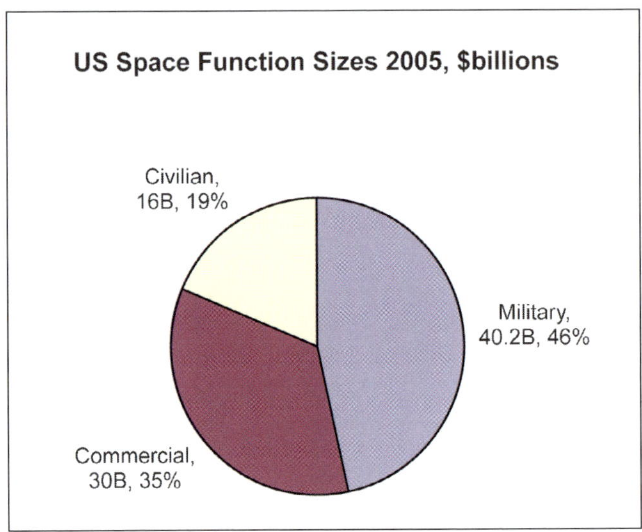

Figure 9.3—Relative size of U.S. space functions, 2005.[18]

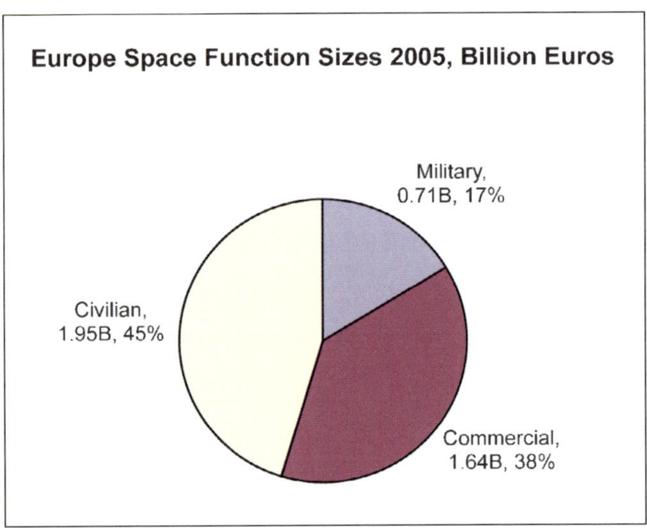

Figure 9.4—Relative size of European space functions, 2005.[19]

18. These civilian and military figures come from the 2005 U.S. national budget, Budget of the United States Government, Washington, DC: Government Printing Office. It can be found online at *http://www.gpoaccess.gov/usbudget/* (accessed January 2007). The commercial figures is a rough estimate derived from several sources, including Space Foundation, *The Space Report 2006* (Colorado Springs, CO: Space Foundation, 2006), the International Space Business Council, and Department of Commerce publications.

19. ASD-Eurospace, *Facts and Figures, the European Space Industry in 2005* (Paris: ASD-Eurospace, 2006), p. 12.

The major trend over the nearly 50 years of spaceflight has been the growth of commercial activities relative to civilian and military expenditures. From essentially zero in the 1950s, by the early twenty-first century the commercial aspects of space have grown to significance, with revenues in the tens of billions of dollars each year. Military spending has remained relatively constant or growing over the years, whereas civilian space spending has remained essentially fixed after the initial bump of the 1960s space race.

INSTITUTIONAL STRUCTURE

A typical economic view of an industry subdivides that industry into the major suppliers and then lower-tier companies down the supply chain. The lowest tier provides small components to the next higher tier, which integrates them into larger units, and then ultimately to the top-tier provider, which provides the full system or service to the consumer. This type of analysis can be applied to the space industry or, more accurately, to the aerospace industry.[20]

While some "pure" space companies exist, many space products are supplied by larger companies with diversified product lines. In most, though not all, cases, these are aerospace companies that provide a host of aviation, space, and defense products to governments and to other companies. Boeing is one of the best known, being most famous as a commercial airline manufacturer. However, in 2006 Boeing also manufactured the Delta launcher, a variety of commercial, civilian, and military satellites, and many other defense-related products, along with a few other smaller lines of business. The EADS is the largest European aerospace manufacturer, with a similarly diversified portfolio.

The largest companies, which supply entire launch vehicles, satellites, aircraft, and missiles, are known as prime contractors. After decades of consolidation, in the U.S. and Europe these had been reduced to a very small number: Lockheed Martin, Boeing, Northrop Grumman, Raytheon, EADS, and Alcatel. These gigantic companies consisted of what had formerly been many smaller companies, though in their time these had been considered large prime contractors in their own right, such as Douglas, Martin, Messerschmidt, Hawker Siddeley, McDonnell, Grumman, Sud-Aviation, Dornier, North American, and so on.[21] The massive prime contractors

20. Definitions of prime, subcontractors, and component suppliers can be found in U.S. General Accounting Office, "Best Practices: DoD Can Help Suppliers Contribute More to Weapon System Programs," (March 1998) GAO/NSIAD-98-87. See also Keith Hayward, *The World Aerospace Industry: Collaboration and Competition* (London: Duckworth & RUSI, 1994), p. 6.

21. Roger E. Bilstein, *The American Aerospace Industry* (New York: Twayne Publishers, 1996), pp. 117, 176–177, 215–218; Bromberg, *NASA and the Space Industry*, pp. 12–13; European Aeronautic Defence and Space Company, *On the Wings of Time: A Chronology of EADS* (2003), found online at *http://www.eads.com/1024/en/eads/history/wings_of_time/wings_of_time.html* (accessed January 2007); Alcatel-Lucent Web site, *http://www.alcatel-lucent.com* (accessed July 2007).

engaged in horizontal consolidation (merging with their peer prime contractors) and vertical consolidation (acquiring lower-tier companies that supplied components, subsystems, and services to them). One can see Marxist capitalist dynamics behind many of these mergers. During economic slowdowns, less profitable companies exit the market, go out of business, or are acquired by more successful firms.

These mergers could not have occurred unless the governments of the relevant nations approved, or at least declined to intervene to prevent them.[22] Antitrust regulations in the U.S. and Europe meant that mergers were reviewed to ensure they did not unduly hamper competition. In Europe, additional complications ensued because the mergers needed to create companies on the scale of the American behemoths had to cross national lines. No single European nation had a large enough aerospace industry to compare with the American domestic market. In both the U.S. and Europe, many mergers occurred from the late 1940s through the 1960s, and then again in the 1990s and early 2000s. The truly massive companies that exist in 2006 came into being only after the DOD—no doubt in consultation with other parts of the U.S. government—told the major aerospace companies in 1993 that now that the cold war had ended there was not enough business to support them all at the levels to which they were accustomed, that there would be fewer of them after a few years, and that the U.S. government would look favorably on consolidation.[23]

Observing events in the U.S., European aerospace officials in both private industry and government concluded that they could not compete with the U.S. unless they, too, allowed the formation of truly massive companies. By the early 2000s, the European consortia and so-called national champion companies merged into the multinational aerospace companies EADS and Alcatel, on scales nearly the size of Boeing and Lockheed Martin.[24] One could argue that Marxist dynamics apply—governments, responding to or in submission to the large companies, allow them to merge, though many of the conditions attached to various mergers somewhat weaken the argument.

Below the giants are second-tier companies. These are often very large companies in their own right, often with significant or preponderant business interests outside of aerospace. In the U.S., these include companies such as Honeywell, well known for building thermostats for private homes but also a primary builder of avionic systems in aerospace; International Business Machines (IBM), the world's largest computer manufacturer; Bendix, supplier of a variety of electronic

22. For a general discussion of these mergers prior to 1994, see Keith Hayward, *The World Aerospace Industry: Collaboration and Competition* (London: Duckworth & RUSI, 1994), chapter 3.

23. Walter J. Boyne, *Beyond the Horizons: The Lockheed Story* (New York: St. Martin's Press, 1998); John A. Tirpak, "The Distillation of the Defense," *Air Force Magazine* (July 1998), *http://www.afa.org/magazine/July1998/0798industry.asp* (accessed 13 January 2007).

24. European Aeronautic Defence and Space Company, *On the Wings of Time: A Chronology of EADS* (2003). Alcatel-Lucent Web site, *http://www.alcatel lucent.com*, (accessed July 2007).

components; Litton, manufacturer of gyroscopes and other electronic systems; Pratt & Whitney, builder of jet and rocket engines; rocket engine manufacturer Aerojet; and a number of others.

Some second-tier suppliers also build full systems as well as components. They are considered second-tier mainly because they are much smaller than the giants. Orbital Sciences is a typical example, building small satellites and launchers. Ball Aerospace, which is a part of the larger Ball Corporation known to homemakers as the maker of fruit jars, has built many scientific satellites over the years and is a manufacturer of star trackers.

Below the subsystem suppliers are component suppliers. These consist of hundreds of small companies, each of which typically specializes in a small set of products. Examples include Adcole, a well-known maker of sun sensors; Eagle-Picher, the manufacturer of most space-qualified batteries; and Moog, which manufactures actuators.[25]

One can analyze the relative success of these tiers, as shown in figure 9.5. From 1997 to 2002, second-tier companies in Europe declined dramatically in comparison to first- and third-tier corporations, based on the political-economic dynamics of the mergers during that period.

Political factors influenced the formation of collaborations between private companies and with governments. European companies formed complex alliances, becoming official consortia starting in the late 1960s and 1970s to ensure that each nation that contributed to a space project's funding would benefit by contracts to companies in their nations in return. In 1995, Lockheed Martin and Rockwell International (later Boeing) created a joint venture known as United Space Alliance to operate the Space Shuttle, as the U.S. government pressed for privatization of Shuttle operations. The fall of the Soviet Union presented new opportunities for multinational joint ventures between launcher companies and organizations, such as International Launch Services (Lockheed Martin and Khrunichev State Research and Production Space Center),[26] Starsem (Arianespace, EADS, Russian Federal Space Agency, and Samara Space Center),[27] and Sea Launch (Boeing, Rocket and Space Corporation Energia [RKK Energia], SDO Yuzhnoe [Ukraine], and Kvaerner ASA).[28] These organizations, which had both Russian and Western ownership, made it easier for the U.S. and European governments to purchase Russian launches. Because Western companies profited by the purchase of Russian launch services, purchasing the significantly cheaper Russian launches was made more politically palatable.

---

25. Many examples of various-sized suppliers can be found in, for example, *2004 North American Space Directory,* 7th ed. (Bethesda, MD: Space Publications LLC, 2004), *http://www.SpaceBusiness.com* (accessed January 2007). Similar directories exist for Europe.

26. International Launch Services Web site, *http://www.ilslaunch.com/aboutus/legacy/* (accessed 13 January 2007).

27. Starsem Web site, *http://www.starsem.com/starsem/starsem.html* (accessed 13 January 2007).

28. Sea Launch Web site, *http://www.boeing.com/special/sea-launch/* (accessed January 2007).

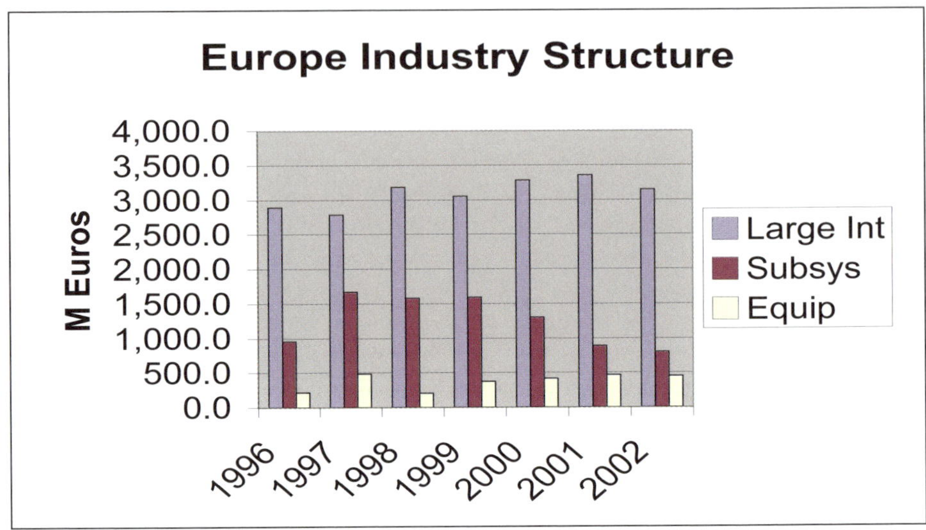

Figure 9.5—Turnover of first-, second-, and third-tier suppliers in Europe from 1996–2002 *(Source: Eurospace).*

Some joint ventures, such as International Launch Services, included government organizations as partners in a commercial enterprise. In other cases, such as Arianespace, SPOT Image, and RKK Energia, governments owned some shares in the otherwise private companies. In many of these cases, the owning governments turned over designs, hardware, or facilities to the company they owned, with the agreement that the private company merely had to cover the operational costs but not the development costs of the systems given to them. Governments also create commercial arms of their organizations, such as Antrix Corporation, the commercial arm of India's Department of Space.[29]

Governments also hired or created nonprofit organizations to perform various functions related to space activities. In the U.S., The Aerospace Corporation and Mitre Corporation were both created with the guidance and approval of the U.S. Air Force (USAF) to provide "systems engineering and technical direction" to private contractors building space systems. The Aerospace Corporation, created in 1960, performed these functions initially for the USAF ballistic missile programs and later for USAF space satellite programs.[30] Mitre did the same for electronics and command and control systems such as the complex built at Cheyenne Mountain, Colorado, for the North American Aerospace Defense Command (NORAD).[31]

29. Antrix Web site, *http://www.antrix.gov.in/* (accessed 13 January 2007).

30. The Aerospace Corporation, *The Aerospace Corporation: Its Work, 1960–1980* (El Segundo, CA: The Aerospace Corporation, 1980).

31. Davis Dyer and Michael Aaron Davis, *Architects of Information Advantage: The MITRE Corporation since 1958* (Newport, RI: Community Communications Corporation, 1998).

Other nonprofit institutions were hired to perform functions normally run by government organizations, such as the California Institute of Technology's Jet Propulsion Laboratory (JPL), which was initially hired by Army Ordnance to build ballistic missiles. It then became, in effect, a NASA field center, building and managing most of NASA's deep space probes and its Deep Space Network. JPL often hired and managed contractors just as other true government organizations did.[32] The Johns Hopkins University Applied Physics Laboratory (APL) became a prime contractor both for the DOD and for NASA, competing against profit-making corporations to integrate and operate satellites.[33] The Charles Stark Draper Laboratories of the Massachusetts Institute of Technology built components, such as the Apollo inertial guidance system, and Stanford University built the Gravity Probe B satellite for NASA.

Government space organizations also vary significantly from country to country. Although many nations, including the U.S. and France, quickly centralized their civilian space programs, others, including Japan, Canada, the Soviet Union, Germany, and the United Kingdom, operated space programs for decades before creating national space agencies. Many smaller nations continue to operate their space programs without a centralized civilian space agency. NASA initially centralized civilian space programs but, since the 1960s, other civilian and commercial institutions have gained access to, and eventually some control of, various space programs, such as the National Oceanic and Atmospheric Administration's (NOAA) operation of weather satellites and some remote sensing satellites.

Military programs also showed a variety of institutional forms. Generally speaking, tension existed between the civilian and military institutions and also between the regular armed services (Army, Navy, Air Force) and other military and intelligence institutions over control of space activities. In the U.S., the USAF was the predominant space organization but the Army and Navy retained significant programs; the intelligence agencies maintained significant interests and their own programs, either individually or shared with the military services. The DOD established its own space organizations at times, such as the Missile Defense Agency, the descendant of the Strategic Defense Initiative Organization. Early on, the Soviet Union created separate organizations to control many space activities as part of its Strategic Rocket Forces. In 2001 these became the Space Forces.

By the nature of the communist state, the Soviet Union developed all of its initial space programs as government institutions. Initially these were "special design bureaus" (OKBs) that performed the research and development activities to create

32. Clayton Koppes, *JPL and the American Space Program: A History of the Jet Propulsion Laboratory* (New Haven, CT: Yale University Press, 1982).

33. William K. Klingaman, *APL—Fifty Years of Service to the Nation: A History of the Johns Hopkins University Applied Physics Laboratory* (Laurel, MD: Applied Physics Laboratory, 1993).

the novel technologies of ballistic missiles and spacecraft. The design bureaus were either newly created or modified from former scientific research institutes (NIIs). Once designed, manufacturing of these systems was assigned to manufacturing facilities, most typically old World War II plants. Over time, alliances between the design bureaus and manufacturing plants stabilized, and in the 1970s they were combined into scientific-production associations (NPOs). With the collapse of the Soviet Union, some of these NPOs were privatized, such as RKK Energia, while others remained government institutions, such as Khrunichev.[34]

Sector View

Perhaps the most useful way to view space activities is by sector, which for the purposes of this paper is defined as an economic division in which the suppliers compete to provide a specific product line that consumers wish to purchase. Space transportation provides examples of how this definition applies. All spacecraft must be launched into space, and those who have spacecraft to place there are consumers of the space transportation product. At this level of analysis, all launch providers are potential suppliers to the spacecraft operators because spacecraft operators will generally investigate all possible suppliers and may ask for bids to launch. In responding to requests for bids, the launch providers must be aware of their competition and set their prices and services based partly on the potential prices and services of their competitors. This self-grouping of suppliers and consumers defines economic sectors.

This seems straightforward, but there are always complications. In the transportation sector, although it is true that all suppliers are potentially competitors for all launches, in practice this is not the case. Some launches, such as U.S. military and intelligence satellites, can by law only be provided only by U.S. suppliers. Similar restrictions exist for military satellites of other nations. Some European civilian launches are restricted to using the Ariane launcher, and so on. After the *Challenger* accident of 1986, the Space Shuttle was prohibited from commercial launches and soon the military pulled its payloads from the Shuttle as well. Other limitations are technical. Although, in principle, small satellites could be placed in orbit on any launcher (in some cases by multiple small satellites on a single launcher), large satellites can be put in orbit only by large launchers. Only the U.S. Space Shuttle, the Russian Soyuz, and the Chinese Long March can place humans in orbit. For some purposes we can consider space transportation a single sector, but in practice one can argue it consists of several subsectors that are semi-independent of each other. Similar reasoning applies to other sectors and subsectors.

34. Siddiqi, *Challenge to Apollo*, Table 3.

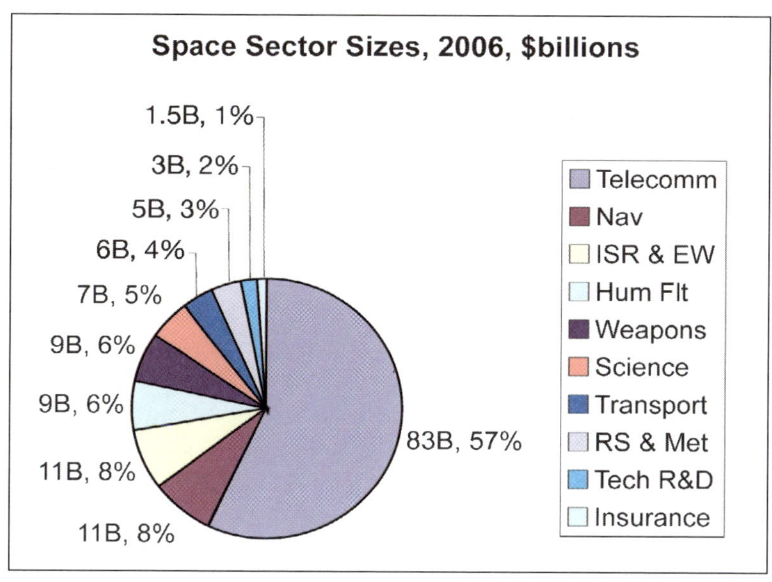

Figure 9.6—Space sector sizes, 2006.[35]

Politics is another reason to group economic activities by sector. In general, national laws and policies affect only one or two sectors. For example, a variety of laws and policies relate to the development of private remote sensing satellites and the regulations that govern sale of imagery from them. These laws have little or no bearing on other sectors. Similar sector-specific policies relate to transportation, navigation, communications satellites, and so on, making the sector view one of the most important ways of understanding economic activity.

A quick tour of the major sectors provides a view of the scope and scale of space endeavors. Figure 9.6 provides an estimate as of 2006 of the relative scale of the major space sectors. The most obvious deduction from this chart is that the communications sector is by far the largest, at roughly $83 billion. It is interesting that historical research into space activities gives little indication that this is the case, with the overwhelming majority of both academic and nonacademic writing focused on the human flight and science programs, with secondary emphases on satellite reconnaissance programs and space politics. The other large sectors are of approximately equal size: science (including both microgravity and physical sciences); navigation; human flight; intelligence, surveillance, and reconnaissance (ISR); and early warning (EW). Remote sensing and meteorology, and transportation, are of roughly equal size, in the range of $5–$11 billion annual expenditures. Technology research and development is in the $2–$3 billion per year range, and the insurance sectors are smaller, from $1–$2 billion annually.

35. Author's estimate, based on assessment of several sources that are described in this chapter.

TRANSPORTATION

No spacecraft arrives in space unless transported there by some launch vehicle. Launchers originally derived from ballistic missile designs and were therefore initially controlled exclusively by the military. By the late 1950s and early 1960s, however, both the Soviet Union and the U.S. realized that the initial ballistic missile designs that used liquid cryogenic propellants were poor choices for weapons, compared to solid and liquid storable designs. Cryogenic systems using liquid oxygen provided greater performance and thus continued to be used to put spacecraft in orbit, whereas ballistic missiles switched to storable technologies that could be launched within minutes or seconds instead of the hours or days required for cryogenic systems.

The connection of ballistic missile with space launcher technologies has remained a primary facet of space transportation politics and largely explains the fact that throughout the twentieth century no privately funded space launch system has succeeded, despite some attempts. Governments have generally sought to develop launch technologies for both military and civilian purposes and have generally prohibited the export or sale of these technologies (or at least current versions of these technologies) to others. The U.S., in particular, has taken a variety of measures to prevent export of its own launcher technologies (with some exceptions, such as the sharing of Delta launcher technology with Japan), and has also attempted to prevent the export of ballistic missile technologies by other nations, such as Russia's sale of KVD-1 rocket engines to India in the 1990s.[36]

National control of launcher technologies is the only way to guarantee national access to space. No nation or group of nations that has long-term space ambitions or military interests in space can afford to rely solely on other nations to put their commercial or military payloads in orbit. Europe developed its integrated launcher programs, first Europa in the 1960s and then Ariane in the 1970s, to ensure that it could launch its own commercial communications satellites in the face of American resistance. China's ballistic missile and launcher programs derived from its desire for military, economic, and cultural independence from both Soviet and American models.

The rise of a commercial space launch industry to provide space transport for private or semiprivate payloads posed problems for the national control idea. Initially, NASA negotiated with other nations to provide launch services for their payloads, though its system of reimbursement was extremely cumbersome—continuing to bill nations for services months and years after their payload was launched, as NASA's government accounting system figured out all of the expenses involved. NASA essentially purchased launchers from corporate manufacturers and then integrated the payload and launched the vehicles itself. In the 1970s, NASA promoted the Space Shuttle as a means to dramatically lower the cost of access to space, which forced European nations (then developing the Ariane launcher through the ESA) to reconsider its organizational and pricing mechanisms. NASA's advertised Shuttle

36. Brian Harvey, *Russia in Space: The Failed Frontier?* (Chichester, U.K.: Springer-Praxis, 2001), pp. 257–260.

prices were based not on actual costs, but on a variety of mathematical models of future Shuttle operations as well as on the prices of expendable launchers such as Delta and Atlas, because the Shuttle had to charge less than these systems to make the Shuttle economically competitive. This, of course, would put American vendors of expendable launch vehicles (Martin Marietta, McDonnell Douglas, and General Dynamics) out of business, while Europe would have to meet the challenge of Shuttle pricing even if it meant selling Ariane launches at a loss. Both sides preemptively set very low initial prices as a means to lure customers to their vehicles.[37]

These policies clashed with the Ronald Reagan administration's ideology of the superiority of free enterprise over government activities. Although NASA aimed to lure all payloads to the Shuttle (hence, putting all other competitors out of business), the Reagan administration and the Europeans worked to develop and promote private industry. The Europeans created Arianespace, the first private space launch company, though it was partially owned by European governments and only manufactured and operated vehicles whose development was fully funded by the ESA. The Reagan administration sanctioned the development of commercial launchers, to be licensed by the Department of Transportation (not NASA) only after the government-funded manufacturing infrastructure was sold back to industry. Commercial launches would not be viable if the Shuttle delivered on its promise of cheap and reliable space access. Unfortunately for NASA, the Shuttle never did achieve these goals, and the Reagan administration pressed NASA to set its prices closer to the actual Shuttle costs. The coup de grâce was the *Challenger* accident of January 1986, which ultimately led the Reagan administration to prohibit NASA from commercial launches. This, along with other regulations that loosened the military's grip on the contractors, paved the way for the emergence of American commercial launch companies to compete with Arianespace.[38]

Unfortunately for all Western nations, in the late 1980s and early 1990s the collapse of the Soviet Union and China's first steps to earn cash from the West meant that Russian and Chinese organizations began to offer launch services as well. Due to the very low wages in both nations, both could offer their launchers at very low prices that could still earn huge profits while remaining far below Western costs. Left unhindered, they would destroy the American and European commercial launch industries (the noncommercial launch industry would still survive by guaranteed launches from domestic markets). American policy makers faced the dilemma of wanting to integrate the Chinese, and especially the Russians, into the capitalist system to stabilize the Russian economy and bind Russia to the West, while simultaneously trying to protect the Western space launch industry and enable cheaper launches for the American communications satellite industry.

37. J. Krige, A. Russo, and L. Sebesta, *History of the European Space Agency 1958–1987, Vol. 2, The Story of ESA 1973-1987* (Noordwijk, The Netherlands: ESA SP-1235, 2000), chapter 11.

38. Joan Lisa Bromberg, *NASA and the Space Industry* (Baltimore, MD: Johns Hopkins University Press, 1999), chapter 6.

Two solutions prevailed. The first was to sign agreements with China (1989), Russia (1992), and Ukraine (1995) to place a quota on the number of Chinese, Russian, and Ukrainian commercial launches and ensure that they charged prices comparable to Western companies. Put bluntly, the U.S. forced Chinese, Russian, and Ukrainian companies to charge high prices and make very large profits so as not to tempt American and other satellite operators to choose exclusively non-Western launchers.[39] For Russia and Ukraine, this policy came to an end in the early 2000s.

The second solution was the integration of Russian, Ukrainian, and Western companies into corporate joint ventures so that Russian companies could bid for Western launches; the Russians could learn how to do business with the West; and to ensure that some of the money going for Russian launches benefited Western companies. This resulted in a number of joint ventures, including International Launch Services (Lockheed Martin and Khrunichev), Sea Launch (Boeing, RSC Energia, Kvaerner, Yuzhnoe), Starsem (EADS, Arianespace, Russian Federal Space Agency, and Samara Space Center), and Eurockot (EADS and Khrunichev).[40]

Spurred by the prospect of hundreds of satellite launches for medium-Earth orbit constellations of communications satellites, a number of purely private companies formed in the middle to late 1990s to build space launchers. Companies such as Kistler, Beal Aerospace, Pioneer Rocketplane, and Rotary Rocket spent tens and hundreds of millions of dollars before the bottom fell out of the prospective satellite boom in 1999 and 2000. Most of the companies went under, and those that remained sought government funding to survive. The November 2005 announcement by NASA Administrator Michael Griffin that private companies could bid for government contracts to deliver water and other bulk materials to space provided some encouragement to these efforts, but as of 2007 the result of this program is yet to be seen.[41]

In the United States, U.S. military funding on the procurement and operation of military launchers exceeded $1.4 billion in 2006.[42] The ESA's 2006 budget

39. "Memorandum of Agreement Between the Government of the United States of America and the Government of the People's Republic of China Regarding International Trade in International Launch Services," 26 January 1989, Document III-24, in *Exploring the Unknown, Volume IV: Accessing Space,* John M. Logsdon, ed., et al. (Washington, DC: NASA SP-4407, 1999), pp. 497–502.

40. Roger Handberg, *International Space Commerce: Building from Scratch* (Gainesville, FL: University Press of Florida, 2006), pp. 106–109.

41. Andrew J. Butrica, "Reusable Launch Vehicles or Expendable Launch Vehicles? A Perennial Debate," in *Critical Issues in the History of Spaceflight,* Steven J. Dick and Roger D. Launius, ed. (Washington, DC: NASA SP-2006-4702, 2006), pp. 319–320; *Reusable Launch Vehicle Programs and Concepts* (Washington, DC: Associate Administrator for Commercial Space Transportation, January 1998). See also various articles in *Space News* in the late 1990s and early 2000s on the various commercial vehicles. Michael D. Griffin, "NASA and the Business of Space," speech to American Astronautical Society 52nd. Annual Conference, 15 November 2005.

42. U.S. Government Budget, 2006. Budget of the United States Government, Washington, DC: Government Printing Office. It can be found online at *http://www.gpoaccess.gov/usbudget/*, accessed January 2007. Evolved Expendable Launch Vehicle Acquisition $864.4; Medium Launch Vehicles acquisition, $111.2 million; Space Launch Operations & Maintenance, $443.4 million.

allocated roughly €531 million ($640 million) for launchers, primarily the Ariane and Vega programs.[43] NASA has invested hundreds of millions of dollars per year on launchers in some years, particularly in the 1990s and early 2000s, such as its $284 million spent in 1997 on reusable launch vehicles, and the later Space Launch Initiative; however, by 2006 its launcher funding was going into the Crew Launch Vehicle, or Ares I program, in support of human flight.

Technology Research and Development

The research and development sector is characterized by a confusing competition among government, academic, corporate, and nonprofit organizations for research and development (R&D) funding. It is a specific sector because these funds are largely dispersed by government organizations with their own separate budgets allocated for the purpose of developing new technologies. The DOD and NASA dominate these funds in the U.S., whereas their equivalents in other nations distribute these moneys according to equivalent procedures. Within the DOD, research appears in the Research, Development, Test, and Evaluation (RDT&E) budgets. Research is allocated in Basic Research and Applied Research funds, whereas development funding appears in Advanced Technology Development, Advanced Component Development and Prototypes, and System Development and Demonstration, with other related support showing up in RDT&E Management Support line items.[44] Within NASA, these funds have historically been in the Office of Aerospace Research and Technology budgets and their predecessors, though as of 2007 these are now folded into Exploration, Science, and Operations budgets.

In the former Soviet Union, competition existed among the scientific research institutes such as Mstislav Keldysh's NII-1, which performed aerodynamic research, and the various development design bureaus. This competition was just as stiff and certainly more devious than in the U.S., due to institutional secrecy and the Byzantine nature of the Soviet state.

Competition for technology R&D funds is fierce, but typically government funding is restricted to domestic suppliers due to the criticality of technology development for national economic and strategic aspirations. Since the bulk of these funds are distributed by governments, government institutions have priority access to these funds when they believe they should run the program. In the U.S., government policy dictates that the majority of funding should go to industry, but because governments must also train their personnel, technology R&D is a prime location to gain experience. University researchers have obviously been strong competitors for technology R&D funding, since their jobs require them to

43. Peter B. de Selding, "ESA Budget Emphasizes Independence, Satcom Technology, *Space News* (23 January 2006).

44. See U.S. Government Budget, Department of Defense, for these categories.

advance the state of the art. Similarly, nonprofit institutions such as The Aerospace Corporation and the MITRE Corporation need to keep their knowledge at the state of the art, and they perform a variety of technology R&D studies. Finally, corporations compete for these same funds, winning a significant proportion of them while also committing their own.

Outside of the U.S., similar dynamics apply, though the separation between government and industry is often less clear. It is notable that in Europe all member states contribute to technology R&D programs and all member states benefit. This is one major exception to the usual policy of restricting technology development funding to domestic suppliers only.[45]

In 2006, DOD Basic and Applied Research Budgets allocated more than $520 million to space technologies, with $223.8 million of that budget coming from the Defense Advanced Research Projects Agency. This does not account for U.S. Army and U.S. Navy basic and applied research into space technologies, which are not readily apparent from the top-level budget figures given by the U.S. government. NASA set aside $693 million for Exploration Systems Research and Technology and $624.1 million for Human Systems Research and Technology in its 2006 budget.[46] The ESA's 2006 budget allocated roughly €126 million ($152 million) for technology development.[47]

ASTRONOMY AND PLANETARY SCIENCE

Astronomy has been a popular topic for centuries, as it has dealt with fascinating questions such as the nature of the universe, its history, and Earth's (and hence humanity's) place in the cosmos. Satellites to observe the cosmos (with the Hubble Space Telescope being the most famous) and probes to visit other planets (such as Voyager's flybys of Jupiter, Saturn, Uranus, and Neptune) have been among the most captivating space programs. As such, they have garnered political support to entice young people into science, engineering, and mathematics, and hence to improve the technological competitiveness of the U.S. NASA's educational programs feature large doses of space science.

Another major use of space science has been as a political diplomatic tool. During the International Geophysical Year of 1957–1958, the early U.S. space program was used explicitly as a means to improve foreign relations with allies,

45. Roger M. Bonnet and Vittorio Manno, *International Cooperation in Space: The Example of the European Space Agency* (Cambridge, MA: Harvard University Press, 1994), p. 25.

46. President's FY 2007 Budget Request for NASA, NASA Web site, *http://www.nasa.gov/about/budget/* (accessed February 2007).

47. Peter B. de Selding, "2006 ESA Budget Emphasizes Independence, Satcom Technology," *Space News* (23 January 2006).

while at the same time paving the way for overflights of reconnaissance satellites over the Soviet Union. During the late 1950s and 1960s, the space race between the U.S. and the Soviet Union included the so-called robotic races—first to the Moon and then to Venus and Mars. The winner of these races accrued benefits in national prestige and perceived technological prowess, a proxy for the long-term viability and strength of the capitalist and socialist visions.

NASA's charter explicitly included international cooperation, and under International Programs Chief Arnold Frutkin, in its early years created dozens of agreements and cooperative programs with nations around the world as an active element of American foreign policy. These initiatives included free launches of foreign scientific payloads; foreign instruments on American scientific satellites and vice versa; establishment of communications ground stations for scientific satellites; and a variety of other scientific information exchanges. Not to be outdone, the Soviet Union ran similar programs for its socialist allies, and soon for capitalist or nonaligned nations to woo them away from American interests. Europe, through the European Space Research Organisation (ESRO), and later the ESA, made space science part of its mandatory program that all member states had to support.[48] From the inception of the Space Age, and even before, space science has been a popular, cooperative diplomatic tool, which helps to explain its significant and stable funding.

Corporations typically (though not always) build the satellites, whereas academic, government, and corporate researchers usually build the instruments placed on-board. With relatively few spacecraft put in orbit, the competition to build the satellites and the instruments has always been fierce. Corporations like to build science-craft because it provides an opportunity to advertise their capabilities in ways not possible with often-secret military programs. For scientists, getting an instrument in space can make or break their careers and reputations.

Though the politics of determining the relative priorities and funding of space science programs has made or broken the hopes of many scientists, total space science funding has remained fairly stable over the years, and by 2006 reached roughly \$6–\$7 billion worldwide. This stability owes much to the use of science as a diplomatic tool, since it is much more difficult to cancel a program that has foreign partners.[49]

48. Roger M. Bonnet and Vittorio Manno, *International Cooperation in Space: The Example of the European Space Agency* (Cambridge, MA: Harvard University Press, 1994), chapter 2.

49. John Krige, "The European Space System," in John Krige and Arturo Russo, *Reflections on Europe in Space* (Noordwijk, The Netherlands: ESA HSR-11, January 1994), pp. 7–8.

COMMUNICATIONS

As early as 1945, when Arthur C. Clarke first published a paper proposing the use of the geosynchronous orbit as promising location for telecommunications satellites, the potential utility and profitability of space communications beckoned. Military, civilian, and commercial organizations all began to develop communications satellites in the U.S. by the early 1960s.

All of the military services had interests in communications satellites (comsats) for command and control, as part of a worldwide military communications system. In part responding to Kennedy administration concerns about the potential that AT&T would extend its communications monopoly into space by developing a purely private system, NASA funded its own experimental satellites. AT&T's launch of Telstar in 1962 fueled these government fears, leading to the passage of the Satellite Telecommunications Act of 1962 that created Communications Satellite Corporation to build and manage the international satellite communications system. Military and NASA funding built several strong competitors to AT&T, including Hughes Aircraft Corporation, Philco Corporation, and Radio Corporation of America. The successful 1963 launch and operation of Hughes's NASA-funded Syncom satellite proved the viability of geosynchronous telecommunications satellites. The U.S. then opened negotiations with European nations, Canada, and Japan regarding the management of the international telecommunications system.[50]

Although the U.S. wanted to negotiate on a bilateral basis with each nation, the Europeans quickly banded together to negotiate with greater strength as a group, forming the European Conference for Satellite Telecommunications (French acronym CETS) in 1963. The resulting negotiations created the International Telecommunications Satellite Organization (Intelsat) the next year, with Communications Satellite Corporation as the organization's manager. Though the Americans dominated Intelsat in the 1960s, the Europeans acquired one major concession—to renegotiate Intelsat's terms in 1971. In the meantime, the United Kingdom, Germany, France, and Italy began development of their own national satellite programs, while working through CETS and ESRO to create an integrated European satellite program.[51]

During this period, the U.S. had monopoly power over spacecraft launches, and U.S. policy dictated that the U.S. would not launch any satellite that threatened Intelsat's international monopoly. "Experimental" satellites were launchable, but

50. David J. Whalen, *The Origins of Satellite Communications 1945-1965* (Washington, DC: Smithsonian Institution Press, 2002).

51. A. Russo, "The Beginning of the Telecommunications Satellite Programme in ESRO," in J. Krige and A. Russo, *A History of the European Space Agency 1958–1987, Volume I, The Story of ESRO and ELDO 1958-1973* (Noordwijk, The Netherlands: ESA SP-1235, 2000), chapter 9.

not regional or international systems. Convinced that the U.S. was attempting to retain control of satellite telecommunications, the Europeans, led by France, pushed to develop their own launcher to ensure that European comsats would get to orbit. Ultimately, this resulted in the development of the Ariane and was a major factor in the negotiations that led to the creation of the ESA. Intelsat's permanent agreement of 1973 reallocated shares in Intelsat based on actual usage of the system, and allowed for regional as well as domestic systems. American shares shrank from more than 50 percent to the range of 25 percent to 30 percent, and hence control of Intelsat passed out of American hands by the late 1970s. ESA began development of the Orbital Test Satellite, the first of several ESA-developed systems from that time forward.[52]

From the 1970s through the 1990s, many nations built or purchased comsats for domestic purposes, while several private companies orbited systems that leased transponders to government and private customers. The growing demand led to many negotiations over the use of the frequency spectrum through the International Telecommunication Union (ITU). Nations own the spectrum rights and geosynchronous orbital slots, and thus this provided a mechanism for governments to control both government and corporate uses of space communications. Many countries ensured that their domestic systems were government-owned or -controlled. The most profitable use of comsats turned out to be direct broadcast television, with companies such as DirecTV, Echostar, and SES selling dozens and eventually hundreds of broadcast channels directly to millions of consumers. Direct-broadcast radio, led by XM, Sirius, and Worldspace, began in 1998 and was growing rapidly by the first decade of the twenty-first century.[53]

Mobile communications for ships, aircraft, and eventually telephones and computers began from military experiments in the U.S. Systems such as Tactical Satellite (TACSAT), launched in 1969, proved the concept of using relatively small ground antennas that could be ported on ships, trucks, and other vehicles. The U.S. and Soviet navies desperately needed such systems for fleet communications, and merchant ships would also be obvious beneficiaries.[54] The International Maritime

52. Ibid., and A. Russo, "The Early Development of ESRO's Telecom Programme and the OTS Project," in Krige and Russo, *A History of the European Space Agency, Volume 1,* chapter 10; John L. McLucas, *Space Commerce* (Cambridge, MA: Harvard University Press, 1991), pp. 40–43.

53. McLucas, pp. 75–80. To see the growth of direct broadcast television, see Office of Space Commercialization, *Trends in Space Commerce,* prepared by Futron Corporation (Washington, DC: Department of Commerce, 2001), chapter 3. On satellite radio market growth, see International Space Business Council, *2004 State of the Space Industry* (Bethesda, MD: Space Publications, 2004), pp. 23, 42–43.

54. Norman Friedman, *Seapower and Space: From the Dawn of the Missile Age to Net-Centric Warfare* (Annapolis, MD: Naval Institute Press, 2000), chapter 5; William W. Ward and Franklin W. Floyd, "Thirty Years of Space Communications Research and Development at Lincoln Laboratory," in *Beyond the Ionosphere: Fifty Years of Satellite Communication,* Andrew J. Butrica, ed. (Washington, DC: NASA SP-4217, 1997), pp. 84–86.

Communications Satellite Organisation (Inmarsat) was established in 1976 to provide worldwide mobile commercial service for shipping. Unlike Intelsat, in which the Soviet Union and other Communist-bloc nations refused to participate, the Soviet Union and other socialist countries became members of Inmarsat.[55]

In the 1990s, Iridium, Globalstar, and ICO developed large constellations of medium-Earth orbit comsats to provide worldwide mobile telephone and data services. By the first years of the twenty-first century, all three systems had been deployed and all three companies had gone bankrupt after raising and spending billions of dollars; they had overestimated the demand for global service and underestimated the competition from ground-based cellular telephone systems.[56]

Both ground-based and space-based private competition put pressure on the international and domestic government systems, making some of them economically obsolete. Both Intelsat and Inmarsat were privatized in response to U.S. demands; if these systems wanted to compete for domestic U.S. business, they could no longer be owned by governments.[57] However, military demand continued to grow rapidly, leading to the leasing of many commercial transponders for military purposes and the development of more-capable military systems that were less susceptible to jamming or radiation effects (such as Milstar), and with greater bandwidth to send imagery and intelligence data from reconnaissance and intelligence satellites and from various sensor systems to military users around the world.[58] In 2006, the DOD was budgeted more than $2.9 billion to develop and acquire communications satellites,[59] and more than $340 million to procure USAF and U.S. Army satellite communications ground terminals.[60] Precise numbers to operate military communications satellites are difficult to estimate from the DOD overview budgets, but are certainly in the hundreds

55. McLucas, *Space Commerce*, pp. 46–50; Edward J. Martin, "The Evolution of Mobile Satellite Communications," in Butrica, ed., *Beyond the Ionosphere*, pp. 265–281.

56. Roger Handberg, *International Space Commerce: Building from Scratch* (Gainesville, FL: University Press of Florida, 2006), pp. 144–145.

57. Intelsat Web site, *http://www.intelsat.com* (accessed January 2007). Inmarsat Web site *http://www.inmarsat.com* (accessed January 2007).

58. David N. Spires and Rick W. Sturdevant, "From Advent to Milstar: The U.S. Air Force and the Challenges of Military Satellite Communications," in Butrica, ed., *Beyond the Ionosphere*, pp. 65–78.

59. From the U.S. Government 2006 budget: $1,194.3 million for Advanced Extremely High Frequency Satellite Acquisition; $835.8 for Transformational Satellite Communications Acquisition; $166.4 million for Wideband Gapfiller Acquisition; $28.7 million for USAF Military Satellite Communications Procurement; $2.185 million for USAF Polar Military Satellite Communications RDT&E; $541.98 million for U.S. Navy Satellite Communications RDT&E; $71.8 million for U.S. Navy Satellite Communications Systems Procurement.

60. From the U.S. Government 2006 budget: U.S. Army, DSCS Ground Systems Program Acquisition, $66.5 million; USAF, Management Support for MILSATCOM Terminals, $273.974 million. U.S. Navy satellite terminal procurements are buried in other budget lines.

of millions of dollars per year.[61] By 2005, the satellite communications sector was more than $80 billion annually—by any estimate, the largest space sector by a wide margin.[62]

RECONNAISSANCE AND REMOTE SENSING

The technologies of reconnaissance and scientific or commercial remote sensing are essentially the same, though the specific characteristics of the sensors vary depending on what the satellite is trying to sense. High-resolution, often black-and-white (panchromatic) imagery tends to be most useful for intelligence gathering, whereas lower-resolution color (multispectral) imagery is useful for environmental monitoring. Monitoring of volcanoes, clouds, ice, ocean, and land features are each optimized with the use of different portions of the spectrum, with infrared bands looking for heat sources best for volcanic activities, and various shades of green for vegetation monitoring, for example. High-resolution imagery is not particularly useful to monitor large-scale weather fronts but it is crucial to differentiate one kind of airplane from another.

Military and intelligence organizations developed the first reconnaissance systems, starting in the mid-1950s with the USAF's WS-117L, which led to the Central Intelligence Agency (CIA)-sponsored Corona and the Air Force SAMOS projects. The Eisenhower administration secretly created the National Reconnaissance Office in 1960 to manage the spy satellite programs.[63] Also in the late 1950s, the U.S. Army began to investigate weather satellites, but these study results and projects were soon transferred to NASA, becoming the Tiros program, first launched in 1960. The USAF, needing weather imagery over the Soviet Union to better utilize Corona, created what was eventually called the Defense Meteorological Satellite Program (DMSP).[64] The Soviet Union created its equivalent systems, Zenit (based on the Vostok capsule) for reconnaissance, and Meteor for weather.[65]

61. These U.S. Government 2006 budget figures are part of the Operations and Maintenance budgets, including the U.S. Navy's $156.8 million for Space Systems and Surveillance; the USAF's C3I and Early Warning's $1.201 billion; and the U.S. Army's O&M budget, which does not have any specific "space" line items. Original numbers, from which I make these estimates, are drawn from International Space Business Council, *2005 State of the Space Industry* (Bethesda, MD: Space Business.com, 2006). SIA/Futron, and the U.S. military communications satellite and operational budgets.

62. This includes approximately $42 billion for direct-broadcast, fixed, and mobile satellite revenues. I estimate some $5 billion for acquisition of government-paid communications satellites worldwide ($16 billion for ground systems), assuming some two-thirds of all ground system are for communications antennas and related equipment, and another $3 billion of government-paid operational (launchers, launch operations, and mission) budgets to launch and operate communications satellites.

63. Dwayne A. Day, John M. Logsdon, and Brian Latell, ed., *Eye in the Sky: The Story of the Corona Spy Satellites* (Washington, DC: Smithsonian Institution Press, 1998).

64. R. Cargill Hall, "A History of the Military Polar Orbiting Meteorological Satellite Program," *Quest: The History of Spaceflight Quarterly* 9, no. 2 (2002): pp. 4–25.

65. Bart Hendrickx, "A History of Soviet/Russian Meteorological Satellites," *Space Chronicle, The Journal of the British Interplanetary Society* 57, Supplement 1 (2004): pp. 56–102.

Classified reconnaissance programs evolved rapidly in both the U.S. and Soviet Union. Early systems generally used film that had to be returned to Earth, which meant that space reconnaissance required many launches every year. In 1977, the first American Keyhole (KH)-11 satellite was launched. This program helped create charged-coupled device (CCD) technology that allowed for digital capture and radio transmission of imagery to Earth.[66] The Soviet Union soon followed this lead, as its Yantar series evolved to include Yantar Terilen, the first Soviet satellite with electro-optical capabilities, first launched in 1982.[67] China began development of its own reconnaissance systems by the late 1970s,[68] and other nations such as France and Israel developed their own reconnaissance systems by the 1990s. Many nations now have access to American high-resolution commercial satellite imagery, which became available in 1999 with the launch of Space Imaging EOSAT Company's Ikonos 1, which provided ~3-foot (1-meter) resolution. These companies have remained viable only through major government imagery purchases and direct subsidies for satellite development. The costs of classified reconnaissance systems remain classified, but some estimate that the funding of the National Reconnaissance Office, which manages U.S. imagery intelligence, exceeds $7 billion per year in the early twenty-first century.[69] The costs of the U.S. reconnaissance program in the 1960s may have ranged in the same order of magnitude as the Apollo program, which totaled roughly $21 billion in then-year dollars.

The development of commercial remote sensing began with NASA's Landsat program (first launch 1972, final launch of Landsat 7 in 1999), a civilian project to provide satellite imagery of landforms and vegetation for scientific purposes. The program soon developed a clientele of users that included mining companies, land use planners, and others who were not scientists.[70] These commercial applications lent weight to arguments to privatize Landsat. Congressional legislation led to the privatization of Landsat operations in 1985 by the Earth Observation Satellite Company (EOSAT). Certain conditions of the legislation hampered EOSAT's business activities, while the large increase in image prices alienated Landsat's scientific clientele. Congress repealed the legislation in 1992, returning control of Landsat to the government but also creating procedures for companies to develop

66. William E. Burrows, *Deep Black: Space Espionage and National Security* (New York: Berkley Books, 1988).

67. Peter A. Gorin, "Black 'Amber': Russian Yantar-Class Optical Reconnaissance Satellites," *Journal of the British Interplanetary Society* 51/8 (August 1998): pp. 309–320; Harvey, *Russia in Space: The Failed Frontier?*

68. Phillip S. Clark, "Development of China's Recoverable Satellites," *Quest: The History of Spaceflight Quarterly* 6, no. 2 (1998): pp. 36–43.

69. Federation of American Scientists Web site, *http://www.fas.org* (accessed January 2007).

70. Pamela E. Mack, *Viewing the Earth: The Social Construction of the Landsat System* (Cambridge, MA: MIT Press, 1990).

commercial high-resolution remote sensing, which EOSAT and others quickly pursued. This was spurred by the success of international competitors, such as France's Satellite pour l'Observation de la Terre (SPOT), first launched in 1986; India's Indian Remote Sensing (IRS) satellites (1988); and others, all of whom sold imagery on the commercial market.[71]

France and India sold SPOT and IRS imagery, respectively, through private companies SPOT Image and Antrix, respectively, but these companies did not have to develop or launch the spacecraft. Other nations sold the imagery directly from government organizations. Commercial remote sensing has thus developed as an international competition among governments and government-supported private companies. None of the organizations makes enough revenues from imagery sales to fund the development of the satellites, but the imagery sales have defrayed some small fraction of government costs. Commercial space imagery sales worldwide in 2004 were less than $600 million annually, while reconnaissance satellites cost several hundred million each.[72]

Despite the unpromising economics, when the U.S. in 1992 created provisions for private companies to build high-resolution remote sensing satellites, several companies jumped at the opportunity. The first high-resolution commercial remote sensing satellite, Ikonos, was orbited in 1999, with two other companies following soon thereafter. However, the revenues from imagery sales never met expectations, and survival of the American commercial companies depended on large government contracts for imagery and also for direct government subsidies to fund their next-generation satellites. The National Geospatial Intelligence Agency obligingly awarded imagery contracts, and then provided $500 million manufacturing subsidies, to DigitalGlobe in 2003 and to Orbimage the next year, but not to EOSAT, leading to Orbimage's purchase of EOSAT to form GeoEye.[73]

Most users are not able to use raw remote sensing images. Instead, the so-called value-added services sector processes raw imagery and adds other information to make it useful. This sector is usually estimated as having revenues an order of magnitude larger than the raw imagery market. Numerous small companies create and market to the many small niche markets providing processed imagery to government, scientific, and commercial users, often through Geographic Information Systems.[74]

---

71. Joanne Gabrynowicz, "The Perils of Landsat from Grassroots to Globalization: A Comprehensive Review of U.S. Remote Sensing Law with a Few Thoughts for the Future," *Chicago Journal of International Law* 6 (2005).

72. *2004 State of the Space Industry*, p. 23.

73. John C. Baker, Kevin M. O'Connell, and Ray A. Williamson, ed., *Commercial Observation Satellites: At the Leading Age of Global Transparency* (Santa Monica, CA: RAND, 2001). See relevant articles by *Space News* and *http://www.space.com* on the NGA contracts.

74. Office of Space Commercialization, *Trends in Space Commerce*, prepared by Futron Corporation (Washington, DC: Department of Commerce, 2001).

Governments have borne the entire costs of weather satellites throughout their history, and thus have provided a steady demand for these systems and several hundred million dollars annually worldwide for their development and operation. In the United States, the U.S. Army started such work but soon transferred it to NASA, where it became the Tiros series, first launched in 1960. That same year, the Corona program finally succeeded in taking images of the Soviet Union, finding out that much of its precious film recorded cloud tops. This led the National Reconnaissance Office and the USAF to develop what eventually became known as the Defense Meteorological Satellite Program (DMSP), initially derived from the Tiros design. Organizational issues plagued the civilian program because NASA specialized in satellite development, but the users of satellites were in the National Weather Service (NWS). NASA preferred to push the technological envelope, whereas NWS forecasters wanted a reliable system that typically used tried-and-true techniques. Thus, in the mid to late 1960s, while NASA developed Nimbus satellites, the NWS purchased systems derived from the military's DMSP due to their perceived greater operational utility and proven design. These same issues continued to characterize relationships between NASA and the user community, and continued through the development of the Geostationary Operational Environmental satellites and Polar Operational Environmental satellites. All the while, the possibility of combining military and civilian systems continued, leading in the 1990s to the decision to create the joint civilian-military National Polar-Orbiting Environmental Satellite System (NPOESS), scheduled to launch in the early twenty-first century.[75]

In 2006, NOAA was funded to spend $782 million to acquire weather satellites, another $97 million to operate them, $53 million to distribute data, and $11 million to integrate Landsat sensors onto NPOESS.[76] That same year, the USAF spent an additional $323 million on NPOESS and $71 million on DMSP.[77]

Other nations recognized the practical utility of weather satellites and developed their own, starting with the Soviet Union's Meteor program, the first of which launched in 1964. Europe, through the ESA, developed its Meteosat series, first launched in 1977, and eventually operated through Eumetsat, established in 1986. India combined weather observation with communications in its Indian National Satellite series, while Japan developed its Geostationary Meteorological Satellites and China its Feng Yun series. The Coordination of Geostationary Meteorological Satellites (CGMS) organization formed in 1972 to coordinate worldwide satellite

75. Frederik Nebeker. *Calculating the Weather: Meteorology in the 20th Century* (New York: Academic Press, 1995).

76. R. Cargill Hall, "A History of the Military Polar Orbiting Meteorological Satellite Program," *Quest* 9/2 (2002): pp. 4–25. Budget data from 2006 NOAA budget, available on NOAA Web site, *http://www.corporateservices.noaa.gov/~nbo/* (accessed January 2007).

77. U.S. Government Budget, 2006. DMSP figures are $67.2 for Procurement and $3.908 million for RDT&E Management Support.

weather monitoring from geostationary satellites. In 1992 the organization expanded its purview to include polar-orbiting satellites, and became the Coordination Group for Meteorological Satellites. Among other activities, the group coordinated the use of one nation's satellite when another nation's satellite had problems.[78]

Electronic Intelligence and Early Warning

Closely related to imagery collection is electronic data collection, which gathers data from various nonvisible portions of the electromagnetic spectrum. These data can be used for strategic assessments of national capabilities or for tactical military purposes. Strategic uses, such as intelligence gathering by the U.S. National Security Agency and its Russian equivalents, remain highly classified; they started in the early 1960s. Tactical uses include monitoring of shipping and naval signals, as well as strategic early warning systems that monitor infrared signatures of ballistic missile launches or actively send radar signals to reflect off targets. Another application is the tracking of objects in space, which is necessary to distinguish between a ballistic missile and a meteor hitting Earth. In all cases, these are purely government-controlled functions. The U.S. contracted to industry for the satellites while the government operated them, whereas the Soviet Union and later Russia developed and operated its systems purely through government organizations.[79]

Electronic intelligence-gathering budgets, like the satellite programs themselves, remain classified, so their costs remain highly speculative. Some analysts estimate their costs in the range of $3 billion per year in the U.S. and presumably significantly lower in Russia.[80] Naval reconnaissance systems, such as the American White Cloud system or the Russian EORSAT, also remain classified, along with their funding. Early warning systems, by contrast, have official published budgets in the U.S. Published figures from the 1960s through the mid-1980s show early warning budgets ranging from a low of $45.6 million in 1967 to $707.2 million in 1984.[81] In 2006, the USAF budgeted $1.21 billion on procurement of early warning and space tracking systems.[82]

---

78. Eumetsat, *The History of Eumetsat—European Organisation for the Exploitation of Meteorological Satellites* (Darmstadt: Eumetsat, 2001); Bart Hendrickx, "A History of Soviet/Russian Meteorological Satellites," *Journal of the British Interplanetary Society* 57/1 (2004): pp. 56–102.

79. Friedman, *Seapower and Space*; William E. Burrows, *Deep Black: Space Espionage and National Security* (New York: Berkley Books, 1988); James Bamford, *Body of Secrets: Anatomy of the Ultra-Secret National Security Agency* (New York: Anchor Books, 2001).

80. Federation of American Scientists, *http://www.fas.org*, accessed January 2007.

81. Paul B. Stares, *The Militarization of Space: U.S. Policy, 1945–1984* (Ithaca, NY: Cornell University Press, 1985), Table 2, pp. 256–257.

82. From the U.S. Government Budget, 2006. It includes $756.6 million to procure the Space-Based Infrared Systems-High; $225.8 million for a Space-Based Radar system; $151 million to support its Space Tracking efforts; $32.8 million on Nuclear Detection; and $42.7 million for RDT&E Management Support to its older Defense Support Program system.

WEAPONS

Space weapons refer on one hand to ballistic missiles, and on the other to weapons to destroy ballistic missiles (antiballistic missile system) and satellites (antisatellite systems). As with intelligence-gathering satellites, space weapons have been the exclusive domain of governments, with the exception of contracting to industry in Western nations. The very first space expenditures were for the development of ballistic missiles, which received major funding first in Nazi Germany in the mid-1930s, and by the mid to late 1940s in the U.S. and Soviet Union, followed quickly by Great Britain and France, and somewhat later by many other nations including China, India, Pakistan, Iran, North Korea, and others. Even though some of the budget figures for ballistic missile, antiballistic missile, and antisatellite programs are unclassified, working through the maze of government budgets to determine actual spending on space-related research, development, facilities, and operations is not trivial.

One significant exception is the budget of the Missile Defense Agency, the descendant of the 1980s Strategic Defense Initiative Organization. From a budget of $1.6 billion in 1985, the agency's budget jumped to $7.8 billion in 2002 and remained in the $7 to $9 billion range through 2006 as the George W. Bush administration began to deploy a national missile defense system in Alaska.

Ballistic missiles consumed enormous amounts of funding in the U.S. and Soviet Union in the 1950s and 1960s in particular, leveling off somewhat in the 1970s and 1980s, and declining significantly after the end of the cold war in the 1990s and early twenty-first century. For example, in 1959, the Atlas, Titan, and Minuteman programs consumed $1.321 billion, which, when converted to 2006 dollars, is equivalent to more than $7 billion.[83] And that is not all, as in that year there were several other ballistic missile programs ongoing, including the Navy's Polaris, the Army's Corporal, Redstone, and Sergeant systems, and the USAF's Thor Intermediate-Range Ballistic Missile. By contrast, in 2006, the USAF spent $791 million on intercontinental ballistic missile acquisition[84] and the U.S. Navy spent $936.1 million on Trident II submarine-launched ballistic missile modifications and another $830 million to operate its fleet ballistic missile systems.[85]

83. Data from Jacob Neufeld, *Ballistic Missiles in the United States Air Force 1945–1960* (Washington, DC: Office of Air Force History, 1990), pp. 198, 229. Atlas: $641.5, Titan $495.4, Minuteman $184 (all in millions).

84. U.S. Government Budget, 2006. USAF Minuteman III Modifications, $672.6; Missile Replacement, $41.6; ICBM RDT&E, $44.67; ICBM Components, $32.42 (all in millions).

85. U.S. Government Budget, 2006.

NAVIGATION

Satellites assist navigation on Earth by providing targets with known positions in space, from which objects on Earth can triangulate their position. This was particularly useful for naval operations, where navigation on featureless oceans has historically been particularly difficult. However, given sufficient reliability, accuracy, and timeliness, it is also useful for movement of vehicles and people on land and in the air. The first application of satellite navigation was for American and Soviet submarines to determine their positions with sufficient accuracy such that their nuclear-tipped ballistic missiles could hit cities in the opposing countries. For this purpose, the U.S. Navy funded the Transit program, the first satellite of which launched in 1960. The Soviet Union countered with its Tsiklon satellites, first launched in 1967. Both systems provided two-dimensional (not vertical) accuracy of several hundred feet, with update rates on the order of tens of minutes in the worst case. This was sufficient for slow-moving naval vessels but not fast-moving aircraft at various altitudes.[86]

Naval success spurred further studies and research, leading both nations to create more capable navigational systems. In the U.S., the tri-service Global Positioning System (GPS) program began in 1973, with its first experimental satellite launch in 1978. The full 24-satellite constellation was not operational until 1995. The Soviet Union's equivalent, the Global Navigation Satellite System (GLONASS), put its first satellite into orbit in 1982, with its full 24-satellite constellation in place in 1996. While the U.S. maintained its GPS constellation, Russia was unable to maintain its GLONASS constellation, with only 14 satellites operating by late 2005.[87] The Transit system was relatively cheap, with acquisition costs at roughly $20–$30 million per year in the 1960s.[88] GPS is much more capable and also much more expensive. In 2006, the USAF budgeted $719.6 million for acquisition of GPS satellites, which primarily paid Boeing for the latest generation of Block 2F satellites.[89]

The military uses of the satellite navigation systems were obvious and crucial, and soon expanded to include aircraft and land force positioning, and precision-guided weapons. The use of GPS has been a key element in the dramatic enhancement of U.S. military capabilities so apparent from the two Iraq wars in 1991 and 2003. Certain civilian and commercial uses were also clear from the beginning,

86. Friedman, *Seapower and Space,* chapter 4.

87. Harvey, *Russia in Space: The Failed Frontier?,* pp. 130–136.

88. Paul B. Stares, *The Militarization of Space: U.S. Policy, 1945–1984* (Ithaca, NY: Cornell University Press, 1985), table 2, pp. 256–257.

89. U.S. Government Budget, 2006.

such as for merchant shipping, air transport, and air traffic control. However, the military did not expect the tremendous surge in commercial applications that blossomed in the 1990s and 2000s. Once in place, the satellite navigation signals were essentially a free resource that individuals and commercial firms could tap into for their own uses simply by purchasing the appropriate receivers. Receiver sales were roughly $800 million in 1994; by 2005, estimates of GPS equipment sales ranged from $5 billion to $22 billion, with American and foreign firms splitting the market roughly 50-50.[90]

Such a wide discrepancy in estimates says much about the difficulty of estimating the size of this market. However, it should be noted that, historically, many if not most estimates of current and future commercial space market sizes have been overestimated. For GPS in particular, the market estimate depends a great deal on what is counted. For example, for a precision weapon, does the entire weapon count or merely the GPS receiver in the weapon? For a handheld cellular phone or a GPS fish finder with a GPS chip, does the entire phone or fish finder count or merely the GPS chip? There are many ways to estimate the entire market size, each of which starts with different assumptions and yields radically different answers.[91]

By 2006, with GPS receivers available for under $100 and packaged in many products, commercial uses included car and boat navigation, surveying, animal and child tracking, cellular telephone positioning, and various recreational purposes such as locating precise positions and distances on a golf course or fish finders on lakes. Civilian purposes grew to include various emergency services such as fire, police, and search and rescue, and scientific research in geodesy and geology. Civilian and commercial applications dwarfed military ones in terms of receiver sales by the early twenty-first century.

Europe and China were the first to develop space-based navigation capabilities independent of the old superpowers. China launched three Beidou satellites in 2000–2003, most likely to support its small intercontinental ballistic missile force. In 2002, the Europeans began development of the Galileo system, but as a civilian commercial

90. Office of Space Commercialization, *Trends in Space Commerce*, prepared by Futron Corporation (Washington, DC: Department of Commerce, 2001). The high estimate, $13 billion, comes from the RNCOS Web site, "GPS Market Update (2006)," April, 2006, *http://www.rncos.com/Report/COM33.htm* (accessed 28 December 2006). The low estimate, $4.8 billion, comes from International Space Business Council, *2004 State of the Space Industry* (Bethesda, MD: International Space Business Council, February 2004), p. 23. The $22 billion figure comes from ABI Research, in Space Foundation, *The Space Report 2006* (Colorado Springs, CO: Space Foundation, 2006), footnote 262.

91. Discussion with Scott Sacknoff, head of the International Space Business Council (ISBC), 28 December 2006. The ISBC methodology is a "conservative" one that will tend to yield lower estimates by parsing out the GPS components from the overall products. It also adds up revenues of major market players, such as Garman and Trimble, to compare with other estimates that add up the total number of cars produced, for example, and estimating the percentage of those cars that have GPS receivers.

venture. The GPS signal is free but potentially interruptible should the U.S. deem it necessary, whereas the Galileo system would provide coarse signals for free but could provide higher-precision signals (for a price) to users. China, Israel, India, Saudi Arabia, Ukraine, and Morocco have decided to participate in the program as well, making it a global venture to ensure independence from the U.S. The project, as of 2006, was projected to cost some €3.8 billion, with European costs split 50-50 between the European Commission Transport Directorate and the ESA. Germany is the largest single national contributor, with some €500 million committed to the project. The funding goes primarily to a European industrial consortium, Galileo Industries, which consists of EADS, Alcatel Space, and Alenia Spazio. The first Galileo experimental satellite went into orbit in December 2005.[92]

Several nations, including the U.S., Europe, Japan, and India, developed systems to augment the capabilities of the GPS. The U.S. Wide Area Augmentation System was developed to aid precision flight approaches. Europe deployed the European Geostationary Navigation Overlay System. Japan placed navigational capability in its Multi-functional Transport Satellite, which is primarily a meteorological satellite, first launched in February, 2005. In parallel, it developed the Quasi-Zenith Satellite System to enhance regional navigation in Asia and the Pacific region.[93] By 2006, the Indian Space Research Organisation was developing its Geo-augmented Navigation (Gagan) Satellite System, which was to augment GPS signals primarily for aircraft navigation in south Asia.[94]

Since the deployment of GPS, the military, civilian, and commercial success of navigational systems has generated global interest in navigational capabilities, generating political and economic opportunities and conundrums. For the U.S., the key issue has become the debate between military and civilian control of GPS signals. The military maintains the authority to shut down certain capabilities in wartime, yet civilian and commercial uses have become so important that it is clear that GPS signals cannot merely be shut down. Russia's major problem is finding the cash to maintain its GLONASS constellation. Other nations debate whether to create very expensive, independent capabilities so as not to depend on the U.S., as

---

92. Peter B. de Selding, "Initial Galileo Satellite Navigation Bid Exceeds ESA's Budget," *Space News Business Report*, *http://www.space.com*, 22 March 2004 (accessed January 2007); "EU Launches First Satellite in Galileo Navigation Program," 28 December 2005, *http://www.galileo-navigationsystem.com/Archiv/galileo-archive1.htm* (accessed January 2007); "Contract Signed for Second Phase of Galileo Navigation Satellites," 19 January 2006, *http://www.galileo-navigationsystem.com/Archiv/galileo-archive1.htm* (accessed January 2007).

93. Japan Meteorological Agency, MTSAT Web site, *http://www.jma.go.jp/jma/jma-eng/satellite/about_mt_index.html* (accessed 28 December 2006).

94. "ISRO, Raytheon complete tests for GAGAN Satellite Navigation System, IndiaDefence Web site, *http://www.india-defence.com/reports/2239* (accessed 28 December 2006).

Europe and China have done, or to enhance existing and forthcoming systems, as Israel, Japan, and India have chosen. European motivations include the desire to make significant profits from European sales of Galileo receivers, but as of 2006 it is not obvious whether this wish will be fulfilled, since American, European, and Japanese commercial vendors already compete to use signals from GPS and GLONASS and will simply expand their repertoires to include Galileo. Commercial vendors will combine signals from several navigational systems to enhance local performance, regardless of who funds and maintains the space-based systems.[95]

HUMAN SPACEFLIGHT

By far the most well-known space activity is human spaceflight, which has been the primary political arm of space programs around the world. The primary initial motivation for human spaceflight in the late 1950s and 1960s was competition for prestige, as the U.S. and Soviet Union sought to project their technological capabilities in an ideological battle for the hearts and minds of people around the world. The Soviet Union gained an early lead in the space race by putting the first man, Yuri Gagarin, into space in 1961, and other firsts such as the first woman in space, the first multi-man mission, and the first spacewalk. However, Soviet efforts to move beyond the R7 launcher and Vostok capsule designs fizzled in bureaucratic infighting, while the American response surged forward. The Gemini program proved U.S. rendezvous and docking capabilities, and Apollo placed the first man on the Moon in 1969. Secret Soviet efforts to put a man on the Moon fizzled as the huge N-1 launcher failed in four test flights, while the political imperative disappeared with the American success.[96]

With the failure of their piloted lunar program, the Soviet Union moved to long-duration space station efforts, with the U.S. shifting its efforts to the Space Shuttle in the 1970s and the beginnings of its own space station program in the mid-1980s. With their *Salyut* space stations and *Soyuz* ferry vehicles, the Soviet Union set record after record in long-duration human spaceflight and proved that military uses of piloted space stations were not effective. The *Mir* station was the culmination of Soviet and Russian efforts. The Soviet Buran program, which mirrored the American Shuttle, was fielded in the late 1980s but was too expensive to survive the Soviet Union's collapse. The American Space Shuttle also proved to be very expensive, though the U.S. was willing to pay that price to keep the American astronaut program alive. It also proved too unreliable, with flight delays the norm and the loss of *Challenger* in 1986 and *Columbia* in 2003 causing the loss of 14 astronauts' lives. The American Space

95. Handberg, *International Space Commerce*, chapter 6.

96. There are many books on these subjects. A good place to start is William E. Burrows, *This New Ocean: The Story of the First Space Age* (New York: The Modern Library, 1998).

Station *Freedom* program also proved problematic, with many design changes as NASA struggled to determine its purpose.[97]

Although its initial motivation was superpower competition, the shuttle and space station programs of the two superpowers provided opportunities for international cooperation, though initially that cooperation remained fixed within ideological boundaries. The Soviet Union took advantage of its regular *Soyuz* crew replacement flights to its *Salyut* and *Mir* stations to offer its Communist-bloc allies the opportunity to put their citizens into space through the Interkosmos program. Similarly, the U.S., which had seven seats on each Shuttle flight, offered rides to its allies. The Space Shuttle program also included some opportunities for allies to build flight hardware, with Canada's robot arm and the Europeans' Spacelab module the most significant. Although partly motivated by the prestige factor, scientific and economic factors provide more significant motivations for these nations, particularly in their interest in developing and using microgravity facilities and developing niche capabilities in space.[98]

These international precedents were the basis for much more intimate and expansive cooperation in the International Space Station (ISS) program. Initially including the U.S., Europe, Japan, and Canada, the ISS came to include Russia as well. On the brink of cancellation in the early 1990s, the American space station program was saved by the opportunity for the U.S. to use it as a means to keep Russian technical talent working on peaceful programs as opposed to being forced by economic necessity to sell their services to nations such as Iran and North Korea. The Russians built ISS modules under contract to NASA and also as part of their own independent contribution. In addition, Russia supplies Progress ferry vehicles for supplies and its reliable *Soyuz* transfer vehicles for crews. With the American Shuttle fleet grounded from early 2003 to mid-2005 due to the *Columbia* accident, Russia provided the only access to the ISS. This situation has shown the wisdom of international partnerships and their alternative means to keep programs alive when one partner has difficulties.[99]

Separate from the U.S. and the ISS program, China started its own Shenzhou program to put Chinese yuhangyuan (astronauts) into space. After a number of test flights, it finally succeeded in this ambition in 2003, with the launch of Yang Liwei in *Shenzhou 5*. China's motives appear quite similar to that of the U.S. and Russia: to garner world prestige and to inspire its own people, particularly young people, to study technical subjects.[100]

---

97. Robert Zimmerman, *Leaving Earth: Space Stations, Rival Superpowers, and the Quest for Interplanetary Travel* (Washington, DC: Joseph Henry Press, 2003); Dennis R. Jenkins, *Space Shuttle: The History of the National Space Transportation System, the First 100 Missions,* 3rd ed. (Cape Canaveral, FL: Dennis R. Jenkins, 2001).

98. Zimmerman, *Leaving Earth*; Michael Cassutt, *Who's Who in Space*, 3rd ed. (New York: Macmillan, 1998).

99. D. M. Harland and J. E. Catchpole, *Creating the International Space Station* (Chichester, U.K.: Springer-Praxis, 2002).

100. Brian Harvey, *China's Space Program: From Conception to Manned Spaceflight* (Chichester, U.K.: Springer-Praxis, 2004), chapters 1, 10.

In early 2004, the George W. Bush administration announced its Vision for Space Exploration, which directed NASA to replace the Space Shuttle, return humans to the Moon, and eventually go to Mars. By 2006 these plans had been translated into the Constellation program, whose first major systems are the Orion Crew Exploration Vehicle and Ares 1 Crew Launch Vehicle. NASA also began consultations with other nations regarding their future participation.[101]

NASA's overall human flight program costs are published, though sometimes the specific details are difficult to interpret, such as estimating the true cost of the Shuttle program. From the beginning of human flight, the U.S. has had by far the largest budget compared with other nations. NASA's historical human flight budget, shown in figure 9.7, shows a huge spike in the 1960s due to the Apollo program, but since that time has remained relatively stable, with a small spike to build the Shuttle *Endeavour* after the *Challenger* accident. The USAF spent significant funds on the Manned Orbiting Laboratory (MOL) program in the 1960s as well.

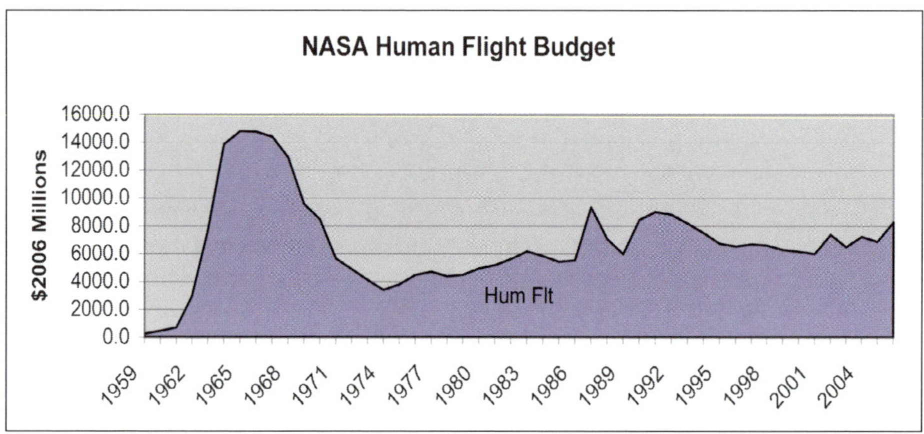

Figure 9.7—NASA human flight budget, 2006 dollars.[102]

Unfortunately, other budget figures are more difficult to acquire. The true costs of the Soviet human flight program may never be known, though all agree that Soviet and Russian costs have been significantly lower than U.S. costs. As compared to Apollo's cost, which has been estimated variously between roughly $21.8 and $25 billion in then-year dollars, the Soviet piloted lunar landing program has been estimated at roughly 4 to 4.5 billion rubles, or approximately $12–$13.5 billion.[103]

101. NASA Web site, *http://www.nasa.gov* (accessed January 2007).

102. Author's compilation. Based on data from NASA annual budget submissions, generally using previous year actuals, and the standard U.S. government deflator schedule.

103. Asif A. Siddiqi, *Challenge to Apollo: The Soviet Union and the Space Race 1945–1974* (Washington, DC: NASA SP-2000-4008, 2000), p. 838. Brian Harvey estimates $4.5 billion in his *Russia in Space: The Failed Frontier?*, p. 13. This figure appears to come from Mishin's writings, though it is not footnoted.

The Energia-Buran program, which was significantly larger than the *Salyut* and *Mir* space station programs, cost some 14 billion rubles over 17 years from 1976 to 1993.[104] China's human flight program, Shenzhou, is reported to have cost ¥18 billion from inception to its first docking flight, or about ¥1.6 billion per year, starting in 1992, which equates roughly to $240 million per year in 2002.[105]

The Soviet and Russian Interkosmos program, which allowed guest flights to *Salyut* and *Mir*, also created the procedures that, with slight modification, allowed for paying customers to travel to *Mir*. In what is apparently the first private booking of a flight to space, the Tokyo Broadcasting System paid $12 million for reporter Toyohiro Akiyama to travel to *Mir* in December 1990.[106] British scientist Helen Sharman flew to *Mir* the next year, paid for by a pool of private companies in the United Kingdom. Starting in the late 1980s, the ESA and European nations paid between $12 and $40 million for European astronauts to visit *Mir*, the price depending on the length and complexity of the mission.[107] In the 1995, the first NASA astronaut arrived at *Mir*, as part of a larger agreement between the U.S. and Russia for the International Space Station. While the U.S. pressed Russia to end the *Mir* program, American entrepreneur Walt Anderson created MirCorp, registered in The Netherlands, to investigate saving *Mir* for commercial purposes. MirCorp delivered a $7 million down payment in January 2000, and Russia flew a MirCorp-paid mission to recommission the station. MirCorp then received funds from American technology investor Denis Tito for two flights, with himself riding on the second. These missions did not take place. *Mir* was deorbited in March 2001; for a reputed $20 million, Tito rode instead as the first "space tourist" on *Soyuz* to the International Space Station, over strenuous American and European objections.[108] Anousheh Ansari became the fourth tourist (and the first woman tourist) in 2006.[109] After losing the battle to keep Tito off the ISS, the U.S. and Europe reluctantly acquiesced in Russia's right to sell tourist seats.

Another route to space tourism opened in June 2004 when Burt Rutan's Scaled Composites Company spacecraft, *SpaceShipOne*, won the $10 million, privately funded X-Prize by reaching suborbital space twice within two weeks, at a cost of roughly $25–$27 million paid by Microsoft cofounder Paul Allen. Richard

104. Siddiqi, *Challenge to Apollo,* p. 841. Brian Harvey puts the cost of the Energia-Buran program at 20 billion rubles. He also translates this to approximately €4.4 billion, which at that time would be roughly equivalent to $4.4 billion.

105. Harvey, *China's Space Program*, p. 293.

106. Wikipedia claims this price was $28 million. *http://en.wikipedia.org/wiki/Space_tourism* (accessed 3 January 2007).

107. Harvey, *Russia, the Failed Frontier?*, p. 31.

108. David M. Harland, *The Story of Space Station Mir* (Chichester, U.K.: Springer-Praxis, 2005).

109. "Female Space Tourist Blasts Off," 18 September 2006, *http://www.CNN.com* (accessed 3 January 2007).

Branson's Virgin Galactic company started a joint venture with Scaled Composites to build *SpaceShipTwo*, which would start carrying tourists for suborbital flights in roughly 2008 for a price of $200,000 each. Founded by motel magnate Robert Bigelow, Bigelow Aerospace began testing inflatable habitats for a space hotel, with its first successful test of Genesis 1 in July 2006.[110] Bigelow created "America's Space Prize," a $50-million award for the first successful reusable spacecraft that could reach his orbiting hotel.[111] The U.S. began to take commercial spaceflight seriously, creating a licensing process through the Federal Aviation Administration's Office of Space Commercialization in 2005.

MICROGRAVITY RESEARCH

Microgravity research differs from space physical sciences in several ways. First, it has historically been closely tied to human spaceflight because one of the major motivations has been to understand how the human body functions in zero gravity. Human spaceflight and the long-term movement of humans into space depend on deep understanding of microgravity effects both on humans and on other biological organisms. Materials research has generally required human-tended experiments in space and usually must return its samples back to Earth; thus, it also has had strong ties to the human spaceflight program because of the requirement to return astronauts and cosmonauts to Earth. Also, both material and biological research have often been portrayed as having commercial potential, such as purer crystals or proteins grown in space as opposed to on Earth. These promises have yet to be fulfilled, as the cost of doing business in space—despite its potential improved materials quality—is far higher than on Earth.[112]

The need for microgravity science to support human spaceflight, and the potential commercial possibilities of materials research, have led to large and direct government funding of this research and certainly to justifications to build various experiment modules in which to perform it, such as the Spacelab module for the Shuttle, the Japanese Experiment Module and Columbus module for the International Space Station, and various experimental stations on *Skylab*, *Mir*, and in the Space Shuttle payload bay. The various space programs provided funds to create microgravity research disciplines which, unlike the physical sciences, essentially did

110. Jeff Foust, "Bigelow Aerospace's Big Day at the Rodeo," *The Space Review* (24 July 2006), *http://www.thespacereview.com/article/667/1*, accessed January 2007.

111. "Space Tourism," Wikipedia, *http://en.wikipedia.org/wiki/Space_tourism* (accessed 3 January 2007).

112. Bromberg, *NASA and the Space Industry*, pp. 107–109, 167–174; Lou Dobbs with H. P. Newquist, *Space: The Next Business Frontier* (New York: ibooks, 2001).

113. A. Russo, "The European Use of Spacelab," in Krige et al., *A History of the European Space Agency, Volume II*, chapter 14.

not exist until human spaceflight programs required their existence. For example, the ESA had difficulties in the late 1970s when it realized that it was building the Spacelab module but had few researchers capable of using it.[113]

Microgravity research has also been a favorite educational tool, with NASA and other organizations providing major funding and subsidies to allow students from elementary school to graduate school to propose and create experiments such as how spiders build webs in space, how peas grow in space, etc. Microgravity science budgets have grown from essentially zero in the 1950s to tens of millions of dollars by the early twenty-first century so as to fill the experimental facilities of the International Space Station. In 2003, NASA allocated $304 million for biological science, $351 million for physical science research, and another $254 million for various research partnerships.[114] ESA allocated $120 million for microgravity science in 2006.[115]

Support Services

The U.S. government places restrictions on the number of civil service positions available in its various agencies, but encourages private enterprise. This combination has led to the fact of far more work being assigned to civil service personnel than they can accomplish. To deal with this situation, support service contractors supply the arms and legs for the government personnel, working under government supervision to perform many of the tasks that the civil servants do not have the time to perform. These companies sprang up around all of the NASA facilities, frequently owned or operated by women, minorities, and small businesses so as to meet government guidelines and quotas in these categories. Although the businesses are often owned by people in these categories, the work itself is usually directly supervised by NASA civil servants, with relatively little direction or guidance from the companies.

These unusual political and technical circumstances lead to unusual dynamics. A common experience among these contractors is that one company that held an engineering service contract in one year would lose it to another company in the next. Many of the personnel from the losing company are immediately hired by the winner, such that the personnel doing the actual work often remain the same while the managerial and financial structures are shifted to the new organization. Some support service contractors develop core competences in certain technologies or capabilities, which allow them to compete for other hardware, software, or service contracts, or even to sell products in the nongovernment commercial marketplace. Others, such as SAIC and Teledyne Brown Engineering, become quite large, with parts of the business continuing with support service contracting and other parts with specialized

114. President's FY 2007 Budget Request for NASA, NASA Web site, *http://www.nasa.gov/about/budget/* (accessed January 2007).

115. Peter B. de Selding, "2006 ESA Budget Emphasizes Independence, Satcom Technology," *Space News* (23 January 2006).

expertise that allowed them to bid on manufacturing contracts. Through several legal rulings, NASA's Marshall Space Flight Center was involved with determining the legality and guidelines for the use of such contractors in the 1960s and 1970s. A lawsuit that challenged the legality of support service contracts was filed in 1968 and dragged on for years, but in 1978 was settled in NASA's favor.[116]

INSURANCE

Governments funded all early spaceflight launches, and if one of the launchers or satellites failed, the government funded a replacement, occasionally in advance. For example, throughout the 1960s and 1970s, NASA typically funded two identical spacecraft on exploration missions to another planet, such as Mariner 1 and 2 in 1962. Targeted to fly by Venus, the first failed while the second completed the mission. The Voyager program of the 1970s built two matching pairs of spacecraft, while Viking had two identical orbiters and landers to ensure that at least one of them succeeded. Liability was another potential concern. Governments could not be sued if a rocket fell on its citizens' property, and could pay for damages through taxes.

Commercial companies were in a different situation. They could not afford to build two copies of each spacecraft to ensure that one worked. In addition, they could be quickly bankrupted should a commercial launch cause damage to private property. In 1965, Communications Satellite Corporation insured its Early Bird satellite for liability and for prelaunch damage to the satellite. Three years later, insurers covered the launch of Intelsat II, and by the 1970s created launch property damage and on-orbit coverage. The 1967 Outer Space Treaty and the 1972 Convention on International Liability for Damage required that nations pay for damage caused to other nations. This spurred the U.S., and later other nations, to set minimum liability insurance requirements on private companies. In the American case, the government requires private companies to obtain insurance to cover damages up to $500 million, and the government will cover any additional damages between $500 million and $2 billion.

Insurers in the U.S. and Europe typically set rates ranging from 15 percent to 20 percent for satellite loss, with several companies generally pooling their efforts for each launch. Liability and on-orbit rates were much lower, reflecting the much lower probability of damage to private property and of on-orbit failures compared to the loss of a satellite during ascent. From revenues of $5 to $10 million in the 1970s, insurance revenues by the 2000s typically ranged from $800 million to $1 billion per year, with loss payouts usually, but not always, smaller.[117]

116. Andrew J. Dunar and Stephen P. Waring, *Power to Explore: A History of Marshall Space Flight Center 1960-1990* (Washington, DC: NASA SP-4313, 1999), pp. 142–143.

117. Alden Richards, "The Early History of the Satellite Insurance Market," *Quest* 9/3 (2002): pp. 54–59.

Insurance companies hate failures that cause them to pay for losses, so when failures do occur they press the launcher and satellite organizations to understand and fix the problems before they would insure another launcher or satellite. The desire to enforce reliability on launcher and satellite organizations led to political problems. In the late 1980s, China entered the commercial launch market, creating Great Wall Industries Corporation to market its Long March rockets. When Long March rockets failed in 1992 and 1995, insurers launched investigations, which forced China (if it wanted to compete commercially) to open up its secretive processes and technologies. In addition, the insurers desired independent investigation committees, which included personnel from Hughes Aircraft, the leading satellite manufacturer, and Space Systems/Loral (SS/L), another major satellite vendor. In the ensuing interactions between the investigators and the Chinese, the U.S. government concluded that Hughes and SS/L had broken American laws regarding International Traffic in Arms Regulations (ITAR), leading to fines of $20 million and $32 million, respectively, on the satellite companies. Insurers in the U.S. and Europe complain that ITAR laws decrease their ability to get the information they need to underwrite policies.[118]

Capital

Space activities require lots of money. In the past, most of it was provided by the government, but private activities have grown dramatically since the 1980s. Aerospace and space enterprises must occasionally acquire venture capital; the means by which they have done so has gone largely unnoticed by space historians. It is also worth mentioning the 2005 creation of the SPADE Defense Index and the Space Foundation's Space Index, the first indices directly tied to financial performance to defense, space, and homeland security companies.

Burial Services

Celestis provides the service of sending a person's cremated ashes, for a fee, into space. To do this, Celestis Group first received a license from the Office of Commercial Space Transportation in 1984 to fly cremated remains aboard the Space Services International Conestoga rocket. Conestoga never flew and in 1994 a new company, Celestis Incorporated, negotiated with Orbital Sciences for a launch on its Pegasus launcher, which eventually did take place in 1997. Since then other flights have occurred, with prices ranging from about $495 to $12,500, depending on the amount of ashes and their ultimate destination.[120]

118. *The United States House of Representatives Select Committee on U.S. National Security and Military/Commercial Concerns with the People's Republic of China* (Cox Report) (Washington, DC: Government Printing Office, 3 January 1999), chapter 8, "Commercial Space Insurance."

119. Space Foundation, *The Space Report 2006* (Colorado Springs, CO: Space Foundation, 2006), pp. 149–150, *http://www.spaceindex.com/* (accessed 14 January 2007).

120. Celestis Web site, *http://www.memorialspaceflights.com* (accessed 14 January 2007).

MANAGEMENT, ECONOMICS, AND POLICY

The assessment of management, economics, and policy issues has spurred a small market of government and industry organizations. Management companies such as Booz Allen Hamilton, Inc. and McKinsey & Company have consulted for businesses and the government for years, occasionally including space organizations, such as McKinsey's support to JPL's reorganization efforts in 1959.[121] Economic issues have been a staple of industry associations, such as the Aerospace Industries Association in the U.S., or Eurospace in Europe. They gather and collate aerospace industry-wide data to assist member company marketing and lobbying.[122]

The Department of Commerce's Office of Space Commerce, founded in 1988, has performed political-economic analyses since that time. In 1998, Congress passed the Technology Administration Act, which founded the Office of Space Commercialization; it was transferred to the National Oceanic and Atmospheric Administration (NOAA) in 2004. It funded private companies like Futron Corporation and The Tauri Group to assess various political-economic issues.[123] The Commercial Space Launch Act of 1984 gave the Department of Transportation's Office of Commercial Space Transportation regulatory authority over commercial space launches. It was transferred in 1995 to the Federal Aviation Administration (FAA) as the Associate Administrator for Commercial Space Transportation (AST). AST publishes quarterly and annual reports of worldwide commercial space launch activities.[124]

EDUCATION

As space activities developed, so too did educational programs to train space industry personnel. These started at institutions involved with early rocketry or satellite projects, such as California Institution of Technology's Jet Propulsion Laboratory (JPL), Massachusetts Institute of Technology's (MIT's) Lincoln Laboratory, and Johns Hopkins University's Applied Physics Laboratory (APL). These institutions created engineering and scientific programs to train their own employees and other government and industry personnel. Some programs and institutions were created based on government or commercial lobbying, such as Wernher von Braun's successful effort to create a research institute at the Huntsville

121. Stephen B. Johnson, *The Secret of Apollo: Systems Management in American and European Space Programs* (Baltimore, MD: Johns Hopkins University Press, 2002), p. 252, footnote 45.

122. See the Aerospace Industry Association, *Aerospace Facts and Figures* (Arlington, VA: Aerospace Industries Association), or Eurospace's *Eurospace Facts and Figures*, both published annually.

123. See the Office of Space Commercialization Web site, *http://www.nesdis.noaa.gov/space/* (accessed January 2007).

124. *Aeronautics and Space Report of the President, Fiscal Year 2004 Activities* (Washington, D.C.: Government Printing Office, 2005).

Extension Center of the University of Alabama, which was a major factor in the 1966 creation of the independent campus, the University of Alabama in Huntsville. Similar relationships occurred elsewhere, such as Stanford's relationship with the USAF and NASA's Ames Research Center, NASA Johnson's relationship with the University of Houston Clear Lake, or NASA Kennedy Space Center's relationship with the University of Central Florida. Some space organizations spun off from educational institutions, such as the Charles Stark Draper Laboratory spun off from MIT, MITRE Corporation created from Lincoln Laboratory, Aerojet from JPL, and so on. Diversified (not merely technical) programs such as the University of North Dakota's Department of Space Studies and the International Space University (Strasbourg, France) were created in 1987. Other specialized programs now exist in space systems management, politics, and other space-related topics. Private companies also provided technical courses and consulting. The political and economic dimensions of space education have gotten little attention to date, except for the creation of economic clusters such as those in Colorado Springs, the greater Los Angeles and Boston areas, and Florida's Space Coast.[125]

With strong Congressional support to ensure educational funding in every state, NASA created its own educational programs. The NASA Space Grant programs provide space educational materials to elementary and high school teachers, provide student scholarships, promote space education to minorities and disadvantaged groups, and support various university programs. NASA also creates educational programming on its own television channel, much of which is targeted to these same audiences. The economic and political effects of these programs has, to date, not been investigated, though tens of millions of dollars are spent annually.

Space camps to educate young people were created specifically to spur science and mathematics education, and hence technical development in their societies. Soviet space camps originated from Young Cosmonauts Groups, which came into being soon after Gagarin's 1961 flight. These spread across the Soviet Union and then into its allied European nations such as East Germany, Poland, and Hungary. The first U.S. space camp took place in 1982 in Huntsville, Alabama, at United States Space and Rocket Center, which had been founded earlier under the direction of Wernher von Braun. By the early twenty-first century, space camps existed worldwide.[126]

125. Ann Markusen, Scott Campbell, Peter Hall, and Sabina Deitrick, *The Rise of the Gunbelt: The Military Remapping of Industrial America* (Oxford, U.K.: Oxford University Press, 1991). William Barnaby Faherty, *Florida's Space Coast: The Impact of NASA on the Sunshine State* (Gainsville, FL: University Press of Florida, 2002).

126. Anne Baird and Robert Koropp, *The Official U.S. Space Camp Book* (1992); Pablo de Leon, "Space Camps," in Space Exploration and Humanity, 2 vols. (Santa Barbara, CA: ABC-CLIO, in press).

PROFESSIONAL AND ADVOCACY GROUPS

As rocketry moved from amateurs to military organizations and corporations in the 1940s, the amateur rocketry groups matured into advocacy and professional groups. Professional groups such as the American Astronautical Society (AAS) and the American Institute for Aeronautics and Astronautics (AIAA), primarily acted as professional networking organizations, sponsoring conferences and publications for engineers and managers working in corporations and government organizations. Others, such as the British Interplanetary Society and the National Space Society (NSS), continued the tradition of space advocacy. Despite their small size, these organizations have significant political and economic impact, as they provide forums for technical and political interchanges between government, industry, and academic organizations.

MEDIA

Space, with its inherent aura of mystery, risk, and danger, has always been a popular media topic, which a variety of media cater to and profit from. Some general media outlets cater to space activities, such as the "New Series in NASA History" by the Johns Hopkins University Press; the Public Broadcasting System's creation and airing of Carl Sagan's *Cosmos* series in the 1980s; Home Box Office's series *From Earth to the Moon* in the early twenty-first century, and *The Space Show*, a San Francisco radio show. Professional space organizations such as the AIAA, the AAS, and the British Interplanetary Society have their own in-house publication capabilities for trade journals, books, and newsletters; so do NASA, ESA, and other government space organizations. Some publishers publish mainly or only space topics, such as Apogee Press, Orbit Press, and Space Publications. Some publications are space-only or aerospace-only, such as *Space News* or *Aviation Week & Space Technology*. Though this sector is small in terms of money, its influence is much larger than its economic size indicates.

CONCLUSION

Space historians select a variety of topics to study based on their own individual interests, but they should also take note of those subjects that humanity as a whole values, and of the wide variety of institutions that support activities in space. The money spent on particular areas is a proxy for the value that humans place on those activities. By this standard, space telecommunications has become, by the early twenty-first century, the most important space topic. Perhaps because its rise to prominence became significant in the 1980s and 1990s, it has yet to rate highly among historians—a fact this paper hopes to elevate. Also, the variety of economic activities that take place to support space endeavors is significantly broader than is typically portrayed. Topics such as space education, insurance, weaponry, and support services are subjects as worthy of analysis as are human flight and space science. Analysis of the interaction of civilian, military, and commercial activities must take place within the context of these many sectors, as these interactions vary greatly from sector to sector.

In one respect, all of these interactions have a common characteristic—economic and political issues are tightly intertwined. A number of studies and analyses try to separate the economic and political realms. Advocates of private enterprise such as those in the so-called alt-space movement try very hard to divide true "commercial" endeavors from government-funded activities.[127] Unfortunately for them, separating government from space activities has been extremely difficult and is likely to remain so for a variety of economic, legal, and political reasons. Indeed, businesses in general seek any profitable markets, and governments are too large a market for space-related businesses to ignore. Dollars, after all, are just as green coming from the government as from other corporations or from individuals.

Space historians have the opposite bias because they have depended largely on government sources, which in turn have led to histories that take a government viewpoint—often with similar priorities to the governments themselves. Instead, we space historians need to provide a much better balance between government and private activities in our studies and analyses. Whether investigating governmental activities (where politics are the priority but economics the grease) or investigating private activities (where economics is often primary and politics unavoidable), a political-economic approach is a useful tool to ensure a proper balance of perspectives.

127. See, for example, Edward L. Hudgins, ed., *Space: The Free-Market Frontier* (Washington, DC: The Cato Institute, 2002); Paula Berinstein, *Making Space Happen: Private Space Ventures and the Visionaries Behind Them* (Medford, NJ: Medford Press, 2002), and Dobbs, *Space: The Next Business Frontier.*

Chapter 10

The Role of Space Development in Globalization

James A. Vedda

Hundreds of books and countless articles have been written since the mid-1990s about the phenomenon of globalization. A Google search on the term "globalization" yielded 18.8 million hits in September 2005—a number that multiplied almost six times (to 109 million) by March 2006. Amazon.com displayed at least 120 new books on the subject that were scheduled to be released between September 2005 and mid-2006. Despite this impressive volume of literature, the author has had little success in finding discussions that directly address the effect of space development on globalization (and vice versa) other than cursory acknowledgments of the role of satellite communications as a key supporting technology.

Globalization has been identified as the dominant trend that has replaced the cold war, although its development overlaps the cold war by at least three decades. For much of that time, U.S. government space activities had fairly well defined roles that were closely associated with the nation's cold war-era interests. If globalization is the successor to the cold war paradigm, then U.S. space efforts, particularly those involving exploration and development, must be redefined appropriately. This is not a simple task, since debates rage as to what globalization means, where it is headed, and whether the net effect will be good or bad. Although globalization debates primarily address economic, social, and environmental issues, the continuing influence of space development cannot be ignored or viewed in isolation from these issues.

This chapter highlights the role of space development in the emergence of the current era of globalization, and briefly discusses how space activities are likely to continue to influence this evolving process. Globalization, in turn, has influenced the course of space development. The implications for the future may be both positive and negative, including the risk of a backlash against space development stemming from anti-globalization movements.

DEFINING GLOBALIZATION

The economic and societal developments that today are labeled globalization have been around for centuries, waxing and waning at least since the sixteenth century. Popular use of the term "globalization" goes back to the late 1980s. The term's popularity is partly due to its ambiguity and ability to assume different connotations depending on who is using it and in which context.[1] For some, it connotes international connectedness; liberation from geographic and nationalistic limits to innovation and growth; leveling of inequalities; improvement of living standards and the human condition; and an avenue for avoiding major conflicts such as the wars of the twentieth century. For others, globalization is just the opposite. They see it as "a project for polarizing and dividing people—along axis of class and economic inequality, axis of religion and culture, axis of gender, axis of geographies and regions . . . a new caste system."[2]

The deep and often heated disagreement over the nature and ramifications of globalization makes it difficult to find a generally accepted definition for the phenomenon that *The New York Times* writer Thomas Friedman calls "the overarching international system shaping the domestic politics and foreign relations of virtually every country."[3] For purposes of this chapter, a good definition that recognizes the contributions of technological development comes from Joseph Stiglitz, Nobel laureate in economics, who describes globalization as "the closer integration of the countries and peoples of the world which has been brought about by the enormous reduction of costs of transportation and communication, and the breaking down of artificial barriers to the flows of goods, services, capital, knowledge, and (to a lesser extent) people across borders."[4]

The literature on globalization tends to mention space technology (if it is mentioned at all) in no more than a sentence or two acknowledging satellite communications as a component of the revolution in telecommunications that enabled the current era of globalization. The following statement is representative:

> The revolution in microelectronics, in information technology and in computers has established virtually instantaneous worldwide links which, when combined with the technologies of the telephone, television, cable and satellite, have dramatically altered the nature of political communication.[5]

---

1. Mathias Koenig-Archibugi, "Globalization and the Challenge to Governance," in *Taming Globalization: Frontiers of Governance*, David Held and Mathias Koenig-Archibugi, ed. (Cambridge, U.K.: Polity Press, 2003).
2. Vandana Shiva, "The Polarised World of Globalisation," International Forum on Globalization, 27 May 2005, *http://www.zmag.org/sustainers/content/2005-05/27shiva.cfm* (accessed June 2006).
3. Thomas L. Friedman, *The Lexus and the Olive Tree* (New York: Farrar, Straus & Giroux, 1999), p. 7.
4. Joseph E. Stiglitz, *Globalization and Its Discontents* (New York: W.W. Norton & Company, 2003), p. 9.
5. David Held, "From Executive to Cosmopolitan Multilateralism" in *Taming Globalization: Frontiers of Governance*, David Held and Mathias Koenig-Archibugi, ed. (Cambridge, U.K.: Polity Press, 2003).

This cursory treatment is understandable, since most of the literature focuses on international economics, implications for developing countries and the world's poor, and potential impacts to the environment. But there is much more to the story of how space development has made the current globalization experience different from previous ones, and will continue to affect its evolution. As former Secretary of Labor Robert Reich noted, technology and globalization are often discussed as separate trends, but they are becoming one and the same.[6]

GLOBALIZATION PAST AND PRESENT

The concept of globalization has been popularized in newspaper articles and best-selling books by Thomas Friedman, who divides its history into three distinct eras:

- Globalization 1.0 (1492–1800). Countries and governments drove global integration. Trade began between the Eastern and Western hemispheres.
- Globalization 2.0 (1800–2000, interrupted by the Great Depression and the two World Wars). The Industrial Revolution and multinational companies were the key agents of change.
- Globalization 3.0 (2000 onward). Individuals have newfound power to collaborate and compete globally.[7]

Friedman's view of globalization history hinges on the increasing empowerment of ever-smaller components of societies. This view, though compelling, is not shared by historians, who perceive three eras of globalization as follows:

- The age of exploration and colonization from the fifteenth century to the early nineteenth century.
- Industrialization and expansion of world trade from the mid-nineteenth century to 1914, at which time globalization was halted by the outbreak of the first World War.
- The current era from the post–World War II recovery of the global economy to the present.[8]

When one looks at the differences between the pre-war and post-war eras, it is clear that this latter view is more accurate—the wars and the Great Depression separated two distinct eras rather than simply being a pause within a single continuous period. This view is also better suited to the analysis of the role of space development in the current era.

6. Robert B. Reich, *The Future of Success* (New York: Vintage Books, 2000), p. 23.

7. Thomas L. Friedman, *The World Is Flat: A Brief History of the Twenty-First Century* (New York: Farrar, Straus & Giroux, 2005), pp. 9–10.

8. Jeffrey A. Frieden, *Global Capitalism: Its Fall and Rise in the Twentieth Century* (New York: W. W. Norton & Company, 2006).

Before looking at the differences between the current and previous experiences with globalization, it is instructive to note characteristics that are similar. For example, nineteenth-century globalization featured the following: unprecedented international movement of capital, raw materials, and people; revolutionary technological innovation, including the telephone, radio, and internal combustion engine; and an ongoing struggle for balance between protectionism and free trade. Allowing for a century of technological advances, these characteristics sound very familiar today. There were also tensions in that era's international order that resemble today's headlines: imperial overstretch, great power rivalry, an unstable alliance system, rogue regimes sponsoring terror, and the rise of a revolutionary terrorist organization (the Bolsheviks) hostile to capitalism.[9]

The end of World War II launched the current era by heralding a revitalization of the economies of the industrialized nations, particularly the U.S. An important legacy of the war that would stimulate the re-emergence of globalization was the new relationship between the U.S. government and the research community. Always a key supporter of infrastructure projects, the U.S. government became the nation's primary patron of science and engineering. An effort to sustain this relationship beyond the war years was spearheaded by President Franklin Roosevelt's director of scientific research and development, Vannevar Bush.[10] The result was what some have called a "social contract with science" that portrays the pursuit of scientific knowledge as intrinsically good and useful: as long as the nation maintains its input into the reservoir of knowledge, the system is working as it should, and application of that knowledge will take care of itself. Institutions created in this image, such as the National Science Foundation and NASA, persist to this day, as does the dominance of government funding in certain fields, such as medical research. However, it remains to be seen whether this social contract is sustainable in an evolving post-cold war political environment.[11] Large science budgets will be increasingly difficult to justify if the scientific enterprise, or at least some part of it, is perceived to be isolated from societal needs.

The output of this government partnership with science was the eventual widespread availability of technologies that could only be dreamed of—or in some cases, were unimaginable—in the prior era of globalization. For example, technologies that allowed more rapid movement of people, goods, and information in the latter half of the twentieth century included the following:

- Jet air transport for passengers and cargo multiplied the speed of long-distance travel, effectively shrinking travel times from days to hours. Equally important, it eventually became affordable to a broad swath of society.

9. Niall Ferguson, "Sinking Globalization," *Foreign Affairs* (March/April 2005).

10. Vannevar Bush, *Science—The Endless Frontier: A Report to the President* (Washington, DC: U.S. Government Printing Office, July 1945).

11. Radford Byerly and Roger Pielke, "The Changing Ecology of United States Science," *Science* 269 (1995): pp. 1531–1532.

- Supertankers and container ships have dramatically reduced the cost of transporting cargo across the oceans, and refrigeration allows perishable products to make their way around the world.
- Near-instantaneous, high-bandwidth communications have evolved so far beyond the telegraph and radio of our great-grandfathers' day that the benefits are beyond our ability to quantify. By the 1960s, telephones became ubiquitous and continents were connected by undersea cables. The fax machine became popular in the mid-1980s, at the same time that so-called microcomputers were maturing. By the 1990s, the expectation was that every desktop would have its own computer, probably linked to a corporate network and the Internet.
- Space technology in its various forms started making its contribution to globalization in the 1960s.

Clearly, the contribution of space technology was not limited to the addition of satellites to an already expanding network of global communications, as the globalization literature seems to imply. The full array of emerging space capabilities had significant influence. For example, numerous business and government activities at the local, regional, national, and international level are dependent on the weather. The improved weather forecasts enabled by satellites beginning in the 1960s enhanced productivity and safety of operations in areas such as agriculture, air transport, shipping, construction, mining, and utilities, to name a few. Over time, these improvements had cumulative effects that altered business cycles and planning to reflect an evolving Information Age economy.

As weather monitoring matured, another form of Earth monitoring known as satellite remote sensing became available to civilian users starting in the 1970s. Able to produce images much more detailed than weather satellites, and in some cases using multiple spectral bands that reveal even more information, remote sensing opened new avenues for industries such as those listed above and others, including urban planners, environmentalists, fossil fuel geologists, and even archeologists. Early in NASA's Landsat series, interest in this new capability spread around the world. Assisted by the U.S. policy of nondiscriminatory access to Landsat data, remote sensing became a new tool for resource exploration, environmental stewardship, and disaster assistance, among other applications. Commercial descendants of Landsat are cultivating global markets in ventures that are very much in keeping with the proliferation of know-how and exchange of data that are characteristic of a globalized world.

The use of satellites for navigation began in the 1960s, primarily to serve the needs of military ships and submarines. Today, GPS has become a household word (or more appropriately, a household acronym) even to those who don't know that it stands for Global Positioning System. By the end of the 1980s, the GPS constellation was taking shape and its services—accurate positioning, navigation, and timing—were being shared at no cost with the world. Essentially, those services provide value to anything that moves, and even some things that don't move but depend on precise timing signals. Though it is often overlooked, this capability is in a class with satellite communications as an enabler of globalization. Its ability to assist the movement of people, goods, capital, and information around the world is widely recognized, as evidenced by several global and regional navigation satellite systems operated or planned by Russia, China, Europe, Japan, and India.

One of the most important but least acknowledged contributions of satellites to globalization is their role in keeping the cold war from turning into a hot war. As noted earlier, the previous era of globalization ended abruptly with the outbreak of major military conflict in 1914. Despite the best efforts of many, it took more than three decades to resurrect globalization. The same thing could have happened in the years following World War II, dramatically worsened by the addition of nuclear weapons into the mix. The "balance of terror" in offensive weapons is generally given credit for the fact that this never happened, but the nuclear arsenal could not have allowed us to achieve this without the support of satellites for surveillance, reconnaissance, and targeting.[12]

Not all observers during the cold war saw the government relationship with science and technology, or space in particular, in a positive light.[13] A noteworthy example that specifically addresses space comes from historian Walter A. McDougall, who received a Pulitzer Prize for his 1985 book on the political history of the early space age.[14] His views on the social consequences of the space program were perhaps more starkly displayed in a 1982 journal article in which he identified the U.S. response to Sputnik as the catalyst that turned the United States into a "full-fledged technocracy" in the 1960s.[15] McDougall proposed his own definition of technocracy (usually taken to mean the management of society by technical experts). He defined it as "the institutionalization of technological change for state purposes." Institutionalized stimulation of science and technology, in his view, is artificial, and in the U.S. it "extended not only to military spending, science, and space, but also to foreign aid, education, welfare, medical care, urban renewal, and more." The "symbol and vanguard" of this movement, he said, was NASA.

---

12. For a striking example of how satellites helped keep the cold war cold, see the discussion of the Cuban missile crisis in William E. Burrows, *Deep Black* (New York: Random House, 1986). A more recent expression of this view can be found in David Kahn, "The Rise of Intelligence," *Foreign Affairs* (September/October 2006).

13. For example, see Amitai Etzioni, *The Moon-Doggle* (Garden City, NY: Doubleday and Co., 1964).

14. Walter A. McDougall, *. . . the Heavens and the Earth: A Political History of the Space Age* (New York: Basic Books, 1985).

15. Walter A. McDougall, "Technocracy and Statecraft in the Space Age: Toward the History of a Saltation," *The American Historical Review* 87 (1982): pp. 1010–1040. For a later reiteration of these views, see Hal Bowser, "How the Space Race Changed America: An Interview with Walter A. McDougall," *Invention & Technology* (Fall 1987): pp. 25–30.

McDougall saw the Space Age as "defined by the discontinuous leap in public stimulation and direction of research and development" and worried that "progress may, at times, undermine the values that make a society worth defending in the first place." He suggested that "[T]he net gains from space technology should be measured not only against the total cost, or the economic cost, of the program itself but also against the continuing loss incurred from misdirected military and social spending encouraged by the same technocratic mentality that inspired Apollo."

McDougall gave NASA and the space program far too much credit for shaping the management of late-twentieth-century society. Many phenomena of equal or greater influence existed, some of which preceded the Sputnik era by many years. The U.S. government turned to centralization of large-scale projects during the New Deal of the 1930s. The war effort continued this trend and increased government influence over technological and industrial development. In the post-war era, as the population and economy grew, the government pursued many macro-projects unrelated to space while exercising its cold war responsibility as the leader of the "free world." These examples put the space program's role into proper perspective: high-profile space research was a product, not the cause, of its socio-political environment.

McDougall's primary focus was on the research and development enterprise, which he saw as a "command economy" approach to "choosing new technology for social (and political) goals ... abandoning the concept of a free society" and reinforcing "the national state as the most efficient agent of technological change." His skeptical assessment of the nation's research and development enterprise (including the space program)—highlighting centralization, nationalism, and direction of technological development by public institutions—is the polar opposite of the globalization concept we know today. However, the key aspects of the nation's space efforts that have helped to enable globalization are not the direct public funding of research but, rather, the dissemination, adoption, and routine use of the resulting space applications.

Present-day globalization is reaping the benefits of space applications created and disseminated in the cold war in an environment that kept major threats at bay and allowed global markets to flourish. Government space efforts aimed at national security, national prestige, and technology development have led us to a point where civil and commercial space applications are fundamental—though often transparent—in a globalizing world.

SPACE DEVELOPMENT UNDER A GLOBALIZATION PARADIGM

Friedman believes that the current system of globalization "has come upon us far faster than our ability to retrain ourselves to see and comprehend it."[16] Certainly

16. Friedman, *The Lexus and the Olive Tree*, p. 22.

this has been the case with space development. As in other societal activities, space-related institutions seek to continue their existence and their traditional priorities despite the fast pace of change in key segments of their environment. Since the end of the Apollo era, for example, U.S. civil space efforts have struggled with questions on the role of government vs. the private sector, made all the more difficult by the fact that the answers are moving targets. Who should finance, build, and operate space infrastructure elements such as launch systems and space stations? To what extent should the government support research projects that have the potential to produce private-sector revenues? In an era of tight federal budgets, should the government shift as much responsibility and expertise as possible to the private sector, or is this a short-sighted strategy that will undermine the nation's continuing need for large-scale, evolving space capabilities? Can the private sector, at the current stage of technical development, always be counted on to choose better space investments and technical approaches than the government?

A significant percentage of the space industry is designed to serve governments, since these constitute much of the customer base in key areas such as space hardware manufacturing and launch services. The relatively small number of competitors and customers in these areas, and the dominance of government customers, yield a space industry that is slower to adapt and innovate than most other high-tech industries. The tendency to protect space technologies as sensitive national assets slows their adoption in the world market and may hinder the competitiveness of nations employing export restrictions and protectionist measures. These circumstances do not bode well for the U.S. space community's ability to rapidly adapt to the globalized environment.

In addition to keeping up with the frenetic pace of the world's economic evolution, the U.S. space community also must adapt to changes in its character. In the globalization era, this must take into account the diffusion (or "democratization") of technology, information, economic power, and international influence. The leveling effect that results will change relationships with international partners, increase competition in space products and services on the world market, and challenge U.S. space leadership across the board. This is already forcing the U.S. civil space program to rethink its post-cold war identity, as demonstrated by the shift away from NASA's flagship programs of the 1970s and 1980s and toward ambitious human space exploration.

As before, geopolitics and economics will drive the search for a new national identity in space. The issues are somewhat different today than they were during the cold war, but the challenges and risks remain. The International Forum on Globalization warns of this when it states: "The world's corporate and political leadership is undertaking a restructuring of global politics and economics that may prove as historically significant as any event since the Industrial Revolution. This restructuring is happening at tremendous speed, with little public disclosure of the profound consequences affecting democracy, human welfare, local economies, and the natural world."[17] Warnings of this type are echoed by many observers of globalization, some of whom see the movement as unsustainable, sparking nonlinear trends that will be impossible to manage.[18] In a recent assessment, the Central Intelligence Agency recognized that sustained financial crises or other prolonged disruptions could occur, but was generally optimistic about the prospects for the global economy by 2015. The implications for geopolitics were also viewed with optimism but were tempered by caution:

> This globalized economy will be a net contributor to increased political stability in the world in 2015, although its reach and benefits will not be universal. In contrast to the Industrial Revolution, the process of globalization is more compressed. Its evolution will be rocky, marked by chronic financial volatility and a widening economic divide . . . Regions, countries, and groups feeling left behind will face deepening economic stagnation, political instability, and cultural alienation. They will foster political, ethnic, ideological, and religious extremism, along with the violence that often accompanies it. They will force the United States and other developed countries to remain focused on "old-world" challenges while concentrating on the implications of "new-world" technologies at the same time.[19]

As noted earlier, supporters of globalization believe it will allow more and more individuals, as consumers and producers, to enjoy the benefits of economic liberalization, competition, and innovation. It is natural to want to see oneself as part of the solution rather than part of the problem, so the space community undoubtedly would like to view itself as an essential tool of globalization for redressing deficiencies and providing solutions for global problems. But general acceptance of this view is not automatic. In fact, there is a risk that the opposite may occur.

Globalization has a dark side, and there is no shortage of critics around the world who are eager to point this out. Among the negative aspects of globalization that have been cited are:

- Exposure of workers and firms to unwelcome competition from abroad, and increased risk that companies will relocate their production elsewhere.
- Competition between locations for mobile capital that may lead to a "race to the bottom" in environmental standards.

---

17. The International Forum on Globalization claims to represent more than 60 organizations in 25 countries and identifies itself as "an alliance of sixty leading activists, scholars, economists, researchers, and writers formed to stimulate new thinking, joint activity, and public education in response to economic globalization." See *http://www.ifg.org/* (accessed June 2006).

18. Ervin Laszlo, *Macroshift: Navigating the Transformation to a Sustainable World* (San Francisco: Berrett-Koehler Publishers Inc., 2001), p. 8.

19. Central Intelligence Agency, *Global Trends 2015: A Dialogue About the Future With Nongovernment Experts*, (New York: Cosimo Inc., 2005; originally released as a U.S. government report in December 2000).

- Diffidence to the outside world or fear that a cherished way of life will disappear as a result of cultural standardization.
- Potential worsening of inequality and injustice and erosion of democratic governance.[20]

The challenge for space development is to continue its role as a key element of globalization without becoming associated with its negative consequences. The same entities that dominate space development—government institutions and transnational corporations—are seen by critics as orchestrating globalization to serve the wealthy at the expense of the poor. In this view, observations of Earth from space might be interpreted as security threats or as a way to spy on economic activities in other parts of the world, rather than being seen as an instrument of environmental protection and disaster relief. Satellite communications might be depicted as a tool for extracting information and capital from unsuspecting regions of the world, rather than as a means of bringing information and capital to them. Even incoming information can be pejoratively portrayed as "cultural contamination" or Western propaganda designed to influence national or regional policies and attitudes.

Space technology could be seen by globalization critics as a tool of transnational corporations that exploit workers, of foreign investors who undermine local businesses, or of wealthy (i.e., spacefaring) countries that economically take advantage of developing nations. The result could be neo-Luddite controls on technology and onerous trade protection schemes that suppress economic dynamism.[21] Therefore, it is critical that government-supported space development be directed at—and perceived as—seeking solutions for the planet in areas such as disaster relief, environmental monitoring, climate research, medical research, and in the long term, the use of extraterrestrial resources and capabilities for the benefit of Earth.

So far, the government institutions criticized most often by globalization opponents are the World Bank, the International Monetary Fund, and the World Trade Organization.[22] Multinational corporations typically are disparaged generically rather than by individual sectors,[23] and mention of aerospace companies is notably absent so far. But there is still the possibility for an anti-technology backlash akin to the Vietnam-era experience.[24]

---

20. Held, "From Executive to Cosmopolitan Multilateralism," pp. 2–3.

21. Reich, *The Future of Success*, p. 247.

22. For example, see Stiglitz, *Globalization and Its Discontents*; Peter Isard, *Globalization and the International Financial System: What's Wrong and What Can Be Done* (New York: Cambridge University Press, 2005).

23. For example, see Robert O. Keohane, "Global Governance and Democratic Accountability" *Taming Globalization: Frontiers of Governance*, David Held and Mathias Koenig-Archibugi, ed. (Cambridge, U.K.: Polity Press, 2003).

24. W. Henry Lambright, "Managing America to the Moon: A Coalition Analysis" in *From Engineering Science to Big Science: The NACA and NASA Collier Trophy Research Project Winners*, Pamela E. Mack, ed. (Washington, DC: NASA SP-4219, 1998), p. 209.

For at least part of the cold war era, large public expenditures on space projects were widely perceived in the U.S. as good investments to counter powerful, unfriendly forces in the world that could wreak nuclear destruction at any moment. Today, the public's perception of U.S. civil space efforts as a counterweight to unfriendly forces appears much weaker (although this factor is not measured directly by public opinion polls). Certainly the national prestige argument for the space program has lost much of its impact, since terrorist networks and rogue nations are not winning hearts and minds around the world by demonstrating their prowess in spaceflight. Even China's human spaceflight program fails to stir fears in the West as it follows a path that was tread by the U.S. four decades earlier.

In the current era, the value of government space efforts needs to be measured by a different yardstick that takes into account the multipolar geopolitical environment and the globalized nature of economics and technology. The space community must recognize the effect of this environment on trade, technology, and leadership in space, and resist the urge to preserve the outdated aspects of institutions, processes, and relationships that insulate it from the evolving "big picture."

Possible Futures

A recent multiphased study by the Organization for Economic Cooperation and Development (OECD)[25] addressed the future of space applications through the 2030s and perhaps provided some guidance on avoiding an anti-globalization backlash. The study did not express itself in globalization terms but, rather, sought "to understand how OECD countries may reap the benefits of civil and commercial space applications for society at large." It used a scenario approach based on "the interaction of three main drivers of social change: geopolitical, economic, and environmental." Viewed from a globalization perspective, the three scenarios presented in the study ranged from successful globalization to its failure to sustain itself in the face of a combination of unfavorable factors.

Although the OECD effort does not speak in terms of globalization, it nonetheless covers similar ground. Significantly, despite its generally optimistic outlook for civil and commercial space, it identifies obstacles to future growth, including:

- Market access restrictions due to incomplete trade liberalization in some countries.
- Procurement policy problems resulting from the unreliability and unpredictability of government customers.

25. Organization for Economic Cooperation and Development, "Space 2030: Exploring the Future of Space Applications," 3 May 2004, and "Space 2030: Tackling Society's Challenges," 31 May 2005. Both titles are available from *http://www.oecdbookshop.org* (accessed June 2006). The OECD is a multinational forum for addressing economic, social, and environmental challenges. It has 30 member countries in North America, Europe, and the Pacific Rim.

- Export controls and investment restrictions.
- Spectrum allocation problems.
- Insufficient government support for the development of new technologies.
- Legal and regulatory constraints that cause uncertainty and delay in the deployment of new applications.[26]

The OECD frames its recommendations in three "blocks" aimed at what governments can do to strengthen the contributions space can make to solving important socioeconomic challenges:

Block I: Implement sustainable space infrastructure that is fully integrated with ground infrastructure and takes into account user needs, especially in the areas of Earth observation, navigation, communications, and access to space.

Block II: Take advantage of productivity gains that space solutions may offer for delivery of public services and development of new ones, particularly through international cooperation, data sharing, disaster and treaty monitoring, emergency management, and economic development.

Block III: Encourage the private sector to contribute fully to the development of new, innovative applications and to the development and operation of space-based infrastructures by making national and international space laws business-friendly, and by encouraging entrepreneurship, open markets, and international standards.[27]

These recommendations align well with a belief that space development should continue to play a significant role—far beyond just communications—in shaping globalization's evolution and keeping it focused on societal needs. In its detailed recommendations, the OECD study suggests ideas on how we can get there from here. At least for the next three decades, the study sees great hope and promise for applied space research and development, relegates basic research to government space agencies, and seems to marginalize human spaceflight.

Today's space community must consider timeframes even beyond 30 years. Given the scope and difficulty of exploring and developing space, it is not too early to ask: Where do we want the United States to be when it reaches its tricentennial in 2076? Will the United States still be one of the leaders in space at that time?

26. OECD, 2004, p. 16.

27. OECD, 2005, pp. 211–215.

Conclusion

Analyses of the circumstances and outcomes of the previous era of globalization, and the similarities to and differences from the current era, are instructive in defining the role space has played and will play in the years ahead. Will globalization continue to flourish in the decades to come, or will it end relatively soon, perhaps suddenly, as it did nearly a century ago at the outbreak of World War I? At that time, Britain was still the world's financial center but the United States had become the world's largest national market and had surpassed Britain in industrial output.

> Between 1870 and 1913 the size of the British economy well more than doubled; even if one takes into account population growth, British output rose by more than 50 percent per person in those years. Yet the gap between Britain and the rest of the world narrowed continually. British manufacturers were being beaten out of export markets, even out of the British market. The United States and Germany were the world's manufacturing dynamos; the United Kingdom maintained its leadership only in such services as banking, insurance, and shipping. It was no longer a given that the next power plant or railroad built in Africa or eastern Europe would be British; it was just as likely to be German, French, or American. Even in international investment, Continental financial centers—as well as New York—were challenging London's supremacy. It could hardly have been imagined that Britain's enormous industrial lead would last forever, but the speed of its erosion led many Britons to ask how this had happened . . . .[28]

After 1914, the hardship of the war and the faltering economic recovery of the following decade shifted financial leadership to the U.S., where it has remained ever since. But go back and read the above quote again, this time substituting the U.S. for all the references to Britain, and emerging economic powers like China and India for the references to Germany, France, and America. The description becomes eerily familiar, mirroring news media reports of the past quarter-century.

Could either continuation or disruption of globalization shift leadership roles the way it did in the early twentieth century? It happened once, so it can happen again. The story of space development in the globalization era, and of U.S. ambitions in this arena, is still being written.

---

28. Frieden, *Global Capitalism*, p. 107.

Chapter 11

Nasa as an Instrument of U.S. Foreign Policy

John Krige

Has space exploration, and NASA's role in it in particular, had an effect on society, and, if so, on what aspects of it? And how do we measure any such impact? These are challenging questions indeed. The stakeholders in the huge American space program are multiple and include scientists; engineers; research, development, and launch facilities; industry; administrators; and many government agencies, not to speak of Congress and the U.S. taxpayer. The impacts of spaceflight vary widely, from adding to the stockpile of knowledge and stimulating innovation and industry, to training, education, and creating jobs and—if we move beyond the civilian sphere—to enhancing national security and intelligence gathering. And then there are the intangible, difficult to quantify cultural effects that range from inspiring a young girl to become an astronaut to building national pride and prestige in what are, after all, spectacular scientific and technological, managerial, and industrial achievements.

This paper briefly considers one small, but I think important and often overlooked, corner of this vast panorama: the place of spaceflight in American foreign policy. I do not simply want to insist that NASA's international programs have had an important impact as instruments of foreign policy. I also want to suggest that today they have a particularly significant political and cultural role to play in projecting a positive image of American power and American democracy abroad. In a world increasingly torn apart by conflicts over values—conflicts which history teaches us can seldom be resolved by force—I believe we overlook the potential of NASA as an instrument for American foreign policy at our peril.

International cooperation for peaceful purposes was one of NASA's important missions from its inception, and those who drafted the Space Act that created the organization in 1958 gave it considerable prominence. The range of international activities covered by NASA is truly vast.[1] These are partly a response to the nature

1. Arnold W. Frutkin, *International Cooperation in Space* (Englewood Cliffs, NJ: Prentice Hall, 1965); Arnold W. Frutkin, *Space and the International Cooperation Year: A National Challenge* (Washington, DC: U.S. Government Printing Office, 1965 OL-779-251). Arnold W. Frutkin "International Cooperation in Space," *Science* 169 (24 July 1970): pp. 333–339 surveys NASA's general philosophy on international cooperation and some of the earlier programs of interest to us here.

of space exploration itself, which transcends national boundaries; whether they are launching sounding rockets or astronauts, communicating with satellites or space shuttles, or measuring the properties of the ionosphere or the trajectory of storms, NASA and its sister agencies have to think globally.

However, those who implemented NASA's mandate had a far broader vision of international cooperation than one that was simply subservient to America's national space needs. From its inception, NASA saw its role as fostering the development of space science and technology in other countries. Its officers, in consultation with other parts of the administration (notably the State Department and the Department of Defense), sought to use American scientific and technological preeminence to kick-start and even mould space activities in other countries, notably those of the Western alliance. NASA's international programs were intended to build a world community dedicated to the peaceful exploration of space with American help, under American leadership, and in line with the general objectives of American foreign policy. In brief, as a NASA Task Force put it in 1987, "[I]nternational cooperation in space from the outset has been motivated primarily by foreign policy objectives."[2]

In what follows I shall substantiate these claims by focusing on three space science programs in which U.S. foreign policy has been interwoven, more or less explicitly, with NASA's international initiatives. What makes these cases interesting is that, a priori, many people tend to believe that science is above politics and that international science is conducted independently of foreign policy concerns. This paper will not simply challenge such views but, by picking what is arguably the most difficult case, scientific collaboration, will alert us to the range of areas—some obvious, some less evident—in which NASA has served as a vector of U.S. foreign policy. My aim is to illustrate NASA's impact on strengthening the Western alliance not simply by promoting international scientific collaboration, but also by using it as a platform to consolidate the political and cultural solidarity of the free world. And although my examples are drawn from the cold war and its immediate aftermath, the lessons of history apply just as much today, when new and even more fundamental divisions threaten to tear apart the fragile fabric of Western democracy.

SPACE SCIENCE WHEN THE COLD WAR WAS HOT

In March 1959, just a few months after NASA officially came into being, the American delegate to a meeting of the Committee on Space Research (COSPAR) announced that the U.S. would be willing to launch scientific experiments proposed by scientists from other countries on American-built satellites.[3] NASA would help

2. Task Force on International Relations in Space, National Aeronautics and Space Administration, *International Space Policy for the 1990's and Beyond*, (Washington, DC: NASA, 1987), p.18.

3. The text is reproduced in H. Massey and M. O. Robins, *History of British Space Science* (Cambridge, U.K.: Cambridge University Press, 1986), Annex 4; see also pp. 67–69.

integrate the experiment into the payload and would even consider launches entirely dedicated to foreign experiments. The organization also offered to host foreign scientists in U.S. laboratories where they would help them design, build, and test their experiments. Informally, NASA also let it be known that, initially at least, the payload would be launched free of charge using an American rocket.

The British enthusiastically took up this initiative, and in April 1962 NASA launched Ariel I containing instruments that had been designed, prepared, and funded by the British National Committee for Space Research. Ariel II followed in 1964. A year later the British were engineering and building the payload for their own satellite with NASA's help. In September 1962, NASA also launched a Canadian satellite, Alouette I, that had been designed, funded, and engineered by the Defense Research Telecommunications Establishment, inaugurating a fruitful joint venture with the U.S. in studies of the ionosphere. French and Italian space researchers also benefited quickly from the offer made at COSPAR. Indeed, by 1965 Arnold W. Frutkin, who had been put in charge of NASA's international programs in September 1959, could boast that the organization had already entered into collaborative arrangements with no fewer than 69 countries. Apart from providing for NASA's own needs, as explained above, these programs, Frutkin pointed out, were affording opportunities to the best brains abroad to contribute and participate in space research, were stimulating technical development abroad so perhaps reducing some of the gaps that were causing political and economic strains between the U.S. and its partrners, and were providing a framework for other countries to join NASA in complementary and cost-sharing programs—like that with Canada.[4]

Frutkin was never sentimental about the benefits of international collaboration; his experience in the International Geophysical Year had taught him just how easily the high ideals of internationalism could be thwarted by the centrifugal pull of national interest. His roadmap for international collaboration was one that demanded there be no exchange of funds between the partners; that there be clean technological interfaces at the level of hardware; that the project be of genuine scientific interest and, if possible, complement the American space science program; and that the results be published and open to all. It was implicit in this roadmap that political considerations did not determine the choice of projects and that NASA's civilian mandate was respected.[5]

On the face of it, these collaborations were of purely scientific interest and have no relevance to my topic. Yet the more we probe, the more we realize how deeply embedded they were in the cold war struggle and the pursuit of America's foreign policy objectives. I shall identify just two very different dimensions of this that are pertinent to these cases.

---

4. Arnold W. Frutkin, *International Cooperation* (Englewood Cliffs, NJ, Prentice Hall, 1966), esp. chapter 2.

5. See Frutkin, *International Cooperation*.

First, the determination to help Britain and then Canada orbit their own satellites quickly was provoked, in part, by fears that a communist country, and not a member of the Western alliance, would be the first to launch a satellite after the USSR and the U.S.[6] There was a space race in space science. As early as September 1958, officials hoped to place British instruments on an American satellite launched from the U.K.'s test range in Woomera, South Australia.[7] It soon became clear that even if America's most important ally was putting a national space program in place, it did not yet have the independent capacity to provide an instrument payload. NASA's proposal made to COSPAR in March 1959 was partly a response to the inherent weakness of this and other European space programs. If the British were quick to capitalize on it, it was not only because they valued American help but also because they realized the urgency of the situation, both in terms of national pride and the opportunities provided by cold war rivalry. Ariel I, launched in April 1962, won the race, though it was something of a Pyrrhic victory. Although the instrumentation was British, the satellite was American. It was Canada's Alouette I satellite, launched on 29 September 1962, that had the honor of being "the first satellite to be designed and built by a nation other than the United States or the Soviet Union."[8] Apart from providing valuable information on the ionosphere, it ensured that a country from the Western alliance and not from the Communist bloc was third into space with its own satellite.[9]

Cultural as well as political spinoffs accrued from the early space race in science. As I mentioned earlier, France also took advantage of America's offer to help build a national space science program. Indeed, in the words of Roger Bonnet, an internationally recognized figure in French and European space science, "[W]ithout the [sic] American cooperation, the French space science programme would not have had any chance to start on a competitive basis." Bonnet's own Ph.D. research on the ultraviolet spectrum of the Sun was made possible thanks to the close contact established between his mentor, Jacques Blamont, and the American

6. John Krige, "'Building a Third Space Power.' Western European Reactions to Sputnik at the Dawn of the Space Age," in *Reconsidering Sputnik: Forty Years Since the First Soviet Satellite*, Roger D. Launius, John M. Logsdon and Robert W. Smith, ed. (Amsterdam: Harwood Academic Publishers, 2000), pp. 289–307.

7. Neil Whyte and Philip Gummett, "Far Beyond the Bounds of Science: The Making of the United Kingdom's First Space Policy," *Minerva* (1997): pp. 139–169.

8. C. A. Franklin, "Alouette/ISIS: How It All Began," IEEE International Engineering Ceremony, Shirley Bay, Ottawa, 13 May 1993, available online at *http://www.ewh.ieee.org/reg/7/millennium/alouette/alouette_home.html* (accessed 15 September 2006).

9. *Alouette's* mission had three components: 1) develop a Canadian space capability; 2) acquire new data for the engineering of high-frequency radio communication links; and 3) acquire a better understanding of the properties of the ionosphere for scattering and deflection of radar beams. See Web site quoted in previous note, and also *http://www.sciencetech.technomuses.ca/francais/collection/space2.cfm* (accessed 15 September 2006). Frutkin glosses over the military origins of Alouette.

space science community, notably at the Goddard Space Flight Center. In fact, Bonnet's first launch in the Sahara desert in 1963 used a French Véronique sounding rocket enhanced with two pointing systems developed for the U.S. military by the University of Colorado. Why is this pertinent? Because Roger Bonnet was raised in a French communist family and as a young man it had been Soviet firsts that inspired him to enter space research. Working with the U.S. forced him to revise his political perspective. As he put it in an interview with me recently:

> . . . We were all impressed by the frantic competition which developed between the Russians and the Americans in the race to space. It was fascinating as far as I was concerned. I was listening to the radio each time the Soviets were launching something new and witnessed vividly all their first steps into space: the first intercontinental ballistic missile, the first Sputnik, and all that followed after. It was fantastic! But very soon we realized that the Americans adopted an open policy of information which we could not always get from the Russians. So, ultimately there was a greater appeal to cooperate with the Americans.[10]

Collaboration with France did not simply kick-start the national space science community. It could also pull French space scientists out of a pro-Soviet or neutralist orbit, thereby strengthening the ideological cohesion of the Western alliance.

All of us remember President Kennedy's commitment in May 1961 ". . . to achieving the goal, before this decade is out, of landing a man on the Moon and returning him safely to Earth." We can still be inspired by another speech at Rice University a year later, in which he proclaimed "We choose to go to the Moon in this decade and do the other things not because they are easy but because they are hard, and because that goal will serve to organize and measure the best of our energies and skills." For Kennedy, success in space was a barometer of the capacity of America to mobilize its resources and its dynamism to achieve any goal it wanted. The Apollo program was a direct response to the increasing credibility of communism as a viable alternative to capitalism, of which successes on the ground in Indochina and in space with flights such as that of Yuri Gagarin were simply the most manifest examples. Taking on the challenge of putting a person on the Moon was a deliberate effort to regain the initiative by identifying national prestige and good government with a major scientific and technological achievement which tested the mettle of astronauts, engineers, administrators, and industry alike.[11]

---

10. Roger Bonnet interview, Geneva, Switzerland, by John Krige, 10 February 2005, Historical Archives of the European Economic Community, European University Institute, Villa Il Pioggiolo, Florence, Italy.

11. This paragraph owes much to John M. Logsdon, *The Decision to Go to the Moon: Project Apollo and the National Interest* (Cambridge, MA: MIT Press, 1970), and Howard E. McCurdy, *Space and the American Imagination* (Washington, DC: Smithsonian Institution Press, 1997), chapter 4.

One of Kennedy's main concerns in taking that initiative was the many countries that had recently been decolonized; accordingly, he introduced the section of his speech to Congress in May 1961 that gave birth to the Apollo program by identifying the conquest of space with ". . . the battle that is going on around the world between freedom and tyranny, . . . the battle for men's minds, . . . the minds of men everywhere who are attempting to make a determination of which road they should take."[12] Many in Western Europe were also grappling with that choice. According to space historian Walter McDougall, an April 1960 poll revealed that a majority of Europeans in every country expected the USSR to be stronger than the U.S. after 20 years of "competition without war." More to the point, according to a report from the U.S. Information Agency, only one Frenchman in 14, or about 7 percent of those polled, thought the U.S. would prevail over its communist rival in the long run.[13] Collaborating with Blamont, Bonnet, and their colleagues in space science promoted U.S. foreign policy objectives at a cultural level by tangibly demonstrating the values of an open, democratic system over a closed, communist society.

HELIOS: A PLACE IN THE SUN FOR GERMANY

In December 1974 and in January 1976, two German spacecraft, Helios 1 and Helios 2, weighing about 452 lbs (205 kg) each, were launched by American rockets into elliptical orbits about the Sun. They were designed to fly closer to the Sun than any previous spacecraft (approaching to within 25 million miles) and to provide invaluable scientific information about solar processes and solar–terrestrial relationships. This was the most ambitious bilateral scientific project that NASA had undertaken to date. Its estimated cost in 1970 was $100 million, paid by the (West) German Ministry for Science and Education. Germany designed, manufactured, and integrated the two spacecraft, provided the majority of the payload (which also included some experiments from the U.S., Australia, and Italy), and operated and controlled the spacecraft from a national facility. NASA provided the deep space tracking network to support the mission and participated in the Joint Working Group which was responsible for technical implementation. The Helios spacecraft imposed advanced technical requirements on German industry, particularly for the development of the on-board power system, on-board data processing system, and thermal controls which had to survive high levels of solar radiation. It also introduced German engineers and project managers in the Joint Working Group to the way space projects were implemented in the U.S.[14]

---

12. Logsdon, p. 128. See also Jeremi Suri, *Power and Protest: Global Revolution and the Rise of Détente* (Cambridge, MA: Harvard University Press, 2003), pp. 15–25.

13. Walter A. McDougall, *. . . the Heavens and the Earth: A Political History of the Space Age* (New York: Basic Books, 1985), p. 241.

14. Frutkin, *International Cooperation*. For an overview, see Niklas Reinke, *A History of German Space Policy: Ideas, Influences and Interdependence*, 1923–2002, translated from the German (Noordwijk, The Netherlands: ESA, forthcoming).

Helios was the crowning achievement of U.S.-German space science collaboration that began in the mid-1960s with sounding rocket experiments and graduated through various smaller satellite projects.[15] It was heavily charged with political content and foreign policy concerns. Just before Christmas 1965, Chancellor Ludwig Erhard made an official visit to Washington. In a brief exchange of toasts with his guest in the state dining room, President Johnson took time to mention their ongoing "mutual adventure in space" and said he looked forward to discussing "more ambitious plans to permit us to do together what we cannot do so well alone," including a probe to the Sun, the eventual Helios.[16] Erhard visited Johnson again in October 1966. Despite the fact that he only came for two days, he was taken down to Cape Kennedy to see the progress there. In an official address in the as-yet incomplete Vehicle Assembly Building, Johnson assured Erhard that the Apollo program was progressing as expected and reaffirmed his commitment to mutual space projects.[17] On the way back to Washington, NASA Administrator James Webb took the opportunity to spend an hour with the German chancellor. The "large on-going effort [at the Cape] made a deep impression" on Erhard, Webb wrote Secretary of State Dean Rusk. He went on: "[I]t seems to me that Erhard had a different attitude when we left the Cape than when we arrived. In fact, he did say that it was impossible to learn from pictures, television, and documents the true scope and magnitude of what was being done and that he had a much better appreciation of its importance."[18]

There are many reasons why the American president and his top advisors went to such pains to publicly and personally promote space collaboration with Germany, and Chancellor Erhard, at this particular moment. I shall mention just a few here.[19]

First, it was an attempt to meet European objections that a "technological gap" had opened up between the two sides of the Atlantic that made it impossible

---

15. "German-NASA cooperation, 9/28/66," record no. 14620, International Cooperation and Foreign Countries, Foreign Countries, West Germany, Folder Germany-U.S., 1963–1984, NASA Historical Reference Collection, Washington, DC.

16. "The White House. Exchange of Toasts" between Johnson and Erhard, 20 December 1965, NSF, Country File Europe & USSR, Germany/Erhard Visit [12-65], folder 12/19-21/65, Johnson Presidential Archives, University of Austin, Austin, Texas.

17. "The White House. Remarks by the President at Vehicle Assembly Building, Cape Kennedy, Florida," 20 September 1966, NSF, Country File Europe & USSR, Germany 9/66, Erhard Visit, folder Papers, Cables, Memos [9/66], Johnson Presidential Archives, University of Austin, Austin, Texas.

18. James E. Webb to Dean Rusk, 14 October 1966, record no. 14465, International Cooperation and Foreign Countries, Foreign Countries, West Germany, folder Germany (West), 1956–1990, NASA Historical Reference Collection, Washington, DC.

19. Thomas Alan Schwartz, *Lyndon Johnson and Europe: In the Shadow of Vietnam* (Cambridge, MA: Harvard University Press, 2003) is a fine analysis of the Johnson's foreign policy in the region. See also John Krige, "Technology, Foreign Policy and International Cooperation in Space," in *Critical Issues in the History of Spaceflight*, Steven J. Dick and Roger D. Launius, ed. (Washington DC: NASA SP-2006-4702, 2006), pp. 239–260.

for European high-tech firms to compete effectively in the world market against their American rivals. This had deep political ramifications since it bred resentment against what was perceived as American domination, and undermined Washington's demand that Europe assume more of the burden for its own defense. Both parties quickly realized that the fault lay within Europe itself, and that its institutions and managerial practices needed drastic reform. The indigenous development of space technology with American help was seen as one useful way to overcome this situation. Thus Webb assured Erhard on the way back from Cape Kennedy that

> [T]he President was, in fact, offering him more than friendship and more than dollars. In fact he was offering a partnership in the development of technology that could permit Germany to increase its own capability, gain a better understanding of its own needs and opportunities for multilateral and bilateral cooperation, establish a basis for leadership in the direction it felt its leadership could be effective in Western Europe, and could set a pattern of university/industry/government cooperation suited to the needs of Germany, benefiting throughout from our own experience. [20]

A second factor of more symbolic than financial significance was the idea that Germany could purchase space technology and space launches from the U.S. as part of its "offset" obligations. West Germany was required to offset with military purchases the approximate costs to the American government of retaining U.S. forces in its territory. These offset payments had become a major political and financial liability for the German government in 1966. For one thing, they were associated in the public's mind with a series of crashes of the F-104G Starfighter jets—10 in the first half of 1966 alone, giving the impression that the U.S. was selling unreliable and unnecessary military equipment to its ally.[21] For another, the next round of payments was due shortly; Erhard had undertaken to place $1.35 billion in weapons orders by 31 December 1966 and to make an additional $1.4 billion in offset payments by June 1967.[22] Writing to Johnson in July 1966, the chancellor said he was willing to accept his offset obligations but that he hoped to do so by "payments and services other than the mere purchase of weapons and military equipment."[23] Purchasing space technology and services was one such way, even though such alternatives would probably not amount to more than about $25 to $50 million.[24]

20. Webb to Rusk, 14 October 1966.

21. Schwartz, p. 116.

22. Francis M. Bator, "Memorandum for the President. Subject: Erhard Visit, September 26–27, 1966," 25 September 1966, NSF, Country File Europe & USSR, Germany 9/66, Erhard Visit, folder Papers, Cables, Memos [9/66], Johnson Presidential Archives, University of Austin, Austin, Texas.

23. Schwartz, p.117.

24. "Text of Cable from Ambassador McGhee (Bonn 3361), Subject: The Offset and American Troop Level in Germany," 20 September 1965, NSF, Country File Europe & USSR, Germany 9/66, Erhard Visit, folder Papers, Cables, Memos [9/66], Johnson Presidential Archives, University of Austin, Austin, Texas.

Then there was the hope that Germany would take the lead in strengthening European multilateral cooperation that was being threatened by President de Gaulle's increasing resentment of the limitations placed on French sovereignty by NATO and the European Economic Community (EEC). "The United States has a direct interest in the continuation of European integration," wrote George Ball in the State Department. "It is the most realistic means of achieving European political unity with all that that implies for our relations with Eastern Europe and the Soviet Union." de Gaulle's actions were undermining that unity. "The United States hopes therefore," Ball went on, "that the Federal Republic will continue to exert leadership to preserve the unique character of the European institutions . . . ." [25] In particular, if Germany could become a leading space power in Europe it could play a major role not only in developing European high-tech industry but also in reinforcing European multilateral institutions in the face of the threat posed to them by de Gaulle's affirmation of national sovereignty.[26]

Finally, it must be mentioned that Erhard was a staunch supporter of the war in Vietnam. In fact, Johnson went out of his way in his toast to the chancellor in December 1965 to thank him for "the support which your Government has given to the common cause in Viet Nam, and which you may give in the days ahead . . . . The credible commitment of the United States is the foundation stone of freedom all around the world," Johnson added. "If it is not good in Viet Nam who can trust it in the heart of Europe? But America's word, I assure you," Johnson concluded, "is good in Viet Nam, just as it is good in Berlin."[27] The high-profile offer to collaborate with Germany in space was also a public act of gratitude to a faithful ally and a signal to the Soviets that they had best not challenge the now-established divisions between East and West in Europe.

In replying to Johnson's toast that Christmas Eve in 1965, Erhard, while enthusiastically agreeing that such an ambitious project would "fascinate the imagination of the people," also joked that "Of course, we, the Germans, would not like to get too close to the sun because we wouldn't like to burn our wings . . . ."[28] Actually, it was Helios that survived the journey to the Sun and Erhard who burnt his wings. He resigned after returning from his visit to Washington and Cape Kennedy in October 1966. His failure to achieve a major reduction in offset commitments and his unwavering support for Johnson's policies in Viet Nam were two of the main factors leading to the collapse of his government.

25. Cable signed Ball from the Department of State to the American Embassy, Bonn, 18 November 1965, NSF, Country File Europe & USSR, Germany, Erhard Visit [12-65], folder 12/19-21/65 Johnson Presidential Archives, University of Austin, Austin, Texas.

26. Krige in Dick and Launius, *Critical Issues*.

27. Johnson, "The White House. Exchange of Toasts."

28. *Space Daily*, 22 December 1965, 295, record no. 14548, International Cooperation and Foreign Countries, International Cooperation, folder US-Europe 1965-1972, NASA Historical Reference Collection, Washington, DC.

AND THEN THE WALL CAME DOWN: CASSINI-HUYGENS

As Western European countries gradually put their national and multinational space programs onto sounder footing, they expected to be treated as equals by their American partners. The mantra of Reimar Lüst, the Director-General of ESA (European Space Agency) from 1984 to 1990, was that Europe had to be able to compete with the U.S. in order to collaborate with it from a position of strength. This philosophy was exemplified in the magnificent Cassini-Huygens mission to Saturn and Titan in 2004–2005. In this joint venture, the Jet Propulsion Laboratory built and managed the Cassini orbiter that surveyed Saturn; the Italian Space Agency built Cassini's high-gain communications antennae; and ESA built the Huygens probe that plunged through Titan's atmosphere to its surface. The truly spectacular images of Saturn's rings and of its largest moon will have thrilled many a space scientist, be they at high school or an old hand at the game.

This extraordinary scientific achievement not only called for scientific, engineering and managerial expertise, it also called for diplomacy. Early in 1992, Dan Goldin was appointed NASA Administrator. He resolved to shake up the organization and inaugurated his famous policy of "faster, better, cheaper." Cassini-Huygens was anything but that, and it soon caught his eye; late in 1993 he threatened to cancel the program. The American space scientists and engineers and their European colleagues were outraged. "I remember Carl Sagan calling me on the phone from California asking for help because NASA was trying to stop the mission," Roger Bonnet told me recently. "Three times ESA intervened and asked its ambassadors to interact with the State Department in order to make the Americans understand that they could not stop Cassini, with such a big involvement of Europe . . . ."[29] In June, 1994, ESA Director General Jean-Marie Luton wrote a strong letter to Vice President Al Gore, copied to the Secretary of State and to various senior administrators, including Goldin. In it Luton stressed that Europe regarded

> . . . [A]ny prospect of a unilateral withdrawal from the cooperation on the part of the United States as totally unacceptable. Such an action would call into question the reliability of the U.S. as a partner in any future major scientific and technological cooperation.[30]

Goldin had to back down. The Clinton administration wanted an unambiguous European commitment to what was soon to be the International Space Station and could not afford to alienate ESA. This combination of financial and foreign policy concerns saved Cassini-Huygens from being axed by the NASA Administrator, and avoided a major diplomatic incident.

---

29. Bonnet, interview with the author, 10 February 2005.

30. Quoted in Charley Kohlhase, "Return to Saturn's Realm," *The Planetary Report* (March/April 2004), available online at *http://www.planetary.org/saturn/tpr_vol24no2_kohlhase.html* (accessed 16 September 2006).

Luton's unambiguous position was a symptom of the strength of the European space program. It was also fuelled by his determination that joint U.S.–European projects would never again be sacrificed on the altar of NASA's changing national priorities. This had happened some years before, with the International Solar Polar Mission (ISPM). In the late 1970s, NASA and ESA had agreed to launch a pair of satellites out of the ecliptic plane (the plane that contains most of the objects that orbit the Sun) to perform a variety of challenging scientific experiments in domains that included solar physics, cosmic ray studies, and the exploration of the interplanetary environment.[31] NASA canceled its contribution unilaterally in 1982 due to budget constraints caused by the development of the Shuttle and a new, stricter financial regime inaugurated under President Reagan. A political climate dominated by fears that increased economic competition from Japan and Western Europe was undermining American leadership did the rest. Europeans understood the budget difficulties their American colleagues faced, which derived in part from the very different procedures for funding spaceflight on the two sides of the Atlantic.[32] What they bitterly resented was that they were not consulted before the American decision and that NASA was deaf to pleas to reinstate the ISPM program. The huge disparity in space capability between the U.S. and Europe for the first two decades of the Space Age had reduced Europeans to the status of junior partners who could be manipulated almost at will by their dominant ally. The experience with ISPM taught Europeans, in the words of Bonnet and Manno, never again to "accept being considered a subordinate participant" in a joint project.[33]

NASA's and the Reagan's administration's approach were coherent with, and justified locally by, a persistent tendency of the U.S. during the cold war to fail to consult its Western European allies in important foreign policy decisions which affected both parties, the most blatant example being Kennedy's handling of the Cuban missile crisis.[34] Indeed, veteran U.S. diplomat David Bruce described this as "the vicious circle of American predominance, European dependence and mutual resentment [that] operated for half a century,"[35] up to the collapse of the Berlin Wall and the implosion of the Soviet Union. Thereafter, in a global environment no longer dominated by superpower rivalry, and with a new administration in the White

---

31. What follows is based on Russo's analysis in John Krige, Arturo Russo and Lorenza Sebesta, *A History of the European Space Agency 1958–1987. Vol. II. The Story of ESA*, 1973 to 1987 (Noordwijk, The Netherlands: ESA SP-1235, April 2000), chapter 3.

32. In ESA, once a program is agreed to the participating states agree to fund to completion; in the U.S. projects are subject to the vicissitudes of the annual budget voted by Congress for NASA.

33. Roger M. Bonnet and Vittorio Manno, *International Cooperation in Space: The Example of the European Space Agency* (Cambridge, MA: Harvard University Press, 1994).

34. Frank Costigliola, "Kennedy, the European Allies and the Failure to Consult," *Political Science Quarterly* 110 (1995): pp. 105–123.

35. Costigliola, p. 122.

House, Washington was more willing to take seriously the needs of its European allies—and Cassini-Huygens survived as a joint venture. Today, as the pendulum swings back toward U.S. unilateralism, so the prospects for durable international agreements become bleaker.

CONCLUSION: INTERNATIONAL SPACE COLLABORATION, NASA, AND "SOFT POWER"

In 1998 a Commission of the National Academies pointed out that

> . . . [D]uring the Cold War there was significant political goodwill to be gained by the United States through cooperation with Europe vis-à-vis the former Soviet Union . . . . Competition in space (including the space sciences) was part and parcel of concerted efforts made by the superpowers to convince other countries of their technical capabilities, and hence leadership.[36]

This paper has fleshed out these claims. It has illustrated how international scientific and technological collaboration in space were used to promote American interests abroad, and how it has adapted to the changing balance of power between the American and European space programs. Borrowing the language of Joseph Nye, professor of international relations at the Kennedy School of Government, Harvard University, we can say that NASA's international initiatives have served as agents of "soft," or co-optive power, as opposed to "hard," coercive or command power. Nye puts it thus:

> Soft co-optive power is just as important as hard command power. If a state can make its power seem legitimate in the eyes of others, it will encounter less resistance to its wishes. If its culture and ideology are attractive, others will more willingly follow . . . . If it can support institutions that make other states wish to channel or limit their activities in ways that the dominant state prefers, it may be spared the costly exercise of coercive or hard power.[37]

Echoing Nye, we can say that international collaboration in space is one of a repertoire of instruments the U.S. has at its disposal to legitimate its power in the eyes of others, to promote its culture and its democratic ideals, and to channel the scientific and technological efforts of other nations down paths that cohere with American interests.[38] NASA has played an important role in that process in the past and can continue to do so in the future. The cold war may be over but the struggle for hearts and minds is not.[39]

---

36. Committee on International Space Programs, *U.S.–European Collaboration in Space Science* (Washington, DC: National Academies Press, 1998), p. 15, available online at *http://books.nap.edu* (accessed 16 September 2006).

37. Joseph S. Nye, Jr., "Soft Power," *Foreign Policy* 80 (Autumn 1990): pp. 153–171.

38. For more detail, see Krige in Dick and Launius, *Critical Issues*.

39. Joseph S. Nye, Jr., "The Decline of America's Soft Power: Why Washington Should Worry," *Foreign Affairs* 83 (2004): pp. 16–20.

Chapter 12

"From Farm to Fork": How Space Food Standards Impacted the Food Industry and Changed Food Safety Standards

Jennifer Ross-Nazzal

Most Americans give little thought to the safety of their food until they hear of an *E. coli* outbreak or a recall of their favorite item. They may be surprised to learn that Space Age technology designed to protect the astronauts from food poisoning has slowly become the safety standard for the food industry in the U.S. and abroad. Dubbed the Hazard Analysis and Critical Control Point (HACCP) system, this NASA spinoff has been called "the most revolutionary institutional innovation to ensure food safety of the twentieth century."[1]

For more than 30 years, canners who process low-acid foods have relied upon the risk prevention system developed by NASA to safeguard their products. More recently, HACCP regulations have been implemented by the U.S. Food and Drug Administration (FDA) to maintain the integrity of seafood and juice in the United States. The U.S. Department of Agriculture (USDA) also relies on HACCP systems in the nation's meat and poultry plants and slaughterhouses. There is, however, some disagreement over whether some of the more recent HACCP systems put in place by these regulatory agencies truly reflect the principles of an HACCP plan as outlined by food safety experts.

In nearly all cases, a series of food crises forced the regulatory agencies and industries to implement HACCP. In the 1970s, two well-publicized incidents and a growing consumer movement compelled industry and its trade representatives to adopt and lobby for the implementation of a preventive and comprehensive safety plan. The first occurrence happened in the spring of 1971.

1. John Spriggs and Grant Isaac, *Food Safety and International Competitiveness: The Case of Beef* (New York: CABI Publishing, 2001), p. 11.

Dr. Howard E. Bauman

A woman from Connecticut found glass in her baby's cereal. Soon after, Americans awoke to the news, hearing: "Good morning America, there's glass in your baby food." Pillsbury Company's farina, a creamy wheat cereal for infants, had been contaminated when shards of glass fell into a storage bin at one of Pillsbury's plants, forcing the company to recall the cereal. Upon hearing the news, Robert J. Keith, chief executive officer of Pillsbury, called Dr. Howard E. Bauman, a microbiologist and one of the company's research directors, into his office.[2]

Keith had worked for Pillsbury for more than 30 years and made his way up the corporate ladder, becoming CEO and chairman of the Board in 1967. During the five years he served in this position, he championed many popular causes, one of which included the growing consumer movement. Led by advocate Ralph Nader, consumers increasingly demanded safe food. Keith, sympathetic to such demands, told Bauman this incident would not happen again. Customers needed to know that the company's products were safe.[3]

Publicly, Pillsbury comforted customers by announcing a "considerable change" in the company's manufacturing processes, but this was not a PR campaign designed to halt fading consumer confidence.[4] Significant changes were underway at Pillsbury. In response to the recall, Keith pushed Pillsbury to implement a secure product safety system to minimize the likelihood of another recall of the company's food products.[5] For his part, Bauman saw to it that no food would be recalled under his watch. He planned to implement procedures he had helped develop years earlier while working with NASA, an idea he later pursued with HACCP.

Bauman began working at Pillsbury in 1953, when he completed his doctoral degree at the University of Wisconsin. He started out as head of research in the bacteriology section at Pillsbury and later assisted NASA, the U.S. Air Force Space

2. Dr. William H. Sperber worked with Dr. Bauman at Pillsbury, and he and his colleagues recall hearing this anecdote from Bauman. Dr. William H. Sperber, telephone conversation with author, 21 June 2006.

3. William J. Powell, *Pillsbury's Best: A Company History from 1869* (Minneapolis: The Pillsbury Company, 1985), pp. 190–191; Pillsbury Company, *Annual Report for the Year Ended 1973* (Minneapolis: The Pillsbury Company, 1973), p. 1; Sperber, telephone conversation, 21 June 2006.

4. "Pillsbury Recalls Cereal; Boxes May Contain Glass," *The New York Times*, 24 March 1971; Carole Shifrin, "Warning on Farina Cereal," *The Washington Post-Times Herald*, 25 March 1971.

5. Powell, *Pillsbury's Best,* p. 190.

Laboratory Project Group, and the U.S. Army Natick Laboratories with the food systems for the human spaceflight programs.[6]

Some of the other key individuals involved with the development and testing of the early space food systems included Herbert A. Hollender, Mary V. Klicka, and Hamed El-Bisi of the U.S. Army Natick Laboratories. Paul A. Lachance of NASA's Manned Spacecraft Center in Houston, Texas, rounded out the group.

Pillsbury became involved in the space program in 1959 when the Quartermaster Food and Container Institute of the United States Armed Forces (later called the U.S. Army Natick Laboratories) phoned Bauman and asked for Pillsbury's assistance. Would the Pillsbury Company be interested in producing space food? After some discussion, the company accepted and began working on cube-sized foods for the flight crews.[7]

Concerned about safety, NASA engineers specified that the food could not crumble, thereby floating into instrument panels or contaminating the capsule's atmosphere. To meet the outlined specifications, food technologists at Pillsbury developed a compressed food bar with an edible coating to prevent the food from breaking apart. In addition to processing food that would not damage the capsule's electronics, the food also had to be safe for the astronauts to consume.

Almost immediately food scientists and microbiologists determined that the assurance of food safety was a problem. Bauman recalled that it was nearly impossible for companies to guarantee that the food manufactured for the astronauts was uncontaminated. "We quickly found by using standard methods of quality control there was absolutely no way we could be assured there wouldn't be a problem," he said.[8] To determine food safety for the flight crews, manufacturers had to test a large percentage of their finished products, which involved a great deal of expense and left little for the flights.[9]

A survey conducted among experts in the field indicated there was no single standard quality control program for the food industry. Control programs were numerous and varied widely, according to Bauman: "Our surveys indicated that there were about as many variations of control progammes as there were quality control managers or Government inspectors."[10] Thus, there was no program already in place that could readily be used to provide a 100 percent guarantee of food safety.

6. Wolfgang Saxon, "Howard Bauman, 76, Expert Who Kept Food Safe in Space," *The New York Times*, 12 August 2001.

7. Howard E. Bauman, "The Origin of the HACCP System and Subsequent Evolution," *Food Science and Technology Today* 8, no. 2 (June 1994): p. 67.

8. "A Dividend in Food Safety," [NASA] *Spinoff*, (1991): p. 52.

9. Bauman, "The Origin of the HACCP System," p. 67; Howard Bauman, "HACCP: Concept, Development, and Application," *Food Technology* 44, no. 5 (May 1990): p. 156.

10. Bauman, "The Origin of the HACCP System," p. 68.

Paul Lachance

While Pillsbury was dealing with issues of food contamination, Paul Lachance completed a tour of duty with the U.S. Air Force Aeromedical Research Laboratories at Wright-Patterson Air Force Base in Dayton, Ohio. He was well aware of the issues concerning astronaut food as the Air Force laboratory provided support for the preflight feeding of the Mercury astronauts. Given his experience with Project Mercury, NASA recruited him and offered him the position of Flight Food and Nutrition Coordinator at the Manned Spacecraft Center in Houston.[11]

When Lachance arrived in September 1963, he began evaluating the Gemini and Apollo food systems, which were not very far along in development. Food safety for astronauts became an overriding concern for Lachance, who did not want a late night telephone call from Charles A. Berry "who was the Chief Medical Officer of NASA, telling me that his astronaut or astronauts were sick and had stomach problems and were having a hard time holding things down." Lachance also wanted to avoid putting the crews in jeopardy, and he began thinking about the potential microbiological, physical, and chemical dangers space foods might pose. Microbiological hazards became an overriding concern after NASA found that many of the ingredients they purchased were contaminated with viral or bacterial pathogens. There had to be some way to minimize or eliminate these hazards, Lachance explained.[12]

But no one was sure how to conduct a thorough hazard analysis, Bauman recalled. Eventually a suitable model, called the "modes of failure," was located, adopted, and utilized. Microbiologists began examining each food item and analyzed the potential areas of concern during the manufacturing process. Armed with this information, scientists then scoured publications to determine ingredients that were potentially dangerous—possibly containing viral or bacterial pathogens, heavy metals, other hazardous chemicals, or physical hazards. A list of hazards was then compiled.[13]

11. Paul A. Lachance interview, Houston, TX, 4 May 2006, JSC Oral History Project, JSC History Collection, University of Houston-Clear Lake.

12. Ibid.

13. Natick used "modes of failure" to analyze medical supplies. Bauman, "The Origin of the HACCP System," p. 68; Bauman, "HACCP," p. 156.

For their part, the Natick Labs established the microbiological standards for food that would be flown on piloted missions.[14] Requirements were stringent because scientific research had indicated that stress might weaken an astronaut's ability to fight infection. Even the smallest amount of a relatively harmless microorganism on Earth could potentially cause an astronaut in orbit to become ill. Thus, microbiologist Hamed El-Bisi of the Natick Labs concluded, "All possible measures must thus be taken to eliminate all pathogens and to minimize the microbial load in all food intake." He placed the total aerobic plate count at less than 10,000 per gram, meaning that the food was more likely to be safe for consumption by flight crews.[15]

This was a substantial change for the food manufacturers contracted to develop the Gemini food system. Previously, food processors had not measured pathogens unless they encountered bouts of food poisoning. By contrast, a hazard analysis required contractors to conduct pre- and in-process microbiology tests of food ingredients to ensure the health of the astronauts. Manufacturers had to assure NASA that their foods conformed to the microbiological standards outlined by Natick Laboratories. Food manufacturing conditions were strict; there were rigid temperature and humidity controls. Some foods were even processed in clean rooms, similar to the environment in which McDonnell Aircraft Corporation built the Gemini spacecraft.[16] If the food producers did not meet the microbiological standards, food technologists discarded the food.[17]

Bauman, who was assigned to the Gemini and Apollo Programs, was well suited for the position of ensuring the microbiological safety of astronaut food. Dr. Lachance recalled, Bauman was a microbiologist, " . . . and so he really knew his microbiology. So he was an ideal person, in some ways, to develop a laboratory where microbiology had to be paid attention to."[18]

As work on the Gemini program proceeded, Lachance turned his attention to the Apollo food system. The Apollo Spacecraft Program Office (ASPO) required

---

14. Robert A. Nanz, Edward L. Michel, and Paul A. Lachance, "Evolution of Space Feeding Concepts During the Mercury and Gemini Space Programs," *Food Technology* 21 (December 1967): p. 53; Space Food Systems Contract NAS 9-9032 Final Report, December 1970, Space Food Systems: Mercury Through Apollo (December 1970), Rita Rapp Files, Center Series, JSC History Collection, University of Houston-Clear Lake.

15. The microbiological standards were established in 1964. Hamed M. El-Bisi, "Microbiological Requirements of Space Food Prototypes," *Research and Development Associates for Military Food and Packaging Systems, Inc.* 17 (1965): pp. 55, 57; Charles T. Bourland interview, Houston, TX, 7 April 2006, JSC Oral History Project, JSC History Collection, University of Houston-Clear Lake; Edmund M. Powers, et al., "Bacteriology of Dehydrated Space Foods," *Applied Microbiology* 22, no. 3 (September 1971): p. 441.

16. Lachance interview, 4 May 2006.

17. Space Food Systems Contract NAS 9-9032 Final Report; Powers, "Bacteriology of Dehydrated Space Foods," p. 444.

18. Lachance interview, 4 May 2006.

its contractors to comply with certain reliability standards. Lachance had previously implemented reliability requirements for the Gemini food system but the ASPO required all contractors to develop prediction models for their systems to determine "critical failure areas" and then eliminate those hazards from the system.[19] Food contractors were not exempt from this requirement and had to sketch out their critical control points—places in the manufacturing process where the system could break down and put the hardware at risk.

Writing these blueprints forced Pillsbury to think logically about the steps in their process and identify critical control points. As the Apollo program matured, Pillsbury continued to revise the list of critical control points as they went along. Bauman explained what they learned along the way: "[A]s we worked along in this system, we found certain critical control points like telephones in the room. They are a good source of bacteria, unless you sterilize the receiver. That's something that you really don't always think of."[20]

Even though NASA required its food contractors to identify critical control points, NASA also determined them. In the specifications for most Apollo foods, NASA located 17 quality control stations in the production process; stations had acceptance and rejection standards for the inspectors, or in NASA-ese, "go" or "no go."[21]

Aside from monitoring the critical control points, contractors also had to keep records that documented the history of a food product. Records were kept from the moment the raw foods reached the plant. Logs indicated where the raw materials came from or, if the product had been processed, the name of the plant that produced the item and the names of people who worked in the manufacturing of that item. Strict recordkeeping allowed product tracking. "We knew the latitude and longitude where the salmon used in the salmon loaf were caught," Bauman joked.[22]

As a result of his NASA experience, Bauman became one of the biggest proponents of the HACCP concept, which was introduced to the food industry at the first National Conference on Food Protection in April 1971, just a few days after Pillsbury recalled packages of its farina cereal. The conference, sponsored by the American Public Health Association, opened on April 4 in Denver, Colorado. The main purpose of the conference was "to develop a comprehensive, integrated attack on the problem of microbial contamination of foods."[23]

---

19. NASA, "Reliability Program Provisions for Space System Contractors," NPC 250-1 (Washington, DC: U.S. Government Printing Office, 1963), pp. 3-1, 3-2; Lachance interview, 4 May 2006.

20. The Pillsbury Company, Research and Development Department, *Development of a Food Quality Assurance Program and the Training of FDA Personnel in Hazard Analysis Techniques* (Minneapolis: The Pillsbury Company, 1973), p. 507.

21. Malcolm C. Smith, et al., "Apollo Experience Report—Food Systems" NASA TN D-7720 (Washington, DC: NASA, 1974), p. 9.

22. Bauman, "The Origin of HACCP," p. 68.

23. National Conference on Food Protection, *Proceedings* (Washington, DC: FDA, 1972), p. III.

Bauman served as vice chairman of Panel Number Two, which focused on the prevention of contamination of commercially processed foods. Other panel members included L. Atkin of Arthur D. Little, Inc., James J. Jezeski from Montana State University, and John H. Silliker of Silliker Laboratories, Inc., a food testing laboratory.[24] Convinced of the benefits of HACCP, Bauman encouraged his colleagues to consider the system as a plausible option for the food industry as a whole. The idea, however, was not immediately embraced, and a second incident occurred in the summer of 1971.

On a sweltering June night, Grace Cochran cracked open a can of Bon Vivant's vichyssoise (cold potato soup) for dinner. Samuel, her husband, ate a bite or two but then stopped, noting that the soup tasted spoiled. Grace had a spoonful and agreed. The next morning while driving to work in Manhattan, Samuel's vision began to blur. The condition continued to worsen, and a few hours after arriving at work he scheduled an appointment with doctors at the Eye Institute of the Columbia Presbyterian Medical Center. When he walked into the Center, Samuel's condition had deteriorated and doctors directed him to his personal physician. By the time he arrived at the hospital and met with his doctor, he had difficulty talking, could not turn his eyes left and right, could not swallow, and when he held his arms straight in front of his body, they shook. By 11 p.m., less than 8 hours after being admitted to the hospital, Samuel died.

A few hours later, Grace became ill. Dr. Henry P. Colmore, the internist who had treated her husband, visited Grace at her home. She told the doctor, "I'm doing just like Sam did. You don't suppose it was that soup we had last night? It tasted so bad we couldn't finish it." On the advice of Dr. Colmore, Grace's sons located the soup can while Colmore arranged for Grace to go to the hospital. Colmore was certain that Grace was suffering from botulism poisoning and that her husband had died from botulism. As a result of his findings, he notified the Westchester County Health Department.[25]

His phone call began a series of events leading to a national recall of Bon Vivant soups. Fearing a public health epidemic, Jack Goldman, county health commissioner, concluded that the department had to document the case. They needed the soup's lot number from the recovered can, but they needed additional evidence and found it. Cans of soup on the shelves at the local grocery store, bearing the same lot number, V-141/USA-71, were bulging. Armed with this information, Goldman contacted the state health department and the FDA, relaying the knowledge he had gathered. Eventually recalls for all Bon Vivant soups and other products made by the same manufacturer were issued and the factory shut down.[26]

24. Ibid., pp. 56–83.

25. Boyce Rensberger, "Grim Detective Case: Search for Vichyssoise," *The New York Times*, 18 July 1971.

26. Ibid; Nancy L. Ross, "Tracking Down the Soup Can Killer," *The Washington Post-Times Herald*, 18 July 1971.

This outbreak of botulism cast doubt over food safety in the U.S. and whether the FDA could protect citizens from contaminated food. Doubt had surfaced many times before this incident. In 1968 and 1969, for example, a Ralph Nader summer study group issued critical reports about the agency. In 1970, James S. Turner, the project director, revised the reports and published *The Chemical Feast: The Ralph Nader Study Group Report on Food Protection and the Food and Drug Administration.* In a chapter about the food industry, the bestseller detailed the FDA's friendship with food conglomerates and called upon the FDA to "enforce the law" rather than apologize on behalf of food processors who placed profits over consumer safety. "It is time the FDA set about *its* assigned task of insuring [*sic*] that profits made by the food industry are not the result of fraud, deception, adulteration, or misbranding."[27]

Employees of the FDA recognized the agency had problems. In July 1969, the FDA released the "Kinslow Report," commissioned by FDA Commissioner Dr. Herbert L. Ley, Jr. The study concluded, "The American public's principal consumer protection is provided by the Food and Drug Administration, and we are currently not equipped to cope with the challenge." In total, the panel submitted 45 recommendations to the Commissioner. Ley did not have time to implement any suggestions. In an attempt to overhaul the agency, Robert H. Finch, the Secretary of Health, Education, and Welfare, named Dr. Charles C. Edwards to the position of FDA Commissioner in December.[28]

After being removed from his post, Ley warned the public about the FDA's inability to safeguard consumers. People were being misled, he believed. "The thing that bugs me is that the people think the FDA is protecting them—it isn't. What the FDA is doing and what the public thinks it's doing are as different as night and day," he said. The agency, in his opinion, did not have the motivation to protect consumers, faced budget shortfalls, and lacked support from the Department of Health, Education, and Welfare.[29]

A year and a half later, when Samuel Cochran died from botulism and his wife suffered the ill effects of the disease, the FDA, its leaders, and food inspection processes continued to be under the microscope. Newspapers reported that the FDA had not inspected the Bon Vivant plant for four years. The last inspection took place in May 1967. Reporters asked about the lack of inspections and were told that workforce shortages often resulted in infrequent plant inspections. In some cases, the FDA had not inspected certain food plants for periods of up to 10 years.[30]

---

27. James S. Turner, *The Chemical Feast* (New York: Grossman Publishers, 1970), pp. 85–86, 106.

28. Richard D. Lyons, "F.D.A. Shake-Up Will Start with Naming of New Chief," *The New York Times*, 10 December 1969; "Trouble Over Drugs on the Market," *The New York Times*, 4 January 1970; Andrew Hamilton, "FDA: New Pressures, Old Habits Bring a Change at the Top," *Science* (16 January 1970): pp. 268–270.

29. Richard D. Lyons, "Ousted F.D.A. Chief Charges 'Pressure' from Drug Industry," *The New York Times*, 31 December 1969.

30. Grace Lichtenstein, "Bon Vivant's Soup Plant Not Inspected for 4 Years," *The New York Times*, 21 July 1971.

Later that summer, as the recall of Bon Vivant soups was underway, another soup manufacturer—Campbell's—was recalling a batch of contaminated chicken vegetable soup. Fearing a public outcry, the company tried to quietly recall the canned soup. Testing later indicated that a few cans contained botulinum toxin.[31]

Later that summer, a Congressional investigation of the failure of federal food inspections began. When asked how the FDA was able to protect consumers from food poisoning, FDA Commissioner Edwards admitted that the agency's 250 food inspectors were overextended and the agency was short of funds. "We are daily falling farther and farther behind in our routine inspection activities," he said. Generally the FDA inspected plants once every six years. To inspect plants more frequently and bring them back to normal levels, the FDA needed to hire 1,500 inspectors and have its inspection budget raised from $18 million to $85 million a year. The Bon Vivant investigation had swamped the already overburdened FDA, and the FDA canceled more than 2,000 plant inspections in 1971.[32]

The Bon Vivant case continued to make headlines that fall. A government inspection of the Bon Vivant plant in Newark, New Jersey, indicated that the plant neglected food safety. Two problems in particular stood out: the company regularly undercooked its canned products and kept incomplete records. A government inspector summed up the review by saying, "[N]one [of the firm's products] are considered . . . to be safe for consumption by man or animal." For example, non-soup products suffered from poor quality control, as investigators found that more than half of all spaghetti sauce cans were defective—swollen, leaking, or had imperfect seams.[33]

Records indicate that Bon Vivant knew they had canning problems before this incident. As early as 1959, the corporation was aware of sealing problems, which led to leaking cans and defective seams. In 1962, the American Can Company warned Bon Vivant that the length of time that the company cooked batches of soups and sauces was insufficient.[34]

Newspapers continued to run stories about botulism as other cases became known. For the third time in 1971, the FDA issued a warning about botulism in canned foods when they learned that a batch of Stokley-Van Camp canned green beans might have contained the deadly toxin. The consequences were less deadly than the Bon Vivant case. An 8-year-old boy and his father, who ate beans from a swollen can, developed no symptoms but when the Centers for Disease Control and Prevention (CDC) injected mice with liquid from the can, they died.[35]

---

31. Richard D. Lyons, "Campbell Was Quietly Recalling Contaminated Soup Before It Learned of Botulin," *The New York Times*, 24 August 1971.

32. House Subcommittee on Public Health and Environment of the Committee on Interstate and Foreign Commerce, *FDA Oversight—Food Inspection—1971*, 92nd. Cong., 1st. sess., 1971, pp. 4, 11, 13.

33. Boyce Rensberger, "Federal Inquiry Charges Bon Vivant Soup Factory Had Wide Sanitary Violations and Faulty Records," *The New York Times*, 16 September 1971.

34. Ibid.

35. "Warning Is Issued on Stokley Beans," *The New York Times*, 30 October 1971.

The National Canners Association (NCA), fearful of a public backlash against canned foods as well as lack of consumer confidence in their products, petitioned the FDA for more government regulation to prevent the spread of botulism and other food-borne illnesses. Although only four botulism-related deaths had been linked to commercially canned food since 1925, the NCA hoped that by taking such action they could circumvent any negative press. Dr. Ira I. Somers, the research director for the NCA, explained, "We just don't think the canning industry can tolerate any more bad publicity. From a statistical standpoint our record is good but we want to tighten every screw we can."[36] By pushing for additional regulations, the NCA hoped to prove to consumers that they were committed to food safety practices.

The FDA published the NCA proposal, which reflected many of the principles of HACCP, in the *Federal Register* in November 1971. In the proposal, all canners manufacturing low-acid canned foods had to register with the FDA, listing the type of low-acid canned food processed at the plant. In addition, food processors would have to explain their processes as well as the equipment they employed in the manufacturing of such food. Other requirements included coding for containers, recordkeeping requirements, and training for retort operators and can seam inspectors. If companies failed to follow the outlined requirements, the FDA could invoke emergency permit controls whereby the cannery could not distribute its products until the owner had met specific conditions listed in the permit. Industry had 60 days to respond.[37]

Not all food processors agreed with the steps taken by the NCA and some challenged the association's actions. The American Shrimp Canners Association, for example, asked the NCA to withdraw its proposal. In response to their request, the NCA's Executive Vice President J. E. Countryman explained that their idea, while not a panacea, was "a significant constructive step toward providing increased safeguards in the processing of canned foods," and he added, "There can be no question that the whole canning industry benefits if this proposal begins the renewal of the public's faith in the safety and integrity of canned foods. For this reason alone, NCA has no choice but to allow the proposal to go forward."[38]

A dark cloud continued to follow the food industry and the FDA in the spring of 1972. In April, the U.S. Government Accountability Office (GAO) issued a damning

---

36. "Canners Petition F.D.A. for Stiffer Regulation," *The New York Times*, 26 October 1971.

37. Edward Dunkelberger, "The Statutory Basis for the FDA's Food Safety Assurance Programs: From GMP, to Emergency Permit Control, to HACCP," *Food and Drug Law Journal* 50, no. 3 (1995): p. 365; *Federal Register*, 12 November 1971, General Subject Files (1938-1974), FDA Records, Record Group 88, National Archives II, College Park, MD [henceforth: RG 88, Archives II]; House Committee, FDA *Oversight*, pp. 453–456.

38. J. E. Countryman to H. R. Robinson, 23 December 1971, General Subject Files (1938-1974), FDA Records, RG 88, Archives II.

report about unsanitary conditions in food manufacturing plants. The GAO's study of 97 plants found that standards of cleanliness in food plants had deteriorated from 1969 to 1972. Even worse, the "FDA did not know how extensive these insanitary conditions were and therefore could not provide the assurance of consumer protection required by the law."[39] To alleviate such conditions, the FDA had to take action.

The agency, which had provided some funds for the first National Conference on Food Protection, had learned of HACCP at the meeting. Searching for a "better, more comprehensive food protection [program] for the consuming public," the FDA asked the Pillsbury Company to provide HACCP training for its supervisors and investigators. In September of 1972, 16 inspectors attended the first class offered in Gull Lake, Minnesota.[40] Pillsbury's three-week course included 11 days of lectures and 10 days of field work in Minnesota canning plants.[41] Upon completing the training, the inspectors returned to their posts, and later the following year the FDA established permanent low-acid canned food regulations. This represented the first regulatory use of HACCP in the food industry.

The implementation of HACCP regulations had a tremendous impact on canners of low-acid foods and their quality control programs. Joseph P. Hile, Executive Director of Regional Operations for the FDA, explained, "Some firms had no real quality control program until after FDA made its HACCP inspection and identified the crucial needs." Other food plants, Hile stated, "ceased operations as a result of these inspections until major equipment improvements are made and meaningful plant quality control procedures instituted."[42]

This was the case for Western Natural Growers, Inc., of Ulysses, Kansas. In the fall of 1973, an inspector reported that "Processing procedures, equipment and the firm's general knowledge of retort operations are so grossly inadequate that the production of low acid canned foods from this firm could represent a threat to consumer safety." The plant's retort operators had not attended any FDA- or NCA-approved schools and the plant failed to maintain any processing and production records with the exception of temperature recording charts. Following the inspection, the FDA Bureau of Foods requested that the plant cease operation until the agency believed that they understood and could comply with low-acid canned food regulations. On November 1, the plant was voluntarily shut down. Notes from a December inspection indicate that conditions

39. U.S. Government Accountability Office, *Dimensions of Insanitary Conditions in the Food Manufacturing Industry; Report to the Congress [on the] Food and Drug Administration, Department of Health, Education, and Welfare* (Washington, DC: GAO, 1972), p. 2.

40. Pillsbury Company, *Development of a Food Quality Assurance Program,* iii, 733; F. Leo Kauffman, "How FDA Uses HACCP," *Food Technology* 28, no. 9 (September 1974): p. 51.

41. Sperber, telephone conversation, 21 June 2006.

42. Joseph P. Hile, "HACCP—A New Approach to FDA Inspections," *Food Product Development* 8, no. 1 (February 1974): p. 52.

at Western Natural Growers, Inc., had substantially improved; retort operators, for instance, were scheduled to attend an FDA/NCA school at the University of Arkansas and were maintaining processing and production records.[43] Naturally, these equipment and operation changes resulted in some increased costs for the company.

Other smaller canners were not as fortunate as the Western Natural Growers. Some went out of business as a result of the adoption of these regulations. The rules also had significant impact upon the canned seafood industry, where many smaller plants closed.[44]

Aside from the impact on quality assurance in canneries, plant inspections also changed as a result of the FDA's use of the HACCP concept. Hile, who had at one time worked as an inspector for the agency, recalled that the inspections previously conducted by the FDA varied; some were brief while others were in-depth, and the length of inspections was determined at the local level. HACCP guidelines, by contrast, laid out the details by which all plants across the country would be inspected by the agency and, in general, HACCP inspections followed a nationwide, uniform model.[45]

Another key difference between traditional factory inspections and the HACCP inspections was the approach taken by the investigator. Customarily, canning plant inspections were limited in scope by the time the inspector spent at factory. HACCP inspections, by contrast, entailed the examination of records, thereby giving inspectors a broader picture of how the plant operated over the course of the year, not just the hours the investigator spent at the plant.[46]

FDA records indicate that canning safety programs improved over a period of four years from 1973 to 1977. During this time, FDA inspectors found fewer factories processing food that had either major or critical deviations from low-acid canned food regulations. Most companies complied with FDA requirements and approximately 10,000 people attended about 100 FDA-approved canning courses.[47]

In 1980 the FDA commissioned a study to determine the total costs of the low-acid canned food regulations on plants. Arthur D. Little, Inc., of Cambridge, Massachusetts, conducted the study, and more than 800 plants participated in the review. Arthur D. Little calculated that the industry spent $85 million to comply with the regulations, with an average cost of $102,000 per factory. Compared to

43. Establishment Inspection Endorsement, 18–19 September 1973, General Subject Files (1938–1974), FDA Records, RG 88, Archives II; Establishment Inspection Endorsement, 9 November 1973, General Subject Files (1938–1974), FDA Records, RG 88, Archives II; HACCP-EIR Evaluation Critical Factors-Potential Hazard to Public Health, 19 December 1973, General Subject Files (1938–1974), FDA Records, RG 88, Archives II.

44. U.S. Food and Drug Administration, *Total Industry Cost of Compliance with Low-Acid Canned Food Regulation Report to the Food and Drug Administration*, prepared by Arthur D. Little, Inc., 21 November 1980 (Rockville, MD: U.S. Food and Drug Administration, Office of Planning and Evaluation) p. IV-6.

45. Hile, "HACCP," p. 50.

46. Ibid.

47. "Canning Safety Practices Improve," *FDA Consumer* 13, no. 1 (February 1979): p. 23.

smaller facilities, larger plants tended to spend less on compliance. Overall, however, the burdens of compliance were insignificant, amounting to less than 1 percent of the low-acid canned food's shipment value.[48]

By 1974 Pillsbury had achieved its objective of implementing a new product safety standard at its facilities. The company's annual report boasted that the HACCP system was in use in the Pillsbury food plants and in its Burger King restaurants. The concept employed three principles: 1) conduct a hazard analysis, 2) determine critical control points, and 3) establish monitoring procedures.[49] Soon the concept would be employed in its more recent acquisitions, the Souverain wineries and Wilton plants.[50]

The attainment of Keith's goal represented a significant accomplishment for the company and a distinct turning point in the history of food safety. Instead of relying solely on end-product testing to ensure the safety of their products, Pillsbury had implemented a total safety system which affected not only their quality assurance programs but all phases of production. Bauman contrasted the old and new safety systems in an FDA training seminar. Under the old system, product development, testing, and marketing were quick and relatively easy; all of the Pillsbury offices conducted their work in relative isolation. By contrast, the total safety system integrated the research and development work to involve all employees. Where the company once viewed quality control as a final, isolated step, Pillsbury now viewed all stages of development as interrelated. Conducting a hazard analysis and identifying critical control points involved not only the quality control employees but individuals from all parts of the company—engineers, scientists, marketers, and attorneys. In addition, the company organized a number of offices to ensure product safety, such as the Product Systems Safety Office which verified that all new products had undergone an HACCP assessment. Aside from processing modifications, the culture of the company's middle management also changed.[51]

For his part, Bauman kept his word to the CEO of Pillsbury. Under his watch, the company did not have a major recall.[52] Pillsbury was pleased with their implementation of HACCP, saying, "There have been more than 130 food safety-related recalls of product from the marketplace from 1983 to 1991. None were Pillsbury products. HACCP works!"[53]

48. FDA, *Total Industry Cost of Compliance*, pp. I-4, I-5, I-6.

49. William H. Sperber, "HACCP and Transparency," *Food Control* 16 (2005): p. 506.

50. The Pillsbury Company, *Annual Report for the Year Ended 1974* (Minneapolis: The Pillsbury Company, 1974), p. 14.

51. Pillsbury Company, *Development of a Food Quality Assurance Program*, pp. 315–328, 514.

52. Bauman, "The Origin of the HACCP System," p. 69.

53. "A Dividend in Food Safety," p. 53.

Even with Pillsbury's successful implementation of an HACCP program in the early 1970s, interest in the system dwindled until the 1980s, when HACCP began to be revisited. In 1980, at the request of the National Marine Fisheries Service, the USDA, the FDA, the U.S. Army Natick Research and Development Center, and the National Research Council's Food and Nutrition Board Subcommittee on Microbiological Criteria formulated microbiological standards for food and drafted a plan of action for regulatory agencies to implement an HACCP system. Two members of the committee, James J. Jezeski and John H. Silliker, had previously served as panel members at the first National Conference on Food Protection where the idea had been unveiled. The committee's final report made mention of the historic event, noting that HACCP inspections provided a better approach than traditional inspections. As an example, the committee noted that the HACCP system helped the low-acid canned food industry control microbiological hazards. The group concluded that the HACCP concept was a valuable approach to securing the food system, and members urged regulatory agencies and the food industry to adopt the system.[54]

The Food and Safety Inspection Service (FSIS, a USDA agency) made a similar request of the Food and Nutrition Board of the National Research Council in 1983. They asked the board, which coincidentally included Norman D. Heidelbaugh, a veterinarian who had worked on the food systems at NASA, to evaluate the agency's meat and poultry inspection system. Upon completing its study, the board recommended that FSIS adopt HACCP principles in slaughterhouses and processing plants; in addition, the board encouraged the agency to train inspectors in the HACCP concept.[55] Together, these two reports rekindled widespread interest in HACCP in the U.S.

In response to the recommendations, the FSIS established the National Advisory Committee on Microbiological Criteria for Foods (NACMCF) in 1988. Cosponsored by the FDA, the CDC, the National Marine Fisheries Service (NMFS), and the Department of Defense Veterinary Service Activity, the committee provided an interagency look at microbiological standards for food. Bauman, who still worked at Pillsbury as vice president for science and regulatory affairs, served on the first NACMCF and remained on the committee until 1992. His colleague, William Sperber, joined in 1990.[56] In 1992, the committee recommended HACCP as an effective food protection system. A number of experts came out in favor of HACCP and another key report, "Cattle Inspection," encouraged the U.S. federal regulatory agencies to adopt HACCP-based systems.

54. National Research Council (U.S.) Food Protection Committee Subcommittee on Microbiological Criteria, *An Evaluation of the Role of Microbiological Criteria for Foods and Food Ingredients* (Washington, DC: National Academies Press, 1985), pp. 50–51, 308.

55. National Research Council Committee on the Scientific Basis of the Nation's Meat and Poultry Inspection Program, *Meat and Poultry Inspection: The Scientific Basis of the Nation's Program* (Washington, DC: National Academies Press, 1985), pp. 8, 134–136.

56. Sperber, telephone conversation, 21 June 2006.

Pressure to adopt HACCP systems also came from international governing bodies. In the summer of 1993, the Codex Alimentarius Commission (CAC), a joint program of the United Nation's World Health Organization and Food and Agriculture Organization, adopted *Guidelines for the Application of the Hazard Analysis Critical Control Point System*. But, in spite of the urging of experts, no change came about in the meat and poultry industry.[57]

The impetus came when hundreds of people fell sick and four children died after eating *E. coli*-contaminated hamburgers from a Jack in the Box fast food restaurant in the winter of 1993. The incident might have been avoided if the beef industry, food service establishments, or the USDA had implemented HACCP inspections. Eight years earlier, the National Research Council's Subcommittee on Microbiological Criteria had encouraged restaurants to adopt HACCP systems in their operations because research had overwhelmingly linked such establishments to most outbreaks of food-borne illness.[58]

The deaths of several small children from this incident led many to question the safety of the nation's meat. In a televised PBS *Frontline* interview, Carol Tucker Foreman, director of the Food Policy Institute at the Consumer Federation of America, explained how the deaths altered America's view of safety and the role of the USDA in preventing food crises. The *E. coli* outbreak indicated that the USDA inspections had not kept pace with America's increasing dependence on prepared and processed foods. "[I]t exposed the fact that the meat inspection system has not changed a bit since 1906. We were using methods that were essentially a century old in an industry that had changed radically," she said.[59] For instance, USDA inspectors continued to use the "sniff and poke" method to determine whether carcasses were safe for consumption, rather than rely on microbiological testing.

As Jack in the Box saw its sales slip, the fast food giant hired food scientist David M. Theno to prevent another disaster. Theno was a proponent of the HACCP system and he had previously used such methods to eliminate nearly all traces of *Salmonella* in the poultry at Foster Farms, the largest poultry producer in the western U.S. After reviewing Jack in the Box records, he laid out a plan to implement an HACCP program in the chain's restaurants. Jack in the Box was the first fast food chain to implement the system and require its suppliers to implement such plans. The standards were strict. For instance, meatpackers selling to Jack in the Box had to conduct microbiological tests on their beef every 15 minutes during processing, and managers were required to attend food safety courses. The implementation of the HACCP system increased beef costs by a mere penny per pound.[60]

57. Instead, FSIS conducted an HACCP study to determine how to implement HACCP procedures in the meat and poultry industry. "A Dividend in Food Safety," p. 54.

58. National Research Council, *An Evaluation of the Role of Microbiological Criteria*, pp. 314–315, 326.

59. PBS, *Frontline: Modern Meat, http://www.pbs.org/wgbh/pages/frontline/shows/meat/interviews/foreman.html* (accessed 3 August 2006).

60. Eric Schlosser, *Fast Food Nation: The Dark Side of the All-American Meal* (New York: Perennial, 2002), pp. 208–210.

The Jack in the Box incident proved that USDA inspection methods were antiquated, as inspectors could not necessarily see microbiological hazards. In response, FSIS issued a proposed a Pathogen Reduction/HACCP (PR/HACCP) rule in the *Federal Register* in February 1995. The proposal had three parts: the first section required the meat and poultry industry to develop and implement sanitation standard operating procedures (steps taken to prevent food contamination); second, the agency aimed to reduce *Salmonella* in meat and poultry plants and proposed daily microbiological testing at slaughterhouses and at facilities grinding meat; and third, the proposal would require all meat packing plants, slaughterhouses, and food processors handling meat and poultry to adopt HACCP plans.[61] Industry had 120 days to comment. The proposal pleased those who hoped to modernize the inspection process. "It may not be *Star Trek the Next Generation*, but it gets the USDA out of the horse and buggy era," said Foreman.[62]

When the rule was finalized in 1996, the press touted the achievement as a landmark in food safety. In a Saturday morning radio address, President Bill Clinton proclaimed that the new rules strengthened regulations, protecting families and those most vulnerable to pathogens—children. Recalling the Jack in the Box incident, he said, "Parents should know that when they serve a chicken dinner, they are not putting their children at risk."[63]

Experts, however, disagreed with Clinton's assessment. William Sperber, a food safety expert now with Cargill, believed that this rule, known more commonly as the "Megareg," and the additional HACCP regulations passed by the FDA in 1997 and 2001 did not follow the principles of HACCP as outlined by the NACMCF and later by the CAC. As an example, Sperber explained that sometimes meatpacking plants failed to meet the *Salmonella* performance standards as outlined by the USDA regulation. The rule gave the USDA the authority to close the plant if a packer failed the *Salmonella* monitoring plan three times in a row. FSIS rarely employed such drastic measures, however. Instead, the USDA waited to conduct another round of samples that consumed several months, and the meatpacking plants continued shipping meat until the results came back. This process sometimes took two years to complete. Very rarely did FSIS proceed to close a plant. The hesitancy with which the agency took action is not reflected in the HACCP principles outlined by the NACMCF. "Several hallmarks of a valid HACCP plan are that monitoring procedures and corrective actions, insofar as possible, should be taken in real time, and should be as continuous as possible," Sperber noted.[64] In other words, the USDA failed to implement a true HACCP

61. *Federal Register*, 25 July 1996.

62. Marian Burros, "Sweeping Changes Proposed on Meat Safety," *The New York Times*, 1 February 1995.

63. Todd S. Purdum, "Meat Inspections Facing Overhaul, First in 90 Years," *The New York Times*, 7 July 1996.

64. Sperber, "HACCP and Transparency," p. 507.

system because the agency allowed certain meatpackers to ship inferior and potentially unsafe meat, and because it relied on product testing for *Salmonella* rather than more practical process controls.

Likewise, food inspectors voiced concern about the rule, which, they argued, put the public at greater risk for food-borne illness. The regulation had taken away their authority to check contaminated meat. Instead of visually examining carcasses, inspectors had to ensure that companies followed the HACCP system they had drawn up. The acronym, which had once outlined the steps to ensure food safety—Hazard Analysis and Critical Control Points—was now dubbed "Have a Cup of Coffee and Pray" by those inspectors opposed to the Megareg.[65]

In spite of the criticism leveled against the PR/HACCP regulation, the Economic Research Service (ERS) of the USDA linked the implementation of the rule to a 20 percent reduction in food-borne illness and lower medical costs.[66] Similar trends were noted by another federal agency: the CDC cited HACCP as one factor contributing to the decrease in the number of *Salmonella* infections over a five-year period.[67]

After further review, the costs of developing and implementing an HACCP plan were higher than previously assumed. FSIS had estimated that the costs of PR/HACCP would be relatively insignificant, about 0.12 cents per pound. The ERS found that the actual overhead was higher than anticipated, 0.4 cents a pound for poultry and 1.2 cents for beef. This amounted to a 1.1 percent increase for plant operators. For cattle slaughterhouses the rates were higher, about 5.5 percent of all costs.[68] Although costs have been higher than expected, this has not hindered the adoption of HACCP systems in the U.S. and abroad.

The emergence of new pathogens in foods, as well as consumer demands for safe food, has driven the use of HACCP in other nations. In Australia, for instance, an *E. coli* outbreak sickened more than 100 people and killed one child, forcing changes in food safety requirements. The passage of the Australian Standard for Hygienic Production of Meat for Human Consumption required plants to implement HACCP systems in their meatpacking plants.[69] Scotland required its butchers to employ HACCP

65. Schlosser, *Fast Food Nation*, p. 215.

66. Michael Ollinger and Valerie Mueller, *Managing for Safer Food: The Economics of Sanitation and Process Controls in Meat and Poultry Plants*, Agricultural Economic Report No. AER817 (April 2003): pp. vi, 59, *http://www.ers.usda.gov/publications/aer817/* (accessed 3 August 2006).

67. Centers for Disease Control and Prevention, "Preliminary FoodNet Data on the Incidence of Foodborne Illness—Selected Sites, United States, 2001," *Morbidity and Mortality Weekly Report*, 19 April 2002, *http://www.cdc.gov/mmwr/preview/mmwrhtml/mm5115a3.htm* (accessed 3 August 2006).

68. Ollinger and Mueller, *Managing for Safer Food*, p. 58.

69. Spriggs and Isaac, *Food Safety and International Competitiveness*, pp. 111, 115.

procedures after 21 people died from eating tainted meat at a butcher shop.[70]

Throughout the past three decades, the widespread use of HACCP in the U.S. and abroad indicates the impact NASA has had on the food industry and food safety regulations. Originally implemented on a small scale for NASA's Gemini and Apollo astronauts, the HACCP system is essentially utilized worldwide by many multinational food conglomerates to ensure food safety for billions of consumers. In addition to the tremendous growth of the HACCP approach, many regulatory agencies require certain sectors of the industry to design and utilize systems in their processing plants that can be linked to the techniques first developed to comply with NASA food safety regulations.

Perhaps more important, HACCP has changed the manner in which food manufacturers and regulators look at the issue of food safety. Just 20 years ago many food manufacturers believed that the issues of food safety belong solely in the hands of quality control and quality assurance engineers in food processing plants. Today this is not the case. William Sperber explains this shift: "We now realize that some food safety practices can be applied at each step of the global food chain; from the growing of crops and the raising of animals, to the processing of these commodities, and through the production, distribution, and consumption of consumer food products."[71]

70. Food and Agriculture Organization of the United Nations (FAO), "Escherichia Coli 0157: H7 Outbreak in Scotland in 1996/97," FAO/WHO Global Forum of Food Safety Regulators, Marrakesh, Morocco, 28–30 January 2002, *http://www.fao.org/DOCREP/MEETING/004/X6925E.HTM* (accessed 3 August 2006).

71. William H. Sperber, "Opening Remarks," *Food Control* 14 (2003): p. 73.

CHAPTER 13

THE SOCIAL AND ECONOMIC IMPACT OF EARTH OBSERVING SATELLITES

Henry R. Hertzfeld and Ray A. Williamson

Research, development, and operational investments of the U.S. government and NASA for Earth observations have had a large impact on the economy of the U.S. and on the world. With the participation of other federal agencies such as the National Oceanic and Atmospheric Administration (NOAA), the U.S. Department of Agriculture (USDA), and the Department of Interior, United States Geological Survey (USGS), new industries have been created. New technologies have been advanced from the laboratory to the marketplace more quickly than if there had been no space program. Not only have jobs and income been created, but new ways of viewing the world now exist and other innovations that can be traced to U.S. government requirements and investments have improved the quality of life. Describing these technological advances is relatively easy; measuring their economic and social benefits is difficult. This paper reviews economic and other measures that have been applied to the benefits that Earth observation satellites have brought. The process of measuring benefits faces two major difficulties: 1) economists do not agree on the best approach to measurement because each issue and problem as well as each application focuses on a menu of different approaches and 2) no single measure exists that provides a comprehensive indicator of Earth observations impacts and benefits.

This essay will review developments in two areas: 1) a summary and evaluation of selected studies that have attempted to quantify and describe the various observable and measurable impacts of using Earth observation satellites and 2) a focus on the value of information used in the economy that can be attributed to satellite observations. This is different from just trying to measure historical impacts because it looks to the marginal value of additional information that can be derived from improved forecasts of many variables, which include weather and climate, river flow, soil moisture, snow cover, and land use. In general, information about these variables is derived from ground, aerial, and satellite sources and is combined in models that predict near- and long-term conditions. Since much is already known and published about these variables and the predictions are currently in use by many economic sectors, with each sector making large contributions to the gross domestic product (GDP), improved forecasts—though constituting only small-percentage gains—can add up to equal a large impact.

Over the past 30 years since the launching of the first civilian Earth observation satellite, ERTS-1 (later renamed Landsat-1), much has been written about the potential impacts of the data on various economic sectors. This paper will only briefly summarize these studies as a background for a more comprehensive analysis of the impacts of the value of improved forecasts in three areas of growing importance: natural hazards (mainly from weather and climate), electric energy production, and the management of freshwater resources.

Following the discussion of results, this paper also summarizes the assumptions, methodology, and analytical problems in developing consistent, accurate, and reliable results with the tools currently available.

THE SOCIOECONOMIC BENEFITS OF EARTH OBSERVATION SATELLITE SYSTEMS

Beginning with the polar-orbiting Television Infrared Observing Satellite (TIROS)-1 in 1960, NASA's first Earth observation satellites focused on providing a view from space of Earth's cloud patterns. These were extremely useful in both civilian weather forecasting and in planning military maneuvers. Later versions of these satellites added microwave instruments that could probe layers of the atmosphere to provide estimates of temperature, pressure, and humidity—measurements that are required as inputs to weather forecast models. Because many terrestrial weather measurements were available in North America, early TIROS satellite data made only modest contributions to forecasting accuracy within this continent. However, they made a sharp improvement in weather forecasting in other parts of the world, where weather data were only sparsely available. Steady improvements in the TIROS series (now called POES, the Polar-orbiting Operational Environmental Satellites, and operated by NOAA) have made these polar orbiting satellite measurements indispensable around the world. The U.S. allows all nations, indeed any entity with the appropriate receiver, to download data from these satellites without cost.

In the late 1960s NASA's geostationary Applications Technology Satellites (ATSs) demonstrated the utility of making frequent observations of weather conditions over North America for tracking severe storms, such as hurricanes and tornadoes. In the 1970s, NASA and NOAA agreed to build an upgraded version of ATS called GOES—Geostationary Operational Environmental Satellite. Built and launched by NASA and operated by NOAA, the POES and GOES systems now provide some 95 percent of the data that NOAA's National Weather Service (NWS) uses in its weather forecast models. Information provided by the data from these satellites is credited with saving many lives and reducing property damage from severe storms.[1] Every day, weather forecasts provide information used by

1. The National Climatic Data Center, *http://www.ncdc.noaa.gov/oa/climateresearch.html* (accessed 30 November 2006).

industry, government, and the average citizen to lower the risks they face from adverse weather conditions. Many of these uses can be quantified; others can only be described qualitatively. Later sections discuss some of these results.

In the early 1970s NASA created the Landsat program to gather multispectral data about Earth's surface features.[2] The first of this series of satellites was launched in July 1972. The satellite proved a technical success, returning images of large areas of Earth's surface in four different color bands that could be probed for information about geology, biology, snow and ice cover, and human settlement patterns. Even before the launch, NASA had started an effort to involve other agencies in experimenting with the data for assisting in their applications. It also encouraged countries around the world to establish data receiving stations to collect data over their territories. Although the data showed immediate potential utility for use in resource management, mineral prospecting, and agriculture, NASA found it difficult move the system into operational status. The software and computer hardware were cumbersome to use and, at first, analysis was largely carried out using paper imagery.

Even today, with the benefit of extremely powerful Geographic Information System (GIS) and sophisticated imaging processing software, as well as the Global Positioning System (GPS) to place landscape details into a geographic reference system, incorporating Earth observations into routine agency resource management operations continues to be difficult for several reasons. First, potential users do not feel confident that research sensors will be made operational and data from them will be available in the future. Second, government budgets often do not include enough funds for processing and operations once the hardware has been built, flown, and tested. Third, there are cultural and communications barriers between space researchers and data users that are often difficult to overcome.

Nevertheless, Landsat data have been used widely for such scientific studies as the examination of the state of the world's forests and estimates of the amount of carbon sequestered by them. They have also contributed to a better understanding of the rates of deforestation and reforestation around the world. For agencies that have incorporated Landsat data into their routine operations, the data provide a broad, synoptic view of the landscape and an enhanced ability to manage the natural and cultural resources of the lands they manage.

NASA's Landsat effort was sufficiently successful for it to obtain the financial backing from Congress to build three satellites that carried similar, 80-meter-resolution MultiSpectral Scanners (MSSs) and to extend the Landsat mission to Landsats 4 and 5, which carried an enhanced 30-meter sensor called the Thematic Mapper (TM). These were launched in 1982 and 1984, respectively; the 25-year-old Landsat 5 still returns imagery from orbit, though at a reduced capability. The

2. Pamela E. Mack, *Viewing the Earth: The Social Construction of the Landsat Satellite System* (Cambridge, MA: MIT Press, 1990).

TM sensor carries seven spectral bands, six of which operate between the blue and near-infrared parts of the spectrum. As detailed below, Landsat 6 failed and Landsat 7 is now in orbit and supplying data.

LACIE and NASA's Applications Program

The 1974 to 1978 Large Area Crop Inventory Experiment (LACIE) and its 1978 to 1983 successor program, Agriculture and Resources Inventory Surveys Through Remote Sensing (AgRISTARS) were designed to develop uses for the Landsat data to measure crop production, first in the U.S. and then in other parts of the world. They were a joint program of NASA, NOAA, and the USDA. (AgRISTARS was primarily a USDA program to make operational the results of LACIE, but budget pressures and changing priorities of the Reagan administration greatly reduced the spending for AgRISTARS.)[3]

In one interesting experiment in the LACIE program, Landsat and other data were used to estimate the Soviet Union's wheat crop. Crops yields are heavily dependent on weather conditions. The lack of knowledge and good forecasts of the Soviet crop led to significant increases in the price of wheat in 1972. As described by Dr. Forrest Hall, senior research scientist at NASA's Goddard Space Flight Center,

> In 1972, the Soviet Union experienced a major wheat crop failure. The U.S. had sold large quantities of wheat to the Soviet Union at low prices before the crop failure was announced. The failure drove up wheat prices, and the U.S. ended up buying wheat back from the Soviets at a loss . . . . When we started selling it, we were selling it for $1.92 a bushel, and we ended up buying some of it back at $4 or so a bushel . . . . That really made us realize that our conventional (crop-estimation) systems at that point were not very accurate.

In order to add an element of stability to the world's agricultural markets, NASA and the USDA began a program to see if Landsat data could be used to estimate global crop production.[4] The resulting improvements in wheat crop forecasting from LACIE were documented to be within 6 percent of the final Soviet figures, which were released more than six months later than the LACIE estimates.[5]

---

3. "United States Civilian Space Programs, Volume II, Applications Satellites," Report Prepared for the Subcommittee on Space Science and Applications of the Committee on Science and Technology, U.S. House of Representatives, 98th Congress, May 1983, pp. 231–237.

4. *http://science.nasa.gov/headlines/y2000/ast04dec_1.htm* (accessed 24 August 2007).

5. R. B. MacDonald and F. G. Hall, "Global Crop Forecasting, *Science* (New Series) 208, no. 4445 (16 May 1980): p. 670. In spite of the overall accuracy, it is noted by others that the value of the accuracy was somewhat diminished by offsetting errors in area and yield estimates (U.S. General Accounting Office, "Crop Forecasting by Satellite: Progress and Problems:" PSAD-78-52, 7 April 1978, as quoted in "United States Civilian Space Programs, Volume II, Applications Satellites," p. 233.

An economic methodology was constructed to measure the value of these information improvements in forecasting, which was based on the premise that more accurate observations affect the commodity-price distribution. By reducing the variation in prices of highly volatile commodities, consumers receive direct economic benefits (a Marshallian surplus) in the form of more stable prices.[6] Over time and with improved satellite resolution and additional data, the same type of analyses have more than adequately demonstrated the value of better information in forecasting in many other areas, as described in other sections of this paper. Although the Landsat satellites proved technically successful, NASA did not want to operate Landsat indefinitely and the Office of Management and Budget was not keen to approve continued funding for the system.

In the late 1970s, influential members of the Carter administration also felt that the private sector should assume operation of the Landsat system and provide the data commercially. In order to move the Landsat system to a private operator, the administration crafted a plan that would first transfer operational control over Landsats 4 and 5 to NOAA and then later to the private sector. Several years later, the Reagan administration pushed hard to move the system into private hands as soon as possible. Congress supported this decision by passing the Land-Remote Sensing Commercialization Act of 1984 (P.L. 98-365), which also included important provisions allowing the commercial development and operations of land remote sensing satellites.

As soon as NOAA took over Landsat operation, it raised the price of data from near zero to several thousand dollars per scene, which caused the volume of data sales to plummet. RCA, Inc. and Lockheed Martin Corp. formed EOSAT, Inc. to operate the Landsat system and to increase the small market for the data. However, EOSAT was unable to build a commercial market to a size that would support fully private development of the Landsat satellites. For one thing, the federal government remained by far the largest user of Landsat data. In addition, the price of data continued to be prohibitive for many users. Further, Landsat 6 incurred a complete launch failure and did not achieve orbit.

Congress in 1992 decided that Landsat did indeed provide sufficient benefit to the country to be continued, and drafted legislation to ease the restrictions on private operation of land remote sensing that were in the 1984 law and to bring future Landsat satellites back under government operation.[7] NASA was instructed to build and launch Landsat 7, which is still orbiting[8] and supplying data to users

---

6. David F. Bradford and Harry H. Kelejian, "The Value of Information for Crop Forecasting with Bayesian Speculators: Theory and Empirical Results," *The Bell Journal of Economics* 9 (Spring 1978): pp. 123–144.

7. The Land Remote-Sensing Policy Act of 1992 (P.L. 102-555).

8. Landsat 7, however, suffers from a failure in its Land-scan corrector that makes part of the data in each scene unusable.

around the world. USGS now operates Landsats 5 and 7. NASA will build a successor to this satellite, which will be operated by the USGS. Long-term continuity of this capability is still in doubt, however.[9]

This abbreviated summary of the trials and tribulations of the Landsat system illustrates that even though a satellite system may prove technically successful for the economy, finding the will to move it into operational use can be fraught with difficulties. This state of affairs has come about because Earth observation data have both public and private uses. Governments have invested in expensive space systems because the information obtained fulfills various mission purposes, ranging from national security to planning and monitoring natural resources. At the same time, commercial and private for-profit uses of the very same data provide opportunities for economic growth and benefits.

These dual capabilities have fueled many policy debates over the years that have led to some very odd compromises. In the U.S., for example, Congress has declared that satellite (and other) weather information is a public good,[10] while at the same time leaving all other remote sensing data products undefined and therefore sometimes treated as public goods and sometimes as private goods.[11] To further confuse the policy debate, civil space activities fall under two other legislative mandates: 1) they are "for the benefit of mankind"[12] and 2) the information obtained from space should be openly and widely disseminated.[13] In addition, government policy also calls for government-collected information to be disseminated with user charges set no higher than the costs of dissemination.[14] Also, unlike many other nations, the U.S. government also prohibits the copyright of government publications.

Therefore, all remote sensing data are mixed public-private goods, making market pricing and measuring benefits on an economic basis extremely difficult. Clearly, private sector value-added firms (those taking and/or purchasing government information from satellites and processing the images for commercial

---

9. A forthcoming report by the Office of Science and Technology Policy will reportedly recommend the development by the government of long-term operation of a Landsat-type system to ensure moderate resolution data continuity for the long term.

10. A public good is one that is non-excludable (nobody can be denied the use of the good) and non-rival (one person's consumption of it does not affect another's). Public goods often reflect the intervention of governments into the marketplace where market systems have failed to provide either competition and/or service to everyone for necessary services at an equitable price. Public goods are also collective goods where a profit-motivated firm would not provide the service (e.g., national defense).

11. We need a citation for the mid-1980's legislation.

12. "National Aeronautics and Space Act of 1958," Public Law 85-568, 72 Stat., 426. Signed by the President on 29 July 1958, §102(a).

13. Ibid.

14. Executive Office of the President, Office of Management and Budget, Circular No. A-130, Revised, Management of Federal Information Resources, Washington, DC: 28 November 2000, §8(a)7(c).

purposes) can provide useful, measurable benefits. However, the prices do not reflect the total costs to society of producing the information since the space component is often heavily subsidized by government programs. As policy has evolved over time, the private sector is now developing its own satellites, with data products sold both to governments and to the private sector. However, the legislative mandates still allow for a significant amount of competition between government-subsidized information and for-profit systems, which makes a true economic and benefit analysis of Earth observation data very complex.

Although some of the technology in the U.S. commercial satellites derives from systems developed for classified satellites, much of the hardware and the associated supporting image processing software sprang from NASA's efforts to make Landsat data useful for operational applications. Further, government investments in GPS and GIS software have made Landsat data truly useful for a wide variety of scientific and applied purposes. These ancillary government inputs can also be considered as benefits, though extremely difficulty to quantify.

Scientific research on climate has also proved highly beneficial. By the early 1980s, NASA began to focus on developing a view of Earth as an integrated, interdependent system. NASA scientists reasoned that global satellite observations would be essential in developing global climate models. In 1987, at a time when NASA was reappraising its role in space research and development following the loss of Space Shuttle *Challenger*, NASA published a report entitled *Leadership and America's Future in Space*.[15] Among other things, this report included a proposal for a "Mission to Planet Earth" that would study and characterize Earth on a global scale. This report was rescoped, rebaselined, and reshaped (NASA's terminology) in steps over several years to fit a much smaller $7-billion budget profile in which the original large, polar-orbiting platforms were replaced by several smaller, less capable versions, but which have been highly successful in producing excellent scientific data.

Some of the instruments from the Earth Observing System (EOS) satellites could provide the basis for operational instruments operated by another federal agency. Nevertheless, NASA and these agencies will still face the difficulty of making the transition from research to operations unless the relevant agencies (NOAA and USGS) are able to take over the development and operation of their own satellites or find ways to involve the private sector in supplying such information commercially.

The issue becomes clear in an examination of the longevity of the Tropical Rainfall Measuring Mission (TRMM), launched in 1997. TRMM is a joint Japanese-U.S. mission to study the effects of tropical rainfall, which orbits in an inclined orbit between +/- 35 degrees latitude. Data from the TRMM Microwave Imager (TMI)

15. Ride, Sally K., "NASA LEADERSHIP and America's Future in Space," A Report to the Administrator, August 1987.

and the Precipitation Radar (PR) instruments on this satellite have not only led to enhanced understanding of the role of tropical rainfall in Earth's system but have also proven extremely capable of providing improved estimates of rainfall amounts in tropical cyclones. Data from the PR have also led to much-improved estimates of tropical cyclone path and intensity.[16]

However, by 2004, although the satellite was still in excellent operating condition (well beyond its planned scientific mission), for budget reasons NASA decided to stop collecting data from the satellite and to deorbit it. That move would have saved the agency about $4 million per year (though much of that savings would be consumed by deorbit maneuvers over several years). The outcries of dismay from scientists and from weather forecasters in Japan and in the U.S., and a National Research Council Report on TRMM, caused NASA to rethink its approach. Some weather forecasters, especially in Japan and Europe, were already using the data for operational purposes in measuring rainfall and in tropical cyclone warnings.

Therefore, in May 2005 NASA reversed its earlier decision and extended the operation of TRMM either until it fails or its fuel runs out sometime in 2010 or 2011. TRMM is still operating and in 2005 contributed to improved observations of Hurricanes Katrina and Rita as well as other tropical cyclones. Although its benefits to society have not been quantified, the National Research Council report enumerated many of its contributions to weather and climate prediction models.[17] These successes make it clear that a satellite system, if extended to the globe, would provide continuing and improved data which would result in benefits to all nations.

The following sections summarize results of several studies carried out by the Space Policy Institute on the benefits of EOS systems.

NATURAL HAZARDS, MITIGATION, AND RESPONSE

Some of the most familiar and recognizable Earth observation images in the U.S. popular mind are the dramatic pictures of major hurricanes headed for the U.S. coast. The pictures, captured by the NOAA GOES satellites, serve to illustrate the danger these enormous storms pose for the affected coastline and assist in urging citizens reluctant to evacuate the area that they should leave. To the extent that the satellite systems that produce images of these and other weather-related natural disasters save lives and allow affected communities to prepare their homes and businesses to withstand the storms' onslaughts, they bring a clear benefit to the U.S.

In general, more accurate prediction of severe weather can help to reduce substantially the economic and social costs of weather-related disasters. Better

16. National Research Council, *The Future of the Tropical Rainfall Measuring Mission: Interim Report* (Washington, DC: National Academies Press, 2005).

17. 4.3.1.2 Operational Assimilation of TMI and PR Data for Weather and Climate 21 Prediction Models, p. 61.

information induces governments, businesses, and individuals to invest in loss-reduction activities; it can also reduce economic costs from unnecessary loss-reduction activities that derive from uncertainty about adverse weather (e.g., evacuations during hurricanes). This section summarizes what is known about these types of benefits as applied to weather-related natural hazards such as hurricanes.

A few economists have attempted to quantify the economic impacts of severe storms in specific industries. For example, research by Timothy Considine et al. on the costs of evacuating energy production platforms in the Gulf of Mexico estimated that achieving a 50 percent reduction in hurricane and tropical storm forecast error would save producers about $18 million annually. According to his analysis, a perfect forecast could lead to savings between $225 million and $275 million, illustrating the nonlinear nature of forecast value in this case. However, for energy producers in the Gulf, averting the risk of losing lives is generally far more important than saving short-run operations costs. The costs of evacuation from a platform are much lower than the perceived costs of loss of life. If "losses are perceived to be very substantial, producers will always take preventive action regardless of evacuation costs."[18]

Preparing for and Responding to Hurricanes

Satellite data from several instruments can contribute to the delivery of more accurate, timely hurricane forecasts (table 13.1). Satellite data also have a role in mitigating the damaging effects of hurricanes and in responding to and recovering from hurricane damage (table 13.2). For example, digital elevation models, coupled with land cover information and estimates of storm force, allow modelers to estimate the force and extent of storm surge along the coast.[19]

TABLE 13.1–SATELLITE CONTRIBUTIONS TO MORE ACCURATE, TIMELY HURRICANE FORECASTS

Satellite Instrument	Measurement	Utility
TMI-TRMM Microwave Instrument	Precipitation rate and distribution	Rain estimates, flood warnings
Precipitation Radar	Storm track, rain rate in storm	Increased accuracy of evacuation warnings
QuikSCAT	Surface winds speed and direction	Storm force, track predictions
GOES, POES	Imagery, atmospheric soundings	Storm track, rain estimates, force

18. T. J. Considine, C. Jablonowski, B. Posner, and C. H. Bishop, "The Value of Hurricane Forecasts to Oil and Gas Producers in the Gulf of Mexico," *Journal of Applied Meteorology* 43 (2003): pp. 1270–1281.

19. LIDAR-derived digital elevation models (from aircraft instruments) of Broward County, FL, have allowed the county to avoid significant evacuation costs during severe hurricanes by reducing the required evacuation area. Ray A. Williamson, Henry R. Hertzfeld, Joseph Cordes, and John M. Logsdon, "The Socioeconomic Benefits of Earth Science and Applications Research: Reducing the Risks and Costs of Natural Disasters in the USA," *Space Policy* 18 (2002): pp. 57–65.

TABLE 13.2–SATELLITE CONTRIBUTIONS TO MITIGATION, RESPONSE, AND RECOVERY OF HURRICANE DAMAGE

Satellite Instrument	Measurement	Utility
Landsat Thematic Mapper, SPOT	Land cover, flooding extent	Flood modeling, recovery planning
QuickBird Ikonos Orbview	Damage type, extent; detailed digital elevation model	Insurance estimates, detailed cleanup planning
Shuttle Radar Topographic Mapper	Digital elevation model	Storm, flood modeling
Radarsat	Flooding	Flood extent, disaster designation

The most destructive tropical cyclone in recent years to strike the U.S. was Hurricane Katrina, which made landfall southeast of New Orleans in late August 2005 and quickly moved northeast, spreading death and destruction across southern Louisiana and western Mississippi. Much of the storm damage was directly related to the massive amount of rain. The heavy rainfall and storm surge destroyed parts of New Orleans levees, flooding the city and displacing much of the city's population. Less than a month later, this storm was followed by Hurricane Rita, which made landfall near the Texas-Louisiana border. The storm surge it caused led to extensive damage along the Louisiana and southeastern Texas coasts. Both storms were among the most well-forecast storms in U.S. history because of the early concern they raised among forecasters and the public. Despite highly accurate forecasts for both storms, they caused at least 2,000 deaths directly or indirectly, massive short- and long-term population displacement, and thousands of destroyed homes and businesses.

Earth observation satellites had a major role in tracking the storms and in response, recovery, and rebuilding efforts immediately afterwards. Information derived from NOAA's GOES and POES satellites was used to estimate storm intensity with considerable accuracy. NASA contributed data from the TRMM satellite, which had led to improved hurricane path and rainfall predictions. However, even though the information was highly accurate, response at all levels of government was slow and halting, which led to a much higher death rate—demonstrating that better information does not always lead to better decision making in times of crisis.

During response and recovery after the storm, NOAA, NASA, and private companies contributed time and considerable effort to acquiring both aerial and satellite imagery of the damaged areas. This helped citizens, some of whom were several hundreds of miles from their homes, view the damage to their neighborhoods and houses and decide how to respond appropriately. In addition, the International Disaster Charter was activated to assist.[20] The Charter is an international consortium

20. The formal name is Charter on Cooperation To Achieve the Coordinated Use of Space Facilities in the Event of Natural or Technological Disasters; see *http://www.disasterscharter.org/main_e.html* (accessed 1 November 2006).

of space-capable nations, including the U.S., that have pledged to provide imagery to countries afflicted by major disasters.

Potential International Benefits of Improved Weather and Climate Information

The international benefits of improved weather and climate information involve virtually the same list that we would put together for the U.S., with the important difference that for developing countries, especially, these improvements could have even greater primary economic and social benefits. As one example, table 13.3 summarizes the immediate economic damage and recorded deaths for the 1998 Hurricane Mitch, which swept across Central America in November 1998. However, these figures do not reveal the costs associated with damaged agricultural production or the long-term displacement of residents.

TABLE 13.3–ESTIMATED DEATHS AND DAMAGE COSTS FROM HURRICANE MITCH

Country	Deaths	Damage Costs
Honduras	6,500	$4 billion
Nicaragua	3,800	$1 billion
El Salvador	239	Not available
Guatemala	256	Not available
Mexico	9	Not available
Other	14	Not available

Source: http://lwf.ncdc.noaa.gov/oa/climate/severeweather/extremes.html (accessed 24 August 2007).

El Niño and the Southern Oscillation (ENSO)

Recent development of forecast models of the short-term climate variation of the El Niño and La Niña cycle has proved a significant success. These models could not have been developed without the data from global satellite observations. The temperature and precipitation changes caused by the ENSO phenomenon have led both to significant losses and benefits, depending on which region of the world is studied. Among other things, this interannual climate swing is responsible for a significant level of uncertainty in the prediction of long-term weather patterns. Hence, U.S. and global climate research has focused considerable attention on not only a deeper understanding of the biophysical mechanisms behind ENSO, but also on the ability to predict ENSO effects. Scientists have also focused on the economic and social effects of ENSO in order to reduce the level of uncertainty and risk faced by agriculture, fisheries, and the general public throughout the world.

The climate research community has made significant progress in the past decade in understanding the physical relationships between the warming or cooling of the ocean along the western coast of South America and changes in weather patterns elsewhere in the world. This understanding, coupled with data from several satellites, has led to an improved ability to predict the return of El Niño, which can then be used to alert weather-sensitive industries around the world that they may face increased risk of experiencing abnormal weather phenomena in their regions.[21]

Learning to predict the onset of El Niño and its sister phenomenon La Niña with sufficient accuracy, can have a major impact on the U.S. economy. Table 13.4 summarizes one analyst's estimates[22] of the socioeconomic gains and losses from the El Niño of 1997–1998. Note that, contrary to popular belief, in this case the gains vastly outweigh the losses for North America. Similar charts for other regions for the same incident would probably show a different picture, with greater losses than gains. Whether gains or losses are at stake, however, better knowledge of the timing and strength of the ENSO cycle would assist governmental policy makers and private sector investors to capitalize on the benefits of this climate cycle and reduce the risk of loss.

TABLE 13.4–1997–1998 ENSO LOSSES AND BENEFITS IN THE UNITED STATES

Source	**Losses**
Property losses	$2.8 billion (insured losses were $1.7 billion)
Federal government relief costs	$410 million
State costs	$125 million
Agricultural losses	$650–$700 million
Lost sales in housing and snow-related equipment	$60–$80 million
Losses in the tourist industry	$180–$200 million
Source	**Savings/Benefits**
Reduced heating costs	$6.7 billion
Increased sales of merchandise, homes, and other goods	$5.6 billion
Reduction in costs for snow/ice removal from roads	$350–$400 million
Reduction in normal losses because of the lack of snowmelt flood and Atlantic hurricanes	$6.9 billion
Income from increased construction and related employment	$450–$500 million
Reduced costs to airline and trucking industry	$160–$175 million
Cost/Benefit Summary	
Costs	**Benefits**
Human lives lost: 189	Human lives not lost: 850
Economic loss: $4.2 to $4.5 billion	Economic benefit: $19.6 to $19.9 billion

21. Richard A. Kerr, "Signs of Success in Forecasting El Niño," *Science*, 297 (26 July 2002): pp. 497–498.

22. Stanley A. Changnon, *El Nino 1997–1998: The Climate Event of the Century* (New York: Oxford University Press, 2000), pp. 144, 149, 152

THE BENEFITS OF WEATHER AND CLIMATE INFORMATION FOR THE ELECTRIC ENERGY INDUSTRY

Weather is an important component in the analysis and operational components of both the demand and supply of electricity. It strongly affects electricity demand via heating and cooling needs in businesses and residences. Accurate weather forecasts are extremely valuable for accurate electricity demand forecasts, which are used to determine the load carried by the electric infrastructure, conduct transactions on the electricity market, and manage electricity flows across the power grid. On the supply side of the electric power industry, weather data have applications in both electricity transport and generation. High temperatures and severe weather events (such as hurricanes, lightning, and ice storms) can damage transmission and distribution systems and interrupt electricity supply. Weather also affects the capacity to generate electricity from fossil fuels and renewable sources; the latter, which represents an ever-growing portion of the electricity supply, is particularly sensitive to weather conditions. NOAA's operational environmental satellites, augmented by land- and sea-based systems, gather meteorological data that lead to valuable information inputs on both the demand and supply sides of electricity production. Innovation in space meteorological technology, as well as more extensive understanding and utilization of current capabilities, will provide the electric power industry with more sophisticated weather information of even greater economic value than that available today.

Accurate weather forecasts are crucial in maintaining the reliability of the supply of electricity to users through management of the power grid (especially, close monitoring of overload conditions) and the prediction of severe weather. Partly as the result of the increasing deregulation of the electricity industry, electric utilities have installed more efficient transmission technologies to compete effectively. Yet these measures have often also introduced new vulnerabilities to the grid from weather by making it more sensitive. In the U.S. the aging of the hardware and equipment in the electric power grid has also led to reduced reliability. These changes have increased the need for the industry to make more efficient use of weather and climate data than ever before.

Temperature is the most important weather factor influencing electricity demand. People use more energy on hot days to cool indoor environments and on cold days to warm them. Heating degree days (HDDs) and cooling degree days (CDDs) are commonly used measures of energy demand. They indicate the variation of daily temperatures from a temperature that would require no external energy inputs for heating or cooling.[23] Differences in temperature above or below 65°F determine the need for heating and cooling, the largest component of electricity use.

23. Daily temperature is calculated as the average between the daily minimum and maximum, and the HDD/CDD is the absolute value of the difference between this and 65°F. Energy Information Administration, "Short-Term Energy Outlook," July 2007, *http://www.eia.doe.gov/emeu/steo/pub/a2tab.html* (accessed 24 August 2007).

Decreasing forecast errors can reduce the costs of unnecessarily buying and selling electricity on the open market. Error grows more costly as the time between purchase and consumption diminishes, which becomes apparent in high spot-market prices. Commercial weather information vendors such as Itron, Inc., Weather Bank, Inc., and Weather Services International specialize in providing load forecasts and forecasting software to energy utilities and independent system operators (ISOs). They obtain raw data from the National Weather Service and other data providers and then turn this information into forecasts tailored to the specific needs of each customer in the electric power industry. The most common electric power applications are for the very-short-term (minutes to hours ahead) to the short-term (1 to 10 days ahead).[24]

The costs to utilities of an inaccurate forecast can be very high, especially for day-ahead or hour-ahead forecasts. Hourly changes in weather can result in over- and underestimating demand and costly decisions regarding the operation of electricity generation units. Improved forecasts from the use of satellite weather information have resulted in direct economic payoffs to the electric utility industry. Electric load forecasts are valuable to utilities and ISOs for allocating power over different parts of the electric grid and for optimizing purchases on the spot and day-ahead markets.

Electric utilities derive the greatest economic benefit from weather forecasts that are accurate over the 2- to 4-day time frame. Improved 7- to 10-day weather forecasts would also provide some economic benefit for utilities. The companies use monthly and seasonal weather forecasts for scheduling maintenance and for meeting U.S. Environmental Protection Agency (EPA)-set yearly emission allotments. Long-term forecasts assist in planning for new power generation facilities.

The many studies of the value of better terrestrial weather forecasts all indicate that benefits to the electric utility industry are significant, often reaching millions of dollars. However, the studies have been made in an uncoordinated way—each one measuring the benefits at one point in time for one region and often for one particular application. As enumerated below, economic benefits from better weather information are measured in many ways with a variety of methodologies. Each methodology may be particularly relevant to specific case studies and situations. Yet measures derived from different methodologies cannot easily be added together, making it impossible at present to calculate a single, aggregate measure of the economic value of improving weather forecasts and other information. However, more accurate forecasts coupled with intelligent and timely use of those forecasts by the industry is already yielding benefits in the tens of millions of dollars annually. As weather forecasts improve with new satellite-based information, and improved data assimilation into forecast models, these benefits will increase.

---

24. Frank A. Monforte, vice president of forecasting, Itron, Inc., personal communication. See Henry C. Hertzfeld and Ray A. Williamson, "Weather, Climate Satellite Data and Socioeconomic Value in the Electric Utility Industry," A Report to NOAA (Washington, DC: Space Policy Institute, George Washington University, December 2004), p.13.

The social benefits of supplying better weather forecasts to the public and government agencies are equally robust but even more difficult to measure accurately. Nevertheless, because the entire modern infrastructure depends in some way on the availability of electrical power, it is clear that when the U.S. electrical power grid operates reliably, the general public and public services benefit substantially from reduced uncertainty in the supply of electricity.

Satellite information can also provide significant benefits in planning, locating, and operating electric production dependent on renewable sources of energy such as wind, sunlight, and water. As of 2007, at least 17 states have mandated the use of renewable energy sources in generating electrical power; other states are rapidly adding similar regulatory requirements. Some have followed the federal example and instituted tax incentives to assist the development of this component of the industry. Satellite-based remote sensing can aid in realizing the potential of exploiting renewable energy resources by aiding in the optimal siting of generating facilities as well as in the operational decisions of generating facilities and electric power grid management. State and federal governments may wish to consider increased investment in the research and development of environmental satellites to support sound and sustainable economic and environmental policies, both in the energy and space industries. The increasing global demand for energy resources makes this particular use of satellites very significant and immediately practical. There are clear economic and social benefits to the use of satellite data for locating sites and for routine operations of renewable-source generating stations, yet the magnitude of the economic benefits that satellite data can provide have not yet been quantified.

SATELLITE INFORMATION IN QUANTIFYING AND MANAGING WATER RESOURCES

Clean, fresh water, so crucial in supporting life and national economies, is becoming increasingly difficult to obtain, especially in arid and semi-arid climates. Freshwater, with less than 0.5 parts per thousand dissolved salts, may be found in lakes, rivers, and bodies of groundwater. Only 3 percent of water on Earth is freshwater, and more than two-thirds of this is frozen in glaciers and ice caps.

In the near future, ensuring adequate supplies of freshwater to support all the competitive water needs of the world will likely become one of the most contentious issues facing global society. Improving water resource management (supply and distribution) has clearly become one of the most important challenges of modern life. As noted in a recent report, "Earth's water resources can no longer be taken for granted. Water is an issue that cannot be ignored, if we want the world to sail safely through the century ahead."[25]

---

25. ITT Industries, *ITT Guidebook to Global Water Issues, http://www.itt.com/waterbook* (accessed September 2006).

Information derived from Earth observation satellites could improve knowledge of the supply of freshwater and assist in managing its distribution to water users. However, so far, few researchers have attempted to assess the value of space systems in addressing the challenges of improved water management.

As noted in the preceding section, electricity generation and transmission derives significant benefits from satellite data. Socioeconomic analysis of this industry is aided by the fact that electricity is a commodity and the prices and markets that exist are very important in the allocation of electric power among users, despite the distortions created by the significant amount of government regulation that is also involved.

Electricity and water are both treated as public utilities in the U.S., but that is where the direct comparisons end. Electricity is a uniform commodity, being transmitted to users by wires from power plants. Water stems from many sources, is transferred to users by different means, and cannot efficiently be transported over long distances. Table 13.5 summarizes some of these differences.

TABLE 13.5–ELECTRICITY AND WATER COMPARED

	Electricity	**Water Resources**
Sensitivity to price changes	Yes	No
Distribution system	National/regional	Regional/local
Sources	Coal, hydroelectric, nuclear, oil, alternatives	Water cycle
Originates (distribution)	Power plants	Rivers; ground
Reusability/recovery	None	In some applications
Social/cultural approaches	Commodity, becoming a necessity	Basic need, but some uses are "commodities"
Markets	Sophisticated trading: spot, day-ahead, long-term markets	No large-scale, organized economic markets
Legal impediments	Regulatory, but relatively consistent across the nation	Many different local systems; different treatment of ground and river water
Measures of value	Sales, usage, cost savings, hedging on prices	Gross usage (not $), cost savings, scarcity
Direct benefits	Industry (profit incentive) Consumers (price effects)	Agriculture, hydroelectric, nuclear plant cooling
Indirect benefits	Quality of life	Recreation
Age of industry	Approximately 100 years	Ancient

Additional theoretical problems create even more uncertainties and issues in trying to grasp the aggregate value of satellite information for uses of water resources. Overall, valuing information is quite difficult because information only has value if it is used or expected to be used. Thus, measuring the benefits of access to information depends on the ability to be able to measure the expected use of the information rather than the information itself. When the uses are diffused among different users and markets, the measurement problem is greater. Further, when no true price-responsive markets exist for the commodity, the problem is many times harder to evaluate. Finally, when the supply of water and the raw information are not precise or even affect the user in a direct buyer/seller market, yet another difficult variable is introduced

Therefore, we face a multipart problem: valuing weather and moisture information from proxy measures created by satellites; valuing a commodity that is not a market commodity; and valuing a commodity that, for many high-value uses, is not consumed but is replaced after its use. Figure 13.1 illustrates these issues and problems.

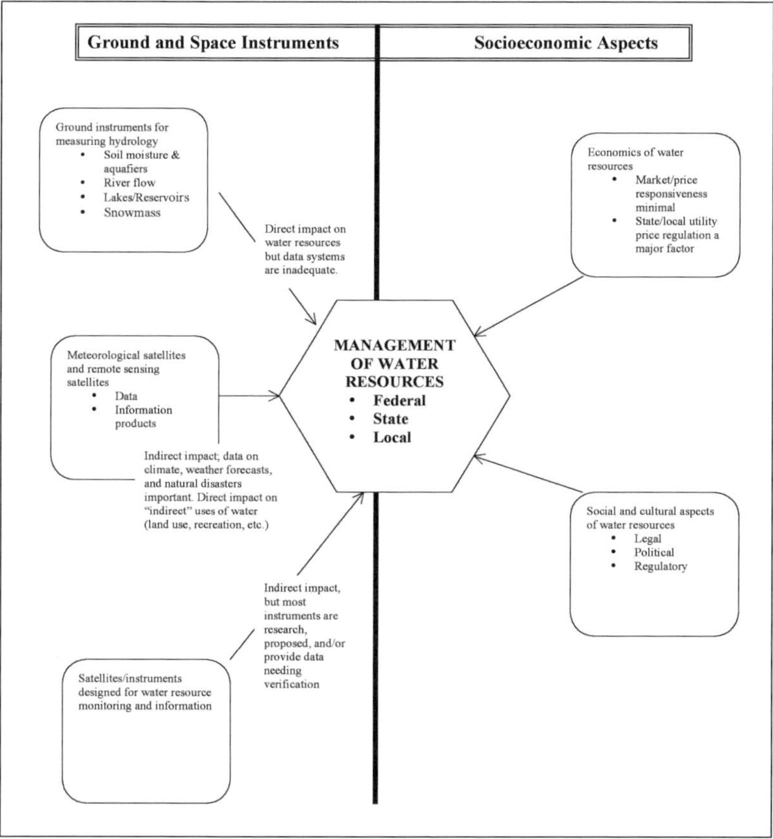

Figure 13.1—Data sources for water resource management and the use of economic models to measure the impact of weather resource data.

Despite the difficulties of actually measuring the socioeconomic benefits of satellite data for water resource management, satellite data can contribute numerous benefits for specific economic sectors of the economy. For example, table 13.6 illustrates the potential use of satellite data for irrigated agriculture.

TABLE 13.6–REMOTE SENSING IN IRRIGATED AGRICULTURE

Agricultural use of water for irrigation is commonly one of the highest-volume usage categories for freshwater; in the United States nearly 40 percent of freshwater withdrawals in 2002 were used for irrigation purposes.[26] Worldwide, on average 70 percent of water use is in the agriculture sector.[27] However, agriculture is not typically the highest-value use of water; municipal and industrial usages generally have greater social and economic value. As development occurs and populations grow, more pressure is felt to provide greater amounts of water to the higher-valued usages. Therefore, increasing efficiency and productivity in agricultural usage of water may provide benefit in allowing increased availability of water for higher-valued usages. Remote sensing applications are able to measure several indicators of performance related to irrigated agriculture. Applying remote sensing techniques in measuring these indicators may provide avenues to increased productivity and efficiency. A selection of these indicators is presented in the following chart.

Indicator	Remote Sensing Principle	Potential Satellites/Instruments
Crop water stress index	Surface energy balance	Landsat (Thematic Mapper)
Evaporative fraction	Surface energy balance	Landsat (Thematic Mapper)
Water deficit index	Surface energy balance	Landsat (Thematic Mapper)
Evapo-transpiration	Surface energy balance	ASTER, AVHRR
Spatial geometry of crop yield	Vegetation index	Landsat, IRS-LISS (Indian Space Research Organization–Linear Imaging Self-Scanner)
Irrigation intensity	Multispectral classification	Landsat, IRS-LISS
Crop intensity	Multispectral classification	Landsat, IRS-LISS
Irrigated area	Multispectral classification	Landsat, IRS-LISS
Soil salinity	Microwave	SMOS (planned)
Soil moisture	Microwave	SMOS (planned)

Some of these measurements are relatively robust, and some require additional research and/or the development of new sensors and new, sophisticated algorithms and modeling to make operational use of the data they provide.

The preceding sections have illustrated some of the measurable and nonmeasurable benefits from the use of Earth observations satellite data. They also pointed out some of the practical problems in measuring these benefits. The following sections elaborate these issues from a methodological standpoint.

26. S. S. Hutson, N. L. Barber, J. F. Kenny, K. S. Linsey, D. S. Lumia, and M. A. Maupin, *Estimated Use of Water in the United States in 2000* (Reston, VA: U.S. Geological Survey Circular 1268, 2004).

27. Chris Perry, "Irrigation Reliability and the Productivity of Water: A Proposed Methodology Using Evapotranspiration Mapping," *Irrigation and Drainage Systems* 19 (2005): pp. 211–221.

MEASURING SOCIOECONOMIC BENEFITS

Economic Measures

There are several approaches to measuring economic benefits:

- A macroeconomic approach that attempts to gauge the impact on the entire economy through measuring changes in GDP, employment, income, or other economy-wide parameters.
- A microeconomic approach that focuses on the impact of consumer welfare through the price mechanism; that is, with new technology and new products, the relative price of a particular good or service will decrease, which, in turn, makes consumers better off.
- An approach that focuses on reducing uncertainty in decision making, which can be evaluated in a number of ways, including assessing consumer preferences through surveys, hedonic measures (parameters associated with the attributes or use of a good or service), avoidance of a particular externality (e.g., costs of cleanup from pollution), or the value-added by using one method over alternatives (e.g., irrigated vs. non-irrigated land for agriculture).
- Other indirect or proxy measures such as counting the number of patentable inventions, number of users of a good or service, or other measures where the actual value or affect on the market is indeterminate.

The following two tables illustrate some of the many economic sectors and applications in which the impact of Earth observation data is very important. Table 13.7 summarizes some of the uses of weather and climate data in the public sector and table 13.8 in the private sector and by individuals. Often, the economic and social values of these uses are very difficult to estimate because they are spread throughout the economy and through a wide variety of entities, including local communities, families, and diverse businesses.

TABLE 13.7–REPRESENTATIVE INDUSTRIES FOR WHICH WEATHER PREDICTIONS HAVE AN IMPORTANT FINANCIAL IMPACT

Major Industry	Examples of Specific Applications
Agriculture	Crop management Irrigation decisions Prevention of weather-related diseases
Energy	Planning purchases of gas and electric power Managing responses in emergency situations Managing capacity and resources
Aviation/Transportation	Optimizing flight patterns Reducing wait times on runways Avoidance of sudden volcanic plumes
Tourism/Recreation	Improving ski slope demand/production of artificial snow Marine forecasts/warnings

TABLE 13.8–USES OF WEATHER AND CLIMATE DATA IN THE PUBLIC SECTOR AND BY INDIVIDUALS	
Entity	**Examples**
Federal, state, local government	Managing public resources
	Managing assistance programs
	Managing disasters, emergencies
	More efficient emergency evacuation
	Reducing operational costs
	Improving operational capacity and safety of U.S. military forces
Citizens	Improving safety
	Managing daily choice of activities
	Improving quality of life
	Reducing lives lost

Macroeconomic Measures of Socioeconomic Impacts/Benefits of Space Programs

A macroeconomic production function model can be used to estimate impacts of technological change attributed to R&D spending on the GDP and derivative measures such as employment and earnings. The results of using this type of model are expressed as a rate of return to a given investment or as a total value.

This economic formulation is best used to develop estimates of a very large program or agency because it focuses on the totality of the economic impact on the entire national economy. Even for an agency such as NASA, the $16 billion annual budget (not all of which is either space- or R&D-related) is comparatively small in relation to the more than $10 trillion GDP for the United States. Trying to ferret out the technology improvements attributable to a specific program such as Earth observations, attribute these improvements to a particular budget line, and then estimate their impact on the economy is extremely speculative on a macroeconomic level.

Another approach to a consolidated measure of benefits is to add up the measured benefits in particular industrial sectors from microeconomic studies to a U.S. or national total. This also yields inappropriate and unreliable estimates. The reason for this is not the logic of the addition, but simply that different point estimates for different products and sectors, coupled with somewhat different econometric methodologies and different time periods, amount to adding up apples and oranges and fails virtually all tests of validity and reliability.

Microeconomic Measures of Space Applications

As appealing as a single number to aggregate all of the benefits from space R&D (or even just from the R&D spent on Earth observations) may be, economists have turned to using more focused examples to measure impacts and benefits through the tools of microeconomics. Analyses at the industry, firm, and product or service levels have been able to provide a useful window on the benefits derived from Earth observations. Several different tools are used for these analyses.

The first is based on the benefits consumers realize from lower prices and greater capability from innovations. If the results of R&D can be translated into goods and services that are less expensive, then the benefits can be measured by the amount "saved" from not having to pay as much as without the new products or services. Conversely, producers can also benefit from being able to offer more products and services at lower prices. The distribution of benefits between consumers and producers depends on the market structure of the industry and products. For Earth observations, these types of benefits can be analyzed by comparing the costs of obtaining weather and land use data from nonspace sources (airplanes, ground measures, etc.) with the costs of buying satellite imagery. Clearly, the greater the area that needs to be observed in detail, the larger the benefits from using space imagery. In many cases, space imagery provides new services that were not available through more traditional methods.

Although cost/benefit analyses are derived from this general framework, the results of those analyses are inherently inaccurate. The costs involved are largely from government expenditures for mission-oriented, dual-use, public goods and are very difficult to isolate program-by-program and mission-to-mission. One should note that cost/benefit analysis was developed to analyze the impacts of regulatory measures (a true before-after situation) rather than on the impacts of new technologies, many of which have no comparable "before" market uses.

Another microeconomic method is the examination of data that provide evidence of the direct transfer of technology from federal space R&D programs to the private sector. The results of these analyses are reported as actual numbers measured (number of patents or inventions, value of royalties, value of sales, etc.). They are rarely compared to associated government expenditures, again because of the difficulty of linking general government funding to specific products or patents.

Qualitative analyses of the benefits of Earth observations, which range from monitoring vegetation and tracking hurricanes, to national security operations, focus on descriptive case studies. Although these activities may have a clear positive effect on human life, valuing the information in a market/price format does not fully describe their impacts.

In summary, there are numerous methods for valuing the impacts of new and better goods and services from Earth observations. Each is useful for particular purposes and in particular situations. No measure can capture the entire impact. The best that can be done is to take particular uses of Earth observation data that have been studied in detail and report on the benefits from those uses.

Reducing Uncertainty

The creation and distribution of accurate weather forecasts involves several elements, beginning with scientific research and continuing through to the delivery of information to government agencies, businesses, and consumers. The process can be viewed both over time (i.e., research results may precede actual use of information by end-users) and at a given point in time (the institutional system structure of information delivery). Measuring the value of information therefore requires evaluating a complex process and has typically only been attempted through studies of specific, isolated examples.

Benefits to society derive from public investment in increasing the amount and the quality of information about natural processes such as weather and climate. Increased scientific knowledge per se generates real benefits. For example, better observations of the geophysical processes that influence weather and climate help advance scientific knowledge directly, or indirectly, by providing better data for calibrating scientific models and/or testing scientific hypotheses. However, despite considerable research on the topic, no accurate metrics exist that enable economists to determine both the quality and the future monetary value of economic benefits that may arise from acquiring new knowledge. Indeed, even the use of peer review and other methods of selecting future scientific missions cannot predict with accuracy the success of such scientific pursuits in operations.

Nevertheless, better information about weather and climate provides tangible socioeconomic payoffs that, at least in principle, lend themselves to quantification. These benefits derive from the fact that weather and climate information can help reduce uncertainty in several ways, as illustrated in the following sections.

IMPROVED CIVIL GOVERNMENT AND MILITARY PLANNING

Weather conditions have a major role in government planning for such tasks as administering forests, grasslands, and other lands under federal management. The 2000 fire in Los Alamos, New Mexico, provides an instructive example. In that case, a fire that was deliberately set by federal officials to reduce the load of dry underbrush raged out of control when the winds turned unfavorable. Better local weather forecasts of wind conditions[28] might have prevented the devastating effects of that fire—reducing or eliminating the severe social and economic effects of that experience. Also, weather forecasts at airports can reduce operational costs. A 1995 Australian study found savings of $6–$7 million per year from improved fueling decisions.[29]

28. Keith Easthouse, "Park Service Unfairly Scapegoated for Los Alamos Fire," *Forest Magazine*, April 2001, *http://www.forestmag.com/losalamosfire-update.cfm* (accessed 24 August 2007).

29. Roy J. Leigh, "Economic Benefits of Terminal Aerodrome Forecasts (TAFs) for Sydney Airport, Australia," Meteorological Applications, Volume 2, (Royal Meteorological Society, 1995), pp. 239–247.

Military operations, whether in war or peacetime, are affected by weather conditions. The military services need accurate weather information in order to increase personnel safety and to gain an information edge over adversaries. Accurate weather forecasts can reduce operational costs by allowing commanders to make better decisions regarding movements and deployments of troops. For example, accurate information regarding winds, sea state, and ocean currents can enable ships to follow more cost-effective courses than would be possible without such information.

Responding to Natural Hazards

The unexpected and severe flooding of the many major rivers in Europe and China in the summer of 2002 and the 1998 devastation in Central America from Hurricane Mitch serve as reminders of the potentially huge economic costs of natural hazards. Better prediction of weather and climate cannot reduce the likelihood that severe weather events will occur but can help substantially lower the costs to society of such events. These cost savings come in two forms: 1) people are more likely to invest in loss-reduction activities when better information is available and 2) better information can also reduce economic costs that arise when uncertainty about adverse weather causes government authorities, people, and business to "err on the side of caution" and undertake what later turn out to be unnecessary loss-reduction activities.

Improved Industrial Planning

Reducing uncertainty about weather and climate facilitates the process of planning in a variety of industrial sectors. More accurate predictions about future weather and climate enable farmers and agribusinesses to estimate future crop yields, leading to reduced uncertainty about yields and prices. In economic terms, such reduced uncertainty translates directly into better use of scarce productive resources, as well as dampening the fluctuations in prices of agricultural products. Similarly in the energy generation industry, improving the predictive ability of forecasts by an average of only one degree can result in more efficient use of power generating resources and can mean hundreds of thousands of dollars saved each year for electric utilities.[30] Many utilities employ their own forecasters at a high annual cost because of these potential large savings. Weather forecasts are also critical for airline operations since better forecasts will reduce operational costs (mainly by saving fuel and improving safety) at airports and in-flight.

30. National Oceanic and Atmospheric Administration (NOAA), Geostationary Operational Environmental Satellite System (GOES), "GOES-R Sounder and Imager Cost/Benefit Analysis (CBA)," prepared for the GOES Users Conference, 1–3 October 2002, Boulder, CO, *http://www.osd.noaa.gov/goes_R/goesrconf.htm* (accessed October 2002); Del Jones, "Forecast: 1 Degree Is Worth $1B In Power Savings," *USA Today*, 19 June 2001. Note that other factors, including political and regulatory actions, can overshadow any savings from forecasts. For example, the wild fluctuations in price and energy availability in California over the past several years resulting from a policy of deregulation would make an economic analysis of separating out the price and efficiency effects of better forecasts very difficult.

Insurance and Hedging against Uncertainty

Finally, providing better information concerning the probabilities of weather-related events also enables the emergence of markets that help mitigate the economic and financial consequences of uncertainty. These markets, which allow the consequences of uncertainties to be "priced" in the form of insurance and hedge contracts, are able to function because information about weather and climate makes it possible to attach probabilities to uncertain events.

In each of these instances, however, new information has value only to the extent that more scientific information reduces uncertainty in ways that are economically valuable. In the case of planning for and responding to natural hazards, information about weather and climate will be valuable to the extent that 1) having more information provides a measurable or significant reduction in uncertainty and 2) reducing uncertainty "matters" in the sense that having more reliable information has the potential to affect choices made by individuals, businesses, and government. Similarly, increased scientific knowledge about weather and climate, by itself, does not facilitate pricing in insurance and/or hedge markets if this information cannot be translated into the probability distribution of future weather events and then efficiently distributed to users.

The value of information has particularly interesting qualities. Before information is released to potential buyers (ex ante), the value to a potential user of the information is not known. Information has economic value only when it is actually used. The transmission of information gained from analysis of data from the environmental satellites to end-users is complex and much information is ignored, lost, or not used. Even if information is disseminated in a timely fashion, sometimes the interpretation may not be clear and potential benefits will disappear. Who will ultimately pay for the information, how much they will pay for it, and what is the actual value of the information are all difficult to evaluate until after the information is obtained and actually used.

Derivatives

Virtually all companies face financial risks from unexpected variations in temperature, precipitation, and other weather-related events. In order to reduce the financial risk that unexpected weather variations might cause, companies whose income depends significantly on the weather are likely to make use of use financial instruments such as weather derivatives to hedge against major losses from unpredicted weather. Whether it is a ski resort protecting itself from a warm winter or an electric utility hedging against price increases in fuels from a cold winter, actual market transactions can provide a window on the value of these natural events to businesses.

Weather derivatives are financial instruments that act very much like puts and calls in the stock and futures markets, and are specific to each company, location, and type of weather condition (temperature, precipitation, wind speed, snowfall, etc.). They tend to cover short periods of time (typically, two weeks to one season in length) and the contracts are usually written to limit the seller's financial exposure. Since they are traded on markets developed for this purpose, the makers of the markets charge a fee (premium) for this service. Since derivatives are especially relevant to business market transactions and are not well understood outside of the industry, they are useful in providing a view of an often-overlooked indicator of the value of weather forecasts.

Purchasing derivatives reveals one facet of the economic value of information on weather and business activity. In March 2003 an analysis of the weather derivative market reported a total of 7,239 contracts (from both a survey of the industry and the contracts reported from the Chicago Mercantile Exchange) with a notional value of nearly $4.2 billion. More than 98 percent of these weather derivative contracts have been based on temperature (the rest were based on precipitation).[31]

Although satellites play a long-term role in improving the accuracy of forecasts and of historical data, the information from satellites tends not to affect the short-term assessment of risks for weather derivatives since these risk assessments are based on history, not on predictions. Nevertheless, future improvements in predictive capabilities (particularly from improvements in satellite instrumentation and data distribution) may well influence the derivative market.

Clearly, as weather prediction capabilities improve, the *potential* for directly using satellite data for derivatives (along with other weather information) will become economically and financially more feasible. As real-time data become more accurate, the historical time series in future years will improve. Satellite weather data will have a great influence on the market and price volatility of weather derivatives.

SUMMARY AND CONCLUSIONS

The preceding short descriptions of socioeconomic benefits from satellite Earth observations data illustrate some of the existing and potential contributions that these systems make to the economy and to societal well-being. It is clear in examining such cases in more detail that numerous impediments in U.S. institutions and in organizational culture prevent government agencies and private companies from taking full advantage of the benefits these data supply. Impediments include the mixed record of NASA and NOAA in moving research findings to operational use; lack of knowledge within companies and local communities about the benefits satellite data can bring to them; institutional inertia and reluctance to make investments in new ways of conducting operations; and the necessary costs of training and equipment to upgrade operations.

31. PricewaterhouseCoopers, "The Weather Risk Management Industry: Survey Findings for November 1997 to March 2001," prepared for The Weather Risk Management Association, Washington, DC, June 2001 (updated in 2003).

Further, it is apparent that we cannot develop a reliable overall estimate of what we know intuitively must be true—that the benefits from Earth observations from space have had a huge and significant impact on the economy. The quality of life, the ability to protect our nation, and the ability to manage environmental and natural resources are very much improved by the use of space-based instruments.

The considerations in this paper suggest that increases in *scientific* information about weather and climate do not automatically or immediately create information that is of *economic* value. A direct implication is that the mix of government-funded projects could change over time depending on how policy makers take into consideration the balance between the economic and commercial value of Earth observations and the research, scientific, and qualitative (social) value of Earth-sensing activities.[32]

The value of weather and climate information itself has been shown to be relatively small as a percentage of the economy.[33] However, when dealing with weather and climate where each year billions of dollars of property is damaged and many lives are lost as a result of severe weather events, even a small improvement in predictive capability can add up to major savings.[34]

Despite these concerns and the methodological measurement difficulties we have enumerated, government agencies and private companies derive sufficient benefit from many of the systems to justify continued and expanded government investment in them, especially when their utility for nonquantifiable international and national security operations is taken into account. Nevertheless, especially in an era of substantial pressure on the discretionary portion of the federal budget, decision makers will continue to press for hard evidence that the investments are worth the cost. Part of this presents a dilemma. Increases in technological capabilities will advance the potential of benefits; however, without a corresponding increase in providing incentives to users and in moving the research results to operational capabilities, it will be very difficult to achieve greater economic benefits, particularly those that can be measured quantitatively.

---

32. It should be clearly recognized that these two goals and not mutually exclusive, due to the dual-use nature of most Earth observation data. In other words, providing data that has social value also contributes to economic and commercial uses.

33. A good review of some of the economic issues in measuring the value of weather information can be found in Molly K. Macauley, "Some Dimensions of the Value of Weather Information: General Principles and a Taxonomy of Empirical Approaches," *http://sciencepolicy.Colorado.edu/socasp/weather1/macauley.html* (accessed March 2002).

34. For a review of the magnitude of losses from extreme weather, see Tom Ross and Neal Lott, The National Climatic Data Center, *http://www.ncdc.noaa.gov/oa/climateresearch.html* (accessed 30 November 2006); "Extreme Weather Sourcebook 2001," *http://sciencepolicy.Colorado.edu/sourcebook/data.html* (accessed 28 July 2007). Other compilations can be found in "Natural Disasters 2000, Annual Review," (Munich, Germany: Munich Re; also available as a CD-ROM); A. Arguez and J. Elsner, "Trends in U.S. Tropical Cyclone Mortality During the Past Century," (Florida State University: Tallahassee, FL, 11 April 2001).

Hence, additional research on socioeconomic benefits will be essential, quantifying where possible the economic benefits satellite systems provide to the U.S. economy. Our research so far demonstrates, among other things, that too little effort has been put into this important task. The recent National Research Council study, *Earth Science and Applications from Space: National Imperatives for the Next Decade and Beyond*, underscores this need in its Chapter 5: Earth Science Applications and Societal Benefits.[35]

35. National Research Council, Committee on Earth Science and Applications from Space, *Earth Science and Applications from Space: National Imperatives for the Next Decade and Beyond* (Washington, DC: National Academies Press, 2007).

Section IV

Applications Satellites, the Environment, and National Security

Chapter 14

Satellites and Security: Space in Service to Humanity

Erik M. Conway

In his classic political history of the early Space Age, Walter McDougall explained the cold war competition between the Soviet Union and the United States as a competition between two increasingly technocratic states.[1] In the American case, his narrative represents a cry for restraint. The technocratic imperative is not a democratic one. He saw in the coming of technocracy the rise of a narrow, technologically focused elite to power. He apparently did not like the technocratic vision of the future this gave him; not surprising, I think, given his experiences in one of American technocracy's great Apollo-era disasters, the Vietnam War.

But one of the key traits of technocracy is the state's effort to use "technology," often defined very amorphously, to improve lives. This is, I think, what Steve Dick meant when he asked me to discuss the societal impact of military, applications, and science satellites. The United States government has financed these in the general belief that they would result in progress of some sort. In many cases, the hoped-for outcomes were obvious. Everyone thought weather satellites would result in longer-range weather forecasts, an economic, as well as a social, good. Military satellites, too, had obvious uses. The earliest military satellites were oriented toward surveillance and intelligence gathering, developed to supplement intelligence aircraft.

The same can be said about the literature on applications satellites. To date, only communications satellites have a significant literature, as David Whalen will discuss in his paper. Only military satellites, my first topic, have received extensive study, although to date imaging satellites have drawn most attention, leaving several areas untouched. It is also the case that many "big picture" questions about the impact of these vehicles on the military, and on our political culture, have yet to be asked.

Even scientific satellites were often utilitarian in nature. Asif Siddiqi wrote in his essay for the 2005 Critical Issues conference that "[T]he literature on the history

1. Walter A. McDougall, . . . *the Heavens and the Earth: A Political History of the Space Age* (New York: Basic Books, 1985).

of space-based science has, however, not been significant."[2] Indeed it has not! Very few efforts in space science have drawn scholarly attention. But scientific satellites and solar system probes have had profound impacts on the scientific community and, as is the case with military satellites, on our national politics.

SATELLITES AND THE MILITARY

One of the fundamental historical realities exposed in all of the papers in this session is that satellites have been seen, since Sputnik, as militarily useful. The first weather satellite, TIROS, started its road to existence as an Army reconnaissance satellite project at Fort Monmouth, New Jersey, a couple of years prior to Sputnik's launch. Army leaders thought, quite rightly, that just as airborne reconnaissance had transformed knowledge of the battlefield in World War II, spaceborne reconnaissance would cement that tactical advantage and make it a global, and strategic, asset. It would also place that asset under Army control, relieving the institution's dependence on its chief rival, the U.S. Air Force, for intelligence. They were thwarted by President Eisenhower, who gave imaging reconnaissance to the Central Intelligence Agency and the Air Force, weather satellites to NASA and the Weather Bureau in an unsuccessful partnership arrangement, and to his own former service, the Army, communications.

There is already some good literature on reconnaissance satellites and their impact on the cold war. Jeffrey Richelson has examined two important systems in a pair of books, the Defense Support Satellite Series, which served to provide early warning of Soviet ballistic missile launches, and the Keyhole series of imagery satellites (also known as Corona and Discoverer.) The Corona satellites have also drawn attention from Dwayne Day, from a policy and technology perspective. There is also at least one popular treatment of space surveillance.[3]

---

2. Asif A. Siddiqi, "American Space History: Legacies, Questions, and Opportunities for Future Research," *Critical Issues in the History of Spaceflight*, Steven J. Dick and Roger D. Launius, ed. (Washington, DC: NASA SP-2006-4702, 2006), p. 440.

3. Jeffrey Richelson, *America's Secret Eyes in Space: The U.S. Keyhole Spy Satellite Program* (New York: Harper & Row, 1990); Richelson, *America's Space Sentinels: DSP Satellites and National Security*, Modern War Studies (Lawrence, KS: University Press of Kansas, 1999); Dwayne A. Day, "A Strategy for Reconnaissance: Dwight D. Eisenhower and Freedom of Space," in *Eye in the Sky: The Story of the Corona Spy Satellites*, Dwayne A. Day, John M. Logsdon, and Brian Latell, ed. (Washington, DC: Smithsonian Institution Press, 1998), pp. 119–142; William Burrows, *Deep Black: Space Espionage and National Security* (New York: Random House, 1986). Also see R. Cargill Hall, "Missile Defense Alarm: The Genesis of Space-Based Infrared Early Warning," *Quest: The History of Spaceflight Quarterly* 7, no. 1 (Spring 1999): pp. 5–17.

A thorough overview of what's often called "MILSPACE" can be found in Stephen Johnson's article in the *Critical Issues in the History of Spaceflight* volume, so instead of repeating his work I want to highlight a few areas that seem particularly interesting.[4]

From a technological standpoint, several types of intelligence satellites have existed since the late 1960s that haven't been examined. Since the early 1980s, the United States has flown a series of synthetic aperture radar satellites known as Lacrosse (and sometimes Onyx). These satellites are intended to detect vehicles of all types and can also detect motion and change. The first civilian synthetic aperture radar (SAR) flew in 1977 and, as far as I can determine, the first intelligence SAR was initiated in 1976. The Jet Propulsion Laboratory (JPL) built the first civilian SAR but does not seem to have been the source for the military version.[5] Are these truly independent developments? Or is there a relationship between the two? Our current understanding of space history would pretty strongly suggest that there must have been a relationship between the two. John Cloud has described the relationship between classified imaging satellites built for intelligence purposes and their unclassified counterparts (i.e., Landsat) as a shuttered lamp—a relationship controlled by the intelligence community.[6] The radar case might show us that the relationship between civilian and military technology is more complex than this model.

There are also other kinds of intelligence satellites. As was revealed at least as far back as the early 1980s, the United States operates "signals intelligence" (sigint) satellites. These satellites capture millions of electronic signals every day that are analyzed by an ever-changing array of supercomputers. There are many possible stories to tell buried in this topic. As Stephen Johnson also pointed out, the technologies underlying sigint are in need of historical exegesis.[7] The political, military, and intelligence utility of these kinds of systems is in need of study, too. In the early 1990s, the U.S. National Security Agency apparently considered using it for industrial espionage against our European allies. This briefly produced diplomatic problems for the United States (whose governmental allies were, apparently, quite content with the system's existence until their publics found out about it).[8]

---

4. Stephen Johnson, "The History and Historiography of National Security Space," in *Critical Issues in the History of Spaceflight*, Steven J. Dick and Roger D. Launius, ed. (Washington, DC: NASA SP-2006-4702, 2006), pp. 481–548.

5. The Jet Propulsion Lab's SAR flew on Seasat-A; see Peter Westwick, *Into the Black: A History of the Jet Propulsion Laboratory* (New Haven, CT: Yale University Press, forthcoming); for Lacrosse, see *http://www.fas.org/spp/military/program/imint/lacrosse.htm* (accessed 5 September 2006).

6. John Cloud, "Imaging the World in a Barrel: CORONA and the Clandestine Convergence of the Earth Sciences," *Social Studies of Science* 31:2 (April 2001): pp. 231–251.

7. Johnson, "History and Historiography," note 4.

8. Ron Suskind, *The One Percent Doctrine: Inside America's Pursuit of its Enemies Since 9/11* (New York: Simon and Schuster, 2006), p. 85.

The history of this technical capability is also directly relevant to the current administration's attempts to discredit revelations that it is using signals intelligence to spy on terrorists (and everyone else, too.)[9] The historical fact of the matter is that this is old news.[10] The technical capacity to do this has been a matter of public record since the public unmasking of the National Security Agency in the Ford administration and its role in illegal domestic surveillance; wise leaders should assume enemies have a basic awareness of our capabilities—particularly old, well established ones. The subject of signals intelligence and its use and abuse deserves further historical exegesis.

Better covered historically than signals intelligence has been the development of military communications satellites. So far, the literature suggests that although the military saw the utility of surveillance satellites well before the ability to build them existed, the armed services rather missed the boat with communications satellites. The first communications satellites that we're aware of were commercial. AT&T immediately sought a monopoly in this new business and was denied by government policy; the Department of Defense (DOD) seems to have bought access to the technology in the early years. It became an innovator later in the Space Age. One author places responsibility for this in the decision to place communications satellite development in the Army and in the Advanced Research Projects Agency, delaying development until after private companies had already flown their satellites. NASA also played a significant role in communications satellite development in the 1960s and early 1970s, although apparently not after that.[11] To make this picture of innovation still more complicated, I was recently told by a retired NASA manager that the Air Force paid for a significant fraction of the last of NASA's Advanced Technology Satellite comsat series, ATS-6, launched in 1974.

9. Liberal columnist Paul Krugman discusses the treason charge in Paul Krugman, "The Treason Card," *The New York Times*, 7 July 2006; also see Frank Rich, "Will the Real Traitors Please Stand Up," *The New York Times*, 14 May 2006.

10. The technical capability to carry out large-scale eavesdropping on electronic communication was revealed in the 1970s, when the National Security Agency's illegal activities in spying on Americans was investigated by Congress. See James Bamford, *The Puzzle Palace: A Report on America's Most Secret Agency* (Boston: Houghton Mifflin, 1982), esp. pp. 280–308. The role of satellites in this system is not clear; much of what was revealed during the 1970s involved wiretapping and interception of radio communications from ground-based "listening stations." As Dwayne Day has shown recently, signals intelligence satellites first flew in 1960, although these seem to have been radar intercept satellites for characterizing air defenses, not communications intercept satellites. See Dwayne A. Day, "Early American Ferret and Radar Satellites," *Spaceflight* (July 2001): pp. 288–293.

11. See David N. Spires and Rick W. Sturdevant, "From Advent to Milstar: The U.S. Air Force and the Challenges of Military Satellite Communications," in *Beyond the Ionosphere: Fifty Years of Satellite Communication*, ed. Andrew Butrica, (Washington, DC: NASA SP-4217, 1997), pp. 65–70; David Whalen, *The Origins of Satellite Communications, 1945–1965* (Washington, DC: Smithsonian Institution Press, 2002).

Communications satellites have had a profound impact on the armed services. My own former service, the U.S. Navy, evolved command traditions over 200 years that were based on the unreliability and often non-existence of communications with the National Command Authority back in Washington. Even radio did not fundamentally change this reality, because the high-frequency radios that were the basis of World War II communications were "environmentally unreliable." As radio engineers all knew, sometimes one could get global range from a radio set—and sometimes you couldn't reach a ship 50 miles away. The Navy continued to operate under the expectation that senior commanders were unlikely to be immediately available.

Satellite communications changed that. Essentially instantaneous communications were possible from virtually any spot on Earth or in the air above it after the mid-1960s. In addition to the culture shock this imposed on the Navy (and which it is still dealing with), I rather suspect it accounts for some of the severe cultural differences between the relatively new Air Force—which has only existed in the era of electronic communications—and the Navy. But in any case, buried in the history of military communications satellites are stories of cultural change as well as technological change, deserving of documentation and analysis. There are also, of course, changes in operations and in operational capacity due to this capability in need of examination.

Although there is at least some literature on communications satellites, largely missing from the historical record is an examination of the first military navigation satellite series, Transit. Operational in 1962, Transit was a Navy project implemented by the Applied Physics Laboratory of the Johns Hopkins University. Transit operated until 1996. The technical literature contains articles on its genesis, but I can identify no examination of its impact on Navy operations, capabilities, etc.[12] Yet Transit was simple and inexpensive, and it was part of the justification for belief in "faster, better, cheaper" in the 1990s within NASA. It also paved the way for the NAVSTAR Global Positioning System, which one paper in this book addresses. So it had important ramifications but it hasn't found its historian yet.

The last kind of military space technology that I want to discuss is weather satellites. The United States has maintained a civilian weather satellite series since 1960 and a military weather satellite series since 1961.[13] President Eisenhower had intended that there be a single system operated by the Weather Bureau, but the resulting agreement collapsed very quickly. The prime justification for splitting into two satellite systems was that the DOD and the Weather Bureau wanted different overpass times, with DOD officials wanting an early morning orbit that would

12. See, for example, Robert J. Danchik, "An Overview of Transit Development," *Johns Hopkins APL Technical Digest*, vol. 19, no. 1 (1998): pp. 18–26.

13. See the essay by Purdom and Menzel in James R. Fleming, *Historical Essays on Meteorology, 1919–1955* (Boston: American Meteorological Society, 1996), for basic information on the civilian weather satellite program.

permit planning of reconnaissance and aerial refueling missions. One source also states that the civilian system would not overfly Moscow, increasing the Pentagon's desire for a different orbit choice.[14] I rather suspect that the Pentagon's desire for control also had a lot to do with this decision.

Initially, the two organizations used two variants of the TIROS satellite, with very similar instrumentation. This changed over time and, although the current civilian and military satellites use the same "bus," they have different instrumentation. It is the military weather satellites that have provided strong evidence that the Arctic ice cap is shrinking rapidly, for example—a capacity the civilian satellites do not have. Other than R. Cargill Hall's recent short, and very technically oriented, history of the Defense Meteorological Satellite Program, however, there are no studies of DOD's weather satellite system or its evolution.[15] My own forthcoming work deals only with the civilian system.

A Larger View of Military Impacts

Paul Edwards expounded a thesis a few years ago that in constructing its network of surveillance and communications assets, the DOD sought nothing less than a "closed world"—an Earth in which nothing could happen without the Pentagon's knowledge. His argument is more sociocultural than historical, and I am not convinced. The sheer arrogance of such a goal (not to mention its technological unlikelihood!) gives me pause. Yet he might be right. The very name of the DOD's communications architecture, The World Wide Military Command and Control System, certainly suggests that he is right.[16] Rigorous investigation of his claim would tell us a great deal about the DOD and its leaders' faith in technology if someone were willing to dig into it.

All in all, the extant historical literature on military use of space pretty strongly suggests that surveillance and communications have been stabilizing influences in terms of the strategic deterrence and the cold war contest with the Soviet Union. That's the point made by Glenn Hastedt in his paper in this book. It's far less clear that this is true in terms of the myriad "small wars," "regional conflicts," and "operations short of war" that the United States has engaged in since 1945. It's likely true that these space assets have expanded the United States' ability to engage in so-called conventional war, the sort fought by large formations of men and machinery.

14. Dwayne Day comments on an earlier draft of this paper, September 2006.

15. R. Cargill Hall, "A History of the Military Polar Orbiting Meteorological Satellite Program," (Washington, DC: National Reconnaissance Office, 2001).

16. Paul N. Edwards, *The Closed World: Computers and the Politics of Discourse in Cold War America* (Cambridge, MA: MIT Press, 1996), pp. 1–42. On WMMCS, see David E. Person, *The World Wide Military Command and Control System: Evolution and Effectiveness* (Maxwell AFB, AL: Air University Press, 2000).

One recent analysis describes the very conventional first Gulf War as the first "space war."[17] Yet satellites have not similarly improved the nation's capability to wage "guerilla war"—we remain poor at it, as the current wars in Iraq and Afghanistan show, and as was also true in Vietnam. The impact of space technology has been profound, in other words, but not limitless.

Further, one should wonder whether the global view offered by our space assets have encouraged the belief prevalent among neoconservatives that we can construct and maintain a perpetual Pax Americana. Based on a presumed "Revolution in Military Affairs" that derives from the integration of space-based communications, intelligence, and space- and airborne weapons, they contend that America can maintain and expand its dominion over Earth forever. In their 1998 program for America, for example, the movement's leaders called for the development of "global missile defenses to defend the American homeland and American allies, and to provide a secure basis for U. S. power projection around the world," and for establishment of control over "the new 'international commons' of space and cyberspace."[18] Their faith in technology and the utility of the space weaponry seems to me founded directly on a heavily distorted view of space capabilities.[19]

In fact, they seem to believe in exactly what Edwards claims the Pentagon did in the 1960s—that a closed world is possible and desirable. There is a great subject here for someone to dig into. Satellites may very well have encouraged the rapid rise of militarism in the United States as well as encouraging our political classes to undertake global wars of choice.[20]

Civilian Applications Satellites

So, I claim, military satellites of many descriptions and functions have had significant impacts on the American military, and may have had far larger impacts on American political culture. What about their civilian twins?

NASA launched the first civilian weather satellite in April 1960. This was TIROS 1, the direct descendant of the U.S. Army's work toward an imaging

17. Peter Anson and Dennis Cummings, "The First Space War: The Contribution of Satellites to the Gulf War," *RUSI Journal* 136 (1991): pp. 45–53.

18. Thomas Donnelly, et al., "Rebuilding America's Defenses: A Report of the Project for the New American Century," pp. iv, 54–64, published online at *http://www.newamericancentury.org* (accessed 5 February 2007).

19. Frances Fitzgerald, *Way Out There in the Blue: Reagan, Star Wars, and the End of the Cold War* (New York: Simon and Schuster, 2000) is the best work on Star Wars to date, although it is nontechnological. A good summary of this "Revolution in Military Affairs" thinking is Andrew Bacevich, *The New American Militarism* (New York: Oxford University Press, 2005), pp. 165–170.

20. On American militarism, see Andrew Bacevich, *The New American Militarism*. The Project for a New American Century calls such small, voluntary wars "constabulary duties." Donnelly, Rebuilding America's Defenses, p. 10.

surveillance satellite. It was hugely popular, drawing a large, front-page *New York Times* article, complete with "cloud pictures" from space. NASA officials used it to sell the agency to Congress and the public. Better weather forecasting was the first obvious and direct benefit to non-astronaut, ground-dwelling humans (i.e., 100 percent of the American population).

Yet there is no published history of the weather satellite program. The last attempt at a history was Chapman's study of the conflict between NASA and the Weather Bureau over control of the technology completed back in 1967.[21] It was rather heavily redacted due to DOD concerns. Since then, no one has found the subject worthy of historical study. Not even the Weather Bureau's descendant, NOAA, has bothered. The agency doesn't even publish data on economic benefits from weather satellites. There is a single essay touching on weather satellites in the *Exploring the Unknown* series.[22] What gives here?

My speculation on the subject goes like this: Weather satellites did not live up to the grand promises upon which they were sold. Using them, meteorologists of the 1960s thought they could produce monthly forecasts of great accuracy. As I show in my forthcoming work, however, by the late 1960s simulation studies done at the Goddard Institute for Space Studies showed that forecasts of more than a dozen days were completely impossible and forecasts of more than five days were impossible with current technologies. This was confirmed by the Global Atmospheric Research Program in 1978, a hugely demoralizing outcome to NOAA and to the meteorological profession. Satellite data emerged from this in such low regard that NOAA didn't use it in routine operational numerical forecasting until 1998.[23] So I rather suspect NOAA leaders have not wanted to undermine themselves with a historical study.

I do not want to leave the impression, however, that the weather satellites have been without impact. The geosynchronous satellites provide the hurricane and typhoon warnings and imagery and track forecasts that we've been accustomed to since the early 1980s. I suspect that they've also improved severe storm forecasting in the Midwest. Meteorologist friends of mine say this is true, anyway. Both of these are significant economic and human goods, although the lack of research on these topics available makes it impossible to make the case with any rigor. These satellites have also transformed the meteorological profession. Although operational forecasts in the United States have not used the satellite data, researchers and research-oriented forecast models have been using it routinely since the early 1980s.

---

21. Richard Leroy Chapman, "A Case Study of the U.S. Weather Satellite Program: The Interaction of Science and Politics," Ph.D. dissertation, Syracuse University, 1967.

22. Pamela E. Mack and Ray A. Williamson, "Observing the Earth from Space," in *Exploring the Unknown: Selected Documents in the History of the U.S. Civil Space Program Volume III: Using Space*, John M. Logsdon et al., ed. (Washington, DC: NASA SP-4407, 1998), pp. 155–176.

23. National Research Council, *From Research to Operations in Weather Satellites and Numerical Weather Prediction: Crossing the Valley of Death* (Washington, DC: National Academies Press, 2000).

It has led to much greater understanding of the structure of the atmosphere and its organization, particularly over the otherwise data-sparse oceans. Satellites permitted identification of the region of the Atlantic where hurricanes typically form, for example, something that hadn't been known prior to the Space Age.

There is also a policy issue to be discussed. The United States has not flown a new instrument on its civilian, polar-orbiting weather satellites since 1978. This is not a failure of innovation; NASA is flying relevant research instruments and NOAA has developed airplane-based demonstration instruments as well. But NOAA hasn't been able to get any of these instruments into space. This problem has been exacerbated by the 1994 decision to merge the civilian and military weather satellite programs into the National Polar Orbiting Environmental Satellite System, whose 1998 launch date has been "slipped" to somewhere around 2013–2014.[24] There's a gross policy failure here, and possibly several policy failures all working together to produce an exceptionally bad outcome.

Finally, in his recent movie, *An Inconvenient Truth*, former Vice President Al Gore, Jr., shows an image of Earth from space, taken by Apollo 17 astronauts from lunar orbit in 1973.[25] By showing the "Big Blue Marble" in all its beauty and fragility and free of arbitrary, human-drawn political borders, it provided a potent symbol for environmentalists to rally around.

Although I certainly agree that images have great power, I have always wondered a bit about the specifics of this story. The first Earthrise image was sent back by a robot, Lunar Orbiter 1, in 1966. The first full-disk color image of Earth was actually sent back by Vern Suomi's experimental spin-scan camera on the communications satellite ATS-3 in 1967. Yet only meteorologists talk about this image. In 1968, Apollo 8 astronauts took the first color Earthrise image. All of these images appeared in public media before the first Earth Day in 1970; the Apollo 17 images did not. The space program as a whole produced a series of stunning images that placed Earth in a context far different than Americans' daily experience. I think the steady stream of dramatic imagery mattered more to the spread of environmental consciousness than any single picture.

I have touched on some of these issues in my own forthcoming history of atmospheric science, but since my focus is on NASA science, a lot of the above ground is essentially untilled. In terms of "societal impact," I suspect that the weather satellites have been enormously important—but neither I nor any other scholar has made the case with any rigor.

---

24. A bare hint of NPOESS's long, messy history is contained in Eli Kintisch, "Stormy Skies for Polar Satellite Program," *Science* 312 (2 June 2006): pp. 1296–1297.

25. Sheila Jasanoff, "Image and Imagination: The Formation of Global Environmental Consciousness," in *Changing the Atmosphere*, ed. Clark Miller and Paul N. Edwards (Cambridge, MA: MIT Press, 2001). The book companion to Gore's movie is Al Gore, *An Inconvenient Truth* (New York: Rodale Books, 2006).

Commercial communications satellites do not have quite the historical lacuna that the weather satellites do. Because David Whalen addresses them in his paper, I am not going to say much about this subject other than to reiterate what I said in the above Military Impacts section: although we have some good treatments of the evolution of comsat technology and of the policy issues surrounding them, no one has made the leap yet to societal impact. The economic value of the industry seems clear from Whalen's work but there's more to society than the pure social construct we call "money." We often believe that our technologies make our lives "better" in some meaningful way, although one should always consult Ruth Schwartz Cowen's book *More Work for Mother* before making this claim too strongly! Have communications satellites done so? I am reminded of Bruce Springsteen's lyric "57 channels and nothing on."[26] Satellites have brought me hundreds of channels; I watch two regularly.

But let me dig a bit further into this question of "impact." In one of his exploration essays, Steve Dick threw out the idea that communications satellites have brought the world "closer together."[27] The airplane was similarly touted as having the potential to "bring the world closer together" and other happy things—bring peace on Earth through greater understanding among peoples, etc. Joe Corn wrote a wonderful book on this.[28] But the airplane's legacy is far more mixed. It is true that I can take a relatively short flight to Tokyo to visit the land of sushi. It's also true that the airplane allowed us to nearly wipe Tokyo off the map two generations ago, without even bothering with nuclear weapons. The airplane's legacy is so mixed and contested that the nation's shrine to aviation, the National Air and Space Museum, is not *allowed* to construct interpretive exhibits about it. Interpreting the airplane accurately would offend powerful political and economic interests.

My personal sense is that the same will turn out to be true for communications satellites: the legacy will be heavily mixed. It is certainly true that satellites facilitate emergency communications and they provide the opportunity for vastly more programming. At the same time, we have seen the growth of targeted media—channels that exist to promote specific interests or specific politics. The same capability that enables dedicated sports networks (e.g., ESPN, NESN) also permits Al Jazeera, the first independent Arab media organization (despised by the American

26. Ruth Schwartz Cowan, *More Work for Mother: The Ironies of Household Technology from the Open Hearth to the Microwave Oven* (New York: Basic Books, 1985); Bruce Springsteen, "57 Channels (and Nothin' On)," *Human Touch*, Sony BMG Records, 1992.

27. Steven J. Dick, "Societal Impact of the Space Age," 4 April 2005, *http://www.nasa.gov/mission_pages/exploration/whyweexplore/Why_We_09.html* (accessed 11 September 2006).

28. Joseph J. Corn, *The Winged Gospel: America's Romance with Aviation, 1900–1950* (New York: Oxford University Press, 1983), esp. pp. 29–50.

political right), and News Corp's Fox News channel, derided by American liberals as "Faux News" for its blatant political bias.[29]

George Orwell feared that the end of the twentieth century would bring a perfect surveillance state into existence. Big Brother would be watching all of us, all of the time. We are not quite there yet, but we *have* achieved a world full of specialized propaganda bubbles—hidden behind the euphemisms of "targeted media" or "narrowcasting." One can go through one's daily life and never be subjected to uncomfortable or challenging ideas—or even facts. In short, we have a postmodern version of Orwell's Ministry of Truth.

This is not the fault of the technology exclusively, of course. The Reagan administration's elimination of the old Fairness Doctrine that required the airing of multiple viewpoints is another key enabler of this unfortunate outcome. It is also not new. Most major European cities have had partisan newspapers for centuries and the United States has its share of political magazines, the famous ones being *National Review* on the right and *The Nation* on the left, although there are others. But there are differences. Television is more immediate and immersive than print and has much greater emotional power. Whether these differences are sufficient to make a difference in how the newly partisan electronic media affects the nation is a subject worthy of study. To wrap up this brief discussion of communications satellites, I am not at all sure Dick's optimistic appraisal is warranted.

Finally, the last applications satellite type I'll discuss in this extended editorial essay is land use. As Pam Mack has shown in her book on Landsat, there were many possible and interested users of satellite-based land imaging during its developmental period, and their competing goals and interests made development of the system very difficult.[30] And partly because of this, and also partly due to Pentagon restrictions on allowable spatial resolution (because of ill-conceived Congressional efforts to force "privatization" of Landsat), Landsat has never achieved a large enough user base to pay for itself. Instead, its imagery has been used by researchers, not by the economic interests that might be able to afford it on a commercial basis.[31] Indeed, the primary

---

29. *http://www.sourcewatch.org/index.php?title=Fox_News* (accessed 22 October 2006); Steven Kull, "Misperceptions, the Media and the Iraq War," 2 October 2003, *http://65.109.167.118/pipa/pdf/oct03/IraqMedia_Oct03_rpt.pdf* (accessed 22 October 2006). Renegade conservative David Brock has written a detailed account of the creation of the right-wing media establishment in America. David Brock, *The Republican Noise Machine: Right Wing Media and How it Corrupts Democracy* (New York: Basic Books, 2004); also see Neil Hickey, "Is Fox News Fair?" *Columbia Journalism Review* 36:6 (March-April 1998): pp. 30–36. Also see Daphne Eviatar, "Murdoch's Fox News," *The Nation* (22 February 2001), online edition, *http://www.thenation.com/doc/20010312/eviatar* (accessed 5 February 2007).

30. Pamela E. Mack, *Viewing the Earth: The Social Construction of the Landsat Satellite System* (Cambridge, MA: MIT Press, 1990).

31. There is a single historical study of Landsat use for scientific purposes to date: Peter Leimgruber, Catherine A. Christen, and Alison Laborderie, "The Impact of Landsat Satellite Monitoring on Conservation Biology," *Environmental Monitoring and Assessment* 106 (2005): pp. 81–101.

buyer of the data has been the intelligence community, which apparently finds that Landsat data serves as an effective supplement to its own classified imagery sources.

The satellite series itself has lurched from one crisis to the next, with each administration since Reagan willing to commit to only one more mission prior to commercialization; with commercialization never succeeding, each new administration has had to cope with the question of how to continue the series. The fundamental policy issues of what agency should maintain the capability and who should pay for it have not been resolved. So Landsat has been a technical success, but programmatically its history has been tortured. At the very least, there's a good policy study here for someone interested in the subject.

EARTH SCIENCE FROM SPACE

This is the subject of my own recently completed history of atmospheric science at NASA, so what follows is an essay on my own findings in the context of what very little else has been done on this subject. NASA's scientific satellites, and—a very important point—its planetary probes, have revolutionized our understanding of Earth, its processes, and our place on it. They have also radically altered our beliefs about the solar system and the universe around us. Finally, they have fundamentally changed our national politics.

There is, to date, not a single history of any NASA Earth science program. The only work that even comes close is Henry Lambright's monograph.[32] Yet as several (non-NASA) studies show, the Agency's stratospheric ozone research program initiated in the early 1970s led directly to the worldwide banning of a class of highly profitable chemicals. NASA research provided the first conclusive evidence that human activity was capable of causing global-scale damage. Yet these studies are not aimed at NASA's science program. They discuss the politics and policies surrounding the 1987 Montreal Protocol.[33] The political angle is important to these studies because ozone science was politically controversial, with leaders of the American political right claiming for many years that the idea of ozone depletion was an environmentalist hoax. Congressional hearings in 1995 underscored this view. But these earlier studies leave the reader wondering how NASA wound up leading this research field.

This is particularly germane as NASA's role in the ozone wars led to repeated and long-running attacks on the agency on the editorial page of the *Wall Street Journal*, by Rush Limbaugh on his radio show and in his best-selling books, and by a

32. Henry Lambright, *NASA and the Environment: The Case of Ozone Depletion* (Washington, DC: NASA SP-2005-4538, 2005).

33. The best of these works is Edward A. Parson, *Protecting the Ozone Layer: Science and Strategy* (New York: Oxford University Press, 2003); see also Maureen Christie, *The Ozone Layer: A Philosophy of Science Perspective* (New York: Cambridge University Press, 2000).

small constellation of other lesser actors in the right-wing political journals: *Reason, Commentary*, and *National Review*. The current administration made a brief attempt to quash NASA's James Hansen, the agency's most prominent climate modeler. But he was not the first NASA scientist to be attacked by what journalist Bill Moyers accurately terms the "radical right."[34] That honor actually belongs to Robert T. Watson, head of NASA's ozone research program in the 1980s.[35]

In *High Speed Dreams*, my history of supersonic transport research, I suggested in my conclusion that NASA's entry into political controversy came with its decision to embrace stratospheric ozone as a research program in hopes of absolving the Space Shuttle of claims that its solid rocket boosters would damage the ozone layer.[36] But in fact, the Agency's conflict with the New Right political movement that came to power with Ronald Reagan has its origins much earlier, in the Agency's planetary missions of the 1960s.[37] The JPL's Mariner Venus and Mars missions showed NASA's scientists—those employed by the Agency as well as university-based scientists involved in them—that relatively small initial differences between the three "terrestrial" planets (Venus, Earth, and Mars) had led to huge differences among these planets as they exist today. This fact, obvious by 1965, forced NASA's scientific constituency to start to come to grips with the relationship between *chemistry* and *climate*.

Probably the best known expression of this is in James Lovelock's Gaia hypothesis. Lovelock, who consulted briefly at JPL in the early 1960s, argued that biological activity regulated Earth's climate via its impact on atmospheric chemistry. But he was not the only person making chemical claims about climate by the end of the 1960s. Carl Sagan, never a NASA employee but always associated with the

---

34. Amanda Griscom, "Now Hear This: Bill Moyers Speaks his Mind on Bush-Brand Environmental Destruction and More," Grist Magazine (26 August 2003), online edition, *http://www.grist.org/news/maindish/2003/08/26/griscom-moyers/* (accessed 22 February 2007).

35. Andrew Revkin, "Climate Expert says NASA Tried to Silence Him," *The New York Times* (29 January 2006): p. 1; Donald Kennedy, "The New Gag Rules," *Science* (17 February 2006): p. 917; Rush Limbaugh, *The Way Things Ought to Be* (New York: Pocket Books, 1993), pp. 154–157. A few citations from the ozone depletion denial effort will suffice: Robert W. Pease, "Ozone Chicken Littles Are at It Again," *Wall Street Journal* (23 March 1989): p. 24; S. F. Singer, "My Adventures in the Ozone Layer," *National Review* 41 (1989): pp. 34–38; R. Bailey, "The Hole Story: The Science Behind the Scare," *Reason* 24 (1992): pp. 24–31; R. S. Bennett, and Robert W. Clack, "Ozone, CFCs and Science Fiction," *Wall Street Journal* (24 March 1993): p. 15; Kent Jeffreys, "Too Many Holes," *Wall Street Journal* (11 February 1993): p. 15.

36. Erik M. Conway, *High Speed Dreams: The Technopolitics of Supersonic Transportation* (Baltimore, MD: Johns Hopkins University Press, 2005), p. 303.

37. There is already a substantial body of literature on the New Right political movement. For examples, see Sara Diamond, *Roads to Dominion: Right Wing Movements and Political Power in the United States* (New York: Guildford Press, 1995), esp. pp. 108–128 and 205–227; for surveys placing the New Right in the context of the New Left of the 1960s, see Maurice Isserman and Michael Kazin, *America Divided: The Civil War of the 1960s* (New York: Oxford University Press, 2000), esp. pp. 205–220, and Rebecca E. Klatch, *A Generation Divided: the New Left, the New Right, and the 1960s* (Berkeley, CA: University of California Press, 1999).

Agency anyway, did a relatively speculative study for JPL on Venus's climate in 1960 and did a series of comparative studies between Earth, Mars, and Venus over the course of his career. His first graduate student, James Pollack, built NASA's planetary climate modeling program at the NASA Ames Research Center. [38]

It was also Pollack's group at Ames that produced the "nuclear winter" hypothesis in 1984, bringing down right-wing opprobrium on NASA while also stimulating tropospheric aerosol research.[39] The nuclear winter hypothesis helped trigger the foundation of the George C. Marshall Institute by Robert Jastrow. Jastrow, horrified by what he saw as a deliberate political attack on the Reagan administration disguised as science by Sagan, Pollack, and other scientists (including Donald Kennedy, the current editor of *Science*), appealed to conservative foundations for funds to finance a pro-nuclear "scientific" organization. He envisioned it as the conservative response to the Union of Concerned Scientists.[40]

Planetary climate studies also landed on the East coast, NASA's Goddard Institute for Space Studies (GISS). This organization, founded (somewhat ironically) by Jastrow in 1960, was intended to be the Agency's center for theoretical study of planetary atmospheres. That's still what it does. Unlike Goddard, Ames, and JPL (all of which also retain experimental and hardware programs), GISS is primarily theoretical. In modern space science, theory and data are both examined through the use of models. GISS actually performed the simulation studies that suggested weather satellites of the 1970s would not produce a great improvement in daily weather forecasts, for example.

GISS, of course, is also the home of NASA climate modeler James E. Hansen. Hired there in the early 1970s, Hansen initially worked on a scattering model for Venus' planetary cloud layers. Jule Charney, founder of numerical weather forecasting, performed some regional (Earth) climate studies using a GISS weather model in the middle of the decade; these, and Pollack's comparative planetary studies, triggered Hansen's interest in Earth's climate. Hansen and his colleague at GISS, Andrew Lacis,

---

38. Carl Sagan, "The Radiation Balance of Venus," *JPL Technical Report* no. 32-34 (15 September 1960); D. R. Hitchcock and J. E. Lovelock, "Life Detection by Atmospheric Analysis," *Icarus* 7:2 (1967): pp. 149+; Lynn Margulis and J. E. Lovelock, "Biological Regulation of the Earth's Atmosphere," *Icarus* 21 (1974): pp. 471–489; James Lovelock, *Gaia: A New Look at Life on Earth* (New York: Oxford University Press, 1974); Carl Sagan and George Mullen, "Earth and Mars: Evolution of Atmospheres and Surface Temperatures," *Science* 177 (7 July 1972): pp. 52–56.

39. Paul Ehrlich et al., *The Cold and the Dark: The World after Nuclear War* (New York: W. W. Norton, 1984); Lawrence Badash, "Nuclear Winter: Scientists in the Political Arena," *Physics in Perspective* (2001): pp. 76–105.

40. On the Marshall Institute's founding, see Robert Jastrow to Robert Walker, 1 December 1986, box 21 file "George C. Marshall Institute, accession 20001-01, William Aaron Nierenberg papers, Scripps Institution of Oceanography Archives, University of California–San Diego; Draft Proposal for the George C. Marshall Institute, sent to Bill Nierenberg, 12 December 1984, MC 13, William Aaron Nierenberg Papers box 75, folder 6, Scripps Institution of Oceanography Archives, University of California–San Diego.

came from the University of Iowa. Hansen considers his intellectual forebears to be the late James van Allen of Iowa and Jule Charney.[41] He began building GISS's climate modeling capabilities in mid-decade.

Hansen's climate model was one of two that formed the basis of the first declarative study by the National Academy of Science on the subject of global warming, the 1979 "Charney Report." In its preface, the University of Wisconsin's Verner Suomi stated that there was "no reason to doubt that climate changes will result and no reason to believe that these changes will be negligible."[42] Every study by the National Academy of Science since has ratified its conclusions, as has the Intergovernmental Panel on Climate Change in each of its three assessments of climate science since its foundation in 1988.[43]

Global warming is caused by human emissions of carbon dioxide, which exceed volcanic output by more than 150 times. Carbon dioxide is a chemical, of course, although one that is currently difficult to measure accurately from space. Sagan was the first to blame Venus's extremely hot climate on a surplus of this gas; Venus, American and Soviet robotic spacecraft found in the 1960s and 1970s, has about 300 times as much atmospheric carbon dioxide as Earth. And these facts bring us back to my claims about chemistry, climate, and NASA's role in fostering political controversy.

To be blunt about it: NASA has shown conclusively that humans cannot continue to change the chemistry of Earth's atmosphere without enormous, and negative, consequences. One of these consequences is ozone depletion, a cancer risk. The larger consequence is global warming. A 1983 National Academy study laid out some of the problems for the United States: much of the irrigated area of the southwest would have to be abandoned unless sufficient additional water could be imported. The highest forecasts of sea level rise would leave most of coastal America underwater unless defended by dikes of 15 to 20 feet (5 to 6 meters)—quite an expensive undertaking along several thousand miles of coast.[44] Further, this committee pointed out, "[I]ncreasing carbon dioxide is expected to produce changes in global mean temperature that, in both magnitude and rate of change, have few or no precedents in the Earth's recent history."[45] This had led much of the

---

41. Author interview with Hansen, January 2006; J. Hansen et al., "Climate Modeling in the Global Warming Debate," *General Circulation Model Development* (New York: Academic Press, 2000), pp. 127–164.

42. National Research Council, *Carbon Dioxide and Climate* (Washington, DC: National Academies Press, 1979), p. vii.

43. Naomi Oreskes, "The Scientific Consensus on Climate Change," *Science* (3 December 2004): 1686, DOI 10.1126/science.1103618; also see Oreskes, "The Scientific Consensus on Climate Change: How Do We Know We're Not Wrong?" in *Climate Change: What It Means for You, Your Children, and Your Grandchildren*, Joseph DiMento and Pamela Doughman, ed. (Cambridge, MA: MIT Press, 2007), forthcoming.

44. NRC, Carbon Dioxide and Climate, 1983, p. 46.

45. NRC, Carbon Dioxide and Climate, 1983, p. 51.

committee to a state of "unease," as they put it, because they could not assume that the effects would appear in a gradual, linear fashion.

Yet these facts are a direct affront to the belief system that dominates the American political right. Beginning with the popular work of famed economist Milton Friedman, the American right has adopted a worldview that equates economic freedom with political freedom.[46] Although this is historical fantasy—we have never been a free-market nation—this belief system forces its adherents to deny the reality of the scientific community's fact claims. True believers seek to protect their faith by denying inconvenient facts, and they believe that no scientific agency of our government has done more to undermine what George Soros has called "free market fundamentalism" than NASA has.[47] The accepted facts are that unregulated carbon dioxide emissions are the cause of global warming; solving the problem will require regulation of some kind. Such regulations would violate fundamental precepts of free-market theology, so the rightists have decided to reject the facts. Instead, they have formulated a convenient set of conspiracies. And, perhaps further reinforcing my point about "targeted media" above, a recent Pew Research Center for the People and the Press poll indicates that 43 percent of college-educated Republicans reject the fact of global warming, whereas only 25 percent of Democrats do.[48]

Hence, if one reads what's published in the rightist political literature mentioned above, one will find environmentalism equated with communism throughout the 1980s and 1990s, and with terrorism after 2001. One will also see scientists, including NASA scientists, derided as dupes and frauds for promoting "environmentalist nonsense." The "hidden agenda" of these scientists, in the words of one writer, is "against business, the free market, and the capitalistic system."[49] In the words of

46. Milton Friedman, *Capitalism and Freedom* (Chicago: University of Chicago Press, 1962). In his obituary of Friedman, economist Paul Krugman delineates carefully between Friedman's careful and meticulous academic work and his role as a public intellectual, promoting free markets with not entirely honest zeal. See Krugman, "Who Was Milton Friedman? *The New York Review of Books* 54/2 (15 February 2007).

47. George Soros, "The Capitalist Threat," *The Atlantic Monthly* 279 (February 1997): pp. 45–58; Myanna Lahsen has done signal work in digging up some of this material; see "Climate Rhetoric: Constructions of Climate Science in the Age of Environmentalism," Ph.D. dissertation, Rice University, 1998). The clearest expression of the belief that environmentalism is merely a plot to destroy capitalism I have found to date is in S. Fred Singer, "Global Warming: Do We Know Enough to Act?" in *Environmental Protection: Regulating for Results*, ed. Kenneth Chilton and Melinda Warren (Boulder, CO: Westview Press, 1991), pp. 29–50. My favorite exegesis of the behavior of fanatics is Eric Hoffer, *The True Believer: Thoughts on the Nature of Mass Movements* (New York: Harper and Row, 1951; Perennial Classic reprint, 2002), esp. pp. 79–80.

48. Pew Research Center for People and the Press, "Global Warming: A Divide on Causes and Solutions," 24 January 2007, *http://people-press.org/reports/display.php3?ReportID=303* (accessed 9 April 2007).

49. S. Fred Singer, "Global Warming: Do We Know Enough to Act?," p. 45. For a larger discussion of the political affiliations of the global warming denial effort see Lahsen, "Climate Rhetoric: Constructions of Climate Science in the Age of Environmentalism."

another, global warming is merely "a means of achieving an egalitarian society."[50] It is this literature that Senator James Inhofe (R-Oklahoma) and his colleagues draw on when they denounce global warming as "the greatest fraud ever perpetrated on the American people."[51] A new book from the Competitive Enterprise Institute puts it still more bluntly, calling them anti-American, anti-capitalist, and anti-human; to make clear their linkage to communism, they are "green on the outside, red to the core."[52] "Commies," as it were, in green makeup. To be clear, these political actors are not merely attacking leftist politicians. They attack practicing scientists for the results of their research—the very content of science. One journalist has labeled this assault "The Republican War on Science."[53]

One can argue, as physicist William Nierenberg did in the National Academy of Science's 1983 study of global warming, that the phenomenon is real and is likely to have severe consequences, but we don't need to do anything about it. Human civilizations have come and gone as climate changed around them; survivors simply migrate and rebuild elsewhere.[54] This is scientifically, and historically, entirely correct. One can argue about the ethics of such an approach—I would not subscribe to it—but it is correct and honest. One does not have to deny the legitimacy of a science to defend capitalists' right to pollute; in a pluralist, democratic society, one can simply accept Nierenberg's argument that we adapt to the new, warmer world. Yet this not the argument being made by the global warming deniers. Instead, they have chosen to demonize working scientists by applying the McCarthyite tactic of linking them to communism and accusing them of a global conspiracy.

This extreme reaction against climate science by the American right is not merely a quibble over interpretations of data. And it is not happening simply because ExxonMobil has spent millions of dollars a year supporting the denial industry.[55]

---

50. Aaron Wildavsky, "Global Warming as a Means of Achieving an Egalitarian Society: An Introduction," in Robert C. Balling, *The Heated Debate: Greenhouse Predictions versus Climate Reality* (San Francisco: Pacific Research Institute for Public Policy, 1992), p. xv.

51. Inhofe's speech is available at *http://inhofe.senate.gov/pressreleases/climateupdate.htm* (accessed 10 July 2005); also see Chris C. Mooney, *The Republican War on Science* (New York: Basic Books, 2005), p. 78.

52. Christopher C. Horner, *The Politically Incorrect Guide to Global Warming* (and Environmentalism) (New York: Regnery Publishing, 2007).

53. Chris Mooney, *The Republican War on Science* (New York: Basic Books, 2005), esp. pp. 78–101. Mooney attributes this "war" to the business interests of the afflicted corporations and their ability to buy political protection, ignoring the ideological issues at the root of the conflict.

54. See Carbon Dioxide Assessment Committee, *Changing Climate* (Washington, DC: National Academies Press, 1983), pp. 48–55, 61–63; and physicist Alvin Weinberg's review, "Comments on NRC Draft Report of the Carbon Dioxide Assessment Committee," folder 7, box 86, William Aaron Nierenberg Collection, MC 13, Scripps Institution of Oceanography Archives, University of California–San Diego.

55. Union of Concerned Scientists, *Smoke, Mirrors and Hot Air: How ExxonMobil Uses Big Tobacco's Tactics to Manufacture Uncertainty on Climate Science* (Washington, DC: Union of Concerned Scientists, 2007), pp. 3–6.

It is a defense of a political ideology. Nothing else adequately explains the rightists' reaction to a half-century of scientific research.

I do not wish to make too much of NASA's role in the science of global warming. As Spencer Weart's recent history of global warming shows, there are many other threads to that story. But since the late 1970s, NASA funding for climate science has overwhelmed all other sources. It is by far the dominant funder of the current Climate Change Research Program, a fact in need of explication. Of the $1.86 billion dollars spent on climate science in fiscal year 2005, NASA spent $1.24 billion. The next largest funder, the National Science Foundation, spent $198 million that year.[56] NASA's interest in planetary climates stemming from the Mariner missions of the 1960s, and the seeming need for a global view of global warming, made its entry into this research field an obvious choice.

It's also true that these controversial fields were not originally as politically charged as they are now. During the 1970s, both American political parties accepted the reality of environmental damage and argued over policy details: the most efficient form of regulation, standards of evidence, how to determine when a sufficiency of evidence existed to base regulation upon. The modern anti-environmental movement—often it is called "pro-business" to disguise its true nature—started in the Western states at the same time and achieved its first national expression with Reagan's election.[57] When NASA officials decided to enter these fields, the technocratic impulse still ruled: once one understood the science, science-based regulation would follow. It made sense to them to try to lead in fields relevant to the agency's technological capacities. As technocratic managers, they also didn't expect that anti-environmentalism would wind up dominant 20 years later. The magnitude of the controversy they entered came as a surprise.

Inside the scientific community, the intersection of space science and Earth science also engendered controversy. Lovelock's Gaia hypothesis was widely criticized for its depiction of Earth as a self-regulating *organism*. In one paper he used a metaphor of a planetary engineer to describe how the integrated Earth system worked. This had obvious metaphysical implications that he later regretted. In short, he was attacked for Gaia's religious implications. But his view of planetary climate as a system of

56. Spencer Weart, *The Discovery of Global Warming* (Cambridge, MA: Harvard University Press, 2003); also see the expanded online edition at *http://www.aip.org/history/climate* (accessed 3 July 2006). Current funding for the Climate Change Science Program is given in CCSP-4Budget14Jan2006.pdf, available at *http://www.climatescience.gov/infosheets/factsheet4/default.htm* (accessed 12 September 2006).

57. On the anti-environmental policies of the Reagan administration, see Samuel P. Hays, *Beauty, Health and Permanence: Environmental Politics in the United States 1945–1985* (New York: Cambridge University Press, 1985), pp. 490–526; on the origin of anti-environmentalism in the West, see Hal K. Rothman, *The Greening of a Nation? Environmentalism since 1945* (San Diego: Harcourt Brace Publishers, 1998), pp. 169–181.

nested feedback-control systems involving what came in the 1980s to be called "biogeochemistry" was ultimately highly influential. In 1986, a group of scientists working under a NASA charter created a new, less offensive label for this integrated view of Earth, "Earth Systems Science." One can find textbooks bearing this name in university bookstores now and some universities have integrated their separate geology/geophysics/atmospheric/ocean science programs into a single department. The California Institute of Technology's integrated department is Earth and Planetary Sciences, while the University of California, Irvine called its program Earth System Science (it was organized and named by a former NASA scientist, Michael Prather). Earth sciences are in the midst of a sea change in organization and, I think, their intellectual structure, inspired by the availability of planetary-scale data.

Satellites seem to have disappeared from my narrative, so let me bring some back in. I have ignored solar physics and the tightly linked field of space weather. Our friendly local star's radiation and particulate output affects Earth's upper atmosphere, and solar radiation trapped by Earth's magnetic field does as well. These also affect satellites in Earth orbit and deep space probes. Because satellites have significant economic and military value, NASA, the DOD, and NOAA have all spent quite a bit of money over the last several decades on satellites and model studies aimed at understanding and predicting these effects. Scientists are also interested for the intrinsic scientific questions involved, of course—the motivation isn't solely utilitarian!

Three recent works discuss the evolution of solar science and space weather—the border between these two issues being, like space itself, rather tenuous. Karl Hufbauer's book *Exploring the Sun* focuses exclusively on the history of solar physics and the solar wind. He does not discuss their influence on Earth. Two other works make attempts at this. In *Storms from the Sun*, Michael Carlowicz and Ramon Lopez focus on the impact of solar eruptions on telecommunications and electrical power distribution networks. But their work is weak on the science and on the historical evolution of the field.[58]

A more useful treatment of the subject is Barbara Poppe's and Kristen P. Jorden's *Sentinels of the Sun*. Written largely from NOAA's perspective, this work focuses on the evolution of space weather forecasting with that agency. It gives the reader a good sense of the bureaucratic politics of the issue. Unfortunately, it is completely undocumented and, while it leaves the reader with the understanding that NASA and the Air Force have always been parallel actors in the field of space weather, it tells us little about their respective roles.[59] There is plenty of room left for new research in this area.

58. Karl Hufbauer, *Exploring the Sun: Solar Science since Galileo* (Baltimore, MD: Johns Hopkins University Press, 1991); Michael J. Carlowicz and Ramon E. Lopez, *Storms from the Sun: The Emerging Science of Space Weather* (Washington, DC: Joseph Henry Press, 2002).

59. Barbara Poppe and Kristen P. Jorden, *Sentinels of the Sun: Forecasting Space Weather* (Boulder, CO: Johnson Books, 2006).

The second type of satellite to discuss is oceanographic. NASA has flown satellite sensors aimed at both physical and biological oceanography. Neither type has been addressed in an historical study. The most significant of these has probably been TOPEX/Poseidon, a joint US-France mission that performed sea surface altimetry. Altimetry directly measures the height of the ocean surface; indirectly, it can tell us a great many other things about the oceans. Its most public outcomes have probably been in El Niño forecasting and in measuring sea level rise.

Scientifically, however, it may turn out to be revolutionary for its impact on ocean modeling. JPL and several other institutions are involved in a joint modeling effort aimed at transforming data from TOPEX/Poseidon and NOAA's Argo buoy network into a new, four-dimensional view of the world ocean. They are not done yet but, from what this group has published so far, they're building a radically new interpretation of how the ocean absorbs and distributes heat.[60] In turn, this will affect scientific understanding of how the geographic distribution of heat and precipitation will change under global warming.

The final type of satellite I will discuss is geodetic. These measure the shape of Earth and its gravity field. They were among the first kinds of scientific satellites flown, as they are very simple by their nature and militarily useful. Earth's gravity field is not perfectly spherical—it has "bumps" due to local concentrations of higher-density material within Earth—and the DOD wished to know where these were for more accurate targeting of nuclear missiles. Dwayne Day has published a set of articles on military geodetic satellites, the first historical studies on this subject.[61]

At NASA, the Goddard Space Flight Center has led the development of this technology until very recently. In collaboration with the DOD, NASA has flown a series of these satellites. These haven't been part of the Agency's "controversy portfolio," however, perhaps because most of the public has no idea what they do. But that's likely to change because the latest geodetic satellite, the JPL's GRACE mission, is accurate enough to produce mass estimates for ice sheets and aquifers. Recently, the mission scientists have pronounced that their data shows mass loss from both Greenland and Antarctic ice sheets, thus sticking themselves directly into

---

60. Carl Wunsch and Detelef Stammer, "Satellite Altimetry, the Marine Geoid, and the Oceanic General Circulation," *Annual Review of the Earth and Planetary Sciences* 26 (1998):219–253; D. Stammer et al., "The Global Ocean State during 1992–1997, Estimated from Ocean Observations and a General Circulation Model. Part III: Volume, Heat and Freshwater Transports," ECCO Report no. 6, 24 August 2001, *http://www.ecco-group.org* (accessed 1 July 2006).

61. Dwayne Day, "Mapping the Dark Side of the World, Part 1: The KH-5 ARGON Geodetic Satellite," *Spaceflight* 40/7 (July 1998): pp. 264–269; Dwayne A. Day, "Mapping the Dark Side of the World, Part 2: Secret Geodetic Programmes after ARGON," *Spaceflight* 40/8 (August 1998): pp. 303–310.

the climate wars.[62] There is a lot more historical work on the subject of geodetic satellites and their impacts on science, however. To date, very little examination or interpretation of this field has been done.

Space science, then—not merely Earth science but aspects of planetary science as well—has radically altered our beliefs about Earth and its processes. It has forced scientists, often against their own political preferences, to come to grips with the very uncomfortable notion that humans have become geological agents. We humans have the ability to change the basic conditions of life on Earth. The dominant belief from the nineteenth century—that humans were too puny to have any significant impact on Earth—can no longer be sustained in the face of NASA's research. This is the root of the political problems faced by Earth scientists in the 2000s.

Conclusion

Much of the historical literature on the Space Age to date has focused on human spaceflight. To borrow the analytical terminology of feminist history of science, this privileges the narrative of a handful of transient males—the professional space tourists we call astronauts—over that of most of the species. But the largest impacts of space technologies to nonastronaut humans have come from robotic spaceflight. These impacts have been positive and negative, economic, political, scientific, military, and ideological.

Unlike humans (to date, at any rate), robot spacecraft can be engineered to live in space, providing routine data for sometimes decades. They can also be sent to places inhospitable or deadly to humans. This has been the source of their success as scientific explorers, routine storm monitors, communications facilitators, and intelligence gatherers. It is also probably why they have virtually no place in the narrative of space history to date. As nonhumans, they're ignored. It's far easier to write a compelling heroic narrative about human actors than robotic ones, and space history has often been little more than advocacy written in heroic prose.[63]

62. I. Velicogna and J. Wahr, "Greenland Mass Balance from GRACE," *Geophysical Research Letters* 32/18 (30 September 2005); I. Velicogna and J. Wahr, "Measurements of Time-Variable Gravity Show Mass Loss in Antarctica" *Science* 311 (24 March 2006): pp. 1754–1756; J. L. Chen et al., "Antarctic Mass Rates from GRACE," *Geophysical Research Letters* 33 (9 June 2006). On geodesy more generally, see Paul D. Lowman, *Exploring space, Exploring Earth: New Understanding of the Earth from Space Research* (New York: Cambridge University Press, 2002). The title of this book is deceptive. It deals only with geodesy from space, ignoring all other fields of Earth remote sensing.

63. Roger D. Launius, "The Historical Dimension of Space Exploration: Reflections and Possibilities," *Space Policy* 16 (2000): pp. 23–38, esp. pp. 24–26.

But engineering is a human endeavor as well. As the above discussion should suggest, spacecraft engineering is wrapped up in the politics of science and government—both human enterprises. The New Space History, as Roger Launius has called it, has plenty of room in it for this narrative as well.[64] Precisely because there is so little already written, it's also a rich area for research.

64. Launius, "Historical Dimension of Space Exploration."

CHAPTER 15

FOR ALL MANKIND: SOCIETAL IMPACT OF APPLICATION SATELLITES

David J. Whalen

> *First, I believe that this nation should commit itself to achieving the goal, before this decade is out, of landing a man on the moon and returning him safely to the earth . . . .*
>
> *Secondly, an additional 23 million dollars, together with 7 million dollars already available, will accelerate development of the Rover nuclear rocket . . . .*
>
> *Third, an additional 50 million dollars will make the most of our present leadership, by accelerating the use of space satellites for world-wide* communications.
>
> *Fourth, an additional 75 million dollars—of which 53 million dollars is for the Weather Bureau—will help give us at the earliest possible time a satellite system for world-wide* weather *observation.*
>
> JOHN F. KENNEDY
> Special Message to the Congress on Urgent National Needs
> 25 May 1961

The Apollo 11 plaque states that the U.S. astronauts "came in peace for all mankind." But long before the Apollo landing President Kennedy wanted the United States to be seen as running a high-technology program whose *practical benefits* would be *for all mankind*. Kennedy's principal audience was meant to be the third world—as well as the U.S. public—but his message also provided "cover" for the Congressmen who would have to authorize billions of dollars for the Apollo program. The U.S. program would also provide practical, nonmilitary applications that would benefit all Americans as well as all mankind.

The military had already begun their program of applications satellites. By the time of Kennedy's speech, the Department of Defense (DOD) had launched a whole generation of reconnaissance, navigation, and weather satellites. These systems were operational and new generations were in the works. Strangely, this DOD litany of successful applications did not include communications satellites.

Much of the military thinking about space originated with RAND, a Douglas Aircraft Company Research and Development (R&D) unit at the time. On 2 May 1946, RAND published their famous "Preliminary Design of an Experimental World-Circling Spaceship."[1] Chapter 2 of this document, drafted by Louis Ridenour, was titled "The Significance of a Satellite Vehicle." Greatest significance is given to the use of rockets (satellite vehicles) as bombardment vehicles, but next in importance was the observation capability of a satellite over enemy territory. This observational capability would allow weather observation before the raids and accurate bomb damage assessment after raids. The document also discusses the advantages of satellites as communications relay stations. The simplicity of operations if satellites are in geostationary orbits is addressed in passing. The value of then-current communications through the ionosphere is given as $10 billion.

In 1947 RAND published the first of many follow-ups to the 1946 report. These reports, prepared under the direction of James E. Lipp, covered a variety of topics. One report, "Communication and Observation Problems of a Satellite," continued the discussion of satellite communications and brought up the issue of a "spy satellite" for the first time.[2] RAND continued its studies of reconnaissance and weather satellites, in 1951 publishing a report titled "Inquiry into the Feasibility of Weather Reconnaissance from a Satellite Vehicle" by William Kellogg and Stanley Greenfield, and another report on the "Utility of a Satellite Vehicle for Reconnaissance" by James E. Lipp, Stanley M. Greenfield, and R. S. Wehner.[3] Perhaps more important for the space race was an earlier RAND report entitled "The Satellite Rocket Vehicle: Political and Psychological Problems."[4] This document was considered by space historian Walter McDougall as "the birth certificate of American space policy."[5]

Communications satellites had their origins in science fiction and their first serious exposition by an AT&T engineering manager. Arthur C. Clarke (1917–), then a Royal Air Force radar officer, published an article in the October 1945 issue

1. RAND Corp., "Preliminary Design of an Experimental World-Circling Spaceship," SM-11827, 2 May 1946.

2. RAND Corp., "Communication and Observation Problems of a Satellite," RA-15028, 1 February 1947.

3. William Kellogg and Stanley Greenfield, "Inquiry into the Feasibility of Weather Reconnaissance from a Satellite Vehicle," RAND R-218, 1947; James E. Lipp, Stanley M. Greenfield and R. S. Wehner, "Utility of a Satellite Vehicle for Reconnaissance," RAND R-217, 1947.

4. Paul Kecskemeti, "The Satellite Rocket Vehicle: Political and Psychological Problems," RAND RM-567, 4 October 1950.

5. Walter A. McDougall, . . . *the Heavens and the Earth: A Political History of the Space Age* (New York: Basic Books, 1985), p. 108.

of *Wireless World* entitled "Extra-terrestrial Relays."[6] In this article he discussed the advantages of 24-hour geostationary orbits that would allow a satellite to maintain position over the same portion of the equator indefinitely. Clarke foresaw use of space stations at this altitude for radio and television broadcasting. John R. Pierce (1910–2002) was, like Clarke, a science fiction writer, but he was also an engineering manager at Bell Telephone Laboratories. In an article "Don't Write, Telegraph," published in *Astounding Science Fiction* in 1952, Pierce discussed some possibilities regarding communications satellites.[7] In 1954 he was asked to give a space talk to the Princeton section of the IRE (the Institute of Radio Engineers, now the IEEE).[8] According to Pierce, "The idea of communication satellites came to me. I didn't think of this as my idea, it was just in the air. Somehow, I had missed Arthur Clarke's paper on the use of manned synchronous satellites for communication." In 1958 Pierce and his colleague Rudolf Kompfner prepared a presentation on satellite communications for a conference. This was later published in the *Proceedings of the IRE* in March 1959.[9]

Thus, all four applications areas—reconnaissance, navigation, weather, and communications—had been discussed at some length in the immediate post-World War II period. Interest had accelerated in response to the U.S. Air Force spy satellite Request for Proposal in 1954, but the real push for applications satellites came as a response to Sputnik.

Reconnaissance Satellites

All of the early RAND recommendations had been for a "direct readout" satellite—one that transmitted pictures to the ground electronically. Many of the studies assumed a standard television camera. The Air Force (aided by its RAND think tank) had begun development of a reconnaissance satellite, Weapon System 117L (WS-117L), on 16 March 1955. The program, initially called Advanced Reconnaissance Satellite (ARS), then SENTRY, and finally Satellite and Missile Observation System (SAMOS), was slow to mature. By 1957, members of the Presidential Science Advisory Committee (PSAC) were dissatisfied with the Air Force program; they wanted a "film return" satellite and they wanted the program managed by the Central Intelligence Agency (CIA). The success of the U-2 seemed to indicate that the CIA was better at bringing new technology into operation in a short period of time. On 7 February 1958, President Eisenhower authorized the CIA to proceed with CORONA.

---

6. Arthur C. Clarke, "Extra-terrestrial Relays," *Wireless World* 51 (October 1945): pp. 303–308.

7. John R. Pierce, "Don't Write: Telegraph," *Astounding* (March 1952). Pierce wrote at least 20 articles for *Astounding* under his pen name, J. J. Coupling, and at least one under his real name.

8. Later published: John R. Pierce, "Orbital Radio Relays," *Jet Propulsion* 25 (April 1955): pp. 153–157.

9. John Robinson Pierce and Rudolf Kompfner, "Transoceanic Communication by Means of Satellites," *Proceedings of the IRE* 47 (March 1959): pp. 372–380.

CORONA/GAMBIT

Within a few months, the WS-117L program had been reoriented to include CORONA (film return), MIDAS (early warning), and SAMOS (direct readout—later to include film return). The first Discoverer (CORONA) launch was on 28 February 1959; it was a failure, as were most launches over the next two years. The first successes were in August 1960 when reentry vehicles (SRVs) were recovered from the ocean and in mid-air. From 1959 to 1972, almost 150 CORONA (KH-1 through KH-4B GAMBIT) satellites were launched on Thor-Agena vehicles. After August 1960, most were successful.[10]

From 1959 to 1971, CORONA was the principal U.S. reconnaissance satellite (along with a few ARGON and LANYARD special-purpose satellites). SAMOS was eventually cancelled. In 1960 a joint program office was formed and designated the National Reconnaissance Office (NRO). NRO was staffed by the CIA and Air Force. The existence of NRO was "revealed" on 18 September 1992. The entire CORONA program was declassified on 24 February 1995. Later programs are still classified, making accurate descriptions difficult.

The KH-7 and KH-8 GAMBIT satellites provided increased resolution (~0.5 m) over the CORONA satellites (~3 m). The CORONA satellites had grown in size from 800 kg to about 2,000 kg (two SRVs), but were all launched by Thor-Agena launch vehicles. The KH-7 satellites were launched on Atlas-Agenas. The heavier (3,000 kg) KH-8 satellites were launched on Titans. About 100 GAMBITs were launched between 1963 and 1984 with about a 95 percent success rate. Early GAMBITs had lifetimes of days but, over time, lifetimes grew to weeks.

The Rest of the "Spysats"

The Air Force had always wanted to put men in space; the Manned Orbiting Laboratory (MOL) was their great opportunity to do so. Although MOL had many goals, its primary purpose was the KH-10 (DORIAN) reconnaissance system. The vehicle would have weighed about 15,000 kg. First authorized in 1962–1963, MOL would eventually be cancelled in 1969 after an expenditure of billions of dollars.

Starting in the late 1960s, the United States and the USSR began discussing arms limitation. The Soviet Union had established—or was establishing—rough parity in nuclear weapons and intercontinental ballistic missiles (ICBMs). Both sides were developing anti-ballistic missile (ABM) systems. In the process of negotiating the Strategic Arms Limitation Treaty I (SALT I, the ABM Treaty), it was agreed

10. Much of the information on CORONA comes from Merton E. Davies and William R. Harris, "RAND's Role in the Evolution of Balloon and Satellite Observation Systems and Related U.S. Space Technology," RAND, R-3692-RC, September, 1988; and Dwayne A. Day, John M. Logsdon, and Brian Latell, *Eye in the Sky: The Story of the CORONA Spy Satellites* (Washington, DC: Smithsonian Institution Press, 1998). Also of interest is Robert A. McDonald ed., *CORONA—Between the Earth and the Sun: The First NRO Reconnaissance Eye in Space* (Bethesda, MD: American Society for Photogrammetry and Remote Sensing, 1997).

that "national technical means" would be used to verify compliance and that no interference with these means would be allowed. Spy satellites were legal!

It has been argued that the KH-9 was developed as a backup to MOL. The vehicle weighed more than 11,000 kg—almost four times the weight of a KH-8 GAMBIT and not much less than the MOL. Known as "Big Bird," the KH-9 HEXAGON carried a television camera as well as film cameras and four SRVs. It was launched by Titan 3D rockets. Big Birds increased satellite lifetimes to months. Of 20 KH-9 launches, only one failed, the last launch in 1986. Declassification was progressing until the fall of 1997. There were even plans to place a KH-9 in the new Smithsonian hangar at Dulles Airport. According to Dwayne Day, a military space historian who has written about CORONA and other spy satellites, Big Bird probably "gathers dust in a classified warehouse . . . only a few yards down from the Lost Ark of the Covenant."[11]

Perhaps the biggest improvement in spy satellites was the all-electronic, direct-readout KH-11 Kennan/Crystal and its successor, the KH-12/KH-11B Improved Crystal. These satellites finally provide the capability that SENTRY/SAMOS hoped for: real-time direct readout—facilitated by communications relay satellites including SDS, TDRSS, and MILSTAR. About two dozen of these satellites have been launched. The Improved Crystal weighs almost 20,000 kg and can only be launched by the Space Shuttle or Titan 4. Lifetimes are now measured in years. The KH-12 carries about 7,000 kg of fuel and its lifetime is more than 10 years. In addition to the visible and infrared capabilities of the KH satellites, at least a half-dozen Lacrosse radar satellites have been launched.

Image intelligence (imint) and human (spy) intelligence (humint) have been supported by various forms of signal intelligence (sigint)—including satellite sigint. These include Navy systems from the 1960s (GRAB, DYNO, POPPY), ferrets launched with KH-9 satellites, Air Force systems (CANYON, VORTEX, MERCURY), and the CIA's AQUACADE. Many of these satellites are now in geosynchronous or Molniya orbits—and are all but invisible.

Societal Impact of Reconnaissance Satellites

Both the United States and the Soviet Union had been brought into World War II as the result of surprise attacks. Each of these two countries suspected its rival of planning a surprise attack. Perhaps the greatest impact of reconnaissance satellites was reducing that threat. In the words of President Lyndon B. Johnson,

> I wouldn't want to be quoted on this, but we've spent thirty-five or forty billion dollars on the space program. And if nothing else had come out of it except the knowledge we've obtained from space photography, it would be worth ten times what the program has

11. Dwayne A. Day, "The Invisible Big Bird: Why There Is No KH-9 Spy Satellite in the Smithsonian," *The Space Review* (8 November 2004), *http://www.thespacereview.com* (accessed 24 August 2007).

> cost. Because tonight we know how many missiles the enemy has and, it turned out, our guesses were way off. We were doing things we didn't need to do. We were building things we didn't need to build. We were harboring fears we didn't need to harbor.[12]

NAVIGATION SATELLITES

In the days immediately following the launch of Sputnik in October 1957, scientists and engineers worked to analyze the spacecraft's signal and its orbit. Bill Guier and George Weiffenbach of the Johns Hopkins University Applied Physics Laboratory (APL) listened to the satellite's signal and monitored the change in its frequency due to the Doppler effect. They used this Doppler shift to compute an orbit for the Russian satellite.

Another APL engineer, Frank McLure (1916–1973), realized that if the orbit were known, the Doppler information could be used to determine the position of the radio receiver on the ground. In early 1958, McLure described the potential for developing a space-based navigation system. Within a few weeks, APL proposed a navigation system to the Navy.

Transit Satellites

The earliest Transits were launched on the Thor-Able and Thor Able-Star rockets from Cape Canaveral. The very first of these occurred on 17 September 1959. The last two experimental Transit satellites demonstrated that precise navigation was possible using two frequency beacons broadcasting the satellite ephemerides (orbits). This system was so robust that it was capable of determining the harmonics of Earth's gravitational field and the effects of propagation through the ionosphere. The last satellites were also able to demonstrate the availability of the satellites when in a near-circular orbit at about 1,000 km and inclined about 66 degrees.

After the poor reliability of the Naval Avionics Facility Indianapolis (NAFI)-built Transit satellites, RCA built the rest. It was always clear that Transit had significant limitations. The accuracy was good enough for nuclear weapons (<1 km) but not good enough for conventional weapons. The Transit position fixes took some time to obtain, making Transit almost useless for moving objects. The Navy continued research—especially at Naval Research Laboratory (NRL)—on improvements. In 1964, the Air Force started a new navigation satellite program, Project 621B.

NAVSTAR/GPS

By 1972, DOD wanted just one program: a system that could be used to navigate fast-moving aircraft and even to deliver conventional weapons. In 1973, the NAVSTAR Global Positioning System (GPS) program was approved. Over the next few years, arguments and tradeoffs between the Navy and Air Force were

12. Lyndon B. Johnson, Nashville, Tennessee, March 1967.

adjudicated—mostly on their merits—and GPS Block I launches began in 1978. Transit was kept in operation until 1996.[13]

GPS satellites are in 12-hour circular orbits inclined 55° to the equator—much higher than Transit. There are six orbital planes, each containing four satellites. Two L-band frequencies are broadcast (L1: 1575.42 MHz and L2: 1227.60 MHz) containing the time (Universal Time Coordinated and a Pseudo Random Noise code) and satellite position. Differences between the satellite time and the vehicle time provide range measurements—three range measurements allow a position to be determined within 10 to 100 m. The civilian Standard Positioning Service (SPS) signal has a conditional access code (CA) that degrades accuracy. This selective access means that civilian signal is not good enough to use for weapons delivery. Military users can get position to within a few meters.

A contract for eight Block I GPS satellites was awarded to North American (Rockwell) in 1974. A contract modification for four additional Block I satellites was awarded in 1981. NAVSTAR 12 was produced as the Block II qualification model. There was only 1 failure out of 11 launches. Block II/IIA added nuclear detonation detectors and many improvements. The satellite was still manufactured by Rockwell, but the launch vehicle was now a Delta 6925 for Block II and a Delta 7925 for Block IIA. There was only 1 failure in 33 launches.

After building 44 GPS satellites, Rockwell lost the "replacement" contract to General Electric Astro Space (formerly RCA Astro Electronics, currently Lockheed Martin Commercial Space Systems). The Block IIR satellites were based on the Astro Space series 4000 geosynchronous communications satellite. The 1989 contract was for 21 satellites. Many improvements in cost, lifetime, autonomy, and improved precision were made on the IIR series.

In 1997 the Air Force awarded a contract for six GPS satellites and 27 options to Rockwell (now Boeing). In 2000 the decision was made to rebid the contract. A series of studies for a "generation after next" system, Block III, was begun in 2000. This was revised in 2003 and again in 2005. The competitors are Lockheed Martin (Block IIR) and Rockwell (Block I, II, IIA).

Civilian Use

The decision to allow Transit use by commercial ships seems to have been made at an early date. This was given extra urgency when the supertanker *Torrey Canyon* ran aground on the Cornish coast in March 1967, spilling 120,000 tons of oil.[14]

---

13. Much of the material in this section was obtained from National Research Council, *The Global Positioning System* (Washington, DC: National Academies Press, 1995), pp. 145–276 and *http://www.astronautix.com* (accessed 24 August 2007).

14. Abraham Hyatt, memo to Deputy Administrator and Associate Administrator, "Informal Discussions Regarding Navigation Satellites," 7 September 1961; Alton B. Moody, "Navigation Satellite Progress," National Electronics Congress, 9 October 1962; "NAV/TRAF SAT program," *Space Daily* (21 April 1964): p. 118; Walter Sullivan, "How to Navigate with Satellites," *The New York Times*, 2 April 1967: p. E7.

By 1995, civilian use of GPS exceeded military use. Ten years later, GPS was the established navigation system—an "international utility." Most other navigation systems were in the process of shutting down. But GPS remained a military system: use could be denied during a military emergency. The U.S. Federal Aviation Administration (FAA) started a program in 1995 called the Wide Area Augmentation System (WAAS) that would facilitate use of GPS for instrument landings. This would obviate the need to build the Microwave Landing System (MLS) scheduled to replace the old Instrument Landing System (ILS). Most MLS systems in the United States have been turned off and replaced by GPS. [15]

As GPS replaced all previous navigation (and instrument landing) systems, many foreign countries became quite concerned that transportation safety was dependent on an American military system. In part to assuage these fears and to increase the precision of GPS, selective availability (SA)—a system that ensured lower civilian accuracy—was turned off on 1 May 2000. Somewhat earlier in 1982, the Soviet Union had launched its own navigation system, GLONASS (Global Orbiting Navigation Satellite System). The system was fully operational in 1995. Unfortunately, the collapse of the Soviet economy left Russia unable to maintain the system for several years. A planned replenishment will be complete by 2010.[16]

Both GPS and GLONASS are military systems that allow civilian use. The European Galileo system will be completely civilian-run by a private consortium. The four GIOVE (Galileo In-Orbit Validation Experiment) satellites will be launched by 2008. The 30-satellite operational system will be complete by 2010. Galileo will provide greater accuracy (~1 m) and will work in buildings and under trees. Galileo and GPS will be compatible.[17]

Societal Impact of Navigation Satellites

The original purpose of navigation satellites was to maintain the so-called balance of terror. Even if the Soviet Union had launched a first strike, the submarine-launched ICBMs (SLBMs) would have enough navigational accuracy to level most of the cities of the Soviet Union—whose positions were now well-known thanks to reconnaissance satellites. NAVSTAR/GPS gave aircraft the same navigational assurance—and accuracy to within meters, not kilometers. This improved accuracy led to GPS-guided munitions used in the Gulf wars.

Most surprising is the ubiquity of GPS receivers in the civilian world. These are now the *primary* means of navigating ships and aircraft. They are also widely used

15. David Field, "U.S. to Let Airliners Navigate by Its Satellites," *Washington Times*, 28 March 1995: p. B7; Warren E. Leary, "Civilian Uses Are Proposed for Satellites," *The New York Times*, 1 June 1995: p. A23.

16. *http://www.spacetoday.com* (accessed 24 August 2007).

17. Ibid.

in cars and trucks, and by hikers. Of the three applications pioneered by the military, this is by far the greatest success story. Commercial sales of GPS receivers are now a $9 billion industry.[18]

Weather

By the 1950s, the idea of weather satellites was beginning to surface. In 1951, RAND published "Inquiry into the Feasibility of Weather Reconnaissance from a Satellite Vehicle" and Arthur C. Clarke depicted polar and geosynchronous "metsats" in the endpapers of *The Exploration of Space*.[19] In 1954, a tropical storm was discovered accidentally when pictures taken from an Aerobee sounding rocket were analyzed. Also in 1954, Dr. Harry Wexler, the Weather Bureau's chief scientist, presented a paper on "Observing the Weather from a Satellite Vehicle" at the Third Symposium on Space Travel.[20] In 1955, when the decision was made to launch a satellite during the upcoming International Geophysical Year (IGY), weather observation and radiation balance payloads were considered and eventually were flown on Vanguard and Explorer satellites.

Polar Satellites/TIROS

In spite of the influence of scientists such as Wexler and Verner Suomi, the first weather satellite was a product of the military. TIROS (Television Infra-Red Observation Satellite) was RCA's losing entry in the Air Force WS-117L competition won by Lockheed in 1956. The Army was persuaded to support development of TIROS as a polar-orbiting weather satellite. The project was transferred to ARPA and eventually to NASA in 1958. The first launch was on a Delta on 1 April 1960. The satellite had two television cameras: one wide-angle and one narrow-angle (high-resolution) on TIROS-1 and -2, and both wide-angle on succeeding TIROS satellites. TIROS-8 pioneered the Automatic Picture Transmission (APT) camera system. TIROS satellites had the cameras mounted on the base of the satellite, aligned with the spin axis. This meant that the cameras were Earth-pointing for only a small fraction of their orbits. TIROS-9 pioneered the "cartwheel" configuration wherein the cameras were mounted on the sides of the spacecraft; the spacecraft spin axis was aligned with orbit normal and pictures were taken continuously. All launches were from Cape Canaveral into high-inclination (481)° orbits until TIROS-9 and -10 were launched into Sun-synchronous (SS) polar orbits. Sun-synchronous orbits allowed pictures to be taken at the same local time every day (usually early morning).

---

18. Satellite Industry Association (SIA)/Futron Corp., *State of the Satellite Industry Report*, (San Diego, CA: SIA/Futron, June, 2006).

19. Arthur C. Clarke, *The Exploration of Space* (New York: Harper, 1951).

20. Harry Wexler, "Observing the Weather from a Satellite Vehicle," Third Symposium on Space Travel, 4 May 1954.

The Pentagon recognized the disadvantages of the TIROS baseplate-mounted cameras and the advantages of Sun-synchronous orbits. Joseph V. Charyk, director of the National Reconnaissance Office (NRO), concluded that NASA development of a better weather satellite (Nimbus) would be delayed and expensive. He also was uncomfortable with the international commitments NASA had made to share TIROS weather pictures. Weather information was critical to NRO—too many spy satellite pictures showed nothing but clouds. In 1961, Charyk sponsored what was to become the Defense Meteorological Satellite Program (DMSP). The program envisioned an improved RCA TIROS launched on a Scout launch vehicle from Vandenberg Air Force Base. The satellite was much lighter than TIROS and carried a single television camera that would "snap" pictures of Earth when the horizon sensors indicated that the camera was pointed in the appropriate direction.

At about this time, it was decided that the next-generation civilian weather satellite, TOS (TIROS Operational System, also known as ESSA) would be a copy of DOD's DMSP Block 4A rather than the NASA Nimbus, which became a research vehicle and later the model for ERTS/Landsat. The Block 5 satellites were three-axis-stabilized rather than spin-stabilized. A variant of Block 5 became the civilian ITOS (Improved TOS, also known as National Oceanic and Atmospheric Administration [NOAA]). The DMSP program remained classified until late 1972, when DMSP data were routinely delivered to the Weather Bureau.[21]

The military and the intelligence community were the initial sponsors of weather satellites. NASA took over part of this remit, and by the mid-1960s the Weather Bureau (ESSA, then NOAA) took responsibility for polar weather satellites. Geostationary weather satellites seem to have been championed by NASA and transferred to the Weather Bureau almost immediately after their launch in the 1970s. The military and the intelligence community retained a separate polar weather satellite program into the twenty-first century, but a single National Polar Orbit Environmental Satellite System (NPOESS) is in the works—amid much disarray.

Geosynchronous Weather Observation

In spite of Arthur C. Clarke's work, NASA (and Hughes) first looked at geosynchronous orbit (GEO) as a place for weather satellites, not communications satellites. A consequence of the stationary orbit over the equator was that a GEO weather satellite could take continuous pictures of about one-third of Earth's surface. The polar weather satellites only took one picture (two if we include night-time IR pictures) of a given location each day.

21. Most of the material on DMSP is taken from R. Cargill Hall (NRO Historian), "Chapter Three: Weather Reconnaissance," n.d. (1988?), original classified TOPSECRET/TALENT KEYHOLE. Other sources include *http://earth.nasa.gov/history and http://www.astronautix.com* Web sites (accessed 24 August 2007).

In 1965, Verner Suomi and Robert Parent started the Space Science and Engineering Center (SSEC) at the University of Wisconsin-Madison with funding from NASA and the National Science Foundation. While at SSEC, Suomi developed the spin-scan camera. SSEC's spin-scan camera was launched on ATS-1 in 1966. The camera scanned an east-west strip of Earth with each rotation of the spinning satellite. By tilting a mirror in the camera slightly at each rotation, a multi-strip image of Earth could be created in less than 30 minutes.

The NASA Applications Technology Satellites (ATSs) had originally been conceived as Advanced Syncom satellites. The creation of the Communications Satellite Corporation (Comsat) following the Communications Satellite Act of 1962 led to suggestions that communications satellite R&D by NASA was inappropriate as Comsat was a private entity. NASA was more than willing to add a meteorological payload, and DOD asked that gravity gradient stabilization and medium Earth orbits (MEOs) experiments also be conducted.

The first five ATS satellites were all built by Hughes. None of the gravity gradient experiments worked. All of the cameras worked, as did all of the C-band transponders. ATS-1 (7 December 1966) and ATS-3 (6 April 1967) were complete successes, taking the first black-and-white (ATS-1) and first color (ATS-3) pictures of Earth from geosynchronous orbit. The first three ATS launches were on Atlas-Agenas (A-As), the second two on Atlas-Centaurs (A-Cs), and the sixth on a Titan. ATS-1 and -3 were deactivated in 1978. ATS-5—a nominal failure—provided communications services for many years.

The experimental/operational Synchronous Meteorological Satellites (SMSs) were built by Ford Aerospace. Their lighter weight allowed use of the cheaper, more reliable Delta launch vehicle. All carried a Visible Infrared Spin Scan Radiometer (VISSR) built by Hughes Santa Barbara Research Center. SMS-1 was placed over the Atlantic (17 May 1974) and SMS-2 was placed over the eastern Pacific (6 February 1975).

The Geostationary Operational Environmental Satellites (GOES)-1 through -3 were identical to the SMS-1 and -2 satellites. GOES-4 through GOES-7 were built by Hughes. The more advanced GOES-8 through GOES-12, (again built by Ford, now Space Systems Loral) have an imager much like the Advanced TIROS-N polar-orbiting satellites. Attitude control is three-axis and detailed position and pointing are obtained using the VHRR imager. As with the polar satellites, GOES now combines operational capabilities and research capabilities. One satellite is usually positioned over the Atlantic at 75°W (GOES-East) and another at 135°W (GOES-West). Early in the program satellites were placed over the Indian Ocean to provide "global" coverage for the Global Atmospheric Research Program (GARP). European, Indian, and Russian satellites now provide Indian Ocean coverage while Japan provides coverage of the western Pacific. Any "extra" GOES satellites are stored at 105°W—ready to replace GOES-East or GOES-West.

International Cooperation

Meteorological data have always been shared with other countries. In 1977 both Europe and Japan launched geosynchronous orbit weather satellites. When the GOES-NEXT program was delayed, the Europeans loaned NOAA a Meteosat. When the Japanese MTSAT was delayed, NOAA loaned the Japanese Meteorological Agency (JMA) a GOES. Starting from the prime meridian, Eumetsat covers the eastern Atlantic from 0°E and the Indian Ocean from 62°E. JMA covers the western Pacific from 140°E (and 155°E). NOAA covers the eastern Pacific from 135°W and the western Atlantic from 75°W. The five satellites of these three agencies continuously monitor Earth's weather, except for polar latitudes. These countries also cooperate by sharing polar weather data; Russia also supports this activity.

Societal Impact of Weather Satellites

The major U.S. hurricanes of 1900 and 1938 came from nowhere and killed people on the shoreline who had no idea a major storm was coming. On 8 September 1900, Galveston, Texas, had a population of about 36,000; by nightfall, one in six would be dead. The 1938 New England hurricane completely wiped out several vacation areas and flooded sea-level Providence, Rhode Island, and interior Hartford, Connecticut. The Galveston Hurricane of 1900 may have been the deadliest natural disaster in the United States but it does not even appear on any list of storms sorted by damage cost. In contrast, Galveston was evacuated over the single bridge linking it to the mainland before Hurricane Rita hit in September 2005. Evacuation was probably easier to enforce after the Hurricane Katrina disaster a month earlier. Katrina was among the most costly hurricanes to hit the United States but the death toll—in spite of poor evacuation plans—was much lower than it might have been. The inflation-adjusted cost of Katrina damage was 100 times the cost of the 1900 Galveston hurricane damage but the death toll was one-third. NOAA predicted landfall at New Orleans more than two days in advance, and the day before landfall the local NOAA office recommended immediate evacuation. Weather satellites don't just provide cloud pictures and warnings of hurricanes. They also detect forest fires, volcanic activity, and severe storms, and provide measures of rainfall and winds. Somewhat surprisingly, the value of weather satellites for forecasting is much less clear than the value for severe weather detection and monitoring.[22]

The commercial value of weather satellites was demonstrated in the mid-1980s when the Reagan administration was trying to privatize Landsat. Comsat offered to take over Landsat only if it was also given the weather satellites. NOAA has provided a compendium of economic statistics in which the costs of weather and climate events are summarized along with some estimates of the benefits of weather forecasting. Severe weather causes damages well in excess of $10 billion every year. Total benefits

22. Wikipedia, *http://www2.sunysuffolk.edu/mandias/38hurricane*, *http://www.1990storm.com*, *http://www.dailycomet.com* (accessed 24 August 2007); discussions with Erik Conway.

to the householders are estimated at over $10 billion per year. Benefits to agriculture, construction, and transportation would presumably increase this total.[23]

Communications

Although several pre–World War II mentions of satellite communications have been found, the first well-known discussion was Arthur C. Clarke's 1945 article in *Wireless World*.[24] Perhaps of greater importance were later articles by John R. Pierce in *Jet Propulsion* (1955) and *Proceedings of the IRE* (1959).[25] Clarke was a member of the British Interplanetary Society and a budding science-fiction author. Pierce was also a science-fiction author but, more importantly, he was the director of communications research at AT&T's Bell Telephone Laboratories (BTL).

In early March 1958,[26] John R. Pierce and Rudolf Kompfner of AT&T (independent inventors of the traveling-wave tube) saw a picture of the shiny, 100-foot sphere that William J. O'Sullivan of NACA Langley Research Center was proposing to launch into space for atmospheric research. It reminded Pierce of the 100-foot communications reflector he had envisioned in 1954. Pierce persuaded William H. Pickering of JPL to provide a West coast antenna for the experiment. To support this plan, Kompfner and Pierce wrote a paper[27] that they presented at an IRE conference on "Extended Range Communications" at the Lisner Auditorium of George Washington University in Washington, DC, on 6–7 October 1958.[28]

Echo I was launched into a 1,000-mile circular orbit on 12 August 1960. During the first orbit of the 100-foot sphere, a recording of President Eisenhower speaking was transmitted from JPL's Goldstone, California, Earth station to AT&T's Holmdel, New Jersey, Earth station. In spite of Echo's success, it was clear that active, rather than passive, satellites were the technology to develop.[29] In a 13 May 1960 letter to Leonard Jaffe at NASA Headquarters, Rudolf Kompfner had described the current AT&T/BTL research program as shifting to *active* satellites. In this letter Kompfner reviews the active satellite component/subsystem studies that had been underway since late 1959.[30]

23. "Economic Statistics for NOAA," (Washington, DC: U.S. Department of Commerce, NOAA, May 2005), pp. 10, 38.

24. Arthur C. Clarke, "Extra-terrestrial Relays."

25. John R. Pierce, "Orbital Radio Relays," *Jet Propulsion* (April 1955): p. 44; John R. Pierce and Rudolf Kompfner, "Transoceanic Communications by Means of Satellites," *Proceedings of the IRE* (March 1959): pp. 372–380.

26. Donald C. Elder, *Out From Behind the Eight-Ball: A History of Project Echo* (San Diego, CA: American Astronautical Society, 1995), p. 25.

27. Pierce and Kompfner, "Transoceanic Communications by Means of Satellites."

28. J. R. Pierce, *The Beginnings of Satellite Communications* (San Francisco: San Francisco Press, 1968), pp. 9–12.

29. NASA, *Fourth Semi-Annual Report to Congress* (Washington, DC: U.S. Government Printing Office, 1961), pp. 10–17.

30. A. C. Dickieson, *TELSTAR–Management* (Bell Telephone Laboratories, July 1970), pp. 34–39.

By 1960, Pierce had convinced AT&T management to build and launch a medium-Earth-orbit (MEO) satellite system. Even earlier, the Pierce and Kompfner paper had also energized a group of young engineers—Harold Rosen and Donald D. Williams—at Hughes Aircraft Company to prove wrong their 1959 argument that geosynchronous Earth orbit (GEO) satellites were beyond the state of the art. AT&T's plan to launch a satellite system was put on hold when NASA refused to provide launch services. NASA argued that launch vehicles were in short supply and must be rationed. The rationing mechanism would be a competition to design a MEO communications satellite. Proposals were submitted to NASA by seven companies—including Hughes (Syncom) and AT&T (Telstar). RCA won the competition for the Relay satellite in May 1961, but AT&T was allowed to purchase launch services and the Hughes satellite was jointly funded by NASA and DOD by the end of that summer.

On 10 July 1962, at 8:25 UT (4:25 a.m. EDT) the Delta carrying Telstar 1 lifted off from its pad at Cape Canaveral. AT&T had placed a commercially funded communications satellite in orbit before the government-funded projects, but the Communications Satellite Act would be passed less than two months later. AT&T, after expending more than $100 million (in 1960 USD) was out of the satellite manufacturing business for good. The NASA/RCA Relay (MEO) and the Hughes/NASA Syncom (GEO) satellites would be launched over the next two years.

The Communications Satellite Act of 1962

Just after NASA's announcement of the Relay communications satellite program award to RCA, President John F. Kennedy delivered a speech to Congress on "Urgent National Needs." In this famous 25 May 1961 speech, Kennedy promised to land a man on the Moon and also asked the Congress to provide the funds that "will make the most of our present leadership, by accelerating the use of space satellites for world-wide communications." This speech has been characterized as being driven by the unfortunate events of April 1961—Gagarin's orbital flight and the Bay of Pigs—but his comments echo the Wiesner Committee's "Report to the President-Elect of the Ad Hoc Committee on Space," delivered to Kennedy on 10 January 1961.[31] It is also consistent with his State of the Union message of 30 January 1961:

> Finally, this Administration intends to explore promptly all possible areas of cooperation with the Soviet Union and other nations "to invoke the wonders of science instead of its terrors." Specifically, I now invite all nations—including the Soviet Union—to join with us in developing a weather prediction program, in a new communications satellite program and in preparation for probing the distant planets of Mars and Venus, probes which may someday unlock the deepest secrets of the universe.[32]

31. *http://www.hq.nasa.gov/office/pao/History/report61.html* (accessed 24 August 2007).

32. *http://www.infoplease.com/t/hist/state-of-the-union/174.html* (accessed 24 August 2007).

In any case, politics—cold war politics—would be a driver in deployment of communications satellites.

The Space Council drafted an Administration Bill in November 1961 providing for a public-private corporation directly regulated by the president. Before it was submitted, Senator Robert Kerr (D-Oklahoma) submitted a similar bill that gave more control to the international communications carriers—as the FCC recommended. Another bill was introduced by Senator Estes Kefauver (D-Tennessee) that advocated government ownership. The Administration Bill passed the House 354 to 9 and, after a liberal filibuster, passed the Senate 66 to 11. On 31 August 1962, President Kennedy signed the bill into law.

The Interim Agreements

Comsat had been advised by the common carriers—especially AT&T—that bilateral arrangements between Comsat and each of the foreign post, telegraph, and telephone (PTT) organizations were preferable. Even as bilateral negotiations were being considered and before the incorporators had met, a U.K., Canada, and U.S. (Foreign Ministry/State Department) conference on satellite communications took place in Washington. In October of 1963, the International Telecommunications Union (ITU) held an Extraordinary Administrative Radio Conference (EARC) in Geneva to discuss frequencies for satellite communications. Somewhat to their surprise, Comsat got almost everything they wanted out of the conference.[33]

Comsat, and later Intelsat, had a major problem: Were they "commercial" entities, in the limited sense that government-owned PTTs were "commercial," or were they instruments of foreign policy? If they were commercial entities, then their purpose was to earn a profit for their owners by providing global satellite communications. If they were profit-oriented, then decisions should be based on costs and profits. For a long period, purchase of American satellites by Comsat and Intelsat was based on the cost-benefit analysis that these showed these satellites would provide the best service—and hence greatest profits—at the lowest cost. If Intelsat and Comsat were instruments of foreign policy, however, then profits were irrelevant. If these organizations were instruments of technological advance, then each country should obtain "work" (manufacturing contracts) in proportion to their contribution of funding. This later became the European Space Agency's (ESA's) principle of *juste retour.*[34]

In early 1964, the United States (the State Department and Comsat) met with the Europeans in Rome. It was clear at this meeting that the Europeans would insist on some amount of control over satellite communications. The next meeting was in London with additional participation. At this meeting it became obvious that

33. Joseph McConnell interviewed by Frederick Durant III, 18 July 1985; William Gilbert Carter interviewed by Nina Seavey, 15 July 1985.

34. Ibid.

there would be two agreements: a government-to-government agreement and a PTT-to-PTT agreement, with Comsat as the American PTT. The final version of the interim agreements was presented to the world on 20 August 1964 in Vatican City, where 14 countries immediately signed it. It is interesting to note that during this negotiation process, Comsat contracted for the geosynchronous Early Bird and raised $200 million in an initial public offering (IPO).[35]

The three most important consequences of this interim agreement were: (1) Comsat would not go it alone, but it would manage the interim system under an Interim Communications Satellite Committee (ICSC); (2) the organization would have both Foreign Office and PTT representation; and (3) a new definitive agreement would be negotiated in five years.[36]

Early Bird and Intelsat-II

Early Bird was launched in April 1965 and entered service in June. A few months later, the interim organization adopted the name Intelsat. Four Intelsat II series satellites were launched in 1966–1967; three were successful. The Intelsat II series was launched to support NASA's Apollo program. Early Bird covered only the northern hemisphere over the Atlantic Ocean Region (AOR). The Intelsat II series covered the globe and were located over both the AOR and Pacific Ocean Region (POR).

The Intelsat Definitive Agreements

When it came time to meet in February 1969 to discuss the definitive arrangements, the old disagreements were still present. The ICSC, representing the Intelsat consortium, had been dominated by Comsat. The first Plenipotentiary Conference was held from 24 February to 21 March 1969 in Washington. All but one of Intelsat's 68 member states sent a delegation. Some indication of where things might be headed was the reception that Katherine Johnsen of *Aviation Week & Space Technology* got when she tried to interview the members of the ICSC in 1967: 17 agreed to be interviewed; only John A. Johnson refused. Similarly, at an ICSC meeting in December 1968, a vote was taken as to whether Comsat should remain as manager: the result was 17 to 1 against.[37]

The third and final Conference on Definitive Agreements began on 14 April 1971. On 20 August 1971, the Agreements were opened for signature and by 14 December 1972 two-thirds of the members had signed. Sixty days later, on 12 February 1973, the interim agreements were terminated and the new agreement entered into force.[38]

35. Ibid.

36. A dry but fairly complete discussion of both the Interim and Permanent (definitive) Agreements can be found in Marcellus S. Snow, *The International Telecommunications Satellite Organization* (Baden-Baden, Germany: Momos Verlagsgesellschaft, 1987).

37. Katherine Johnsen, "France Backs UN Intelsat Control," *Aviation Week & Space Technology* (13 February 1967); Maddox, Babel, pp. 103-104.

38. Richard R. Colino, *The INTELSAT Definitive Arrangements* (Geneva: European Broadcasting Union, 1973), pp. 19–21.

The strange public-private, commercial-political nature of Comsat was also reflected in the structure of Intelsat. Intelsat had two "governing bodies": nations signed the Intelsat Agreement (also referred to as the Intelsat Treaty), but telecommunications entities (signatories) signed the Intelsat Operating Agreement. The Intelsat Assembly of Parties consisted of the sovereign governments that signed the Intelsat Agreement. Voting in the Assembly was by country: one nation–one vote. Its powers were limited. The Meeting of Signatories consisted of all the telecommunications entities that signed the Intelsat Operating Agreement. Voting in the meeting was on the basis of shares and the shares were allocated (and paid for) on the basis of usage. This has been referred to as "one telephone call–one vote." A board of governors had functions similar to the ICSC, or to the functions of a commercial board of directors. The Board consisted of about 20 members, each having a minimum specified investment, and individual representatives of member-groups whose total investment met the minimum specified (about 2 percent). Finally, there was a manager (Comsat for six years after the agreements enter into force, terminating on 12 February 1979), reporting to a secretary general until 31 December 1976 and to a director general thereafter.

The major antagonists, the United States and the Franco-Europeans, each compromised in some way but the result was both semi-commercial and semi-political. It could be argued that the State Department got what it wanted because the third-world countries wanted international communications at reasonable rates and with some national control—at least control of their own Earth stations. The Europeans continued to complain that satellite contracts went exclusively to the United States (until the 1990s), but the third-world countries preferred cheaper, higher-quality American satellites and launch vehicles.

Intelsat III and IV

Comsat had not officially chosen geosynchronous orbit when the Intelsat III contract was put out for bids. Hughes decided not to bid an MEO option. This allowed TRW to sneak in a winning bid. These satellites were the first to provide coverage of the Indian Ocean Region (IOR). Almost in parallel to the Intelsat III program was the Intelsat IV program. The first three generations had relatively limited capacity. Intelsat IV would be a significant increase in power, number of transponders, mass, and coverage options. The first three generations had Earth coverage only. Intelsat III was considered a big advance because it had a despun antenna that always pointed at Earth—dramatically increasing equivalent isotropically radiated power (EIRP). Intelsat IV had two narrowbeam antennas covering the East and West hemispheres. Because of its size, this series would use the Atlas-Centaur launch vehicle instead of the Delta. About 20 percent of the content was provided by international manufacturers.

It was no surprise when Intelsat discovered that North Atlantic traffic (AOR) was greater than POR or IOR. Intelsat IVA F6 was the last satellite launched by Comsat as Intelsat "manager." The definitive agreements left them without a major role in satellite development. Comsat still monitored construction under contract to Intelsat, but executive decisions were made by Intelsat.

Domestic Satellite Communications

Neither the Communications Satellite Act of 1962 nor the Interim Intelsat Agreements precluded domestic communications satellites (domsats). It was assumed—and later made explicit—that domsats should not interfere with Intelsat. A small working group was put together in 1969 to formulate Nixon administration domsat policy. Among the members of the group was Clay T. Whitehead. Whitehead's boss, Peter Flanigan, sent a memo to Dean Burch at the Federal Communications Commission (FCC) recommending open entry—"open skies"—on 23 January 1970. Thirteen entities had already filed for authorization to launch domsats. In March 1972, the FCC released a proposed Second Report and Order on DOMSATs, requesting that the filers consolidate their filings. Nobody liked this. The actual Second Report and Order was released on 16 June 1972 after a 4 to 3 vote by the commissioners. The dissenters objected to the restrictions on AT&T and Comsat. A final Report on DOMSAT was issued 22 December 1972 modifying (but retaining) these restrictions.

Meanwhile, Canada had quickly decided to launch a Canadian satellite to service the Far North. In 1967, the Chapman Report recommended that a satellite system be developed. In 1969, Telesat Canada was established. On 9 November 1972, Anik A1—a Hughes HS-333—was launched on an American Delta launch vehicle. RCA Global Communications began service to Alaska on Anik. RCA may have been first into service, but Western Union (the telegraph company) was the first U.S. company to launch its own satellite (13 April 1974).

RCA built its own satellite using the services of RCA Astro-Electronics in East Windsor, New Jersey and RCA Canada (later Spar) in Montreal. The RCA satellites had twice the number of transponders and twice the power of the HS-333. They were the first operational (as opposed to the experimental ATS-6 and Symphonie) three-axis-stabilized communications satellites.

AT&T had built its own experimental Telstar satellites but opted to buy Hughes satellites for its operational program. More accurately, it leased satellites from Comsat. AT&T was constrained to provide only point-to-point services—they could not offer television distribution services.

Indonesia was the third nation to launch a commercial geosynchronous communications satellite business. The Palapa series, like Anik and Westar, was based on the HS-333. Within a few years of the first launch, the tens of thousands of Indonesian islands were connected via satellite.

The Television Revolution

The original U.S. filing for a domestic communications satellite had been made in 1965 by ABC with encouragement from Hughes. Comsat and Intelsat had never been much interested in television—some at Comsat argued that only four television transponders were necessary, one for each network (ABC, CBS, NBC, and educational television). Some at RCA (owners of NBC) argued that at least 20 transponders were needed—one for each of the four networks in each of the three time zones plus one for each NFL game. Although a few had seen the future, the explosion in television, especially non-network cable television, was a shock. Within a few years there were a dozen satellites carrying more than 200 transponders. Two-thirds of the traffic was television—a ratio that persists to this day. By the 1990s in the U.S. (earlier in Europe) dedicated direct-to-home broadcast satellites had revenues in the tens of billions of dollars.

Comsat and Intelsat: Competition and Other Problems

Comsat was looking for a mission after 1979. The company tried domestic satellites (Comstar with AT&T, SBS with IBM), broadcast satellites (STC), software, ground systems, and especially Earth stations (RSI). None of them worked. By the mid-1980s the company's profits were bouncing up and down. It seemed that every other year the company lost tens of millions of dollars. Intelsat did well in these years, and it was still the largest and most profitable satellite company. It had few business barriers because its owners were the national PTTs. In the late 1980s competition did begin to affect Intelsat. More dangerous was the fact that the U.S. government was seeking to destroy its monopoly.

There had long been concerns that Comsat/Intelsat was a monopoly and monopolies are "bad." On 28 November 1984 President Reagan announced that "separate systems" were required in the national interest. Based on the president's decision, the FCC began granting conditional licenses. On 25 July 1985 the FCC issued its Separate Systems Report and Order. On 1 June 1988 PanAmSat's PAS-1 was launched and eventually drifted to a longitude of 45°W to provide trans-Atlantic service.

That same year, the fiber-optic TAT-8 cable began to provide service across the Atlantic—service that was cheaper than Intelsat. The trans-Atlantic telephone cables had always competed with satellite transmission across the Atlantic but the savings, if any, were minimal. Fiber-optic cables were cheaper, provided higher-quality transmissions, and *much* higher data rates.

Comsat and Lockheed

How did it all begin? In 1995 Lockheed and Martin Marietta merged. The earlier combination of RCA Astro and GE Space that had been purchased by Martin Marietta became Lockheed Martin Commercial Space Systems (LMCSS). Martin Marietta had been looking at getting into communications satellite operations rather than (or in addition to) manufacturing. Profit margins in manufacturing looked slim compared to operations.

By early 1998, it seemed clear that Comsat was on the market. Early in the year, Comsat denied that it was being acquired by Loral.[39] In July, fallout from the Cox Report caused aerospace stocks to tumble: Comsat fell to $28.75 from $42

39. "COMSAT Denies Acquisition Report," *Washington Business Journal*, 18 February 1998.

over a "couple months."[40] Comsat claimed it was in the process of "unlocking the value of investments in Intelsat and Inmarsat." On September 20, Lockheed (LMT) announced plans to purchase Comsat after failing to buy Northrup Grumman for $8.3 billion. The Lockheed offer was for 49 percent of the Comsat stock at $45.50 per share with the last 51 percent to be purchased, with one share of Lockheed for two shares of Comsat. The total value of the purchase was about $2.7 billion. The Lockheed offering of $45.50 was about one-third higher than the market price $341/8. Comsat was apparently vulnerable to takeover due to its small size. Lockheed shares fell while Comsat shares climbed.[41]

On August 20, 1999 Comsat shareholders voted to accept LMT's offer (99 percent of votes, 74 percent of shares). The plunge in value of Lockheed shares reduced the value of the deal from $2.7 billion to $2.2 billion.[42] On September 15 the FCC authorized Lockheed to purchase 49 percent of Comsat. Lockheed was also authorized to buy Comsat Government Services, Inc. (CGSI).[43] In addition to buying 49 percent of the shares at $45.50 per share, the remaining 51 percent would be a share-for-share deal (Lockheed had split). Lockheed would also assume $455 million in Comsat debt.[44] On September 16, the Department of Justice authorized the Lockheed-Comsat merger.[45]

The Orbit Act of 2000

In 1996 the U.S. Government Accountability Office (GAO) made a report to Congress, requested by Thomas J. Bliley, Jr. (R-Virginia), on Intelsat restructuring. [46] In 1997 Bliley and Edward J. Markey (D-Massachusetts) submitted a bill (H.R. 1872) to privatize Intelsat and Inmarsat.[47] On 30 July 1997 Senator Conrad Burns (R-Montana) and his Communications subcommittee held hearings on "Satellites and the Telecommunications Act." The FCC, NTIA, and State Departments testified, as did Intelsat, PanAmSat, and Comsat.[48] The claim was made that Comsat's markup on

40. Jerry Knight, "A Probe of US-China Policy Pummels Satellite Stocks," *Washington Business Journal*, 13 July 1998: p. 7.

41. "LockMart Swallows COMSAT," *Space Daily* (21 September 1998).

42. Tim Dobbyn, "COMSAT Shareholders Back Lockheed Sale," 20 August 1999, *http://www.space.com* (accessed 24 August 2007).

43. "FCC Authorizes Lockheed Martin to Purchase up to 49 Percent of COMSAT," *FCC News* (15 September 1999).

44. Tim Smart, "FCC OKs Lockheed-COMSAT Deal," 15 September 1999, *http://www.washingtonpost.com* (accessed 24 August 2007).

45. "Receives Department of Justice Clearance for COMSAT Corp. Merger," *S&P Daily News* (5 October 1999).

46. "Telecommunications: Competitive Impact of Restructuring the International Satellite Organizations," GAO Report GAO/RCED-96-204, July 1996.

47. Doug Abrahms, "Bearing Weight of 141 Nations, Intelsat Tries to Change Itself," *Washington Business Times*, 23 June 1997: pp. D13–D14.

48. U.S. Senate Commerce, Science, and Transportation Committee, Subcommittee on Communications, Hearing on Satellites and the Telecommunications Act, 30 July 1997.

Intelsat pricing was as much as 86 percent.[49] On 25 March 1998, the House committee passed the Bliley bill (H.R. 1872), including "direct access" and a variant of "fresh look." "Direct access" allowed customers to deal directly with Intelsat—bypassing Comsat. "Fresh look" afforded customers an opportunity to renegotiate all Comsat-mediated contracts for Intelsat bandwidth. Intelsat's Tony Trujillo commented that the bill was fatally flawed.[50] Nevertheless, H.R. 1872 passed the entire House on 6 May 1998.

The House bill provided for both "direct access" and a variant of "fresh look." Sen. Burns introduced a different bill (S. 2365) on July 29, a bill seen as more favorable to Comsat.[51] PanAmSat claimed that S. 376, the Open-Market Reorganization for the Betterment of International Telecommunications (ORBIT) bill did not go far enough. On January 21, Tom Bliley asked the FCC to reject any ownership greater than 10 percent until after the passage of reform legislation. On 5 May 1999, the full Senate committee approved the ORBIT bill (S. 376).[52] On July 1, the ORBIT bill was passed unanimously by the Senate. "Fresh look" was not allowed, but "direct access" was. Tony Trujillo of Intelsat described the bill as "the heavy hand of the U.S. Congress."[53] Senator Lott insisted that Burns allow "direct access" by 1 July 2002. The Bliley House bill passed in 1998 had removed almost all Comsat privileges.[54]

On February 17 the House and Senate conference committee reached a compromise. Direct access was allowed and the Intelsat IPO was delayed to 1 January 2003.[55] On April 4, an FCC Public Notice was published to the effect that LMT had applied to transfer control of Comsat to LMT/CGS.[56] On July 31 the FCC authorized Lockheed to merge with Comsat based on the provisions of the 17 March 2000 ORBIT Act.[57]

The End of Comsat

Lockheed Martin Global Telecommunications (LMGT) did not last long. On 7 December 2001 (the 60th anniversary of Pearl Harbor), LMGT announced

---

49. "Ending a High-Flying Monopoly," *Washington Post*, 20 March 1998: p. A10.

50. "House Commerce Committee Passes Bliley Bill, But Hurdles Remain," *Satellite News* (30 March 1998): p. 1.

51. Sam Silverstein, "COMSAT Prefers Senate Bill," *Space News*, 3–9 August 1998): p. 3.

52. "International Satellite Reform Bill Approved by Committee," U.S. Senate Committee on Commerce, Science, and Transportation Committee Press Release, 5 May 1999.

53. Daniel Sood, "Senate Unanimously Passes Satellite Reform," 1 July 1999, *http://www.space.com* (accessed 24 August 2007).

54. "Senate Satellite Bill Faces Tough Fight in the House," *Aerospace Daily* (6 July 1999).

55. Greg Schneider, "Accord Reached on Bid for COMSAT," *Washington Post*, 18 February 2000: p. E01; Mary Motta, "Lockheed Gets Good News on COMSAT Deal," 19 February 2000, *http://www.space.com* (accessed 24 August 2007).

56. FCC Public Notice, Report No. SAT-00040, 4 April 2000.

57. "Commission Authorizes Lockheed Martin to Take Control of COMSAT," *FCC News* (31 July 2000). Yuki Noguchi, "Intelsat Agrees To Merger Deal With PanAmSat," *Washington Post*, 29 August 2005: p. A7.

that it was shutting down its operations. Some units would be absorbed by other Lockheed divisions, some units would be sold off, and some units would simply disappear. Earlier that year, much of Comsat Laboratories was sold to ViaSat. In January 2002 the sale of Comsat Mobile Communications to Telenor had been finalized. In March 2002 Lockheed sold its remaining Comsat World Systems facilities to Intelsat. Finally, in 2004 Comsat General's remaining facilities were also sold to Intelsat by Lockheed. The public-private experiment was over.

Merger with PanAmSat

After the dot-com crash and general telecom meltdown of the early twenty-first century, market share and profitability became critical. On 28 August 2005 Intelsat and PanAmSat agreed to merge. The merger of the second- and fourth-largest fixed service satellite (FSS) companies would produce a giant that owned between a quarter and a third of all FSS satellites. What makes this merger particularly strange is that PanAmSat was formed as "the non-Intelsat" by René Anselmo in 1984. The PanAmSat motto was "Truth and technology will triumph over bullshit and bureaucracy." René despised the Comsat-Intelsat monopoly. The PanAmSat mascot was the dog Spot—usually seen urinating on the leg of a representative of bullshit and bureaucracy.

Intelsat agreed to buy PanAmSat Holding Corp. for $3.2 billion in cash. The merged companies would form the world's largest satellite company and give the companies a more diversified set of businesses. The new company would own 53 satellites spanning the globe and generate annual revenues of more than $1.9 billion.

On 7 July 2006, the $6.4-billion purchase of PanAmSat was completed, creating a merged company carrying one-quarter of the world's commercial satellite-delivered television programming. The acquisition leaves PanAmSat a wholly owned subsidiary of Intelsat. The combined company would initially lose money. PanAmSat earned $72.7 million in 2005 but Intelsat lost about $325 million. Intelsat chief executive David McGlade told *The Washington Post* that, given the level of debt and interest payments, the company did not expect to become profitable in the foreseeable future. He said the company's investors have been pleased with Intelsat's positive cash flow and its heavy backlog of orders. The traditional core of Intelsat's business has been telephony, a difficult market in recent years. The combined company will be more diverse with the addition of PanAmSat's television customers.[58]

Societal Impact of Communications Satellites

Revenues for commercial satellite applications have been dominated by satellite communications.[59] This industry has yet to attain the $100 billion estimated in the

58. *http://broadcastengineering.com/beyond_the_headlines/intelsat_mergers_panamsat/?r=1* (accessed 24 August 2007).

59. Much of the material for this section comes from the Satellite Industry Association and Futron. SIA/Futron, *State of the Satellite Industry Report* (San Diego, CA: SIA/Futron, June 2006).

early days, but total revenues are approaching this figure ($88.8 billion according to SIA/Futron). The fixed (FSS), mobile (MSS), and broadcast (BSS) satellite service sectors had revenues exceeding $50 billion in 2005. Direct broadcast satellite (DBS) revenues led at $41.3 billion, followed by FSS revenues at $9.8 billion and MSS revenues at $1.7 billion.

Some Final Thoughts

The first space application to be operationalized was remote sensing with the launch of the first Discover on 28 February 1959. One could argue that "success" wasn't achieved until August 1960 but, in any case, this was the first application where significant funds were expended. These funds came from DOD (Air Force) and the intelligence community (CIA). Neither NASA nor commercial firms were involved except as manufacturers or other contractors. Although NASA launched the first Earth Resources Technology (ERT)/Landsat satellite in 1972, it is not clear that remote sensing has ever been truly commercialized, although one can argue that by the twenty-first century it was possible to buy fairly high-resolution imaging on the open market.

The second space application to be operationalized was navigation with the launch of the first Transit on 17 September 1959. Transit funding came from DOD (Navy), as did funding for GPS/NAVSTAR (Air Force) later. In 1967, Transit use by the civilian maritime industry was allowed. Although funding for the satellites has come exclusively from DOD, this application has definitely been commercialized, as evidenced by the billions of dollars expended every year for GPS receivers.

The third space application to be operationalized was weather with the launch of the first TIROS on 1 April 1960. TIROS was based on the RCA proposal for a reconnaissance satellite. Initial funding came from the Army but the project was transferred to NASA. NASA funded TIROS and many of its upgrades, although many of these upgrades were initially funded by DOD on the DMSP program. The Weather Bureau eventually began funding operation of the satellites and, somewhat later, satellite procurement. NASA seems to have taken the lead on geosynchronous weather satellites, launching the first Synchronous Meteorological Satellite (SMS-1) in 1974. Commercialization of this application probably started with the launch of TIROS-1, but transfer to the Weather Bureau (ESSA) didn't formally occur until TOS in 1966.

The fourth space application to be operationalized was communications with the first launch of Telstar on 10 July 1962. Earlier dates (Courier and Echo in 1960) and later dates (Syncom 2 in 1963) can be proposed, but it is fascinating to observe that the most commercial of all space applications was the last to be actually launched. Not surprisingly, it was the first to be commercialized—in every sense of that word—when Early Bird was launched on 6 April 1965. Funding for the earliest communications satellites is complicated. By far the largest investor was AT&T but much of that investment was for manufacturing capability. NASA was the second-largest investor, funding most of Syncom and all of Relay. Hughes was the third-largest investor and may have gained the greatest profit by building proto-Syncom with its own funds.

Communications satellites showed the most interesting behaviors, possibly because they are so commercial. The failures—Aerosat, SBS, STC, the LEOs, and the MEOs—all seemed to have misread the market for their offerings. The international projects (Inmarsat and Intelsat) seemed to have generated geopolitical hassles. Others have had trouble generating profits, such as DARS and, to a lesser extent, DTH TV.

It is interesting to note that the earliest space applications are the ones developed by or for DOD (reconnaissance and navigation). It should be no surprise that the most commercial of all the applications (communications) shows the greatest commercial funding and the earliest commercialization. The role of NASA is hard to evaluate. NASA seems to have been more of a facilitator than anything else. NASA had no real part in reconnaissance and navigation but certainly "facilitated" the development of weather satellites. It is also possible to claim that NASA "facilitated" the development of communications satellites. If NASA had not been involved, AT&T would have gone ahead with its MEO Telstar system. This might have made it very difficult for the Hughes "better idea" to make it into a marketplace that was dominated by AT&T. It would be interesting to examine the effect on NASA priorities of its R&D agency status. Any application developed by NASA would have to be given away. Perhaps one measure of NASA's influence would be to examine what would have happened without NASA. Reconnaissance, navigation, and communications satellites would have been developed by DOD and industry, but the weather story is more complicated. DOD did not want their weather pictures circulated; DMSP is proof of this. It is not clear that the Weather Bureau would have invested the funds that NASA made available. NASA is *still* supporting development of weather satellites.

Applications satellites are not as glamorous as Moon landings—or Mars landings—but they have made a huge difference in the world we live in: financially, culturally, and in the areas of safety and security. They have created the global village. It is a feisty, angry, violent village, but there are fewer unknowns and a greater chance for peace and prosperity.

Chapter 16

NASA and the Environment: Science in a Political Context

W. Henry Lambright[1]

The advent of the Space Age has paralleled the rise of the environmental movement. NASA was born in 1958 and Rachel Carson wrote *Silent Spring* in 1962; that book is generally seen as marking the onset of modern environmentalism.[2] NASA has intersected with the environmental movement—a set of values and interest groups concerned about the need to protect our natural setting for the current and future generations—in many ways over the years. How did NASA do so? How did it evolve an environmental mission? What did it do with that mission? What were the consequences for society—and NASA—of its environmental role? To answer these questions, this paper will discuss two of the most important ways NASA and the environmental movement related.

First, NASA has had direct impacts through the images of Earth taken by Apollo astronauts as well as by satellites in Earth orbit. Those satellite images and theories about Earth as a system evolved into an organized NASA program, initially called Mission To Planet Earth (MTPE), later the Earth Sciences Program. Second, there was an indirect relation through NASA's mission *from* Earth. Comparative planetology came into existence as a new field; learning about other planets stimulated better understanding of Earth.

There are many other issues in the NASA–environment relation, such as space debris and the contamination of other planets, but these two themes—Earth monitoring and comparative planetology—are especially salient in NASA's history, present, and likely future. The first theme focused on the use of space-based remote sensing and became the dominant emphasis in NASA's environmental history.

1. The author wishes to thank NASA for support for this paper's research and also Sara Pesek for research assistance. Ms. Pesek was a Master of Public Administration graduate student at the Maxwell School of Syracuse University.

2. There are those who would date the beginning of the modern environmental movement from an earlier Rachel Carson book, *The Sea Around Us* (1951). *Silent Spring* came out in 1962. Carson worked for the Bureau of Fisheries, which became the National Marine Fisheries Service within NOAA. Personal correspondence, John Cloud, NOAA, to author, 25 September 2006.

There evolved a set of satellites that can be seen as environmental satellites. They are the centerpiece of NASA's mission to Earth. That mission has had an uncertain, somewhat uneasy relationship with other parts of NASA and other agencies. Some people think it is NASA's most societally relevant mission whereas others think it is extraneous to NASA and belongs somewhere else—the National Oceanic and Atmospheric Administration (NOAA) is the usual candidate. Virtually every observer has found it a controversial mission, one in need of high-level policy attention and improvement for the sake of NASA, the nation, and the world.

The second theme, comparative planetology, has indirectly influenced the main debate—a reminder of the fact that Earth is an island home in the vast sea of space and is the only planet (so far) supporting life. NASA's long association with the environmental and Earth sciences has been fruitful but it has also been contested and even tortuous. The history of NASA and its environmental mission is one of science in a political context.

Beginnings in the Apollo Years

In the 1960s, as NASA concentrated on the Moon project and environmentalism emerged as a conscious political and philosophical movement, NASA began to monitor aspects of the environment in the name of an "applications" program, part of NASA's Office of Space Science and Applications (OSSA). Its chief activity was initially weather satellites. Early on, NASA negotiated a relationship with the predecessor of NOAA, the Weather Bureau, such that NASA developed weather satellite technology and then transferred it to the weather forecasting agency for operational use. The weather satellite program was clearly one of the great successes of the 1960s. It was obvious to all that weather satellites improved forecasts and aided early warnings of approaching hurricanes. There was a technology push from NASA and a pull from a user agency.[3]

Comparative planetology also began in the 1960s, with the Mariner spacecraft flybys of Venus and Mars. Venus revealed a runaway greenhouse effect that heated it into an inferno, providing an early example to some scientists of what could happen here. Mars also seemed inhospitable to life. But the most important impacts on the environmental movement of the early space program were pictures of Earth taken by Apollo astronauts, beginning with Apollo 8 in 1968—the Christmas Eve flight around the Moon. For the first time, humanity saw a blue Earth in the desolate blackness of space. As the environmental movement emerged, it used these images

3. Homer Newell, "Space Science and Practical Applications" in *Beyond the Atmosphere: Early Years of Space Science* (Washington, DC: NASA, 1980).

of our planet as a symbol for the first Earth Day in 1970. Without a doubt, the space program helped catalyze the environmental movement, especially insofar as getting it to think about the *global* environment.[4]

NASA in the 1970s

The 1970s are often called an environmental decade. One reason is that Environmental Protection Agency (EPA) was born in 1970 and Congress enacted a sequence of laws to deal with water, air, and other forms of pollution. Another reason was the energy crisis of the decade, and writings on "limits to growth."[5] The Carter administration, in particular, integrated environment and energy conservation and conveyed the notion that "small is beautiful." This notion was applied to technology, which environmentalists argued had to be "appropriate" to the user.[6]

The space program in the 1970s mirrored the decade's political setting. NASA devolved from Apollo, whose Moon flights ended in 1972, to the Space Shuttle. NASA diminished in size and exploratory capacity. The NASA Administrator, James Fletcher, who served from 1972 to 1977 (the first of two terms at NASA's helm) was personally interested in environmental stewardship, a characteristic Launius has linked in part to his Mormon roots. However, the situation he faced was problematic. The environmental movement of the early 1970s had a distinctly anti-technological flavor. It contributed to the termination of the supersonic transport (SST) in 1971. There were concerns in NASA that environmentalists might attack the Space Shuttle as they had attacked SST, on environmental factors. NASA realized that it needed to research environmental-atmospheric impacts of the shuttle to defend itself, if necessary, against possible opposition[7] More positively, Fletcher sought to align his Agency with environmental values. In 1973, he told Congress that NASA should be considered "an environmental agency." He declared: "Everything we do ... helps in some practical way to improve the environment of our planet and helps us understand the

---

4. Comparative planetology helped crystallize global warming as an issue. See Spencer Weart, *The Discovery of Global Warming* (Cambridge, MA: Harvard University Press, 2003), pp. 87–89. See also Steven Dick and James Strick, *The Living Universe: NASA and the Development of Astrobiology* (New Brunswick, NJ: Rutgers University Press, 2004); Marina Benjamin, *Rocket Dreams* (New York: Free Press, 2003); and Neil Maher, "Gallery: Shooting the Moon," *Environmental History* (July 2004).

5. *Limits to Growth* was the name of a best-selling book authored by Donella Meadows and others and published in 1972 under the auspices of a business group called the Club of Rome. It modeled the consequences of a world with a growing population and limited resources. It was republished in 2002 by Chelsea Green Publishing, White River Junction, Vermont.

6. For views of appropriate technology, see E. F. Schumacher, "Buddhist Economics," and Paul Goodman, "Can Technology be Humane?" in *Technology and the Future*, 7th ed., Albert Teich, ed. (New York: St. Martin's Press, 1997).

7. Kim McQuaid illuminates NASA's early ambivalence about environmentalism in "Selling the Space Age: NASA and Earth's Environment, 1958–90," *Environment and History* 12 (2006), pp. 127–163.

forces that affect it. Perhaps that is our essential task, to study and understand the Earth and its environment."[8]

Fletcher promoted a new Earth-oriented satellite, called Landsat, launched in 1972, that was capable of helping to forecast world food harvests and other resource issues. Conscious of environmentalist concern about pollution of the stratosphere from high-flying planes that could extend to the shuttle, he sought legislation to undergird NASA's emergent environmental (and shuttle) interests. In 1975 Congress authorized NASA "to conduct a comprehensive program of research, technology, and monitoring of the phenomena of the upper atmosphere." In 1977, Congress required NASA to issue biennial reports to Congress on the status of ozone depletion, an issue beginning to worry some scientists and environmentalists. This legislation (1975 and 1977) was important; it gave NASA legitimacy not only to do environmental research but also to link it with policy.[9] It was not a "given" that NASA would have this mission rather than another agency. It reflected NASA's administrative entrepreneurship and Congressional support relative to that of potential rivals.

The 1970s also featured the growth of comparative planetology, as a field, energized by the 1976 Viking mission to Mars and James Lovelock's Gaia Hypothesis. Lovelock, who published his ideas in 1974, held that Earth was a living system in which physical and biological components worked together to enable life. As it developed, the Gaia Hypothesis drew on studies of other planets, including Viking's apparent failure to find life on Mars. It made a number of scientists and environmentalists better appreciate Earth as a precious and vulnerable home and the role of human beings in altering it. There even were those such as Gerard O'Neill, a Princeton physicist, who speculated that man might need to migrate beyond Earth and establish colonies in space—a new, better place, a utopia where a more eco-friendly existence could be practiced.[10]

THE RISE OF NASA'S ENVIRONMENTAL ROLE IN THE 1980S

The 1970s produced a set of ideas about Earth as an interacting system. Also, some scientists felt that techniques used to study Mars could be applied on Earth.[11] The decade also gave rise to NASA's thinking more strategically about crafting an

8. Roger Launius, "A Western Mormon in Washington, DC: James C. Fletcher, NASA and the Final Frontier," *Pacific Historical Review* (1995), p. 236.

9. W. Henry Lambright, *NASA and the Environment: The Case of Ozone Depletion* (Washington, DC: NASA, 2005), pp. 7–8.

10. Dick and Strick, p. 49; Benjamin, pp. 51, 121–138.

11. In 1976, when NASA sent Viking to Mars, it found the intended landing spot to be unsuitable. NASA and leading Mars specialists searched for three weeks to find a good place, surveying virtually the entire planet. At the end, Michael McElroy of Harvard declared: "You know, we've never done anything like this for the Earth." Burton Edelson, "Mission to Planet Earth," *Science*, 227, no. 4185 (25 January 1985), p. 6.

environmental mission and how better to use the legislation it had obtained for its atmospheric research. In the 1980s, NASA charted a larger and broader Earth observation program that built on its work in weather, land satellite monitoring, and initial attempts at ocean surveys. What NASA contributed, through its satellites and comparative planetology studies, was a perspective different in scale from other agencies—literally a global view. There were certain environmental issues that were indeed global in scale and NASA, in the view of some of its officials and external supporters, was particularly suited to address these.

In 1982, NASA Administrator James Beggs, responding to overtures from OSSA, went to the United Nations Conference on the Peaceful Uses of Outer Space, where he called for "an international cooperative project to use space technology to address natural and manmade changes affecting habitability of Earth." The reaction to his "Global Habitability" overture was overwhelmingly negative—not to the idea of global habitability but to NASA as leader. "It came across like NASA was trying to take over the world," Burton Edelson, associate administrator of OSSA, recalled.[12] There had been no spadework ahead of time to build a coalition of support for the proposed endeavor. Beggs told Edelson to build that support base with other agencies, the White House, the scientific community, and the public before surfacing the proposal again.

This he began to do, starting in 1983 with a broad-gauged Earth System Science Committee. Edelson worked outside and inside NASA—outside with the scientific community, the National Science Foundation (NSF), and the National Oceanic and Atmospheric Administration (NOAA) in particular, and inside by establishing an Earth Sciences and Applications Division within OSSA to augment the Earth science personnel and presence within NASA. In 1985, Edelson wrote an editorial in *Science* magazine. Dropping the "Global Habitability" name, he proclaimed the need for a new "Mission To Planet Earth."[13] While this planning and support-building was underway, events provided NASA an opportunity to demonstrate how it could lead in a new global environmental mission.

The issue was ozone depletion. It had risen and declined as an issue in the 1970s but in the mid-1980s had returned with a vengeance. In 1985, British scientists using ground-based studies discovered extraordinary and shocking ozone depletion over Antarctica. NASA quickly followed-up with satellite observations confirming what became known as the "ozone hole."[14]

The media, using NASA satellite images, conveyed an eerie and graphic display of a gigantic hole that seemed to grow like an organism over Antarctica. The images alarmed the public, as they were accompanied with reports of how skin cancer could be caused by ozone depletion if the hole spread to more populated places. Scientists,

12. W. Henry Lambright, "Entrepreneurship and Space Technology: The Ups and Downs of 'Mission to Planet Earth,'" *Public Administration Review* (March/April 1994), pp. 97–104.

13. Edelson, "Mission To Planet Earth."

14. Lambright, *NASA and the Environment*.

media, environmentalists and politicians sounded the alarm. Industry defended itself against over-hasty regulations, but was on the defensive.

NASA seized the initiative. It had the legislation from the 1970s that gave it legitimacy to take a "lead agency" role on the science side of this issue. It organized a scientific expedition to Antarctica. Enlisting NOAA, NSF, academic scientists from the United States and abroad, and even industry researchers, NASA and its allies sought to determine the cause of the hole. Robert Watson, an energetic OSSA manager, was the prime mover in the Antarctic expedition. He struck a close alliance with a key scientist-administrator of NOAA, Dan Albritton, and practiced what Albritton called "ecumenical" leadership. One Antarctic expedition was soon followed by another, more extensive one. It became increasingly clear to the scientists that the prime suspects behind the hole were common chemicals called chlorofluorocarbons, or CFCs, found in a host of everyday products. This growing consensus extended to industry scientists involved in the expedition and was endorsed by an independent group of scientists Watson set up to review the expeditions' findings.

Watson and Albritton played dual roles—scientist-administrators of two agencies collaborating in an ad hoc program to determine causality in ozone depletion, and science advisors to the State Department and the EPA. These "policy agencies" were users, in real time, of information from the science agencies. There was again a push and pull on the part of providers and users. The technical information—what was known and what was not known—was conveyed to diplomats in the field. The diplomats were meeting under international political pressure to act, and did so in 1987, producing the pathbreaking Montreal Protocol, an agreement which set deadlines for replacing CFCs with less harmful chemicals. The Montreal Protocol was remarkable in calling for amendments as science produced more precise knowledge. Ozone depletion in many ways became the model for subsequent international environmental policy.

It was also seen as a potential model for NASA. NASA leaders contemplated ozone depletion as the first step in the broader global environmental initiative it had sought since 1982. The ozone experience, highly positive for NASA in a public relations sense, took place in a period when the Agency otherwise suffered severe criticism for the *Challenger* Space Shuttle disaster. That disaster, in 1986, had brought James Fletcher back to the Agency. While presiding over NASA's return to flight, he also looked ahead to new missions. He asked Sally Ride, America's first woman in space, to study possible initiatives beyond the space station, NASA's existing flagship project, that would give long-term direction to the Agency. Her 1987 report listed four options, without giving any priority. The first she listed, however, was a Mission To Planet Earth. Calling the mission of "fundamental importance to humanity's future," Ride said NASA was "uniquely suited to lead the effort." In the wake of the ozone experience, NASA was now in a far better position to make the case politically for a new mission than it had been in 1982.[15]

15. Lambright, "Entrepreneurship and Space Technology."

ACHIEVING POLICY ADOPTION

Advocates of the new mission, largely in OSSA, began pushing harder for policy adoption of a new program. Drawing on the work of the Earth System Science Committee, which completed its studies in 1986, NASA set as the new program's centerpiece an Earth Observation System (EOS). EOS would feature two 13-ton, bus-sized platforms launched by the Space Shuttle and linked with the Space Station *Freedom*. They would have multiple sensors, enough to permit comprehensive and simultaneous views of land, atmosphere, and sea interactions. The estimated cost of a system that delivered 15 years of observations was $30 billion. The full system would be launched by 2001. A series of specialized "precursory" missions in the 1990s would lead up to the main event, an operational two-platform EOS replete with an unprecedented data-handling and disbursing system. The EOS vision included potential additional systems contributed by other nations.

A program this big could only be justified if applied to a very big problem. The problem that had emerged over the 1980s in parallel with EOS planning was called "global change." Global change included, but was not limited to, climate change. Climate change was a politically neutral way of referring to global warming. Global warming was controversial among scientists and even more so among politicians. Climate change was a much more complicated problem than ozone, both scientifically and politically. Climate change had even been connected with scientific opposition to the Reagan administration's nuclear weapons policy. Scientists, led by space scientist Carl Sagan, warned against a "nuclear winter" if atomic weapons were ever used. Whether hot or cold, Earth's climate was complex and many scientists felt it premature to forecast a dire future, for whatever reason, given existing understanding. Moreover, there was not an ozone hole to focus scientific work and trigger policy action. Instead, there was a slow accumulation of contentious scientific information. Although NASA's lead role in investigating ozone depletion might be seen by the Agency to be a model for a Mission To Planet Earth, the transfer of the model to climate and global change would not be straightforward. NASA had no special legislation that gave it legitimacy to assert its claims, but at least one man within NASA was asserting his own views.

In 1988, James Hansen, director of NASA's Institute for Space Studies, proclaimed before Congress that global warming was almost certainly a reality now. He declared to reporters afterward that it was time to "stop waffling and say that the evidence is pretty strong that the greenhouse effect is here."[16] His view was not popular with the Reagan White House. The Reagan administration established an interagency committee, whose dominant members were NASA, NOAA, and NSF, to get at the facts. From this committee, initially called Committee on Earth Sciences (CES),

16. Weart, p. 155.

came the proposal for a U.S. Global Change Research Program (USGCRP). When George H. W. Bush became president in 1989, CES had a report waiting for him and his science advisor, Alan Bromley. Bush said he wished to be "the environmental president" and, on the advice of Bromley, made global change his first "presidential priority" in science and technology. Such a designation slated global change for policy adoption and budget support. The USGCRP combined space and ground-based observations. The largest single item in the interagency effort would be EOS, but NASA was not a "lead" agency. Leadership was vested in the USGCRP interagency committee, whose name was broadened, along with its membership, to Committee on Earth and Environmental Sciences (CEES). In 1990, Bush officially adopted Mission To Planet Earth as a NASA priority. Congress provided endorsements and an initial appropriation to start EOS. The advocates of a greater NASA role in global environmental monitoring thereby achieved a victory after almost a decade of planning and strategizing. NASA was poised to use space technology in a grand effort to develop an "Earth System Science." Such a science would pave the way for a "predictive capability" in global change that would guide policy makers, and thereby protect planet Earth.[17]

WOUNDED AT THE OUTSET

In theory, a new government program moves from adoption to implementation, at which point it has the opportunity to show its mettle before later undergoing evaluation and possible mid-course correction. Not so MTPE/EOS. NASA's environmental initiative faced drastic reorientation almost as soon as it got underway, before implementation could commence in a serious way. There were five obstacles to stable growth it confronted in the years of George H. W. Bush. The first was a larger, more compelling national policy imperative concerning the overall federal budget. With the cold war waning and the federal budget deficit soaring, politicians increasingly wanted to cap spending on discretionary programs as much as possible to lower the deficit. This put all "big science" programs in danger, and EOS was clearly big science at a projected $30 billion. The Bush White House and Congress pressed NASA for redesign to lower costs.

OSSA's associate administrator, Len Fisk, argued in 1991 for staying the course in plans and budget. Stating that a "comprehensive" approach was essential, he declared that Earth was "too complicated, and its workings too interrelated" for a piecemeal strategy. "It's a big Earth," he stated. "and there are big consequences for getting the wrong answers."[18] Unfortunately for EOS, there was a second problem: lack

17. Lambright, "Entrepreneurship and Space Technology."

18. Lennard Fisk, "Mission to Planet Earth," address to Maryland Space Business Round Table, 26 February 1991, NASA History Office Files.

of cohesion in support on the part of the scientific-environmentalist constituency. NASA's own Hansen, clearly alarmed about the global warming issue, said that the comprehensive approach of EOS would take too long to show results. While saying he supported EOS, he proposed a short-term, smaller-satellite strategy specifically geared to resolving the climate change debate.[19] The Hansen view was shared by various environmentalists and their prime legislative spokesman, Senator Al Gore (D-Tennessee). Many environmentalists in fact regarded EOS as an administration delaying tactic—researching instead of responding to climate change with emissions controls. At a time when Fisk needed solid backing for EOS, there were fissures in his potential scientific and environmental constituency.

The third problem Fisk faced was that NASA's flagship program, Space Station *Freedom*, was running into serious opposition in Congress. In her 1987 report, Ride had mentioned MTPE as one of four new initiatives that would build on the space station. However, EOS seemed now to be competing with the space station for funds. Both programs would cost billions. Congress told Bush and NASA: prioritize! When Congressional threats to terminate the space station forced Bush and NASA Administrator Richard Truly to choose, they made it clear that the space station was their priority. Congress in 1991 directed NASA to cap EOS expenditures at $11 billion through the year 2000. This was $6 billion less than NASA had projected during the 1990s for the two-platform design. Reorientation of the program became essential.[20]

The fourth obstacle to the program adopted in 1990 came in April 1992, when Dan Goldin arrived as NASA Administrator. By this time, the two-platform design was giving way to a fleet of six satellites, carrying fewer instruments. Goldin removed Fisk and reorganized OSSA into three smaller units. MTPE/EOS survived, but Goldin demanded it adhere to his philosophy of faster, better, cheaper (FBC). He called the original EOS design a "Battlestar Galactica," and even the six-satellite fleet failed his criteria.[21]

The fifth obstacle to implementing the original vision was the end of the relative political bipartisanship that had made adoption of MTPE/EOS possible in the first place. In the view of some conservative Republican lawmakers, NASA became politicized in the early 1990s. Ironically, NASA gave ammunition to its critics. As it shifted emphasis from ozone depletion to global climate change, it continued its ozone work, using the first satellite deployed under the MTPE rubric, the Upper Atmosphere Research Satellite (UARS). The focus of NASA's ozone research moved from the South Pole to the Arctic. NASA researchers in early 1992 detected an Arctic ozone hole opening up and publicly announced that fact with a sense of alarm.

19. James Hansen et al., "The Missing Data on Global Climate Change," *Issues in Science and Technology*, vii, no. 1 (Fall 1990), pp. 62–69.

20. Lambright, "Entrepreneurship and Space Technology," p. 102.

21. Stephanie Roy, "The Origin of the Smaller, Faster, Cheaper Approach in NASA's Solar System Exploration Program," *Space Policy* 14 (1998), p. 165–167.

Environmentalists, the media, and various legislators demanded action. Senator Gore, whose book, *Earth in the Balance*, came out in 1992, pointed out the danger of an Arctic ozone hole spreading over populated territory in the North. Critics of Bush said that there would soon be an ozone hole over Bush's head in his summer home in Kennebunkport, Maine. Bush did in fact act, using powers he had under domestic legislation related to the Montreal Protocol. He ordered a speed-up in the phasing out of ozone-depleting chemicals. There was thus again a swift progress from science to policy. The only problem this time was that NASA subsequently admitted that it had sounded the alarm prematurely and that there was no Arctic ozone hole. Nor, said a chastened NASA spokesman, was there a hole over Kennebunkport.[22]

The consequence was that conservative lawmakers and media seized the moment to criticize NASA's credibility and link the space agency to the environmentalists' agenda on global warming. As Bush left office, NASA's environmental mission was flayed by critics as politicized, and opposition in Congress loomed.

Thus, in the course of just four years—one presidential administration—NASA's environmental mission went seemingly from a high priority of NASA and the nation, with bipartisan support, to an embattled program struggling to survive, technically and politically.

More Turbulence Under Clinton

The inauguration of Bill Clinton as president, and especially Al Gore as his vice president, in 1993 guaranteed that NASA would continue to have an environmental role. The issue remained one of content and scale. Clinton retained Goldin as NASA Administrator, and NASA's Robert Watson moved up to the White House Office of Science and Technology to a new post of associate director of the Office of Science and Technology Policy (OSTP) for Environment.

The problem for MTPE/EOS was that Clinton, not Gore, was president. Also, both men had many other priorities and were far more interested in environmental action—including CO2 emissions reduction—than a long-term research program like EOS. Moreover, the larger enterprise of which EOS was the centerpiece, USGCRP, continued but lost momentum as a federal coordinating vehicle. The interagency committee that led USGCRP (and to some extent MTPE/EOS) found itself engaged in turf battles with the OSTP Environmental office. "Who was in charge of what" became an issue that caused confusion in global change policy generally.[23]

Most importantly, Clinton made deficit reduction and economic growth his top priority and looked to NASA for budget savings. There was at first no guarantee that NASA's major program, Space Station *Freedom*, would survive the transition

22. Lambright, *NASA and the Environment.*

23. W. Henry Lambright, *The Challenge of Coordinating "Big Science"* (Washington, DC: IBM, 2003).

to the Clinton White House and the new Congress. Following Clinton's decision to reduce the space station in scale and then his decision to merge Space Station *Freedom* with Russia's *Mir-2* space station effort, the NASA flagship project was secured within the NASA budget. The president and Congress agreed to stabilize the budget of the newly rechristened International Space Station (ISS) in succeeding years. However, the overall NASA budget would hold steady or even decrease, thus putting strain on other programs.

Goldin applied his faster, better, cheaper formula to space science and MTPE, but his heart was clearly more with missions beyond Earth, especially Mars, than Mission To Planet Earth. He cut the MTPE/EOS budget further and brought Charles Kennell, an astrophysicist from UCLA, aboard to head MTPE. Shelby Tilford, architect of the EOS program under Fisk, departed, complaining that faster, better, cheaper was not better in EOS's case. The budget for EOS was now down to $7.25 billion for the period from 1993 to 2000. Kennell's job was to reorient EOS further and adapt MTPE generally to Goldin and administration priorities. Adaptation included Landsat's return. Landsat, a 1970s NASA initiative, had spun off in the 1980s to NOAA in what proved a fruitless effort to privatize the program. It now came back to NASA to become part of an effort to bring space remote sensing applications "down to Earth." The aim was to apply space remote sensing to agriculture, city planning, fisheries, and other practical concerns. The new emphasis won friends for the program in Congress and the states. The flagship MTPE program, EOS, continued in development but its ambition was lowered in accord with the budget.

Although MTPE won friends because of its new "down to Earth" approach, EOS continued to be a target of critics. Because of its association with global change and especially global warming, EOS continued to be controversial. When the Republicans gained control of Congress in 1995, they sought substantial cuts in EOS. Rep. Dana Rohrabacher (R-California), chair of the House Science Committee's Energy and Environmental Subcommittee, derided EOS as concerned with "scientific nonsense." Global warming, he charged, was at best "unproven and at worst . . . liberal claptrap." Goldin strongly defended MTPE/EOS and, in the end, the effort to make draconian cuts failed.[24] The budget projection did go down, however, to $6.8 billion, along with NASA's budget generally.[25] Defenders of EOS commented sardonically that it was "restructured," "rescoped," "rebaselined," and "reshaped." In the end, it stabilized around the development of three intermediate-sized, multi-sensor satellites, one oriented to land, another to water, and the third to the atmosphere.

24. W. Henry Lambright, "The Rise and Fall of Interagency Coordination: The US Global Change Research Program," *Public Administration Review* (January, February 1997), pp. 41–42.

25. "Senators Rally Around NASA, Mission to Planet Earth," *The American Institute of Physics Bulletin of Science Policy News* (17 May 1996).

MTPE/EOS IN THE SECOND CLINTON ADMINISTRATION

There was relative stability for MTPE/EOS in Clinton's second term. To be sure, every year was a budget struggle. However, the precursor and applications missions showed results and the three EOS satellites made progress in development. Some of the partisan heat surrounding the program cooled, aided perhaps by NASA's changing the name of MTPE in 1998 to the more innocuous "Earth Sciences." The aim was to emphasize it as a policy-relevant, not policy-driven, science program—a subtle shift. Gore did not help NASA's effort to convey a neutral-science approach, however, when he proposed a new environmental satellite, called Triana, largely as an educational/inspirational tool. Detractors called it "Goresat" and Congress postponed its introduction, thereby taking some of the political heat off NASA.[26]

Meanwhile, positive results for the environment did flow from NASA precursor (i.e., pre-EOS) missions. The Agency's work in ozone was continuing to bear fruit. The UARS satellite helped NASA monitor ozone depletion and thus global compliance with the Montreal Protocol. Although it would take years for the ozone depletion problem to be fully alleviated, there was evidence that the depletion situation was beginning to improve.

Other important results in precursor missions lay with El Niño. Working with France, NASA developed a specialized satellite, called Topex-Poseidon, for measuring sea-surface temperatures. In March 1997, Topex-Poseidon detected a significant rise in sea level that spread across the Pacific toward the South American coast. In April, it found a rise in sea-level temperatures off the coast of Ecuador and Peru and monitored its spread north and south. This early discovery of a developing El Niño indicated that space technology, linked with other technologies of communication, could provide early warnings for natural disasters associated with the El Niño phenomenon.[27] Similarly, in 1998, NASA used the Tropical Rainfall Measuring Mission (TRMM) satellite to model more precisely the formation of hurricanes and other storms off the coast of Florida. Topex-Poseidon and TRMM again pointed up the practical value of environmental satellites. NASA's work in these "near-term" disasters was generally applauded. It still ran into controversy when it came to global warming, however. This was because it could not avoid what was happening in its political setting, where the controversy over the issue intensified.

---

26. Office of the Vice President, "Vice President Gore Challenges NASA to Build a New Satellite to Provide Live Images of Earth from Outer Space," The White House, Press Release (13 March 1998); "Scripps Institution, GSFC, Picked to Put Goresat at L-1 Point," *Aerospace Daily* (29 October 1998), article 117803. The satellite was eventually terminated. See Andrew Lawler, "NASA Terminates Gore's Eye on Earth," *Science* (6 January 2006), p. 26.

27. Madeline Nash, *El Nino* (New York: Warner Books, 2002), pp. 91, 114.

In late 1997, Vice President Gore defied Congressional sentiment when he went to Kyoto, Japan, and agreed to a protocol with other nations to cut back greenhouse gas emissions by a specific amount over a designated span of time. Clinton said he agreed with Gore but that he would not submit the Kyoto Protocol to the Senate for confirmation. He would instead emphasize voluntary measures to reduce emissions to Kyoto standards. The political consensus for binding emissions reduction and thus the regulation of the energy and automobile industries was not present, in Clinton's view. The political debate over Kyoto spilled over to make the setting for research on climate change more conflictual. [28]

In 1999, NASA launched the first of its climate-related EOS satellites, Terra. This satellite emphasized land masses but also had sensors relevant to other aspects of the physical environment. At $1.3 billion, Terra was intermediate in scale. Although far from the original concept, it was bigger than the faster, better, cheaper model Goldin had initially promoted. The remaining two satellites in the three-satellite series were scheduled to go up in the first term of Clinton's successor.

Environment at NASA Under George W. Bush

When George W. Bush became president in January 2001, the major policy question for the new administration involving NASA and the environment was: What would be NASA's role in environmental research after EOS? The answer has come slowly, tortuously, and is still emerging.

In March, Bush withdrew the United States from agreements under Kyoto. The negative reaction to that decision in the United States and Europe caused him to subsequently announce he would support research on climate change and take policy action if it proved necessary. The USGCRP was maintained under a new label, Climate Change Research Initiative, with the assistant secretary of commerce in charge of the interagency effort. For a year, NASA's Earth Science program was essentially on hold while the administration determined what it would do. Meanwhile, at the beginning of January 2002, Sean O'Keefe came aboard as new NASA Administrator. O'Keefe's priority was the ISS cost overrun, not Earth Sciences, but he provided support for EOS to complete what had begun under Goldin. The first EOS satellite was performing well, and the second, Aqua (costing $1 billion), was launched May 4. As its name implied, Aqua's mission was to study the global water cycle, including precipitation and evaporation.[29] O'Keefe also symbolized NASA's environmental concern by including in its mission statement the phrase "To understand and protect our home planet."

---

28. Lamont Hempel, "Climate Policy on the Installment Plan," in *Environmental Policy*, 6th ed., Norman Vig and Michael Kraft, ed. (Washington, DC: CQ Press, 2006), pp. 294–297.

29. Brian Berger, "NASA Aqua Mission to Study Global Water Cycle," *Space News* (29 April 2005), p. 18.

On 1 February 2003 the *Columbia* Space Shuttle disintegrated, killing all astronauts aboard. Once again, momentum for NASA's future, Earth Sciences included, was interrupted. That there would be some kind of future for Earth Sciences seemed clear. In the summer 2003, the State Department held an International Earth Observations Summit. The United States pledged to provide leadership in climate change research through an international "system of systems" in which satellite capabilities of many nations would be coordinated and linked. Following the conference, NASA announced it would develop on an accelerated pace a new, post-EOS, climate change-oriented satellite called Glory.[30]

On 14 January 2004, President Bush announced his Vision for Space Exploration (VSE) initiative, "To the Moon, Mars, and Beyond." In line with the new priority, O'Keefe reorganized the Agency, creating an Exploration Systems Division and merging NASA's separate Earth Science and Space Science programs into a single organization, the Science Mission Directorate. Proponents of Earth Science viewed the reorganization as a downgrading in priority. Advocates of reorganization argued that the moving of Earth and Space Science together promoted synergy and enhanced learning about planetary change in general from work on Earth, Mars, and other planets. They gave a comparative planetology perspective, one historically always present as part of NASA's environmental mission but usually as a background element. Now it was moved to the foreground to emphasize mutual learning from missions to and from Earth.

In July, NASA launched Aura, the third and final leg of EOS. Another billion-dollar satellite, Aura would make comprehensive measurements of the atmosphere and take over for the aging UARS in monitoring ozone trends.[31] The completion of EOS after so many years was a matter of relief for some Earth Science observers, but frustrating for others. Valuable though they would be, the EOS satellites were far short of the program creators' vision: large dual platforms providing comprehensive and simultaneous air, land, and sea interactive images over a 15-year time frame. There were many gaps in knowledge to be filled.

The more pressing problem continued to be: What next for Earth Sciences? Especially: What next in an era when Moon-Mars would increasingly shape all priorities for NASA? That is, the closer a program was to VSE, the more relevant it could be seen to be to NASA's future. The further away, the less relevant. There was a great deal of concern among scientists and administrators involved with NASA's environmental mission that, when push came to shove, they would suffer. Indeed, budget projections seemed to indicate that fact, as did a decision to cancel the Glory climate satellite. Moreover, there were rumors that the administration was considering moving much of NASA's Earth observation work to NOAA.

---

30. Brian Berger, "Reversing Course," *Space News* (25 July 2005), p. 10.

31. Tariq Malik, "Last EOS Satellite to Study Air Quality at Earth's Surface," *Space News* (14 June 2004), p. 8.

The future remained murky when O'Keefe departed and Michael Griffin came aboard as NASA's leader in April 2005. Griffin reversed a number of O'Keefe's decisions, including the Glory termination. Earth Science had friends in Congress, especially Sherwood Boehlert (R-NY), chair of the House Science Committee.[32] However, Griffin had to deal with the reality of increased shuttle repair costs and a relatively static overall NASA budget. If VSE were to move forward, money would have to come from lower-priority programs, including Earth Sciences. At the same time, the cross-pressured Griffin found himself dealing with a highly public dispute involving NASA's outspoken James Hansen. The NASA scientist complained to the media that administration appointees in NASA's public affairs office were censoring his statements about global warming. Griffin declared that NASA scientists could comment freely about science, and even its implications, but they should make clear that they spoke for themselves, not NASA, when they crossed the line into policy. A NASA public affairs official who had allegedly censored Hansen was fired when it was revealed he had lied about his educational credentials.[33]

In 2006 Griffin presented a budget that delayed various post-EOS missions but reinstated Glory.[34] EOS orbited and provided considerable information, but there was worry about whether there was enough money to analyze the information properly. There were also various older specialized satellites, including Landsat, which needed replacement in some way. NASA was a junior partner in a NOAA-Department of Defense operationally-oriented system called the National Polar Orbiting Operational Environmental Satellite System (NPOESS). For a while, the Bush administration seemed intent on linking NASA's post-EOS climate science future to NPOESS, but when NPOESS ran into budget problems of its own, connecting NASA's climate role to NPOESS lost a great deal of its rationale.[35]

What NASA needed was an overarching strategy for its future in Earth Science, one like the one it developed in the 1980s prior to the birth of MTPE and EOS. To help provide advice, the National Research Council undertook an Earth Science decadal survey on priorities in the next 10 years. To be completed by November 2006, the survey was intended to help Griffin determine what made sense for the Earth Sciences program. Griffin, in turn, would require major decisions from the White House and Congress to set NASA on a course fruitful both for science and climate policy.

32. Brian Berger, "NASA's Exploration Focus Blamed for Earth Science Cuts," *Space News* (2 May 2005), p. 4.

33. Rick Weiss, "NASA Sets New Rules on Media," *The Washington Post*, 31 March 2006, p. A10; Warren Leary, "New NASA Policy Backs Free Discussion by Scientists," *The New York Times*, 31 March 2006, *http://www.nytimes.com*.

34. Berger, "Reversing Course."

35. Jeffrey Mervis, "Climate Sensors Dropped from US Weather Satellite Package," *Science* (16 June 2006), p. 1580.

Without question, the Earth Science program was at a crossroad. NASA's environmental role was not explicitly mentioned in the overarching NASA mission statement of 2006. The phrase O'Keefe had used in his mission statement—"To understand and protect our home planet"—was gone. The new Griffin mission statement read "To pioneer the future in space exploration, scientific discovery, and aeronautics research."[36] Whether the omission was symbolic or substantive remained to be seen. The basic problem NASA Administrator Griffin has faced is that he has too many programs on NASA's plate and not enough money for all.

CONCLUSION

How did NASA evolve an environmental mission? The answer is that it did so gradually and somewhat tortuously, with ups and downs along the way. What did it do with that mission? It did a great deal that was positive in a host of ways, but with special significance in regard to ozone depletion and the Montreal Protocol. What has this meant for society? It has meant benefits in knowledge with spillover into policy. What has the environmental mission meant for NASA? It has meant enmeshment in a political context that has helped it survive and grow at some points in history and made it extremely controversial at other times.

Taking the long view of history, NASA has had an enormous impact on environmental policy since its founding in 1958. The weather satellite and its successors, developed by NASA and transferred to operational use, have saved lives and money through early warnings of hurricanes and improved weather forecasting. The first pictures of Earth from the Apollo program crystallized the vulnerability and beauty of Earth and were used by environmentalists to mobilize support for the first Earth Day in 1970. As the environmental movement gathered momentum, NASA in the 1970s aligned itself with the new imperative and extended its weather satellite effort to land- and sea-oriented images. This activity took place as NASA's planetary program stimulated a comparative planetology consciousness in which Earth seemed increasingly unique as a home to life. The Gaia Hypothesis, which emerged from comparative planetology, stressed thinking about Earth as a living system of interacting and interdependent parts.

In 1972, with Landsat, NASA initiated a long series of Earth monitoring satellites. In doing so, it became more and more, as NASA administrator Fletcher said, "an environmental agency." Becoming "an environmental agency," through the science it provided, immersed NASA in the controversies and political emotions surrounding the environmental question. Sometimes the political winds were favorable to NASA; at other times they were not so. In the 1980s NASA, armed with

36. Andrew Revkin, "NASA's Goals Delete Mention of Home Planet," *The New York Times,* 22 July 2006.

legislative authority gained in the 1970s, played a lead role within the government in responding to the ozone depletion threat. It was the ozone crisis that provided a proactive model for NASA as an environmental agency, one that could conduct and transfer policy-relevant science. Advocates of a larger global environmental mission sought to build on the ozone success to position NASA for an expanded role. After *Challenger*, in 1987 Sally Ride listed Mission To Planet Earth as first among new initiatives that could galvanize a shaken agency and make it more relevant to American society and the world's needs.

Such a mission seemed ripe indeed in the wake of the Montreal Protocol as the new concern about climate change rose on the nation's agenda, propelled in part as NASA's James Hansen. President George H. W. Bush and Congress in 1990 provided authority and initial funding for MTPE and its centerpiece, EOS. However, in the 1990s, MTPE ran into a budgetary crunch and competition with other NASA programs—human spaceflight and space science—and was never funded as its founders expected. EOS was downsized and split up, subject not only to budget constraints but also pressures to conform to Goldin's faster, better, cheaper doctrine. What finally was launched as EOS in 1999, 2002, and 2004 was a much circumscribed version of what its architects had planned in the late 1980s. The system could not realize the promise of a 15-year comprehensive data set from the simultaneous monitoring of land, sea, and atmosphere interaction. A new Earth System Science got started but did not get as far as expected. The "predictive capability" remained to be developed for global change.

With ozone depletion, NASA had legitimacy to lead and a crisis context in which there was science push and policy pull. With climate change, however, NASA had no special mandate to lead and the science push has come gradually in a societal context where the perception of crisis is lacking. Rather than policy pull, there has been policy discord and even "push back."

A strategy for the post-EOS era has yet to be fully charted and adopted, much less implemented. Agreement on next steps has to be forged. It is easy today for advocates of global environmental satellites to see a glass half empty. However, a more positive, glass half-full view surely would say that it is remarkable how much a space agency has done in directing attention to Earth. NASA is the space research and development arm of the federal government when it comes to global environment. In addition to EOS, there have been numerous specialized satellites that have monitored ice melting, El Niño, storms, and others ills. Ozone provides a genuine success story in science and public policy. There was science-push and policy-pull. Climate change is not a success story yet. If one were giving grades, NASA would get an "A" for ozone depletion and "Incomplete" for global climate change. Having said that, the problem of climate change is more complex than ozone for many reasons, and there has been consensus-building progress, at least on the science side. There is a limit to what NASA (and science generally) can do in eliciting policy response, however.

The dilemma NASA now faces in designing a post-EOS future is shared with other agencies associated with the global change initiative of the early 1990s. This interagency initiative never was fully implemented, coordinated, or led. There is a need to re-energize the vision many of the early advocates of EOS and USGCRP had—a strong Earth system science and a capacity to predict global change (especially climate change)—that can be put to policy use. Achieving such a vision requires a planetary perspective and that is NASA's distinctive environmental competence. It is based on NASA's mission to the home planet *and* the comparative approach derived from its work beyond Earth. That perspective needs renewal and advocacy for a twenty-first century setting. That setting almost surely will be influenced, perhaps dramatically, by events involving climate change. Remaking NASA's environmental mission, with resources to match, and connecting that role to other agencies and nations is a challenge. It is less a problem in science and technology, however, and much more a challenge of political will.

Chapter 17

NAVSTAR, the Global Positioning System: A Sampling of Its Military, Civil, and Commercial Impact

Rick W. Sturdevant

Historical Background

The NAVSTAR[1] Global Positioning System (GPS), the first satellite navigation system that enabled users to determine precisely their location in three dimensions and time within billionths of a second, grew from a concept into a fully operational system in slightly more than two decades. This is not to suggest, however, that selling the idea was easy. As early as 1969–1970, Aerospace Corporation president and GPS pioneer Ivan Getting had suggested to Lee DuBridge, President Richard Nixon's science advisor, that a presidential commission be created to review how satellite navigation ought to proceed, because there were so many potential users. After thinking about it for several weeks, DuBridge concluded that execution of Getting's proposal would be too difficult. He told Getting, "There are too many people, too many bureaucracies, too much politics, and too many agencies involved. Why don't you just have the Air Force develop it the way we always did?"[2]

By 1972, both the U.S. Air Force (USAF) and the U.S. Navy had been studying for several years the possibility of improved satellite-based radio navigation. Three earlier space-based navigation systems or programs contributed to GPS: The Johns Hopkins University Applied Physics Laboratory (APL) Transit, otherwise known as the Naval Navigation Satellite System; the Naval Research Laboratory's Timation satellite program, led by Roger Easton; and USAF Project 621B. Colonel Bradford

1. Over time, some people identified "NAVSTAR" as an acronym derived from either "Navigation Signal Timing and Ranging" or "Navigation Satellite Timing and Ranging." Apparently, TRW had once advocated a navigational system for which NAVSTAR was an acronym (NAVigation System Timing And Ranging). Bradford W. Parkinson, "GPS Eyewitness: The Early Years," *GPS World* 5 (9 September 1994): pp. 32–45, explained that his team at the Joint Program Office never considered "NAVSTAR" an acronym but "simply a nice-sounding name."

2. Jacob Neufeld, ed., *Research and Development in the United States Air Force* (Washington, DC: Center for Air Force History, 1993), pp. 91–92.

Parkinson, USAF director of the newly formed multi-service Joint Program Office (JPO) for GPS, assembled about a dozen members of the JPO over Labor Day weekend in 1973 and directed them to synthesize the design for a new satellite navigation system by drawing the best ideas and technology from everything then available.[3]

From that point, GPS developed at a reasonably steady pace. By June 1974, the JPO had selected Rockwell International as the satellite contractor. The JPO oversaw deployment of the initial control segment at the Army's Yuma Proving Ground in Arizona, followed by the launch of the first operational prototype in February 1978. Eleven years later, in February 1989, the USAF launched the first fully operational Block 2 version. Although a complete constellation of 24 Block 2 satellites existed in December 1993, GPS did not officially achieve full operational status until April 1995. It had cost $10–$12 billion to field the system, and the USAF estimated the cost annually of sustaining minimal GPS services at $400 million.[4]

The main reasons for GPS development were the need to deliver weapons precisely on target and to reverse the proliferation of navigation systems in the U.S. military. From the beginning, however, the Department of Defense (DOD) recognized the usefulness of GPS to the worldwide civilian community. To withhold full accuracy from enemies but provide GPS service to civilian users, the USAF designed the system with a protective feature called "selective availability" (SA) that, when used, gave the U.S. military and its allies significantly more precise satellite signals than what other users received. After Korean Airline Flight 007 went astray in September 1983 and Soviet fighters shot it down, President Ronald Reagan reassured the world that the coarser signal would remain continually and universally available at no cost once GPS became fully operational. As GPS approached that status in the early 1990s, civilian and commercial users, who already had 10 times as many GPS receivers as the military, mounted an increasingly vocal campaign for unrestricted access to the more precise satellite signals. Many GPS equipment manufacturers, anticipating a multitude of applications, had begun forming strategic alliances with outside companies in such fields as communications, Geographic Information Systems (GIS), computing, and transportation. Finally, in May 2000, President William Clinton acknowledged the global utility of GPS and directed immediate discontinuation of SA, thereby giving millions of nonmilitary users access to the more precise GPS signals.[5]

3. Bradford W. Parkinson, "Introduction and Heritage of NAVSTAR, the Global Positioning System" in *Global Positioning System: Theory and Applications, Volume I*, Bradford W. Parkinson and James J. Spiker, Jr., ed. (Washington, DC: American Institute of Aeronautics and Astronautics, 1996), pp. 3–28; Ivan A. Getting, *All in a Lifetime: Science in the Defense of Democracy* (New York: Vantage Press, 1989), pp. 574–599; Richard Easton, "Who Invented the Global Positioning System?" *The Space Review* (22 May 2006), *http://www.thespacereview.com/article/626/1* (accessed 12 September 2006).

4. Michael Russell Rip and James M. Hasik, *The Precision Revolution: GPS and the Future of Aerial Warfare* (Annapolis, MD: Naval Institute Press, 2002), pp. 68–79.

5. Parkinson, "GPS Eyewitness," pp. 42–44; Rip and Hasik, *The Precision Revolution*, pp. 9–10 and 429–441; Anthony R. Foster, "GPS Strategic Alliances, Part I: Setting Them Up," *GPS World* 5 (5 May 1994): pp. 34–42; National Research Council, *The Global Positioning System—A Shared National Asset: Recommendations for Technical Improvements and Enhancements* (Washington, DC: National Academies Press, 1995), pp. 1–12.

With increasing demand for accuracy beyond what GPS alone yielded, users found ways to augment it. For small areas, those included pseudolites (ground-based transmitters that could be configured to emit GPS-like signals) and differential GPS (DGPS), which required a high-quality GPS "reference receiver" at a known, surveyed location. Wide Area DGPS (WADGPS), involving reference receivers at multiple monitor stations and a master-control hub, would achieve similar results over a broader region. An example of WADGPS was the Wide Area Augmentation System (WAAS), which the Federal Aviation Administration (FAA) believed would eventually provide the integrity, reliability, time availability, and accuracy to permit pilots and air traffic controllers to rely confidently and safely on GPS as their primary navigation system.[6] By 2006, continuously operating reference stations (CORS), coordinated across the United States by the National Geodetic Survey, enhanced GPS services to millions of users.[7]

Meanwhile, other countries and geographic regions had begun developing their own GPS augmentation systems: the European Geostationary Navigation Overlay System (EGNOS); India's GPS and Geostationary Augmentation Network (GAGAN); Australia's Ground-based Regional Augmentation System (GRAS); and Japan's Multi-transport Satellite Augmentation System (MSAS). In addition, Russia operated its own Global Navigation Satellite System (GLONASS); China experimented with its Beidou navigation satellites; and Europe pressed hard toward launching the Galileo satellite navigation system. Whether all those capabilities could be melded into a fully integrated Global Navigation Satellite System (GNSS) remained a question without an immediate answer, but the military, civil, and commercial utility of GPS was unquestionable. Since a full accounting of the GPS's societal impact would require hundreds of pages, what follows is merely suggestive. [8]

6. Bradford W. Parkinson and James J. Spilker, Jr., eds., *Global Positioning System: Theory and Applications, Volume II* (Washington, DC: American Institute of Aeronautics and Astronautics, 1996), pp. 3–142; Robert O. DeBolt et al., *A Technical Report to the Secretary of Transportation on a National Approach to Augmented GPS Services* (U.S. Department of Commerce, NTIA Special Publication 94-30, December 1994), *http://www.navcen.uscg.gov/pubs/gps/gpsaug/auggps.doc* (accessed 13 September 2006).

7. James Stowell, "GPS Reference Stations: 24/7, High Accuracy, Differential Data in Your Own Backyard," CE News (May 2002), *http://www.leica-geosystems.com/us/articles/2002/CE%20News_May%202002_GPS%20Reference%20Stations.pdf* (accessed 12 September 2006); "Continuously Operating Reference Stations," National Geodetic Survey—CORS, *http://www.ngs.noaa.gov/CORS/* (accessed 12 September 2006).

8. Rosalind Lewis et al., *Building a Multinational Global Navigation Satellite System: An Initial Look* (Santa Monica, CA: RAND Corporation, 2005); "New U.S. Policy for Positioning, Timing and Navigation (PNT) Services," Embassy of the United States, Spain (12 January 2005), *http://barcelona.usconsulate.gov/emba/pntpolicyb.html* (accessed 12 September 2006); Brian Evans, "Setting a Course: Beyond GPS," Aviation Today (1 May 2006), *http://www.aviationtoday.com/cgi/av/show_mag.cgi?pub=av&mon=0506&file=beyondgps.htm* (accessed 12 September 2006).

MILITARY APPLICATIONS

The first weapon in the U.S. arsenal to become operational using GPS navigation was the Conventional Air-Launched Cruise Missile (CALCM) or AGM-86C. Its development began in June 1986 when Boeing Company received a 12-month contract for rapid conversion of existing ALCMs, the inaccuracy of which had resulted in accidentally bombing the French Embassy during Operation Eldorado Canyon, a night attack against Libya two months earlier. A loosely coupled integration of GPS capabilities with the missile's existing inertial navigation system (INS) permitted Boeing to deliver the first CALCMs to the U.S. Air Force in June 1987. During the initial hours of the air campaign for Operation Desert Storm in January 1991, seven B-52G bombers flying from Barksdale Air Force Base, Louisiana, delivered a total of 35 CALCMs against eight high-value targets in Iraq and achieved 85 percent to 91 percent success, including several exact hits.[9]

In the 1990s, the U.S. military fielded a host of air-dropped munitions that used GPS to one degree or another. Foremost among them, the Joint Direct Attack Munition (JDAM) turned fundamentally "dumb" bombs into high-precision ordnance capable of destroying multiple targets in a single sortie any time of day or night, in any kind of weather, no matter how adverse. During Operation Allied Force, the NATO air campaign against Serbia in 1999, U.S. B-2 Spirit bombers dropped more than 500 relatively inexpensive JDAMs with such great success that U.S. military strategists foresaw the possibility of unguided bombs completely disappearing from the arsenal. Other GPS-aided U.S. aerial weapons that debuted in the 1990s included the AGM-154 Joint Stand-Off Weapon (JSOW), the AGM-130 air-to-surface missile, the BGM-109 Tomahawk cruise missile, and the SLAM-ER (Stand-off Land Attack Missile—Enhanced Response).[10]

A suite of technologies known as Advanced Spinning-Vehicle Navigation (ASVN) permitted GPS/INS guidance in smaller and smaller munitions. By 2001, the U.S. Army planned to use it in artillery shells; the U.S. Navy had similar plans for rocket-assisted projectiles fired from its deck guns. The Army's howitzer-fired, 155-mm XM982 Excalibur round underwent a demonstration at Yuma Proving Ground, Arizona, on 15 September 2005 in which its accuracy was better than 33 ft (10 m) at a distance of 9 mi (15 km). Excalibur tests continued into February 2007, with operational fielding scheduled for later that year. A Navy contract with Raytheon Missile Systems, primary designer of the Army's Excalibur, called for development of the 5-in (13-cm) MK-171 Extended-Range Guided Munition (ERGM). Although achieving satisfactory ERGM performance proved harder than

---

9. John Tirpak, "The Secret Squirrels," *Air Force Magazine* 77, no. 4 (April 1994), *http://www.afa.org/magazine/perspectives/desert_storm/0494squirrels.asp* (accessed 5 September 2006); John T. Nielson, "The Untold Story of the CALCM: The Secret GPS Weapon Used in the Gulf War," *GPS World* 6, no. 1 (January 1995): pp. 26–32.

10. Rip and Hasik, *The Precision Revolution*, pp. 233–264.

for Excalibur and delayed its operational deployment, flight demonstrations at White Sands Missile Range, New Mexico, on 16 February 2005 proved the ERGM could achieve great accuracy at a distance of more than 40 nautical miles. In both cases, these new projectiles offered greater lethality and lower collateral damage, with increased range and a considerably reduced logistical burden for deployed forces.[11]

In early 2003, the public became aware of GPS guidance for precision airdrops. The U.S. Army Operational Test Command employed GPS in two different instrument packages—one to verify the optimal parachute rigging for heavy cargo pallets, and the other to evaluate new troop-parachute designs. Within months, the U.S. military operationally tested Onyx, an autonomously guided parachute system developed by Atair Aerospace, for delivering payloads ranging from 75 to 2,200 lbs (34 to 998 kg) within 246 ft (75 m) circular error probable from altitudes up the 35,000 ft (10,668 m) in darkness and other extreme conditions. On 9 August 2004, U.S. Marines near Camp Korean Village, Iraq, witnessed the first operational use of a GPS-assisted Sherpa parafoil cargo delivery system, a key component of the Joint Precision Airdrop System (JPADS) technology demonstration program, in a combat zone. Two years later, the first joint Air Force-Army operational drop using JPADS in Southwest Asia supplied ammunition and water to troops in Afghanistan. Meanwhile, design and testing of GPS-equipped navigation units for paratroopers proceeded. In 2005, the French Military Agency (DGA) and Army Special Forces in Singapore were using more than 200 GPS-assisted Operational Paratroopers Navigation System (OPANAS)

11. James H. Doty, "Revolution in GPS: Advanced Spinning-Vehicle Navigation," *GPS World* (September 2004), *http://www.gpsworld.com/gpsworld/article/articleDetail.jsp?id=120802* (accessed 14 August 2006); Lawrence L. Wells, "GPS Guidance for Projectiles," *GPS World* (October 2001), *http://www.gpsworld.com/gpsworld/article/articleDetail.jsp?id=1806* (accessed 14 August 2006); Raymond Sicignano, "Picatinny Support the Warfighter: New Picatinny-Developed Smart Artillery Munition Could Reach Troops by March," *The Voice* 18, no. 18 (7 October 2005), *http://w4.pica.army.mil/Voice2005/051007/051007%20Smart%20artillery.htm* (accessed 14 August 2006); Raytheon News Release, "Successful Flight Test of GPS-guided Artillery Projectile Puts Raytheon-Bofors Excalibur Closer to Fielding," 26 September 2005, *http://www.prnewswire.com/cgi-bin/micro_stories.pl?ACCT=683934&TICK=RTN7&STORY=/www/story/09-26-2005/0004131807&EDATE=Sep+26,+2005* (accessed 14 August 2006); Staff Writers, "Testing Of GPS-Guided Projectile Puts Raytheon-BAE Excalibur Closer To Fielding," *GPS Daily* (21 August 2006), *http://www.gpsdaily.com/reports/Testing_Of_GPS_Guided_Projectile_Puts_Raytheon_BAE_Excalibur_Closer_To_Fielding_999.html* (accessed 21 August 2006); Raytheon News Release, "Raytheon's Excalibur Successfully Completes Final Testing, Clearing the Path for Early Fielding," *http://www.prnewswire.com/cgi-bin/micro_stories.pl?ACCT=910473&TICK=RTNB12&STORY=/www/story/03-08-2007/0004542432&EDATE=Mar+8,+200* (accessed 31 July 2007); Raytheon News Release, "Raytheon's ERGM Guides into Success," 17 February 2005, *http://www.prnewswire.com/cgi-bin/micro_stories.pl?ACCT=683934&TICK=RTN7&STORY=/www/story/02-17-2005/0003027460&EDATE=Feb+17,+2005* (accessed 14 August 2006); Dan Coskren, Tim Easterly, and Robert Polutchko, "More Bang, Less Buck: Low-Cost GPS/INS Guidance for Navy Munitions Launches," GPS World (September 2005), *http://www.gpsworld.com/gpsworld/article/articleDetail.jsp?id=180139* (accessed 14 August 2006); John Pike, "EX-171 ERGM Extended-Range Guided Munition," *http://www.globalsecurity.org/military/systems/munitions/ergm.htm* (accessed 14 August 2006); Rupert Pengelley, "Guided Artillery Projectiles Turn the Corner," Jane's International Defence Review 39 (February 2006): pp. 57–61.

Onyx autonomously guided parachute system using GPS navigation.

units produced by SSK Industries, and North Atlantic Treaty Organization (NATO) countries were testing OPANAS for high-altitude, high-opening (HAHO) jumps. These new capabilities ensured more accurate landings for both cargo and troops in all kinds of conditions and, by permitting releases from altitudes above 25,000 ft (7,620 m), protected aircraft and personnel from hostile forces armed with inexpensive surface-to-air missiles.[12]

After 2000, a rapidly expanding family of autonomous or remotely controlled robotic military systems relied on GPS navigation to perform a wide variety of tasks on land, at sea, underwater, and in the air or outer space. Unpiloted aerial vehicles (UAVs) equipped with GPS, which had been under development since the 1980s, flew reconnaissance and surveillance missions over Bosnia and Kosovo during the late 1990s and, thereafter, proliferated among military organizations worldwide. In 2001–2002, Predator drones carrying AGM-114 Hellfire missiles attacked Taliban and Al Qaeda forces in Afghanistan.[13] On 25 January 2002, a GPS-Aided Inertial Navigation System (GAINS) on the third stage of an SM-3 interceptor launched from a U.S. Navy Aegis cruiser helped successfully demonstrate the exo-

12. Joseph Strus et al., "15 Tons. 1500 Feet. 4 Gs.—Airdrop Behavior of Parachuted Cargo Pallets," *GPS World* (April 2003), *http://www.gpsworld.com/gpsworld/article/articleDetail.jsp?id=53305* (accessed 17 August 2006); Nanker Phelge, "Take the Plunge—Navigating from 35,000 Feet," *GPS World* (August 2003), *http://www.gpsworld.com/gpsworld/article/articleDetail.jsp?id=65593* (accessed 17 August 2006); Bill Lisbon, "GPS-Guided Paradrop," *Special Operations Technology 2*, no. 6 (2004): pp. 29–31; Tom Vanden Brook, "'Smart' Airdrops May Save Lives of U.S. Troops," *USA Today*, 8 January 2007; Joseph Strus et al., "Stand Up. Hook Up. GO!—Instrumenting Paratroopers," *GPS World* (April 2003), *http://www.gpsworld.com/gpsworld/article/articleDetail.jsp?id=53449* (accessed 17 August 2006); Jason I. Thompson, "A Three Dimensional Helmet Mounted Primary Flight Reference for Paratroopers," Master's thesis, Air Force Institute of Technology, March 2005), pp. 1–8.

13. Defense Airborne Reconnaissance Office, "UAV Annual Report FY 1997," (6 November 1997), *http://www.fas.org/irp/agency/daro/uav97/toc.html* (accessed 17 August 2006); John Pike, "RQ-1 Predator MAE UAV," 6 November 2002, *http://www.fas.org/irp/program/collect/predator.htm* (accessed 17 August 2006); Greg Loegering, "The Global Hawk Advantage," *GeoIntelligence Magazine* (November 2004), *http://www.geointelmag.com/geointelligence/article/articleDetail.jsp?id=134277* (accessed 17 August 2006).

atmospheric interception of an incoming target missile.[14] During Operation Iraqi Freedom in 2003, the U.S. Navy ordered the first wartime deployment of the GPS-equipped, Remote Environmental Monitoring Units (REMUS) autonomous underwater vehicle (AUV) to locate suspected mines in the port of Umm Qasr.[15] By November 2004, the DOD had contracted with Applied Perception, Inc. (API) to build two interrelated, GPS-enabled robotic prototypes—a 3,500-lb (1,588-kg) Robotic Evacuation Vehicle (REV) and a 600-lb (272-kg) Robotic Extraction Vehicle (REX)—for the safe recovery and transport of wounded soldiers from the battlefield to a nearby field hospital. Clearly, such vehicles could have numerous other military and nonmilitary applications ranging from logistical supply, ordnance disposal, surveillance, and assault to fire fighting, crowd control, soil sampling, pest abatement, and exploration of hazardous environments.[16]

Perhaps the most pervasive U.S. military application of GPS by 2006 involved the tracking and coordination of various battle elements in real time using the GPS-enabled Force XXI Battle Command, Brigade-and-Below (FBCB2) satellite-based tracking system and its Blue Force Tracking (BFT) variant. American soldiers had suffered roughly a quarter-million casualties in the twentieth century due to so-called friendly fire, primarily because of the difficulty during intense conflict of discriminating quickly between friend and foe. As early as springtime of 1987, engineers and scouts from the U.S. Army's 4th Infantry Division used two 17.5-lb (8-kg) GPS manpacks during exercises in the Pinõn Canyon training area of southern Colorado to maneuver through "enemy" lines in snow, rain, fog, and darkness to accomplish their objective. Four years later in Operation Desert Storm, with only 16 satellites in the constellation, GPS aided positioning and maneuvering of large troop formations, plus precision bombing, artillery fire support, and special operations in relatively featureless desert terrain. Coalition forces relying primarily on more than 12,000 personal receivers, each costing about $3,500, prevented countless casualties by reducing the so-called fog of battle.[17]

---

14. James L. Anders et al., "From View to Kill—Integrated System Guides Ballistic Missile Intercept," *GPS World* (May 2003), *http://www.gpsworld.com/gpsworld/article/articleDetail.jsp?id=57971* (accessed 17 August 2006).

15. Ken Jordan, "REMUS AUV Plays Key Role in Iraq War," UnderWater Magazine (July/August 2003), *http://www.diveweb.com/rovs/features/034.01.htm* (accessed 17 August 2006); Kevin McCarthy, "REMUS—A Role Model for AUV Technology Transfer," *International Ocean Systems* 7, no. 6 (November/December 2003), *http://www.hydroidinc.com/pdfs/Hydroid_REMUS.pdf* (accessed 17 August 2006); Marty Whitford, "In the Swim—GPS Guides Autonomous Underwater Vehicles," *GPS World* (April 2005), *http://www.gpsworld.com/gpsworld/article/articleDetail.jsp?id=154866* (accessed 17 August 2006).

16. Marty Whitford, "Robotic Rescue—No One Gets Left Behind," *GPS World* (November 2004), *http://www.gpsworld.com/gpsworld/article/articleDetail.jsp?id=131750* (accessed 17 August 2006); Paul J. Lewis et al., "Applications Suitable for Unmanned and Autonomous Missions Utilizing the Tactical Amphibious Ground Support (TAGS) Platform," *http://www.autonomoussolutions.com/Press/SPIE%20TAGS.html* (accessed 17 August 2006).

17. "U.S. Army Trains on New System—Exercise in Piñon Canyon Teaches Usefulness of GPS," *Space Trace* (29 May 1987): p. 2.

A decade after Desert Storm, U.S. forces conducting war games in California and Florida located and tracked troops, aircraft, and assorted equipment in real time using an Inexpensive Range Instrumentation System (IRIS). That experimentation demonstrated that low-cost, commercial off-the-shelf hardware could be used to coordinate ground and air activities, thereby enhancing the safety of friendly forces. By 2005, U.S. and allied forces in Kosovo, Afghanistan, and Iraq relied on more than 8,000 GPS-enabled FBCB2 units and another 2,000 FBCB2-BFT units to track their own positions, neighboring friendly forces, and spotted enemy forces along with the location of bridges, mine fields, and other potentially dangerous geo-points. In 2006, Globecomm Systems won a $7.8 million contract to provide NATO forces with a similar BFT capability. The enhanced situational awareness provided by FBCB2 and its BFT variant also allowed battlefield commanders to plan and coordinate maneuvers—offensive and defensive—with unprecedented precision. [18]

NAVIGATION

The most rapidly expanding area of GPS use for civil, commercial, and personal purposes was probably location-based services (LBS)—positioning and navigation. Land-based users include automobile drivers, railroads, fleet managers of trucks, delivery vehicles, and public transportation; emergency responders such as fire, ambulance, and police; and recreational activities such as hiking, hunting, skiing, biking, and golfing. According to Alan A. Varghese from ABI Research in Oyster Bay, New York, shipments of recreational GPS devices alone rose from 3.2 million in 2002 to 5 million in 2003, with a predicted annual growth of 31 percent until 2009. Sea-based applications ranged from recreational sailing, fishing, and managing shipping fleets, to assisted steering, risk assessment, and hazard warning. Pilots of all varieties—airplane, helicopter, hot-air balloon—relied increasingly on GPS for monitoring their flight path, for collision avoidance, and for landing. Search-and-rescue personnel on land, at sea, and in the air all came to view GPS as indispensable. Ultimately, scientists and engineers experimented with using GPS for launch and on-orbit operation of spacecraft.[19]

---

18. James Madeiros, "New Tracking System Keeps Eye on Troops," *Air Force Link* (25 October 2001), *http://www.af.mil/news/Oct2001/n20011025 1525.shtml* (accessed 26 October 2001); Marty Whitford, "Friend or Foe?—FBCB2 Enhances Battle Planning, Reduces 'Friendly Fire'" *GPS World* (February 2005), *http://www.gpsworld.com/gpsworld/article/articleDetail.jsp?id=146087* (accessed 17 August 2006); Tony Skinner, "NATO Orders GPS-Based Blue Force Tracking System," *Jane's Defence Weekly* 43, no. 33 (16 August 2006): p. 31.

19. Joe Mehaffey, "Recreational and Car Navigator GPS Sales and Sales Projections" (15 May 2005), *http://gpsinformation.us/joe/gpssalesprojections.html* (accessed 30 August 2006); "GPS Applications" (20 May 2003), MSU Global Positioning System (GPS) Laboratory, *http://www.montana.edu/places/gps/* (accessed 21 August 2006); Martin J. Unwin, "Experience with the Use of GPS in a Low Earth Orbit," *Satellite Communications* 13, no. 1 (1995): pp. 25–34.

When GPS receiver technology and cellular phones started to become more affordable during the mid-1990s, industry analysts and entrepreneurs perceived LBS as an emerging multibillion-dollar market. By 2002, however, only two automotive companies—General Motors (OnStar) and Mercedes (TeleAid)—offered consumers telematic LBS; such factors as cost, technological drawbacks of early-generation systems, and privacy concerns had led to slower than expected market growth. Nonetheless, an April 2002 nationwide survey of 20,000 U.S. households revealed strong consumer interest in using GPS for security-related purposes, especially for stolen vehicle tracking, with more moderate interest in services such as real-time traffic alerts, navigation assistance, and monitoring family vehicles. By 2006, North American sales of GPS-equipped navigation systems built into many different makes of automobiles, in portable devices, and in cell phones totaled 4.5 million units, with an increase of nearly 50 percent anticipated for 2007. Demand for up-to-date digital maps, derived in large part using GPS, of 7.3 million miles of North American highways caused the revenues of Tele Atlas and Navteq Corporation, the two principal competitors, to grow by 57 percent and 26 percent, respectively, in 2005. Meanwhile, after a decade of work, developers were ready to begin marketing in the United States and Canada an in-vehicle navigation system (IVNS) that promised to save private motorists billions (in terms of gallons, hours, and dollars) by integrating real-time data on traffic incidents, construction, and traffic flow to suggest alternate routes.[20]

Use of GPS-aided technology for management of vehicle fleets has saved governments and businesses hundreds of millions of dollars by enabling more efficient planning of routes, monitoring misuse by employees, or locating stolen vehicles.[21] The proliferation of organizations and conferences promoting development of automatic vehicle location (AVL) systems and intelligent vehicle/highway systems (IVHS) began in the late 1980s and fostered a veritable parade of specific projects in the early 1990s. By 1993, cities including Denver, Colorado, and Dallas, Texas, had begun installing GPS-based AVL systems in their buses and other city vehicles to keep riders informed of projected arrival times, to assist mobility-impaired riders, and to expedite response time in emergency situations.[22] Meanwhile, Netherlands-based

20. Clement Driscoll, "What Do Consumers Really Think?" GPS World (1 July 2002), *http://www.gpsworld.com/gpsworld/content/printContentPopup.jsp?id=25975* accessed 30 August 2006); Rafe Needleman, "Navigation PDAs Look for Their Market," *Business 2.0* (19 June 2003), *http://www.business2.com/subscribers/articles/web/0,1653,50392,00.html?cnn=yes* (accessed 24 October 2003); "The New Navigation," *The Gazette* (Colorado Springs, CO), 11 June 2006; Business 2.0 Staff, "Nowhere to Hide: GPS is Spreading," *Business 2.0* (March 2002), *http://www.business2.com/subscribers/articles/mag/0,1640,37777,00.html?cnn=yes* (accessed 24 October 2003); Marty Whitford, "Unstuck in Traffic," *GPS World* (1 June 2005), *http://www.gpsworld.com/gpsworld/content/printContentPopup.jsp?id=165057* (accessed 30 August 2006).

21. W. Eric Martin, "Mobilizing the Fleet," *Mobile Government* (May 2003): pp. 6–9.

22. Glen Gibbons et al., "Automatic Vehicle Location: GPS Meets IVHS," *GPS World* 4, no. 4 (April 1993): pp. 22–26; Paul Ledwitz, "DART AVL Hits Transit Bull's-Eye with GPS," *GPS World* 4, no. 4 (April 1993): pp. 29–34; Steve Blacksher and Tom Foley, "Boulder HOPs Aboard GPS Tracking," *GPS World* (1 January 2002), *http://www.gpsworld.com/gpsworld/content/printContentPopup.jsp?id=7782* (accessed 5 September 2006).

Büchner Transport, which owned 50 trucks and specialized in delivery of house plants, fruits, and vegetables under controlled temperatures throughout Europe, had experimented with a GPS-assisted AVL system in two of its trucks from December 1991 to June 1992 and decided, despite the high price of the hardware, to begin equipping its entire fleet over the next five years.[23] During 1988–2004, San Diego-based Qualcomm, the largest AVL supplier worldwide, installed its position-reporting system on more than 500,000 commercial vehicles belonging to more than 1,500 trucking firms. By autumn of 2004, the Defense Transportation Tracking System used GPS-derived location reports to monitor annually more than 47,000 arms, ammunition, and explosives shipments by commercial motor carriers in the continental United States.[24]

Emergency responders found GPS capabilities invaluable. In January–May 1992, Amoco tested a GPS-aided response system in its large Crossfield natural gas field north of Calgary, Alberta, and concluded that it offered noteworthy cost and safety improvements over earlier systems by "providing nearly immediate identification of an alarm site and the nearest field personnel, as well as detailed maps that show the best route to the scene of an alarm."[25] The San Francisco Bay Area's Freeway Service Patrol (FSP), a special team of approximately 40 tow-truck drivers who patrolled the most congested freeways during peak commute hours, began using a GPS-supported AVL system in August 1993 to assist stranded motorists. On average, FSP trucks arrived at the scene of a breakdown within seven minutes, almost six times faster than regular tow services.[26] Also, in 1993, Doug Baker, founder and president of LaSalle Ambulance Service of Buffalo, New York, adopted a GPS-equipped tracking and dispatch system for his fleet of 42 vehicles, two aero-medical helicopters, and one fixed-wing aircraft to ensure speedier response times and, consequently, save more lives.[27]

Recovery of stolen vehicles became much more likely with GPS. Founded in 1998 by Pakistani crime fighter and businessman Jameel Yusuf, Trakker Pvt. Limited used GPS technology to track and recover more than 1,000 stolen rental cars and private vehicles in Karachi by October 2003. By then, the Trakker system had been installed in 12,000 Pakistani vehicles and attracted roughly 500 new customers

---

23. J. A. J. Beerens, "Fleet Monitoring with GPS and Satellite Communications," *GPS World* 4, no. 4 (April 1993): pp. 42–46.

24. Clement Driscoll, "Finding the Fleet: Vehicle Location Systems and Technologies," *GPS World* 5, no. 4 (April 1994): pp. 66–70; Glen Gibbons, "HAZMAT Keeps on Truckin'," GPS World (1 October 2004), *http://www.gpsworld.com/gpsworld/content/printContentPopup.jsp?id=126157* (accessed 31 August 2006).

25. Edward J. Krakiwsky and Donald Chalmers, "Emergency Response with GPS in Oil and Gas Fields," *GPS World* 4, no. 4 (April 1993): pp. 48–51.

26. Michelle Morris, "High-Tech White Knights on the Freeway," *GPS World* 5, no. 4 (April 1994): pp. 54–59.

27. Margaret Ferrentino, "Code Red: GPS and Emergency Medical Response," *GPS World* 5, no. 4 (April 1994): pp. 28–38.

per month, each paying up to $750 for installation and $17 monthly for service.[28] Meanwhile, in January 2002 a thief stole a taxicab in Colorado Springs, Colorado, when the driver left the engine running while he went inside a 7-Eleven. In the first incident of its kind locally, a Yellow Cab dispatcher used the vehicle's on-board tracking system to update police on the car's location, and they arrested the culprit a short while later.[29] During less than a year with integrated GPS/cellular tracking systems, Lieutenant Tim Stewart of the North Texas Auto Theft Task Force reported in 2004 that his agency had recovered more than $6 million worth of property, including more than 50 truck tractors and 75 trailers—many filled with stolen equipment or merchandise.[30]

Among the recreational uses for GPS, golfing became prominent. In 1997, Darryl Sharp's Geodetic Services, Inc., began using GPS technology for three-dimensional mapping of premier golf courses in the United States and, by May 2002, had mapped 55 U.S. courses. Sharp estimated he would have lost $5,000–$6,000 per job without GPS but, with advances in GPS equipment, he had tripled his production. Under a five-year contract, videogame manufacturer EA SPORTS used Sharp's data to make playing simulated golf on some of the country's top-ranked courses stimulating and amazingly realistic for anyone with a personal computer or PlayStation. PGA TOUR later contracted with Geodetic Services to map courses for its new ShotLink scoring and statistics system that brought television or cyberspace fans closer to the game by reporting real-time information on every shot by every player in every tournament.[31]

Professional, amateur, and casual golfers all latched onto GPS-aided technology to improve their performance, and course managers enthusiastically supported the demand by purchasing GPS-equipped golf carts. In addition to recording exact yardage with each stroke and displaying actual yardage to the green, which aided club selection, personal GPS systems helped golfers maintain the pace of play. After courting by the electronics industry and a couple of years testing equipment such as the handheld SkyCaddie, the United States Golf Association finally permitted "distance-measuring devices, including GPS-based systems and laser range finders" by local rule starting in 2006. As demand for GPS-equipped golf carts surged in early 2006, UpLink Corporation, one of the three largest suppliers of golf-related GPS equipment, signed up 26 new courses with its system after a single trade show,

28. Reuters, "Satellites Help Slash Karachi Car Thefts, Kidnaps," CNN.com/Technology (24 October 2003), *http://www.cnn.com/2003/TECH/ptech/10/24/satellite.tracking.reut/index.html* (accessed 24 October 2003).

29. Anslee Willett, "Global Satellite System Tracks Down Suspect in Theft of Taxicab," *The Gazette* (Colorado Springs, CO), 18 January 2002.

30. Marty Whitford, "Thief Relief—GPS/Cellular Combo Acquires Abducted Assets," *GPS World* (1 October 2004), *http://www.gpsworld.com/gpsworld/content/printContentPopup.jsp?id=128322* (accessed 30 August 2006).

31. Krista Stevens, "GPS Fore the Golf Course," *Point of Beginning* (30 April 2002), *http://www.pobonline.com/CDA/Archives/6c07817cac0f6010VgnVCM100000f932a8c0* (accessed 30 August 2006).

thereby generating $7.4 million in sales. UpLink technology allowed course managers to track every golf cart in their fleet, a financially important capability considering that golf-cart rentals yielded the largest source of revenue outside green fees at most courses. For example, a 2005 progress report from the mayor of South Bend, Indiana, mentioned that golf-cart revenue from GPS was more than $70,000.[32]

AGRICULTURE

Use of GPS for precision farming (i.e., site-specific management) commenced in the early 1990s and rapidly took a variety of forms. These included planting or cultivating crops at night; locating weed, insect, and disease infestations; applying fertilizer or pesticides at a variable rate; preventing skips or overlaps when fertilizing; monitoring and mapping crop yield; and pinpointing crop damage due to hail or drought. John Ruth, chief executive officer of Amana Farms in Iowa, explained in a 1992 interview, "With GPS, we'll be able to determine crop yield by the square foot and not by the traditional bushels per acre . . . . This will dramatically change the way farmers plant, fertilize, apply weed killers and harvest crops."[33] To further improve efficiency and increase the profitability of their operations, farmers also used GPS for detailed base-mapping of physical features such as borders, fence lines, wells, buildings, landscape features, irrigation canals or pipelines, and wetlands.[34]

Montana State University (MSU) began research and planning in 1986 for a GPS-assisted navigation system that would enable agricultural producers to apply variable amounts of seed and fertilizer exactly where needed to maximize crop yield in the most cost-effective manner. In August 1990 MSU researchers used the first GPS-guided agricultural fertilizer application system in a field trial near Power, Montana.[35] Two years later, agronomist Mitch Schefcik and electrical engineer William Bauer

---

32. Doug Pike, "High Tech Takes a Big Step Forward," *Houston Chronicle*, 18 October 2005, *http://www.uplinkgolf.com/cms/website/mediacoverage/HoustonChron10_18_05.pdf* (accessed 30 August 2006); Mayor Stephen J. Luecke, "South Bend, Indiana—Progress Report, 2005," *http://www.southbendin.gov/doc/Comm_ProgressReport2005.pdf* (accessed 30 August 2006); Roger Graves, "From the Dailies—Golf Cars Drive up Course Revenues with New Models, Technology," *PGA Magazine* (26 January 2006), *http://www.pgamagazine.com/article.aspx$id=2750* (accessed 30 August 2006); Tim Schooley, "Tracking a New Market—Global Positioning Systems Find Their Way into Golf Carts, Providing Much More than Distance Readings," *Pittsburgh Business Times* (12 June 2006), *http://pittsburgh.bizjournals.com/pittsburgh/stories/2006/06/12/focus1.html* (accessed 30 August 2006); Pat Dailey, "GPS System Welcome Addition," *Branson Daily News*, 29 August 2006, *http://www.bransondailynews.com/story_print.php?storyID=1540* (accessed 30 August 2006); SkyCaddie advertisement, Sky Mall (Early Spring 2006): p. 31; GolfPS Personal GPS Golf System Web site, *http://www.golfps.com/index.htm* (accessed 30 August 2006).

33. Mark W. Brown, "GPS Helps Farmers Reap Higher Crop Yields," *Space Trace* (November 1992): p. 4.

34. "GPS Applications" (20 May 2003), MSU Global Positioning System (GPS) Laboratory, *http://www.montana.edu/places/gps/* (accessed 21 August 2006); Robert "Bobby" Grisso et al., "Precision Farming Tools: Global Positioning System (GPS)," July 2003, Virginia Cooperative Extension, *http://www.ext.vt.edu/pubs/bse/442-503/442-503.html* (accessed 21 August 2006).

35. Carolyn Peterson, "Precision GPS Navigation for Improving Agricultural Productivity," *GPS World* 2, no. 1 (January 1991): pp. 38–44.

combined their expertise to devise a DGPS/Geographic Information System (GIS) system to vary the herbicide application rate on sugar-beet fields in western Nebraska, thereby increasing crop yield while satisfying government chemical application regulations. Experimentation by the Agricultural Research Service of the U.S. Department of Agriculture (USDA) found DGPS extremely useful for mapping initial, variable applications of nitrogen fertilizer and evaluating the effect on crop yields in irrigated central Nebraska cornfields during 1995. Improvement in nitrogen fertilizer management also offered the prospect of improved groundwater (i.e., drinking water) quality, something much desired by the U.S. Environmental Protection Agency.[36]

One expert estimated in 1994 that only 5 percent of U.S. farms, mostly in the corn-producing states of the Midwest, used GPS-derived reference points for soil-specific management, but the technology was "booming." A study of soil-specific management in Stoddard County, Missouri, during that period showed reductions in fertilizer costs of $18.70 per acre with no loss of crop yields.[37] By 2003 University of Florida researchers touted GPS/GIS technology's great value for locating and managing site-specific crop losses (more than $77 billion worldwide in 1987) due to plant-parasitic nematodes.[38] According to an Associated Press story in late November 2004, up to 15 percent of U.S. farmers had GPS-controlled tractors or combines and were saving as much as 5 percent in fertilizers and pesticides by using precision guidance systems. Farmers in Kentucky, where rolling terrain led to both high and low production in the same field, realized a cost saving of $30/acre by using GPS-enabled yield monitors. Meanwhile, under a NASA grant, researchers at Ohio State University worked to perfect a GPS-enabled tomato-picking robot that could significantly reduce labor costs on large corporate farms.[39]

To meet demand for greater accuracy, lower labor costs, and documentation of spraying, the USDA began in the early 1990s to explore GPS-aided navigation for precision aerial applications to control insects on cropland or rangeland. During the autumn of 1992, the Plains Cotton Growers Diapause Program in Texas had used a GPS flight guidance system from Satloc, Inc., to spray more than 450,000 acres of cotton for boll weevils. That application offered savings by eliminating the need for approximately 170 ground flaggers. Based on that program's success, the USDA's Grasshopper Integrated Pest Management (GHIPM) Project near Watford City in northwestern North Dakota tested a DGPS system in 1993 on a 6,400-acre (2,590-hectare) section of rangeland. Chief pilot Tim Roland for the USDA's Animal and Plant Health Inspection Service, Plant Protection and Quarantine (APHIS/PPQ)

---

36. William D. Bauer and Mitch Schefcik, "Using Differential GPS to Improve Crop Yields," *GPS World* 5, no. 2 (February 1994): pp. 38–41; Tracy M. Blackmer and James S. Schepers, "Using DGPS to Improve Corn Production and Water Quality," *GPS World* 7, no. 3 (March 1996): pp. 44–52.

37. Hale Montgomery, "GPS Down on the Farm," GPS World 5, no. 9 (September 1994): p. 18.

38. Jim Rich et al., "Using GIS/GPS for Variable-Rate Nematicide Application in Row Crops," University of Florida IFAS Extension, *http://edis.ifas.ufl.edu/IN464* (accessed 21 August 2006).

39. Associated Press, "GPS Goes Down on the Farm—Technologies Can Help Some Farms Cut Costs," 29 November 2004, MSNBC, *http://www.msnbc.msn.com/id/6608881/* (accessed 21 August 2006).

division subsequently remarked, "We're just getting our foot in the door with GPS, but I'm so confident [about the] equipment that I'd recommend it to support any of our programs."[40] When the accuracy of Loran-C navigation proved inadequate to meet the parallel-swathing requirements of California's Cooperative Medfly Project during 1993, Roland recommended DGPS guidance. In California's war against the Mediterranean fruit fly, which could cause more than $1 billion in damage annually to fruit, nut, and vegetable crops, not to mention the loss of thousands of jobs, DGPS brought significant savings and provided a previously unavailable element of quality control to aerial treatments.[41]

Disaster Relief And Recovery

Use of GPS in conjunction with GIS, cartographic mapping, and other technologies proved beneficial in disaster relief and recovery efforts. After Hurricane Andrew devastated Florida in 1992, the Federal Emergency Management Agency (FEMA) contracted with survey crews to experimentally inventory the damage using GPS/GIS technology instead of the traditional, manual assessment that involved house-by-house interviews. Based on encouraging results from that experiment, FEMA, the U.S. Army Corps of Engineers, and a private contractor with GPS/GIS expertise formed a team in July 1993 to produce maps for disaster response, recovery efforts, and risk mitigation in the wake of severe Mississippi River floods that inundated more than 13 million acres, destroyed billions of dollars in crops, and left hundreds of people homeless. Following a GPS-equipped helicopter survey, a pair of two-person ground observer teams with GPS/GIS handheld receivers inspected and inventoried structures in approximately 75 communities south of Quincy, Illinois. More than 1,500 maps/data sheets were produced within a week of the teams' initial transfer of data to the Corps of Engineers' Rock Island, Illinois, base station. Prior to GPS/GIS, it would have taken a team of 50 people years to complete the same task. With the maps quickly delivered to FEMA decision makers, they began meeting with local officials and citizens to discuss assistance and requirements to rebuild above the 100-year flood elevation.[42]

When New York City officials faced the daunting cleanup of "Ground Zero" after the 11 September 2001 terrorist attacks on the World Trade Center, they found multiple applications for GPS. As Fire Department teams and others began the monumental task of sifting through the rubble to recover and manually catalog thousands of pieces of evidence at the crash site, which was also a crime scene, it became obvious the process was too time-consuming and too error-prone. To

---

40. Mike W. Sampson, "Getting the Bugs Out: GPS-Guided Aerial Spraying," *GPS World* 4, no. 9 (September 1993): pp. 28–34.

41. Mike W. Sampson, "No Small Affair: DGPS Battle the California Medfly," *GPS World* 6, no. 2 (February 1995): pp. 32–38.

42. Harry Bottorff, "Rapid Relief: GPS Helps Assess Mississippi River Flood Damage," *GPS World* 5, no. 3 (March 1994): pp. 22–26.

automate it, the city gave a handheld device equipped with a GPS receiver and an attached bar-code reader to one person on each of the eight recovery units working at the 16-acre site at any given time. Searchers put a bar-coded tag on each significant piece of recovered evidence; scanned the tag to create an electronic record that included the exact location, date, and time; and selected an item description from a scroll-down list programmed into the handheld device. At the end of each shift, searchers placed their handhelds into a cradle to download their information to a central database, from which the city produced maps that helped investigators pinpoint where various kinds of vehicles, equipment, and personnel were when the buildings collapsed. The fire and police departments could analyze that information to improve their response to future emergencies.[43]

A second challenge related to the September 11 disaster involved learning how various buildings in a 10-square-block area surrounding the World Trade Center survived, because that knowledge would permit the design and construction of more damage-resistant structures. The solution lay in modification of a GPS/GIS software application for a Palm Pilot developed by Georgia Institute of Technology civil engineering professor David Frost and graduate student Scott Deaton. This handheld damage assessment technology, which Frost had conceived while surveying damage from a 1999 earthquake in Izmit, Turkey, allowed researchers to capture digital images and select an option from a list of damage descriptions, at which time the information was automatically linked to GPS coordinates. The collected data could be uploaded periodically to a single, spatial database. In approximately four days during mid-October 2001, Frost, Deaton, and master's student Prateek Goel documented and mapped in nearly real time most of the structures in the study area. A team using more traditional methods would have taken at least a month just to enter all the data into a computer for mapping and analysis.[44]

Yet another GPS application in disaster recovery efforts at Ground Zero involved debris removal. Hauling away 1.8 million tons of debris efficiently became a top priority, because experience suggested this would be the most expensive aspect of the recovery effort. New York City's Department of Design and Construction employed a team of contractors led by Criticom International Corporation of Minneapolis, Minnesota, to devise a GPS-based management system for debris removal. The contract team quickly assembled and installed commercial off-the-shelf components in 235 trucks and a control center, thereby creating a management system that integrated GPS-based positioning with communications, camera monitoring, and Internet data. This first use of GPS-based AVL to manage debris removal in a disaster

---

43. John Rendleman, "GPS Aids Recovery Effort," *InformationWeek* (12 November 2001), *http://www.informationweek.com/showArticle.jhtml;jsessionid=YH2WRBQHCAKCCQSNDLPSKH0CJUNN2JVN?articleID=6507967* (accessed 12 September 2006).

44. Beth Bacheldor, "Damage Assessment: Lessons from Ground Zero," *InformationWeek* (18 March 2002), *http://www.informationweek.com/showArticle.jhtml;jsessionid=YH2WRBQHCAKCCQSNDLPSKH0CJUNN2JVN?articleID=6501464* (accessed 12 September 2006).

Artistic rendition of GPS Block 3 satellite.

recovery operation contributed substantially to finishing the removal project four months earlier than originally predicted and at a cost of $750 million, far less than the $7 billion city officials initially estimated.[45]

During and after Hurricane Katrina in August 2005, experts found further applications for GPS. Military pilots flew into the approaching hurricane to deploy GPS-enabled dropsondes, first used in 1996–1997, that helped scientists predict the strength, speed, and direction of the storm with greater precision than allowed by previous systems. After Katrina hit the Gulf Coast, the Urban and Regional Information Systems Association (URISA) GISCorps, a volunteer organization, went there to support U.S. Coast Guard operations by translating distressed survivors' addresses or locations into GPS coordinates, which enabled dispatch of rescue helicopters. Within two weeks after the storm, major ports in the Gulf region reopened, due in large measure to Nationwide DGPS (NDGPS) allowing the Coast Guard to precisely reposition 1,800 buoys and fixed aids to navigation; the Army Corps of Engineers to survey and dredge 38 critical waterways; and commercial pilots to navigate even when they could not "read" silt-clouded waterways. The National Park Service used NDGPS to map areas rendered unsafe by hazardous materials and to mark safe passageways; the Department of Agriculture used it help locate disposal sites for animal carcasses and to map areas where blocked drainage demanded future clearing. Despite problems with keeping the NDGPS signal available in the aftermath of Hurricane Katrina, it proved crucial to relief and recovery efforts.[46]

45. Raymond J. Menard and Jocelyn L. Knieff, "GPS at Ground Zero: Tracking World Trade Center Recovery," GPS World (2 September 2002), *http://www.gpsworld.com/gpsworld/content/printContentPopup.jsp?id=30686* (accessed 12 September 2006).

46. Marty Whitford, "Hurricane Hunters: GPS Dropsondes Trace Katrina's Course," *GPS World* (1 October 2005), *http://mg.gpsworld.com/gpsmg/content/printContentPopup.jsp?id=262187* (accessed 13 September 2006); Bob Henson, ed., "Hurricane!" *SN Monthly* (October 1998), *http://www.ucar.edu/communications/staffnotes/9810/hurricane.html* (accessed 13 September 2006); "The Nationwide Differential Global Positioning System's Role with Hurricane Katrina's Recovery," U.S. Coast Guard Navigation Center, January 2006, *http://www.navcenter.org/misc/HurricaneKatrina.pdf* (accessed 13 September 2006).

TIMING

People generally think of GPS as a navigation system and, consequently, do not fully comprehend its usefulness in the dissemination of precise time, time intervals, and frequency. Because a GPS receiver location is established through simultaneous ranging of signals from several GPS satellites, every signal from each satellite includes time of transmission. Atomic clocks on every satellite ensure that system time is closely synchronized through cesium and rubidium frequency standards. By providing timing and frequency accuracies in the range of 100 nanoseconds and a few picoseconds, respectively, to receivers worldwide 24 hours a day, GPS quickly became a boon to specialists in a wide variety of fields ranging from science, satellite tracking, and industrial plant operations to power distribution, television broadcasting, and banking.[47]

The Bonneville Power Administration (BPA), which supplies about half of all electrical power used in the Pacific Northwest, began gradually integrating GPS technology into its various operations in 1988. Its ultimate goal was a GPS-based infrastructure for measuring voltage and current at selected generation and transmission sites and for time-tagging that information with microsecond precision to more effectively and efficiently stabilize the electrical grid, thereby minimizing system disturbances that potentially could cascade into major blackouts. In 1989, BPA began using GPS in its power-line fault location system and in solving problems associated with standard time distribution across the grid. Meanwhile, it started investigating GPS as a source of precise time for its phasor measurement units (PMUs), which computed the phase angle between points on various circuits to provide a good representation of the power system's "health." During 1991–1992, BPA tested two GPS-synchronized prototype PMUs on a 260-mile section of a main transmission link, the resulting data being more accurate and less noisy than comparable analog information. A new GPS-based Central Time System (CTS) with triple redundancy went into the BPA Dittmer Control Center in 1994. By 2005, BPA was touting its Wide Area Control System (WACS), a demonstration project that BPA engineers predicted could route $7.2 million worth of additional power—enough for 20,000–60,000 typical homes—from the Pacific Northwest to California and could prevent a system blackout costing more than $1 billion.[48]

---

47. Peter H. Dana and Bruce M. Penrod, "The Role of GPS in Precise Time and Frequency Dissemination," *GPS World* 1, no. 4 (July/August 1990): pp. 38–43; Harold Hough, "A GPS Precise Timing Sampler," *GPS World* 2, no. 9 (October 1991): pp. 33–36; Edgar W. Butterline, "Reach Out and Time Someone," *GPS World* 4, no. 1 (January 1993): pp. 32–40; Hewlett-Packard Company, "GPS and Precision Timing Applications," Application Note 1272 (1996), *http://www.fgg.uni-lj.si/~/mkuhar/Zalozba/HP1272.pdf* (accessed 10 September 2006).

48. Kenneth E. Martin, "Powerful Connections: High-Energy Transmission with High-Precision GPS Time," *GPS World* 7, no. 3 (March 1996): pp. 20–36; Dennis C. Erickson and Carson Taylor, "Pacify the Power: GPS Harness for Large-Area Electrical Grid," GPS World (1 April 2005), *http://www.gpsworld.com/gpsworld/content/printContentPopup.jsp?id=154868* (accessed 8 September 2006).

The explosive growth of computer networks from the 1980s onward made high-quality time synchronization essential for proper functioning and for rendering failures more manageable. GPS-enabled time synchronization directly affected network operations in several key areas ranging from file time stamping and directory services to access security, log file accuracy, and fault diagnosis and recovery; it also affected applications in numerous areas from transaction processing and e-mail to software development and legal or regulatory requirements.[49]

At the beginning of the twenty-first century, financial institutions relied on millions of networked computers to execute billions of transactions per second involving extremely rapid changes in value. On 14 November 2001, for example, during the closing minutes of trading on the New York Stock Exchange, a single stock—Intel—averaged six transactions/second and as high as 20/second during intense trading. With traders frequently calling their brokering institution to dispute the recorded value of a particular transaction among countless others, computer-system clock resolution had to be less than the minimum transaction composition and transmission time—hence, 5- to 20-millisecond resolution. An international investment banking firm already had begun using a Palisade network time protocol (NTP) synchronization kit, introduced by Trimble Navigation Ltd. in 1999, to ensure simultaneous recording of transactions at its London, New York, and Tokyo offices. By mid-2002, the New York Stock Exchange, the World Bank, and other major financial institutions used Symmetricomm (formerly TrueTime), Spectracom, or other corporations' GPS-based NTP products to synchronize their computer systems at the required level of accuracy. Experimenters at Ireland's University of Galway had installed Trimble's newer Acutime 2000 GPS synchronization kit to create what they called a Stratum-1 (defined as micro-second accurate) NTP server that other European servers and clients used for time synchronization. So important had GPS become as a time-reference standard for the world's aggregate financial network that the Heritage Foundation advised designating it a critical infrastructure.[50]

As wireless communication networks expanded at the end of the twentieth century, GPS receivers replaced less reliable timing technologies. So-called smart GPS antennas in base stations synchronized transmitter sites to precisely the same time, which prevented a

49. Paul Skoog, "The Importance of Network Time Synchronization," TrueTime (2001), *http://www.truetime.net/pdf/imp_netsync.pdf* (accessed 8 September 2006).

50. Timble Navigation Ltd., "Trimble Launches GPS Clock for Micro-Second Synchronization of Network Computers and Internet Applications," Directions Magazine (27 July 1999), *http://www.directionsmag.com/press.releases/index.php?duty=Show&id=742&trv=1* (accessed 10 September 2006); "Trimble Offers Internet GPS Time Standard," *Space Daily*, *http://www.spacedaily.com/news/gps-99j.html* (accessed 10 September 2006); Fact Sheet, "Acutime 2000 GPS Smart Antenna for Precise Timing and Synchronization," Trimble Navigation Ltd. (2000), *http://www.navgeocom.ru/oem/download/main/sync/a2000/acutime2k_info.pdf* (accessed 10 September 2006); Alan Cameron, "Billions per Second: Timing Financial Transactions," *GPS World* (31 July 2002), *http://www.gpsworld.com/gpsworld/content/printContentPopup.jsp?id=30096* (accessed 10 September 2006); "GPS Tutorial—Timing," Trimble Navigation Ltd. (2006), *http://www.trimble.com/gps/gpswork-timing.shtml* (accessed 10 September 2006).

user from receiving the same cellular telephone, pager, or other wireless signal multiple times. In August 1996 at the Republican National Convention in San Diego, California, Pacific Bell Mobile Services (PBMS) implemented a new type of all-digital, wireless telephone network called personal communications service (PCS), which relied on relatively low-cost, GPS-based time synchronization. By early 1997, PBMS had begun building on that success by installing an identical capability in Las Vegas, Nevada.[51]

Wireless positioning and tracking also benefit immeasurably from the accuracy of GPS timing. A caller's handset, for example, can be located by calculating and triangulating the time differences in arrival of its signal at cell towers whose positions are accurately known. In 1993, the Federal Communications Commission (FCC) took regulatory interest in extending enhanced-911 (E911) emergency service, including automatic location of a caller, to wireless mobile subscribers; in October 1999 Congress passed the Wireless Communications and Public Safety Act mandating that manufacturers incorporate E911 features in cell phones and, thereafter, the FCC stipulated that 95 percent of all new digital cell phones must have the automatic location identification (ALI) feature by 1 October 2002. The FCC later extended that deadline three years when wireless carriers appealed for more implementation time. At the beginning of 2005, Nextel Communications, Inc., was the only national wireless provider offering GPS services, and nearly all carriers doubted they could meet the end-of-year deadline for having 95 percent of their handsets upgraded to GPS-capable models. Although wireless E911 offered the prospect of improving the personal safety of millions of people nationwide, indications were that its availability would be piecemeal for years to come, and some citizens questioned whether the government, employers, spouses, or parents might misuse the ALI feature to infringe upon personal privacy and constitutional rights.[52]

---

51. "Wireless Profile," Wireless Magazine (2006), *http://www.trimble.com/tmg_wirelessprofile.shtml* (accessed 10 September 2006); Don Britten, "Wireless and Seamless: GPS-Timed Mobile Communications," *GPS World* 8, no. 3 (March 1997): pp. 32–39.

52. Peter Kuykendall and Peter V.W. Loomis, "In Sync with GPS: GPS Clocks for the Wireless Infrastructure," Trimble Navigation (2006), *http://trl.trimble.com/docushare/dsweb/Get/Document-8439* (accessed 10 September 2006); Dale N. Hatfield, "A Report on Technical and Operational Issues Impacting the Provision of Wireless Enhanced 911 Services," Federal Communications Commission, 24 November 2004, *http://gullfoss2.fcc.gov/prod/ecfs/retrieve.cgi?native_or_pdf=pdf&id_document=6513296239* (accessed 10 September 2006); U.S. Government Accountability Office, "Telecommunications: Uneven Implementation of Wireless Enhanced 911 Raises Prospect of Piecemeal Availability for Years to Come," GAO-04-55, November 2003, *http://www.gao.gov/new.items/d0455.pdf* (accessed 10 September 2006); David H. Williams, "Wireless Carriers are Making Great Progress Implementing Wireless E911 . . . So What's the Problem?" Directions Magazine (22 September 2004), *http://www.directionsmag.com/article.php?article_id=665* (accessed 10 September 2006); David Colker, "Parents, Bosses Call on GPS Tracking Cell Phones," *Los Angeles Times*, 9 January 2005; Sascha Segan, "Inexpensive Emergency Phones for Children, Elders Could Be Used As GPS Tracking Device," *PS Magazine* (8 January 2005), *http://www.infowars.com/articles/bb/parents_bosses_gps_track_cellphones.htm* (accessed 10 September 2006); David H. Williams, "The Deadline for the E911 Mandate Approaches . . . Where Do Things Stand?" *Directions Magazine* (30 November 2005), *http://www.directionsmag.com/article.php?article_id=2014&trv=1* (accessed 10 September 2006); Associated Press, "Satellite Navigation Finds Its Way to Phones: Latest Technology Allows You to Track People's Whereabouts," MSNBC, 7 April 2006, *http://www.msnbc.msn.com/id/12206651* (accessed 10 September 2006); "Sprint service tracks children: Parent can locate kid's cell with GPS," *The Gazette* (Colorado Springs, Colorado), 14 April 2006.

SUMMARY

Nothing about GPS could be more obvious than the rapid expansion of entrepreneurial activity and the veritable explosion of applications from the early 1990s into the first decade of the twenty-first century. One can measure that growth, at least notionally, in several ways: patent activity, manufacturers, employment, revenues, and sales. The number of GPS-related patent families (a family being a collection of related patents and published patent applications) grew internationally from less than 20 in 1988 to nearly 80 in 1991, with U.S. and Japanese corporations holding the most.[53] Meanwhile, the DOD procured only 7,253 GPS receivers during 1986–1992 but zoomed to 19,086 receivers in 1993 alone.[54] Firms that listed themselves as providing GPS-related goods or services totaled 301 in 1997 compared to 109 in 1992. North American revenues from GPS products in 1999 grew 21.1 percent from 1998, indicating that "an increasing number of end users in previously untapped markets" had begun accepting the new technology. Revenues worldwide from GPS user equipment sales totaled $3.39 billion in 1996 and rose to $6.22 billion in 1999; during the same period, the number of GPS industry employees worldwide increased from 16,688 to 30,622. With average prices for GPS products expected to decrease 7.4 percent annually through 2006, according to strategic research by the consulting firm Frost & Sullivan, the number of products sold for nontraditional applications in commercial and consumer markets would steadily increase. Among the five basic GPS market segments—land, aviation, marine, military, and timing—examined by Frost & Sullivan analyst Ron Stearns, the land-based segment alone accounted for 61.8 percent of total North American revenues in 1999, a statistic unlikely to change appreciably through 2006.[55]

While the average price of GPS products continued downward in the new century, the number of units sold and total revenues advanced according to predictions. One estimate in 2000 put the number of GPS users worldwide at 1.5 million, with an economic impact of $6.2 billion. According to Frost & Sullivan, total revenues from the North American GPS equipment market alone amounted to $3.46 billion in 2003. A breakdown of that figure by end-user group—consumer, commercial, and military—revealed 52 percent, 40 percent, and 8 percent, respectively.

53. Scott Pace, et al., *The Global Positioning System: Assessing National Policies* (Santa Monica, CA: RAND Corp., 1995), pp. 114–118.

54. National Academy of Public Administration, *The Global Positioning System: Charting the Future* (Washington, DC: NAPA, 1995), pp. 62–68.

55. Scott Pace and James E. Wilson, *Global Positioning System: Market Projections and Trends in the Newest Global Information Utility*, International Trade Administration, Office of Telecommunications, U.S. Department of Commerce, September 1998, *http://www.nesdis.noaa.gov/space/library/reports/1998-09-gps.pdf* (accessed 13 September 2006); Frost & Sullivan, "Price Points, Product Integration Fuel GPS Revenue Growth," *Directions Magazine* (26 April 2000), *http://www.directionsmag.com/press.releases/index.php?duty=Show&id=1694&PRSID* (accessed 11 September 2006); Futron Corporation, *Trends in Space Commerce*, Office of Space Commercialization, U.S. Department of Commerce, May 2001, *http://www.nesdis.noaa.gov/space/library/reports/2001-06-trends.pdf* (accessed 13 September 2006).

Recreational GPS devices constituted a huge market, with shipments rising from 3.2 million in 2002 to 5 million in 2003.[56] Sales of portable navigation devices, used primarily for in-vehicle navigation, rose faster than projected in the United States from 300,000 units in 2003 to 550,000 in 2004 and about one million in 2005. Analyst Ron Stearns of Frost & Sullivan calculated the automotive portion of the consumer GPS business at $922 million, with estimated unit sales of 1.2 million for 2006; the addition of outdoor units for recreational users would send sales to $1.8 billion and more than 4 million units for 2006. Stearns expected combined annual sales for automotive and outdoor GPS units to reach 8.3 million, worth $2.7 billion in revenues, by 2010.[57] As Bradford Parkinson, one of the founders of GPS, said in 1980, the "potential uses [of GPS] are limited only by our imaginations."[58]

56. "Keeping Pace with Consumer Applications Vital to GPS Market Expansion," Frost & Sullivan, 9 February 2004, *http://www.frost.com/prod/servlet/press-release.pag?mode=open&docid=10127615* (accessed 11 September 2006); "GPS Market," Google Answers, 28 February 2005, *http://answers.google.com/answers/threadview?id=481415* (accessed 11 September 2006); "U.S. GPS Market Size," Google Answers, 27 August 2005, *http://answers.google.com/answers/threadview?id=561093* (accessed 11 September 2006).

57. "Taiwan GPS Sector Update," MasterLink Securities, 30 December 2005, *http://web6.masterlink.com.tw/project/english/main_page/files/company_reports/GPS_DEC_2005_20051230.pdf* (accessed 11 September 2006); "Wireless Tracking: One GPS Giant Too Many?" *BPM Today* (21 June 2006), *http://www.bpm-today.com/story.xhtml?story_id=012000EP08VC* (accessed 11 September 2006); Clem Driscoll, "Strong Consumer Interest in GPS for Traffic, Nav, LBS," *GPS World* (1 August 2006), *http://www.gpsworld.com/gpsworld/content/printContentPopup.jsp?id=360934* (accessed 11 September 2006).

58. Bradford W. Parkinson, "Overview," in *Global Position System: Papers Published in Navigation, Volume I*, P. M. Janiczek, ed. (Washington, DC: The Institute of Navigation, 1980), pp. 1–2.

Chapter 18

Dual-Use as Unintended Policy Driver: The American Bubble

Roger Handberg

In the earliest days of the Space Age, the U.S. government, for diverse and pressing policy reasons, elaborated the dual-use distinction. Their view became that space activities could in fact easily be delineated into peaceful civil purposes and clearly military purposes. This distinction in practice ultimately proved unsustainable but was especially convenient for arms control purposes beginning with the 1960s nuclear arms race. The dual-use concept has proven to embody several unanticipated effects which decisively and negatively impact future U.S. engagement in space commerce. This policy arose when the United States was effectively a monopolist with regard to space applications, but has different implications in a globalizing economy.

This paper analyzes, first, the rise of the dual-use concept and its general impact on civil/commercial space applications; second, how that situation changed with the cold war's end and the lessening of security restrictions; and, third, the destabilizing economic effects that have arisen for the United States. This analysis focuses mostly on the American experience, but the dual-use concept proved particularly useful internationally with regard to slowing nuclear proliferation (the Nonproliferation Treaty)[1]; it was extended specifically in 1987 to space launch technologies with the Missile Technology Control Regime (MTCR),[2] and more broadly with the "Wassenear Arrangement on Export Controls for Conventional Arms and Dual-Use Goods and Technologies."[3] The latter is more wide-sweeping in its implications for transfer of dual-use technologies.

---

1. Treaty on the Non-Proliferation of Nuclear Weapons (entered into force 5 March 1970), *http://disarmament.un.org/TreatyStatus.nsf* (accessed 15 October 2006).

2. The original signatories were allies of the United States: Canada, France, Germany, Italy, Japan, and the United Kingdom. *http://www.mtcr.info/english/index.html* (accessed 15 October 2006).

3. The Arrangement "complements and reinforces, without duplication, the existing regimes for non-proliferation of weapons of mass destruction and their delivery systems, by focusing on the threats to international and regional peace and security which may arise from transfers of armaments and sensitive dual-use goods and technologies where the risks are judged greatest. This arrangement is also intended to enhance co-operation to prevent the acquisition of armaments and sensitive dual-use items for military end-uses, if the situation in a region or the behaviour of a state is, or becomes, a cause for serious concern to the Participating States." *http://www.wassenaar.org/publicdocuments/Basic%20documents%202006%20-%20January.doc* (accessed 15 October 2006).

The application areas focused on here include remote sensing (including weather), navigation, and communications satellite policies as the most obvious areas. Launch vehicle restrictions arose later than the others due to other factors. The changes occurring are not merely technological (i.e., improvement in the scale of images provided or accuracy of positioning information or enhanced communications) but are due to the fact that these largely military-initiated and -dominated sectors are becoming engines for economic growth and improvements in productivity. These applications are driving the commercial space sector to become more truly international in scope and operations. Early space visionaries often envisioned a world economically and politically integrated through the use of space applications, but that has not occurred because of national security restraints. Those security restraints have not vanished in the American case, but the global spread of technological competence regarding space technologies has removed the capacity of any single state to control these applications. The image projected by these changes is a cooperative, peaceful world but, for the United States, the political focus remains upon these applications' potential to disrupt the U.S. economy and security operations.

DUAL-USE AS A CONCEPT—BEGINNINGS

From the perspective of the late 1940s and early 1950s, at the cusp of the first Space Age, the dual-use concept was largely irrelevant because the operative assumption was that national space programs would be controlled and led by their military, with whatever civilian presence that developed being clearly subordinate one.[4] The historical U.S. model for nonmilitary participation in space activities was the scientific expedition, such as Lewis and Clark in 1804, led by the military. Wernher von Braun's famous series of articles in *Collier's* assumed that the expeditionary model would continue.[5] This concept faded in the 1960s but has been resurrected by U.S. Air Force space power advocates as a means by which to recapture their control over human spaceflight.[6]

The purely civilian (especially commercial) aspect of space activities existed initially as a theoretical concept, despite Robert Goddard's pioneering research on launch vehicles and early speculation by Arthur Clarke about communications

4. Howard E. McCurdy, *Space and the American Imagination* (Washington, DC: Smithsonian Institution Press, 1997), chapter 1.

5. William E. Burrows, *This New Ocean: The Story of the First Space Age* (New York: Modern Library, 1998), pp. 142–146. For excerpts of the original articles, see John M. Logsdon, ed., *Exploring the Unknown: Selected Documents in the History of the U.S. Civil Space Program, Volume I: Organizing for Exploration* (Washington, DC: U.S. Government Printing Office, 1995), pp. 176–200.

6. Simon P. Worden and John E. Shaw, *Whither Space Power? Forging a Strategy for the New Century* (Maxwell Air Force Base, AL: Air University Press, September 2002), pp. 110–112.

using Earth-orbiting satellites. The dominant reality became that the military (the Nazis and Soviets first, later the Americans) controlled the space technology development process, including funding of whatever launch technology was deemed useful. The military's initial and primary interest focused on building more effective and farther-reaching weapons carriers.[7] The wider possibilities for space activities were understood to exist and were the subject of preliminary analysis but all were considered within the paradigm of military control over any space-related technologies that might be developed.[8] In fact, the original thought was that the military itself, through an arsenal system, would control production of such technologies. In the American case, however, the U.S. Air Force had extensive experience with contractors as technology producers under military supervision.

This contractor approach fit better with American ideological proclivities and provided greater flexibility for expansion and contraction of production, it being easier to lay off contractor employees than civil servants employed at an arsenal. One direct consequence of this approach was creation of an aerospace industry that was in place if and when military control loosened. The aeronautical side of the industry was an excellent prototype of a dual-use capability, although it was not thought of in those terms.

Embedded in the beginnings of the Space Age was the shadow of a dual-use concept, but even there the military remained dominant. The rise of space scientists was in part built around the reality that their initial value was as payload providers whose results had direct military relevance, especially improvement in communications and scanning capabilities through better understanding of the ionosphere and other environmental forces impacting radio wave transmissions at different frequencies and thus, by extension, military operations.[9] Beginning with sounding rockets (high-altitude balloons were already in use), scientists found that the ability to leave the atmosphere even briefly to observe atmospheric and celestial events was truly liberating, opening up new vistas of scientific information and understanding. This original relationship explains why the first successful U.S. satellite launch carried a scientific experiment on-board whereas the Soviets' first satellite was basically a transmitter in space, famously annoying the Americans with its repetitive beeping.[10]

---

7. Walter A. McDougall, . . . *the Heavens and the Earth: A Political History of the Space Age* (New York: Basic Books, 1985), pp. 100–111.

8. Ibid., pp. 116–124.

9. David H. DeVorkin, Science with a Vengeance: How the Military Created the U.S. Space Sciences after World War II (New York: Springer-Verlag, 1992).

10. Paul Dickson, *Sputnik: Shock of the Century* (New York: Berkeley Publishing, 2001), pp. 108–109. The original satellite payload for the Sputnik was downsized in order to ensure that success would occur when the attempt was made to orbit a satellite. Cf. Asif A. Siddiqi, *Sputnik and the Soviet Space Challenge* (Gainesville, FL: University Press of Florida, 2000), pp. 152–170.

Concurrent with these military-dominated development efforts were research efforts within the communications industry exploring the use of satellites for facilitating global communications.[11] These efforts were less publicly visible since the corporations involved, including AT&T ("Ma Bell") and RCA, focused more on securing corporate economic advantage rather than the publicity sought by military services in their quest for Congressional attention. The 1940s and 1950s saw intense public campaigns by the military services to gain appropriations advantages relative to other services. In one sense, these private efforts were more realistically the harbingers of the dual-use concept because they were truly nonmilitary although communications satellites (comsats) possessed obvious military usefulness regardless of their origin.

Therefore, the dual-use concept is premised on a distinction that technologically had no reality but was considered politically critical if weapons of mass destruction (WMD) proliferation threats to world security were to be controlled. This nonproliferation effort focused heavily on missiles or rockets that could deliver weapons with no effective means of defense, although other applications came under similar constraints. However, the biggest political booster of this distinction, President Dwight Eisenhower, saw the dual-use concept's value first with regard to his Herculean struggle to keep the U.S. budget under control (i.e., fiscally balanced).[12] Eisenhower's greatest political nightmare was that the United States would rush off in pursuit of glory in the heavens—a quest that he considered to be of little relevance security-wise.

On 4 October 1957, the first Sputnik launch led to an immediate and vociferous public demand for a U.S. space effort commensurate with that being apparently mounted by the Soviets. The public's demand was fed by a Congress controlled by the president's political adversaries.[13] This situation further inflamed existing rivalries among the three U.S. military services, especially between the Air Force and Army; each eagerly sought to seize outer space as its new area of operations, just as they had previously squabbled over missile systems.[14] With visions of large but separate competing military space programs dancing in his head, Eisenhower moved at once

11. David J. Whalen, *The Origins of Satellite Communications*, 1945–1965 (Washington, DC: Smithsonian Institution Press, 2002). Whalen's argument is that these private efforts got lost in the publicity given to NASA for forays into communications satellite development.

12. David Callahan and Fred I. Greenstein, "The Reluctant Racer: Eisenhower and U.S. Space Policy," in *Spaceflight and the Myth of Presidential Leadership*, Roger D. Launius and Howard E. McCurdy, eds. (Urbana, IL: University of Illinois Press, 1997), pp. 31–39.

13. For example, Senate Majority Leader Lyndon Johnson opened committee hearings on the question of the American space program and what was to be done in the fall of 1957. Robert A. Divine, *The Sputnik Challenge: Eisenhower's Response to the Soviet Satellite* (New York: Oxford University Press, 1993), pp. 41–76; Eugene M. Emme, "Presidents and Space," in *Between Sputnik and the Shuttle: New Perspectives on American Astronautics*, Frederick C. Durant, III, ed. (San Diego, CA: American Astronautical Society, 1981), pp. 16–23.

14. David N. Spires, *Beyond Horizons: A Half Century of Air Force Space Leadership* (Peterson Air Force Base, CO: Air Force Space Command, U.S. Air Force, 1997), pp. 21–38.

to first, stop the interservice rivalry, and second, to cut off any military aspirations for pursuing human spaceflight, the most expensive and difficult of space activities. All of this domestic activity occurred against an international background where the global nuclear arms race loomed as an enormously expensive contest with no end in sight, except possibly Armageddon. Eisenhower was definitely not interested in having this nuclear arms race spread to the heavens, raining death from space. Plus, Eisenhower had severe doubts as to the military usefulness of outer space in terms of weapons.

Cutting off a potential interservice rivalry was politically easier than stopping the political rush to start a human spaceflight program. The Air Force had earlier been awarded control over land-based, long-range ballistic missiles; the competitive Army missile effort had been reduced to developing short-range or tactical missiles.[15] This decision, however, had come only after years of bitter service infighting but was only put in place just prior to the first satellite launch. The president further confronted the political reality that his original effort to stymie the Army's growing space efforts had come a cropper when the first Vanguard launch (a civilian program run by the Naval Research Lab) failed to leave the pad in December 1957—a perfectly dismal response to two Soviet successes with their larger and physically more impressive Sputniks. Prior to the Vanguard launch, the administration had hedged its bets by authorizing the Army Ballistic Missile Agency team, led by Wernher von Braun, to build a back-up launcher and satellite. This cobbled-together effort flew to orbit on 31 January 1958—Explorer I became the first U.S. satellite. The Army had argued in 1956 that it could orbit a satellite immediately but that option was rejected by the administration whose interests were different and more internationally focused.[16] After the very public Vanguard failure, Eisenhower needed the political success symbolized by a successful Explorer launch, but von Braun's and the Army's larger space ambitions, including manned spaceflight, were not encouraged and were finally terminated.

Eisenhower's resistance was premised on two views he held regarding a potential manned spaceflight race: first, the cost was too high given available national resources (remember, balanced budgets), and second, the Army could not be allowed to build up its space efforts in competition with the Air Force. The latter goal was comparatively easily achieved with the Army space program being shut down. That decision was bitterly resisted (almost to the point of active insubordination) but the president was adamant. That left the other piece—the question of a manned space effort run by the Air Force, the victor in the interservice space wars, with visions of a vast space effort to compete with the Soviets across the spectrum of activities.

15. Raymond H. Dawson, "Congressional Innovation and Intervention in Defense Policy: Legislative Authorization of Weapons Systems," *American Political Science Review* 56 (March 1962): pp. 49–50. The dispute began over air defense systems and escalated.

16. Rip Bulkeley, *The Sputniks Crisis and Early United States Space Policy* (Bloomington, IN: Indiana University Press, 1991), pp. 95–101.

The answer to Eisenhower's latter problem came in the form of NASA, killing the proverbial two birds with one stone. NASA became the president's stalking horse for removing the Air Force from the manned space arena. His view was that a manned space program run by a military service would be impossible to control in terms of budget growth because the military could always invoke military necessity in order to stymie any presidential efforts at budget control. The problem of military services end-running around the president to Congress regarding their relative budget share had been a continuing feature of domestic politics in the 1950s, especially since the other party, the Democrats, controlled Congress.[17] Eisenhower's great prestige as the commander of Allied victory in Europe helped him beat back some efforts, but military space activities were literally completely new—the president was considered no more expert than many others. Also, equally relevant was the U.S. desire to slow down the proliferation of nuclear weapons.

NASA became the stalking horse for achieving the president's efforts—first to deflect Air Force ambitions and second to initiate successful arms control. The Air Force part was somewhat easier because, as a military service, it had to have an assigned mission in order to pursue a particular technology or approach. The decision was to transfer all manned or crewed spaceflight operations to the new agency, NASA, established 1 October 1958, leaving the Air Force out of the picture. The decision to make this move had two bases.

First, Eisenhower thought that a civilian agency would be more easily managed in terms of budget, lacking the political clout of the military services. In his assumption of NASA's political weakness, Eisenhower was ultimately correct but mistaken in the short term—the Southern paladins within Congress (especially at the committee level) saw great opportunities for constituency service in terms of creating constituent jobs.[18] When the Mercury program was approved despite Eisenhower's misgivings, the door was opened for Congressional pork. With President Kennedy's announcement of the Apollo program in 1961, NASA entered its Golden Age. Even in its relative political eclipse later, NASA has retained its usefulness for Congress as a source of constituent jobs.[19]

17. Samuel P. Huntington, *The Common Defense: Strategic Programs in National Politics* (New York: Columbia University Press, 1961), pp. 374–378.

18. Robert A. Divine, "Lyndon B. Johnson and the Politics of Space," in *The Johnson Years, Volume 2: Vietnam, the Environment and Science*, Robert A. Divine, ed. (Lawrence, KS: University Press of Kansas, 1990).

19. Eisenhower's resistance to government growth stymied Congressional leaders' desire to create jobs. The creation of NASA presented the golden goose since Eisenhower was constrained by the political pressures to "do something about space." New NASA Centers—Goddard Space Flight Center, Johnson Space Center, Stennis Space Center, and Kennedy Space Center—were established, while former NACA Centers—Langley Research Center, Glenn Research Center, Dryden Flight Research Center, and Ames Research Center—were upgraded and former Army space assets—Marshall Space Flight Center and the Jet Propulsion Lab—were acquired. All kinds of economic opportunities were brought to the southern United States, which historically had a severe deficit of such facilities and industries.

Second and more critical, by transferring manned spaceflight to a nonmilitary agency, the political opportunity existed for establishing outer space as an international sanctuary devoid of space-based weapons. This latter was extremely important for nuclear arms control purposes since, for the first time, the parties were not attempting to remove or restrict weapons already fielded; rather, they were attempting to deny weapons' initial entry into a location. The realm of outer space was defined as the common heritage of mankind, as stated in Article 1 of the so-called Outer Space Treaty:

> The exploration and use of outer space, including the moon and other celestial bodies, shall be carried out for the benefit and in the interests of all countries, irrespective of their degree of economic or scientific development, and shall be the province of all mankind.

It was not to be an arena for direct military confrontation (Article 4):

> States Parties to the Treaty undertake not to place in orbit around the Earth any objects carrying nuclear weapons or any other kinds of weapons of mass destruction, install such weapons on celestial bodies, or station such weapons in outer space in any other manner. [20]

This perspective allowed an intense public space race competition between the two nuclear space powers to occur without necessarily leading to a military confrontation.[21] After the Cuban missile crisis in October 1962, the Soviets and Americans were both aware of how fragile the nuclear peace was—a fact which heightened their interest in decreasing confrontation potential. Thus, the race for the lunar surface could end with a clear winner, the United States, but without the hazardous outcome implied by an arms race.

Out of this mishmash of goals and motivations, the concept of dual-use arose as one primary methodology by which all space-related technologies could be evaluated as to whether they possessed significant military implications. This concept created a truly artificial distinction since the only real difference between military and civilian or commercial uses was, at its essence, user intent. The technology remained basically the same but its purposes varied. Military technologies were often more robust in terms of their survivability (i.e., military specifications or "mil-specs"), but the central application remained the same for both.

---

20. Treaty on Principles Governing the Activities of States in the Exploration and Use of Outer Space, Including the Moon and Other Celestial Bodies (entered into force 10 October 1967), *http://www.state.gov/t/ac/trt/5181.htm* (accessed 15 October 2006).

21. In the interest of this nonmilitary space race, President Richard Nixon has been identified as reshuffling the Apollo crew schedule in order to ensure that a civilian was first to set foot on the lunar surface. Worden and Shaw, pp. 112.

Those with a more Machiavellian orientation might think that another important motivation was economic in that the dual-use concept permitted tightened political control over the dissemination of such technologies. The United States, for example, aggressively protected its monopoly over other Western nations with regard to space lift and any other space technologies they might develop that potentially competed with U.S. economic interests.[22] For example, the launching of experimental communications satellites built by potential economic competitors was resisted until sufficient political pressure was brought to bear, one example being the French-German Symphonie comsat in 1974. Once that political barrier was broken, space communications technologies could now be sent to orbit by other nations, increasing their competitiveness with the United States. Success there, however, did not change the reality that military security-imposed limitations still affected their usefulness. In effect, these externally imposed technological disabilities distinguished the nonmilitary usefulness of the same technologies from the military—the essence of dual-use.

The impact of the dual-use distinction was very real economically because it imposed restrictions upon the usefulness of several space applications in competition with terrestrial-based competitors. Often discussions of space-based commercial applications ignore the existence of robust and established economic competitors. In fact, those competitors either directly or indirectly have impacted development of space applications. Early U.S. policy regarding comsats was driven by awareness that AT&T (the Bell system) and IT&T sought to dominate the new field of space-based communications. Controlling their monopolistic tendencies was a major factor in U.S. policy. Ironically, the pathway chosen solidified their critical role in the field's future development since the initial satellite linkages were to their phone lines.

DOMESTIC IMPLICATIONS OF DUAL-USE

Dual-use space applications are, by definition, useful for civilians but their military potential renders their dissemination problematic. Simply put, during the cold war the domestic economic usefulness of such technologies was, as a matter of policy, subordinated to their potential as a threat enhancer for other nations. Therefore, strict constraints were imposed regarding how useful the application could be made or how widely it was disseminated. The universe of dual-use applications has become large and comprehensive, as can be seen in the various lists generated under the Wassenear Arrangement.[23]

---

22. Roger Handberg, *International Space Commerce: Building from Scratch* (Gainesville, FL: University Press of Florida, 2006), pp. 52–58; Lorenza Sebesta, *The Availability of American Launchers and Europe's Decision "To Go It Alone,"* (Noordwijk, The Netherlands: ESA Publications Division, 1996).

23. The Control Lists generated by the Arrangement include: "List of Dual-Use Goods and Munitions List," which is publicly available on their Web site: *http://www.wassenaar.org* (accessed 15 October 2006).

The reality is that the United States has adhered to a much more restrictive view regarding dual-use technologies. Efforts at loosening those restrictions were underway in the 1990s but were partially reversed with regard to exports when allegations were made that China was stealing U.S. secrets.[24] Congressional action in 1998 led to stricter enforcement of existing International Traffic in Arms Regulations (ITAR) by the Department of State rather than the Department of Commerce, which was deemed to be too willing to facilitate expanded international trade by loosening security restrictions. As a result, the United States returned to a policy similar to that during the cold war, when the trend was toward liberalized global trade of such technologies. Global trade in space applications expanded, but with less and less U.S. participation. American fears were that certain nations were stealing U.S. secrets—many with military implications, since the technologies involved were dual-use.

Launch technologies are the obvious dual-use technology; however, policy makers at first did not consider them critical simply because governments, through their militaries, controlled all the missile launch vehicle derivatives. In fact, not until after the Space Shuttle *Challenger* accident in January 1986 did a private launch sector come into view in the United States.[25] This privatized sector flew legacy launch systems received from the government. Internationally, all major launch vehicles have been government-developed and -owned, even when they were spun off in the Arianespace context as a commercial corporation. Development of new launchers or upgrades of existing ones have thus far always been government-funded and ultimately government-controlled.

Efforts at purely private launchers have been more disappointing than successful, and even the successes (or near-successes such as SpaceX's Falcon series) get sucked into the government orbit.[26] As a result, launch technologies did not pose an issue—the U.S. had an effective monopoly over space launch in the West while the Soviets controlled the rest. As launch technology spread in the form of ballistic missiles rather than space transportation, the United States and other nations concerned with weapons proliferation became alarmed at the fact that rogue and other unsavory states could readily acquire such militarily useful technology. Out of that concern arose the Missile Technology Control Regime (MTCR) in 1987 as the mechanism by which the dissemination of such dual-use technology could be regulated. MTCR attracted only limited support initially but its existence has

24. Christopher Cox, *Report of the Select Committee on U.S. National Security and Military/Commercial Concerns with the People's Republic of China*. House Report 105-851, 25 May 1999 (Washington, DC: U.S. House of Representatives); Joan Johnson-Freese, "Alice in Licenseland: U.S. Satellite Export Controls since 1990," *Space Policy* 16 (July 2000): pp. 195–204.

25. Handberg, International Space Commerce, pp. 82–83.

26. The Commercial Orbital Transportation Services (COTS) demonstration program involves up to $500 million for support of the International Space Station using either the Rocketplane Kistler or Space Exploration Technologies (SpaceX) as the supply vehicles.

been used to leverage other nations into compliance. India's attempted purchase of cryogenic upper-stage engine technologies from Russia was derailed because of Russia's need for Western support and investment to support the newly established Federation. The MTCR has not prevented the dissemination of missile technologies but, similar to nonproliferation treaties, has slowed the process. [27]

Comsats

Communications satellites remain the lodestar of commercial space activity since their applications can generate significant revenues while also being militarily useful. Development of the field was dominated at first by the United States, both technologically and organizationally. Establishment of Intelsat was orchestrated to put the United States in the dominant position—it was defined as the monopoly over international satellite-based communications. In the early years, U.S. satellite manufacturers were favored, but once that monopoly was broken by the Europeans the United States argued that no comsat could be launched that operated outside the purview of Intelsat or later Inmarsat. In time, that broke down with the Canadian Anik satellites and later with regional systems such as Eutelsat and Arabsat, and finally in the commercial comsat vendors, the first being PanAmSat. The opening up of the comsat market made direct broadcast service (DBS) available, which individual consumers or groups can access directly without going through the gateways of the Intelsat system.[28] In addition, comsat companies became increasingly international in their ownership, which restricted U.S. ability to control communications in and out of certain nations. The international nature of these corporations made them less responsive to U.S. demands, although access to U.S. markets could be denied in retaliation.

In dual-use terms, comsats are available to an even wider group of users, including nations that the United States does not wish to have such access. For comsats, the dual-use argument was weaker and has largely been discarded except by the United States, since the idea of globalization implies and translates into more and more access to the worldwide communications net (never mind the World Wide Web, which is in fact a minor part of it). Proliferation of comsats and their methods of operation have completely undermined U.S. efforts at control since the United States no longer controls the manufacture or launch of such spacecraft. The Europeans are the strongest competitors, but other nations such as Japan and China are developing their manufacturing capabilities.

---

27. The standard treatments of U.S. policy regarding Intelsat in the early years can be seen in Joseph N. Pelton, *Global Communications Satellite Policy: INTELSAT, Politics, and Functionalism* (Mt. Airy, MD: Lomond Books, 1974); and Jonathan F. Galloway, *The Politics and Technology of Satellite Communications* (Lexington, MA: Lexington Books, 1972).

28. Ironically, PanAmSat was later acquired by a privatized Intelsat in 2005—a move symbolic of the changed policy environment.

Remote Sensing

The two areas most directly impacted by the dual-use concept are remote sensing and navigation. Clear restrictions were placed on all nonmilitary remote sensing satellites; weather satellites were the major exception, but even there the level of resolution was kept large (i.e., the aperture was kept at the half-mile [kilometer] level or higher rather than at the 3-foot [meter] or less level in order to deny any military usefulness to the images produced). In the late 1960s NASA began looking for space applications that would have social utility; remote sensing was one obvious application because the images produced had great social potential for social purposes, including environmental monitoring in particular.

With the empowerment of the environmental movement in the early 1970s, such a satellite became an obvious route to pursue. The Earth Resources Technology Satellite (renamed Landsat 1) was the result. The fact was that the satellite was deliberately kept less accurate than it could have been, ranging from approximately 131 feet (40 meters) to 246 feet (75 meters) with an approximately 115-mile (185-kilometer) swath depending upon the imager used. The effect was to thwart any attempt at commercialization of the Landsat or equivalent systems. The security-related fear was that a truly commercialized remote sensing approach would allow potential enemies to acquire detailed images at little cost even though they did not own space assets capable of doing the job or any space assets at all.

A series of struggles ensued over the next two decades, reaching their crescendo in the Reagan administration's efforts to commercialize Landsat.[29] That effort failed because, in reality, there was no large commercial market for the images produced—the images had usefulness but their large scale limited what could be observed. On a macro level, the images produced were useful but further advances in Landsat technology were effectively stymied by dual-use considerations. The pictures taken could be manipulated to improve the view, but that was of limited utility despite significant increase in resolution. The reality was that for two decades the remote sensing field was effectively a dead-end in terms of civilian applications. Efforts to open the field to commercial or other players took form in the passage of the Land Remote Sensing Act of 1992.[30] This allowed the entry of commercial interests but did not change the security restrictions, which meant it remained a dead-end until that last part was put in place. The French, with their SPOT Image satellite, were the only international competitor; its images had somewhat greater resolution but also were not considered to be of military significance.

29. For an abbreviated summary of the controversy, see Roger Handberg, *The Future of Space Industry: Private Enterprise and Public Policy* (Westport, CT: Quorum Books, 1995), chapter 4.

30. Pamela E. Mack and Ray A. Williamson, "Observing the Earth from Space," in *Exploring the Unknown: Selected Documents in the History of the U.S. Civil Space Program, Volume III: Using Space*, John Logsdon, ed. (Washington, DC: U.S. Government Printing Office, 1998), pp. 173–176.

This national security dimension weakened in the 1990s as a consequence of two factors, the first Gulf War and the rising U.S. concern about international economic competitors. During the run-up to the 1991 coalition attack on Iraqi forces in Kuwait and Iraq, images from both Landsat and SPOT Image satellites were incorporated into military planning. Their large scale was in fact more useful for some purposes than the images acquired from intelligence satellites, which provided detail but no larger perspective. That usage meant that the security restrictions imposed over the years were in fact less useful than originally thought.

Subsequently, in 1994, the Clinton administration effectively removed any restrictions on the image resolution being sought. That led to an explosion of applications for remote sensing licenses, although economic reality proved much harsher since most applicants lacked the financial resources to make their satellites happen.[31] Ironically, the Department of Defense (DOD) now found this loosening-up to its advantage because the existence of such commercial options meant that the military did not have to build as large a remote sensing fleet as was earlier projected. This became particularly important after the Soviet Union's collapse and the general decline in defense spending (partially reversed after 9/11).

The 11 September 2001 terrorist attack provided evidence as to this new operating environment: for example, in October–December 2001 DOD purchased all Ikonos images of Afghanistan so that others could not gain access. However, the commercial sector remains dependent upon the DOD for its survival given the competition with the aerial surveying industry. Improvements have occurred but delivery of space-based images is still too slow for many customers. Also, the problem of satellite revisit times to take subsequent images still advantages the aerial surveying industry. Further complicating the situation is the existence of a number of international competitors, including the Russians, using military-grade remote sensing data, the French, and, interestingly, the Canadians with their Radarsat system. Clearly, the technology has spread beyond the control of one nation.

Navigation

Navigation represents the clearest example of dual-use applications, since its pedigree was entirely military. What occurs is fairly straightforward; the military are obsessed with knowing their forces' exact location (regardless of the enemy's location). This is particularly true for the Navy, which operates beyond sight of landmarks. Establishing one's position at sea awaited the development of the sextant where one shoots the stars to determine location and accurate clocks. Like a communications satellite, a navigation satellite provides a signal which, when

31. For a broader overview of the global remote sensing market, see John C. Baker, Kevin M. O'Connell, and Ray Williamson, eds., *Commercial Observation Satellites: At the Leading Edge of Global Transparency* (Santa Monica, CA: RAND Corp., 2001).

combined with signals from other navsats, gives one their exact location on Earth's surface. The U.S. Navy pursued such a navsat system first with its Transit system in the 1960s, until the Air Force and Navy combined efforts to build what became known as the NAV GPS system.[32]

Using this combination of radio signals and atomic clocks, the receiver can determine a location on the surface and/or in the air with great exactitude. This application was developed for positioning and search and rescue, although the civilian applications became quickly obvious. For the military, GPS reduced "friendly fire" incidents and other blunders due to the fog of war and facilitated development of GPS-guided munitions, greatly enhancing weapon effectiveness and lethality. The essence of the military transformation hinges on global GPS access.

The degree of precision provided was particularly sensitive because the U.S. military did not want to enhance the ability of America's enemies to find targets employing the same GPS signal. Initially, the United States established two signals (now more)—one very accurate and precise for the military and other authorized users, and a second signal with a deliberate distortion (selective availability, SA) built into that signal.

In response to the shooting down of Korean Airline Flight 007 in 1983 when it strayed into Soviet air space, President Ronald Reagan had ordered the DOD to allow civilian access to the GPS signal. Opening the door to civilian use proved the equivalent to opening Pandora's Box as the military, in time, lost control. Another aspect was that the DOD retained the capacity to completely deny the civilian signal under some conditions due to threat, imminence of war, or actual conflict. This SA function was controversial for non-U.S. users of GPS. Despite the U.S. military's resistance, American and especially Japanese commercial vendors were relatively quickly able to create software that effectively negated the built-in distortion. In fact, the first large military conflict employment of GPS was during the first Iraq war in 1991 when, due to a shortage of military GPS receivers, the United States turned off SA so that commercial receivers given to the troops would work accurately. This decision further dramatized the DOD's waning control over what was becoming a major commercial sector.

Afterwards, the navigation business exploded as more applications were developed that provided even greater precision. One powerful application was the timing function of the navigation satellites (each carries an atomic clock), which is used to control computer networks to allow greater efficiency in moving data across the globe. This enormously increased economic efficiency across large distances in moving information and money transactions. Absolute U.S. control over this critical business resource became a major controversy between the United States and other nations.

32. Handberg, International Space Commerce, chapter 6.

The Europeans, especially, saw SA and DOD control over the turnkey not as security-driven but as further attempts by the American monopolists to protect their economic dominance. The result, after some acrimony, was development of the Galileo satellite radio navigation system as an alternative to the GPS system. This Galileo system, the Europeans say, will not be turned off in a time of military conflict or imminent conflict, although that may not prove out in the long term as the Europeans come to see themselves as becoming more proactive globally, partially replacing the Americans. The DOD's response was initially to reject any change but that view was overridden by President Clinton with the 2 May 2000 removal of distortion from the GPS signal. That decision attempted to forestall the Galileo program or at least make its development proceed more slowly. All such efforts failed, although splits within Europe over various Galileo issues have slowed development. In addition, the Europeans have solicited Galileo participation by non-European nations, including China, a fact that further feeds American concerns.

Dual-Use and Its Implications

The point being made here is simple: dual-use considerations directly and heavily impacted American domestic and international commercial space policy—international considerations are what drive the system even though the major economic impacts occur domestically. Those impacts are the unintended consequences of an American policy generated in an earlier period. No nation has been totally deprived of the capacity to acquire needed space applications because of the U.S. prohibitions, especially in the cases of communications, remote sensing, and navigation (and arguably rocket technology). The reality is that these restrictions have had more of an adverse impact on the U.S. economy than elsewhere. The United States is creating a bubble around its space commerce efforts by imposing security-driven restrictions that significantly blunt any U.S. efforts at competing economically in the global marketplace.[33]

Over time, the global spread of space technologies has eliminated U.S. capacity to determine to whom and for what uses the technologies will be available. That raises some interesting implications for the broader question of U.S. security policy. As a general policy concept, dual-use embodies several implications, the most significant of which is keeping the United States secure from its enemies by denying them improved militarily useful technology. What has slipped out of U.S. hands

33. The "bubble analogy" arose during the discussions of the original papers at the Societal Impact conference in September 2006; I regret not writing down who the author of the phrase was at that time. It is used here as shorthand for the isolation of American space industry from the global marketplace due to security restrictions. In other contexts, Joan Johnson-Freese has written and spoken extensively on the effects of the 1998 decision to reimpose security restrictions on U.S. technology exports, including space applications.

is the ability to control dissemination due to the multiplicity of players. Take any major space technology and you can find multiple providers outside the United States. This issue arose first in the 1980s when Japan, for example, was deemed more advanced in a number of militarily relevant technologies than the United States. That situation has grown worse. That translates into a situation in which U.S. space industry runs the risk of becoming less competitive and, by extension, falling behind possible military competitors in terms of application quality.

During the cold war there were two central players with their alliances; now there exist multiple combinations of players who may coalesce in opposition to the United States regarding various issues. The key problem is that the instruments through which America military power is exercised are now exceedingly vulnerable. One of the critical lessons learned, fortunately by analogy and not by attack, is that Earth-orbiting satellites are vulnerable to interference either directly through physical attack, or indirectly through manipulation of their operating programs or disruption of their signals. Proceeding along orbital paths, easily predictable by observation, means that satellites cannot hide. In fact, in orbital space, commercial satellites are growing even more vulnerable to disruption. Growing U.S. military dependence on commercial comsats and remote sensing satellites for critical tasks increases their vulnerability to disruption since commercial vendors find no reason to harden satellites or provide other means of protection. The costs are not justifiable for a commercial venture. The GPS system itself is capable of being jammed by any combatant more sophisticated than Iraq—their efforts in 2003 were not successful, but the way is clear.

The dual-use concept represented one effort at delaying American vulnerability to nations equipped with equivalent military technologies; the economic motivation has persisted even though the security justification has lost its potency. Both have lost their persuasiveness, especially the latter; American companies are now effectively excluded from many international economic opportunities regarding use of space technologies—strong, technologically sophisticated competitors are taking increasing market share in the different sectors of space commerce. The difficulty is that for the United States, change demands a rethinking of what is to be accomplished using space applications. Previously, U.S. policy was to bind others to the United States through security and economic ties, with the latter thought to be the more lasting. Those days are gone. Space history has seen the relative (not absolute) decline of American dominance with the entry of the Russians and the formerly excluded Chinese into the field, along with the rise of the Europeans and others to increasing prominence.

CONCLUSION

Dwight Eisenhower, along with others, effectively fabricated the dual-use concept to solve certain political problems that he otherwise felt would spin out of control. To that end, Eisenhower was successful and dual-use became embedded in U.S. policy thinking, although initially its implications were not entirely clear if the international environment were to change. Over time, the earlier question of security faded in intensity but not out of existence. The emergence of China as a potential rival led to an intensification of dual-use concerns, with the economic component much more publicly muted. Ironically, the result is that China has not been delayed by U.S. actions; the larger effect has been to severely cripple American space industry competitiveness more by inadvertence than by design. The two are now not mutually supportive, as in the beginnings when dual-use allowed the pursuit of American security and economic interests simultaneously. In a world of global economic competitions, balancing the values of economic competitiveness and security is no longer as simple or clear as before.

Chapter 19

Reconnaissance Satellites, Intelligence, and National Security

Glenn Hastedt

Policy arenas do not arrive on the scene full-blown, nor do they remain static over time. They grow and evolve. We are witnessing this today with homeland security. More than a half-century ago we saw it with national security. One way to conceptualize the dynamics involved in the development of a policy arena is as a stream of activity. Much like the origins of a river are found in the merging of smaller tributaries, a policy arena is the product of several different forces coming together. Typically they involve a definition of a problem, the emergence of a collection of institutions designated to address that problem, and the identification of a strategy set to solve the problem. Once under way, a river reinvents itself daily. The changes are not necessarily visible at the outset, but over time they become clear. External events, both man-made and natural occurrences, play their part in this evolution, but so, too, do the currents of the river and the life its waters sustain within it. International crises, accidents, bureaucratic politics, personalities, as well as new ideas and technologies are such driving and shaping forces in policy arenas. Finally, given enough time, rivers themselves disappear by either merging into larger bodies of water or vanishing into the ground as their water flow is reduced to a trickle. Changing perceptions of a problem or the proper way to address it may cause the first phenomenon to occur in a policy arena, whereas shrinking budgets and public apathy may bring about the second.

A key element of the policy stream that is national security was the development of reconnaissance satellites. They were not present in 1947 at the formal founding of national security as a policy arena—with the passage of the National Security Act—but they became an important force in its subsequent downstream development.[1] Its effects can be seen in the identity and influence of the government agencies that make up the intelligence community; the manner in which intelligence was

1. See R. Cargill Hall, "Clandestine Victory: Eisenhower and Overhead Reconnaissance in the Cold War," in *Forging the Shield: Eisenhower and National Security for the 21st Century*, Dennis Showalter, ed. (Chicago: Imprints Publications, 2005) for a discussion of the technological and political streams from which reconnaissance satellites emerged.

thought about by policy makers; and the problems to which intelligence was put. In none of these cases were reconnaissance satellites the sole factor in producing these changes but in each case they played a major role.

FORMATIVE CURRENTS

Four forces can be seen as having a formative influence on the development of the national security policy stream into which reconnaissance satellites would enter. The first was the problem of strategic surprise as symbolized by Pearl Harbor. This was the event that, in the minds of many, national security policy had to make sure was not repeated. The second force was the solution of greater centralization and cooperation at the national level among bureaucracies involved in foreign diplomatic, military, and economic policy. Pearl Harbor occurred in spite of warning; intelligence was present, but it was not recognized or acted upon. The inherent validity of this solution was reinforced by the wartime experience of ad hoc military centralization that came about out of the need to cooperate with the British. To bring this about, the 1947 National Security Act created the Central Intelligence Agency (CIA), the National Security Council (NSC), and unified the military services under a Secretary of Defense in a national defense establishment that would soon become the Department of Defense (DOD).

The third force was the de facto establishment of an intelligence community that was to work together to prevent another Pearl Harbor. Along with the newly created CIA the other founding members were the Bureau of Intelligence and Research (INR) at the State Department and U.S. Army, Navy, and Air Force intelligence. The final force that exerted great influence on the origins of American national security policy was the advent of the cold war. It presented the United States—and national security policy—with a clearly identifiable enemy in the Soviet Union and then a strategy—containment—around which policy makers could unite.

GROWTH AND DEVELOPMENT OF THE INTELLIGENCE COMMUNITY

Today there are 16 organizations that officially constitute the intelligence community. Reconnaissance satellites played a central role in the formation of one organization and had a substantial impact on the development of two others. Reconnaissance satellites can be most directly linked to the establishment of the National Reconnaissance Office (NRO). President Eisenhower established the NRO by executive order in August 1960. It became operational on 6 September 1961, following an agreement between the CIA and the Air Force setting it up as a joint operation. The Air Force was placed in charge of launching the satellites and recovering the film capsules; the CIA was charged with developing the satellites. The director of the NRO was to be the undersecretary of the Air Force and the deputy director was to be drawn from the CIA. Under terms of the initial agreement, neither the CIA nor the Air Force had to give up control over any of its reconnaissance satellite programs to the NRO. Instead, they would be merged at a higher level into a National Reconnaissance Program.

Reconnaissance satellites also played a role in creating the National Photographic Interpretation Center (NPIC), the predecessor of one of the newest members is the National Geospatial Intelligence Agency (NGIA). The original impetus for creating NPIC lay in a March 1960 suggestion by Secretary of Defense Thomas Gates that Eisenhower commission a study of the defense intelligence bureaucracy, describing it as an inefficient, huge conglomerate. Gates's proposal languished until Francis Gary Powers's U-2 reconnaissance aircraft was shot down. After this incident a Joint Study Group was formed that reported out just prior to the end of Eisenhower's presidency. Among its conclusions were that the military was playing too prominent a role in the intelligence process, and it called for increased efficiency through the creation of NPIC. Both the CIA and DOD sought to run NPIC, with the DOD proposing the creation of a new unit and the CIA calling for the expansion of its already existing Photographic Intelligence Center. Secretary of Defense Robert McNamara acted on this recommendation and NPIC came into existence in 1961 as a community-wide asset in the interpretation of aerial photos. He also followed Eisenhower's inclination to place NPIC within the CIA.

The third member of the intelligence community whose existence and development is tied to reconnaissance satellites is the National Security Agency (NSA). It was established by a secret executive order, National Security Council Intelligence Directive (NSCID) No. 6, entitled "Communications Intelligence and Electronics Intelligence," on 15 September 1952. It formally came into existence on 4 November 1952. NSA is the successor organization to the Armed Forces Security Agency (AFSA). It was set up as the result of a Joint Chiefs of Staff Directive signed by Secretary of Defense Louis Johnson on 20 May 1949. Located within the DOD, the AFSA was assigned responsibility for directing the communications intelligence and electronic intelligence of the three military services signals intelligence units. In spite of this broad mandate, the AFSA had little power. For the most, part its activities consisted of tasks not being performed by the Army Agency, the Naval Security Group, and the Air Force Security Service—the units whose work it was to direct.

Walter Bedell Smith, President Harry S. Truman's executive director of the National Security Council, found this state of affairs to be unsatisfactory. Particularly troubling was the failure of the AFSA's performance during the Korean War when it was unable to break the Chinese and North Korean codes. His view was shared by General James Van Fleet, commander of the U.S. Eighth Army who complained that "[W]e have lost, through neglect, disinterest and possibly jealousy, much of the effectiveness in intelligence work we acquired so painfully in World War II."[2]

2. James Bamford, *Body of Secrets: Anatomy of the Ultra-Secret National Security Agency* (New York: Anchor Books, 2002), p. 30.

Smith wrote a memo in December 1951 calling for a review of communications intelligence activities, describing the current system for collecting and processing communications intelligence as "ineffective." Three days later, on 13 December 1951, the National Security Council set up a committee (commonly referred to as the Brownell Committee after its chair, Herbert Brownell) to examine the matter. The Brownell Committee recommended strengthening the national-level coordination and direction of communications intelligence activities. The NSA was created as a result of these recommendations.

NSA got off to an inauspicious start. Although it successfully engaged in overflights of the Soviet Union, it lacked a capacity to provide intelligence on events elsewhere, such as the Suez crisis. Moreover, its efforts to break Soviet codes repeatedly met with failure. James Bamford goes so far as to speculate that in the 1950s NSA faced the prospect of going out of business and that a "produce or else" atmosphere had settled over the agency.[3] Salvation came in two forms. First, there was support from President Eisenhower and his Board of Consultants along with an influx of funds in an effort to strengthen its code-breaking abilities. Second, there arose the perceived necessity of obtaining signals intelligence from Soviet missiles as a result of the launching of Sputnik in 1957. The initial solution to this need was the construction of Earth-based receiving dishes. The second-generation solution was the deployment of space-based satellite receivers. President Eisenhower gave his approval for the first launching of an ELINT satellite five days after Gary Francis Powers's U-2 was shot down.

The story of the creation and growth of these national security organizations is more complicated than a straight-line response of policy makers to the development of reconnaissance satellites. It is one in which the existing currents of the national security policy stream heavily influenced organizational design. This comes through most vividly in the development of the National Reconnaissance Office.

The decision to create the NRO came years after explorations into the feasibility of space reconnaissance satellites had already begun. Not surprisingly, the Air Force was first to move in this direction. Officials were attracted by the potential power of long-range missiles and tasked the RAND Corporation to study whether they might be used to launch space reconnaissance satellites. Its report, "Preliminary Design of an Experimental Earth-Circling Spaceship," was delivered in May 1946. Three others followed in 1947, 1952, and 1954. The last study was cosponsored by the CIA and recommended that the Air Force begin at "the earliest possible date completion and use of an efficient satellite reconnaissance vehicle."[4] RAND's report formed the basis for General Operational Requirement No. 80, issued by the Air Force in March 1955, requesting proposals from the private sector for the development of a photographic reconnaissance satellite.

3. Ibid., p. 355.

4. William Burrows, *Deep Black: Space Espionage and National Security* (New York: Random House, 1986), p. 83.

Discussions were also under way at the presidential level. On 27 March 1954, President Eisenhower held a meeting with James B. Conant, James R. Killian, Jr., and other scientists that led to the formation of a study group under Killian's direction to develop solutions to the problem of surprise attack. Its report, "Meeting the Threat of Surprise Attack," was completed on 14 February 1955. Project 3 dealt with intelligence and was chaired by Edwin Land. In their briefings to Eisenhower, Land and Killian identified satellites as a promising system for collecting intelligence that would provide warning to the United States of an impending Soviet surprise attack. They also noted that the technology to realize this collection platform would take time to develop and suggested an interim technology: a high-flying reconnaissance aircraft. It would become the U-2.

Interest in space reconnaissance satellites led in multiple directions in the search for a technology to accomplish this mission. The favored Air Force option was to transmit photographs through a radio downlink. First known as WS-117L/Pied Piper and then Sentry, it would ultimately be known as the Satellite and Missile Observation System (SAMOS). The CIA advocated the mid-air capsule recovery system that had been rejected by the Air Force. It became known to the world as Discoverer and to those involved in intelligence collection as CORONA. Competition continued even after a February 1958 Eisenhower meeting with Killian and Land that reviewed the difficulties the Air Force was having in developing its reconnaissance satellite. As a result of that meeting, Eisenhower decided to give the CIA primary responsibility for developing a reconnaissance satellite. The Air Force did not, however, stop work on its preferred option. It proceeded with its Sentry system.

The search for an appropriate technology to use for satellite reconnaissance went hand-in-hand with efforts to devise an organizational structure within which to house it. The search did not begin with a clean slate. A stream of activity was already in place and had left a legacy into which organizational thinking would enter. When the earlier decision had been made to build the U-2, Eisenhower determined that the CIA—not the Air Force—would be in charge of the operation. There was nothing automatic about this. Director of Central Intelligence (DCI) Allen Dulles was a firm believer that human intelligence gathering should be at the core of the CIA's covert operations. He had shown little interest in the project earlier in 1954 and is described as "accepting the inevitable" in later accepting CIA jurisdiction over it.[5] This came after Edwin Land wrote a letter to him strongly urging the CIA to take the lead in the CL-282 Project that would become the U-2. Land wrote "I am not sure that we have made clear that we feel there are many reasons why this activity is appropriate for the CIA . . . . We told you that this seems to us the kind of action and technique that is right for the contemporary version of the CIA: a modern and scientific way for an Agency that is always supposed to be looking, to do its looking."[6]

---

5. Jeffrey Richelson, *The Wizards of Langley: Inside the CIA's Directorate of Science and Technology* (Boulder, CO: Westview, 2002), p. 13.

6. *The CIA and the U-2 Program, 1954-1974* (Washington, DC: Center for the Study of Intelligence, Central Intelligence Agency, 1998), p. 33.

Richard Bissell, a special assistant to Dulles, was given the assignment to develop the U-2, and a special standalone unit within the Directorate of Plans, the Development Projects Staff, was created to manage it. This decision did not put an end to bureaucratic jockeying for control over the U-2 program. Air Force Chief of Staff General Nathan Twining believed that the Strategic Air Command (SAC) under the direction of General Curtis LeMay should be in charge of the U-2. In the spring and summer of 1955, he lobbied for such a change, only to have SAC settle for limited participation in the U-2 program.

Given the speed with which the U-2 was developed and became operational, it is not surprising that Eisenhower again turned to Bissell and his Development Projects Staff in 1958 as the lead organization after the decision was made to go ahead with the Discoverer/CORONA reconnaissance satellite program. Beneath them, the CIA and Air Force continued to go their separate ways. Each encountered internal organizational problems. Within the Air Force, responsibility for space satellite reconnaissance shifted from unit to unit with dizzying frequency. At different times it was the responsibility of the Air Force and the Advanced Research Projects Agency, sometimes leaving the Air Force with responsibility for little more than supervising global surveillance studies and at other times being in charge of total control over satellite reconnaissance programs.

At the CIA a different sort of organizational problem arose. Bissell was appointed Deputy Director for Plans, putting him in charge of all CIA covert operations. In moving into this new position in 1959, he took with him control over U-2 and CORONA, effectively removing them from the Development Projects Staff. This move alarmed Killian and Land, who saw covert action and human espionage as very distinct from espionage based on science and technology. Moreover, they were concerned that within the CIA there was now insufficient attention being given to science and technology issues which were now found in virtually all quarters of the CIA.

The establishment of the NRO as an operational unit in 1961 did not end the conflict between the CIA and Air Force over control over reconnaissance satellites. Where the CIA saw the Air Force and the NRO as one and the same, and as together trying to force it out of the satellite reconnaissance business by taking over its successful CORONA program, the NRO saw itself as a truly national intelligence agency having a small Air Force component. Over the next several years, each body recommended that the other all but go out of the satellite reconnaissance business. In November 1962 Air Force officials proposed that many (if not all) CIA reconnaissance projects should be transferred to the Air Force and that all program functions should be consolidated within the NRO. The CIA would later counter with a proposal to eliminate the NRO with "all research, preliminary design, system development, engineering, and operational employment" going to the CIA. It was not until April 1965 that a truce was achieved with DCI Admiral William Raborn and Secretary of Defense Cyrus Vance agreeing to a formula whereby the Secretary of Defense had ultimate responsibility for managing the NRO, including its budget and choosing the director. The DCI was to have responsibility for determining collection priorities and the CIA was to continue to be responsible for CORONA and the development of new systems once the concept was selected.

A New Decision-Making Environment

Reconnaissance satellites not only helped to create new organizations or transform existing ones; they also altered the shape of the national security policy decision-making environment into which they flowed in three ways. A first change was to complicate and accentuate the managerial challenge facing the DCI. From the outset this individual was simultaneously the head of the CIA and the head of the intelligence community. And from the outset the DCI struggled to transform this grant of authority into something meaningful. Not only was the CIA a new organization but also the other founding members of the intelligence community were located in existing organizations. This would also be true of all others who later joined the intelligence community. This created an immediate point of contention between a DCI trying to forge a community-wide policy and intelligence officials in these agencies who were part of organizations that did not always agree with this policy.

This problem was noted by the First Hoover Commission in its 1948 report. Its subcommittee on national security policy, the Eberstadt Committee, wrote that "[T]he Central Intelligence Agency deserves and must have a greater degree of acceptance and support from old-line intelligence services than it has had in the past."[7] Singled out as still unsatisfactory were relations between the CIA and G-2 (Army intelligence), the FBI, the Atomic Energy Commission, and the State Department.

As we have seen, the development of reconnaissance satellites quickly elevated the Air Force to the position of the CIA's primary antagonist. Conflicts of interest also developed between intelligence agencies involved with the operation and development of reconnaissance satellites and the analysis of their products. NSA and NRO have quarreled over the proper mix of space-based systems, with NRO consistently supporting a more costly systems mix. In addition to competing with NSA, NRO also solicits funds directly from the military services through the promise of tactical intelligence that will support their missions. In an effort to resolve these conflicts the National Imagery and Mapping Agency (NIMA) was created in 1996 by bringing together several offices including NPIC. The creation of NIMA did not end bureaucratic disputes over imagery intelligence. The CIA's Directorate of Science and Technology sought to regain control over functions lost to NIMA, while the NRO continued to have program and budgetary control over ground station and mission control elements of space-based imagery.

7. Frank Gervasi, *Big Government: The Meaning and Purpose of the Hoover Commission Report* (New York: Whittlesey House, 1949), p. 279.

Second, the development of reconnaissance satellites also contributed to changing the balance of power among the members of the intelligence community by directing spending toward some agencies and away from others. Current estimates are that by 9/11, 85 percent of the intelligence budget lay beyond the control of the CIA. The overwhelming portion of this money went to DOD intelligence agencies, most notably NSA and NRO. The inability to control intelligence budgets beyond the CIA, and especially those in the DOD, became a constant issue in studies of intelligence reorganization and a point of debate in the creation of the Director of National Intelligence position. The 9/11 Commission's proposal to create a Director of National Intelligence gave this individual significant budgetary powers over all intelligence community funds. As passed, the legislation accepted the view put forward by the DOD that this power should be limited.

Third, reconnaissance satellites contributed to the development of collection "silos." Under ideal conditions, the relationship between analysts and collectors is one where analysts identify intelligence needs and collectors translate those needs into specific targets. Instead, a system has developed that is driven by collectors and the technology they control. Additionally, the information gathered by these collection systems, more often than not, is treated in a proprietary fashion. Its distribution is controlled and limited. As a consequence, intelligence from different collection sources tends not to merge together in a constructive fashion so that analysts can provide policy makers with answers their questions; instead, it comes forward in competing streams from different collection silos.

Even in their early stages, the managerial impact of these changes was recognized. In 1971 the Schlesinger Report, an inquiry into the operation of the intelligence community, began by stating what it saw as two disturbing trends in the operation of the intelligence community. The first was the "impressive" rise in cost and size. The second was the inability to translate those two features into improved intelligence products. Among the factors it cited as responsible for this state of affairs were competition between collection units that has led to unproductive duplication and unplanned growth, which has led to a series of compromise solutions. It concluded that the main hope for realizing any such improvement lay in a "fundamental reform of the intelligence community's decision making bodies and procedures." What was needed were "governing institutions."[8]

VIEW OF INTELLIGENCE

Reconnaissance satellites fit uneasily into the ongoing thinking about the role of intelligence in the national security process. This relationship was anchored in two guiding assumptions. First, the purpose of intelligence was to prevent strategic

8. The Schlesinger Report, "A Review of the Intelligence Community," can be found at George Washington University's National Security Archive Web site, *http://www.gwu.edu/~nsarchiv/NSAEBB/NSAEBB144/document%204.pdf* (accessed 15 February 2006).

surprise. The development of reconnaissance satellites fit comfortably here. Second, the collection of intelligence was thought about primarily in human terms: covert action and espionage. Reconnaissance satellites ran counter to this assumption and would ultimately undermine this tendency to equate intelligence collection with human intelligence. In the process it would lend an aura of legitimacy to espionage that had never existed.

Reconnaissance satellites, along with their predecessor, the U-2 reconnaissance aircraft, were quick to demonstrate their value as instruments for preventing strategic surprise. Beginning in the mid 1950s, political forces within the U.S. intelligence community (led by the Air Force) raised the specter of a bomber gap in which the Soviet Union held a decided and threatening lead over the United States in the development of a large strategic bomber force, creating an American vulnerability to a surprise attack. U-2 overflights in 1956 provided visual evidence that this gap did not exist. Satellite reconnaissance photographs would do the same just a few years later when they provided visual evidence that led to a repudiation of the charge that a missile gap now existed.

The change in emphasis from human to technological intelligence collection can be traced both to failures of the former and successes in the field of photographic reconnaissance. The late 1940s and 1950s were the heyday of covert action against procommunist regimes around the world and efforts to place agents inside the Soviet Union and behind the iron curtain. The failed 1961 Bay of Pigs invasion marked the end of that period, calling into question the credibility of the CIA and its top leadership in the area of covert action. Eighteen months later, the Cuban missile crisis cast doubt upon relying on human intelligence to prevent strategic surprise. U-2 photographs provided the conclusive proof needed by the Kennedy administration, confirming that Soviet missiles were being installed in Cuba. Classical human espionage in Cuba had been unable to provide such intelligence.

There is, however, one aspect of intelligence that the increased prominence of information (especially photographic images) gathered from U-2 overflights or reconnaissance satellites did not change. It did not provide a silver bullet that ended policy debates over how to interpret intelligence. Army and Navy intelligence, along with the CIA, saw in early U-2 photographs evidence that future Soviet ICBM launching sites would resemble the testing site at Tyuratam, Kazakhstan. The Air Force disagreed and argued that no particular configuration could be assumed.[9] In fact, the arrival of the U-2 photographs may have accentuated the problem of interpretation. Photographs were compelling and easily understood by policy makers. At the same time, they were too compelling and lent themselves to self-deception and wishful thinking.[10]

---

9. Lawrence Freedman, *U.S. Intelligence and the Soviet Strategic Threat*, 2nd ed. (Princeton: Princeton University Press, 1986), p. 71.

10. Mark Lowenthal, *Intelligence from Secrets to Policy*, 2nd ed. (Washington, DC: CQ Press, 2003), p. 64.

EMERGENCE OF A NEW POLICY AREA

The success of first aerial reconnaissance and then satellite reconnaissance in determining Soviet weapons capabilities helped to usher in a shift to thinking about the fundamental purpose of intelligence and national security policy more broadly. Although a concern for preventing a surprise attack never totally disappeared, it was now joined by a concern for developing a framework for managing U.S.–Soviet superpower relations.

Historically, attempts at reducing international tensions were predicated on two assumptions and frustrated by one overriding concern. First, military cooperation among enemies had to be preceded by some form of political accommodation. Second, such cooperation was negotiated into existence by a treaty or similar international agreement. Even when these were realized, a reduction in tensions could be frustrated by the fear of cheating.

By the late 1950s and early 1960s, the cold war had become recognized by all as an international fact of life. No abatement was in sight. Efforts to formally negotiate cooperation, such as the Open Skies proposal, had not met with success. The development of huge nuclear inventories also made it clear that neither side in this struggle could hope for a military triumph over the other at anything except a tremendous cost. The Cuban missile crisis reminded policy makers and the public that conflicts between the two superpowers were not a thing of the past and that they held the real potential for leading to war. Together, large nuclear inventories and the danger of accidental war made it increasingly clear to policy makers in both countries that even though they were enemies they had an interest in reducing tensions.

Steps such as the hot line linking Moscow and Washington were post–Cuban missile crisis moves in the direction of seeking to have a more peaceful and stable relationship without a formal treaty of any kind. This was followed later by interest in negotiating a reduction in the number of nuclear weapons through the Strategic Arms Limitations Talks (SALT I). Still, the problem of cheating remained and it was accentuated by the recognition that the United States and Soviet Union remained enemies. Where political considerations made on-site inspections impossible, reconnaissance satellites offered a more reliable and politically acceptable method for ensuring that each side lived up to the SALT I agreement. They did not infringe on state sovereignty in a traditional sense because the principle of nonterritorial spaceflights had been established in 1955, when the Eisenhower administration and the Soviet Union both announced plans for launching of International Geophysical Year satellites. Reconnaissance satellites also operated unilaterally; they did not require the formal cooperation of other states.

Their ability to verify behavior and act as a stabilizing force in world politics was dependent, however, on three conditions being met. First, both sides had to possess this satellite capability. The first Soviet reconnaissance satellite was launched in April 1962 and it appears that the Soviet Union reached this capability in 1963 when Khrushchev began to publicly refer to such a capability. Second, both the United States and Soviet Union had to agree not to try and shield information from the reconnaissance satellites. This was accomplished in SALT I with the agreement on noninterference with national technical means of verification. Third, neither side could have a serious anti-satellite capability.[11] The U.S. moved its policy in this direction in late 1962 when the DOD "reoriented" or canceled the Air Force's Satellite Interception Program. The Soviet Union ceased its anti-satellite testing in 1971, on the verge of the SALT I treaty.

As we asserted at the outset, policy arenas are constantly evolving and changing as they move forward in time. So it was with conflict management. The ability of reconnaissance satellites to perform their verification function was dependent upon more than technology. As the political foundations of détente began to crumble in the late 1970s, unilateral actions on the part of the Soviet Union and United States began to undermine this verification function. Talk of winning nuclear wars appeared in official pronouncements; definitions and standards of verification were now openly debated; and both sides moved once again to test and develop anti-satellite capabilities. Currently the development of a strategy of preemption threatens to reduce the stabilizing influence of space reconnaissance satellites by making them early targets for military action.

A spinoff from the employing space reconnaissance satellites as a key element in the development of a conflict management framework for stabilizing U.S.–Soviet relations was the need to obtain ground stations to receive the information they collected. Here again, reconnaissance satellites did not create a new national security issue area as much as they added a new element with its own unique dynamics into an ongoing policy stream. The early 1950s saw the United States establish ground stations to support the gathering of electronic intelligence and to monitor Soviet nuclear tests from Great Britain and Norway. More politically sensitive ground stations were set up in Turkey and Iran.

The 1966 decision to rely upon satellites in geostationary orbit to detect Soviet missile launches required the construction of ground stations outside of the United States. The most advantageous site for such a station was Australia. Two would be constructed, one at Nurrangar and the other at Pine Gap. Over the next few decades this decision became the focal point of conflict within Australia and consequently an occasional issue in U.S.–Australian relations.[12] One point of contention was the potential such a station created for making Australia the target of a nuclear attack. This issue was raised in Australia's Parliament by the opposition Labor Party. Another

11. John Gaddis, "The Evolution of a Satellite Reconnaissance Regime," in *U.S.-Soviet Security Cooperation: Achievements, Failures, Lessons*, Alexander George et al., eds. (New York: Oxford University Press, 1988), pp. 353–374.

12. Jeffrey Richelson, *America's Space Sentinels: DSP Satellites and National Security* (Lawrence, KS: University of Kansas Press, 1999), pp. 137–156.

issue (and one that resonated well in Australia) was the secrecy surrounding the project. Gough Whitlam, the head of the Labor Party, raised this point in Parliament, asserting that while it was right for Australia to cooperate with the United States, it was wrong for the Australian government to withhold information on the project from Parliament and the public. Relations between the United States and Australia became particularly touchy during and immediately after Whitlam's short-lived Labor government of 1973–1975. Whitlam had made it known that he wanted to review the future of American bases on Australian territory. In 1975, with his government locked in a budget crisis, Governor-General Sir John Kerr removed Whitlam from office and replaced him with Conservative leader Malcolm Fraser. Left-wing forces in Australia asserted that Whitlam's removal from office was a CIA-engineered coup.

The Post–Cold War Era

The impact of reconnaissance satellites on intelligence and national security policy did not end with the passing of the cold war. Instead, the policy stream in which reconnaissance satellites now operate has altered course. In most cases the changes now evident were present as ripples in the latter part of the cold war, and subsequently have gained in strength. As was the case with the cold war national security policy stream, we can expect reconnaissance satellites and the content of this policy to affect one another. Several indicators already point in directions where this interactive effect is likely to be most pronounced over time.

One notable and already evident area of impact on national security policy is the increased use of reconnaissance satellite imagery for tactical military purposes. Satellites had provided support for military operations on a limited scale prior to the end of the cold war, in the 1986 bombing campaign against Libya, and Operation Just Cause in 1989. A quantum leap in the reliance on satellites took place in 1991 with the Persian Gulf War. Satellite intelligence was used to provide warning of SCUD attacks, target Patriot anti-tactical ballistic missile rockets, provide weather data, aid with land navigation and aerial bombardment, and serve as a communication channel. The war against terrorism also has seen a heavy reliance upon satellite imagery and electronic intelligence in efforts to trace the movements of key terrorist leaders and identify targets.

This changed role for reconnaissance satellites (away from strategic intelligence to tactical intelligence) has brought with it the necessity for adjustments by both the providers of this intelligence and its recipients. In the 1991 Persian Gulf War, postmortems noted that distribution of intelligence was a significant problem and that some senior commanders were unfamiliar with the capabilities and limitations of U.S. intelligence systems.[13]

13. House Committee on Armed Services, *Intelligence Successes and Failures in Operations Desert Shield/Storm*, Committee Print 5, 16 August 1993.

A second emerging change is that the United States and Russia no longer have a monopoly over satellite reconnaissance. Two sets of competitors with different interests have emerged. One group is made of states that have begun to pursue a reconnaissance satellite capability for national security reasons.[14] Foremost among them is Israel, which in the past chafed at the inability or unwillingness of the United States to provide it with satellite images. It launched its first reconnaissance satellite in April 1995. Japan has also acted on regional security concerns to launch reconnaissance satellites; its first launch was in March 2003. Germany, Italy, Spain, India, and Pakistan are also moving in this direction. Down the road, the ability of other states to use satellite reconnaissance for self-defense or offensive maneuvers reduces the ability of the United States to use reconnaissance satellites as a conflict management tool.

Also emerging as competitors are commercial reconnaissance or observation satellites.[15] The first U.S. commercial observation satellite was launched in 1972. France followed in 1986, as did the Soviet Union one year later. The increased technological sophistication of commercial reconnaissance or observation satellites effectively makes them dual-use systems with commercial and military applications. They have the potential for revealing sensitive information that states would otherwise like to keep secret, such as when SPOT and Landsat photos showed the level of devastation at Chernobyl. They also have the ability to allow states to challenge the United States' interpretation of events. This occurred in 1996 when France did not support a cruise missile strike on Iraq because imagery at its disposal did not show a significant Iraqi troop movement into Kurdish areas. Commercial observation satellites may also serve as auxiliary or adjunct intelligence services for states. This occurred in the war against terrorism in Afghanistan when NIMA bought exclusive and perpetual rights to all imagery taken of Afghanistan by the Ikonos satellite.[16]

A third emerging trend is a broadened demand for the services of reconnaissance satellites that transcend the traditional dividing line between national security and non-national security policy issues. These include monitoring of environmental conditions such as droughts and natural disasters, and narcotics trafficking and potential terrorist targets. NIMA was tasked to provide support to the 2002 Winter Olympics in Utah and the 2004 Summer Olympics in Greece. The National Geospatial Intelligence Agency supported Hurricane Katrina relief efforts by providing information to the Federal Emergency Management Agency on affected areas from U.S. government satellites, commercial satellites, and airborne reconnaissance platforms.

14. Jeffrey Richelson, "The Whole World is Watching," *The Bulletin of the Atomic Scientists* 62 (January/February 2006): pp. 26–35.

15. Michael Krepon et al., eds., *Commercial Observation Satellites and International Security* (New York: Carnegie Endowment for International Peace, 1990).

16. Lowenthal, *Intelligence from Secrets to Policy,* p. 68.

In looking to the future impact of reconnaissance satellites on national security policy, the key questions may not be technological in nature as much as they will be contextual. Three features related to intelligence are of particular importance. First, during the cold war the intelligence challenge largely was defined in terms of unearthing secrets. Today, the challenge is more to unravel mysteries.[17] Searching for secrets involves searching for something that is knowable but being denied to you. Mysteries are open-ended and evolving. Reconnaissance satellites proved their worth in revealing secrets; solving mysteries may be a different matter. Photographs could not establish whether Saddam Hussein had weapons of mass destruction or what his intentions were. Years of electronic and communication intercepts have not clarified the nature of the global war on terrorism.

Second, a strong current of reform in intelligence circles pertains to the need to integrate open-source intelligence into both collection and analytic processes.[18] During the cold war, secret intelligence gathered by reconnaissance satellites was combined with secret intelligence collected by human sources to provide the basis for intelligence analysis. The possibility now exists that secret information gathered by reconnaissance satellites and other sources will be combined with information from public sources to produce intelligence products. One consequence may be a reorientation among intelligence agencies and a shift in the balance of power among them.

Finally, should terrorism continue to provide the main context for American national security policy, a readjustment in the place of reconnaissance satellites in the strategy to fight it may come about. Conceptualizing the conflict with terrorism as a war works to place military action at the core of an anti-terrorism strategy. It also favors reconnaissance satellites over other means of intelligence collection for bureaucratic and historical reasons. Should the conflict with terrorism come to be viewed in a criminal justice context, then intelligence from reconnaissance satellites is not so favored. Instead, policing strategies will be most heavily relied upon, and intelligence gathered by reconnaissance satellites will be read by analysts and consumers with different notions about how to use intelligence and what type of intelligence is most valuable.

CONCLUSION

Reconnaissance satellites have contributed in a number of ways to the changing face of intelligence within the national security policy arena. Their influence has been considered not so much as an isolated variable forcing change, but as one force

17. Gregory Treverton, *Reshaping National Intelligence in an Age of Information* (New York: Cambridge University Press, 2001).

18. Jennifer Sims and Barton Gerber, eds. *Transforming U.S. Intelligence* (Washington, DC: Georgetown University Press, 2005), pp. 63–78.

of many. This is because the national security policy arena into which reconnaissance satellites entered already existed as a stream of activity. Reconnaissance satellites entered this stream and helped change it. The impact of reconnaissance satellites on intelligence and national security policy does not end because the cold war is over. They will continue to shape intelligence and national security policy as this policy arena moves further downstream.

In the cold war period, two particular areas of impact were, first, on the changing fortunes of the CIA within the intelligence community, and second, on the development of a framework for managing superpower cold war relations. From the outset, the CIA faced challenges in establishing a position of leadership within the intelligence community. The advent of reconnaissance satellites, combined with the CIA's own failings in the areas of covert action and human espionage, helped bring into existence an intelligence community whose key organizational players lay beyond its effective control and whose key intelligence collection methodologies were rooted in science and technology. The resulting situation proved to be a mixed blessing. On one hand, reconnaissance satellites produced unprecedented insight into the national security policies of the Soviet Union. On the other hand, collection silos arose, human intelligence capabilities declined, costs rose dramatically, and managerial problems festered.

Reconnaissance satellites also helped usher in an era of conflict management between the United States and Soviet Union. They were instrumental in transforming an area of competition into one of conflict management by providing each side with a largely unilateral means of verifying the behavior of the other. Students of international relations have long commented that it is the absence of trust and the fear of cheating that makes cooperation so difficult in world politics. Reconnaissance satellites showed that, with proper motivation, technology can provide a mechanism allowing states to cooperate in the absence of trust. The changed atmosphere of the cold war during the Reagan administration also showed the limits of technology as a proxy for trust.

Section V

Social Impact

Chapter 20

Space History from the Bottom Up: Using Social History To Interpret the Societal Impact of Spaceflight

Glen Asner

The methods and concerns of social history, or what is often referred to as The "New Social History,"[1] came to dominate the study of U.S. history during the 1960s and 1970s. To this day, practitioners in the field continue to shape the history curricula at major universities and the content of leading professional journals. Despite the popularity of social history, it has had almost no influence on the practice of space history. Social historians have ignored spaceflight as a topic of study and space historians have looked infrequently to social history for ideas and inspiration. Perhaps the reason for this divide rests on how each field addresses a fundamental methodological question, namely: What activities, events, and individuals are worthy of inclusion in the historical record? Whereas social historians see matters of historical importance in the lives and experiences of ordinary people, space historians, following a tradition established by nineteenth-century Scottish historian Thomas Carlyle, tend to look to the other end of the social ladder for material and inspiration—to the actions and accomplishments of cultural elites.[2]

Although ordinary people remain bit players in space history, the field has not been a staid backwater of historical analysis. With the emergence in the past two decades of what former NASA chief historian Roger Launius has labeled the

1. The description of the field as "new" is to distinguish it from an older body of social history, often derided as "pots and pans history," that chronicled domestic practices and local cultural life without paying attention to larger socioeconomic issues, such as the influence of race and gender on the organization of work and social life. Alice Kessler-Harris, "Social History," in *The New American History,* Eric Foner, ed. (Philadelphia: Temple University Press, 1990), pp. 163–165.

2. For Carlyle's 1840 lectures on "great man" history, see Thomas Carlyle, with introduction by Michael K. Goldberg, *On Heroes and Hero Worship and the Heroic in History* (Berkeley, CA: University of California Press, 1993).

"New Aerospace History,"[3] space historians have moved away from hagiographic approaches that uncritically relate tales of inspirational individuals, revolutionary technologies, and momentous political decisions. The New Aerospace History includes an impressive range of topics, from studies that explore the role of spaceflight in American culture to those that illuminate the broader political, diplomatic, and cultural context in which political leaders have made major decisions regarding federal space policy. The field, nonetheless, remains focused on elites and artifacts.[4]

The central question of this conference—the impact of spaceflight on society—is itself an indication of the gulf separating space history from both social history and mainstream historical practice. The concept of societal impact is problematic to the extent that it is based on an assumption that the influence of spaceflight on society is more worthy of analysis than other conceptualizations of the relationship, such as the influence of society on spaceflight or the mutual shaping of spaceflight and society. By framing the conference with a unidirectional model of historical change (from government to society; from technology to social change), the conference organizers (among whom I was one) unwittingly encouraged presenters to focus on the production end of spaceflight (the accomplishments and products of government space activities) and to ignore the individuals and social groups who ostensibly were "impacted" by space activities. As presented in the program for this conference, society appears as a monolithic, homogenous blob that reacts as a singular entity to new capabilities and ideas generated from federal space activities. Society does not

3. Roger D. Launius, "The Historical Dimension of Space Exploration: Reflections and Possibilities," *Space Policy* 16 (2000): pp. 23–38. The ideas in this paper fit squarely in the realm of the New Aerospace History. Launius, as well as Margaret A. Weitekamp and Asif A. Siddiqi, have noted the absence of social history ideas in the history of spaceflight and have suggested interesting avenues for filling the gap in the literature. This paper differs from those of Siddiqi, Launius, and Weitekamp primarily in that it focuses almost exclusively on social history and extends the discussion on avenues for future research. Margaret A. Weitekamp, "Critical Theory as a Toolbox: Suggestions for Space History's Relationship to the History Subdisciplines," in *Critical Issues in the History of Spaceflight,* Steven J. Dick and Roger D. Launius, eds. (Washington, DC: NASA SP-2006-4702, 2006), pp. 549–572; and Asif A. Siddiqi, "American Space History: Legacies, Questions, and Opportunities for Future Research," in Dick and Launius, eds., *Critical Issues*, pp. 433–480.

4. Historians who consider the role of spaceflight in popular culture—a topic of recent and growing interest—tend to focus more on the spectacle of space technology and its presentation (in museums, movies, books, television, etc.) than on the spectators. The emphasis on ideas and meaning is welcome, but to the extent that such works substitute the perspectives of spaceflight elites and the culture industry for broader popular sentiments, they serve as a crutch that allows yet another generation of historians to avoid the difficult work of explaining how spaceflight has influenced the material conditions of everyday life. For examples, see the essays in the culture and ideology sections of this volume.

act but is acted upon.[5] In this context, conference presenters made strong assertions about the influence of space on society with very little, if any, evidence about changes in the lives of ordinary people to support those assertions.

By bringing ordinary people and social groups into our analysis, we can avoid reifying the concept of society and relegating masses of people to passive subjects of historical forces. To assess the role of spaceflight in society, we need to know the extent to which such activities altered work patterns, social practices, value systems, and how people relate to one another. We also need to know whether and how ordinary people resisted, reshaped, and adapted to changes brought on or supported in some way by space-related activities.

This paper suggests possibilities for examining the history of spaceflight through the lens of social history, focusing primarily on the relationship between the national space program and conceptions of status, race, and gender in the context of work, the community, and education. The most basic step for incorporating the concerns of social history into the history of spaceflight is to recognize as viable subjects for historical analysis all individuals and social groups involved in space endeavors regardless of their social standing.[6] Adopting a social history perspective also requires dispensing with the notion of one-way impacts and instead understanding the space program as a site of human activity in which individuals, social networks, communities, and institutions are all participants in the process of change and in the attribution of meaning to historical changes. A third concept to borrow from social history is the notion that individual and group identities—how we see ourselves

5. Several schools of thought within the historical profession, from both social history and the history of technology, call into question the conceptual foundation of this conference. To historians of technology, for example, the conference theme may appear to have been conceived without consideration of longstanding discussions about technological determinism, the social shaping of technology, and the uneven benefits of new technologies. For a brief introduction to these streams of research, see Donald MacKenzie, *Knowing Machines: Essays on Technical Change* (Cambridge, MA: MIT Press, 1996); Ruth Schwartz Cowan, *More Work for Mother: The Ironies of Household Technology from the Open Hearth to the Microwave* (New York: Basic Books, 1983); Merritt Roe Smith and Leo Marx, eds., *Does Technology Drive History? The Dilemma of Technological Determinism* (Cambridge, MA: MIT Press, 1994); and Wiebe E. Bijker, Thomas P. Hughes, and Trevor J. Pinch, eds., *The Social Construction of Technological Systems: New Directions in the Sociology and History of Technology* (Cambridge, MA: MIT Press, 1987). Given that historical changes are typically the result of multiple forces, attempting to single out the contributions of spaceflight/NASA to specific changes (whether they are technological, social, political, or economic) may be an exercise in futility. A more coherent and balanced approach would be to reverse the search process—to start with specific changes and then attempt to identify contributing factors, focusing on the relative significance of spaceflight in comparison to the other factors.

6. Peter N. Stearns, "Social History Present and Future," *Journal of Social History* 37/1 (Fall 2003): pp. 9–19.

and how others see us—are social creations that vary across time and place.[7] With these basic concepts and the literature of social history as a foundation, this paper raises questions for further exploration regarding the role of the space program in creating the conditions for social and economic mobility and providing a forum for the negotiation of social norms and conceptions of race, gender, and status for the millions of Americans who benefited from NASA's educational programs, lived in close proximity to NASA Centers, or worked directly for the Agency or one of its contractors over the past 50 years.

WHAT IS SOCIAL HISTORY AND HOW MIGHT IT CONTRIBUTE TO SPACE HISTORY?

Social history was a product of the 1960s, but not the 1960s the space community knows as the Apollo Era. Tensions over civil rights, gender relations, and an unpopular war in Vietnam inspired historians that entered the profession in the tumultuous decade to focus attention on groups that previously had been on the sidelines of American history, including women, the poor and working classes, and ethnic and racial minorities. Studies of migration, social movements, women, labor, and slavery dominated the field in these early years. As social history expanded, its practitioners borrowed methodologies from other academic disciplines. One group turned to quantitative methods to interpret changes in the material conditions of daily life, while others laid a foundation for cultural history by employing ethnographic techniques to explore the belief systems, rituals, language, and symbolic behavior of discrete communities. The field also broadened its topical scope and provided new insights on a variety of themes in American and world history, including immigration, social movements, demographic changes, education, leisure, ritual, social networks, sexuality, globalization, and consumerism.[8]

Social history has also influenced or bolstered similar trends in other history subdisciplines, including military history, urban history, the history of technology, Holocaust studies, and environmental history. Social history's greatest contribution to the study of history has been its challenge to the assumption that formal politics, wars, and great men alone shape history and are the only subjects worthy of study. From the start, social history was based on the premise that everyday people are worthy subjects of analysis, not for sentimental reasons but because their actions, behaviors, ideas, and experiences contribute to historical change and offer a broader perspective on the past.

---

7. Barbara Jeanne Fields, "Ideology and Race in American History," in *Region, Race, and Reconstruction: Essays in Honor of C. Vann Woodward*, J. Morgan Kousser and James M. McPherson, eds. (New York: Oxford University Press, 1982); Joan W. Scott, "Gender: A Useful Category of Historical Analysis," *American Historical Review* 91/5 (December 1986): pp. 1053–1075; David R. Roediger, *The Wages of Whiteness: Race and the Making of the American Working Class* (New York: Verso, 1991); and Roger Horowitz, ed., *Boys and Their Toys? Masculinity, Class, and Technology in America* (New York: Routledge, 2001).

8. For a brief overview of the field, see Kessler-Harris, "Social History," pp. 163–184.

The willingness of social historians to experiment with a wide range of methodologies has complemented the field's topical diversity. Over the years, social historians have borrowed freely from other academic disciplines, including anthropology, economics, sociology, psychology, philosophy, and literary studies. Grand narratives of societal change sit comfortably alongside quantitative studies of slavery's effects and thick descriptions of local cultures on the shelves of social historians. While this interdisciplinarity has at times factionalized the field, it is indicative of a pluralist ethos that puts resolving questions about historical change over narrow methodological commitments.[9]

The two greatest weaknesses of social history have been the tendency of some writers to exaggerate the influence of the oppressed on historical change and the unwillingness of a large portion of the field to grapple with politics and political institutions.[10] The tendency to portray minority groups as more influential in social and political change came in response to earlier histories that completely ignored the contributions of such social groups or portrayed them merely as victims. Although individual authors may still exhibit a bias one way or the other, the field as a whole has come to recognize the importance of capturing the contributions of all relevant social actors to political and social change.[11] Despite a widely held assumption that social history is allergic to politics, furthermore, the earliest practitioners in the field never intended to abandon formal politics and political institutions completely. Some topics in social history have proven resistant to the incorporation of discussion of politics, but many of the most influential studies have explained changes in the lives of ordinary people within the context of larger political and socioeconomic trends. The best scholarship of the past 20 years, while not always identified explicitly as social history, has explored societal transformations through grand narratives that integrate politics, economics, demographics, and the experiences of ordinary people.[12]

---

9. Stearns, "Social History Present and Future," pp. 9–12.

10. Ibid., p. 13; Jürgen Kocka, "Losses, Gains and Opportunities: Social History Today," *Journal of Social History* 37/1 (Fall 2003): pp. 21–28; and Nicole Eustace, "When Fish Walk on Land: Social History in a Postmodern World," *Journal of Social History* 37/1 (Fall 2003): pp. 77–91.

11. Social history works that explore groups other than the disenfranchised include Nancy MacLean, *Behind the Mask of Chivalry: The Making of the Second Ku Klux Klan* (New York: Oxford University Press, 1994); James Oakes, *The Ruling Race: A History of American Slaveholders* (New York: Alfred A. Knopf, 1982); and Elaine Tyler May, *Homeward Bound: American Families in the Cold War Era* (New York: Basic Books, Inc., 1988). Among one of the most profound explorations of relationships between social groups is Eugene D. Genovese, *Roll, Jordan, Roll: The World the Slaves Made* (New York: Random House, 1972).

12. Pathbreaking works that explain changes in daily life within the context of larger political and socioeconomic trends include Alice Kessler-Harris, *Out to Work: A History of Wage-Earning Women in the United States* (New York: Oxford University Press, 1982); Kenneth T. Jackson, *Crabgrass Frontier: The Suburbanization of the United States* (New York: Oxford University Press, 1985); May, *Homeward Bound*; Lizabeth Cohen, *Making a New Deal: Industrial Workers in Chicago, 1919–1939* (Cambridge, U.K.: Cambridge University Press, 1990); William Cronon, *Nature's Metropolis, Chicago and the Great West* (New York: W. W. Norton and Company, 1991); and Stephen Kotkin, *Magnetic Mountain: Stalinism as a Civilization* (Berkeley, CA: University of California Press, 1995).

To reiterate, the best of social history is driven by the central concerns of historical practice: understanding causation, change, and continuity, and providing a basis for comparison of similarities and differences across cultures and societies. Social history is also malleable and expansive. Rather than a rigid commitment to a single methodology or theory, social history has been receptive to new approaches. Social history is also expansive in the sense that it attempts to understand the contributions of all relevant actors to the process of historical change, and its practitioners have always had a hunger for new topics (except those related to space history, of course). As with the New Aerospace History, furthermore, the best social history studies are situated within a larger historical and political context.

Rather than a side note in space history, the social history of NASA can be understood as a central element of U.S. history if we discuss it in the context of the federal government's expanded role in science, education, and technological development in the twentieth century, and with reference to the individuals, communities, and social groups that defined their own existence and their ambitions in the shadow of the space program. The mundane aspects of NASA's existence, including its relationship with employees, contractors, grantees, and the communities in which it has facilities, become more central to twentieth-century history when we recognize that they embody the tensions of a capitalist state attempting to retain its superiority over an anti-capitalist foe through technological advance.[13] Discussing NASA as an arm of an expanded cold war state, rather than as it has defined itself—by its civilian spaceflight mission—allows us to deal more coherently with the Agency's role in society.

With these ideas in mind, this essay will now suggest ways for incorporating the concerns of social history into the history of spaceflight in three contexts: the context of work at the NASA Centers and among space contractors; the context of NASA's educational programs; and the context of the relationship between the Centers and the communities in which they are located.

WORK

Although the topic of work has received a great deal of attention among social historians, studies that explore the labor process in the aerospace industry and the lives or aerospace workers are sparse. We have some sense of the perspectives of aerospace workers through Sylvia Fries's study of Apollo Era engineers, and Howard McCurdy's *Inside NASA* surveys NASA's changing "confederation of cultures" throughout the decades.[14] Some of the Center histories touch upon the social

13. On the political context of the space race, see Walter A. McDougall, *. . . the Heavens and the Earth: A Political History of the Space Age* (New York: Basic Books, 1985).

14. Howard E. McCurdy, *Inside NASA: High Technology and Organizational Change in the U.S. Space Program* (Baltimore, MD: Johns Hopkins University Press, 1993); and Sylvia Fries, *NASA Engineers and the Age of Apollo* (Washington, DC: NASA SP 4104, 1992).

lives of NASA workers and episodes of conflict between workers and managers,[15] but our overall historical knowledge of the experiences of government and space industry workers remains shallow. No significant works in the field of space history explore the daily lives and work environment of the construction workers who built the Centers; the engineers, machinists, and production workers who created the technologies of exploration; the workers involved in assuring safe and continuous space operations; or the individuals at the lowest rungs of the employment ladder.

David Onkst's dissertation, which focuses on the work lives of Grumman engineers, stands alone in dealing with questions of race, ethnicity, and workplace relations in the aerospace industry. It raises a few basic questions that deserve emphasis and should be explored in other settings. For example, what motivated workers involved in the space race? Were they motivated by national objectives, simple financial gain, personal pride, or a commitment to their peers? Onkst finds that the desire to do a good job and to attract positive recognition for themselves, their work unit, and their company, rather than patriotism or competition with the Soviet Union, served as the central motivation for Grumman aerospace engineers.[16]

From this starting point, we can ask a whole new set of questions. How did the workplace culture and the approach of managers influence the motivations of workers and how they perceived their contributions to the space race? Were relations between workers and managers paternalistic, fraternal, or conflict-ridden? Did employees and managers share a common perspective on the quality of workplace conditions and the performance of work units? A few Center histories discuss labor conflict at NASA, but mostly in the context of the Apollo era. We have little sense of how management–employee relations have played out over time and among different segments of the workplace.[17]

---

15. Mack R. Herring, *Way Station to Space: A History of the John C. Stennis Space Center* (Washington, DC: NASA SP-4310, 1997), pp. 56–62; and Andrew J. Dunar and Stephen P. Waring, *Power to Explore: A History of Marshall Space Flight Center, 1960–1990* (Washington, DC: , NASA SP-4313, 1999), pp. 142–178.

16. David H. Onkst, "The Triumph and the Decline of the 'Squares': Grumman Engineers and Workers in the Apollo Era," working paper for presentation to the Historical Seminar on Contemporary Science and Technology at the National Air and Space Museum, 15 February 2001, in author's possession.

17. Benson and Faherty's brief treatment of conflicts at Cape Canaveral in the 1950s and 1960s between the government and its employees, between craft workers and contractors, and between union and nonunion employees, suggests that a rich labor history of KSC workers remains to be told. Although President Kennedy took measures to prevent the rampant work stoppages that plagued government missile sites in the late 1950s and early 1960s from continuing in the 1960s and infecting the civilian space program, NASA's relations with the construction trade unions remained contentious throughout the building of KSC. Charles D. Benson and William Barnaby Faherty, *Moonport: A History of Apollo Launch Facilities and Operations* (Washington, DC: NASA SP-4204, 1978), pp. 299–308. See also William Barnaby Faherty, *Florida's Space Coast: The Impact of NASA on the Sunshine State* (Gainesville, FL: University Press of Florida, 2002).

Despite the weakness of unions in the space industry, they were not entirely absent. Grumman was somewhat unique in that it successfully prevented unionization through a combination of good compensation and paternalistic labor practices.[18] What techniques and incentives did NASA and other space industry employers use to reduce the attractiveness of unionization? How did unionized corners of the space program differ from nonunionized corners in terms of the conditions of work and the attitudes of workers and managers?

Did workers throughout NASA and the space industry share the sentiments of one Grumman employee who described the company as a "classless work environment?"[19] Benson and Faherty similarly described the atmosphere at the Kennedy Space Center (KSC) as devoid of "snobbishness."[20] What did "classless" mean in the high-tech workplace of the 1960s? Did differences exist across the Centers? Was the sense of classlessness greater at KSC than at the Jet Propulsion Lab (JPL), for example? My sense is that despite such claims, the notion of meritocracy was undergoing a transformation in the post–WWII period. Scientific and technical knowledge replaced pedigree and other forms of knowledge as markers of status.[21] In this context, how did such meritocratic ideals shape how workers viewed themselves and others? Was a "classless environment" fostered, intentionally or unconsciously, to mute very real differences in power and status? If so, why was a putatively egalitarian rather than a strictly hierarchical atmosphere seemingly more appropriate for the aerospace workplace?

The contention that the technologies and organization of work structure both the workplace and broader social relations has its roots in nineteenth-century philosophy and remains influential in some corners of the historical profession today. In the context of NASA and the space industry, a softer version of the argument is worth considering—that certain production regimes privilege certain types of social

18. Onkst, "The Triumph and the Decline," pp. 23–28.

19. Ibid., p. 18.

20. Benson and Faherty, *Moonport*, p. 316.

21. On the increased influence of the scientific community and its postwar efforts to strengthen the association between social privilege and scientific and technical knowledge, see Daniel J. Kevles, *The Physicists: The History of a Scientific Community in Modern America* (Cambridge, MA: Harvard University Press, 1971; reprint 1995); Patrick McGrath, *Scientists, Business, and the State, 1890–1960* (Chapel Hill, NC: University of North Carolina Press, 2002); H. L. Nieburg, *In the Name of Science* (Chicago: Quadrangle, 1966; revised edition 1970); and Nicholas Lemann, *The Big Test: The Secret History of the American Meritocracy* (New York: Farrar, Straus and Giroux, 2000).

relations and behaviors within the workplace.[22] How did the format of the space complex—batch production, heavy design and testing elements, heavily reliance on networks of contractors and subcontractors, heavy government oversight, etc.—influence relationships between managers and workers?[23] How did employees get around the imperfections and limitations of formalized management and reporting systems? What informal practices facilitated the completion of projects?

According to the historian of technology Eugene Ferguson, engineering education became based more on mathematical modeling in the 1950s and engineering schools stopped teaching design.[24] Did trained engineers from the best universities need to go through a process of re-education to learn how to use slide rules when they entered the aerospace workplace, or did the new methods of the engineering schools come to dominate? To what extent did different educational experiences generate conflicts between older and newer generations of engineers? Sylvia Fries's discussion of the reaction of Apollo engineers confirms both the dominance of computing over time and the concerns of older engineers with what might have been lost with the turn to computers.[25] In terms of what was lost, Ferguson suggests that the Space Shuttle *Challenger* explosion and the problems the Agency faced with the Hubble space telescope were products of the engineering profession's turn away from older

22. See, for example, Harry Braverman, *Labor and Monopoly Capital: The Degradation of Work in the Twentieth Century* (New York: Monthly Review Press, 1974); Allen J. Scott and Michael Storper, *Production, Work, Territory: The Geographical Anatomy of Industrial Capitalism* (Boston: Allen & Unwin, 1986); Nelson Lichtenstein, "Auto Worker Militancy and the Structure of Factory Life, 1937–1955," *Journal of American History* 67/2 (1980): pp. 335–353; Simon Marcson, ed., *Automation, Alienation, and Anomie* (New York: Harper & Row, 1970); and Shoshana Zuboff, *In the Age of the Smart Machine: The Future of Work and Power* (New York: Basic Books, 1988).

23. On the history of batch production and other flexible production formats, see Philip Scranton, *Endless Novelty: Specialty Production and American Industrialization, 1865–1925* (Princeton, NJ: Princeton University Press, 1997); Philip Scranton, *Proprietary Capitalism: The Textile Manufacture at Philadelphia, 1800–1885* (New York: Cambridge University Press, 1983); Philip Scranton, *Figured Tapestry: Production, Markets and Power in Philadelphia Textiles, 1855–1941* (New York: Cambridge University Press, 1989).

24. Eugene S. Ferguson, *Engineering and the Mind's Eye* (Cambridge, MA: The MIT Press, 1992), pp. 153–194.

25. Sylvia Fries, *NASA Engineers and the Age of Apollo* (Washington, DC: NASA SP 4104, 1992), pp.154–159.

practices, such as visualization and "bottom-up design."[26] Although determining the veracity of such claims should be left to technical experts, historians can contribute to such debates by explaining how changes in techniques and knowledge have shaped relations across generations of NASA engineers.

How did conceptions of status and hierarchy relate to the opportunities for economic mobility that existed in the space complex? Social historians have long been interested in documenting both the reality and the ideology of economic mobility in the United States.[27] To what extent did space work create a stepping-stone into the middle class? How important was economic mobility to workers in the space complex? For people who did move from lesser circumstances, what did their newfound status and wealth mean for them? What were the visible signs of their economic mobility? How did they relate to those who had come from better financial circumstances? Did the economically mobile retain ties to their friends and family who were not so fortunate? Finally, to what extent did conceptions of economic mobility, class, and social status change over time among space workers?[28]

Social historians and labor historians have also studied the lives of workers outside of the workplace.[29] Iconic images of risk-taking astronauts and "geeky"

26. Bottom-up design involves the design, testing, and modification of components before stabilizing the design of a technological system. In top-down design (also known as concurrency), all components of a system are designed at once and then tested as an integrated system, which makes identifying problematic components more difficult and fixing them more costly and time-consuming. Ferguson's critique of NASA was based, in part, on the ideas of Richard Feynman. Eugene S. Ferguson, *Engineering and the Mind's Eye* (Cambridge, MA: The MIT Press, 1992), pp. 186–189; and Richard P. Feynman, *What Do You Care What Other People Think?* (New York: W. W. Norton, 1988).

27. Stephan Thernstrom, *Poverty and Progress: Social Mobility in a Nineteenth Century City* (Cambridge, MA: Harvard University Press, 1964); and John Bodnar, *The Transplanted: A History of Immigrants in Urban America* (Bloomington, IN: Indiana University Press, 1985).

28. The theme of economic and social mobility resonates throughout the memoir of Homer Hickam, a former NASA aerospace engineer. The memoir, of course, formed the basis of the popular movie, *October Sky*. Homer Hickam, *Rocket Boys* (New York: Random House, 1998).

29. Historians first began to study leisure as an offshoot of labor history in the 1960s. From studies of working-class cultural traditions, the field expanded to include a diversity of topics, including the role of mass entertainment in the lives of immigrants, the dating and social rituals of working women, African American travel experiences in the Jim Crow era, and the relationship between sports and social identity. See, for example, E. P. Thompson, *The Making of the English Working Class* (New York: Pantheon, 1964); Keith Thomas, "Work and Leisure in Pre-industrial Society," *Past and Present*, no. 29 (December 1964): pp. 50–66; Roy Rosenzweig, *Eight Hours for What We Will: Workers and Leisure in an Industrial City, 1870–1920* (New York: Cambridge University Press, 1983); Kathy Peiss, *Cheap Amusements: Working Women and Leisure in Turn-of-the-Century New York* (Philadelphia: Temple University Press, 1986); Cohen, *Making a New Deal*; Eliot Gorn, *The Manly Art: Bare-Knuckle Prize Fighting in America* (Ithaca, NY: Cornell University Press, 1986); Gail Bederman, *Manliness and Civilization: A Cultural History of Gender and Race in the United States, 1880–1917* (Chicago: Chicago University Press, 1995); and Cotton Seiler, "So That We as a Race Might Have Something Authentic to Travel By: African American Automobility and Cold-War Liberalism," *American Quarterly* 58/4 (December 2006): pp. 1091–1117.

engineers notwithstanding,[30] we know little about the social lives of the men and women who built and operated the equipment that made spaceflight possible. Few historians who have studied the NASA Centers have provided insight into the social lives of NASA contractors and employees, perhaps considering it a pedestrian topic or fearing intruding too much into the personal lives of individual workers. How did space workers deal with the stresses of work?[31] Did managers and workers socialize? Was it common for engineers and scientists to intermingle? Did occupational categories factor into socializing? How important were bars or sports in social life? What role did formal and informal rituals play in work and social life? To what extent were sexually integrated or sexually segregated activities important to the social lives of workers? Were class, status, and gender differences muted or accentuated by such activities? How have formal NASA social events differed from informal gatherings among employees? What were the unarticulated rules of engaging in such activities? What were the implications of socializing for workplace camaraderie? Did such activities create a sense of common purposes and positive sentiments toward NASA? Did they serve as a safety valve? Did they help to homogenize social, political, or ethical values?

Historians have begun to explore questions about women and gender in the context of the space program. Yet, as Margaret Weitekamp notes in her excellent essay in the *Critical Issues* volume, much of this literature falls in the realm of compensatory history, adding the stories of women to the historical record rather than contextualizing their stories. Weitekamp's analysis and proscriptions for dealing with gender stand on their own as essential reading for anyone interested in the topic.[32] The point that the literature to date on women focuses almost entirely on

30. Roger D. Launius, "Heroes in a Vacuum: The Apollo Astronaut as Cultural Icon," AIAA 2005-702, paper delivered at the 43rd AIAA Aerospace Sciences Meeting and Exhibit, 10–13 January 2005, Reno, Nevada; and M. G. Lord, *Astro Turf: The Private Life of Rocket Science* (New York: Walker & Company, 2005).

31. Bowles provides a few pages on the topic of social outlets at Plum Brook, which included playing music, formal social functions, dances, athletics, games, and practical jokes, some of which were loaded with gendered assumptions. Mark D. Bowles, *Science in Flux: NASA's Nuclear Program at Plum Brook Station, 1955–2005* (Washington, DC: NASA SP-4317, 2006), pp. 188–192. Benson and Faherty discuss the stresses of male engineers at KSC during the Moon race. Although KSC employees enjoyed opportunities for recreation along the Banana River and the waters of the Atlantic Ocean, work life was so consuming that the stresses of work commonly spilled over into the home. The children of space workers, according to a Titusville physician, had unusually high rates of ulcers. Benson and Faherty, *Moonport*, pp. 314–316.

32. Weitekamp, "Critical Theory as a Toolbox," pp. 549–572.

women astronauts deserves emphasis.[33] With the exception of a couple of studies that deal with women engineers, women workers and gender conceptions in the context of work have been ignored almost completely. Sylvia Fries's study of engineers during the Apollo era includes mention of women engineers, as does Sheryll Goecke Powers's short monograph on *Women in Flight Research at Dryden*. From these two works we get a sense of the hurdles that women scientists and engineers faced, as well as the nature of their work and changing opportunities for women within NASA over time.[34]

M. G. Lord's memoir, *Astro Turf*, is exemplary in its treatment of historically situated gender constructions and could be applied to an analysis of masculinity at the Centers, as Weitekamp suggests. Following Lord's lead, historians could explore changes over time in gender conceptions, as well as differences across the Centers. Conceptions of gender, including topics such as manliness, domesticity, motherhood, and sexuality, deserve a great deal more discussion within the context of the workplace of the NASA Centers and the space industry.[35] But we must not allow the discussion of gender and gender conceptions to remain isolated to those at the top of the work pyramid. Women in nonprofessional roles, including secretaries and support staff, deserve to be explored as legitimate historical subjects, as do the men at the lowest rungs of the work hierarchy, such as entry-level technicians and janitors.[36]

The story of race is similar to that of gender in the sense that we have some sense of the experiences of black astronauts[37] and a few of the histories of NASA Centers located in the south mention racial issues, primarily in the context of the civil rights movement and from a broad, top-down perspective. Dunar and Waring's history of Marshall Space Flight Center and Steven Moss's excellent master's thesis on NASA's implementation of federal civil rights policy in the South indicate that the

---

33. Margaret A. Weitekamp, *Right Stuff, Wrong Sex: America's First Women in Space Program* (Baltimore, MD: Johns Hopkins University Press, 2004); Joseph D. Atkinson, Jr., and Jay M. Shafritz, *The Real Stuff: A History of NASA's Astronaut Recruitment Program* (New York: Praeger, 1985); and Stephanie Nolen, *Promised the Moon: The Untold Story of the First Women in the Space Race* (New York: Four Walls Eight Windows, 2002).

34. Sylvia Fries, *NASA Engineers and the Age of Apollo*; and Sheryll Goecke Powers, *Women in Flight Research at NASA Dryden Flight Research Center from 1946 to 1995*, Monograph in Aerospace History No. 6 (Washington, DC: National Aeronautics and Space Administration, 1997).

35. Weitekamp, "Critical Theory," p. 569; and Lord, *Astro Turf*.

36. For historical work dealing with gender in the workplace, see Horowitz, ed., *Boys and Their Toys?*; Ava Baron, ed., *Word Engendered: Toward a New History of American Labor* (Ithaca, NY: Cornell University Press, 1991); Arwen Mohun, *Steam Laundries: Gender, Technology, and Work in the United States and Great Britain, 1880–1940* (Baltimore, MD: John Hopkins University Press, 1999).

37. See, for example, Joseph D. Atkinson, Jr., and Jay M. Shafritz, *The Real Stuff: A History of NASA's Astronaut Recruitment Program* (New York: Praeger, 1985); J. Alfred Phelps, *They Had a Dream: The Story of African-American Astronauts* (Novato, CA: Presidio, 1994); and James Haskins and Kathleen Benson, *Space Challenger: The Story of Guion Bluford* (Minneapolis, MN: Carolrhoda Books, 1984).

Agency as a whole took a strong stance on equal rights—promoting desegregation among contractors, universities, and local businesses near the centers—but NASA fell short of its goals in recruiting black engineers and skilled workers.[38]

Without taking anything away from these studies, the literature to date falls short in several respects. First, it focuses exclusively on the South. The struggle for equality and improved work opportunities for African Americans was not confined to the South. Second, it tells little of the post-1960s story. The struggle to integrate the workplace did not end with the first Apollo lunar landing, nor did our habit of seeing ourselves and one another through a racial lens. Third, these studies leave the perspectives and actions of African American workers out of the story entirely. As David Onkst's excellent dissertation chapter on race relations at Grumman on Long Island shows, African American workers and civil rights groups that supported them engaged in a wide variety of confrontational tactics to secure promotions and better-paying jobs. African Americans were not merely passive recipients of the actions of well-intentioned employers.[39]

Onkst's dissertation points to numerous opportunities for exploring race throughout NASA and among its contractors. In addition to further exploration of conditions in the South, historians might wish to look at changes in job opportunities and conceptions of race across all of the Centers and beyond the 1960s. If the history is not readily apparent, we might wish to search for instances of conflict in lawsuits, complaints from workers filed with the Equal Employment Opportunity Commission, and Congressional hearings. Onkst points out the startling figure of 125,000 cases in the EEOC's case backlog in the 1970s.[40]

In addition to cases of conflict, we need to gain a better sense of the less contentious ways in which minorities created advancement opportunities for themselves and in which the Agency aided their advancement. We know little about how formal groups (i.e., Hispanics in Government, Blacks in Government) and informal social networks have operated to provide career advancement opportunities, a sense of community, and have aided the quest for full workplace equality.

---

38. Andrew J. Dunar and Stephen P. Waring, *Power to Explore: A History of Marshall Space Flight Center, 1960–1990* (Washington, DC: NASA SP-4313, 1999), pp. 115–135; and Steven L. Moss, "NASA and Racial Equality in the South, 1961–1968," M.S. thesis, Texas Tech University, 1997.

39. David H. Onkst, "David and Goliath? Race Relations at Grumman," working paper, 31 January 2005, in author's possession.

40. Ibid., p. 101.

Local Communities

Despite numerous studies of NASA Centers and some solid quantitative data on the local economic impact of NASA Centers,[41] our knowledge of the role of the Centers, their employees, and NASA contractors in those communities remains shallow.[42] Studies of the Centers focus primarily on activities within their gates. Some do not deal with the surrounding community at all,[43] while others deal with it only in a cursory fashion, focusing primarily on quantifiable demographic and economic changes.[44] One study discusses the NASA's relationship with the local community from the perspective of the Agency—as a problem to be managed rather than a subject to be explored.[45] Yet a few valiant attempts to explain the role of NASA Centers in generating and resolving local conflicts, and to characterize the contributions of NASA Centers to local communities, raise possibilities for writing the history of NASA from a local or multi-local perspective.

Despite the general lack of interest in understanding the local context of NASA Centers, at least two histories have explored specific conflicts between the NASA Centers and their communities. In *Science in Flux*, Mark Bowles skillfully examines the conflict between the government and local farmers over the World War II War Department's forcible acquisition of the land that would eventually become Plum Brook. Bowles also remains sensitive to the viewpoints of members

41. See, for example, Annie Mary Hartsfield, Mary Alice Griffin, and Charles M. Grigg, *NASA Impact on Brevard County* (Tallahassee, FL: Florida State University, 1966); and Richard J. Keegan, "Economic and Social Impact of the National Space Program," a report of the Stanford-Sloan Program, Graduate School of Business, Stanford University, 1971.

42. Studies that document the social, economic, and demographic patterns of geographically defined areas have been a staple of social history and other related fields, such as sociology and the history of technology, for decades. Works that deal with industrialization and social change provide a foundation for examining the role of NASA and the space community in the socioeconomic transformation of towns and regions. See, for example, Anthony F. C. Wallace, *Rockdale: The Growth of an American Village in the Early Industrial Revolution* (New York: Norton, 1978); Judith A. McGaw, *Most Wonderful Machine: Mechanization and Social Change in Berkshire Paper Making, 1801–1885* (Princeton: Princeton University Press, 1987); David F. Crew, *Town in the Ruhr: A Social History of Bochum, 1860–1914* (New York: Columbia University Press, 1979); John Demos, *A Little Commonwealth: Family Life in Plymouth Colony* (New York: Oxford University Press, 1970); and Philip J. Greven, Jr., *Four Generations: Population, Land, and Family in Colonial Andover, Massachusetts* (Ithaca, NY: Cornell University Press, 1970).

43. Alfred Rosenthal, *Venture into Space: Early Years of Goddard Space Flight Center* (Washington, DC: NASA SP-4301, 1968); Edwin P. Hartmann, *Adventures in Research: A History of Ames Research Center, 1940–1965* (Washington, DC: NASA SP-4302, 1970); and Clayton R. Koppes, *JPL and the American Space Program: A History of the Jet Propulsion Laboratory* (New Haven, CT: Yale University Press, 1982).

44. Benson and Faherty, *Moonport*, pp. 309–313; and Henry C. Dethloff, *Suddenly Tomorrow Came . . . A History of the Johnson Space Center* (Washington, DC: NASA SP-4307, 1993), pp. 257–271.

45. Mack R. Herring, *Way Station to Space: A History of the John C. Stennis Space Center* (Washington, DC: NASA SP-4310, 1997), pp. 27–137.

of the community surrounding Plum Brook who were concerned about living near a nuclear reactor in the 1960s and early 1970s. In *Power to Explore*, Dunar and Waring shed light on the relationship between the Marshall Space Flight Center and the city of Huntsville, Alabama, particularly in regard to civil rights, education, and the local economy. Despite their meticulous research and thoughtful presentation of what some might consider sensitive issues, Dunar and Waring provide only a top-level overview and leave readers with the sense that much more could be written on Marshall's place in the rise of Huntsville as one of the most important small cities in the United States.[46]

Elizabeth Muenger's brief description of the changing relationship between the Ames Research Center and Silicon Valley suggests the possibility of discussing change over time in the character of the relationship between NASA Centers and their communities rather than merely providing snapshots of controversies and collaboration or listing quantitative data on changing demographics and local economic conditions. Muenger explains the shift at Ames from isolation in relation to its surrounding community to stronger public relations after the formation of NASA. Under the National Advisory Committee for Aeronautics (NACA), Ames's leader, Smith J. De France, discouraged employees from interacting in a formal capacity with members of the surrounding community and denied the broader public the privilege to attend triennial open houses. However, in the 1960s Ames became more active in promoting public education through teacher education and prize competitions, providing employment opportunities for local youth, and assisting local, small businesses.[47]

One theme that comes through in some of these studies is an insider/outside divide between employees at the Centers and the local community. Although based on limited evidence, the concept of insider and outsider deserves further consideration. In reality, we know very little about how well NASA employees integrated into the communities they inhabited.[48] One would expect that they became regular members of the local community in every respect but, at least for the first waves of employees at newly built Centers after the formation of NASA, the question of initial contact is an important one. How did the entrance of highly educated outsiders into the community influence local social and political dynamics? How did local elites deal with sharing influence with newcomers, particularly newcomers who held meritocratic conceptions of status based on educational achievement? Did

46. Mark D. Bowles, *Science in Flux*, pp. 1–34; and Dunar and Waring, *Power to Explore*, pp. 116–130.

47. Elizabeth Muenger, *Searching the Horizon: A History of Ames Research Center, 1940–1976* (Washington, DC: NASA SP-4304, 1985), pp. 183–195.

48. On the movement of outsiders into new communities, see Kotkin, *Magnetic Mountain*; Bodnar, *The Transplanted*; and Clyde and Sally Griffen, *Natives and Newcomers: The Ordering of Opportunity in Mid-Nineteenth Century Poughkeepsie* (Cambridge, MA: Harvard University Press, 1978).

newcomers integrate into existing social institutions, such as churches and social clubs, or did they create their own? Did local residents warmly accept the outsiders, or did conflict exist at the local political level or interpersonal level? What was the basis of such conflicts? Class? Culture? Racial or gender attitudes? How were such conflicts muted? To what extent did outsiders adopt local attitudes? To what extent did they simply grin and bear existing social and racial relations?

Or did the Centers have a significant influence on local perspectives? Did individuals who worked at the Centers, both outsiders and locals, learn anything from working in an environment that promoted equal opportunity that they carried with them to the local community? Did, for example, work at the Centers provide a context for reconfiguring conceptions of race and gender? Did the presence of a large federal employer mute racial antipathy? Did increased social standing in the workplace, for example, translate into increased social power outside of the workplace for African Americans and women? Did NASA employees draw upon their equal status within a federal institution to reshape local institutions and conceptions of race?

Works that hail the local economic impact of NASA Centers have attempted to put a positive spin on the negative consequences of NASA's arrival in new locations. In his study of the Stennis Space Center, for example, Herring includes the perspectives of residents of Hancock County, Mississippi, who were forcibly removed from their properties to make way for the Center. Yet, he adopts the perspective of political leaders in framing the involuntary sacrifices of discontented residents as a necessary step in the path toward progress.[49] It would be nice to know how the arrival of NASA facilities altered local economic patterns in each community and what the consequences were for native residents. Who, in particular, benefited? Existing elites? The poor? The middle classes? Did the Centers in any sense facilitate the opening up of local communities to broader national or international trade or social networks? Did the local economy become dependent on government largesse or did the NASA Centers facilitate the growth of self-sustaining companies that generated wealth from commercial ventures?[50] To what extent did the children

49. Herring, *Way Station to Space*, pp. 27–137. Faherty's compact history on the impact of KSC on the Space Coast is well written, well conceived, and filled with excellent statistics and economic data, but it breezes too quickly over topics that would be of interest to social historians. Faherty, *Florida's Space Coast*.

50. In this regard, studies of the communities surrounding the NASA Centers should engage with the literature on high-tech regional development, the growth of western cities, and suburbanization. See, for example, Carl Abbott, *The Metropolitan Frontier: Cities in the Modern American West* (Tucson, AZ: University of Arizona Press, 1993); Annalee Saxenian, *Regional Advantage: Culture and Competition in Silicon Valley and Route 128* (Cambridge, MA: Harvard University Press, 1994); Richard Florida, *Cities and the Creative Class* (New York: Routledge, 2005); and Margaret Pugh O'Mara, *Cities of Knowledge: Cold War Science and the Search for the Next Silicon Valley* (Princeton, NJ: Princeton University Press, 2004).

of locals and newcomers, especially in previously isolated rural locations such as KSC and Stennis, follow similar life paths after graduating from high school? Did locals and outsiders come to share a common destiny and culture?

Education

NASA's role in the education of millions of Americans is a topic barely examined except in the context of scientific and engineering programs.[51] NASA has a long history of sponsoring education programs for elementary and high school teachers and students, providing grants to universities for research and training, and serving as a site of education through internships, temporary work positions, on-the-job training, and summer fellowships for university professors. A 1988 inventory of education programs listed 59 separate elementary and secondary education programs, 37 university programs, 30 minority outreach programs, 19 employment programs, and 17 public education programs—162 in total. While the combined programs reached millions of teachers and students, most programs supported a more limited number of individuals, often under 100, and a large number, at least at that time, were run out of the Field Centers.[52] The implications of such activities for economic mobility are significant.

Who benefited from these programs? Can we identify any common patterns in their personal backgrounds? To what extent has NASA served as a conduit for social and economic mobility? In the course of such educational programs, what values unrelated to science and engineering have NASA conveyed? To what extent, for example, have educational programs succeeded in indoctrinating students toward the goals of the space program? How have such values conflicted with or reinforced other values and goals, such as religious values or requirements for diversity? To what extent, furthermore, have educators altered the space curriculum to accommodate competing social values?

One of the early directors of NASA's Educational Programs Division claimed that many of the programs were established in response to external demands.[53] This statement raises interesting questions about whether NASA was pulled into

51. One important exception is Virginia Dawson's book on the Lewis Laboratory (now Glenn Research Center) in which she discusses the laboratory's role as a locus of knowledge production and education. Through their research programs, most of the NASA Centers have served an educational function. Virginia P. Dawson, *Engines and Innovation: Lewis Laboratory and American Propulsion Technology* (Washington, DC: National Aeronautics and Space Administration, 1991), pp. 65–123.

52. Education Affairs Division, Office of External Relations, "NASA and the Educational Community: An Inventory of Programs," August 1988 (Washington, DC: NASA, 1988).

53. Aaron P. Seamster, "The Nature of NASA's Educational Programs and Services Office, and its Interest in Science Education," in *Science Education in the Space Age* (Washington, DC: NASA, November 1964), Proceedings of a National Conference conducted at the NASA Western Operations Office, 1–4 June 1964, pp. 19–20.

supporting education for reasons apart from the Agency's mission, or whether internal needs, or even presumed future needs, were the primary impetus for establishing educational programs. While the Space Act contains provisions for the widest practical dissemination of information possible, and NASA has been active in promoting education, to what extent has demand for knowledge, educational materials, and educational opportunities stimulated the Agency's programs? Did the Agency ultimately benefit in cases in which the demand was largely without an identifiable NASA requirement or need? To what extent has the adoption of non-space-related responsibilities contributed to or detracted from the Agency's ability to fulfill its central mission?

Women, minorities, and economically disadvantaged youth have been a focus of many of the educational programs. The only historian that I am aware of who has ventured into this territory is Amy Slaton, who studied the history of one university, Prairie View College in Texas, which received research and graduate training funding through NASA's Historically Black Colleges and Universities program.[54] Beyond the practical benefits of such programs, both for economic mobility for participants and workforce training and recruitment for NASA, what messages did such programs send to society at large regarding the federal commitment to equality? In what sense did such programs generate loyalty to the Agency? To what extent were such programs inspired by or tailored to local needs and socioeconomic patterns? How have external factors, such as the civil rights and women's rights movements, impinged on both the focus and the content of NASA's educational programs?

CONCLUSION

By whatever name—the Space Age, the Atomic Age, the Information Age—the period from achievement of the first self-sustained nuclear chain reaction on 2 December 1942 to at least the end of the cold war in 1989 represented a cohesive era in American history. Although many in the space community believe that the Apollo 11 landing on the Moon was the single most significant event of the century, we must not forget that the technologies of space exploration—not just the space race itself—were products of the cold war. The post-atomic military imperative to quickly and efficiently deliver nuclear weapons to precise locations spurred the federal government to invest massive sums of money in nuclear technology, rocketry, computers, and materials research. While the specific technological systems that made the Mercury, Gemini, and Apollo missions possible represented spectacular

54. Amy Slaton, "NASA Funding for Historically Black Universities: Diversity in Scale and Scope," paper presented 28 April 2006 at the Fourth Laboratory History Conference at Green College, University of British Columbia, Vancouver, British Columbia.

engineering achievements, the likelihood that the space program would have existed at all without the discovery of the destructive potential of the atom or the military's cold war space requirements is low. Just as significant portions of the technologies of exploration were built upon a foundation of cold war military technologies, so, too, NASA was built largely upon a foundation of military organizations.

Rather than isolated entities, the NASA Centers represented the civilian tip of a vast, interrelated complex of facilities and bases that served the cold war national interest. Social and cultural differences undoubtedly separated the people who worked from the communities that surrounded military bases, intelligence agencies, Department of Energy production and research facilities, and the NASA Centers, as well as the networks or electronics, nuclear, defense, research and development, and aerospace contractors that supported the government programs. We have almost no basis, however, upon which to make generalizations (let alone comparisons) about the social lives and culture of the communities that were most directly shaped by the vast expansion and geographical dispersion of U.S. government activities.[55] Assumptions of cultural homogeneity should not prevent historians from studying military and space communities as unique cultural formations and considering whether the occupants of those communities represented or diverged from American cultural traditions and how they compare with other elements of American society.

Several generations from now, historians looking back at our time are likely to be just as interested in understanding how Americans of the Space/Atomic Age lived as they will be in understanding our technological achievements and changes in economic, political, and international dynamics. Robust social histories of the cold war and post–cold war may help to guard against the possibility that the cultural productions of the television and movie industry will substitute for actual knowledge of how Americans lived in this era.

55. Notwithstanding the excellent work of historians and geographers on regional transformations as a result of cold war imperatives and Elaine Tyler May's exegesis on cold war families. Abbott, *The Metropolitan Frontier*; Ann R. Markusen, Peter Hall, Sabina Deitrick, and Scott Campbell, *The Rise of the Gunbelt: The Military Remapping of Industrial America* (New York: Oxford University Press, 1991); Bruce J. Schulman, *From Cotton Belt to Sun Belt: Federal Policy, Economic Development, and the Transformation of the South, 1938–1980* (New York: Oxford University Press, 1991); and May, *Homeward Bound*.

Chapter 21

Space Science Education in the United States: The Good, the Bad, and the Ugly

Andrew Fraknoi

As an astronomer and astronomy educator, I do not have any original historical research to report for this volume. Nevertheless, my hope is that this brief summary of some issues in the uneasy relationship between space science and education in the United States may be of use in getting historians and general readers interested, concerned, and perhaps even upset. A much more detailed overview of the state of astronomy education can be found on the World Wide Web.[1]

The history and societal impact of space science education in the United States is difficult to sort out from the complexities of education in general. If it were an experiment, any scientist would throw up his hands in despair and say, "There are simply too many variables!" For example, what was more important in determining the attitudes of today's U.S. adults about the space program:

- The various science education reform curricula in the 1960s that grew partly in response to Sputnik, or the growth of television as a pervasive, pernicious, and pseudo-educational medium?
- The Shuttle accidents and subsequent revelations, or the disintegration of the U.S. family due to the rise of divorce rates and single-parent families?
- The success of the Mars Exploration Rovers, or the consistently low salaries we have paid science teachers when compared to other professions requiring similar training?

These are very hard questions to answer.

To prepare this essay, I wrote to about two dozen experts in the field of space science education to see if they could help me find the most important parts of the literature on the impact of spaceflight on education. With one exception, they all wrote back quickly and courteously—to inform me that they knew of no such literature but that they were looking forward to our conference starting one.

1. See Andrew Fraknoi, "Astronomy Education in the United States, 1998; available online at *http://www.astrosociety.org/education/resources/useduc.html* (accessed 20 October 2006).

Has the Space Age affected education in the United States? It is a romantic notion that it has. Everyone cites great anecdotes, memoirs, and TV shows such as *How William Shatner Changed the World*. But no one has been able to tease out the effect of the space program from the myriad other factors that affect our educational system. And, as educational researchers are fond of pointing out, "The plural of anecdote is not data!"

A recent paper in *Astronomy Education Review* reported that no research could be found that looked at teacher knowledge of astronomy across grade levels and multiple astronomy concepts.[2] In the absence of relevant data, what I would therefore like to do, in the brief space I have, is to get the reader thinking just a bit about a few issues in the relationship between space science and education.

A Brief History of Space Science Education in the United States

Astronomy was a subject taught in U.S. high schools toward the end of the nineteenth century because it fit with the *mental discipline model* of curriculum.[3] Like Latin and Greek, astronomy was supposed to be good training for student thinking, whether students needed it later in life or not. This approach soon began to unravel as the schools became less elite and more practical, and colleges demanded more practical knowledge from high school graduates.

Also, school curriculum under the mental discipline approach was hard, repetitive, and boring. In 1913, Helen Todd, whose job was to inspect factories, surveyed 500 children who labored there under difficult conditions. She asked if, economic circumstances permitting, they would not rather be back in school. Four hundred twelve out of 500 said no. They preferred sweatshops to what she called the "monotony, humiliation, and cruelty" of school.[4]

In 1893, a Committee of Ten, set up by the National Education Association, was asked to report on a uniform curriculum for high schools that a variety of colleges could accept. (Before that, all colleges had set their own requirements and the situation was a mess, much like many of the local K–12 standards we have across the United States today.) The committee, chaired by Harvard President Charles Eliot, had many suggestions and many critics. But for us, one consequence was that they defined the main science topics as biology, chemistry, and physics—and

2. Eric Brunsell and Jason Marcks, "Identifying a Baseline for Teachers' Astronomy Contents Knowledge," *Astronomy Education Review* 3, issue 2 (October 2004–April 2005): pp. 38–46, available online at *http://aer.noao.edu/cgi-bin/article.pl?id=119* (accessed 27 March 2007).

3. Herbert Kliebard, *The Struggle for the American Curriculum*, 2nd ed. (New York: Routledge, 1995). See Chapter 1 for a nice introduction to the curriculum in the nineteenth century.

4. Ibid., p. 6.

astronomy only as an elective. It soon began to slip into the noise of the demands on American curriculum, and today, few schools teach a separate astronomy class (and then only as an elective).[5]

It is true, as those who follow the politics of the space race know, that the National Defense Education Act of 1958 (the reaction to Sputnik) increased the amount of science and math being taught and led to some interesting curriculum studies and reform, especially in elite schools.[6] But it is not clear that things in science education got better in the long run. Other factors besides curriculum play an ever-increasing role in the education of our children. Teacher training or lack of it, latch-key kids, peer and media pressure against seeming smart (think of films such as *Dumb and Dumber),* MTV—all have combined to "beat us" far better than the Russians ever could have.

What space science students do get in our schools today is mostly found in general science, Earth science, and physical science classes in middle school and high school. Numbers are hard to come by, but many factors are contributing to a decline in the amount of space science being taught. Let's consider just three of them.

First is "the less is more" movement among science educators and those who train them. This movement, which says we should teach fewer topics in science but in greater depth and with more hands-on activities, is generally a positive development in getting students to learn what real science is about. But there are already very few hours devoted to science in the curriculum. Generally, when time is short, basic biology, chemistry, and physics take precedence.

Next we have the lack of trained teachers to teach science. During the 1999–2000 school year, only 55 percent of high school students received physical science instruction (chemistry, Earth science, and physics) from a teacher with a major or minor in physical sciences. At the middle school level, only 18 percent of students received physical science instruction from a teacher with a major or minor in physical sciences. Nearly 50 percent of middle school students received physical science instruction from a teacher without a major or minor in *any* science or science education field.[7]

---

5. Kliebard, *Struggle for the American Curriculum*, discusses the Committee of Ten in his first chapter. For more on the effect of this group on astronomy teaching, see Jeanne Bishop, "U.S. Astronomy Education: Past, Present and Future," *Science Education* 61, no. 3 (1977): p. 295.

6. Arthur S. Flemming, "Government Science and the Universities: The Philosophy and Objectives of the National Defense Education Act," *Annals of the American Academy of Political and Social Science* 327 (January 1960): pp. 132–138.

7. Statistics in this paper about the state of the U.S. science education system come for the most part from the National Science Board, *Science and Engineering Indicators 2004* (Arlington, VA: National Science Foundation, 2004). Chapter 1 contains information about K–12 education, Chapter 7 about public attitudes toward science.

Finally, consider the effects of the No Child Left Behind Act (and the general tendency in the United States toward simple-minded, multiple-choice testing as a measure of teaching and learning). Right now, schools are required to test all students in math and English, but by the 2007–2008 school year, states must test all students in science three times: in grades 3–5, 6–9, and 10–12.

The good news is that districts that had shifted emphasis away from science in the last few years because the testing for English and math was first and science was still in the future, will now have to start putting more emphasis again on science, since science testing is going to be a reality! The bad news is that science exams will not be used to determine whether schools are making "adequate yearly progress," so the pressure from those exams is not as great. Still, science test results will have to be publicly announced (to the parents), so they will be part of how parents and the public judge the schools.

But the far greater issue will be that each state will be allowed to set its own standards and to decide just what kind of science tests to offer. And, make no mistake, once the tests are set up, science teachers will teach to the tests! Anything not being fully tested will be much less likely to be taught. In my home state of California, to take an example, far more high school students take biology than physics. As you can imagine, there is pressure to make the grade 10–12 tests about biology only. If that is so, it will tremendously accelerate the trend to take biology over other sciences.

How will space science do in all this? Unless its teaching and testing are mandated by the states, what teachers will have time to teach astronomy with the short school days and short school years and all the pressure to do well on the standard tests that are increasingly the hallmarks of the U.S. educational system?

How many of you know how much space science is in your state standards? How many of you have participated in the fashioning and review of those standards? How many of you have talked with a local science teacher and tried to encourage increased teaching of space science? Yet if we, who have a deep interest in space science, are not engaged in this process, how can we expect others to carry the torch for us?

SCIENCE EDUCATION AND SCIENCE LITERACY

Lest you think that space science education is being uniquely singled out for failure, I note that things are not great throughout the arena of science education. In 2000, the National Assessment of Educational Progress indicated that in the United States only 29 percent of 4th graders, 32 percent of 8th graders, and a depressing 18 percent of 12th graders performed at or above the level termed *proficient* in science for their grade.[8] Roughly half of all freshmen entering California state colleges

8. Ibid., Chapter 1.

cannot understand English or math at the college level and need remedial courses.[9] It is also an interesting statistic that in 1956, the year *before* Sputnik, there were twice as many B.A.'s in physics in the United States than there were in 2004.[10]

The results of the lack of good science education in this country is that adult Americans know very little about science. Jon Miller of Northwestern University, the foremost science pollster in the United States, has come to the conclusion that fewer than 20 percent of adult Americans know enough science for minimal civic literacy. For example, 50 percent of adult Americans believe that humans lived at the same time as dinosaurs. Only 22 percent of adults in the United States can correctly define a molecule.[11]

At the same time that American science literacy is declining, the U.S. Department of Labor reports that in the next decade jobs requiring science, engineering, or technical training will increase by 51 percent—four times higher than general job growth. Where will all the trained people to hold those jobs come from? Clearly, the reports warning that the competitiveness of our country may be undercut by the lack of adequate education in science and engineering are worthy of far greater political attention than they have so far received.[12]

Now that I have thoroughly depressed you in general, let me at least mention one comparatively positive trend. Others in these conference proceedings are focusing their papers mostly on the programs involving human spaceflight. But these are only one part of how NASA has transformed public perception; the other part is the array of robotic missions and telescopes whose images are now part of the visual and verbal vocabulary of our times.

Even the lowest-level science textbooks are full of the images these missions have returned to us. Just think about the visual impact and drama of these:

- The Hubble space telescope views of star birth in the great clouds of cosmic raw material, black holes at the centers of galaxies, and the depths of galactic space.
- The Voyager mission images of Jupiter, Saturn, and their moons.
- The Galileo spacecraft's first close-up views of asteroids (and an asteroid with a moon).
- The continuing pictures from Spirit and Opportunity, the intrepid little Mars Exploration Rovers.

9. This statistic has been cited by California newspaper reports for years. It is confirmed by the information provided by the California State University system, such as *http://www.calstate.edu/pa/news/2004/proficiency.shtml* (accessed 27 March 2007).

10. National Academies of Science, *Rising above the Gathering Storm: Energizing and Employing America for a Brighter Economic Future* (Washington, DC: The National Academies Press, 2004), available online at *http://www.nap.edu/catalog/11463.html* (accessed 27 March 2007)

11. Miller's work is cited extensively in National Science Board, *Science and Engineering Indicators 2004*. See also Jon D. Miller, "The Measurement of Civic Scientific Literacy," *Public Understanding of Science* 7 (1998): pp. 203–223.

12. National Academies of Science, *Rising above the Gathering Storm: Energizing and Employing America for a Brighter Economic Future* (Washington, DC: The National Academies Press, 2004), available online at *http://www.nap.edu/catalog/11463.html* (accessed 27 March 2007).

Planetary science, as Carl Sagan stressed, has in one lifetime converted the planets from dots in the sky to worlds with tourist attractions. And Hubble images adorn musical CD covers, advertising, popular books and magazines, and even toys.[13]

One benefit of all this is that people seem to have an almost proprietary interest in the solar system these days. Witness the public and media uproar that followed the perceived downgrading of Pluto's planetary status.[14] NASA officials have also told me that the millions of hits on the Mars Rover Web sites and images have far exceeded everyone's expectations. In this way, the vision of robotic spaceflight has succeeded beyond our wildest dreams. It is part of the cultural landscape—video games, movies, television, comic books, and amusement parks all take a public familiarity with space probes and the planets for granted.

A marvelous example of this was the television program *The West Wing,* created by Aaron Sorkin. Although the show portrayed mostly Washington politics and the lives of those working in the White House, it was also one of the best "stealth space science education" vehicles I have ever seen. (And, not coincidentally, it was one of the few shows on television that made being smart an object of pride rather than ridicule.) Throughout the series, Sorkin introduced NASA missions and astronomical ideas. (Even the program I directed at the Astronomical Society of the Pacific, Project ASTRO, was mentioned once.) And the president of the United States was shown as loving space science and robotic planetary exploration and eager to show off about it to school kids.[15]

We need to enlist other friends in the media and public relations world to do more stealth science—science that becomes part of our popular culture and can insinuate itself into the public awareness. We should be putting stealth science into our TV shows, malls, parks, airport waiting areas, fast food places, and anywhere else we can.

THE MIXED BLESSING OF ASTRONOMICAL PSEUDOSCIENCE

One negative result of this widespread but suffused public interest in space travel and astronomical ideas is that some of our cultural need for the supernatural has recently become translated into space-related terms. Whereas people in past

13. Images from planetary exploration can be found at NASA's Planetary Photojournal Web site, *http://photojournal.jpl.nasa.gov/index.html.* Hubble images are best organized at the Space Telescope Science Institute public information site, *http://hubblesite.org/newscenter/newsdesk/archive/releases/image_category/* (both accessed 27 March 2007).

14. See, for example, Andrew Fraknoi, "*Teaching What a Planet Is: A Roundtable on the Educational Implications of the New Definition of a Planet,*" Astronomy Education Review 5, no. 2 (2006), available online at *http://aer.noao.edu/cgi-bin/article.pl?id=207* (accessed 27 March 2007). References at the end of this roundtable discussion can lead you further into other responses.

15. R. Lance Holbert, David A. Tschida, Maria Dixon, Kristin Cherry, Keli Steuber, David Airne, "*The West Wing* and Depictions of the American Presidency: Expanding the Domains of Framing in Political Communication," *Communication Quarterly* 53 (October 2005): pp. 505–522; Staci L. Beavers, "*The West Wing* as a Pedagogical Tool," *PS: Political Science & Politics* 23 (2002): pp. 213–216.

centuries saw supernatural visions in terms of guardian angels or leprechauns or haunted houses, today these same feelings are expressed through UFO and alien abduction experiences.[16] A 2001 Gallup poll, for example, revealed that fully one-third of adult Americans believe that extraterrestrials are visiting Earth. And 31 percent accept the tenets of astrology, that the position of the Sun, Moon, and planets at the time of our birth can affect our personality and destiny.[17]

In a way, those of us concerned with education should be grateful for the growth of tabloid-level interest in alien life-forms or completely misinterpreted "faces" on Mars. Although some students or adults may become interested in space science through such nonsense, they can sometimes go on to learn or read about the real world of space and astronomy once their interest is piqued. Still, without better information about the scientific method in our schools, our media, and in our public consciousness, too few people have the tools to go beyond a fascination with pseudoscience to the real science that would await them.[18]

Still more disturbing, a 2004 Gallup poll said that 45 percent of Americans believe that God created humans in their present form at some time in the last 10,000 years. When asked about the origin of humanity and being given several choices, only 13 percent of Americans chose an answer that did not include something about God's role. A December 2004 *Newsweek* poll asked whether respondents favored teaching creation science in addition to evolution: 60 percent favored it, 12 percent were undecided, and only 28 percent were opposed. For many people, this seems to be only fair, in a loose, democratic sort of sense. Many of these people are the ones who have a problem understanding the scientific method and ways of deciding on evidence in science.[19]

16. See, for example, Carl Sagan, *The Demon-Haunted World: Science as a Candle in the Dark* (New York: Random House, 1995). A listing of resources about astronomical pseudoscience by the present author is kept at *http://www.astrosociety.org/education/resources/pseudobib.html* (accessed 27 March 2007).

17. National Science Board, *Science and Engineering Indicators 2004*, Chapter 7. Belief in pseudoscience is discussed on pp. 7–21 to 7–23. See also Martin Gardner, *Fads and Fallacies in the Name of Science* (New York: Dover Publications, 1957); Michael Shermer, *Why People Believe Weird Things: Pseudoscience, Superstition, and Other Confusions of Our Time* (New York: Owl Books, 2002); Robert L. Park, *Voodoo Science: The Road from Foolishness to Fraud* (New York: Oxford University Press, 2001).

18. This is the approach of Phil Plait, who uses these crackpot ideas to teach about science. See Philip C. Plait, *Bad Astronomy: Misconceptions and Misuses Revealed, from Astrology to the Moon Landing "Hoax"* (Englewood Cliffs, NJ: Wiley, 2002).

19. A Web site that keeps tabs on polls related to science and follows public attitude surveys about evolution is *http://www.pollingreport.com/science.htm* (accessed 27 March 2007). Useful introductions to this subject may be found in Mark Perakh, *Unintelligent Design* (Amherst, NY: Prometheus Books, 2004); Niles Eldredge, *The Triumph of Evolution and the Failure of Creationism* (New York: W. H. Freeman and Co., 2000); Tim M. Berra, *Evolution and the Myth of Creationism: A Basic Guide to the Facts in the Evolution Debate* (Stanford, CA: Stanford University Press, 1990).

This ambiguity about (or ignorance of) the scientific method spills over into public views of the teaching of astronomy as well. The legislatures of a number of states (including Kansas) have begun to pass laws, at the behest of fundamentalist religious groups, to include anything that contradicts a young Earth and young universe in the subjects that should be taught "only as a theory" or taught with "alternative theories" or de-emphasized. This includes radioactive dating and Big Bang cosmology—ideas essential to understanding the longevity and formation of the universe. (The key issue is that the time scales of our modern understanding are too long to satisfy those who seek to promulgate a very literal interpretation of the Bible as science.)[20]

In response, scientists have begun to speak up and write about these issues. The Education Board of the American Astronomical Society has produced a booklet and Web site called *An Ancient Universe* to help teachers and school board members understand how we know that the natural world is old.[21] Other members of the space science community are now writing and talking more about the need to answer such claims, particularly to help beleaguered teachers in those states where the fundamentalist pressure is the greatest.[22]

We might note, as one example of why science educators are concerned, that religious radio stations now number almost 2,000—all but a few controlled by evangelical Christian organizations. They outnumber every other radio format except for country music and news talk.[23] In general, the political and media clout of evangelical Christian organizations is growing, while the amount of time devoted to serious science on the radio and many TV networks is shrinking. There are two syndicated short radio features about space science, *StarDate* and *Earth & Sky*, but their brief spots are hardly enough to overcome the relentless radio proselytizing of the religious right.

CASE STUDY: A NASA EDUCATION AND OUTREACH "ECOSYSTEM"

I hope it is clear even from such a brief introduction that those interested in the expansion (or even continuation) of astronomy and space science education in the United States have a huge task before them. Most teachers, especially at the elementary level, have little or no background in the field and are unlikely to teach

20. An excellent discussion of the history of the Big Bang theory may be found in Simon Singh, *Big Bang: The Origin of the Universe* (New York: Fourth Estate, 2004).

21. Andrew Fraknoi et al., *An Ancient Universe* (Washington, DC: American Astronomical Society, 2004), *http://www.aas.org/education/ancientuniverse.html*.

22. See, for example, Matt Bobrowski, "Dealing with Disbelieving Students on Issues of Evolutionary Processes and Long Time Scales," *Astronomy Education Review* 4, issue 1 (2005): pp. 95–118, on the Web at *http://aer.noao.edu/cgi-bin/article.pl?id=143* (accessed 27 March 2007).

23. Mariah Blake, "Stations of the Cross," *Columbia Journalism Review* (May/June 2005), available o the Web at *http://www.cjr.org/issues/2005/3/blake-evangelist.asp* (accessed 27 March 2007).

much of it without requirements to do so, or active training or assistance. Even many upper-level teachers see it only as possible appetizer for the main dish of the "regular" sciences.

Since there are only on the order of about 10,000 space scientists in the country, it is hard for a community that small to have a major effect. What is needed is either a larger community of interpreters to help us, or a "leveraged" effect where the work we do can translate into a cascade of influence. The first thrust is being pursued by the Astronomical Society of the Pacific, among others. For example, Project ASTRO is a program that trains amateur as well as professional astronomers (and college students) to partner with teachers in grades 4–9 to bring effective, hands-on astronomy activities into classrooms. It is presently operating in 13 regional sites around the country.[24]

The second thrust is far more difficult and requires the kinds of effort that only a major government initiative might have the resources to undertake. In the brief space I have, I would like to conclude with one specific case study where NASA's educational efforts have had a disproportionately important and salutary effect.

Today, about $40 million per year is flowing from the NASA's division formerly known as the Office of Space Science (now part of the oddly named Science Mission Directorate) to a whole range of educational and public outreach (EPO) programs. It has been, over the last few years, the largest investment space science EPO in the history of our country.[25] It's a good story about a little-known effect of spaceflight on education that I hope will be written up in more detail by future historians.

In the early 1990s Congress began to ask the science enterprise in the United States to document how it was benefiting the nation and it asked the science funding agencies, such as the National Science Foundation and NASA, to be more explicit in requiring such justification. At the same time, NASA's Office of Space Science also had some concerns about how effectively NASA's Office of Education was conveying the excitement of fundamental science, as opposed to piloted spaceflight done for political and other reasons (that sometimes wound up having little connection to science).

So the Office of Space Science hired (in a quiet way) a series of dynamic leaders in the field of education to create its own EPO program—focused much more strongly on space science and not on what astronauts ate for lunch. Another aim

24. Andrew Fraknoi and S. Lalor, "Project ASTRO: How to Survive a Visit to a 6th Grade Classroom and Come Back for More," in *Amateur-Professional Partnerships in Astronomy,* J. Percy and J. Wilson, eds. (San Francisco, CA: A.S.P. Conference Series, 2000), vol. 220, pp. 260–263. Also see Andrew Fraknoi and D. Zevin, "Ten Years of Project ASTRO," *Mercury* (September/October 2003): p. 12.

25. See, for example, Jeffrey Rosendhal et al., "The NASA Office of Space Science Education and Public Outreach Program," available online at *http://science.hq.nasa.gov/research/Cospar_Manuscript.pdf* (accessed 27 March 2007). See also the publications listed at *http://science.hq.nasa.gov/research/epo.htm* (accessed 27 March 2007).

of the program was to involve space scientists in an active and leveraged way in the education effort. Eventually, the program would find its influence by requiring each of its instruments (missions) and research programs to devote 1 percent to 2 percent of its budget to EPO. When larger programs, such as the Hubble space telescope, were involved, those small percentages generated a considerable amount of money.

Dr. Cheri Morrow began to organize this community in smart ways and to write material to facilitate better EPO programs, but the real progress came with the appointment of Dr. Jeffrey Rosendhal, a veteran Washington and NASA hand, who in 1993 took this effort in a novel direction. After getting advice from education experts around the country, Rosendhal set up what he called an "ecosystem" for EPO. Rather than each scientist, each institution, and each mission pursuing EPO on its own, the ecosystem would coordinate and amplify the individual efforts so that the overall effect could be greater than the sum of its component parts.[26] There were two distinct parts to the organizing ecosystem:

- A series of topical forums to coordinate EPO activities among like missions (for example, among those exploring the solar system), and
- A set of regional brokers/facilitators to link missions and scientists to the real world of education.

The brokers were supposed to act like old-fashioned marriage brokers did—setting up relationships between the scientists and professional staff doing a NASA mission and local school systems, museums, educational publishers, nonprofit societies, etc. Although not all parts of this system worked right away, the coordination these strategically placed elements provided did help take the NASA space science EPO program to a new level, going way beyond the usual bookmarks, key chains, and pretty lithographs that were the hallmark of earlier mission outreach.[27]

Rosendhal and his staff also set up a series of small educational start-up grants (the IDEAS program) and a way for scientists who already had research grants from the Office of Space Science to request an EPO supplemental grant for some kind of educational project involving scientists or graduate students.

When all these vehicles for doing education and outreach were added together, by the year 2003, 400,000 people around the country had participated in 5,000 educational programs or events; 6 million had visited some Internet site sponsored under the program; and a whole host of curriculum modules and learning materials had been created and infused into the educational system.

26. See, for example, "Partners in Education," *http://spacescience.nasa.gov/admin/pubs/edu/educov.htm*; and "Office of Space Science Education Implementation Plan," *http://spacescience.nasa.gov/admin/pubs/edu/imp_plan.htm* (both accessed 27 March 2007).

27. The system's Web site can be found at *http://science.hq.nasa.gov/research/ecosystem.htm* (accessed 27 March 2007).

As someone who served on the review committee that was set up to see how well the system was doing after the first few years, I can personally attest to the fact that the leaders were open to criticism, able to make changes and course corrections, and had the best interests of education in mind.[28] I was also impressed by the fact that the system worked not only by itself, as NASA education has often tended to do, but was willing to partner with respected institutions and programs in the outside world.

An excellent example of this is that the program, instead of merely inventing more curriculum modules of its own, worked with such curriculum developers as the GEMS program at the Lawrence Hall of Science. And instead of developing curricula focused solely on NASA missions, the investment was in modules that would increase students' general background in space science (including elementary ideas that are often taught incorrectly in the schools.)

The Growth of the Community of EPO Professionals

Now don't get me wrong. NASA's Office of Space Science was not alone in making investments in and becoming more sophisticated in its understanding of the real worlds of education and outreach. The National Science Foundation, the national observatories, the American Astronomical Society, and the Astronomical Society of the Pacific were also increasing their involvement in these areas around the same time. Universities, research labs, and observatories also found it to their political advantage to become more involved in public education. But there is no question that the NASA investment and involvement were the most significant, both in terms of the size of the budget and the sophistication of the approach.

As a result, there was suddenly a critical mass of education and outreach workers in the United States, all doing work outside the traditional classroom—and a new profession may have been born.[29] Think about how a job category becomes a "profession" in this country, whether we are thinking of lawyers (who have done a great job with this), podiatrists, teachers, or NASA historians. Some of the factors that make a profession include professional organizations, journals and newsletters, awards, and a shared literature. Soon, most professions develop a mode of certification—a way of training candidates for the profession which guarantees that certain shared goals and standards will be observed (and that those who do not share those goals and standards will be excluded). Such modes of certification can include undergraduate or graduate degree programs, apprenticeships, examinations,

---

28. The review can be found on the Web at *http://newfrontiers.larc.nasa.gov/PDF_FILES/NASA_OSS_EPO_TF_Report_FINAL.pdf* (accessed 27 March 2007).

29. For more on the specifics of this development, see Andrew Fraknoi, "Astronomy Education and Public Outreach: Steps and Missteps Toward an Emerging Profession" *Mercury* (September/October 2005): p. 19, available online at *http://www.astrosociety.org/pubs/mercury/34_05/epo.pdf* (accessed 27 March 2007).

certifying boards, professional development programs, and (of course) fancy certificates printed on parchment to put on your wall.

The new profession is just beginning to make its way through this list. There is now a journal, called *Astronomy Education Review*;[30] there is an annual conference, held by the Astronomical Society of the Pacific; and several awards (including the American Astronomical Society Education Prize, the Klumpke-Robert Prize, the Las Cumbres Outreach Award, and a new education prize from the Astronomical League) are now given specifically for education and outreach work. At least one group, at the Center for Educational Technology in Wheeling, West Virginia, is beginning to think about a master's degree program in EPO work. Several of these profession-building steps are happening with NASA support but there is still a way to go before the community of such workers becomes a profession.

For example, there are still no agreed upon standards for entering the profession, so that some people come with extensive background in science and/or education whereas others are trained on the job. Many scientists dabble in the field without feeling the responsibility to get to know the literature of astronomy education, in the way that they would if they dabbled in a field of space science research. To some degree the new EPO profession is still what physics education researcher Joe Redish calls "a community of weakly interacting individuals." Nevertheless, the seeds of a new profession seem to have been planted and NASA's efforts have been a major cause of why some of those seeds may be starting to bloom.

Although some disturbing recent changes in the NASA funding picture may profoundly impact the support this fledgling profession that NASA has had such a strong part in creating, I suspect the sense of community and the movement toward professionalization could well continue if the funding is diminished.[31]

CONCLUSION

So what can we conclude about the impact of the space program on education in the United States from this brief review? There were certainly strong influences on small subgroups, which are recorded mostly anecdotally and are worthy of more serious study. There are cases, such as the NASA Office of Space Science ecosystem, where a small, leveraged infusion of funds can make a significant difference in the

30. The journal is available only online at *http://aer.noao.edu* (accessed 27 March 2007).

31. The new Administration at NASA is indicating that they do not believe that education should be a primary mission of the Agency. They are also being quite honest in letting Congress and the public know that they simply cannot accomplish all the goals set for NASA with their present funding. By giving higher priority to piloted spaceflight, replacing the Shuttle, continuing the International Space Station, and other (often politically motivated) programs, they are consigning science and science education to lower priorities. Many commentators are predicting this will soon translate into lower funding and thus to less money being available for the EPO system.

effectiveness of EPO programs. But changing the educational system in the United States as whole is a monumental undertaking, one completely beyond the resources of a small government agency, a single industry, and a group of enthusiasts. Whether space science can play a pivotal role (or even a supporting role) in bringing such a change about remains to be seen.

The educational system in our country reflects and enforces our best and worst values. Pockets of entrepreneurial effectiveness can be found among vast bureaucratic morasses of lazy and lackluster learning and teaching. Having an influence on this system will require a concentrated national effort whose expense and difficulty might well put the Apollo program to shame.

Chapter 22

"Racism, Sexism, and Space Ventures": Civil Rights at NASA in the Nixon Era and Beyond

Kim McQuaid

Race and gender are almost invisible aspects of the early Space Age. The civil rights movement, the women's movement, and early spaceflight occurred simultaneously, but they are normally analyzed as if they occurred in separate universes. Realities were different; there was a social history of the Space Age. Exclusions of women and racial minorities from key portions of America's civilian space effort have had major effects on the political credibility of spaceflight.

Recent books on how the U.S. Astronaut Corps stayed closed to women in the late 1950s and early 1960s, and about how it opened to women of all races and minority men after 1978, tell part of this story. But the political struggles within and around NASA that made such openings a political necessity remain unknown. This essay addresses the fight to make America's space program—like America itself—more diverse and inclusive.

NASA's Troubles

Late in 1973, NASA was an agency in budgetary, policy, and staffing trouble. Efforts to gain public or political support for a human mission to Mars had failed resoundingly in 1969 and 1970. As the Apollo lunar program ended in December 1972, NASA was losing no less than one-third of its civil service workforce. Between 1965 and 1975, the Agency also lost half of its budget, in purchasing power terms. In national opinion polls, only foreign aid had less support than space exploration; even welfare spending fared better.[1]

1. George M. Low, personal notes, "Fiscal Year 1972 . . . Discussions," pp. 13–14, 18 July 1970, Box 27, Rensselaer Polytechnic Institute Archives, Troy, New York (hereafter designated "Low/RPI"); Howard E. McCurdy, *Inside NASA: High Technology and Organizational Change in the U.S. Space Program* (Baltimore, MD: Johns Hopkins, 1993), esp. p. 124ff; Mark E. Byrnes, *Politics and Space: Image-Making by NASA* (Westport, CT: Praeger, 1994), p. 115.

A major reason for NASA's difficulties was that the society around it was changing. An era of external cold war military and diplomatic concerns was giving way to an era dominated by domestic social and economic issues. Chief among these were the fast-paced rise of environmentalism, the women's movement, and the African American civil rights movement. Every recent poll, NASA's George M. Low noted privately as early as July 1970, showed that space and national defense had little appeal. American society was opening up to major new political constituencies in a fashion not seen in almost 40 years.[2]

By October of 1973, Washington, DC, and the nation beyond it were also in the midst of the wrenching Constitutional crisis called Watergate. In June, former White House Counsel John W. Dean implicated Republican President Richard M. Nixon in a conspiracy to obstruct justice regarding political spying and sabotage carried out against his Democratic and other enemies. In July, investigators for a special prosecutor's office created by Congress uncovered the existence of a White House taping system that could confirm or disprove Dean's charges. Political stonewalling and judicial guerilla warfare then commenced in earnest after Nixon refused to provide access to unsanitized versions of the official records of this presidency.[3]

Finally, accumulated pressures caused multiple political faults. First, on 10 October 1973, Vice President Spiro T. Agnew resigned over a bribery scandal. As one of Nixon's key partisan warriors against "radical liberals" surrendered, Nixon himself counterattacked. From 15 to 19 October, Nixon demanded that Special Prosecutor Archibald M. Cox cease requiring unedited White House tapes. After Cox refused, Nixon fired him in the "Saturday Night Massacre" of the 20th of October. In a pre-Internet world, 250,000 to 300,000 telegrams then cascaded into Washington, demanding Nixon's impeachment for high political crimes and misdeeds. The mood in Congress was grim. *The New York Times* and major regional papers editorialized that Nixon should resign, and *The Washington Post* called for impeachment. *Time* magazine, long a Nixon supporter, sadly noted he had "irredeemably lost his moral authority, the confidence of most of the country, and therefore his ability to govern effectively."[4]

Until precisely this point, NASA had avoided any connection with the dangerous and polarizing politics of Watergate. But, on the 11th of October, one day after Agnew's resignation and as the battle between Nixon and Congress, the courts, and the special prosecutor's office peaked, NASA performed a "Nixonesque purge"

2. Low, "Fiscal Year 1972 . . . Discussions," pp. 13–14; Samuel Lubell, *The Future of American Politics* (New York: Doubleday, 1965 ed.) and Samuel Lubell, *The Hidden Crisis of American Politics* (New York: W. W. Norton & Co., 1970) chart the sometimes explosive politics of race.

3. Kim McQuaid, *Anxious Years* (New York: Basic Books, 1989), pp. 232–246 provides a succinct chronicle. See also Stanley Kutler, *The Wars of Watergate* (New York: Alfred A. Knopf, 1990), esp. p. 390ff.

4. Three good sources for Watergate are James Doyle, *Not Above the Law: The Battles of Watergate Prosecutors Cox and Jaworski* (New York: Morrow, 1977); Elizabeth Drew, *Washington Journal: The Events of 1973-1974* (New York: Random House, 1975); and Gladys and Curt Lang, *The Battle for Public Opinion: The President, the Press, and the Polls During Watergate* (New York: Columbia, 1982), esp. pp. 99–105.

of its own. In doing so, it threw itself into the firestorms of the final year of Nixon's plagued presidency. Ironically, it also began a process of political protest and oversight that opened up the U.S. civilian space program to women and to racial minorities.[5]

The "Nixonesque Purge"

The occupational desegregation process began when NASA fired Mrs. Ruth Bates Harris, the highest-ranking woman in the Agency, for submitting a private report to NASA Administrator James Fletcher stating that NASA's belated equal opportunity program was "a near total failure." NASA still employed fewer racial minorities and women than any other agency in the federal government. What women it did hire were almost all clustered in dead-end clerical jobs, and still nothing had been done to open the Astronaut Corps to anyone but white men. "Without denying the validity of the materials in the report, the agency fired its principal author," noted an indignant *Washington Post* editorial of November 24. Despite Fletcher's statements that NASA understood and was dealing with civil rights in employment problems for previously excluded or discriminated-against groups in its largely Southern installations, *The Washington Post* (located in a city which had grown from half to two-thirds African American in the 1960s) was not impressed. "Institutionalized racism and sexism" existed throughout NASA, it concluded. "[N]either simple pieties nor eloquent declarations of principle" would change that fact. *The Washington Post* illustrated this by quoting a NASA Headquarters spokesman who "made a large point of the fact that the agency official who [most immediately] recommended Harris's dismissal was himself black"—"physically blacker than Mrs. Harris," he said.[6]

NASA was now in major political trouble with no strong president to protect it. Its rocky ride through the new politics of America in the 1970s was about to get rockier still. How had it all begun? Basically, an elite agency began coming to social understandings late; when it did, it proceeded to address them in a half-hearted and ambivalent fashion. Not feeling it had much to learn, it denied that problems—or solutions—existed.[7]

---

5. Recent books about the abortive early involvements of women in the space program include Martha Ackmann, *The Mercury 13: The Untold Story of Thirteen American Women and the Dream of Space Flight* (New York: Random House, 2003); Margaret A. Weitekamp, *Right Stuff, Wrong Sex: America's First Women in Space Program* (Baltimore, MD: Johns Hopkins University Press, 2004); and Bettyann Holtzmann Kevles, *Almost Heaven: The Story of Women in Space* (New York: Basic Books, 2003) and the revised second edition (MIT Press, 2006), which extends the story on into the era of the first female astronauts of the 1980s and 1990s; see also David J. Shayler and Ian A. Moule, *Women in Space—Following Valentina* (Chichester, U.K.: Springer-Praxis, 2005).

6. "Racism, Sexism, and Space Ventures," *The Washington Post*, 24 November 1973: p. A14. For the increase in African American population in Washington, DC, in the 1960s, see *Historical Statistics of the United States: Colonial Times to 1970, Volume 1* (Washington, DC: U.S. Government Printing Office, 1975), p. 26.

7. One way to understand how visible and important was the case started by the Ruth Bates Harris firing is to see how many stories exist about her, in whole or part, in the online index to *The Washington Post*. From 1965 to 1985, this number totals 222. See *http://pqasb.parchiver.com/washingtonpost/search.html/?nav=left* (accessed 18 January 2005).

NASA ignored social issues such as affirmative action (compensatory activity for those previously excluded from occupations or training by rules or custom) for as long as it could. It was not alone. *The New York Times*, *Newsweek*, and ABC News all discriminated against and excluded women from many jobs, just as television news stations generally refused to hire female reporters of any hue throughout the early 1970s. 1968 through1971 nevertheless marked a watershed in enforcement of civil rights in employment laws for women and racial minorities in both the private and public sectors.

Further, in 1971 the Supreme Court, in the *Griggs v. Duke Power Company* case, enunciated a clear compensatory action and preferential treatment argument. A previously segregated Southern utility was using competency tests that had a "disparate impact" on minority groups. The court ruled it could not do so unless the tests had a very clear relationship to advertised jobs. "Business necessity" claims for minimum educational attainments or literacy levels were not enough. No intent to discriminate against individual applicants needed to be proved if a pattern of group exclusion was demonstrated. The burden of legal proof, bluntly, was shifted from the historically excluded worker to the employer. Given the very long history of occupational exclusions and low funding for segregated African American schools in Southern states, the utility had to make up for the effects of past discrimination in the present and the future.[8]

Most NASA installations were located in the apartheid American South. The aerospace sector of the economy was also no stranger to job discrimination. A Wharton School report of 1966 noted that African Americans were simply excluded from the aviation industry until World War II and very rarely achieved higher occupational standing thereafter. Just as there were no African American aviators until the wartime Tuskegee Airmen proved to segregationists that African Americans were mentally capable of flight, pre-1941 anti–African American policies were not disguised in corporations. African Americans pushed brooms; whites built airplanes. In the Apollo era of the 1960s, one-third of 1 percent of managers and seven-tenths of 1 percent of professionals in the 60 percent of firms supplying data were African American. The largest wartime and cold war changes, most pushed by unions such as the United Auto Workers, were in semi- and unskilled labor ranks, where 8 percent and 14 percent of workers, respectively, were African American. Office and clerical staff (2 percent) and skilled workers (3 percent) lagged badly, compared with the 20 percent of broom-pushers who were African American in 1966. Aerospace executives generally stated that "direct experience" was essential to success in their industry, while making few, if any, efforts to upgrade such minorities or women as they had already hired.[9]

8. By far the best survey of affirmative action is Nancy MacLean, *Freedom Is Not Enough: The Opening of the America n Workplace* (New York: Russell Sage Foundation and Cambridge, MA: Harvard University Press, 2006), esp. pp. 108–127, 140–141. For *Griggs*, see Herman Belz, *Equality Transformed: A Quarter Century of Affirmative Action* (New Brunswick, NJ: Transaction, 1990), pp. 52–54, 81–85.

9. Herbert R. Northrup, *The Negro in the Aerospace Industry* (Industrial Research Unit, Wharton School, University of Pennsylvania Press, 1966), p. 17ff, esp. n. 29.

NASA reflected the mindsets and occupational patterns of the sector of the economy from which most of its engineers, scientists, and technical managers came. Despite a brief "Rosie the Riveter" interlude during World War II, (white) women disappeared from aerospace—except as clerks and typists—postwar. Women of any race were also not normally admitted to technical schools or to undergraduate or graduate training in engineering and the physical sciences until the late 1960s without intense personal effort. Female engineering Ph.D.'s, accordingly, were less than 4 percent of the total even 25 years after Sputnik. Levels of 10 percent weren't reached until 1990. In that latter year, by comparison, 40 percent of Ph.D.'s in biology and 50 percent in the social sciences and humanities were women. African Americans, meanwhile, still only earned 2 percent of all doctorates in all fields of science and engineering in the 1990s, and only 5 percent of the bachelor's degrees in aerospace, electrical, and mechanical engineering specialties of prime interest to NASA in 2002, while Native Americans and Hispanics were as low and lower.[10]

The effects of the lags in both the aerospace sector and in physical sciences and engineering education could be seen in NASA's approach to race and gender issues. Until September 1971 (in the wake of *Griggs*), NASA had no systematic civil rights element in its employment program at all, even though three-quarters of its facilities were located in Southern states, including Virginia, Alabama, Texas, Louisiana, Florida, and Mississippi. Instead, NASA's director of personnel in Washington carried out tasks "on a part-time basis." The labs where most of NASA's people worked, meanwhile, were generally devoid of any organizational structure, lines of responsibility, or policy guidelines regarding affirmative action.[11]

The ad hoc approach, however, was no longer enough because, also in 1971, groups such as the Women's Equity Action League, the NAACP Legal Defense Fund, and the National Organization of Women banded together to bring successful suits against for-profit and nonprofit organizations accepting federal money in any form. Suddenly and decisively, elite meritocracies such as Harvard University and

10. House Committee on Education and Labor, *A Report on Equal Employment Opportunity and Affirmative Action in the Southern California Aerospace Industry*, 100th Cong., 2d sess., June 1988, pp. 46–52. The first investigations of hiring and promotion of racial minorities and women in the aerospace industry by this House committee began in 1975, immediately after the Harris case was concluded. See U.S. Department of Education, *Digest of Education Statistics—1997* (Washington, DC: U.S. Government Printing Office, 1999), pp. 318–320; National Science Foundation, *Science and Engineering Degrees by Race and Ethnicity [and Gender] of Recipients* (Arlington, VA: NSF, 2000).

11. George M. Low, "Corrected Draft of Forthcoming Congressional Testimony for EEO Head Dr. Dudley McConnell Dated January 10, 1974," Box 67, file 4, Low/RPI; W. Henry Lambright, *Powering Apollo: James E. Webb of NASA* (Baltimore, MD: Johns Hopkins University Press, 1995), p. 133, and an accompanying file of clippings from *The Huntsville* [Alabama] *Times* at the NASA History Office, Washington, DC (hereafter "NHO"). Lambright's brief discussion of race relations and NASA is almost unique. Normally, a single anecdote suffices. See, for example, Walter McDougall, . . . *the Heavens and the Earth: A Political History of the Space Age* (New York: Basic Books, 1985), p. 412.

the University of Michigan were in the same legal position as any of the 10 major integrated aerospace contractors at that time. By 1972, Congress further extended the scope of affirmative action when it ordered all executive-branch agencies, including NASA, to obey the same civil rights employment rules now mandated on private firms and state governments.[12]

A "Harlem Princess" Comes to NASA

Accordingly, NASA's newly installed Administrator James Fletcher started off his tenure as the fourth leader of the U.S. civilian space program with bold moves on the civil rights front. On 24 August 1971, several months after he assumed office, a press release signed by Fletcher announced that a 52-year-old African American woman, Ruth Bates Harris, would become NASA Headquarters's new Director of Equal Opportunity. Harris would "provide direction" to all civil rights in employment programs for all the "approximately 29,000 NASA Civil Service employees." This would include the top managers at all of NASA's far-flung labs. She would also oversee "contract compliance": the hiring of women and minorities by the many private firms providing products and services to NASA facilities. The new head of an agency whose "social center of gravity was exceedingly conservative" had just acted decisively.[13]

Ruth Bates Harris, who died in 2004, "Courtesy of University of Kansas Libraries"

12. Margaret W. Rossiter, *Women Scientists in America: Before Affirmative Action, 1940–1972* (Baltimore, MD: Johns Hopkins University Press, 1994), pp. 377–380; Becky Thompson, "Multicultural Feminism: Recasting the Chronology of Second Wave Feminism," *Feminist Studies* (Summer 2002), p. 337ff; Maren L. Carden, *The New Feminist Movement* (New York: Russell Sage Foundation, 1974), esp. pp. 135, 141; Jo Freeman, *The Politics of Women's Liberation: A Case Study of an Emerging Social Movement and Its Relation to the Policy Process* (New York: McKay, 1975), esp. pp. 49–67. Freeman and Carden make clear differentiations between feminist generations. For a good survey of how African American women related to the civil rights and women's movements, see Paula Giddings, *When and Where I Enter: The Impact of Black Women on Race and Sex in America* (New York: Morrow, 1984).

13. James C. Fletcher, "Director of Equal Employment Opportunity Memo Dated August 24, 1971," EEO (Equal Employment Opportunity)—Very Sensitive file, George M. Low papers; alphabetical files, NHO. (Hereafter "Low/NHO"); Kevles (2003), p. 204; Ruth Bates Harris, *Harlem Princess: The Story of Harry Delaney's Daughter* (New York: Vantage Press, 1991), p. 104.

The woman Fletcher hired was a self-described "Harlem Princess" whose first marriage had been to a Tuskegee Airman. An honors graduate of Florida A&M University, she had gone on to earn an M.B.A. with a specialization in personnel and industrial relations from New York University. What Fletcher called her "distinguished career in human relations" included service as the executive director of the District of Columbia Commission in Human Relations, a civil rights oversight and implementation group. Her nine-year tenure at the DC Commission began with a successful push to get *The Washington Post* to stop carrying racially restricted housing ads, and moved on to an increasing variety of housing, community–police relations, and other work. Through several "long hot summers" of racial discontent in the late 1960s, Harris was among those who exercised front-line leadership in restoring peace and stopping (or avoiding) riots. Because inhabitants of the nation's capital had only gotten the right to vote for local government in 1967, Harris not only became a de facto affirmative action officer for city government in a majority-African American metropolis, she also learned to work well with the Congress and senators of all political persuasions who were the overseers of DC government. As the Congressional Black Caucus was formed (in 1969), as *Ms.* Magazine first began publishing (in 1971), and as the Equal Rights Amendment first passed both houses of Congress and Congresswoman Shirley Chisolm became the first woman—and first African American—to seek the nomination of a major party for the presidency (in 1972), Harris worked hard to understand and guide quiet revolutions in race and gender relations. "I'm never just talking about people being nice to each other. I'm talking about changing the system," she told a civil rights oral history interviewer for Howard University in 1971. "We ought to have one big coalition... [of African Americans and] ... all our minorities [with which] we could change anything in this system."[14]

Ruth Bates Harris was a bridge-builder, not a wild-eyed or other radical. In 1969, for example, she left the DC Commission to become the human relations director for the public school system of Montgomery County, Maryland. In this very affluent county, only 4 percent of the populace was then African American and only 7 percent nonwhite. Her constituents included top policy makers in Washington

14. Harris, *Harlem Princess*, pp. 247–248. Two oral history transcripts of Harris's career at this time exist. One is dated 16 March 1971 and the other is undated, circa 1971. The 44-page and 36-page transcripts are both available in the Ralph Bunche Civil Rights Collection at the Moorland-Spingarn Research Center of Howard University, Washington, DC. The data regarding Montgomery County's economic status and racial composition come from the first Harris oral history, p. 36. Hereafter, the transcripts will be designated "Harris (1971-1)" and "Harris (1971-2)." For profiles of Harris in this period, see Carolyn Lewis, "Her Job: To Pass the Peacepipe," *The Washington Post*, 10 October 1967: pp. C1, C4; Paul Richard, "Ruth Harris is Named Head of Human Relations Council," *The Washington Post*, 21 May 1966: p. B1; "Harris (1971-1)," pp. 31, 36; "Harris (1971-2)," pp. 20, 40; Harris, *Harlem Princess*, pp. 142–220, 183ff. *The Washington Post* coverage of Harris began in 1965 and was regular by the 1968–1970 period, and Harris was honored by the NAACP on several occasions.

and NASA staff from the agency's Goddard Space Flight Center in neighboring Prince Georges County, Maryland. It was via her connections with Goddard that NASA Headquarters heard of her and hired her.[15]

Once she was vetted, security cleared, and hired, however, problems with Harris's posting immediately developed. In fact, all this started before she had even officially left her post at the Montgomery County school system to report to work at NASA. Only one week after publicly announcing Harris's hiring as a change agent with power, Fletcher demoted Harris and rehired her as an assistant deputy director rather than a director. Instead of enforcing change in an era of *Griggs* within a largely Southern agency, Harris reported to a white male overseeing private companies. Instead of regulating NASA, and in particular directors and deputy directors of NASA labs, Harris was sidetracked into an office dealing only with NASA contract employees and firms. Enforcing virtue upon itself had very quickly ceased being a NASA priority.[16]

Harris's immediate demotion before she even arrived for work at NASA Headquarters on 4 October 1971 was a warning signal. It was also what NASA Deputy Administrator George Low later privately admitted was "the weakest part" of NASA's official position. Nobody at NASA Headquarters ever accepted responsibility for the action. Lower echelons passed the buck to a man who had died.[17]

What had happened? Here, organizational culture and cultural context apparently interwove. NASA knew it had to do something about including hitherto largely unrepresented groups, but it also wasn't comfortable getting started. NASA Headquarters staff also might have expected Harris to be a patient schoolteacher rather than the street-smart civil rights implementer that she was, networked into official Washington and to organizations such as the NAACP. One organizational statistic was eloquent: eight of NASA's dozen major facilities had created equal employment/affirmative action offices just before Harris was hired. Four of the eight Center affirmative action people only worked part-time; all eight Center people were under the administrative control of lower-level procurement officers; and no fewer than six of the eight Civil Rights in Employment teams at NASA's labs were all-white.[18]

---

15. Harris, *Harlem Princess,* pp. 247–248.

16. Constance Holden, "NASA: Sacking of Top Black Woman Stirs Concern for Equal Employment" *Science* (23 November 1973): pp. 804–807, esp. 804. The bureaucratic paper trail is in Senate Appropriations Committee, Subcommittee on HUD, NASA and other Independent Agencies, *Hearings before a Subcommittee of the Committee on Appropriations . . . on H.R. 15572,* [the NASA Appropriation Act], Part 1, 93rd Cong., 2d sess., 1971, pp. 6–33, esp. pp. 26 and 33 (regarding how Harris's job description was changed in August of 1971). (Hereafter "Senate Appropriations Committee [1974]".)

17. George M. Low, Box 108, personal notes, "EEO," 25 November 1973, pp. 2–3, Low/RPI.

18. See "Director, Equal Opportunity Programs to Mr. Harnett," 5 February 1971, in Senate Appropriations Committee (1974), pp. 20–21.

NASA, then, talked about wanting "the best equal opportunity program in the federal government," but using part-time and all-white organizations to do it was naïve. NASA employed fewer minorities and women than any other agency in government. It claimed this was because of its elite and expert technical structure, but far from everyone at NASA was a rocket scientist. This disparity between NASA and other federal agencies also grew even as African American professionals sought out government agencies as employers because those agencies also most often obeyed federal civil rights laws. NASA's own statistics showed that it did as well as private corporations in employing minorities and women in the technical half of its operations (at 3.5 percent), but NASA's leaders did not go on to ask why NASA employed only 6 percent of racial minorities in the nontechnical half of its operations. People like Harris were about to pose such uncomfortable questions.[19]

Culture Shock

As the newly arrived Ruth Bates Harris began investigating the human side of space exploration, she also almost immediately transgressed the unwritten folkways of a high-technology agency. At NASA, technical "missions" mattered; personal (and personnel) issues did not. Number Two man in NASA's hierarchy and Chief "inside" Administrator George Low, for example, was so private a man that he never had a listing in *Who's Who in America* during his NASA years, never spoke with reporters, and never let his closest aides at the Agency know keys parts of his background (which included being an Austrian Jewish refugee from Hitlerism).[20]

One reason for Low's cloak of secrecy about himself shared offices at NASA Headquarters with him when Ruth Bates Harris arrived in 1971. Wernher von Braun, NASA's premier missile man, had provided weaponry to a regime that had deemed the Lowensteins of Austria subhuman and deprived them of livelihood, homeland, and loved ones. Low, presented by von Braun's most recent biographer as personally and professionally resentful of von Braun, said nothing publicly against him. His favorite managerial advice was for people to "put their emotional hang-ups aside." To Low, like von Braun, "identity" issues were off-limits.[21]

---

19. "Harris (1971-1)," pp. 31, 36; "Harris (1971-2)," pp. 20, 40; Harris, *Harlem Princess*, pp. 142–220.

20. George M. Low, personal notes, "Meeting Record on February 6-7, 1974: TSC Deputy Directors Council Meeting," pp. 2–4, 19, Box 67, Low/RPI; Shirley Molloy and Professor Stephen E. Wiberley interview with the author, 5 October 2004, Troy, New York.

21. Bob Ward, *Dr. Space: The Life of Wernher von Braun* (Annapolis, MD: Naval Institute Press, 2005), esp. pp. 202–203; "The Groundling Who Won," *Time* (3 January 1969): p. 13. The rare *New York Times* coverage of Low contains minimal information. The NASA History Office files on Low are similarly sparse. Low appeared in *Who's Who* only once, shortly before his death in 1985. Ernst Stuhlinger and Frederick I. Ordway, *Wernher von Braun: Crusader for Space: A Biographical Memoir* (Malabar, FL: Krieger Publishing Company, 1994), pp. 300–303; Francis T. Hoban, *Where Do You Go After You've Been to the Moon?* (Malabar, FL, Krieger, 1997) esp. pp. 53, 57, 58–59, 155, 192; Hoban uses the phrase "catechism, not reformation" to describe Low's approach to *people*-centered change within NASA (p. 192).

Ruth Bates Harris's approach to von Braun's past was quite different. She didn't intimidate easily and she was forceful. So, after reading in a newspaper story in 1971 or 1972 that von Braun had used slave labor to build rockets for the Nazis, she went straight to von Braun's office to ask him whether the story was true. "The silence between us," she recalled, "was deafening and awesome." Von Braun finally replied that there were journalistic distortions. Harris said it was not her role to judge but that she was going to refer the story to an "appropriate office" at Headquarters. Von Braun, Harris then recalled, was both sad and understanding. Neither he nor Harris, after all, needed any instruction about how racism was not restricted to Nazis. Indeed, according to Harris, von Braun pushed for affirmative action "with courage and conviction" at the Center he had previously headed in Huntsville, Alabama. No other top NASA official of the era earned such praises from her. Yet, in officially bringing up von Braun's past, Harris demonstrated more forthrightness than many of her contemporaries in NASA's administrative hierarchy were comfortable with.[22]

Pressure Builds

Harris's problems involved more than frankness regarding prohibited subjects. She also brought uncustomary issues into policy making, via her connections with the Congressional Black Caucus and the NAACP. NASA, for instance, had hired Africans over racist apartheid government opposition at South African tracking stations in the 1960s. But apartheid customs such as Africans eating outside while whites monopolized dining facilities continued. Given that the large majority of Earth's population was not white, NASA might have to consider (as it eventually did) closing the stations.[23]

"Political" ideas like this did not endear Harris to those who saw space exploration as an obvious good that did not require modification. But the major problem with Harris was that she wanted her original job back. She also wanted to do what she had earlier done at the DC Council on Human Relations: transform a junior administrative post into a prominent leadership role after networking within NASA and between NASA and civil rights and women's organizations.[24]

In early 1973, 18 months after her arrival, Harris clearly began pushing top NASA leaders beyond their comfort zones as she tried to implement policy regarding all executive-branch agencies obeying the same affirmative action laws as private industry. New civil service implementation rules also required affirmative action directors within agencies to report *directly* to top administrators of government departments to accomplish this.

---

22. Harris, *Harlem Princess*, pp. 257–258.

23. Transcript of "An Interview with Diane Graham," Harambee program, WTOP TV, Washington, DC, 12 November 1973, 9:00 a.m., pp. 1, 4, Low/NHO; Harris, Hogan and Lynn to James Fletcher, 20 September 1973, Low/NHO, alphabetical files, "EEO—Very Sensitive" folder.

24. Belinda Robnett, *How Long? How Long?: African American Women in the Struggle for Civil Rights* (New York: Oxford University Press, 1997), esp. pp. 19–20, 190ff.

NASA could not just simply ignore Harris. It had 5.6 percent minority and 18 percent female employees in 1973, versus a government average of 20 percent minority and 34 percent female. But of the 4,432 women that NASA employed, only 310 were in science and engineering and just 4—including Harris—were in the highest civil service grades. NASA's technical culture could and did claim that female and minority engineers were scarce. But NASA did no better at hiring more numerous female and minority lawyers or nontechnical professionals (3.7 percent) than scientists and engineers (3.6 percent) in this period. NASA hired more minority male janitors (69 percent) than the government-wide average of 56 percent. But it hired no women at all to do this work and it trailed in *all* other occupational categories, from pilots to guards. What Harris, NASA's highest-ranking woman, was saying about opening up NASA jobs and occupational hierarchies was legitimate and principled.[25]

NASA leaders, however, still didn't want Harris to formulate or implement standards for the Agency. Instead, Fletcher and Low offered Harris a deal. They would raise her a level above that to which they had demoted her even before she arrived in office. She would be Number Two in a new equal employment opportunity (affirmative action) office and also Number Two in an office overseeing NASA contractors. She still would not report directly to either Fletcher or Low, nor would she be able to upgrade the inadequate affirmative action operations out in NASA's Centers created just before her arrival. Harris didn't like the offer, may have threatened to quit, and wanted her original job back.[26]

To Harris, NASA began to look like the uninterested at Headquarters leading the uncommitted in the Centers. To George Low, Harris was an administrative lightweight lacking the qualifications expected of a nontechnical assistant (rather than deputy assistant) administrator. This made NASA's original hiring of her doubly curious. Harris's qualifications were also equivalent to those of the men who were the assistant administrators for public affairs and international affairs at Headquarters at the time.[27]

By March of 1973, bureaucratic knives got whetted. Low privately commented that Harris's affirmative action operation was a "dumping ground for poor people" who "could only say yes [to complaints] and not no." Affirmative action, however, was "too low in the administrative hierarchy" and there was a "lack of management support and no leadership." Low basically agreed with Harris's administrative reasoning while arguing that she was a bad administrator.[28]

---

25. "EEO Statistics," 18 January 1974, Box 107, Low/RPI; "Senate Space Committee Hearings" (1974), pp. 64–65.

26. Low, personal note no. 105, 21 October 1973, pp. 1–2; Fletcher to all NASA employees (2 November 1973), pp. 2–3, both in Box 68, Low/RPI; Holden, "NASA: Sacking of Top Black Woman," p. 805.

27. George Low, "Comments in November 14, 1973 Draft Letter to Senator Moss," 19 November 1973, Box 35, Low/RPI (these men were, respectively, a Chicago businessman and a lawyer with the International Labor Organization).

28. Low to Fletcher, 15 March 1973, 3, Box 68; Low/RPI; Low, personal note no. 91, 14 April 1973, pp. 1–2, Low/RPI; Low, "EEO Contr Compl," notes on the reverse of Low's appointment book for 7 March 1973, Low/NHO; Low, personal note no. 91, 14 April 1973, p. 1, Low/RPI.

This meant Low had to find someone else to get more women and minorities hired, trained, or promoted within NASA. His eyes lighted on NASA's only African American in a high level (or "super-grade") scientific and technical position: Dr. Dudley McConnell of NASA's Scientific and Technical Information Office.[29] Two of Low's aides told him and Fletcher that McConnell "would not want the job," but Low persisted. Low, in fact, had hired McConnell into NASA as an aeronautical research engineer in Cleveland, Ohio, in October 1957. McConnell's personnel or civil rights experience was far less extensive than Harris's. He was also taken out of a senior science and technology position in which there were no other African Americans and put into an office where African Americans were not rare. But, in the end, McConnell agreed to accept the job Harris had originally been hired to do a year and a half earlier.[30]

McConnell Versus Harris

Crude racism played no part in George Low's decision, but the organizational culture of NASA clearly did. Low knew and trusted 16-year NASA man Dudley McConnell far more than he did M.B.A. and personnel/civil rights person Ruth Bates Harris. McConnell, wrote Low, understood the "problems of technical management" in aerospace. McConnell impressed a reporter for *Science* magazine who interviewed him as "soft-spoken, ingratiating, exceedingly articulate, and strong willed." McConnell also compared himself to a ship captain with "unruly crew members."[31]

Turbulence was inevitable, given that Harris was a popular administrator and McConnell definitely was not. Only four months after McConnell took over in April of 1973, Low received two strong oral and written criticisms of McConnell's managerial style from two departing staffers. McConnell didn't relate well to women, "went for appearances rather than substance," and was not communicating affirmative action "concerns to top management." Because "seven or eight" other staffers "might soon be leaving," Fletcher and Low both met privately with McConnell, after which Low privately concluded McConnell might indeed have problems with women and communicating concerns upward. Some of McConnell's difficulties were unintentional. For example, he used a little bell to summon secretaries, who then lampooned him as "Mr. Ding-a-ling" (a silly, affected person). But McConnell's major problem was that for the second time Ruth Bates Harris had now been denied the job for which she was originally hired. NASA's new civil rights in employment chief, moreover, refused some of Harris's early offers of cooperation.[32]

29. For the best biography on McConnell, see Senate Committee on Aeronautical and Space Sciences, *Review of NASA's Equal Opportunity Program*, 93rd Cong., 2d sess., 24 January 1974.

30. Low, personal note no. 91, 14 April 1973, p. 2, Box 68, Low/RPI.

31. Holden, "NASA: Sacking of Top Black Woman," pp. 804, 806.

32. Low, personal note no. 100, 11 August 1973, pp. 1–2, Box 68, Low/RPI; Holden, "NASA: Sacking of Top Black Woman," p. 805.

Discussion therefore soon became argument. Harris and her supporters began to believe NASA would enforce nothing on balky Southern labs. They also knew that organizations such as the National Urban League and the National Organization of Women were displeased by longstanding exclusions of women and minorities from the Houston-based Astronaut Corps. What Harris would later call NASA's "pasteurized and insulated from the real world" aspects, moreover, made it harder to interest the non-male and non-white in space programs. McConnell, for his part, moved slowly, avoided confrontation, and ordered his staff not to undermine him.[33]

This is precisely what Ruth Bates Harris and two of her associates—Joseph M. Hogan and retired Air Force Colonel and Tuskegee Airman Samuel Lynn—did five months after McConnell's too-gradualist tenure began. In September of 1973, the three sent an internal report to Fletcher they had prepared on their own time. In the process, they also did a bureaucratic end-run around McConnell's sponsor, George Low.[34]

The Harris-Lynn report was frank, insistent, and began dramatically. NASA's efforts were "a near-total failure." The agency was denying that problems existed to avoid having to address them. Minorities and women stayed clustered in the lowest civil service job ratings. Thirteen of the 35 total minority hires of fiscal year 1972 were brought in at the lowest pay scale of GS-2. Three NASA Centers had not hired any minorities to do anything in 1972, all Centers in the South—the Kennedy Space Center, the Marshall Space Flight Center, and the Manned Spacecraft Center. Despite cutbacks, these same facilities had hired 10, 24, and 22 people, respectively, that year.[35]

Many considered affirmative action a "sham" at NASA because of uncommitted top management, "insensitive middle management," and "unqualified, uncommitted persons" at NASA Centers. Some NASA installations simply were not going to hire African Americans or women or others until they were forced to. There was a "striking anomaly" between NASA's technical genius and its social insensitivity. The only three females NASA had so far sent into space were two spiders and a monkey. "During an entire generation—from 1958 until the end of this decade—NASA will not have a woman or a minority astronaut in training"; this even as U.S. society was opening up to women and minorities in ways previously seen only during major wars that created labor shortages of white males. Dudley McConnell had demonstrated incapacity and "immaturity in relating to people" and should be removed. Then NASA's human relations policies would begin to match its proven technical excellence.[36]

---

33. Harris, *Harlem Princess*, esp. pp. 267, 262–264 (quote from p. 264); Senate Space Committee Hearings (1974), p. 88ff; Holden, "NASA: Sacking of Top Black Woman," pp. 804–806, passim. For the astronauts issues, see, for instance, Martha Ackmann, *The Mercury 13,* pp. 176, 183, 191; Joseph Atkinson and Jay Shafritz, *The Real Stuff: A History of NASA's Astronaut Recruitment Program* (New York: Praeger, 1985), esp. pp. 134–137.

34. Ruth Bates Harris, Joseph M. Hogan, and Samuel Lynn to Dr. James C. Fletcher, 20 September 1973, Low/NHO (three-page unpaginated cover letter).

35. Ibid., Points II [p. 1] and IV [p. 2].

36. Ibid., the final paragraph, the "Preamble" and the "Special Concerns: No Minority or Female Astronauts" sections.

NASA Administrator Fletcher listened to his frustrated affirmative action staffers. He said vaguely agreeable things about further absorbing their message. Then axes fell. Fletcher fired Harris, transferred Hogan, and told Lynn to work with McConnell or resign. Fletcher claimed the firing and disciplinary actions had nothing to do with policy recommendations, all of which were "already well documented." Instead, in a four-page, single-spaced memo sent to all NASA Headquarters staff, Fletcher presented Harris as a "seriously disruptive force." Though a good advocate, Harris was an uncompromising ideologue who had sabotaged McConnell and NASA. NASA's minority and female hiring record was not one in which it could take pride, but the future would be a marked improvement over the past.[37]

Harris was purged on 11 October 1973. For two weeks, things were quiet. Then the situation changed decisively after the 27th of October. In his private papers, Low made no connection whatsoever between Nixon's purge of the Watergate special prosecutor on the 20th of October and NASA's difficulties after Harris's firing that involved NASA in far wider political struggles roiling all around it.[38]

NASA's timing, nevertheless, was awful. Its leaders could not conceive that anyone would doubt their actions or motives, but plenty did. *The New York Times*, *The Washington Post*, three Senate committees, major African American newspapers, and local Washington radio and TV stations all featured the Ruth Bates Harris story shortly after Fletcher fired her. NASA blithely walked into a journalistic tree-shredder. Its press was so bad that Fletcher's office forwarded to Low a "much more elegant than usual" story about the case from *Science* magazine from 23 November 1973, which Low *should* read.[39]

NASA's Political Beating Begins

The *Science* article showed just how badly NASA had flunked politically. The author, Constance Holden, found the charges that Harris was disruptive or radical to be overblown. "It is difficult to imagine," she concluded, "that it took NASA two years to discover that the woman was a 'divisive' personality." "Fletcher's Nixonesque purge" had only "opened up a can of worms" at NASA, Holden said. Seventy headquarters staff had already protested the firing, and civil rights and women's groups at several NASA Centers were pledging support for Harris. A group called MEAN (Minority Employees at NASA) had been formed to protest employment conditions. Most importantly, the NAACP Legal Defense Fund was about to take on the Harris case to establish legal precedent regarding executive-branch agencies obeying affirmative action laws.[40]

37. Low, personal note no. 105, 21 October 1973, pp. 1–2; Fletcher to all NASA employees (2 November 1973), pp. 2–3, both in Box 68, Low/RPI; Holden, "NASA: Sacking of Top Black Woman," p. 805.

38. Low, personal note no. 107, 13 November 1973, pp. 4–5, Box 67, Low/RPI.

39. "H" to Dr. Low, 20 November 1973, Low/NHO.

40. Holden, "NASA: Sacking of Top Black Woman," p. 806; Holden, "NASA Satellite Project: The Boss Is a Woman," *Science* (5 January 1973): pp. 48–49; "Harris, Mrs. Ruth Bates" biographical file, NHO (hereafter Harris/NHO). Part of the "radical" charge may have come from the dedication of the Harris, Hogan, and Lynn report to "those who have been victims of economic indignities" and "those who are indignant enough to do something about it."

Things only got worse. An eventual 50 national organizations such as the National Conference of Catholic Charities and the National Organization of Women protested. So did prominent members of Congress. When Rep. Fernand St. Germain's (D-Rhode Island) request for an explanation of Harris's firing was simply ignored by Fletcher, St. Germain curtly wrote to Fletcher that his voting on NASA appropriations measures would be "equally unsatisfactory." The Congressional Black Caucus protested and two Senate committees started making serious noise about scheduling hearings.[41]

Accordingly, by early December James Fletcher badly needed proof of Harris's radicalism, so he sent Low and others off on a wild goose chase after he received a brief note stating, "We *Black* People do not want this Subversive person representing us—she is a known 'Trouble Maker.'" Today, NASA handles claims like this the way it handles claims that it is hiding the truth about UFOs—the claims are politely dismissed. Fletcher, however, pressed for evidence of subversive associations but Low—via deniable intermediaries—was able to come up with nothing.[42]

As NASA chiefs belatedly realized the burden of proof about disloyalty might actually be on them, Harris gave interviews to reporters and her NAACP lawyers petitioned the U.S. Civil Service Commission against Harris's dismissal as an illegal reprisal. NASA leaders had presumed that Harris was a political appointee they could fire at will; her NAACP lawyers cogently argued otherwise. In a climate of Constitutional crisis, any presidential agency claiming *lèse majesté* was suicidal. So NASA retreated and started paying Harris her full salary while her status was being determined.[43]

Simultaneously, Low was souring on the man he'd selected only six months earlier to run affirmative action. "Apparently," he noted privately late in December, "McConnell has all the right ideas but he is not pushing very hard insofar as implementing these ideas is concerned." Despite Low's encouragement, McConnell was "still very slow." "I think Fletcher and I will have to 'lead him by the hand' . . . until he can really run the show on his own," Low concluded.[44]

---

41. Holden, "NASA: Sacking of Top Black Woman," p. 806; Ruth Bates Harris, *Harlem Princess*, pp. 261, 263–264; St. Germain to Fletcher, 12 December 1973, Low/NHO. The National Organization of Women (NOW) only belatedly protested Harris's firing. It provided Harris little or nothing in the way of financial or legal assistance. For NOW's "often chaotic" internal organization in its formative years, see Freeman, *The Politics of Women's Liberation*, p. 71ff. For NOW's after-the-fact protest, see Senate Space Committee Hearings (1974), pp. 98–99.

42. Fletcher to Low, 14 December 1973 and M. Johnson to James Fletcher, 30 October 1973, Low/NHO; Low to Fletcher, 14 December 1973, Low/NHO; Unsigned memo from Low, undated (circa December, 1973) regarding report from Bart Fugler regarding "Concerned Citizens for America" group, Low/NHO.

43. Holden, "NASA: Sacking of Top Black Woman," p. 806.

44. Low, personal note no. 110, 23 December 1973, p. 7, Box 67, Low/RPI. For Low's continuing belief that Harris was unqualified and his beginning awareness that he actually had to *prove* that point, see Low to "AD/Deputy Administrator," 19 November 1973, Box 35, Low/RPI.

By January 1974, NASA's yawning credibility gap got wider. NASA lawyers secretly told Low and Fletcher the Agency was probably going to lose its case against Harris and her NAACP Legal Defense Fund attorneys. It also faced the threat of class action suits from women and minorities. Crucial, here, was the unwillingness of those who had attacked Harris verbally to risk public exposure or legal repercussions. Despite requests from Low and Fletcher to repeat their charges to NASA (and NAACP) lawyers or to civil service representatives, they all refused.[45]

Within three months, then, NASA's case against Ruth Bates Harris was in tatters. Blaming Harris wasn't going to work, so the Agency therefore had a "pressing need" to make some progress itself. In January, Low pushed all NASA Center Directors to hire "at least one minority or female at the executive level [apiece] within the next 6 months." Three Southern Center leaders in Texas, Alabama, and Florida refused, saying it was impossible to even begin. NASA had no operational and Agency-wide plan for recruiting or promoting women or minorities. NASA was being flayed in editorials in *The Washington Post* and elsewhere for trying to combat "institutionalized racism and sexism" with "simple pieties." Worse, NASA was by now headed for political appointments on Capitol Hill with three powerful Senate bodies. First on the list was Senator William Proxmire of Wisconsin and the Senate appropriations subcommittee he chaired, meeting on the 11th of January.[46]

Senator Proxmire was trouble, and NASA leaders knew it. He'd worked with Harris often on District of Columbia governance issues in the 1960s. He had also been a notable critic of NASA for a decade and didn't care a whit for vague pieties. Accordingly, Fletcher and Low avoided testifying. Instead, they sent Dr. Dudley McConnell and two other managers in their places. Low, however, ". . . wrote Dudley's [preliminary opening] statement since his was rather weak."[47]

Harris and her associates Hogan and Lynn, meanwhile, argued far more openly. Hogan said NASA's approach to hiring women and minorities was "calculated to provide the appearance of compliance while not doing so." Harris added that NASA was now trying to take credit for proposals, such as recruiting minorities and women into the Astronaut Corps, that she, Hogan, and Lynn had made—this after punishing them and then doing nothing to actually create such a recruitment program. Finally, they said, McConnell had begun his tenure as affirmative action

45. Low, personal note no. 111, 6 January 1974, pp. 6–7 and Low, personal note no. 126, 18 August 1974, p. 8, Box 67, Low/RPI.

46. Low to "AD/Deputy Administrator," 21 January 1974, Box 34, Low RPI; "Center Director's Meeting, December 10-11, 1973 . . ." (minutes dated 18 December 1973), File 5, Box 66, Low/RPI; Senate Space Committee Hearings, (1974), pp. 26–28; "Racism, Sexism, and Space Ventures" [editorial], *The Washington Post*, 24 November 1973; Low, personal note no. 111, 6 January 1974, Box 67, Low/RPI. For the lack of NASA minority or female hiring plans, see Dudley McConnell to Low, 26 November 1973, Low/NHO.

47. Low, personal note no. 112, 20 January 1974, p. 3, Box 67, Low/RPI. For Proxmire and Harris in the 1960s, see "Harris (1971-1)," p. 5ff.

head at NASA observing that forwarding civil rights in employment by "saying it was the law" was not a good idea, ". . . because NASA always breaks the law."[48]

Senator Proxmire then gave the hapless McConnell a political mauling, starting with the "always breaks the law" statement McConnell first claimed he couldn't remember and later publicly apologized for making. No contractor compliance programs existed at key NASA Centers in Houston and Huntsville and no "show cause" orders had ever been issued to any NASA contractor. NASA representatives themselves used phrases like "somewhat poor performance" to describe their efforts. Senator Proxmire refused to accept NASA's contention that it had not excluded minorities and (especially) women from the six groups of all-male and all-white astronauts it had already selected. The color line in professional baseball had fallen in 1947, and the civil rights era was now 15 years old, but NASA was willfully ignoring social change all around it. Dispensing with any pretense at political courtesy, Senator Proxmire said "Congressional monitoring" of NASA's minority and women hiring programs would take place, effective immediately, via his subcommittee because of NASA's "extremely poor record."[49]

Back at NASA, Low was oblivious to NASA's and McConnell's mauling. He used phrases like "not really unfavorable" and "in reasonably good shape" to describe NASA's standing. Selective awareness and denials, however, did not lead either Low or Fletcher to want to personally attend the next hearing, this one by the Senate's space committee. Aides now protested: both leaders absenting themselves looked bad. One or both needed to attend to maintain NASA's fast-fading organizational credibility in the Harris case.[50]

James Fletcher flatly refused. A Utah Mormon, Fletcher feared political and religious embarrassment, and with good reason. Mormonism had major problems with race and gender in the 1970s, and Fletcher wanted to keep himself and NASA removed from them. "Fletcher," Low recorded on 20 January 1974, ". . . is particularly concerned because of his Mormon Church affiliation and the fact that the new 'prophet' of the Mormon Church has made statements that can be interpreted as against women and has also re-emphasized that the Mormon Church will not admit blacks to 'priesthood' [and full membership in the faith]." Low and others argued that Fletcher was not in the Mormon hierarchy and would probably not

---

48. Austin Scott, "NASA Hit on Minority Hiring: Agency Needs Prodding, Proxmire Says," *The Washington Post*, 12 January 1974 (clipping in Low/NHO); Senate Appropriations Subcommittee Hearings (1974), esp. pp. 118–121.

49. *The Washington Post*, "NASA Hit on Minority Hiring"; Senate Appropriations Subcommittee Hearings, (1974), pp. 132–134, 138 (McConnell had approached one aerospace company's central office to get problems solved informally); "NASA Defends Lag on Women's Jobs," *The New York Times*, 12 January 1974 (clipping in Low/NHO). See also Senate Appropriations Subcommittee Hearings (1974), p. 137 and, for audience laughter at NASA expense, pp. 138–140.

50. Low, personal note no. 112, 20 January 1974, p. 2, Box 67, Low/RPI; [Washington, DC] *Afro-American*, 11 January 1974: 107 (clipping in "Harris [1991]").

have any difficulty separating his personal beliefs from those of his faith. Fletcher, however, demurred. Mormonism's racial exclusion practices, meanwhile, were not ended until 1978 (after Mormon leaders received their first divine revelation since polygamy was ended almost a century before).[51]

Low, however, attended the second Senate hearing of January 1974 in Fletcher's place. As he became higher-profile politically, Low also further distanced himself from the physicist he'd so recently and strongly supported over Harris. McConnell, Low observed, was "overwhelmed and overworked," so Low pushed him to hire Dr. Harriett Jenkins, an educator, as a deputy. McConnell bridled, saying Jenkins was not a "professional in EEO [affirmative action]." Given that he wasn't either, Low began wondering whether McConnell wasn't "a little bit afraid" of Jenkins as a woman.[52]

To test this hypothesis, Low had an aide talk at length with two senior African American women at the nearby NASA Goddard Center who had professional interactions with McConnell. Neither woman was impressed. "Apparently," Low concluded, "[McConnell still had to] learn how to communicate effectively with people, especially women." One of the women and a Hispanic section head at Goddard told Low's aide that minorities and women really *did* believe NASA discriminated regarding promotions. One method the Kennedy Space Center then used particularly annoyed the Civil Service Commission: it had 43 different job rankings for secretaries, so women could thus be promoted but kept out of management ranks.[53]

Austrian immigrant and political refugee George Low had remade himself in America. He believed "key people" at NASA's Headquarters and its Centers would do the same. Low's optimism, however, was not widely shared on Capitol Hill, where NASA's political grilling about administrative unwillingness, unmet promises, and belated claims about future virtues continued. By 24 January 1974, NASA and George Low faced frank disbelief at hearings before a normally sympathetic Senate Aeronautical and Space Sciences Committee. Ruth Bates Harris's performance at the hearings was calm and understated. Watergate was now consuming more and more Senatorial energies. Harris said NASA had failed in its duty to enforce laws passed by Congress. She and her supporters had "pricked the conscience" of NASA. "Institutionalized racism and sexism" existed there. The virtues NASA now claimed

---

51. Roger D. Launius, "A Western Mormon in Washington, DC: James C. Fletcher, NASA, and the Final Frontier, "*Pacific Historical Review*, 64 (May 1995): pp. 217–241, esp. 228; Low, personal note no. 112, 20 January 1974, p. 3, Box 67, Low/RPI. The exclusion came from a narrow reading of Mormon scripture, including the Book of Alma in *The Book of Mormon*. Industrious Nephites and slovenly Lamanites fought for control of a New World and the latter lost. Their skins became dark and they became savage by Godly decree. Mormon traditionalists put up a 20-year fight against allowing African Americans full spiritual standing.

52. Low, personal note no. 112, 20 January 1974, p. 4, and Low, personal note no. 113, 3 February 1974, Box 67, Low/RPI.

53. Low, personal note no. 112, 20 January 1974, pp. 4–5, 112; Low, personal note no. 87, "Discussions with the Service Commission," 17 February 1973, p. 4, Low/RPI.

were belated and deceptive. NASA's basic problem was that it was letting some of its Centers evade the law. It had not yet begun internal training and promotion programs for female or minority staff. It claimed it could not do what some unspecified percentage of aerospace contractors had already done: increase percentages of minority and women employees even as overall workforces declined.[54]

Harris's performance earned compliments from the committee's chair; Low and McConnell—who avoided being in the same room with Harris—got rougher treatment. Senators doubted that NASA Centers had any intention of doing what NASA Headquarters said it wanted. It didn't help that the Department of Labor said it looked as if every NASA Center Director had their own "personal view of appropriate compliance policies and procedures." Democratic Senators said NASA was "groping for sympathy" and setting "modest" goals it wasn't really committed to. It didn't help that these goals [of hiring 80 more women and 80 more minorities in professional positions in its 25,000-person workforce in 1974] were decided upon only two weeks before the Senator Proxmire hearing that was held earlier in January. "Where's your sense of urgency?" Senator Howard Metzenbaum of Ohio asked at one point. "He thought our performance was lousy and made no bones about it," Low later recorded. The committee then, per Senator Proxmire, put NASA under legislative oversight. Affirmative action funding (also per Proxmire) was to double; NASA had to report on every major professional position it filled; and minorities and women were to be put on NASA hiring and promotion boards.[55]

NASA Fights On

Still NASA leaders fought on. Harris, NASA representatives told Senators, was "little more than a lobbyist for the cause of minorities and women." Low privately believed Harris and "quite a few" of her supporters within NASA were sabotaging McConnell. The idea that she might be a lightning rod, not a thunderbolt, wasn't considered. NASA's top leaders now knew their case against Harris was legally lost. Nevertheless, in January they decided that she would never be rehired to her former position but, instead, only to a "comparable" post within NASA. If—and when—Harris won, she could be isolated and marginalized.[56]

---

54. Senate Committee on Aeronautical and Space Sciences, "Testimony of Ruth Bates Harris, January 24, 1974," Harris/NHO; Low, personal note no. 112, 20 January 1974, pp. 3–6.

55. Austin Scott, "Senators Eye NASA EEO Goals," *The Washington Post*, 25 January 1974 (Reprinted in NASA *Current News*, 25 January 1974: I); Low, personal note no. 113, 3 February 1974, p. 3, Box 67, Low/RPI.

56. R. Tenney Johnson to "A" [Fletcher and George Low], 3 January 1974, Low/NHO; Low to McConnell, "Decision Statement on Numerical Goals for 1974," 7 January 1974; and Low, personal note no. 113, 3 February 1974, p. 3, Box 67, Low/RPI. The "lobbyist" change is in Senate Appropriations Hearings (1974), p. 143.

Dudley McConnell's days, meanwhile, were numbered. He still didn't want Harriett Jenkins as a deputy, so Low got Fletcher to hire Jenkins himself. McConnell still resisted, offering at least one other female candidate the job after a weekend interview. Low then forced the issue to a final conclusion. Harriett Jenkins's arrival at NASA wasn't quite as messy as Ruth Bates Harris's had been, but it demonstrated precisely the same organizational ambivalence and infighting.[57]

As Harriett Jenkins began working as McConnell's deputy in February of 1974, the drumbeat of external and internal criticisms intensified. The chairman of the Civil Service Commission, for example, told NASA its record of processing racial and sexual discrimination complaints "was not very good." The Department of Labor was similarly distinctly unimpressed with NASA's record on oversight of civil rights in employment standards in firms it contracted with at its Centers. Recruitment teams for minority and female scientists and engineers still weren't in place in at least four NASA Centers. Low's orders to have entry-level jobs offered to likely candidates "on the spot" weren't happening. A disappointed Low told a meeting of deputy Center directors called to jump-start affirmative action that "most of our technical managers were not accustomed to handling human problems."[58]

Meanwhile, protest levels within NASA increased. The first training workshops regarding racial and sexual issues for senior Center leaders were scheduled for April and quickly got postponed. By March, top-level meetings were called to pick an affirmative action officer "more acceptable to women and minorities." A selection panel should probably consist of three of NASA's highest-ranking women, a Hispanic, and the NASA general counsel. They should "probably" pick a minority person as the affirmative action officer. Low was advising McConnell to return to technical work; trouble was brewing at NASA Centers in Virginia and Texas; and Low was addressing all 1,000 NASA Headquarters staff at "communications" (morale-building) sessions. By May, NASA lawyers still advised settling with Harris; an outside consultant told Low and Fletcher they hadn't done well regarding a "generally unstable headquarters situation since Ruth Bates Harris was terminated," and Low wrote that "[O]ur people are still very much concerned . . . and have not seen that real progress has been made." A Government Accountability Office (GAO) investigation based on interviews with NASA employees confirmed Low's

57. Low to Fletcher, "Miscellaneous Items," 21 January 1974; Low to Deputy Associate Administrator for Organization and Management, 5 February 1974; Low to Fletcher, "Harriet Jenkins," 5 February 1974, all in Box 67, Low/RPI.

58. Low to McConnell, "Complaint Processing Time," 4 February 1974, Box 34, Low/RPI; Low to McConnell, 28 January 1974; Low to Assistant Admin. for Institutional Management, "Proposed Recruiting Plan for Minority Scientists and Engineers," 2 February 1974; "Meeting Record, February 6-7, 1974, JSC, Deputy Directors' Council Meeting," all in Box 67, Low/RPI; "Ex-Aide Rehired by NASA," *The Washington Post*, 18 August 1974: p. A20.

statement. No fewer than six Senate and House committees—including NASA's authorization and appropriations committees in both houses—got copies of the GAO findings.[59]

So, after eight months, denials gradually began to end. In June of 1974, as Nixon's presidency entered its final weeks, NASA and NAACP Legal Defense Fund lawyers sat down to hash out a settlement. Two sets of issues especially mattered. The first was organizational. Ruth Bates Harris and her counsel wanted—and got—enhanced training and promotion programs, monitoring of minority and female hiring at middle and senior management levels, and the beginnings of NASA research and development awards to Historically Black colleges and universities. Fletcher and Low refused, however, to give way on a key point. Harris wanted strong Headquarters managerial oversight and control over affirmative action at NASA Centers. This was the centralized job, of course, she had been originally hired by NASA to do. But Fletcher and Low refused, citing NASA's "decentralized management concept." Uncooperative Southern Centers got a breathing space.[60]

On a second set of issues regarding Harris's personal administrative future at NASA, the Agency fought even harder. Harris still wanted what she'd always wanted: Dudley McConnell's job. That meant policy-making power. NASA still didn't want her to have this. She had embarrassed an agency that often saw itself in elite, even transcendent terms. To Low and Fletcher, she was an "advocate," not a "manager." Her skills were ignored. A senior management discrimination case at Headquarters, for example, had become explosive because of what Low called "three levels of poor management" where the white supervisors involved refused to take training in dealing with minority or female workers until Low "urged them to do it six times." Harris, who had handled just such intensive "sensitivity training" programs for Washington, DC, police in the 1960s, was still persona non grata.[61]

Meanwhile, a protesting Dudley McConnell was pushed out of his job in June and July. Low was now de facto affirmative action head. The NAACP Legal Defense Fund threatened class action lawsuits on behalf of groups of employees at NASA Centers. NASA's strongest House ally apparently told NASA the Agency's political status was slipping. NASA could not win but it could not settle, either.[62]

59. Harvey W. Herring, "Meeting Record, March 29, 1974, EEO Officer, Low/NHO; "GAO Investigation of EEO Program," 23 April 1974, Box 107; Low, personal note no. 117, 30 March 1974, p. 4; Low, personal note no. 118, 13 April 1974, p. 2; Low, personal note no. 120, 11 May 1974, pp. 7–8; Low, personal note no. 122, 8 June 1974, p. 11; "Notes for All Hands Meetings [April, 1974]," all in Box 67, Low/RPI.

60. R. Tenney Johnson, "Concerns and Views," circa 5 June 1974, Box 67, Low/RPI.

61. George Low, "Meeting with Headquarters Equal Employment Opportunity Advisory Group," 26 June 1974, esp. p. 3.

62. Low, personal note no. 125, 20 July 1974, pp. 5–6, File 1, Box 67, Low/RPI; Low to Fletcher, "House Staff Study on NASA Public Affairs Activities," 17 July 1974, Box 67, Low/RPI; Constance Holden, "New EEO Leadership at NASA," *Science* (30 August 1974): p. 769.

Richard Nixon's presidential resignation on 9 August 1974 seems to have helped prompt the combatants to settle. McConnell left office after 16 unhappy months; Harriett Jenkins promptly succeeded him. Ruth Bates Harris came back to NASA, but only at deputy assistant rank and only on the PR and community outreach fringes of the affirmative action office. It was what *Science* magazine termed "at least partial vindication." On the 17th of August, Fletcher's official statement welcoming back the woman he had fired complimented her "genuine ability to communicate to members of communities whom NASA has not reached in the past and whom we need to reach." "[S]ome of the forward movement NASA . . . has made . . . ," he continued, "has been stimulated by forces she so eloquently set in motion." An uncompromising and disloyal ideologue now had redeeming importance.[63]

Verbal bouquets, however, did not make reacclimation any easier. Harriett Jenkins did not work with Harris (some of whose efforts involved identifying and recruiting African American astronaut candidates via aviation veterans' organizations such as the Tuskegee Airmen). NASA's rate of accomplishment remained "abysmally low," George Low noted privately. Line managers had "not launched a significant effort" and top management had not made it clear to line and staff administrators what sanctions or rewards would apply to them. "What's wrong?" a discouraged Low wrote on the margin of his first annual performance review with Harriett Jenkins. Pent-up tensions also levied a wider toll. Harris's marriage dissolved and a son was stricken with AIDS. In 1976, Harris suffered a nervous breakdown and returned to New York to get her life back together. Fletcher and Harris smiled for cameras as she departed, but, after her return to Washington in 1978, she never worked for NASA again.[64]

NASA Starts To Change

The Harris deal ended further Senatorial hearings and negative journalistic coverage. All of NASA's internal records on the Harris case were sealed as part of her rehiring agreement. After being legally and politically forced into more active compliance, NASA started congratulating itself on how well it was doing in hiring

63. Dudley McConnell was promoted to Deputy Director for Advanced Programs in the Solar System Exploration Division. He later became Associate Director for Applications in NASA's Earth Science and Applications Division. On 28 October 1991, he died of a heart attack at age 55. R. Tenney to James C. Fletcher, 4 April 1974; Low, personal note no. 126, 18 August 1974, p. I, Box 67, Low/RPI; James C. Fletcher, "Memorandum to All NASA Employees" 16 August 1974—all in EEO—Very Sensitive file in Low/NHO; Gloria Borger, "Woman Fired by NASA Wins Fight to Return," *Washington Star-News*, 17 August 1974: p. A4; "Harris, Mrs. Ruth Bates" biographical file, NHO; Holden, "New EEO Leadership at NASA," p. 769; Lawrence Feinberg, "Ex-Aide Rehired by NASA," *The Washington Post*, 18 August 1974: p. A20; "Ruth Harris Assumes New Post at NASA," 17 August 1974.

64. "Work Planning and Progress Review, Jenkins–Low, June 27, 1975," Box 76, Low/RPI; "Ruth Bates Harris Honored at Headquarters Farewell Party," *Spaceport News* (25 November 1976): p. 4; "Ruth Bates Harris Appointment at Interior," *Weekly Bulletin* 20 (15 May 1978)—all in "Harris, Mrs. Ruth Bates" biographical file, NHO. Harris, *Harlem Princess,* p. 279ff.

and promoting more women and minorities. The basic norms, beliefs, and practices within NASA's "human spaceflight culture," however, only changed slowly. Female and minority supervisors at levels of GS-14 and above stayed rare until the 1980s. A false start at recruiting minority and female astronauts was implemented in 1978. The first female astronauts who "walked through the doors their activist sisters had pried open for them," however, were hardly feminists. Skepticism and distancing only gradually turned into general acceptance by male astronauts. NASA had taken the right path after repeatedly denying it needed to do so; it thus got little credit for belated understanding.[65]

Meanwhile, the new head of Headquarters Civil Rights in Employment, Dr. Harriett Jenkins, had to operate in an organizational landscape littered with fears and grudges. Two vignettes show how her organizational style remained nonconfrontational and gradualist. First, in February of 1975, an affirmative action meeting for top Center heads was finally held, almost a year late. During that meeting a white manager was "quite rude" to Jenkins. The abuse was thorough enough that a normally undemonstrative Jenkins broke into tears and later asked Low for managerial support. Low gave it, but then his initially concerned comments turned flippant. After talking with the all-white deputy Center directors at the meeting, he privately dismissed female and minority objections to Jenkins's treatment as overreaction. Given NASA's "abysmally slow" record since 1971, Low's turnaround was precisely the kind of group denial that had cost Ruth Bates Harris and Dr. Dudley McConnell their jobs. Jenkins was going to have to proceed carefully to survive for long.[66]

In case she had any doubts, another indicator of just how far NASA had to go occurred late in 1975. Jenkins and her associate, Peter Chen, proposed an affirmative action training session "for NASA's senior [Headquarters] management from Fletcher and Low on down." They wanted a one-day session led by two specialists from Arthur D. Little, a well-known management consultancy. Fletcher and Low had a preliminary meeting with the A. D. Little people and came back utterly opposed. Specifically, they did not want any kind of policy discussions or consciousness-raising where disagreements among senior staff might be identified.

65. Kevles, *Almost Heaven*, esp. p. 69ff; Weitekamp, *Right Stuff, Wrong Sex*; and Ackmann, *The Mercury 13*, also briefly cover the opening of the Astronaut Corps to women. Only Kevles very briefly mentions Harris as a change agent (see pp. 58–59). For opposition to female astronauts, see Mike Mullane, *Riding Rockets: The Outrageous Tales of a Shuttle Astronaut* (New York: Charles Scribner's Sons, 2006), esp. pp. 36ff, 346–347, 357.

66. R. Tenney Johnson to Cashdan, 2 August 1974; Low to Frank Zarb of Office of Management and Budget, 18 September 1974; Low, "Telephone Call From Hugh Loweth," 1 October 1974; Low to AC "Center Operations," 20 February 1975; Low, personal note no. 137, 1 February 1975, pp. 3, 8; Low, personal note no. 139, 1 March 1975, p. 3 (all in Box 66, Low/RPI). For "organizational culture" at NASA, see Admiral Harold W. Gehman, Jr., et al., Columbia Accident Investigation Board Report, vol. 1 (August, 2003) (Washington, DC: NASA and the Government Printing Office, 2003), pp. 9, 99–118; McCurdy, *Inside NASA*; Diane T. Vaughan, *The Challenger Launch Decision* (Chicago: University of Chicago Press, 1996); Sylvia Fries, *NASA Engineers in the Age of Apollo* (Washington, DC: NASA, 1977); Charles Murray and Catherine Bly Cox, *Apollo: The Race to the Moon* (New York: Simon & Schuster, 1989).

Because the A. D. Little people "tried to drive a wedge between Fletcher and me, and put Fletcher on the carpet for not knowing as much as I did about the situation in NASA, [they were] not the right kind of people for a NASA training session," Low concluded—especially because they might also drive wedges between Low and Fletcher and their senior staff, which would only be "counterproductive."[67]

Because precisely such differential understandings and actual or contrived ignorance were what the affirmative action mess at NASA was all about, Fletcher's and Low's refusals ensured more policy muddle within Headquarters and Center managerial hierarchies. They also implied that the men leading NASA had nothing significant to learn or agree about. Those, for instance, who only saw racial or sexual bias when white males were excluded from competitions could continue to do so. When Low told Jenkins and Chen this, Jenkins did not protest, but her associate did. Low's unusually graphic notes then describe the scene:

> I said that they must realize the implementation of EEO [affirmative action] activities at NASA is my responsibility in accordance with policies which were established by Fletcher. Peter [Chen] took issue with this and said that if Fletcher were [*sic*] not going to run EEO in NASA and be personally involved in its implementation, it would never succeed. I very deliberately got mad at Peter, told him if he didn't like the way I was running things, he could march right into Fletcher's office and tell him so, but until Fletcher directed me to do differently, I would continue to be in charge. Peter did not take me up on my offer to go see Fletcher, and as a matter of fact backpedaled very nicely.[68]

Given the fate of Ruth Bates Harris, Chen's (or Jenkins's) reluctance to appeal over Low's head to Fletcher was understandable. Low, however, was no cardboard villain. He knew what racism was and had survived Hitlerism. Additionally, his efforts to get NASA to start addressing the problems Harris and her allies identified were relatively strong and sustained as compared with others of his managerial era. On the eve of his departure from the Agency in 1976, for instance, he sponsored an unusual three-day retreat for senior NASA managers. At this meeting, Noel Hinners of the Office of Space Science was unusually frank. "A successful filtering mechanism in the information channel is a major problem," he began. "Many people believe that real problems are being submerged [or dealt with impatiently]." Problems were also "covered-up so top managers won't know." Finally, Hinners added, "[T]here is a failure to face up to personnel problems by many managers who don't like to discuss such issues with their own people." Here was most of NASA's unhappy early experience with minorities and women in a nutshell.[69]

---

67. Low, personal note no. 156, 13 December 1975, p. 9, Box 65, Low/RPI.

68. Ibid.

69. Minutes of Senior Management Conference, Reston, VA, 17–19 March 1976, p. 12, Box 65, Low/RPI; Low, personal note no. 167, 4 June 1976, p. 3, File 1, Box 65, Low/RPI.

The Harriett Jenkins Era, 1974–1992

Affirmative action's new head at NASA, however, possessed key advantages her predecessors Harris and McConnell had not. The most important was timing. Harris had "lanced the boil," as Jenkins later put it. Built-up emotional pressures had begun to decrease as decades of denial began to end. Pioneering affirmative action/EEO managers such as Harris had jarred and infuriated people, and their casualty rates were very high. Succeeding managers such as Jenkins avoided Harris and gradually operated in a less strife-filled environment. Generational change was a factor. By the 1980s, the founding generation at NASA, who grew up in a segregated and apartheid America in which women rarely existed in aerospace and never gave adult males orders about anything, were retiring and dying.[70]

Harriett Jenkins also benefited because she was the patient teacher, in personality and bureaucratic approach, that NASA Headquarters may have thought it was getting when it originally hired Harris in mid-1971. Born in Fort Worth, Texas, Jenkins went to Fisk University in Nashville, Tennessee, on a scholarship. Graduating with a mathematics degree in 1945, she joined her sister in California, where she worked in clerical roles for the Air Force, an insurance company, and the Oakland Police Department from 1948 to 1954.[71]

Jenkins's real occupational journey began in 1954, the year in which the Supreme Court, in *Brown v. Topeka*, ruled that educational segregation should end "with all deliberate speed." Jenkins promptly applied for a high school teaching position in the public schools of neighboring Berkeley, California, a notable university town. Notable or not, Berkeley's public schools also excluded African Americans from relatively senior teaching jobs at that time. However, Jenkins did not give up. If one tactic didn't work, she tried another. Taking several years to earn a certificate in elementary education, she applied again and proved herself in the Berkeley system with a second-grade class which had "sent two previous teachers home" in despair. Rising quickly through the ranks as Berkeley slowly and peacefully addressed varieties of de facto school segregation, she became the city's first female African American vice principal, director of elementary education, and finally assistant superintendent of schools. Along the way, she earned her master's in education at the University of California-Berkeley in 1957 and her doctorate in education there in 1973.[72]

That same year, Jenkins followed her second husband, a career military man, to Washington, where she worked as a consultant before going to NASA. Typically, Jenkins did not confront power; she educated it via alternate routes. While Ruth Bates Harris had charisma, challenging people to energize and motivate them, Harriett

70. Dr. Harriett Jenkins interview with author, 25 August 2005, Bethesda, MD.

71. Carol Harvey, "How to Succeed When It's Trying; Interview with Harriett G. Jenkins . . . of NASA," *The Bureaucrat: the Journal for Public Managers* 20/3 (Fall, 1991), pp. 33–36.

72. Carol Harvey, "How to Succeed," pp. 33–34.

Jenkins emphasized thoroughness and patience. Harris spoke of "changing the system" that made discrimination against many groups possible. Jenkins, in contrast, preferred value-neutral phraseology like "increasing the pool" of qualified applicants. The only NASA Headquarters man who officially honored Harris's efforts at opening space exploration to women and minorities in the 1980s was himself fired for including actress Jane Fonda in a bipartisan group of influential women invited to the launch of first female astronaut Sally Ride in 1982. From then until her death in 2004, Harris was—at best—a footnote in NASA's organizational memory of itself.[73]

As Harris (and McConnell) lapsed into invisibility, Jenkins's 18-year tenure at NASA Headquarters left her with honors from the Agency, from women in aerospace science and engineering organizations, and from the Congressionally chartered National Academy of Public Administration. Moreover, after Jenkins left NASA she became the director of the Office of Fair Employment Practices for the U.S. Senate from 1992 to 1997. She has continued to do consulting work for NASA and other organizations.[74]

Jenkins was honored and Harris was forgotten because Jenkins was a nonconfrontational gradualist. Key, here, was her willingness to work within a decentralized NASA structure and with a Center-oriented Equal Opportunity (affirmative action) Council of deputy directors of NASA labs, joined with staff from Jenkins's Headquarters office. New York-raised Harris had always opposed this approach, seeing it as one more tool for noncompliance by Southern labs and reactionary middle managers who screened matters from upper echelons. Native Southerner Jenkins seems to have assumed that lower levels of compliance were a shorter-term given, one that would be gradually addressed in programs and plans taking 10 to 15 years to come to fruition. Harris and her allies wanted to force Centers to stop stonewalling and start doing things. Jenkins and managers like her, with a process of sometimes reluctant accommodation under way, wanted NASA Center managers to buy into the concept that diversifying NASA's labor force was a key part of their jobs, not some radical, liberal excess that would promptly be ruled unconstitutional when legislators and judges came to their senses. Jenkins's nonconfrontational "What can we do for you?" approach to Center Directors was part of this Fabian strategy. Harris was shocked that a twentieth-century science- and engineering-based Agency had significant numbers of people in it with nineteenth-

73. To illustrate the point about Harris's historical disappearance, no prominent obituaries appeared when Ruth Bates Harris died in 2004.

74. For a Harris challenge to James Fletcher that contributed to her getting fired, see Senate Appropriations Committee hearings (1974), pp. 56–58; Harris, *Harlem Princess*, pp. 303, 311, 333; Thomas S. McFee (and 13 co-authors, including Jenkins), *Final Report and Recommendations: The 21st Century Federal Manager, Human Resources Management Panel* (Washington, DC: National Academy of Public Administration, September 2002) and Ralph C. Bledsoe (and 13 co-authors, including Jenkins), *A Work Experience Second to None: Impelling the Best to Serve* (Washington, DC: National Academy of Public Administration, September 2001).

century habits of mind. Jenkins already knew, from her experience as a teacher and a school administrator in an elite university town, that very smart people could be very ignorant outside of their areas of expertise. Harris was a primary change agent whose firing and rehiring forced NASA to issue quarterly reports on its hiring of women and minorities to Senator Proxmire and Congress generally for three years. Jenkins made it possible for NASA to slowly move on once Senator Proxmire's reporting requirements were removed in 1978.[75]

Conclusion

Thus, the American space program went coed and multiracial. In 1974, women and all minorities combined comprised about 5 percent of the science and engineering workforce of a high-technology agency. As physical sciences and engineering education very slowly opened up to women and nonwhites, these numbers increased, especially for nonminority women and Asian Americans of both sexes. In 1983, Harriett Jenkins wrote Low on NASA's 25th anniversary to tell him about the progress she had made in her first 10 years of integrating women and minorities into NASA's labor force. Jenkins's effort mixed flattery and bureaucratic self-advertisement with a sure hand. She talked about how in a decade nonminority women had risen from 15 percent to just under 18 percent of NASA's total, while minorities had doubled from 6 percent to 12 percent. In the science, engineering, and technical half of NASA, the equivalent percentages were increases from 2.3 percent to 5.5 percent for nonminority women and from 3.9 percent to 8.3 percent for all minorities. Calling this "modest progress," Jenkins went on to tell the man who had done more than anyone else to bring her into NASA that "[T]he greatest challenge still plagues us—the placement of minorities and women in the senior levels of the agency." The 1980s, Jenkins hoped, would accelerate gains because of sympathetic leaders within the Agency in the Reagan era. Complimenting Jenkins's work, Low replied that "the pipeline is filling rapidly" with female candidates qualified for upper managerial ranks, but that the problem of minorities remained "more difficult."[76]

So it was in the nation, in physical sciences and engineering education and occupations generally, and at NASA. The 1980s, as Low expected, saw a gradual increase in nonminority female (and also Asian American) numbers, while progress for other groups remained slower. By 1991, about 12 percent of all NASA science and engineering jobs were held by nonminority women; about 5 percent by Asian

75. Jenkins interview with author, 25 August 2005; "Senate Leaders Appoint a Director of Fair Employment," *Jet* 82 (1 June 1992): p. 22; Nichelle Nichols, *Beyond Uhura: Star Trek and Other Memories* (New York: G. P. Putnam's Sons, 1994), pp. 220–223. Dr. Harriett Jenkins to author, 6 November 2005 (hiring statistics).

76. Harriett Jenkins to George Low, no date (circa November 23), with attached statistical sheets; Low to Jenkins, 14 December 1983 (both in Box 138, Low/RPI).

Americans; and about 4 percent each by African Americans and Hispanics. NASA, like the rest of the U.S. society around it, was gradually opening up to new groups and constituencies. NASA Administrator Dan Goldin flustered many in his agency when he announced in 1994 that the Agency was still too "male, pale, and stale." But important initiatives such as the Science, Engineering, Mathematics and Aerospace Academy program for precollegiate minority and female students that began in 1993 later went national at NASA because of the efforts of NASA officials like Goldin and Jenkins. It also got NASA to establish more serious university research center relationships with Historically Black colleges and universities in 1995. It took until 1999 for Air Force Colonel Eileen Collins to command a Space Shuttle mission but, 42 years after Sputnik, it happened. Problems remained, though. The largest gender gap for any science and technology issue measured by the National Science Foundation in 2000 was in space exploration (14 percent). The under-representation of women and minorities in the physical sciences and in engineering especially mattered.[77]

The bottom line here is that women and minorities comprise nearly two-thirds of the population and a majority of the American labor force, but only a vastly smaller share of its high-technology and science skills pool. In physics (14 percent) and engineering (8 percent), it is still rare to find a woman holding a Ph.D.-level job. The quarter of the nation's science and engineering workforce that is female has also changed little in the last decade. The growth of science and engineering graduates among minorities is still very slow. In a decade when an estimated half of all NASA employees and a quarter of all NASA engineers will be eligible for retirement, and in which noted trade journal *Aviation Week & Space Technology* reports that the average age of aerospace workers in American corporations is the early 50s, recruiting and utilizing historically under-represented groups may be only intelligent selfishness in the longer run, for NASA or for any other organization, which is "an investment in America's future."[78]

A person who makes this latter argument cogently and well is Dr. Shirley Ann Jackson, president of Rensselaer Polytechnic Institute (RPI) in Troy, New York, and the first African American Ph.D. in physics from MIT. That Shirley Ann Jackson

77. National Science Board, National Science Foundation, *Science and Engineering Indicators—2000, Vol. I* (Washington, DC: NSF, 2000), Chapter 8, p. 21; Dan Goldin, "Bullish on Space Symposium: New York, New York, July 22, 1994," p. 6; Goldin, "Welcoming Remarks: Briefing for Dr. Petersen, National Science Foundation, August 9, 1994," p. 1, NHO.

78. "Women, Minorities, and People with Disabilities in Science and Engineering—2002," National Science Foundation, esp. Chapter 6, available at *http://www.nasf.gov/statistics/nsf03312/c6/c6o.htm*; "SEMAA Annual Report: Creating New Horizons, 10 Years," (Washington, DC: NASA, 2003); Margaret Wertheim, "Rational Inequality," *The Los Angeles Times* (30 March 2006); "A Guide to Womanomics," *Economist* (15 April 2006): pp. 73–74; Dorothy Leonard and Walter Swap, "Preserving Deep Smarts at NASA, *ASK: Academy Sharing Knowledge* (Winter 2006) (Washington, DC: NASA, 2006), pp. 9–12; Shirley Ann Jackson, "A Critical Shortage: We Face An Imperative to Tap the Talent of the 'New Majority,'" *Rensselaer* (Spring 2004): p. 2.

assumed her presidency in the same year as Eileen Collins commanded her space mission is no historical accident. The fact that she also succeeded NASA's George Low as a successful and honored president of RPI also has a lot to do with the change era described in this paper. As scholars have recently noted, the early 1970s were a period in which the "woman question" wouldn't go away, inside or outside of NASA. It was also a period during which an all-male and all-white Astronaut Corps came to exclude too many other Americans. NASA's human spaceflight program would have ceased being "manned" and become "human" without Ruth Bates Harris or her supporters, but it would have taken significantly longer than the 20 years it did take. NASA and America's space programs would only have been poorer for it, in terms of public interest, understanding, and regard. Ruth Bates Harris deserves to be remembered as an important actor in the social history of the Space Age.[79]

79. Kevles, *Almost Heaven*, p. 51; Jodi Dean, *Aliens in America* (Ithaca, NY: Cornell University Press, 1998), p. 96.

Chapter 23

NASA Launches Houston into Orbit: The Economic and Social Impact of the Space Agency on Southeast Texas, 1961–1969

Kevin M. Brady

On 19 September 1961 NASA Administrator James E. Webb announced that the Agency's Manned Spacecraft Center, which would serve as the command center for the Apollo missions and future human spaceflight programs, would be located near Clear Lake in southeast Houston, Texas, on 1,020 acres of land donated to the government by Rice University. Following the announcement, Congressman William C. Cramer of Florida and Senator Benjamin A. Smith of Massachusetts cited political pressure from influential Texans (including Vice President Lyndon B. Johnson, who headed the Space Council, and Texas Congressman Albert Thomas, whose district included Houston and who controlled NASA's funds as chairman of the House Appropriations Committee) as the reason why Houston was selected as the site for the NASA Center. During the next year, members of the Space Task Group transferred from the NASA Langley Research Center in Hampton, Virginia, to Houston, where they worked in temporary facilities throughout the city while awaiting the completion of the federal laboratory.[1]

When President John F. Kennedy arrived in Houston on 11 September 1962 to see the construction of NASA's Manned Spacecraft Center (MSC) and visit Rice University, he greeted nearly 200,000 Texans at the city's airport by saying:

> I do not know whether the people of the Southwest [Texas] realize the profound effect the whole space program will have on the economy of this section of the country. The scientists, engineers, and technical people who will be attracted here will

1. *The Houston Chronicle*, 19 September 1961, 20 September 1961; *The Houston Post*, 20 September 1961; "Houston Manned Spacecraft Center," 10 September 1962, NASA 1960s Vertical Files, Houston Metropolitan Research Center, Houston, TX. An additional 600 acres of land were purchased by NASA in 1962.

> really make the Southwest [Texas] a great center of scientific and industrial research as this nation reaches out to the moon. In this place in America [Texas] are going to be laid the plans and designs by which we will reach out in this decade to explore space.[2]

The president's promise was impressive and the prophetic statement came true as the Agency's decision to establish a center near Clear Lake propelled Texas into the Space Age by transforming rural towns in the area into highly visible communities that surrounded a unique facility for the training of astronauts and the control of their spaceflights. During this period of rapid growth, local businesses and nearby schools flourished. The federal installation also brought millions of dollars and enticed highly educated technical specialists into the area. The relocation of national aerospace companies to the Clear Lake area created new jobs and the need for more support services, which primarily added to the diversification of Texas's and the Gulf Coast's economy. The close relationship that the MSC established with local academic institutions expanded their graduate programs, accelerated their research projects, and enlarged their curriculums to meet the needs of the space program. The immense effects that the installation had on Texas ranks alongside other notable events in the state's history, including the discovery of oil at Spindletop, the opening of the Houston ship channel, and the creation of the petrochemical and refining industries following World War II. NASA indeed launched Houston into orbit by contributing to the population, economic, scientific, and technological buildup that the region experienced during the 1960s.[3] This essay focuses on southeast Texas between 1961 and 1969 in an effort to demonstrate how the federal complex represented a catalyst to Houston's economy and enhanced the area's colleges and universities.

Prior to NASA's decision to locate a research facility in Houston, the Clear Lake area had population of 6,520 people. Most of the land in the region, which sold for less than $750 an acre, was devoted to cattle and agriculture. There were also several producing oil and gas fields in the adjacent town of Friendswood. The largest nonagricultural industry was oyster and shrimp fishing, which was centered in the neighboring communities of Seabrook and Kemah.[4]

As a direct result of the MSC location, the Humble Oil and Refining Company and Del W. Webb formed the Friendswood Development Company to create a planned community, known as Clear Lake City, on 15,000 acres of land adjacent to the Center. During the 1960s, the development company oversaw the construction

2. *The Houston Press*, 12 September 1962.

3. *The Houston Chronicle*, 10 November 1986; *The Houston Press*, 23 September 1961; *The Houston Post*, 21 July 1989; "MSC on the Move," *The Houston Magazine* (February 1963): p. 96.

4. *The Houston Post*, 28 August 1977; Dave Lang, "Impact of MSC on the MSC Area," October 1966, Economic Impact Study, Lyndon B. Johnson Space Flight Center, Management Analysis Office, Management Analysis Studies, NASA Record Group 255, National Archives and Records Administration—Southwest Region (Fort Worth, TX); *The Houston Chronicle*, 11 October 1961.

of office buildings, service stations, restaurants, motels, shopping centers, banks, a golf course, a recreation center with swimming pool and baseball field, apartment complexes, and hundreds of homes. With all of these amenities, many of the leading scientists, engineers, and technicians working at the federal laboratory made Clear Lake City their home. By the late 1960s, the area's population exceeded 45,000 people and land values had skyrocketed as half-acre sites sold for more than $6,000. Aside from the rapid growth and booming land prices, the area's economy benefited from the $100-million development project because Houston developers utilized local labor to construct the residential subdivisions and commercial buildings.[5]

Just across the road from the NASA Center were lush pastures owned by Colonel Raymond Pearson. With the need for additional housing and commercial facilities to be built around the MSC, a group of Houston businessmen sought to capitalize on the area's rapid development by purchasing a 570-acre tract of land from Pearson to develop a community known as Nassau Bay. On 15 September 1963 David Bell, who was an aerospace technologist working at the MSC, and his family became the first residents of Nassau Bay. As other NASA employees and workers from the local aerospace companies moved into the development, Nassau Bay's residential sales exceeded $5 million by the fall of 1964. Aside from residential homes, Nassau Bay also included shopping centers, a bank, several apartment units, two motels, a post office, and an office complex which housed the MSC's press corps and more than 25 space-oriented firms. Additionally, the community had its own telephone company and a power plant operated by Thermal Systems Inc., which supplied year-round metered air conditioning and heating to the commercial buildings, townhouses, and apartments.[6]

Established communities near the NASA site also shared in the land boom during the 1960s. Prior to the construction of the MSC in Houston, Hurricane Carla struck the Texas coast on 11 September 1961, which destroyed most of the Seabrook community and discouraged many of the local residents from rebuilding. Thus, the area soon became a ghost town. However, news of the federal government's plans to establish a NASA facility nearby revived the community's hopes and spirits. During the 1960s, Seabrook experienced rapid growth as NASA employees and aerospace company contractors moved to the area. By 1967, the Seabrook Chamber of Commerce reported that the town's population exceeded 5,400 residents. As

5. *The Houston Press*, 19 January 1962, 24 May 1963, 18 June 1963; "Humble-Webb Launch Space," *The Houston Magazine* (March 1962): p. 69; *The Houston Chronicle*, 12 February 1967; Lang, "Impact of MSC on the MSC Area"; *The Houston Chronicle*, 26 May 1963; "One Small Step . . . A Giant Leap," *The Houston Magazine* (September 1969): p. 18.

6. *The Houston Chronicle*, 26 May 1963, 21 June 1962, 15 September 1963, 20 October 1963; *The Houston Post*, 29 September 1963, 22 November 1964, 28 August 1977; *The Houston Press*, 21 June 1962; *Suburban Journal* (Clear Lake City, TX), 5 September 1963; *Spaceland Star* (Webster, TX), 25 February 1965.

Seabrook's population continued to increase, the community needed additional facilities and services to accommodate the new influx of citizens. Therefore, the once-quiet fishing village witnessed the construction of new shopping centers and subdivisions that included 100 new homes and 1,000 apartment units. Road conditions were also improved in Seabrook when a two-lane drawbridge was built over the Clear Creek channel that connected Clear Lake with Galveston Bay.[7]

In 1960, Friendswood, which was 10 miles west of the Center, represented a small rural town of about 75 Quakers who were living off the Texas oil boom with mainly an agricultural lifestyle. When the MSC was built in the Clear Lake area, the small town developed into a suburban community as new residents moved to the area.[8] Dramatic growth led to the construction of two water wells, a new sewage disposal plant, several subdivisions, shopping centers, and a new drugstore. Other notable town improvements included the establishment of a new medical clinic, which housed Friendswood's first doctor and dentist. Many of the MSC employees and aerospace contractors, including Director of Flight Operations Christopher C. Kraft, Jr., Flight Director Glynn S. Lynn, and astronaut Donald K. Slayton, made Friendswood their home. By the late 1960s Friendswood was no longer a quiet country village, but the Quaker influence was still visible within the community because local grocery and convenience stories did not sell alcohol.[9]

The economic impact of the MSC on Houston was felt immediately as national aerospace companies established regional marketing, service, and engineering liaison offices in the area.[10] Robert H. Brewer, who served as manager of the Houston Chamber of Commerce, said this represented "one of the greatest concentrations of national corporate interests brought to focus on a single major metropolitan area in so short a period of time."[11] By 1966, more than 125 firms had established offices in the Clear Lake area. Among the largest space-oriented companies were the Philco Houston Operations of Honeywell Inc., North American Aviation Inc., General Electric, Lockheed Electronics Company, International Business Machines's Federal System Division, Raytheon Company, Thompson Ramo Wooldridge (TRW) Systems Inc., Volt Technologies Corp., the Univac Division of Sperry-Rand,

7. *The Houston Post*, 9 February 1964, 31 May 1964, 15 October 1981; "Character of Seabrook," Seabrook—Texas Vertical Files, Houston Metropolitan Research Center, Houston, TX.

8. *The Houston Chronicle*, 24 September 1979.

9. *The Houston Post*, 1 November 1962; Houston *Chronicle*, 16 June 1963; *The Houston Chronicle*, 26 September 1965; *The Houston Chronicle*, 26 May 1963; Manned Spacecraft Center News Release, MSC 71-85, n.d., Public Affairs Office, NASA Lyndon B. Johnson Space Center, Houston, TX; Manned Spacecraft Center News Release, MSC 69-79, 26 November 1969, Public Affairs Office, NASA Lyndon B. Johnson Space Center, Houston, TX.

10. Letter, Fred Staacke to Kenneth S. Pitzer, 30 January 1962, The Kenneth Pitzer Papers, Box 11, Woodson Research Center, Fondren Library, Rice, University, Houston, TX.

11. *The Houston Post*, 9 February 1964.

Beckman Instruments Inc., Texas Instruments Inc., and Federal Electric Corp. At first, the firms came to Houston to work exclusively with the Agency, but many of them stayed as the Apollo program drew to a conclusion because they viewed the area as a new base for expansion and diversification. During the late 1960s the companies provided their services to the nearby petrochemical industries and engineering businesses in the Houston area. Thus, the space-oriented firms assisted in diversifying Texas's economy. However, a majority of these companies are no longer in business today as budgetary constraints in the space program forced them to close operations or consolidate with other aerospace firms.[12]

Aside from national research and development companies relocating to the Clear Lake area to compete for the Agency's contracts, new businesses were established and others were expanded to provide supplies, equipment, and services to the Center. In January 1962, Gulf Aerospace Corp. was one of the first companies to be organized as a result of the NASA facility. The firm received a contract from the Agency to develop medical equipment designed to monitor an astronaut's heart rate during spaceflights. Another enterprise was the A-OK Business Service, which was established in Houston to provide secretarial and telephone answering services for the institution. By 1967, the firm expanded its operations by providing a placement service for temporary and permanent employees in adjacent Harris and Galveston Counties. Other companies included Test Equipment Corp. and Geo Space Corp., which provided electronics support for the space-oriented firms in the Clear Lake area. By the late 1960s, the total amount of NASA contracts awarded to these companies, along with other Houston-based businesses cooperating with the MSC, was more than $643 billion, which was poured directly into Texas's economy. Thus, the federal funds encouraged and supported future growth and expansion in southeast Texas.[13]

More meaningful to the Houston and Clear Lake area economy was that local firms obtained contracts and awards for space-related projects and construction work at the NASA Center.[14] Among the first Houston companies to receive a

12. Ibid.; "At the Manned Spacecraft Center: Six Years and Spin Office," *The Houston Magazine* (September 1967): p. 62; *The Houston Press*, 15 September 1962; *The Houston Press*, 7 September 1962; "Liaison Branches," *The Houston Magazine* (October 1962): p. 28; "From Cow Pasture To Moon: MSC Grows—Spreads Its Influence Throughout Area," *The Houston Magazine* (September 1966): p. 22; "Working for Industry," *The Houston Magazine* (March 1966): p. 74.

13. *The Houston Post*, 6 June 1964; "Instrument Developers," *The Houston Magazine* (May 1962): p. 24; "Matching People to Jobs," *The Houston Magazine* (December 1967): p. 42; "Lockheed Liaison," *The Houston Magazine* (September 1962): p. 24; "From Cow Pasture to Moon: MSC Grows—Spreads Its Influence Throughout the Area," *The Houston Magazine* (September 1966): p. 22; "NASA's Prime Contractors & Prime Contract Awards as of December 31, 1967," 31 December 1967, Robert C. Eckhardt Papers, Box 95-147/195, Center for American History, University of Texas at Austin, Austin, TX.

14. Letter, Fred Staacke to Kenneth S. Pitzer, 30 January 1962; *The Houston Press*, 2 October 1963.

NASA award was Brown & Root Inc. In February 1962 the construction company was the recipient of a $1,499,280 contract for architectural design work on the MSC.[15] Other local engineering firms to benefit from the Agency's awards was S.I.P. Inc. of Houston, which was awarded a contract for $272,522 to build and install a space environmental chamber at the NASA facility.[16] Following the successful flight of Apollo 8, NASA awarded the Houston-based A-V Corp. a contract to develop a documentary that utilized the first close-up motion pictures of the lunar surface.[17] In 1966, Warrior Constructors Inc. of Houston and the National Electronics Corp. of Houston were the recipients of a NASA contract for $4.3 million to construct and equip the Lunar Receiving Laboratory, which would house the returned lunar samples collected by the Apollo astronauts.[18]

While economic growth and progress was transpiring near the Clear Lake site, colleges and universities throughout the Houston area witnessed these benefits and viewed the establishment of a permanent relationship with federal research complex as a way to enhance their academic prestige and prosperity.[19]

Rice University in Houston was one of the initial educational facilities in the area to be influenced by the arrival of the NASA complex. On 4 January 1963 Dr. Kenneth S. Pitzer, who was president of the 70-year-old university, announced the establishment of the Space Science department at Rice. Prior to the announcement, several institutions around the United States, including the University of California at Berkeley, the University of California at Los Angeles, the University of Michigan, and the University of Maryland offered postgraduate work on the subject, but these programs were usually only part of a physics department or of a college-affiliated space research center.[20] Therefore, Rice represented the only college or university in America to have a department devoted exclusively to space science. MSC Director Robert R. Gilruth welcomed the establishment of the new department at Rice University. "We expect it to be of great assistance in carrying out the mission of the Manned Spacecraft Center," he said.[21] The benefit of the Space Science department would result from being able to train highly specialized professional and technical NASA employees at Rice University,

15. "NASA Groundwork," *The Houston Magazine* (February 1962): p. 26.

16. NASA Manned Spacecraft Center News Release, MSC 63-263, 28 December 1963, Public Affairs Office, NASA Lyndon B. Johnson Space Center, Houston, TX.

17. "Editing Lunar Film Footage," *The Houston Magazine* (June 1969): p. 26.

18. NASA Manned Spacecraft Center News Release, MSC 66-67, 19 August 1966, Public Affairs Office, NASA Lyndon B. Johnson Space Center, Houston, TX.

19. *The Houston Post*, 11 March 1962; Wesley L. Hjornevik, interview by Robert B. Merrifield, 9 March 1967, 1 July 1967, and 9 September 1967, transcript, Center Series Interviews, JSC History Collection, University of Houston-Clear Lake, University Archives, Houston, TX.

20. *The Houston Post*, 5 January 1963; "Space Ph.D.'s at Rice," *The Houston Magazine* (September 1968): p. 56.

21. *The Houston Chronicle*, 5 January 1963; *The Houston Post*, 5 January 1963.

which was less than 40 miles away from the federal site, instead of sending them to institutions on the either the West or East coasts.[22]

Rice's Space Science department provided research and graduate-level instruction in geomagnetism, Van Allen radiation, aurora, atmospheric structures and dynamics, planetary structures, and meteorites.[23] Dr. Alexander J. Dessler, who was appointed chairman of the new department, stated that the establishment of a Space Science department had a large regional value because it expanded space research in the state of Texas and contributed to making Rice University one of the top academic institutions in the nation.[24]

The Space Science department grew remarkably during the 1960s. For instance, the faculty expanded from 4 in the spring of 1963 to 17 by December 1967. The graduate student enrollment increased from 9 in September 1963 to 50 in three years. Many of the graduate students who enrolled in the Space Science department were MSC employees and local aerospace contractors who sought to broaden their understanding of sounding rockets, satellites, astronomy, planetary and meteoritic structures, quantum mathematics, and astrophysics.[25]

Aside from being the recipient of several NASA research grants, Rice also cooperated with the MSC by providing research and teaching facilities for the scientist-astronauts, where they were able to maintain proficiency in their specific fields of study while they were involved in astronaut training at the NASA Center. The close proximity of Rice University and the MSC made the research facilities at the institution convenient to the scientist-astronauts so they could keep up with their scientific work without involving extending amounts of travel time.[26]

22. "Summary," n.d., The Kenneth Pitzer Papers, Box 11, Woodson Research Center, Fondren Library, Rice University, Houston, TX.

23. Memo, K.S. Pitzer to Deans and Department Chairs, "Instruction and Research in Space Science," 2 January 1963, The Kenneth Pitzer Papers, Box 11, Woodson Research Center, Fondren Library, Rice University, Houston, TX.

24. "Summary," n.d., The Kenneth Pitzer Papers, Box 11, Woodson Research Center, Fondren Library, Rice University, Houston, TX; Kenneth Pitzer, "Rice University and Houston in the Space Age," 7 December 1961, The Kenneth Pitzer Papers, Box 49, Woodson Research Center, Fondren Library, Rice University, Houston, TX.

25. Memo, A. J. Dessler to Faculty, "Academic Program," 8 December 1967, The Kenneth Pitzer Papers, Box 11, Woodson Research Center, Fondren Library, Rice University, Houston, TX; A. J. Dessler to President Pitzer, "Current State of Space Science Department," 23 April 1963, The F. Curtis Michel Papers, Box 2, Woodson Research Center, Fondren Library, Rice University, Houston, TX; "Summary," n.d., The Kenneth Pitzer Papers, Box 11, Woodson Research Center, Fondren Library, Rice University, Houston, TX; *The Houston Chronicle*, 26 May 1968.

26. Letter, A. J. Dessler to Dr. Harry Hess, 19 June 1963, The Kenneth Pitzer Papers, Box 11, Woodson Research Center, Fondren Library, Rice University, Houston, TX.

One of the most notable outgrowths of Rice's Space Science Department was that NASA awarded the university a $1.6 million grant to construct a Space Science and Technology Building to house the institution's rapidly expanding research activities. The building consisted of offices, conference rooms, a high-speed computer center, environmental testing facilities, materials research facilities, and a low-level radiation laboratory.[27] During the late 1960s, 18 faculty members, 31 graduate students, and 10 postdoctoral students worked on 35 space-related projects, including meteorite experiments, cosmic rays measurements, solar wind investigations, and aurora studies in the Space Science and Technology Building.[28] The success of these projects enabled Rice to become the first college or university in the United States to receive approval from the Agency to design and development its own satellites under the new University Explorers Program sponsored by NASA.[29] Although Rice University was ranked as one of the top 20 academic institutions in the nation prior to the arrival of the MSC, the close partnership that the university developed with the Center enhanced the school's national and international prestige.[30]

The success of the Space Science department also contributed to Houston developing a new image that never existed before the 1960s. The city had always been viewed as the center of the oil and gas industry, but Rice's program and the MSC caused the nation to recognize southeast Texas as a site for space-oriented research and development.[31]

Probably one of the greatest impacts that NASA had on Rice University was that the national government passed a resolution in July 1963 which stipulated that after 16 August 1964, federal agencies were restricted from awarding contracts to colleges or universities that discriminated because of race, creed, color, or national origins in their admission process.[32] At this time, Rice University was the only

---

27. *The Houston Chronicle*, 1 May 1964; "Space Science Building," 8 February 1965, The Kenneth Pitzer Papers, Box 11, Woodson Research Center, Fondren Library, Rice University, Houston, TX; "Opening Remarks by Dr. Pitzer: Ground-Breaking Ceremonies for the Space Science and Technology Building, Rice University," 12 February 1965, The Kenneth Pitzer Papers, Box 11, Woodson Center, Fondren Library, Rice University, Houston, TX.

28. "Address by James E. Webb—Delivered by Dr. Thomas L. K. Smull, On the Occasion of the Ground-Breaking for the Rice University Space Science and Technology Building," 12 February 1965, The Kenneth Pitzer Papers, Box 11, Woodson Center, Fondren Library, Rice University, Houston, TX; *The Houston Chronicle*, 24 March 1966; *The Houston Post*, 4 June 1965; *The Houston Post*, 8 May 1965.

29. *Austin Statesman*, 10 February 1965; *The Houston Post*, 5 February 1965.

30. *The Houston Post*, 13 September 1969; "First Year: Space-Age Impact on Houston," *The Houston Magazine* (September 1962): p. 30.

31. D. W. Lang, interviewed by Robert B. Merrifield, 3 May 1967 and 9 May 1967, transcript, Box 3, Center Interviews, JSC History Collection, University of Houston-Clear Lake, University Archives, Houston, TX.

32. Warren B. Irons, "Selecting Employees and Facilities for Training," 19 July 1963, The Kenneth Pitzer Papers, Box 16, Woodson Center, Fondren Library, Rice University, Houston, TX; *The Houston Press*, 12 February 1964; *The Houston Press*, 8 January 1964.

nonsectarian, private institution in the South that prohibited the admission of African American students. Thus, President Pitzer and the Board of Governors recognized that if they failed to remove racial restrictions on student admission, the university would suffer the consequences of losing NASA grants.[33] The possibility that Rice could have become ineligible to receive research funds from the Agency represented an ironic twist because one of the main reasons why the NASA installation was centered in Houston was the availability of Rice's scientists and research facilities.[34]

In February 1964 the Rice trustees filed a petition asking for District Court authority to disregard racial restrictions set out by the university's founder, William Marsh Rice, in an 1891 indenture.[35] The trustees claimed that race discrimination prevented the institution from receiving federal research grants and hampered the recruitment of first-rate students and faculty members.[36] A Rice alumni group headed by Congressman Albert Thomas, whose district encompassed Rice University, also filed a petition in support of the trustees' request. On 11 February 1964 District Judge William M. Holland of the 127th District Court heard Rice's case.[37] During the suit, Tom Martin Davis, who was the attorney for the university's trustees, argued in favor of integration by stating, "Rice University today stands at the crossroads—it could go to the moon or it could return to the 19th century."[38] After the jury deliberated, they found that it would be impossible for Rice University to develop as a first-class educational institute if African Americans were barred from enrollment.[39] A month later, District Judge Holland ruled that Rice University's trustees had the authority to remove racial barriers prohibiting the admission of qualified students. With the desegregation of Rice, the institution became eligible once again to receive NASA research grants.[40] In the spring of 1965, Raymond Johnson became the first African American to be admitted to Rice University.[41]

The arrival of the MSC in Houston also led to significant changes at the University of Houston. During the early 1960s the academic institution witnessed the close relationship that existed between Rice University and the Agency, so

33. Memo, K. S. Pitzer to Members of the Board of Governors, "New Non-Discrimination Requirements of Several Government Agencies," 16 October 1963, The Kenneth Pitzer Papers, Box 16, Woodson Center, Fondren Library, Rice University, Houston, TX; *The Houston Press*, 11 February 1964.

34. *The Houston Press*, 8 January 1964.

35. *The Houston Post*, 22 February 1963.

36. *The Houston Chronicle*, 13 February 1964.

37. *The Houston Chronicle*, 9 February 1964; *The Houston Press*, 10 February 1964.

38. *The Houston Press*, 20 February 1964.

39. Ibid., and 9 March 1964.

40. *The Houston Post*, 10 March 1964; *The Houston Chronicle*, 9 March 1964.

41. Memo, K. S. Pitzer to Members of the Board of Governors, "Annual Report: 1963–64," The Kenneth Pitzer Papers, Box 17, Woodson Center, Fondren Library, Rice University, Houston, TX.

the University of Houston expanded its interests and needs to keep pace with the Space Age. For instance, the university inaugurated doctoral programs in physics, mechanical engineering, and electrical engineering after recognizing the need for higher-trained personnel in these fields.[42] Dr. Clark Goodman, a professor of physics, developed a course entitled "Aerospace Science and Engineering," which focused on space environments, temperature controls, heat protection, and launch vehicles. In the fall of 1963 nearly 40 graduate students and seniors enrolled in the course. Aside from Goodman's daily lectures, students had an opportunity to discuss the current activities and operations at the NASA Center with MSC officials, including Maxime A. Faget, who served as Assistant Director of Engineering and Development; Bryan R. Erb, who worked in the Structures and Materials Branch; David M. Hammick from the Space Technology Division; and Chief of the Space Radiation and Space Environment Division Jerry L. Modisette. This new space engineering class at the University of Houston represented one of the first courses of its type in the nation.[43]

Aside from this course, the university jointly sponsored a variety of educational and research programs with the NASA complex, such as the Summer Faculty Fellowship Program, Sustaining University Program, Employee Graduate Training Program, and Pre-doctoral Traineeship Program.[44] Dr. F. M. Tiller, who served as the dean of the College of Engineering, stated, "[T]he establishment of NASA programs aided in the progress and development of the school by accelerating plans that were already underway." Thus, the cooperative efforts assisted the University of Houston in improving its stature as an academic institution in the Southwest.[45]

With the continued expansion of population and industry in the Clear Lake area, the community required additional opportunities for higher education. By 1964, the University of Houston offered undergraduate and graduate courses ranging from elementary Russian, mathematics, physics, political science, and business administration to MSC employees in temporary classrooms at the NASA Center in Houston. However, Center Director Robert Gilruth, seeking more educational opportunities, urged the university to expand its operations in an effort to more effectively serve the interests of NASA employees, residents of the Clear Lake area,

42. *The Houston Post*, 8 September 1963.

43. *The Houston Press*, 23 October 1963; *The Houston Press*, 24 September 1963.

44. "A Study of NASA University Programs," 1968, The Philip G. Hoffman Papers, Box 35, Special Collections and Archives, University of Houston Libraries, Houston, TX; R. W. Hough, "Some Economic Impacts of the National Space Program," (Washington DC: NASA-CR-96809, NASA, 1968), pp. 35–36; "Material Given to Gov. Connally Before His Speech to Law College," n.d., The Philip G. Hoffman Papers, Box 38, Special Collections and Archives, University of Houston Libraries, Houston, TX.

45. *The Houston Post*, 10 February 1963.

and workers at the various aerospace companies. Additionally, he noted that the area's large technical and scientific community made this need more imperative.[46]

On 25 November 1965 the Humble Oil and Refining Company donated 50 acres of land west of the MSC to the University of Houston to establish a branch campus.[47] The president of the university, Dr. Philip G. Hoffman, who accepted the land gift for the board of regents, said, "[S]uch gifts would not only be acceptable, but welcome."[48] Following the announcement, Paul E. Purser, Gilruth's executive assistant, stated that the establishment of a branch campus near the facility would provide a healthy and vigorous university climate in the area.[49] After acquiring the necessary building funds from alumni gifts and private industries, construction began on the property in the early 1970s. Regular scheduled classes at the University of Houston at Clear Lake City began in September 1974 with a total enrollment of 1,069 students and a faculty comprised of 60 professors. Thus, the establishment of the branch campus contributed to the proper development of the state's scientific and industrial strength.[50]

Additional Houston-area educational institutions benefited from specific NASA-related projects, such as the University of St. Thomas, which received funds from the Agency to conduct studies relating to the genetic effects of weightlessness; Baylor University College of Medicine, which was awarded a $258,000 space medicine grant to monitor astronauts' brain waves and blood volume during human spaceflights; the University of Texas Graduate School of Biomedical Sciences, which was awarded a $65,000 grant to collect the astronauts' clinical history data for medical researchers; the Texas Institute for Rehabilitation and Research, which was awarded a $171,183 grant for the development of computer technology for a medical data study relating to the Apollo program; and the Southwest Research Institute of Houston, which was the recipient of a $104,850 grant to assist the Agency in disseminating NASA technological developments to local businesses for

46. *The Houston Post*, 27 November 1965; *The Houston Post*, 24 November 1965; *The Houston Chronicle*, 25 November 1965; *The Houston Post*, 16 March 1963; Letter, Robert R. Gilruth to Dr. Philip G. Hoffman, 10 September 1965, The Philip G. Hoffman Papers, Box 9, Special Collections and Archives, University of Houston Libraries, Houston, TX.

47. *The Houston Post*, 27 November 1965; *The Houston Post*, 24 November 1965; *The Houston Chronicle*, 25 November 1965; *The Houston Post*, 16 March 1963; Robert R. Gilruth to Dr. Philip G. Hoffman, 10 September 1965, The Philip G. Hoffman Papers, Box 9, Special Collections and Archives, University of Houston Libraries, Houston, TX.

48. *The Houston Post*, 24 November 1965.

49. Ibid.

50. "University History," University of Houston-Clear Lake, *http://prtl.uhcl.edu/portal/page?_pageid=353,409744&_dad=portal&_schema=PORTALP* (accessed 4 August 2006).

commercial applications.[51] By the late 1960s, the total amount of Agency funds awarded to these respective educational facilities and research institutions was more than $3.6 million.[52] These NASA grants differed from other research and development funds because they required highly paid and educated workers who demanded higher levels of education and residential accommodations. Houston and local officials responded to these demands by improving the area's quality of education and infrastructure during the 1960s, which resulted in benefits for all members of the community.[53]

The elementary and public schools in the Clear Lake area were also affected by the NASA Center because the children of MSC employees and contractors added to the schools' population.[54] During the 1950s school growth in the Clear Lake area ranged from 90 to 100 additional students each year. However, school growth since the 1960s represented an average increase of 1,021 pupils a year.[55]

During the early 1960s the Clear Creek Independent School District, which was a consolidation of four school districts from the neighboring communities of Webster, Seabrook, Kemah, and League City, also faced a boom in its enrollment resulting from an influx of NASA families to the region. In the fall of 1961 the school district consisted of six schools and had a student population of 1,777 pupils. Within two years the Clear Creek School Board reported a 77 percent rise in its student enrollment. With the school district forced to accommodate an additional 1,046 students, the school board approved the construction of additional buildings and facilities to alleviate the tremendous growth problems, including an immediate four-classroom addition to the Webster Elementary School and a five-classroom addition to the Clear Creek Junior High School.[56]

---

51. "Added Grants," *The Houston Magazine* (August 1966): p. 24; "NASA Research Funds," *The Houston Magazine* (November 1966); p. 30; "NASA Grants," *The Houston Magazine* (February 1967): p. 32; "Space Medicine," *The Houston Magazine* (October 1966); *The Houston Post*, 9 January 1966.

52. "NASA's Prime Contractors & Prime Contract Awards as of December 31, 1967," 31 December 1967, Robert C. Eckhardt Papers, Box 95-147/195, Center for American History, University of Texas at Austin, Austin, TX.

53. "For the Benefit of All Mankind," 14 September 1970, Box 95-147/195, Robert C. Eckhardt Papers, Center for American History, Austin, TX.

54. *The Houston Press*, 29 March 1963; "Pacemaker Schools," *The Houston Magazine* (August 1966): p. 24; "Fantastic Neighbor," *The Houston Magazine* (March 1964): p. 26; "Space For Schools," *The Houston Magazine* (July 1964); "NASA's Prime Contractors & Prime Contract Awards as of December 31, 1967," 31 December 1967, Robert C. Eckhardt Papers, Box 95-147/195, Center for American History, University of Texas at Austin, Austin, TX.

55. "Clear Creek Independent School District Enrollment Figures," Clear Creek Independent School District, League City, TX; Lang, "Impact of MSC on the MSC Area."

56. *The Houston Press*, 29 March 1963; *The Houston Press*, 28 August 1963; *The Houston Press*, 31 July 1963; *The Houston Post*, 7 March 1963; "Clear Creek Independent School District Enrollment Figures," Clear Creek Independent School District, League City, TX.

As children from families working at the NASA site and space-oriented industries in the area continued to overcrowd the Clear Creek Independent School District, the school board applied for federal aid to build additional campuses and expand existing facilities.[57] In January 1965 the U.S. Office of Education awarded the school district a $185,000 grant to construct a new high school campus in League City. The funds were received under a law that grants aid to school districts affected by nearby government installations.[58] During the next three years the school district used bond funds and federal grants to support the construction of three new elementary schools in El Lago, Clear Lake City, and League City, and a junior high school in Seabrook.[59] As the Clear Creek Independent School District expanded to 10 schools and 8,627 pupils in less than five years, the district received $203,079 in federal funds to prepare for additional growth during the late 1960s.[60]

From its inception, the MSC also developed a close relationship with the area's school districts in an effort to heighten students' interests in science and mathematics by disseminating knowledge about the space program to the local elementary and secondary schools.[61] In 1964, Houston Independent School District teachers Erine Baker and Grant Morrison cooperated with Public Relations Specialist Eugene E. Horton, who served as the chief of the educational programs at the MSC, to develop materials designed to inform teachers and students about the Agency's current activities. By 1966, the Houston Independent School District supplied elementary and secondary teachers throughout Texas with educational aids, including film clips, workbooks, and pictures related to the space program. The Houston Independent School District-NASA program demonstrated that educators could effectively work with highly technical information and design materials for classroom instruction. Additionally, the program established a permanent flow of information between Texas schools and the MSC.[62]

The location of the MSC in Houston also changed Texas's image because of the increased space research conducted in the state. As a result, Houstonians were accorded a worldwide identity as "Space City, U.S.A." during the 1960s. Local company owners willingly accepted their new Space Age image by changing the names of their establishments to demonstrate their pride in the space program.

57. *The Houston Chronicle*, 17 November 1964.

58. *The Houston Chronicle*, 6 January 1965.

59. *The Houston Chronicle*, 1 November 1964, 26 May 1965, 13 January 1965.

60. *The Houston Chronicle*, 13 March 1969, 31 March 1968; *The Houston Press*, 29 March 1963; "Clear Creek Independent School District Enrollment Figures," Clear Creek Independent School District, League City, TX.

61. "Year of Apollo," *The Houston Magazine* (September 1968): p. 21.

62. "Pacemaker Schools," *The Houston Magazine* (August 1966): p. 24; "Fantastic Neighbor," *The Houston Magazine* (March 1964): p. 26; "Space For Schools," *The Houston Magazine* (July 1964).

Businesses such as the Space City Bar-B-Q Restaurant, Apollo Broadcasting Company, Space Age Laminating Company, Apollo Paint Company, Astro Nut Company, Space City Hearing Aids Company, Astro Baby Sitters Agency, and the Apollo Restaurant and Lounge surrounded the NASA Center in the Clear Lake area. And naturally the space burger, which most Houstonians claimed "tasted out-of-this-world," was served at the Apollo Restaurant and Lounge.[63]

Texas sports teams also adopted space-oriented names to pay tribute to the nation's space pioneers. Despite the fact that many individuals believe the Houston Rockets basketball team was named after the launch vehicles that carried astronauts and satellites into space, the club was actually christened The Rockets when the organization was founded in San Diego, California, in 1967.[64] However, the Houston Astros baseball team drew their club name from the astronauts.[65] And in October 1965 Houston's new professional ice hockey team was named the Apollos in honor of the lunar landing program.[66]

Local attractions in the Houston area also had Space Age names. For instance, Tranquility Park, which was a tribute to the Apollo 11 mission, was constructed in the city's central business district. The enclosed multipurpose stadium where the Astros played was named the Astrodome, and adjacent to the sports complex was Astroworld, which was a recreational theme park that offered family entertainment.[67]

With Houston serving as the training center for America's astronauts and as the control center for all human spaceflights, astronauts Michael Collins, Edwin E. "Buzz" Aldrin, Jr., and Neil A. Armstrong paid homage to the Lone Star State by flying a Texas flag aboard the command module during the Apollo 11 mission.[68] This flight to the lunar surface was significant to Houstonians in another way because the first word spoken publicly by astronaut Neil Armstrong on the Moon was the name of their city: "Houston . . . Tranquility Base here. The Eagle has landed."[69] In November 1969 astronaut Alan L. Bean, who was a Fort Worth native, also brought worldwide attention to the Lone Star State by carrying a Texas flag to the lunar

63. *The Houston Post*, 21 July 1969.

64. Handbook of Texas Online, "Houston Rockets," *http://www.tsha.utexas.edu/handbook/online/articles/HH/xoh3.html* (accessed 7 August 2006).

65. "Countdown to Blast-Off for 1966 Astros," *The Houston Magazine* (April 1966): p. 63.

66. "Apollos On Ice," *The Houston Magazine* (October 1965): p. 72.

67. "One Small Step . . . A Giant Leap," *The Houston Magazine* (September 1969): p. 18; Handbook of Texas Online, "Astrodome," *http://www.tsha.utexas.edu/handbook/online/articles/AA/xva1.html* (accessed 7 August 2006); "The Beautiful Clear Lake Area," Clear Lake City Vertical File, Center for American History, University of Texas at Austin, Austin, TX.

68. "Texas Flag Flown to the Moon on Apollo 11," The 19th Century Shop, *http://www.19thcenturyshop.com/apps/catalogitem?id=346* (accessed 14 June 2006).

69. "One Small Step . . . A Giant Leap," *The Houston Magazine* (September 1969): p. 18; *The Houston Chronicle*, 10 November 1986; *Austin American Statesman*, 16 July 1989.

surface on the Apollo 12 mission.[70] And although Apollo 16 astronauts Charles M. Duke, Jr., and John W. Young did not carry a Texas flag to the Moon, they did name one of the Moon's craters "Lone Star" in honor of Texas.[71]

Conclusion

Although the Agency did not utilize the MSC as a site to launch rockets into space, the installation did launch southeast Texas into an orbit of growth and prosperity as the federal complex brought international attention to the region and made Houston the focal point of the Space Age fever that was spreading throughout the nation. The NASA facility also served as a catalyst to the state's economy and stimulated enormous development projects in the Clear Lake area. Thousands of new jobs were created in Houston as a result of the relocation of national aerospace firms to the region. Not only did these companies cause a new wave of industrialization in Texas, they also assisted in the diversification of state's economy. The Center also enhanced the prestige and stature of the area's academic institutions. More importantly, the federal complex contributed to the desegregation of Rice University, which recognized that it was necessary for a nationally competitive school to be racially integrated. As NASA placed Texas into orbit, the state entered a new era of technological, scientific, and economic progress during the late twentieth century.[72]

70. *Dallas Morning News*, 1 February 1970; *Dallas Morning News*, 15 November 1969.

71. Randy Attwood and Keith T. Wilson, "Footprints on the Moon—I Can't Believe It! An Interview with Apollo 16 Astronaut Charlie Duke," *Quest: The History of Spaceflight Magazine* (Summer/Fall 1994): p. 55.

72. "The President's Promise for Houston ... 'In This Place Are Going to be Laid Plans to Explore Space,'" *The Houston Magazine* (October 1962): p. 17; "Special to The Houston Post By Governor Price Daniels," n.d., Papers of Price Daniel as Attorney General, U.S. Senator, and Governor, Box 359, Sam Houston Regional Library and Research Center, Liberty, TX; *The Houston Post*, 10 March 1964; *The Houston Chronicle*, 9 March 1964; *The Houston Press*, 2 October 1963; "Houston Spreads Prosperity Throughout Eight-County Houston-Gulf Coast Complex," *The Houston Magazine* (December 1962): p. 45.

Chapter 24

The Jet Propulsion Laboratory and Southern California

Peter J. Westwick

Southern California, as any of its inhabitants will attest, is a rather particular—if not unique—place. Over the course of the twentieth century it evolved from irrigated agriculture, through an oil boom and the emergence of Hollywood, to its present status as a sprawling, high-tech nexus on the Pacific Rim. The central factor in this evolution, especially in the last half-century, has been the aerospace industry. The aerospace complex started with a few aircraft builders around World War I, vastly expanded in the mobilization for World War II, and then continued and broadened in the cold war to encompass a wide range of activities, including military and civilian aircraft, reconnaissance and communications satellites, and strategic missiles in addition to space exploration. By the 1980s about 40 percent of the American missiles and space business resided in southern California, as did about one-third of the aerospace engineers, and the industry as a whole there employed close to a half-million people. Southern California as we know it would not exist without aerospace, and, if one is looking for the societal impact of space exploration, one need look no further than the transformation of southern California in the twentieth century from sunbelt orange groves to high-tech metropolis.[1]

---

1. See Kevin Starr, *Americans and the California Dream, 1850–1915* (New York: Oxford University Press, 2002); Kevin Starr, *Inventing the Dream: California Through the Progressive Era* (New York: Oxford University Press, 1986); Kevin Starr, *Material Dreams: Southern California Through the 1920s* (New York: Oxford University Press, 1991); Kevin Starr, *Endangered Dreams: The Great Depression in California* (New York: Oxford University Press, 1997); Kevin Starr, *The Dream Endures: California Enters the 1940s* (New York: Oxford University Press, 2002); Kevin Starr, *Embattled Dreams: California in War and Peace, 1940–1950* (New York: Oxford University Press, 2003); Mike Davis, *City of Quartz: Excavating the Future in Los Angeles* (New York: Verso, 1990), pp. 120–121; Ann Markusen et al., *Rise of the Gunbelt: The Military Remapping of Industrial America* (New York: Oxford University Press, 1991). Statistics from Allen J. Scott, *Technopolis: High-Technology Industry and Regional Development in Southern California* (Berkeley, CA: University of California Press, 1993), p. 129. The total employment figure includes aircraft, electronics components, search and navigation equipment, and other aerospace-related categories. Scott, *Technopolis*, pp. 12–16. The author is helping launch an initiative to document the history of the regional aerospace industry in a project under the Institute for California and the West at USC and the Huntington Library; see the institute's Web site at *http://www.usc.edu/icw* (accessed 28 March 2007). Throughout the notes, citations to material from the JPL archives follow the form: JPL collection number, box/folder. All interviews by author unless otherwise noted.

Several thousand of these aerospace engineers worked at the Jet Propulsion Laboratory (JPL) in Pasadena. JPL worked with and spun off local aerospace firms, and experienced with them the ups and downs of federal support, but also competed with them in the local labor pool; it attracted publicity to the local community but also a backlash because of environmental pollution; and it intersected the Hollywood movie business in surprising ways. This paper examines a few of these evolving interactions to demonstrate how space programs both reflected and shaped southern California.

ORIGINS AND THE AEROSPACE INDUSTRY

It was no accident that JPL sat at the geographic epicenter of the aerospace industry. Caltech's aeronautics programs had close ties from the 1920s to nearby aircraft firms, and its research and graduates helped fuel the pre-World War II growth of the local aircraft industry. At the time, Caltech was already a center for science and engineering in the United States, fostered by philanthropic funding. This included money from the Guggenheim family for an aeronautics program, which in 1930 helped Caltech lure Theodore von Kármán, a leading authority on aerodynamics in Germany, to become director of the Guggenheim Aeronautical Laboratory at Caltech (GALCIT). In addition to promoting GALCIT's collaboration with local aircraft firms on wind tunnel and other research programs, von Kármán supported research on rockets, which eventually led to Army support and, during World War II, the formal creation of JPL.[2]

The symbiosis continued in the cold war, and JPL's evolution from rockets to spacecraft resonated with the diversification of local industry from aircraft to aerospace. Caltech continued to manage JPL, first under an Army contract and then under NASA. JPL differs from the other NASA Centers in its association with a university—one well-connected with local aerospace industry. Caltech (through GALCIT and JPL) provided to the industry both research underpinning technological advances and also a steady supply of science and engineering graduates, such as Simon Ramo and Dean Wooldridge—the R and W in TRW. These links may be measured by the many aerospace executives on Caltech's board of trustees, several of whom also served on a special committee overseeing JPL and hence injected industrial perspectives. Although Caltech and JPL had exceptionally well-developed connections, they were by no means unique in southern California: Convair engineers helped establish the science and engineering departments at the University of California at San Diego; Boeing similarly cultivated Cal-State Long

2. Judith R. Goodstein, *Millikan's School: A History of the California Institute of Technology* (New York: W. W. Norton & Company, 1991), pp. 156–177; Clayton R. Koppes, *JPL and the American Space Program: A History of the Jet Propulsion Laboratory* (New Haven, CT: Yale University Press, 1982), pp. 1–7.

Beach as a labor source; and Hughes supported the engineering program at the University of Southern California.[3]

JPL also intersected local industry as a competitor in the labor market. JPL managers viewed southern California as a more competitive labor market than that of other NASA Centers, since JPL engineers could jump to any number of industry firms without the hassle of relocating; lab managers consistently worried about losing top talent. As JPL grew to be the largest employer in Pasadena, its evolving mission shaped its ability to recruit and retain staff. After the lab shifted to NASA it gradually relinquished secret defense work, which made it particularly attractive to aerospace engineers who did not want to work in a classified environment. Since many (if not most) local aerospace jobs required a security clearance, this could provide a substantial recruiting inducement, not only for those morally opposed to military work but also those—such as homosexuals—who feared an intrusive clearance process. This became an issue when JPL resumed substantial military work in the 1980s, amid a general remilitarization of the space program.[4]

JPL more generally capitalized (with the rest of local industry) on the southern California climate as a recruiting tool. A number of JPL staff went job hunting in southern California at least in part because of the weather, especially those who started in the early years of the 1950s and 1960s, and even more especially those from colder climes. As one engineer put it, "The Chamber of Commerce sent me this booklet that showed beautiful mountains, no smog, snow on the top, and orange trees," and that was all the inducement necessary to head west.[5]

JPL also reflected the ethnic composition of the local labor pool. Like the rest of NASA, lab staff was very largely white and male, following the national demographic of science and engineering disciplines, but JPL generally had almost twice the proportion of minorities on the staff than did NASA, and Asian ethnic groups made up the largest fraction of minorities, followed by Hispanics and African

3. On Convair engineers and San Diego colleges: Marvin Stern, interview by Finn Aaserud, 1 May 1987, American Institute of Physics.

4. Pete Lyman interview, 1 April 2004; U.S. Government Accountability Office, "Security Clearances: Consideration of Sexual Orientation in the Clearance Process," 24 March 1995 (Washington, DC: GAO, GAO/NSIAD-95-21, 1995); Barbara Spector, "Security Clearance Delays Hamper Gays' Careers," *The Scientist* 6/5 (2 March 1992); *High Tech Gays v. Defense Industrial Security Clearance Office*, U.S. Court of Appeals, Ninth Circuit, 895 F.2d 563, 2 February 1990; Gary Libman, "Scientists Confront Homophobia in their Ranks," *Los Angeles Times*, 2 February 1990; Peter J. Westwick, *Into the Black: JPL and the American Space Program, 1976–2004* (New Haven, CT: Yale University Press, 2006), p. 100. See also M. G. Lord, *Astro-Turf: The Private Life of Rocket Science* (New York: Walker and Co., 2005), pp. 188–195.

5. Charles Kohlhase interview, 2 July 2002; see also John Casani interview, 21 November 2003, and Pete Lyman interview, 26 March 2004.

Americans.[6] Affirmative action policies and the increasing number of minority scientists and engineers nationwide almost doubled minority representation from the 1970s to the 1990s (from 13 percent in 1976 to 25 percent in 1994), although upper management remained a mostly white male province. Hispanics and especially Asian Americans continued to constitute most of the minorities (8 percent and 11 percent, respectively), in particular among scientists and engineers (14 percent Asian), so that whites, Asians, and Hispanics held better views of JPL's diversity than did African Americans.[7]

JPL's connections to local industry extended to spinoffs—from major defense contractor Aerojet, spun off during World War II, to more recent smaller startups such as ViaSpace and Photobit—and of course to local contractors. Many local firms made their living in part off JPL contracts, in what economists would explain as a high-tech, post-Fordist example of earlier concentrations of craft industry (such as Lyon silks or Alsatian calico), where mutually independent producers capitalize on a common labor market and on a network of suppliers and services.[8] About one-third of JPL's subcontracting currently goes to southern California firms, and even for contracts that went to other aerospace hotbeds, such as to Lockheed Martin in Colorado, some of the subcontracts for those programs ended up in southern California as well.[9]

On the other hand, big aerospace firms could view JPL projects as contributing a mere pittance to company balance sheets. For groups such as the Hughes Space Division, which might have sales of several hundred million dollars a year, a JPL contract for $50 million or $100 million spread out over several years provided a very small fraction of business.[10] That did not keep them from competing for contracts or from complaining that JPL kept work in-house instead of contracting it to industry. As firms such as TRW and Hughes developed extensive satellite expertise (most of it through spacecraft for military reconnaissance and commercial communications), JPL faced competition from aerospace firms both near and distant, and NASA and Congress consistently pressured the lab to farm out more work.[11] Periodic downturns in civil and military space spending increased the competition from aerospace firms. The end of the cold war, in particular, marked a turning point for JPL, the aerospace industry, and southern California itself. JPL contemplated

6. T. G. Meikle, "Ethnic Staffing by Job Categories, 28 October 1976 (JPL 142, 16/265); "Advisory Council on Minority Affairs Minutes," 5 October 1977 (JPL 142, 16/258); Employment Comparison in Advisory Council Material," February 1979 (JPL 8, 5/79).

7. JPL, "Response to the NASA Federal Laboratory Task Force Subcommittee Questions," 31 October 1994 (JPL 259, 20/231), and Kirk Dawson, "JPL Institutional Overview," 7 November 1994 (JPL 259, 20/229); JPL Employee Survey, June 1995 (JPL 259, 89/1296).

8. Scott, *Technopolis*, pp. 25–26.

9. Data courtesy of Martin Ramirez in the JPL Legislative Affairs Office.

10. Albert Wheelon interview, 13 June 2002.

11. Westwick, *Into the Black*.

the grim prospect that the lab itself might shut down and, as the aerospace industry hemorrhaged jobs, Caltech's president declared that "This is a time of triage—[the] only question is what is going to die."[12] JPL's response, known as the "employee displacement strategy" or "right-sizing"—or, more bluntly, downsizing—produced staff cuts of 30 percent over the next several years. Though not as deep as the cuts suffered by the aerospace business in general, they represented and contributed to the general malaise of the early 1990s in southern California.[13]

JPL survived because it found new justifications (besides the pork-barrel political support it received as a large local enterprise.) One response to the end of the cold war was to celebrate the triumph of the private sector and, as military threats receded, the federal government shifted emphasis to economic concerns. Clinton's presidential campaign catchphrase of 1992—"It's the economy, stupid"—infected NASA and JPL, which channeled the entrepreneurial buzz into efforts to transfer space technologies to the civilian economy. The usual approach involved licensing and patents, but JPL also spun off people and their ideas in small startup companies. Caltech encouraged the spinoffs by providing startup capital through a "grubstake" grant program, with an equity stake retained by the university. The startups also capitalized, literally, on the thriving venture capital business in California in the 1990s, which quickly learned to target the people and ideas emerging from JPL.

These spinoff companies supported new justifications for the space program based on high technology for economic growth. The clustering of offshoots such as Photobit and ViaSpace around the lab helped the local economy rebound from aerospace cutbacks, but they also demonstrated some resistance to the commercial spirit at JPL. Lab managers worried that engineers constantly on the lookout for business opportunities might not stay focused on the job of building spacecraft. Lab managers also feared the loss of key staff. In 1998, for example, Carl Kukkonen, the architect of JPL's microelectronics program, left to start ViaSpace, and he sought to bring a number of JPL staff with him. Caltech agreed but asked that he keep his recruiting within reasonable bounds—as Kukkonen put it, "retail rather than wholesale."[14] Lab managers also anticipated that NASA auditors, despite the Bayh-Dole Act promoting technology transfer, would charge that individuals were profiting from government-funded research. In some cases the Caltech president and trustees had to intervene to overcome JPL's and NASA's concern about the private–public divide, which continued to dampen the embrace of entrepreneurialism.[15]

---

12. Stone, "Notes on Trustees' Committee on JPL," 10 March 1992 (JPL 259, 32/323).

13. "JPL's Institutional Strategy," ca. January 1993, and "Executive Council Meeting Agenda," 1 March 1993 (JPL 165, 2/11); Stone presentation to Trustees Committee on JPL, 30 October 1998 (JPL 259, 35/343); Stone to Edward Weiler, 7 March 2000 (JPL 259, 7/101).

14. Carl Kukkonen interview, 3 June 2003.

15. Lawrence Gilbert telephone interview, 15 March 2005.

THE ENVIRONMENT AND THE LOCAL COMMUNITY

JPL has always interacted with the local geography and environment. The lab itself sits in the Arroyo Seco, a dry wash three miles above the Rose Bowl in Pasadena. The isolated spot was chosen for initial rocket tests in the 1930s, not so much for safety as because no campus facilities were available. The wisdom of using a remote site was confirmed, however, when subsequent tests on the Caltech campus misfired, one explosively. When the Army Air Corps then stepped up wartime funding, the rocketeers returned to the Arroyo, but only after the city of Pasadena, protecting the interests of nearby high-end homeowners, insisted that the site would serve only for the duration of the war. By war's end, the growth of apparently permanent structures alarmed city managers, but when the Army threatened to invoke eminent domain the city gave in and signed a long-term lease.[16] JPL remains in the Arroyo today, although it had to seek a more remote site for rocket tests in the Mojave desert near Edwards Air Force Base. The California high desert provided isolation of a different sort necessary for deep-space tracking and communication, removed from radio noise, and the cornerstone of what would become the Deep Space Network was laid in the late 1950s on a dry lake bed at Goldstone in the Mojave.[17]

Despite the rocky start with the city of Pasadena, for the local community JPL largely represented a boon. The lab and Caltech came to enjoy substantial influence with local media and city government, by virtue of their size and prestige—enough to squash the occasional protest from neighbors over the construction of new buildings. With more than 4,000 staff by the mid-1970s, JPL at that time was the largest employer in Pasadena.[18] This statistic relied on the fact that JPL staff were Caltech employees; JPL itself was not located in Pasadena. It lay outside the city limits on land annexed by the neighboring town of La Cañada-Flintridge when it incorporated in 1976. After much squabbling between rival city boosters, including the Pasadena city manager's declaration that "to see it end up in some other corporate limits would be like losing the Rose Bowl," the U.S. Postal Service left the decision up to the lab and JPL chose Pasadena. La Cañada-Flintridge residents were still stewing about the episode 20 years later.[19]

16. Koppes, *JPL and the American Space Program,* pp. 11, 21.

17. Ibid., p. 95; Douglas J. Mudgway, *Uplink-Downlink: A History of the Deep Space Network, 1957–1997* (Washington, DC: NASA-SP-2001-4227, 2001), pp. 2–12.

18. JPL Institutional Background and Overview, 3 October 1975 (JPL 142, 27/483).

19. "JPL is in Pasadena" [quote, editorial], *Pasadena Star News*, 19 March 1976; Harold Brown to Mortimer Mathews, 2 December 1975, and D. R. Fowler to A. T. Burke, 2 January 1976 (JPL 142, 24/420); "Pasadena to Ask JPL Land Annexation," *Pasadena Star News*, 1 January 1976; Jackie Knowles, "Pasadena Fails in JPL Annex" *Pasadena Star News*, 11 March 1976; "The Great Battle for JPL," [editorial] *La Canada Valley Sun*, 18 March 1976; Dick Lloyd, "JPL Faces an Identity Crisis Following Incorporation Vote," *Pasadena Star News*, 15 November 1976; see additional clippings in JPL files, folder 4649, NASA History Office.

Pasadena city leaders soon had cause to regret JPL's proximity. Increasing public sensitivity in the United States to environmental pollution resulted, among other things, in the Safe Drinking Water Act of 1976 and testing of water supplies. In 1980 state health authorities began testing water wells in Pasadena and found a number of them contaminated with toxic carcinogens, including four wells in the Arroyo Seco east of JPL. The lab had developed formal control of hazardous waste in the early 1960s; before that, JPL had used cesspools, dumping pits, and an incinerator to dispose of wastes at the east end of the lab.[20]

Pasadena officials assumed the contamination came from JPL's early rocket research. The lab initially denied blame, noting that the contaminants—trichloroethylene, carbon tetrachloride, and tetrachlorethylene—were common industrial solvents; 30 other wells throughout the San Gabriel Valley had also turned up excessive traces. JPL nevertheless agreed in 1985 to fund jointly with Pasadena an engineering study, in the spirit of a "good neighbor" and without admitting liability. An outside consultant subsequently concluded that JPL was the "most probable source" and recommended construction of a water treatment plant.[21] NASA agreed to pay for the plant, built in 1990 at a cost of $1.3 million, and JPL planned to commit about $1 million a year for several years for cleaning up the contamination. In 1992 JPL was named a "Superfund" site, one of more than 1,000 polluted places identified by the federal government for environmental remediation.[22]

In this respect JPL differed little from many other aerospace sites, not to mention nuclear labs and other enterprises, that had raced to beat the Soviets in the cold war—at a time of more cavalier attitudes toward waste disposal and the environment—and then faced the consequences decades later. The concentration of aerospace business in southern California, however, has made the cold war environmental legacy more acute there. This, too, is one of the societal impacts of space exploration.

Environmental liability would be a major issue in contract negotiations with NASA. Caltech feared exposing its endowment to lawsuits, while NASA found itself paying for damage likely incurred under the previous Army contract.[23] This was not just splitting hairs: JPL was shortly slapped with a lawsuit alleging that pollution caused the death of one local woman and Hodgkin's disease in two others;

20. Beverly Place, "Foothill Water Wells Checked for Chemicals," *Montrose Ledger*, 19 January 1980; Fred Felberg to speakers' bureau, with attached fact sheet, 26 November 1986 (JPL 230, 26/246).

21. Felberg to speakers' bureau, 1986; W. E. Rains, memo to the record, 20 August 1986 (JPL 230, 26/241). The "good neighbor" phrase appears in both documents.

22. Management plan for remedial investigation/feasibility study, January 1992 (JPL 239, 3/18); "Environmental Cleanup Review," JPL fact sheet, April 1991 (JPL 239, 1/5); E. C. Stone to senior staff, 22 October 1991 (JPL 239, 2/11); Marla Cone, "Jet Propulsion Lab Added to Superfund list," *Los Angeles Times*, 15 October 1992.

23. Tom Sauret interview, 25 April 2001.

31 other local residents petitioned to join the case. A groundwater study by the federal Health and Human Services Department in 1998 found no current threat and judged past hazards "unlikely," but JPL continued to grapple with lawsuits into the new century.[24]

JPL's response to the groundwater problem is instructive. Despite the lawsuits, JPL avoided the much more active environmental controversies that plagued some other government labs. Brookhaven National Lab, to take one example, encountered an uproar after revelations in the 1990s that a plume of groundwater near a nuclear reactor contained tritium. Hundreds of community activists packed public hearings and vented their anger. Brookhaven scientists could not understand the fuss over what they saw as a negligible hazard, but local residents resented what they viewed as evasions and patronizing reassurances based on statistical risk analysis. Brookhaven learned a hard lesson in community relations: the lab director was effectively pressured out of his job and the contractor, which had run Brookhaven for 50 years, saw its contract summarily terminated by the Department of Energy in 1997.[25]

The JPL case differed in that it dealt with common industrial chemicals, whereas Brookhaven and other nuclear labs involved radioactivity, with its popular associations of danger and secrecy.[26] Another prime difference from labs such as Brookhaven was that JPL, like NASA as a whole, had a finely tuned public relations organization and decades of experience with operating in the media glare, including highly exposed failures. Brookhaven scientists, by contrast, admitted their lack of public relations expertise. At the outset of the controversy at JPL, lab managers emphasized to staff that any public comments "will not underestimate the important nature of the problem. For example, we will not cite statistics in an effort to demonstrate that chances of getting cancer from Pasadena drinking water are low."[27] Another internal memo stressed that JPL and NASA should "maintain a positive, cooperative attitude in dealing with the city. The contribution of a substantial part of the construction cost of a treatment plant is preferable to the consequences of sensational adverse

24. Robin Lloyd, "11 to Join Toxic Suit against JPL," *Pasadena Star-News*, 18 June 1997; "Study Clears NASA Lab of Alleged Threat to Public Health," *San Diego Union-Tribune*, 23 August 1998; Patti Paniccia, "The Devil's Advocate," *Los Angeles Times Magazine* (25 July 2004): p. 16.

25. Colin Macilwain, "Brookhaven Contractor is Sacked over Tritium Leak," *Nature* 387 (8 May 1997): p. 114; Andrew Lawler, "Meltdown on Long Island," *Science* 287 (25 February 2000): pp. 1382–1388; Robert P. Crease, "Anxious History: The High Flux Beam Reactor and Brookhaven National Laboratory," *Historical Studies in the Physical and Biological Sciences*, 32/1 (2001): pp. 41–56; Jack M. Holl, *Argonne National Laboratory, 1946–96* (Urbana, IL: University of Illinois Press, 1997), pp. 485–488; Leland Johnson and Daniel Schaffer, *Oak Ridge National Laboratory: The First Fifty Years* (Knoxville, TN: University of Tennessee Press, 1994), pp. 225–229.

26. Spencer Weart, *Nuclear Fear: A History of Images* (Cambridge, MA: Harvard University Press, 1988).

27. Lawler, "Meltdown"; Felberg to speakers' bureau, 1986.

press coverage of the situation."[28] JPL's public relations experience, well-honed by the civil space program, helped it largely avoid the antagonistic community reaction that could characterize environmental cleanup at other cold war labs.

Hooray for Hollywood

The lab also forged links with another prominent local industry, namely, show business. Hollywood knows a spectacle when it sees one, and celebrities often descended on JPL for planetary encounters. This included not only actors and actresses, but also writers with studio connections—especially local science-fiction writers, who had a natural interest in JPL's work. In the lab's earliest days, two of its founders, Frank Malina and Jack Parsons, tried to sell a story about their rocket group to Metro-Goldwyn-Mayer to raise money for their research; Parsons later hobnobbed with L. Ron Hubbard and Robert Heinlein and lost his life in an explosion while working on pyrotechnic special effects for Hollywood.[29] By the 1970s sci-fi writers were a strong presence around JPL for planetary encounters, including such luminaries as Arthur C. Clarke, Ray Bradbury, and Heinlein. As with aerospace, Caltech helped nourish JPL's Hollywood ties with its own elite connections, including trustees (and JPL overseers) such as mogul Lew Wasserman.

Movie trends revealed Hollywood's periodic fascination with space exploration, such as in the late 1970s, when the first *Star Trek* movie featured a mysterious, nearly sentient object known as V-GER. The trend revived in the late 1990s, when the discovery of a Martian meteorite with possible life-forms, the Shoemaker-Levy comet impact on Jupiter, and Mars Pathfinder were followed by blockbuster movies including *Deep Impact* and *Armageddon*, about meteor impacts on Earth, and *Mission to Mars* and *Red Planet*, about mysterious life-forms on Mars, not to mention several new installments of the *Star Wars* saga. JPL's executive council in 1999 noted the surprising interest in "space movies and the merging of entertainment and realities." The merging included a new mission, approved that year, which planned to fire a 1,000-lb (500-kg) projectile into the comet P/Tempel-1 and observe the resultant crater; JPL's project manager had to insist that the mission's name, Deep Impact, was in fact selected prior to the movie.[30]

There are more substantial connections between JPL and the movie business in the lab's technical program. In the early 1960s JPL engineers had pioneered

28. Rains memo, 20 August 1986.

29. George Pendle, *Strange Angel: The Otherworldly Life of Rocket Scientist John Whiteside Parsons* (New York: Harcourt, 2005), and Davis, *City of Quartz*, pp. 54–62.

30. Gael Squibb notes on EC retreat, 11 March 1999 (JPL 259, 64/758); James Graf quoted in Matthew Fordahl, "Deep Impact," AP newswire, 9 July 1999; Andrew Murr and Jeff Giles, "The Red Planet takes a Bow," *Newsweek* (6 December 1999).

the use of digital image processing, first to clean up pictures returned from the Ranger missions to the Moon, and later on other planetary spacecraft images. The first techniques corrected distortions and removed signal noise, such as a particular frequency superimposed on an image by vibration of the camera, and grew to encompass algorithms for contrast enhancement, cartographic projection, motion compensation, and so on (in short, many of the techniques now available through such software packages as Photoshop).[31] By the mid-1970s, what was known as the Image Processing Laboratory at JPL had perhaps the most advanced capability in the country.

Meanwhile, JPL had also supported important early work on computer animation. In the late 1970s JPL hired a young programmer from Utah named James Blinn, a student of Ivan Sutherland (who was now at Caltech). Blinn was making a name as a guru of computer graphics; Sutherland, himself no slouch, would say, "There are only a dozen great people in computer graphics, and Jim Blinn is six of them." Blinn went to work with Charles Kohlhase on the Voyager project, generating three-dimensional animations simulating Voyager's flight past Saturn, at each point calculating the relative appearance of planets and stars. The three-dimensional (3-D) movies—in which the viewer rode along with Voyager as the spacecraft swooped over Saturn's rings and satellites—proved a hit with TV news editors and viewers. Similar graphics sequences from Blinn enlivened the *Cosmos* television series of Carl Sagan.[32]

Although Blinn was not a part of the Image Processing Lab, his computer animations for Voyager and *Cosmos* used similar techniques, such as reconstructing viewing geometries and surface reflectance.[33] The Image Processing Lab had also produced its own motion pictures by combining still photos, such as those from the Ranger and Mariner spacecraft. Around the time of Blinn's work on Voyager, Kevin Hussey was doing some crude 3-D animation for atmospheric scientists, including modeling the smog distribution in the Los Angeles basin. Hussey heard about Blinn's work and approached him for help, but lacked the budget to hire him. So Hussey used in-house code written by Mike Kobrick in the Image Processing Lab and, with the help of a young programmer named Bob Mortensen, tweaked it for his purposes. The result, in 1987, was *L.A. the Movie*, which projected topographic data from Landsat to simulate the 3-D view from an aircraft swooping through the Los Angeles basin. By that time Hussey's group had grown from 2 to 14 and

---

31. Caltech/JPL Conference on Image Processing Technology, *Proceedings*, 3–5 November 1976 (JPL report SP-43-30).

32. Natalie Angier, "It Was Love at First Byte," *Discover* (March 1981); David Salisbury, "Computer Art Takes Off into Space," *Christian Science Monitor*, 20 July 1979; Frank Colella interview, 26 February 2002; Charles Kohlhase interview, 20 July 2002.

33. William B. Green interview, 12 February 2002. On shared techniques, see Kenneth R. Castleman, *Digital Image Processing* (Englewood Cliffs, NJ: John Wiley & Sons, 1979), pp. 371–377.

had formally organized as the Digital Image Animation Lab (DIAL), alongside and overlapping the original Image Processing Lab.[34]

Blinn had meanwhile gotten several offers from the movie business and in 1981 accepted one from George Lucas to establish a special effects studio for *The Empire Strikes Back*, the first *Star Wars* sequel. At the Lucas studio Blinn teamed up with Alvy Ray Smith, who had himself worked briefly with Blinn at JPL. Blinn's work at JPL had already had some influence: the sequence that persuaded Lucas of the potential of computer graphics was developed by Smith's team for *Start Trek II: The Wrath of Khan*, which showed the view from a spacecraft flying past a dead, Moon-like planet which then is transformed by a fiery cataclysm into an Earth-like environment. In creating this "Genesis" shot, Smith declared, he "was still under the influence of Blinn's Voyager flybys of the planets at the Jet Propulsion Laboratory." After joining Smith for a stint at Lucasfilm, Blinn returned to the Caltech campus and later became a "graphics fellow" at Microsoft; Smith would go on to help found Pixar.[35]

Hussey had also intersected Hollywood through work on digital color correcting, starting with a CBS television special on the *Mona Lisa* and then on color correction of old *Star Trek* backgrounds. He had then helped William Shatner with a zoom shot from the galaxy down to Yosemite Valley for another of the *Star Trek* movies and digitally erased telephone wires from the Egyptian skyline for *Raiders of the Lost Ark*. He had also developed a technique to interpolate between still images, from work for ocean scientists creating continuous animation of ocean circulation from satellite images taken 12 hours apart; he applied this on work for Whitney/Demos Productions, one of the leading graphics studios, to morph images into one another—an early example of the technique popularized by *Terminator 2* and Michael Jackson's *Black and White* video in 1992. Hussey soon had a stream of movie producers coming through his door for what he called "demo after demo" of VICAR (Video Image Communication and Retrieval) and other JPL digital processing software.[36]

Hollywood was hooked, and in the mid to late 1980s began to pour money into digital techniques, aided by the emergence of commercial software packages such as Photoshop. Disney soon hired away Hussey and several others from the DIAL to help build its own digital animation studio, including work with Pixar's movies. By the mid-1990s Hollywood's vastly greater resources allowed it to outstrip the

34. Kevin Hussey phone interview, 11 September 2006; Green interview, 12 February 2002; Kevin Hussey, Bob Mortensen, and Jeff Hall, *L.A. the Movie*, 1987 (JPL audiovisual collection).

35. Natalie Angier, "It Was Love at First Byte"; David Salisbury, "Computer Art Takes Off into Space"; Alvy Ray Smith, "George Lucas Discovers Computer Graphics," *IEEE Annals of the History of Computing* 20/2 (1998): pp. 48–49.

36. Hussey interview, 11 September 2006. For history of Whitney/Demos, see Gary Demos, "My Personal History in the Early Explorations of Computer Graphics," *Visual Computer* 21/12 (2005): pp. 961–978.

capabilities at JPL—even though most of DIAL's funding came from outside NASA, with IMAX as the largest sponsor.[37] As a result, the influence began to run mostly in the opposite direction, from Hollywood to JPL. Hussey himself returned to JPL after several years at Disney, bringing with him the latest from Hollywood studios. These included 3-D gaming techniques, which JPL applied not only for outreach efforts (allowing kids to play online games of planetary exploration), but also to help mission managers visualize what their spacecraft are doing in space, by transforming raw telemetry into real-time animations.[38] Such techniques have closed the loop from JPL to Hollywood, and from popular interest to technical capability.

CONCLUSION

JPL's impact on southern California has ranged from the local aerospace industry and labor pool, to its environmental legacy, to Hollywood studios. But the influence throughout has run both ways. We might ask not only about the societal impact of space programs, but also whether southern California left a particular stamp on the enterprise of space exploration. What were the consequences of the concentration of aerospace for the space program? California historian Kevin Starr has long described the state as a land of dreamers; did the blue-sky California environment encourage dreams of soaring beyond the sky, in the attitude captured by a California-ism from that other local industry, "imagineering"?[39] Or did California's recreation and leisure culture—all that sun and sand and surf—instead distract space engineers from their work, as it can also distract space historians?

Or could one go even further and ask whether there is a distinctive southern California type of spacecraft. Perhaps not, although those Mars Rover airbags look suspiciously like beach balls bouncing across the sand. But one could speculate that southern California attracted or nurtured a particular mindset or sensibility, one attuned to the imaginative possibilities and the public appeal of spaceflight. JPL managers identified two general categories of staff: the benchtop engineers building hardware (or software), and the mission designers.[40] It was the mission designers who were perhaps more susceptible to the context. Consider Bruce Murray, for example, who as JPL director would emphasize missions with popular appeal over strictly scientific goals. In the 1970s Murray released the "purple pigeons," a group of proposed missions with razzle-dazzle for the public (such as a solar sail to Halley's

37. James A. Evans, "The Reimbursable Program," 28 February 1995 (JPL 259, 50/553).

38. Hussey interview, 11 September 2006.

39. See the several volumes in the series by Kevin Starr, *Americans and the California Dream*. See also David Beers, *Blue Sky Dream: A Memoir of America's Fall From Grace* (Garden City, NY: Doubleday and Co., 1996).

40. Felberg et al., "JPL Futures Study," July 1980 (JPL 142, 20/325).

comet); he would continue to push such flights of fancy as an interstellar probe, Mars airplanes, and Venus submarines in the planetary program and in the civil space program in general. Such attitudes were not confined to southern California—but it might be no accident that Murray had grown up just across town in Santa Monica. (It might also be no accident that he has been the only one to wear shorts and sandals behind the JPL Director's desk.)[41]

One might speculate further that JPL, by virtue of its location, attracted more than the usual share of Bruce Murrays. And an attitude like Murray's did in fact influence spacecraft design. Emphasizing popular appeal often meant imaging—that is, spacecraft that returned pictures instead of just reams of numerical data. And imaging cameras in turn meant three-axis-stabilized spacecraft instead of the spin-stabilized spacecraft preferred for particles-and-field experiments. This preference shows up in JPL's long tradition of three-axis-stabilized craft. (For the Galileo spacecraft, JPL review boards and Murray himself recommended dumping a complicated spun/despun design, which combined a spin-stabilized section with a three-axis-stabilized platform, for a straight three-axis spacecraft.) A 1987 proposal from the nearby Planetary Society for a robotic Mars spacecraft captured the people-pleasing sensibility and the sense that planetary missions were all about entertaining the public. The proposal stressed that "Science should be given a low priority on this mission, if it is given any direct participation at all . . . . Imaging, imaging, and more imaging is the name of the game . . . . The aim here is to obtain images that are shameless crowd-pleasers and show Mars from a human perspective. If that's not good science, well then tough."[42] A Hollywood flack couldn't have said it any better. If one considers that the entertainment industry passed aerospace in 1995 as the main employer in California, it may indeed be appropriate to view JPL itself as a subset of the entertainment business.

41. Westwick, *Into the Black*; Bruce Murray, *Journey into Space: The First Three Decades of Space Exploration* (New York: W. W. Norton & Co., 1989).

42. E. J. Gaidos, "Project Precedent," 31 July 1987 (JPL 198, 36/528).

Section VI

Spaceflight, Culture and Ideology

Chapter 25

Overview: Ideology, Advocacy, and Spaceflight—Evolution of a Cultural Narrative

Linda Billings

The ideas of frontier pioneering, continual progress, manifest destiny, free enterprise, and rugged individualism have been prominent in the American national narrative, which has constructed and maintained an ideology of "Americanism"—what it means to be American, and what America is meant to be and do. In exploring the history of U.S. spaceflight, it is useful to consider how U.S. space advocacy movements and initiatives have interpreted and deployed the values and beliefs sustained by this national narrative. The aim here is to illuminate the role and function of ideology and advocacy in the history of spaceflight by examining the rhetoric of spaceflight advocacy.[1] Starting from the premise that spaceflight has played a role in the American national narrative and that this national narrative has played a role in the history of spaceflight, this paper examines the relationship between spaceflight and this narrative.

Examining the history of spaceflight advocacy reveals an ideology of spaceflight that draws deeply on a durable American cultural narrative—a national mythology—of frontier pioneering, continual progress, manifest destiny, free enterprise, rugged individualism, and a right to life without limits. This ideology rests on a number of assumptions, or beliefs, about the role of the United States in the global community, the American national character, and the "right" form of political economy. According to this ideology, the United States is and must remain "Number One" in the world community, playing the role of political, economic, scientific, technological, and moral leader. That is, the United States is and must

1. In this paper, "spaceflight" generally refers to human spaceflight but does not exclude robotic spaceflight. The terms "spaceflight" and "space exploration" are interchangeable in this paper, even though they can convey different meanings in different contexts. General overviews of the history of space advocacy may be found in Tom D. Crouch, *Aiming for the Stars: The Dreamers and Doers of the Space Age* (Washington, DC: Smithsonian Institution Press, 1999); Howard E. McCurdy, *Space and the American Imagination* (Washington, DC: Smithsonian Institution Press, 1997).

be exceptional. This ideology constructs Americans as independent, pioneering, resourceful, inventive, and exceptional, and it establishes that liberal democracy and free-market capitalism (or capitalist democracy) constitute the only viable form of political economy.[2] The rhetoric of space advocacy exalts those enduring American values of pioneering, progress, enterprise, freedom, and rugged individualism, and it advances the cause of capitalist democracy.

Delving into the language or rhetoric of spaceflight is a productive way of exploring the meanings and motives that are embedded in and conveyed by the ideology and advocacy of spaceflight—the cultural narrative of pioneering the space frontier. According to rhetorical critic Thomas Lessl, rhetorical analysis can shed some light on

> . . . [T]he processes of communication that underpin decision making in free societies . . . . Judgments on matters of public policy take their cues from rhetoric, and so an understanding of any society's rhetoric will tell us a lot about its ideas, beliefs, laws, customs and assumptions—especially how and why such social features came into being.[3]

To begin this analysis, some definition of key concepts is warranted, starting with culture and communication. Anthropologist Clifford Geertz's definition of culture is operative in this analysis:

> [Culture is an] historically transmitted pattern of meanings embedded in symbols, a system of inherited conceptions expressed in symbolic forms by means of which men communicate, perpetuate and develop their knowledge about and attitudes toward life. [It is a context within which social action can be] intelligibly—that is, thickly—described.[4]

Building on Geertz's conception, communication theorist James Carey has characterized culture as a predominantly rhetorical construction, "a set of practices, a mode of human activity, a process whereby reality is created, maintained and transformed," primarily by means of communication.[5] Social norms can be constructed, perpetuated, and resisted—and ideologies can be propagated—"through

---

2. For an exposition of this idea, see, for example, Francis Fukuyama, *The End of History and the Last Man* (New York: Free Press, 1992).

3. Paul Newall, "Thomas Lessl: Science and Rhetoric" [interview]. *Galilean Library* [electronic], 2005, *http://www.galilean-library.org/lessl.html* (accessed 28 March 2007).

4. Clifford Geertz, *The Interpretation of Cultures* (New York: Basic Books, 1973), pp. 14, 34.

5. James Carey, *Communication as Culture: Essays on Media and Society* (New York: Routledge, 1992 version, 1988), p. 65.

ritualized communication practices."[6] When advocates speak of advancing scientific and technological progress by exploring and exploiting the space frontier, they are performing ritual incantations of a national myth, repeating a cultural narrative that affirms what America and Americans are like and are meant to do. For the purposes of this analysis, communication is a ritual, culture is communication, and communication is culture.

Standard definitions of ideology and advocacy are operational here. An ideology is a belief system (personal, political, social, cultural). Advocacy is the act of arguing in favor of a cause, idea, or policy.

Ideological Bedrock

The concepts of "progress" and the "frontier" require more extensive explication, as they are bedrock elements of the ideology of spaceflight. The root of "progress" is the Latin word meaning "to go forward." J. B. Bury said progress is movement "in a desirable direction"—but he also noted that "it cannot be proved that the unknown destination towards which man is advancing is desirable."[7] In their histories of the idea of progress, both Bury and Robert Nisbet called progress a dogma. Christopher Lasch contrasted the premodern, Christian idea of progress—"the promise of a secular utopia that would bring history to a happy ending"—with the modern idea representing "the promise of steady improvement with no foreseeable ending."[8] Bury identified progress as an idea originating in the modern era, whereas Nisbet traced its roots to ancient Greek and Roman philosophy, and he documented how it evolved to take on the qualities of destiny and "historical necessity."[9] Nisbet declared progress the most important idea in modern Western history. This modern idea of necessary and inevitable forward movement is deeply embedded in the cultural narrative of U.S. spaceflight.

The idea of progress became the dominant idea in Western thinking in the period 1850–1900, according to Nisbet, serving as "the developmental context for other [key] ideas" such as freedom.[10] Nisbet credited nineteenth-century natural philosopher Herbert Spencer with melding the ideas of progress and freedom, in

---

6. Michele M. Strano, "Ritualized Transmission of Social Norms through Wedding Photography," *Communication Theory* 16/1 (2006): pp. 31–46, esp. 31.

7. J. B. Bury, *The Idea of Progress: An Inquiry into Its Origin and Growth* (New York: Dover Publications, 1932), p. 2.

8. Christopher Lasch, *The True and Only Heaven: Progress and Its Critics* (New York: W. W. Norton & Co., 1991), p. 47. See also Robert Wuthnow, *American Mythos: Why Our Best Efforts to Be a Better Nation Fall Short* (Princeton, NJ: Princeton University Press, 2006); Susan Richards Shreve and Porter Shreve, *How We Want to Live: Narratives on Progress* (Boston: Beacon Press, 1998).

9. Robert Nisbet, *History of the Idea of Progress* (New York: Basic Books, 1980), p. 47.

10. Ibid., p. 179.

declarations of "the rights of life and personal liberty," "the right to use the Earth," "the right of property," and "the right to ignore the state."[11] Spencer's classical liberal thinking is noticeable in the rhetoric of space advocacy.

From the seventeenth through the twentieth century, as Walter McDougall wrote, the Western scientific worldview—itself a cultural narrative of sorts—"elevated technological progress . . . to the level of moral imperative."[12] Science and technology became the means of American progress, and conquest and exploitation became the morally imperative method. Ultimately, progress came to be thought of as the accumulation of material wealth. Robert Wright has said the idea of progress is "a Victorian ideal" of moral advancement that has evolved into an ideal of material improvement.[13] This belief in progress performs the mythic function of providing moral justification for material accumulation. Along those same lines, Kirkpatrick Sale has asserted that the contemporary "myth" of progress advances "the propaganda of capitalism," the idea of continual human improvement by means of resource exploitation and material accumulation.[14]

Author Ishmael Reed has made the link between progress and spaceflight in an essay called "Progress: A Faustian Bargain":

> In order to justify its programs, NASA, in its brochures, describes the Earth as a dying planet, a fact which for them justifies colonizing the universe . . . . You can understand why, in many science fiction movies, the goal of the invaders is to destroy this planet, lest this progress be extended to their neighborhoods.[15]

Historically and presently, the rhetoric of space advocacy advances a conception of outer space as a place of wide-open spaces and limitless resources—a space frontier. The metaphor of the frontier, with its associated images of pioneering, homesteading, claim-staking, and taming, has been persistent in American history. In the rhetoric of spaceflight advocacy, the idea of the frontier is a dominant metaphor. It is worth noting that the root of the word "frontier" is the Old French word for "front." In the English language, that word "front" conveys a complex of meanings, ranging from the most common definition—the part of anything that faces forward—to the definition that probably comes closest to the meaning of "front" in "frontier": an area of activity, conflict, or competition. A common military definition of "front" is also tied up in the meaning of "frontier," that is, the area of contact between opposing

11. Ibid., p. 229.

12. Walter A. McDougall, *. . . the Heavens and the Earth: A Political History of the Space Age* (New York: Basic Books, 1985), p. 11.

13. Robert Wright, *A Short History of Progress* (New York: Carroll & Graf, 2004), p. 3.

14. Kirkpatrick Sale, "Five Facets of a Myth," in *How We Want to Live: Narratives on Progress,* ed. Susan Richards Shreve and Porter Shreve (Boston: Beacon Press, 1998), p. 108.

15. Ishmael Reed, "Progress: A Faustian Bargain," in ibid., quote from p. 102.

combat forces. Other meanings of "front" that should be considered in assessing the meaning of the frontier metaphor are: a façade; a position of leadership or authority; and a person or thing that serves as a cover for secret, disreputable, or illegal activity. What meanings are advocates intending to convey, and what meanings are they in fact conveying, when they talk about the space frontier?[16]

Historian Frederick Jackson Turner's century-old essay, "The Significance of the Frontier in American History," is perhaps the best-known articulation of the frontier metaphor.[17] It is a powerful and evocative piece of writing. In making the case for spaceflight, advocates continue to cite, directly or indirectly, Turner's frontier thesis and the related, potentially dangerous, idea of manifest destiny, seemingly oblivious to a changed cultural context and critiques of Turner's thinking.

As Wright and Sale did with progress, Richard Slotkin, in his trilogy of books about the history of the American West, has deemed the idea of the frontier a myth—a myth in which the United States is "a wide-open land of unlimited opportunity for the strong, ambitious self-reliant individual to thrust his way to the top."[18] Patricia Nelson Limerick has pointed out that space advocates cling to the frontier metaphor, conceiving "American history [as] a straight line, a vector of inevitability and manifest destiny linking the westward expansion of Anglo-Americans directly to the exploration and colonization of space." Limerick has warned that in abusing this metaphor, "[S]pace advocates have built their plans for the future on the foundation of a deeply flawed understanding of the past, [and] the blinders worn to screen the past have proven to be just as effective at distorting the view of the future."[19]

The Cultural Narrative of Spaceflight

According to rhetorical critic Janice Hocker Rushing, "Rhetorical narratives are discourses which explicitly or implicitly advocate moral choices." Rushing has said that the meanings of "definitional [American] cultural myths," such as the myths of the Western frontier and the space frontier, are a source of identity and "moral vision." Rushing noted that the United States "has drawn upon the frontier

16. All definitions were obtained from *http://www.dictionary.com* (accessed June–July 2006).

17. Frederick Jackson Turner, *Rereading Frederick Jackson Turner: The Significance of the Frontier in American History and Other Essays*, with commentary by John Mack Faragher (New York: Henry Holt, 1994, 1947, 1920).

18. Richard Slotkin, *Regeneration through Violence: The Mythology of the American Frontier, 1600–1860* (Middletown, CT: Wesleyan University Press. 1973), p. 5.

19. Patricia Nelson Limerick, remarks, *What is the Value of Space Exploration? A Symposium*, sponsored by the Mission from Planet Earth Study Office, Office of Space Science, NASA Headquarters and the University of Maryland at College Park, Washington, DC, 18–19 July 1994; p. 13.

for its mythic identity,"[20] or moral imperative, as McDougall called it. In this mythic universe, the cultural role of the explorer—the frontier conqueror, as it were—is, as Stephen Pyne has said, to serve as "a moral missionary, telling others and his sustaining civilization who they are and how they ought to behave."[21]

From the start, advocates constructed a narrative of spaceflight that made it a necessary, even biologically driven, enterprise. But, as Pyne has pointed out, spaceflight and other modes of exploration are not in our genes but in our culture. "Exploration cannot be extracted from the historical and cultural context within which it occurs." It is "a specific invention of specific civilizations conducted at specific historical times." Advocates of U.S. spaceflight have created their own frontier mythology, as Limerick has noted, expanding the story of Western American settlement to encompass space exploration. And problems have ensued because, as Pyne has said, "discovery among the planets is qualitatively different from the discovery of continents and seas."[22]

The History of Rocket-Men: Spaceflight as a Belief System

The U.S. and European rocket societies of the 1920s, 1930s, and 1940s were the world's first spaceflight advocacy groups. German advocate Willy Ley documented the exploits of these groups, the titles of his books articulating the ideology of spaceflight: *Engineers' Dreams* (1954), *Across the Space Frontier* (1952, with Wernher von Braun, Fred Whipple, and others), *The Conquest of Space* (1949), *Harnessing Space* (1963), and others.[23] Frank Winter called these early advocates—including Konstantin Tsiolkovskiy, Robert Goddard, and Wernher von Braun—"pioneers."[24] According to William E. Burrows, the Russian Tsiolkovskiy advocated "controlling all of nature—the entire universe," toward enabling human colonization. Tsiolkovskiy was greatly influenced by the late nineteenth-century Russian mystic philosopher Nikolai Fyodorov, who "believed that Earth is not humanity's natural home" and that humanity was intended to live in the cosmos. U.S. rocket developer Goddard

20. Janice Hocker Rushing, "Mythic Evolution of 'The New Frontier' in Mediated Rhetoric," *Critical Studies in Mass Communication* 3/3 (1986): pp. 265–296, quotes from 265, 267.

21. Stephen J. Pyne, "A Third Great Age of Discovery," in Carl Sagan and Stephen J. Pyne, *The Scientific and Historical Rationales for Solar System Exploration*, SPI 88-1 (Washington, DC: Space Policy Institute, George Washington University, July 1988), p. 37.

22. Ibid., pp. 14, 18.

23. His classic statement is Willy Ley, *Rockets, Missiles and Space Travel* (New York: Viking, 1951–1968, multiple editions).

24. Frank Winter, *Prelude to the Space Age: the Rocket Societies, 1924–1940* (Washington, DC: Smithsonian Institution Press, 1983), p. 7.

reportedly shared this belief.[25] An inheritor of this legacy, Wernher von Braun said the aim of the German rocketeers was "to open the planetary world to mankind."[26] In 1930, U.S. science-fiction writers and fans formed the American Interplanetary Society to advocate spaceflight, and in 1934 this group became the American Rocket Society. These early American advocates engaged in "relentless . . . publicizing . . . via newspapers, magazine articles, lectures, demonstrations, exhibits, radio talks, and films" to proselytize for spaceflight.[27]

In July 1958, physicist Freeman Dyson made his contribution to the advocacy campaign for spaceflight with "A Space Traveler's Manifesto":

> From my childhood it has been my conviction that men would reach the planets in my lifetime . . . [T]his conviction . . . rests on two beliefs, one scientific and one political: (1) There are more things in heaven and earth than are dreamed of in our present-day science. And we shall only find out what they are if we go out and look for them. (2) It is in the long run essential to the growth of any new and high civilization that small groups of people can escape from their neighbors and from their governments, to go and live as they please in the wilderness.[28]

As visions of spaceflight advanced toward reality, the rhetoric of spaceflight advocacy continued to promote conquest of the space frontier. Science-fiction author Olaf Stapledon wrote in 1948 for the *Journal of the British Interplanetary Society* that humankind should colonize other planets to exploit their resources for Earth's benefit and to "increase man's power over the environment . . . . The itch to leave a mark is quite wholesome, on condition that, even if it does not actually serve some higher aim, at least it does not positively hinder proper development."[29] Pope Pius XII reportedly told the International Astronautical Federation in 1957, "God who has planted within the heart of man the insatiable desire for knowledge . . . did

25. William E. Burrows, *This New Ocean: The Story of the First Space Age* (New York: Random House, 1998), pp. 37, 42.

26. Wernher von Braun, "German Rocketry," in *The Coming of the Space Age,* ed. Arthur C. Clarke (New York: Meredith Press, 1967), pp. 33–55, quote from p. 55. "There was much regret among us" at Peenemünde, von Braun wrote in this essay, "that the A-4 [a.k.a. the V-2], conceived as it was as a first step to interplanetary rocketry, had joined in the bloody business of war." (p. 54)

27. Winter, *Prelude to the Space Age*, p. 14. Successors to the American Rocket Society include the American Astronautical Society and the American Institute of Aeronautics and Astronautics, now representing professionals rather than amateurs and fans, and more prone to tend to matters of government policy and spending while less prone (though not averse) to proselytizing for spaceflight.

28. Freeman Dyson, *Disturbing the Universe* (New York; Harper & Row, 1979), p. 111.

29. Olaf Stapledon, "Interplanetary Man," in *The Coming of the Space Age*, Clarke, ed., pp. 244–261, quote from p. 255.

not intend to put a limit to man's endeavor."[30] Also in 1957, rocketeer Krafft Ehricke asserted that "the entire solar system, and as much of the universe as he can reach" are humankind's rightful domain: "[B]y expanding through the universe, man fulfills his destiny as an element of life, endowed with the power of reason and the wisdom of . . . moral law."[31] In 1964, space advocate Charles Sheldon wrote that "[M]ankind is destined to step beyond his earthly bonds just as his ancestors once crawled out of the seas." By "colonizing new worlds . . . spread[ing] into new places," Sheldon wrote, "the race will survive."[32]

The rocket-men of earlier decades and the geopolitics of the cold war propelled the U.S. space program into being and kept it going through the Apollo era. In that era, spaceflight advocacy came from two sources, according to William Burrows: political pragmatists and a "hard core of implacable dreamers; the unabashed zealots who shared a religious conviction that it was their race's destiny to explore other worlds and then start colonies on them."[33]

From the end of the Apollo era to the present, the ideology of spaceflight and the rhetoric of spaceflight advocacy have been sustained in public discourse in large part by the so-called grassroots space advocacy groups, such as the National Space Institute and the L5 Society and its successor, the National Space Society; the Space Studies Institute; and the Space Frontier Foundation. Wernher von Braun founded the National Space Institute (NSI) in 1974 to help cultivate public support for the U.S. space program in the post-Apollo era. The L5 Society was founded in 1975 by advocates Carolyn and Keith Henson to promote space colonization, as espoused by Princeton University physics professor Gerard K. O'Neill, who published his first paper on the subject, "The Colonization of Space," in the September 1974 issue of *Physics Today*. Today's National Space Society (NSS) is the product of a merger of the L5 Society and the National Space Institute in 1987. (The NSI was originally named the National Space Association but renamed NSI in 1975.) The

30. Quoted in Eugen Sänger, "Beyond the Solar System," in ibid., p. 216.

31. Krafft Ehricke, "The Anthropology of Spaceflight," in ibid., p. 263.

32. Charles Sheldon, "National Goals in Space," in ibid., pp. 71–93, quote from p. 74. Sheldon's rationale for spaceflight also cited its contributions to national security, technological spinoffs, economic benefits, the advancement of science, and national prestige—values that continue to be cited by advocates of spaceflight and in U.S. space policy.

33. Burrows, *This New Ocean*, p. 332. It should be noted that the frontier metaphor is noticeably absent from other national cultural narratives of spaceflight. Although the word "frontier" does appear from time to time in European space policy discourse, in those contexts it does not appear to mean the same thing as it does in American contexts. See, for example, the European Space Agency (ESA)'s use of the tag line "expanding frontiers" in public documents (*http://www.esa.int*) and the European Commission's reference to space as "a new European frontier." See European Commission, *Space: a New European Frontier for an Expanding Union: An Action Plan for Implementing the European Space Policy* (Luxembourg: European Communities, 2000).

NSI changed its name to the National Space Society shortly before its 1987 merger with the L5 Society.[34]

The NSS says its rationale for promoting space settlement is "survival of the human species." Among the values and beliefs articulated in the Society's "vision" for space exploration and development are "prosperity-unlimited resources," "growth-unlimited room for expansion," individual rights, unrestricted access to space, personal property rights, free-market economics, democratic values—and also enhancement of Earth's ecology and protection of new environments.[35]

Gerard K. O'Neill formed his own advocacy group, the Space Studies Institute (SSI), in 1977 to promote his colonization agenda. The SSI's stated mission is "opening the energy and material resources of space for human benefit . . . to make possible the productive use of the abundant resources in space."[36] Meanwhile, Freeman Dyson updated his spaceflight rationale for the 1970s, writing: "There are three reasons why, quite apart from scientific considerations," human spaceflight is necessary: first, "garbage disposal; we need to transfer industrial processes into space so that the earth may remain a green and pleasant place"; second, "to escape material impoverishment"; and third, "our spiritual need for an open frontier. The ultimate purpose is to bring humanity . . . a real expansion of our spirit."[37]

In the 1980s, the era of NASA's Space Shuttle and space station programs, the space community, as Burrows has noted, heavily promoted human spaceflight: "At the heart of it all, as usual, [were] the core of dreamers . . . who steadfastly believed it was their race's manifest destiny to leave Earth for both adventure and survival."[38] In 1988, some of those believers created the Space Frontier Foundation (SFF) to promote "opening the space frontier to human settlement as rapidly as possible." This group says its "purpose is to unleash the power of free enterprise and lead a united humanity permanently into the Solar System." Like the NSS, the SFF espouses

34. Richard Godwin, "The History of the National Space Society," *Ad Astra* (16 November 2005), *http://www.space.com/adastra/adastra_nss_history_051116.html* (accessed July 14, 2006). Interestingly, the combined membership of NSS and L5 voted after the merger of their groups to continue to use the NSS name instead of renaming their combined group the Space Frontier Society. Aerospace industry groups such as the Aerospace Industries Association, the American Institute of Aeronautics and Astronautics, the American Astronautical Society, and the Space Foundation play an important role in U.S. spaceflight advocacy, but they do so more by pragmatic actions, such as influencing legislation and policy-making, than by sustaining an ideology of spaceflight in public discourse.

35. See the National Space Society Web site for information on this subject, *http://www.nss.org* (accessed 10 June 2006). Interestingly, some of these beliefs appear to be in conflict with others.

36. This statement is available online at *http://www.ssi.org/* (accessed 14 July 2006). For more information on the history of the SSI, see Lee Valentine, "A Space Roadmap: Mine the Sky, Defend the Earth, Settle the Universe," available online at *Http://ssi.org/?page_id=2* (accessed August 2007); Roger D. Launius, "Perfect Worlds, Perfect Societies: The Persistent Goal of Utopia in Human Spaceflight," *Journal of the British Interplanetary Society* 56 (September/October 2003): pp. 338–349.

37. Dyson, *Disturbing the Universe*, pp. 116–117.

38. Burrows, *This New Ocean*, p. 507.

a conflicting set of goals, including "protecting the Earth's fragile biosphere and creating a freer and more prosperous life for each generation by using the unlimited energy and material resources of space." Its stated strategy for achieving these goals is "to wage a war of ideas in the popular culture" and transform U.S. spaceflight "from a government program for the few to an open frontier for everyone."[39]

In a series of essays called "The Frontier Files," SFF founder and director Rick Tumlinson offers his version of the space frontier narrative:

> We . . . see our civilization at a crossroads . . . . Down one path is a future of limits to growth, environmental degradation and ultimately extinction. Down the other path lie limitless growth, an environmentally pristine Earth and an open and free frontier in space.[40]

Regarding the purpose of spaceflight, he asserts:

> The one necessary and sufficient reason we are called to the Space Frontier is buried deep within us. It is a feeling . . . [a] calling to go, to see, to do, to be "there." We believe Homo Sapiens is a frontier creature. It is what we do, it defines what we are.[41]

In 1987, writer Marshall Savage founded the First Millennial Foundation and joined the chorus of advocates claiming humans are destined to colonize the universe:

> Now is the watershed of Cosmic history. We stand at the threshold of the New Millennium. Behind us yawn the chasms of the primordial past . . . before us rise the broad sunlit uplands of a living cosmos . . . . The future of the universe hinges on what we do next. If we . . . stride into space as the torchbearers of Life, this universe will be aborning.

Earth is slipping "into a pit of our own digging," according to Savage, and in order to save itself, humankind must expand into the cosmos: "[I]t is our policy to enliven this sterile universe . . . . If we . . . forsake our cosmic destiny, we will commit a crime of unutterable magnitude."[42]

---

39. "History of the Space Frontier Foundation," *http://www.space-frontier.org/History/* (accessed 14 July 2006). Among the SFF's goals is to keep Gerard K. O'Neill's space colonization manifesto, *The High Frontier*, "in print and available forever." The SFF liberally employs the rhetoric of war in urging people to "join the fight" for its agenda, locating itself on the space "front," characterizing itself as evolving from "a guerrilla band to a professional fighting force," and claiming it has "come down from the hills [to] invade the [U.S.] capital" ("History of the Space Frontier Foundation," subheading "Taking the fight to the nation's capital").

40. Rick Tumlinson, "Preface—Welcome to the Revolution (Message 1 of the Frontier Files)," 1995, *http://www.space-frontier.org/History/frontierfiles.html* (accessed 17 July 2006).

41. Rick Tumlinson, "Introduction—Who We Are (Message 2 of the Frontier Files)," 1995, *http://www.space-frontier.org/History/frontierfiles.html* (accessed 17 July 2006).

42. Marshall T. Savage, *The Millennial Project: Colonizing the Galaxy in Eight Easy Steps* (Boston: Little, Brown & Co., 1992, 1994), pp. 17–18, 230.

The Mars Society, founded in 1998, advocates pioneering the space frontier by the human settlement of Mars. Mars Society founder Robert Zubrin has said he embraces "Turner's belief that the frontier is a crucial part of the American character . . . I would like to see our traditions carried forward."[43] According to the Mars Society's "founding declaration,"

> Civilizations, like people, thrive on challenge and decay without it . . . . As the world moves towards unity, we must join together . . . facing outward to embrace a greater and nobler challenge . . . . Pioneering Mars will provide such a challenge . . . . A humans-to-Mars program would challenge young people everywhere to develop their minds to participate in the pioneering of a new world . . . . The settling of the Martian New World is an opportunity for a noble experiment in which humanity has another chance to shed old baggage and begin the world anew; carrying forward as much of the best of our heritage as possible and leaving the worst behind . . . . [E]xploration and settlement of Mars is one of the greatest human endeavors possible in our time . . . . No nobler cause has ever been.[44]

Government Space Rhetoric

The frontier metaphor, the ideology of progress and the belief in American exceptionalism have been prevalent in government space policy rhetoric as well as the rhetoric of advocacy groups. The National Commission on Space, appointed by President Reagan to develop long-term goals for U.S. civilian space exploration, entitled its final report "Pioneering the Space Frontier" and described in it "a pioneering mission for 21st-century America: to lead the exploration and development of the space frontier." Humankind is "destined to expand to other worlds," the commission said in its report, and "our purpose" is to establish "free societies on new worlds."Toward achieving those goals,"we must stimulate individual initiative and free enterprise in space."[45]

The rhetoric of American exceptionalism remained apparent in space policy documents of the George H. W. Bush administration: "America's space program is what civilization needs . . . America, with its tremendous resources, is uniquely qualified for leadership in space . . . our success will be guaranteed by the American spirit—that same spirit that tamed the North American continent and built enduring

---

43. Bart Leahy, "Save our Planet: Space Advocates see the Bigger Picture," *Ad Astra* (18 May 2006), available online at *http://space.com/adastra/adastra_save_earth_060518.html* (accessed 19 May 2006).

44. Mars Society founding declaration, 1988, *http://www.marssociety.org/about/founding_declaration.asp* (accessed 20 July 2006).

45. National Commission on Space, *Pioneering the Space Frontier* (New York: Bantam Books, 1986), pp. 2–3.

democracy." The "prime objective" of the U.S. space program is "to open the space frontier."[46] NASA declared in its 90-day study of this Space Exploration Initiative, "The imperative to explore" is embedded in our history . . . traditions, and national character," and space is "the frontier" to be explored.[47] "Space is the new frontier," said another space study group of that time, where the United States will find "a future of peace, strength, and prosperity."[48]

In keeping with rhetorical tradition, the Clinton administration declared, "Space exploration has become an integral part of our national character, capturing the spirit of optimism and adventure that has defined this country from its beginnings . . . . Its lineage is part of an ancient heritage of the human race . . . deep in the human psyche and perhaps in our genes."[49]

In the George W. Bush administration, White House Office of Science and Technology Policy Director John Marburger has said the point of the president's so-called vision for space exploration "is to begin preparing now for a future in which the material trapped in the Sun's vicinity is available for incorporation into our way of life."[50] NASA Administrator Michael Griffin has said that the aim of space exploration is "to make the expansion and development of the space frontier an integral part of what it is that human societies do."[51] Griffin has said that when human civilization reaches the point where more people are living off Earth than on it, "we want their culture to be Western." He has asserted that Western civilization is "the best we've seen so far in human history," and that the values space-faring people should take with them into space should be Western values.[52] "We want to be the world's preeminent space-faring nation for all future time," he said on another occasion, "second to none."[53] Griffin has said that space exploration has something to do with "core beliefs" about what societies and civilizations should be doing "on the frontiers of their time . . . North Americans are the way we are because of the challenges of the frontier . . . I believe that Western thought, civilization, and ideals represent a superior set of values," better

46. National Space Council, *Report to the President* (Washington, DC: Office of the President, 1990), p. 17.

47. *Report of the 90-day Study on the Human Exploration of the Moon and Mars* (Houston, TX: NASA Johnson Space Center, November 1989), pp. 1-1, 1-4.

48. Synthesis Group, *America at the Threshold: America's Space Exploration Initiative* (Washington, DC: U.S. Government Printing Office, 1991), pp. iv, 9, 14.

49. "National Apollo Anniversary Observance, A Proclamation by the President of the United States of America, July 19, 1994" (Washington, DC: Office of the President, 1994).

50. John Marburger, keynote address, 44th Robert H. Goddard Memorial Symposium, American Astronautical Society, Greenbelt, MD, 15 March 2006.

51. Griffin made these remarks at a conference sponsored by the Center for Strategic and International Studies, 1 November 2005, Washington, DC. The author attended this event.

52. Griffin made these remarks at a meeting sponsored by Women in Aerospace in Washington, DC, on 2 May 2005. The author attended this event.

53. Griffin made these remarks at a meeting sponsored by Women in Aerospace in Washington, DC, on 5 May 2006. The author attended this event.

than those of civilizations that came before. These values are "irretrievably linked to" expansion, he has said, and now this expansion will continue into the human frontier of space.[54] Most recently, Griffin has said:

> It is in the nature of humans to find, to define, to explore and to push back the frontier. And in our time, the frontier is space and will be for a very long time . . . . The nations that are preeminent in their time are those nations that dominate the frontiers of their time. The failed societies are the ones that pull back from the frontier. I want our society, America, [W]estern society, to be preeminent in the world of the future and I want us not to be a failed society. And the way to do that, universally so, is to push the frontier.[55]

Conclusions

This brief historical review has shown how the rhetoric of space advocacy has sustained an ideology of American exceptionalism and reinforced long-standing beliefs in progress, growth, and capitalist democracy. This rhetoric conveys an ideology of spaceflight that could be described, at its worst, as a sort of space fundamentalism: an exclusive belief system that rejects as unenlightened those who do not advocate the colonization, exploitation, and development of space.[56] The rhetorical strategy of space advocates has tended to rest on the assumption that the values of "believers" are (or should be) shared by others as well.

Although the social, political, economic, and cultural context for space exploration has changed radically since the 1960s, the rhetoric of space advocacy has not. In the twenty-first century, advocates continue to promote spaceflight as a biological imperative and a means of extending U.S. free enterprise, with its private property claims, resource exploitation, and commercial development, into the solar system and beyond. Pyne, among others, has addressed the problematic nature of these arguments: "The theses advanced to promote [solar system] settlement," he noted,

---

54. Griffin made these remarks at a meeting of the NASA Advisory Council's science subcommittees in Washington, DC, on 6 July 2006. The author attended this event.

55. Quoted in William Harwood, "Interview with NASA's Chief: Griffin Defends Budget, Shuttle Plans," *Space Place*, CBS News, 15 August 2006, *http://www.spaceflightnow.com/shuttle/sts115/060815griffin/* (accessed 16 August 2006).

56. In their rhetorical analysis of public statements on foreign policy by Presidents Ronald Reagan and George W. Bush, Kevin Coe and Richard Domke ["Petitioners or Prophets? Presidential Discourse, God, and the Ascendancy of Religious Conservatives," *Journal of Communication* (2006): pp. 309–330] found a common strategy "treading closely to claims regarding a divine vision for U.S. foreign policy, a theological sort of doctrine of Manifest Destiny, which runs the risk of doing what the doctrine of Manifest Destiny has done in the past" (p. 324). See this language in National Aeronautics and Space Administration, *The Vision for Space Exploration* (Washington, DC: NASA NP-2004-010334-HQ, 2004).

"are historical, culturally bound, and selectively anecdotal: that we need to pioneer to be what we are, that new colonies are a means of renewing civilization."[57]

Spaceflight advocacy can be examined as a cultural ritual, performed by means of communication (rhetoric), for the purpose of maintaining the current social order, with its lopsided distribution of power and resources, and perpetuating the values of those in control of that order (materialism, consumerism, technological progress, private property rights, capitalist democracy). Communication research has shown how public discourses—those cultural narratives or national myths—"often function covertly to legitimate the power of elite social classes."[58] And this review has shown how the rhetoric of space advocacy reflects an assumption that these values are worth extending into the solar system.

"Everything now suggests," Nisbet wrote 25 years ago, "that Western faith in the dogma of progress is waning rapidly."[59] This faith appears to have remained alive and well, however, in the ideology of spaceflight. Christopher Lasch wrote 15 years ago, "Almost everyone now agrees that [the idea of] progress—in its utopian form at least," no longer has the power "to explain events or inspire [people] to constructive action."[60] But in the current cultural environment, perhaps it does—at least among space advocates. Progress is, indeed, modern American dogma and a key element of pro-space dogma. But it does not resonate well—as Pyne and others have noted—in the current postmodern (or even post-postmodern) cultural environment, where public discourse is rife with critiques of science, technology, the aims of the military-industrial complex, and the corporate drive for profit.

Pyne observed almost 20 years ago that space exploration was "not yet fully in sync" with its cultural environment.[61] Modern (seventeenth- to twentieth-century) Western (European-American) exploration functioned as "a means of knowing, of creating commercial empires, of outmaneuvering political economic, religious, and military competitors—it was war, diplomacy, proselytizing, scholarship, and trade by other means."[62] But the postmodern exploration of space is different. Outer space is not simply an extension of Earth and the era of space exploration is not simply an extension of the modern era of transoceanic and transcontinental exploration. Its cultural context is different. The modern phenomenon of spaceflight has outlived the

57. Stephen J. Pyne, "Seeking Newer Worlds: The Future of Exploration," Sarton Lecture, American Association for the Advancement of Science, Denver, Colorado, February 2003, *http://www.public.asu.edu/~spyne/FUTURE.pdf* (accessed 10 June 2006), p. 8. See also Gregg Klerkx, *Lost in Space: The Fall of NASA and the Dream of a New Space Age* (New York: Random House, 2004).

58. Janice Hocker Rushing and Thomas S. Frentz, "Integrating Ideology and Archetype in Rhetorical Criticism," *Quarterly Journal of Speech* 77(4) (1991): pp. 385–406, quote from p. 385.

59. Robert Nisbet, *History of the Idea of Progress* (New York: Basic Books, 1980), p. 9.

60. Christopher Lasch, *The True and Only Heaven: Progress and Its Critics* (New York: W. W. Norton & Co., 1991), pp. 41, 21.

61. Pyne, "A Third Great Age of Discovery," p. 32.

62. Pyne, "Seeking Newer Worlds," p. 3.

modern era and its purpose is not clear in a postmodern or even post-postmodern world, characterized by uncertainty, subjectivity, deconstruction, and a rejection of so-called master narratives such as the story of frontier conquest. The moral imperative of the myth of pioneering the space frontier could be interpreted as a narrative that is in tune with its postmodern cultural environment in the sense that it conveys the values of the dominant social order—that is, what communication scholar Herb Schiller has called "the transnational corporate business order" and its ideology of private property ownership, resource exploitation and profit building.[63]

Of course, the idea of the human colonization of space is not publicly compelling in the current cultural environment. Poet Wendell Berry has addressed this dilemma:

> The [space colonization] project is an ideal solution to the moral dilemma of all those in this society who cannot face the necessities of meaningful change. It is superbly attuned to the wishes of the corporation executives, bureaucrats, militarists, political operators, and scientific experts who are the chief beneficiaries of the forces that have produced our crisis . . . .
>
> If it should be implemented, it will be the rebirth of the idea of Progress with all its old lust for unrestrained expansion, its totalitarian concentrations of energy and wealth, its obliviousness to the concerns of character and community, its exclusive reliance on technical and economic criteria, its disinterest in consequence, its contempt for human value, its compulsive salesmanship.

The sales pitch for space colonization goes this way, according to Berry:

> If we will just have the good sense to spend one hundred billion dollars on a space colony, we will thereby produce more money and more jobs, raise the standard of living, help the underdeveloped, increase freedom and opportunity, fulfill the deeper needs of the human spirit etc. etc. . . . Anyone who has listened to the arguments of the Army Corps of Engineers, the strip miners, the Defense Department or any club of boosters will find all this dishearteningly familiar.[64]

63. Herbert Schiller, "Not Yet the Post-Imperialist Era," in *Media and Cultural Studies: Key Works,* rev. edition, M. G. Durham and D. M. Kellner, eds. (Malden, MA: Blackwell, 2006), pp. 295–310, quote from p. 303. Retired Congressional Research Service analyst Eilene Galloway, arguably the oldest living expert on space law and policy at age 100 in 2006, has often said that problems in space policy are due at least in part to the fact that there are too many people in the space community who think that outer space is just like Earth.

64. Wendell Berry, "Mr. Gerard O'Neill's Space Colony Project Is Offered in the Fall 1975 *CoEvolution Quarterly . . .*", in *Space Colonies: a CoEvolution Book,* Stewart Brand, ed. (New York: Penguin, 1977), p. 36

Visions of the human colonization of space present a "moral law of the frontier" that is disturbing, Berry concludes: this law is that "humans are destructive in proportion to their supposition of abundance; if they are faced with an infinite abundance, then they will become infinitely destructive."[65]

Berry wrote his essay about the downside of space colonization in the 1970s. But his views are not necessarily out of date. Environmentalists might argue today that the case Berry made against space colonization is even more relevant today than it was in the 1970s.

In order to survive as a cultural institution, spaceflight needs an ideology. It needs to have some connection to widely held beliefs. It needs a role in a cultural narrative. But as Pyne has noted, "Locating exploration in the human gene or in the human spirit" and not in specific cultures is not viable. Continued reliance on this narrative "only absolves us from making those vital, deliberate choices" we inevitably have to make—about how we should proceed into space, and what values space exploration should embody. "These choices," Pyne has said, "are not intuitive."[66] As a cultural institution, space exploration "has to speak to deeper longings and fears and folk identities." It "is not merely an expression of curiosity but involves the encounter with a world beyond our ken that challenges our sense of who we are. It is a moral act . . . more than adventuring, more than entertainment, more than inquisitiveness." It has to explain "who a people are and how they should behave."[67] And in the current cultural environment, as Pyne has observed, space exploration "will have to base its claim to legitimacy on transnational or ecumenical values."[68]

Unlike the Western American frontier, as Janice Hocker Rushing has pointed out, space is too big to be conquered. The recent focus of space exploration on the search for evidence of extraterrestrial life is a product, she has said, of a widespread understanding that humankind exists in a universe, not only on planet Earth. The narrative of space exploration today might better reflect this understanding by telling a story of "a spiritual humbling of self" rather than "an imperialistic grabbing of territory."[69]

Although she has noted that "the WASP space cowboy version of spaceflight" has persisted from the Apollo era into the present, Constance Penley also has observed that NASA "is still the most popular point of reference for utopian ideas of collective progress." In the popular imagination, "NASA continues to represent . . . perseverance, cooperation, creativity and vision," and these meanings embedded in the narrative of spaceflight "can still be mobilized to rejuvenate the near-moribund idea of a future toward which dedicated people . . . could work together for the common good."[70]

65. Ibid.

66. Pyne, "A Third Great Age of Discovery," p. 54.

67. Pyne, "Seeking Newer Worlds," p. 7.

68. Pyne, "A Third Great Age of Discovery," p. 20.

69. Janice Hocker Rushing, "Mythic Evolution." p. 284.

70. Constance Penley, "Spaced Out: Remembering Christa McAuliffe," *Camera Obscura* 29 (January 1992): pp. 179–213, especially pp. 207–208.

This historical review of the rhetoric of space advocacy reveals competing American cultural narratives, then. The dominant narrative—advancing the values of the dominant culture—upon which the narrative of U.S. spaceflight piggybacks, is a story of American exceptionalism that justifies unilateral action and the globalization of American capitalist democracy and material progress. The story of spaceflight is embedded in this broader narrative. That story is also woven into a competing narrative, a vision of "utopian ideas of collective progress" and "a spiritual humbling of self." This competing narrative may be a site within which the ideology of spaceflight might rejuvenate itself—where the vision of a human future in space becomes a vision of humanity's collective peaceful existence on Spaceship Earth and the need to work together to preserve life here and look for life out there.

Chapter 26

Spaceflight and Popular Culture

Ron Miller

Spaceflight in popular culture has served two important functions—apart from its entertainment value, that is. First, it served to inspire at a time when the entire notion of traveling into space or to other worlds was an idea beneath the contempt of most scientists and engineers. It not only inspired, it carried the torch for the centuries it took for technology to finally make space travel a reality. Second, it acted—and still acts—as a mirror or gauge of both public interest in spaceflight and the state of contemporary astronautical science. For instance, Jules Verne shot his fictional projectile into space by means of a giant cannon[1] simply because his readers would have never believed that contemporary rockets would have been capable of such a feat, while only 50 years later Arthur Train and Robert Wood were able to describe a spaceship propelled by a beam of alpha particles produced by the disintegration of uranium in their remarkable novel, *The Moon-Maker* (1917).[2]

The Dreamers

Astronautics is unique among all the sciences in that it owes its origins to an art form. Long before engineers and scientists took the possibility of spaceflight seriously, virtually all of its aspects were first explored in art and literature, and long before the scientists themselves were taken seriously, the arts kept the torch of interest burning.

Although there had been numerous early fantasies about trips to the Moon, no one really considered the genuine possibility of spaceflight until two important events occurred: first, the discovery that there *were* places in the universe other than Earth, and second, that there might exist the technology for getting there. These events took place about 250 years apart. The first occurred in 1610 when Galileo first turned his telescope toward the heavens and discovered that the planets were not just a special class of wandering stars but real worlds in their own right. Venus

1. Jules Verne, *From the Earth to the Moon* (New York: Thomas Y. Crowell, 1978).

2. Arthur Train and Robert Wood, *The Moon-Maker* (King George, VA: Black Cat Press, 2006).

showed phases just like the Moon; Mars had dusky markings; and Jupiter possessed four tiny moons of its own, like a miniature solar system.

At about the same time that Galileo and the astronomers who followed him were discovering new worlds in our solar system, so were other explorers finding new worlds on the other side of the Atlantic Ocean. That the Moon was now included in geographies such as Peter Heylyn's *Cosmographie* of 1682 is indicative of its new status as an equal among other new worlds such as North and South America.[3]

By the time Galileo had observed the craters of the Moon and Jupiter's satellites, hundreds of ships and thousands of explorers, colonists, soldiers, and adventurers had made the journey to the amazingly rich and fertile lands of the New World. Now that human beings had learned that were there not only new worlds here upon Earth, but unknown lands in the sky as well, it's not very surprising that the discoveries being made by astronomers were quickly followed by a yearning expressed by an unprecedented spate of space travel stories. If these new lands could not be reached in reality, then they would be explored by proxy. Although most of these authors had little or no interest in the realistic depiction of science, their books were nevertheless an accurate barometer of the ever-increasing interest in the possibility of exploring the planets.

Many authors wrote of trips to the Moon in the decades following Galileo's discovery. In 1622, Charles Sorel wrote of "great Engins" that might carry people to the Moon, or that they might get there by means of "all manner of structures, and ladders."[4] In 1638, Francis Godwin published *The Man in the Moone*, in which his hero is carried to that world by a flock of geese who regularly migrate between Earth and the Moon.[5] The great scientist Johannes Kepler—who established the laws of orbiting bodies—wrote a novel called *Somnium* in 1634.[6] Its hero is carried to our satellite by demons along the bridge of darkness that occurs during an eclipse of the Moon. Although his method of getting to the Moon was highly unscientific, Kepler's descriptions of the conditions there were very accurate in light of what was known about the Moon at that time (for instance, Kepler was aware of the fact that most of the journey would have to be made in a vacuum and that the surface of the Moon would be desolate, alternating between extremes of heat and cold).

Cyrano de Bergerac made fun of such fanciful voyages in his *Comic History* (1657), in which he tried to come up with as many utterly ridiculous methods of space travel as he could think of. Ironically, one he thought was funniest was the use of rockets! Although he was probably the first person in history to suggest the use of rockets for launching a spacecraft, he only gets half a point for thinking it was a silly idea.[7]

---

3. Peter Heylyn, *Cosmographie* (London: Thoemmes Press, 2003).

4. Ron Miller, *The Dream Machines* (Malabar, FL: Krieger Publishing Co.,1993), p. 9.

5. Ibid., p.11.

6. Ibid., p.10.

7. Ibid., pp. 12–14.

The second great event in the history of the popular conception of space travel occurred in 1783 when two Frenchmen, brothers Étienne and Joseph Montgolfier, invented the balloon. For the first time, human beings were able to ascend above Earth farther than they could jump. These two events launched avalanches of speculative literature about the possibility of traveling beyond Earth and what might be found on these new worlds. Here at last seemed to be an answer to all those who had been looking at the starry sky with longing. If human inventiveness could devise a way of rising a few thousand feet above the surface of Earth, what was a mere quarter-million miles to the Moon but a matter of degree? Science and technology had conquered the sky—it had to be only a matter of time before they conquered space as well.

Needless to say, writers quickly abandoned their swans and dream-demons and turned to balloons to carry their heroes to the Moon and all over the solar system. But if science was marching ahead, so were the ever-more-knowledgeable readers of these books. The widespread fascination with science that occurred at the end of the eighteenth century also meant that most readers were becoming too knowledgeable about the subject to accept even balloons as a realistic method of getting to the Moon, let alone anything less scientific or technological. Authors were forced to come up with more realistic, believable methods of space travel.

One of the first to do so was the American author George Tucker. Although the spaceship in his 1827 novel, *A Voyage to the Moon*, employed an imaginary antigravitational material, Tucker gave great thought to the actual conditions that might exist beyond Earth's atmosphere and how his vehicle would have to deal with them.[8] What he ended up with might be the very first description of a spaceship that took into account the conditions of outer space as they were known at the time. For instance, the spacecraft—a 6-foot (2-meter), copper-clad cube—is carefully tested to make sure that it is perfectly airtight, while compressed air for breathing is carried in tanks. The walls are even insulated to protect the astronauts from the cold of space.

Tucker was one of Edgar Allan Poe's instructors at the University of Virginia and his novel may have inspired Poe to write his own Moon travel story, "The Unparalleled Adventures of One Hans Pfaal" (1835).[9] Although Poe used a balloon, he still paid more careful attention to science and technology than had any other author before him. For instance, the sealed, spherical gondola of his hero's balloon closely resembles those used by stratosphere balloonists and his descriptions of high-altitude flight and Earth as seen from space might have been written by a present-day astronaut.

Poe rationalized the use of a balloon by assuming that Earth's atmosphere extends as far as the Moon, albeit very tenuously:

8. George Tucker, *A Voyage to the Moon* (King George, VA: Black Cat Press, 2006).

9. Edgar Allen Poe, "The Unparalleled Adventures of One Hans Pfaal," in Richard Adams Locke and Edgar Allan Poe, *Journeys to the Moon* (King George, VA: Black Cat Press, 2006).

> I . . . considered that, provided in my passage I found the *medium* I had imagined, and provided that it should prove to be *essentially* what we denominate atmospheric air, it could make comparatively little difference at what extreme state of rarefaction I should discover it—that is to say, in regard to my power of ascending—for the gas in the balloon would not only be itself subject to similar rarefaction . . . but, *being what it was*, would, at all events, continue specifically lighter than any compound of whatever mere nitrogen and oxygen. Thus, there was a chance—in fact, there was a strong probability—that, *at no epoch of my ascent, I should reach a point where the united weights of my immense balloon, the inconceivably rare gas within it, the car, its contents, should equal the weight of the mass of the surrounding atmosphere displaced.*[10]

To rise into this infinitely thin atmosphere, Poe's balloon is filled with a new gas "37.4 times lighter than hydrogen"—and he also had his hero take advantage of the diminishing gravitational pull of Earth as the balloon traveled farther and farther from it.[11]

Almost all of this was sheer pseudoscientific mumbo-jumbo, of course, but Poe's readers did not know this. No one had ever spoken about a trip to the Moon in such detailed, scientific-sounding terms before. What is important is that for the first time it had become *necessary* to provide an interplanetary story with the trappings of real science. This is what makes Poe's story so significant. He realized that it was no longer sufficient to simply enable a character to reach the planets by fiat. By the time Poe was writing, the conditions at high altitudes were becoming well known and astronomers were becoming more assured about the nature of outer space and the prevailing conditions on our own satellite. An author writing a story set in these places could no longer blithely ignore this knowledge.

Poe himself was perfectly aware of the uniqueness of what he had done. In his postscript to "Hans Pfaal," he reminded his readers that the intent of earlier Moon journeys had been

> satirical; the theme being a description of Lunarian customs as compared with ours. In none, is there any effort at *plausibility* in the details of the voyage itself. The writers seem, in each instance, to be utterly uninformed in respect to astronomy. In *Hans Pfaal* the design is original, inasmuch as regards an attempt at *verisimilitude*, in the application of scientific principles (so far as the whimsical nature of the subject would permit), to the actual passage between the earth and the moon.[12]

10. Poe, "The Unparalleled Adventures of Hans Pfaal," p. 82.

11. Ibid., p. 108.

12. Ibid., p. 112.

By the time Poe was writing, the world had been undergoing an industrial, technological, and scientific revolution. By the time the Montgolfiers had flown their first balloon, the steam engine had already been invented, as had the spinning jenny, the circular saw, and electric generators. Only two years earlier Uranus, a brand-new planet, had been discovered, revealing that the classical solar system known since ancient times was not the limit to the new frontier in the sky. In the half-century following the balloon came the iron plow, the power loom, the first crossing of the Atlantic by a steamship, the first railways, the electric storage battery and motor, the telegraph, cameras, and the revolver. People living in the nineteenth century had good reason for thinking that science could do anything. There seemed to be nothing that scientists or engineers could not understand or conquer. The worlds of the solar system differed only in a matter of scale from Africa or the Poles. Surely, if humans were going to leave Earth and travel to these other planets, then it would be science and technology that would provide the means. By the middle of the century, generals, admirals, and explorers had been replaced by a new hero: the *engineer.*

Still, no one seriously considered the actual technological problems of spaceflight until 1865, when French author Jules Verne wrote his classic novel *From the Earth to the Moon* (its sequel, *Round the Moon*, was published in 1870).[13] Before then, all space travel stories had been fantasies to a greater or lesser degree. But Verne had been a great fan of Edgar Allan Poe and took his lead from Poe's invention of scientific verisimilitude—developing it to a degree Poe never imagined.

In Verne's story, a group of American arms manufacturers find themselves with nothing to do after the close of the Civil War. As an outlet for their energies and creative genius, they propose building an enormous cannon—actually a 900-foot (275-meter) deep, cast iron–lined well with 400,000 lbs (181,000 kg) of explosives at the bottom—and launching a projectile to

Figure 26.1—This illustration of the launch of Jules Verne's projectile accompanied the original edition of From the Earth to the Moon. Although Verne's giant cannon was impossible (at least so far as launching passenger-carrying spacecraft is concerned), he was well aware of this and took great care in making the device at least appear plausible. The result was a novel that inspired generations to take the idea of spaceflight seriously, including many of the seminal founders of astronautics. (*Author's collection.*)

13. Verne, *From the Earth to the Moon*.

the Moon. It eventually dawns on them that it would be much more interesting to not just launch an ordinary projectile but one carrying passengers. Of course, this would never really work—the shock of going from a standing start to 7 miles (11 km) a second in 900 feet (275 meters) would have been instantly fatal to his heroes—but there is compelling internal evidence that Verne realized this and consequently filled his book with so much science and math and engineering that his nineteenth-century readers accepted his story without question. So much so, in fact, that when the novel was originally serialized in France, people wrote to the author volunteering to be passengers in the projectile!

Each chapter of the book is devoted to a different problem facing the space explorers: the method for leaving Earth, the design and construction of the giant cannon, what its explosive charge would consist of, where the cannon would be located, what the projectile would be made of, what its speed and trajectory would need to be, the best time and date for the launch, safety precautions, experimental tests, provisions for life support, methods for tracking the projectile, and how all of this was going to be paid for. All of these questions and more Verne answers in precise detail, even to the point of quoting the actual math used in his calculations (actually performed by his cousin, Henri Garcet of the Lycée Henri-IV).[14]

With the publication of Verne's novel, the possibility of spaceflight was instantly transformed from the realm of the fantastic to a mere exercise in Victorian engineering. For the first time, the problem of space travel had been put on a firm mathematical and technological basis. Indeed, there is some reason to argue for Verne having literally invented the science of astronautics. Verne's method for getting his astronauts into space would not work in reality, but what was important was that he suggested a method that employed nothing but known materials and contemporary technologies. His astronauts did not need to rely upon impossible balloons or imaginary antigravity metals. He demonstrated to his readers one monumentally important fact: the conquest of space was to be a matter of applied mathematics and engineering and nothing else.

Although Verne used a giant cannon instead of rockets to launch his heroes into space, he was perfectly aware of the potential rockets had. But he was also aware that the state of the art of rocketry in the mid-1800s was such that his readers would have never believed that the unreliable, inefficient, and not very powerful rockets then available would ever be capable of speeding a spaceship to the Moon. Still, Verne did have his astronauts carry rockets on board for their eventual (albeit aborted) landing on the Moon and for steering the projectile. In this, Verne was one of the very first to realize that rockets would work in a vacuum and would be the ideal source of propulsion in space—a fact that eluded all but a very few writers and scientists until the early years of the twentieth century.

14. Miller, *The Dream Machines,* pp. 47–54.

Verne's novel was a great success and was translated and published all over the world. It had readers and admirers in almost every nation, and some of these readers were not just admirers—they were inspired to actually *do* what Verne had only *written* about. Konstantin Tsiolkovskiy, Hermann Oberth, Robert Goddard, and many other early pioneers of rocketry and astronautics were directly inspired by the space novels of Jules Verne and others.

By the time the first spacecraft was launched in 1957, most of the problems facing engineers had been solved—or at least dealt with—by the writers of space fiction. This is not because science-fiction writers were more prescient; it's simply because they were forced to solve many of the same problems long before engineers had to, and there are only so many different plausible answers.

The Rocketeers

The period between World War I and World War II was a mixed bag as far as popular interest in space travel was concerned. Aviation was the great technological craze of the time (rivaled only by radio) and rockets seemed something of a trivial side-issue. Robert Goddard had, of course, launched the world's first liquid-fuel rocket in 1926, but the news of this was barely known within scientific circles, let alone the general public. Goddard was much better known as the man who was going to send a rocket to the Moon. In April 1920 he had written an article for *Popular Science* magazine in which he described the possibility of hitting the Moon with a rocket carrying 14 lbs (6 kg) of flash powder—enough, he had calculated, to create a flash visible from Earth.[15] This article immediately made Goddard something he had never had any wish to be: a media celebrity. Scientists and popular-science writers in publications ranging from the *London Graphic* and *Scientific American* to *The New York Times* took Goddard to task for what they considered his utter lack of rudimentary scientific knowledge, while at the same time popular magazines took Goddard dead seriously and published scores of articles on the possibility of spaceflight, often comparing it, rightly, to the advances then occurring in aviation. Many of these were translated reprints of articles originally written by Austrian Max Valier, one of the most enthusiastic and colorful of the rocketry and space travel enthusiasts who were just then beginning to organize in Europe. Valier's ubiquitous writing and spaceship designs (beautifully illustrated by the Römer brothers) influenced science and science-fiction writers, and even appeared in some of the first space movies ever produced, in much the same way that Wernher von Braun's widely distributed designs influenced the public perception of spacecraft in the 1950s and 1960s.[16]

15. Ibid., pp. 123–124.

16. Ibid., pp.144–146.

On the whole, however, the era of the 1920s and 1930s was not a bright one for the depiction of space travel in the media or for its impression on the public at large. Most work on liquid-fuel rockets was being done either in secrecy (by a humiliated and bitter Goddard, who now wanted more than anything else to avoid publicity) or by small groups of amateurs in Europe. "Rocket" to the general public still meant the gunpowder variety familiar to Fourth of July celebrants and Civil War veterans. Although Valier and others had shown that rockets could efficiently propel every sort of vehicle, from cars to sleds to aircraft, these were for the most part seen for the publicity stunts they were and, since almost all of these employed powder rockets, they really did not go very far in educating the public on the possibilities of the liquid-fuel rocket.

The ideas published by Goddard—and the reporting and misreporting of them in the media—combined with the popular writings of Valier and the publicity generated by his experiments, impacted popular culture in only one major way, and it was in a form that was actually a setback for the acceptance of astronautics. This was the introduction of, first, Buck Rogers and then Flash Gordon, two wildly popular comic strip characters whose improbable adventures in space quickly made the transition from the newspaper comic pages to motion picture serials that are still shown today. Inspired by the success of his 1929 novel, *Armageddon 2419 AD*, author Philip Nowlan adapted his hero for a newspaper comic strip, drawn by Dick Calkins. Although Buck Rogers's name has become synonymous with space travel, spaceships did not make an appearance until 1930, after nearly 400 daily strips had been published (although rocket-powered aircraft were introduced in strip 90 in 1929).[17]

Although the early space travel strips had a relatively high standard of scientific accuracy, this was not maintained. By the 1940s, Buck Rogers had descended into utter fantasy. The believable spacecraft of the early 1930s had degenerated into art nouveau monstrosities that would have done credit to Dr. Seuss. This left a lasting impression on a prewar generation who came to derogatorily refer to space travel as "that Buck Rogers thing."

THE GOLDEN AGE OF SPACEFLIGHT

It may seem odd to say so, but the Golden Age of space travel occurred before space travel became a reality. It was a period between the end of World War II and 1961 that is analogous to, say, the 1920s and 1930s in aviation—a time when the public had gone aviation-mad.

Much the same thing occurred in the two decades following World War II. Fueled in large part by postwar optimism combined with a faith in technology and engineering like that of the nineteenth century, the possibility of spaceflight

17. Ibid., pp. 182–183.

took a firm hold on the public imagination. There were references to rockets and space travel everywhere one looked, from television and movies to the pages of the Sunday comics to the hood ornaments of the latest-model Oldsmobile Rocket 88. There were rocket toys and space games, bubblegum cards and gumball machines, cigarette lighters and bedside lamps, breakfast cereal premiums and Buck Rogers Popsicles, and comic books and television serials.[18] A recent collectibles catalog lists more than 300 toys inspired by the most popular children's television space shows of the 1950s: *Space Patrol, Captain Video, Tom Corbett,* and *Rocky Jones*.[19]

Hardly a week went by without an article in at least one national magazine about the coming age of spaceflight. All of this—bad science and good—helped counteract some of the negative effects the depiction of spaceflight in popular culture had created. The popularity of Buck Rogers and Flash Gordon in the 1930s and 1940s—abetted by the series of movie serials they inspired—did little to help the cause. Their emphasis on swashbuckling romance and utter disregard for anything resembling science led too many people—layman and scientist alike—to dismiss the entire concept of space travel as "that crazy Buck Rogers idea."

There has been nothing since the Golden Age that remotely parallels its frenzy about spaceflight, perhaps for the same reason the Golden Age of aviation finally ended: spaceflight became a reality. Just as transatlantic flights stopped making headlines when anyone could get a seat on a regularly scheduled airliner, much of the hold space travel had on the popular imagination faded as flights into space became commonplace.

Why this sudden postwar enthusiasm for space travel, especially since a decade earlier it had been considered a subject fit only for comic strips? A lot of it had to do with advancements in science and technology generally that occurred during the war. Jets and rocket-powered aircraft had shown their mettle as being more than hare-brained publicity stunts. American bomber crews and fighter pilots had brought back stories of incredible German aircraft such as the Messerschmitt Me-109 Komet and Me-262 rocket and jet fighters. Then, of course, there were the giant V-2 rockets that had fallen on England. It was obvious to more than just scientists and space travel enthusiasts that if the V-2 was not a real spaceship, it was mighty close to the real thing.

Given the history of astronautics' close relationship with the arts, perhaps it is only natural that the prime instigators of the Golden Age were a rocket scientist and an artist. The team of Wernher von Braun and Chesley Bonestell were as responsible for the almost universal public enthusiasm for spaceflight as anyone. Bonestell's first book, *The Conquest of Space* (1949), had galvanized its readers. The reaction was much the same as to Galileo's revelations more than 300 years earlier. Although

18. Marek S. Young, Steve Duin, and Mike Richardson, *Blast Off!* (Milwaukie, OR: Dark Horse Books, 2001).

19. T. N. Tumbusch, *Space Adventure* (Radnor, PA: Wallace-Homestead Book Co., 1990).

most people were aware of the planets in a kind of abstract sense, few had thought of them as real worlds like Earth, with landscapes and scenery as fascinating as anything that existed on our own planet. Bonestell's depictions of Mars and the moons of Jupiter and Saturn did not bear the stigma of "artist's impression"—they looked instead like postcards from the future. There was no sensationalism or exaggeration, just a cold-blooded, matter-of-fact representation of places that really existed, presented with no more fanfare than your neighbor's snapshots from their summer vacation in Yosemite.[20]

It was pretty much a given that Bonestell would be invited to illustrate a series of magazine articles being created by von Braun about the near future of spaceflight. The series, written by von Braun and a number of other space experts, repeated Jules Verne's accomplishment by demonstrating that spaceflight could be achieved using only contemporary technology and materials available at the time. It is true that von Braun's giant ferry rockets (the first stage of which would have had 51 separate rocket engines!), lunar landers, and space station might not have been the most practical possible designs, but that was not his point. Anything else would have required speculating on future developments; he wanted to show that spaceflight was possible in 1952, not 2002.[21]

It is difficult to gauge the full effect the subsequent *Collier's* magazine series had on both the public imagination and the contemporary development of America's fledgling space program. Perhaps one way to judge this is by looking at the artifacts of the time. *Collier's* spaceships were literally everywhere. When Bonestell's paintings were not being legally reprinted, they were plagiarized. And where they were not being copied outright, they were being inspirational. One can find *Collier's*-inspired spacecraft

Figure 26.2—This painting by Chesley Bonestell was created for the Wernher von Braun book, *The Exploration of Mars* (1954). von Braun had developed a detailed, consistent, incremental plan for the exploration of space. In much the same way that Jules Verne had, a century earlier, shown that spaceflight was merely a matter of applied mathematics and engineering, von Braun's magazine series and books helped to not only convince a skeptical public that spaceflight was possible using contemporary techniques and materials, but to also make the idea of spaceflight wildly popular. *(Artwork copyright © Bonestell Space Art, used with permission.)*

20. Ron Miller, *The Art of Chesley Bonestell* (London: Paper Tiger, 2000).

21. Frederick I. Ordway and Randy Liebermann, eds., *Blueprint for Space* (Washington, DC: Smithsonian Institution Press, 1992), p. 135.

throughout those decades, from plastic model kits to coloring books, from science-fiction movies to sterling silver money clips! A number of years ago I started a file of Bonestell-inspired artwork and had to quit when it grew to more than two inches in thickness.

In the movie theaters, space travel films had not yet begun to descend into the abyss of alien-ridden space opera. Films such as *Destination Moon* (1950), *Conquest of Space* (1955), and even *Forbidden Planet* (1956) were filled with good science and positive images about the future of spaceflight. On television, series such as *Tom Corbett* employed the advice of experts such as Willy Ley to lend their programs authenticity, while the arrival of a full-size *Space Patrol* "rocket ship" in a town or city was a full-scale media event.[22] Meanwhile, more nonfiction books about rocketry and space travel were being published for young readers than at any time before or since.

One of the most important of the legitimate spinoffs of the *Collier's* articles was a series of programs broadcast as part of Walt Disney's *Disneyland* television show, which had begun airing in 1954. Inspired by the popularity of the recently published magazine series as well as by the increasing public interest in space travel, Disney produced three short films about the future of space exploration: *Man In Space, Man and the Moon* and *Mars and Beyond*. Employing state-of-the-art animation and special effects, combined with the expertise of von Braun, Heinz Haber, Ernst Stuhlinger, and Willy Ley, the shows outlined the history of rocketry and a von Braun-inspired, step-by-step program for the exploration of space—upgraded from the scheme he had earlier proposed in *Collier's*. The series also introduced the American public to the concept of the electric space ship, as proposed by Stuhlinger. The first show was broadcast on 9 March 1955 to an audience of nearly 100 million people—more than half the total population of the United States.[23]

One of these viewers was President Dwight Eisenhower, who borrowed a copy for two weeks to show to officials at the Pentagon. On 29 July, Eisenhower announced that the United States would include the launch of an artificial Earth satellite as part of its role in the coming 1957–1958 International Geophysical Year.[24]

The *Collier's* series and its endless permutations is a perfect example of an event that incorporated the two basic qualities of spaceflight in culture: it not only reflected contemporary fascination with space travel (one that had been gradually growing since the 1930s); it helped fan it into a blaze. Like Verne's stories, the *Collier's* series also served as a self-fulfilling prediction of the future of spaceflight. It was instrumental in raising public awareness of the reality of imminent spaceflight, and it also informed a government that was about to spend many billions of dollars on a space program that it had the enthusiastic support of its taxpayers.

---

22. Miller, *The Dream Machines,* p. 316.

23. Ordway and Liebermann, *Blueprint for Space*, pp. 145–146.

24. Ibid.

TODAY

In spite of the fact that today we are nearly 50 years into the Space Age (if one counts from the launch of Sputnik 1), there appears to be less interest in space travel than there was in 1957, or even 1947, for that matter. In fact, public enthusiasm is virtually nil—at least compared to the frenzy of the 1950s and 1960s. I believe this is largely due to the fact that we *are* in the midst of the Space Age and no longer at its threshold. There are few great "firsts" any more to claim newspaper headlines. There is no race upon which to base national pride. Even an extraordinary accomplishment such as that of *SpaceShipOne* does not result in ticker-tape parades or even children's books and toys. Where are the *SpaceShipOne* table lamps, pillow cases, and cocktail coasters? It took only a few days before the story was moved to the back pages of the newspapers and probably not one person in a thousand today could tell you what the event was all about, what it accomplished, or who accomplished it.

In some ways this is good; it means that spaceflight has been accepted as a commonplace, just as flying nonstop across the Atlantic was eventually accepted. An event does not create headlines when tens of thousands of people routinely do it every day. Making spaceflight an ordinary, everyday occurrence had once been the dream of people like von Braun, but few of them thought about the price that would have to be paid. One cannot have that kind of public acceptance while at the same time maintaining enthusiasm at a high pressure. We have regular transatlantic airline flights but at the expense of not having transatlantic pilots as heroes any longer.

Will space travel ever regain the popularity it had in the 1950s and 1960s? What will it take to regain that kind of support and enthusiasm? What will capture the public imagination in the way that Wernher von Braun and the first astronauts did? Perhaps space travel needs someone as charismatic as von Braun or Carl Sagan had been—an individual upon whom the public can focus and who has the talent and knowledge to maintain and heighten that focus, to say nothing of the traits of leadership and persuasiveness. But such people are vanishingly rare.

Perhaps we need a focused space program, some specific goal that is as exciting conceptually to the general public as it is scientifically important to the scientist. There was a hint that an ember of enthusiasm for space exploration is still alive in the public breast when the Pathfinder Mars rover became one of the media heroes of 1997. For weeks, the official NASA Pathfinder Web site was receiving more hits than any other site in the world. It is no small feat for Pathfinder to have inspired some of the first space-related toys in nearly 40 years; even *SpaceShipOne* did not get a Hot Wheels replica. What was it about that tiny machine that so captured imaginations? Perhaps it was the image of a lone explorer—even if it was only a robot the size of a large toaster—having to fend for itself 30 million miles from home.

CHAPTER 27

MAKING SPACEFLIGHT MODERN: A CULTURAL HISTORY OF THE WORLD'S FIRST SPACE ADVOCACY GROUP

Asif A. Siddiqi

Since the opening of the Space Age in 1957, space exploration has become a powerful embodiment of modernity. This connection between spaceflight and modernity is so fundamental to our vision of space exploration that it no longer requires articulation or elaboration. Like many other technological legacies of the twentieth century, such as aviation, microelectronics, communications, and the Internet, space exploration represents one effective way for nations to assert their arrival on the international stage, i.e., to underscore a nation's commitment to modernity, modernization (especially for developing nations), and ultimately, the future. In the early twenty-first century, this link between spaceflight and modernity is reinforced through cultural imagery and iconography that makes space exploration synonymous with "progress"—through exploration of new frontiers, via accumulation of new knowledge, and by bringing benefits to society. All of these associations are predicated on the fundamental relationship between spaceflight and science and technology. The scientific and technological dimensions of spaceflight—for example, the science of space trajectories and the technology of launch vehicles and spacecraft—have identified spaceflight firmly with a post-Enlightenment view of mastery over nature and "progress" in general.

How did space exploration come to be linked to modernity and, more specifically, with the discourse of science and technology? In other words, how did our collective perceptions of space exploration, as represented through popular culture, become grounded in scientific and technological concerns? This is the question I explore in this paper, focusing particularly on the Russian case.

In Russia, as in other major European nations, space travel was associated for centuries with mythology and mysticism—often in the form of dreams, parables, folk tales, superstition, or mythical tales. A fundamental shift occurred beginning only in the last decades of the nineteenth century, a transformation that culminated in the 1920s. The beginning of this change is rooted in the dissemination of the works of Jules Verne in Russia in the 1860s and 1870s and the general rise in interest of science fiction, in particular the genre of the *astronomicheskii roman* (or the astronomical novel). Partly inspired by such fictional works, the self-educated village school teacher Konstantin Eduardovich Tsiolkovskii (1857–1935) published his first meditations on the mathematics of spaceflight in 1903. Popular interest in space exploration first peaked around the time of the Great War but it was not until the 1920s—after the Russian Revolution—that mass interest in cosmic travel became a cultural phenomenon in Soviet Russia, embodied in the so-called Soviet space fad. It was also at this time that the idea of space travel weakened its link to mythology and superstition; helped by a vast network of scientifically minded amateur space enthusiasts, the language of spaceflight shifted into the domain of science and technology.

During the 1920s Soviet space fad, amateur and technically minded enthusiasts formed short-lived societies to discuss their interests and exchange information on space travel.[1] These societies—the first in the world that were dedicated to space travel—operated largely without material support or encouragement from the state. The men and women who organized the cosmic societies did, of course, absorb official Marxist discourses on the role of technology as a panacea for all social ills. But the record of their actions underscores their own agency in infusing the idea of space travel (and its corollary, rocketry) with the cold, hard power of rationality, science, and mathematics. Moving from fantasy, mysticism, mythology, and lunacy, these enthusiasts appropriated the language of modernity for their case.

In this article, I revisit one specific episode in the cultural history of Soviet space exploration: the formation of the world's *first* active society dedicated to the cause of space travel. In 1924, university students in Moscow formed the Society for the Study of Interplanetary Communications—"interplanetary communications" being a then-common euphemism for "interplanetary travel." The Society sponsored lectures, established networks of enthusiasts, and wrote articles on the value of cosmic flight at a time when few had considered the possibility. Historians in both Russia and the West have described the activities of the Society in very general terms, but a lack of primary sources has prevented a fuller exploration of its significance in the early history of spaceflight.[2] Using archival documents, my goal is not only to reconstruct the Society's activities but also to explore the ways in which the rhetoric and activities of 1920s space enthusiasts showed a distinct appreciation for the language of science and technology. Ultimately, I hope to underscore that the relationship between the culture of space exploration and discourse of modernity was not always a given, but was grounded in historically contingent forces.

1. For a detailed description of the Soviet "space fad" of the 1920s, including an analysis of art (literature, painting, poetry, architecture, etc.) as a site for contesting visions of space exploration, see Asif A. Siddiqi, "Imagining the Cosmos: Utopians, Mystics, and the Popular Culture of Spaceflight in Revolutionary Russia," *Osiris*, forthcoming.

2. See for example, Frank H. Winter, *Prelude to the Space Age: The Rocket Societies: 1924–1940* (Washington, DC: Smithsonian Institution Press, 1983), pp. 27–30; G. Kramarov, *Na zare kosmonavtiki: k 40-letiiu osnovaniia pervogo v mire obshchestva mezhplanetnykh soobshchenii* (Moscow: Znanie, 1965).

FRIDRIKH TSANDER

Historians typically trace the history of the Soviet space program to Tsiolkovskii, who in 1903 published the first mathematical substantiations that spaceflight was possible. In the 1920s and early 1930s, Tsiolkovskii's original ideas on space exploration—which he republished once he discovered that others such as the Romanian-German Hermann Oberth and the American Robert Goddard had also come to the same conclusions—fed enormous popular interest in the Soviet Union in the cause of cosmic travel. Several technology-enraptured (and short-lived) societies coalesced during the period of the space fad. Of these, the most important and influential was the Moscow-based Society for the Study of Interplanetary Communications, formed in 1924. It was not only the first in the world to effectively organize for the cause of space exploration, but was also the first to build a domestic and international network around the idea. The history of the organization—a combination of serendipity, willful devotion, and eventual loss of momentum due to indifference from the state—illustrates the ways in which the technological fascinations of the day inspired a few to bring an esoteric idea to many.

No one played a more important role in spearheading the Society than Fridrikh Arturovich Tsander, who, although not a founding member, gave the Society its heart and soul. Raised in the Latvian capital of Riga, Tsander was an early devotee of the many science fiction novels of the era, and was a convert to the cause of spaceflight by the time he was 20. Unlike Tsiolkovskii, who was concerned only with the theoretical aspects of space travel, Tsander engaged in rudimentary experimentation at an early age. As early as January 1916 he built a lightweight greenhouse that would supply fresh vegetables and absorb excess carbon dioxide for travelers in space. After graduation from the Polytechnic Institute, in 1915 he started work at the Provodnik Factory, a rubber-producing facility, evidently because he believed that rubber could be used as insulating material for spaceships. Soon after the Revolution, in February 1919 he moved to the Aviation Factory No. 4 (the "Motor" Factory) where he spent his free time working on designing a new kind of aircraft equipped with an engine that could breach the atmosphere and fly into space.[3]

Tsander later claimed that soon after the Revolution he had personally met Lenin at a regional conference for inventors, and had told the Soviet leader about his cosmic airplane. Tsander reminisced that "Lenin made a tremendous impression on me: that night I could not sleep. Pacing up and down in my small room, I thought of the greatness of this man—our country is ravaged by war, there is a lack of bread, of coal, and the factories are at a standstill, but this man who controls this huge country

3. For a biography, see L. K. Korneev, "Zhizn', tvorchestvo i deiatel'nost' F. A. Tsandera," in *F. A. Tsander: Problema poleta pri pomoshchi reaktivnykh apparatov. Mezhplanetnyye polety. Sbornik statei*, L. K. Korneev, ed. (Moscow: GNTI Oborongiz, 1961), pp. 5–73.

finds time to listen to interplanetary flights."[4] Like many other stories of the early space years, the veracity of the meeting (which most certainly never happened) is less important than the mythic quality it conveyed. Post-Sputnik Soviet writers repeatedly alluded to the meeting to illustrate the Bol'sheviks' interest in forward-thinking ideas at a time when the very survival of the Soviet state was in doubt.[5]

Within six months of the alleged meeting with Lenin, in June 1922 Tsander quit his job at the Motor Factory to devote all his time to developing a working design of the spaceplane. He survived by the generosity of his former factory workers, who formed a pool of donations to help feed the scientist. Although they might not have shared Tsander's unyielding belief in the possibility of space travel, they were extremely fond of Tsander; in July 1922 the chief of the production section and the chief engineer of the factory jointly wrote a formal review of his interplanetary spaceship.[6] Tsander received a certificate from the Association of Inventors for his "invention" the following month.[7] During this period of self-willed unemployment, Tsander also initiated contact with Tsiolkovskii.[8]

In April 1923, after he quit his factory job to devote his full energy to his interplanetary spaceship, he thanked his coworkers for paying for his living expenses and claimed, "I hope . . . that the money which you have given me will not be in vain, but will make it possible for me to present something of value to your factory." Going beyond pure technical considerations, Tsander genuinely believed that interplanetary flight would be for the benefit of all of humanity, thus underscoring one of the major rationales behind the popular discussions of space travel among

4. Tsander mentioned this meeting in a short autobiography he wrote in 1927 for publication in volume 4 of Nikolai Rynin's famous *Mezhplanetnykh soobshchenii* (*Interplanetary Communications*) encyclopedia published in 1929. He also gave the same account to a future coworker, L. K. Korneev, in 1932. Korneev, "Zhizn'," pp. 23–24.

5. The leading Soviet rocket engine designer of the postwar era, academician Valentin Glushko, requested that the Institute of Marxism-Leninism of the Central Committee of the Communist Party conduct a special investigation into whether Lenin ever met Tsander. The institute found no evidence. G. S. Vetrov, "K voprosu o nauchnom istolkovanii istoricheskikh dokumentov," in *Issledovaniia po istorii i teorii razvitiia aviatsionnoi i raketno-kosmicheskoi nauki i tekhniki: vypusk* 7, B.V. Raushenbakh, ed. (Moscow: Nauka, 1989), p. 191.

6. The review is mentioned in Tsander's letter to Glavnauka (the science agency of the Bol'shevik government) dated 8 October 1926. F. A. Tsander, "Zaiavlenie v Nauchnyi otdel Glavnauki ot 8.X.1926 g.," in *F. A. Tsander: iz nauchnogo nslediya*, F. A. Tsander, ed. (Moscow: Nauka, 1967), p. 82.

7. The certificate is mentioned in Glavnauka's letter to V. P. Vetchinkin dated 15 October 1926. See "Pis'mo Nauchnogo otdela Glavnauki V. P. Vetchinkinu ot 15.X.1926 g.," in *F. A. Tsander: iz nauchnogo naslediia*, p. 85.

8. Tsander first wrote to Tsiolkovskii on 5 March 1923. Archive of the Russian Academy of Sciences (ARAN), f. [collection] 555, op. [unit] 4, d. [file] 670, ll. [pages] 1–8. For a summary of the correspondence between Tsander and Tsiolkovskii, see K. Belyi, "Tsiolkovskii-Tsanderu," *Nauka i zhizn'* no. 10 (1962): pp. 20–21.

Soviet enthusiasts in the early 1920s. As an incurable utopian, he noted, "A flight around the Earth would have tremendous significance; flying like the Moon, we could use telescopes to observe the other planets much better, and could probably construct a habitation in which living conditions would be much better than on the Earth . . . ." To the factory workers, he spoke of "senior citizens [who] will find it much easier to maintain health in [space]," of the "inhabitants of Mars . . . [whose] inventions could help us to a great extent to become happy and well off," and of "[a]stronomy, [which] more than the other sciences, calls upon man to unite for a longer and happier life . . . ."[9]

Tsander was not a scientific dilettante; in fact, he had a masterful grasp of complex mathematics and was full of ground-breaking ideas. In his research (and in his public talks), he compared the properties of various propellant combinations and considered the heat processes, aerodynamics, and engineering behind building rocket engines. To his rapt audience, Tsander also provided details of his new spaceship, a metallic airplane that would literally devour itself as it took to the heavens. The plane would take off from land using a conventional piston engine, use its wings for additional lift until it reached a height of about 17 miles (28 km), at which point the pilot would turn on a powerful rocket engine that would accelerate the vehicle into outer space. Tsander had found that one of the most important weight penalties in designing such a spaceship would be the huge mass of fuels it would have to carry. In order to circumvent this problem, he proposed an idea, radical even today, wherein the space rocket engine would use the melted aluminum parts of its own fuselage as fuel for the engine, in combination with either liquid hydrogen or liquid oxygen. The remaining unused portion of the spaceship would then either go into orbit around Earth or fly to the other planets. He also analyzed the problems of guided reentry into Earth's atmosphere using special wings, and various techniques of landing such a vehicle back on Earth.[10]

The Section on Reactive Motion

Tsander had frequently spoken of trying to organize a group to study space travel. Luckily for him, public discourse on the idea of spaceflight reached a crescendo in early 1924, just as he was looking for a forum to organize. The first intense wave of public fascination with spaceflight was set off by a story "Is Utopia Really Possible?" in the newspaper *Izvestiia* in October 1923, about the recently published meditations on spaceflight by the foreigners Oberth and Goddard.[11] Spurred to

9. F. A. Tsander, "Doklad inzhenera F. A. Tsandera o svoem izobretenii," in *F. A. Tsander: iz nauchnogo naslediia*, pp. 10–13.

10. Ibid., pp. 10–14.

11. "Novosti nauki i tekhniki: neuzheli ne utopiia?," *Izvestiia VTsIK*, 2 October 1923.

promote a Russian source for such ideas, the 66-year-old Tsiolkovskii, ensconced in the provincial town of Kaluga, immediately republished his own pre-Revolutionary works under the title *Raketa v kosmicheskoe prostranstvo* (*The Rocket into Cosmic Space*), and had them distributed to bookstores and "editors and scientists" in Moscow in April of the following year, thus bringing his strange ideas about space exploration to a huge and new audience.[12] Almost simultaneously, the Soviet media began to devote considerable attention to the cosmos. News and rumor of Oberth's and Goddard's exploits, the publication of Aleksei Tolstoi's new space fiction novel *Aelita*, and the "Great Mars Opposition" of August 1924—when Mars and Earth were closer to each other than in hundreds of years—fed an unprecedented explosion public interest in space. On April 15, the newspaper *Pravda* published a long article, "Voyage Into Cosmic Space," in which the author provided a complete history of the idea of space exploration, reaching back to Leonardo da Vinci, Cyrano de Bergerac, Jules Verne, and H. G. Wells, and bringing the story up to date with the works of Tsiolkovskii, Oberth, Goddard, and popular Soviet science writer Iakov Perel'man. Infected by the optimism of the times, the author, Mikhail Lapirov-Skoblo, concluded that "[W]ithin a few years hundreds of heavenly ships will furrow into the starry cosmos."[13] Three days later, *Izvestiia* published yet another article on Tsiolkovskii and Oberth, noting that Goddard had a sensational plan to launch a rocket to the Moon in the near future.[14]

The renewed interest in space in early 1924 might have remained little more than a passing media fad had it not been for the efforts of a few dedicated individuals. Inspired by the recent publications on space travel, in mid-April 1924 students at the prestigious Zhukovskii Academy of the Air Fleet (now the Zhukovskii Military Air Engineering Academy) took action; about a dozen students from the Academy's Military-Science Society (VNO) set up a Section on Reactive Motion to exchange

12. Chizhevskii to Tsiolkovskii, 6 April 1924, ARAN, f. 555, op. 4, d. 689, l. 9. See also Chizhevskii to Tsiolkovskii, 17 November 1925, ARAN, f. 555, op. 4, d. 689, l. 23.

13. M. Ia. Lapirov-Skoblo, "Puteshestviia v mezhplanetnye prostranstva," *Pravda*, 15 April 1924: pp. 5–6.

14. F. Davydov, "Raketa v kosmicheskoe prostranstvo," *Izvestiia VTsIK*, 18 April 1924: p. 7. For Goddard's prominent role in the space fad, see Asif A. Siddiqi, "Deep Impact: Robert Goddard and the Soviet 'Space Fad' of the 1920s," *History and Technology* 20, no. 2 (2004): pp. 97–113.

ideas about rockets.[15] The members of the Section wrote up a short list of objectives that included the following: "(1) to unite all persons, working in the USSR on the given problem; (2) to obtain all possible information on work carried out in the West; (3) to disseminate correct information on the condition of problems on interplanetary communications, and in connection with that, publish on [such] activities; [and] (4) to [conduct] independent scientific-research work and in particular, study questions on the military application of the rocket." Additional goals included scheduling a competition to design a small rocket capable of reaching approximately 60 miles (100 km) up into space; creating a group (*kruzhok*) for more in-depth theoretical study of important problems; organizing a laboratory; opening a book kiosk to "satisfy . . . the wide demand for literature"; and establishing a separate "film group" which would work on sets for films.[16] In other words, the Section touched on all the primary dimensions that would characterize the ensuing space fad, from its technical side (building rockets) to outreach (lectures, publications, and bookstores), to building a community (by involving others interested in the same topics), to opening a channel to the West (by collecting information from overseas), and acknowledging the artistic medium as a possible way to educate and popularize (by branching into film). This Section, although only a student organization, was probably the first organized group in the world dedicated to the cause of space exploration.[17]

The Section may have been small and without any resources, but such considerations did not limit its ambition. In enumerating a list of supplementary goals, it sought to host public reports on astronautics from serious scholars such as Vladimir Vetchinkin and Fridrikh Tsander. At the time, the 35-year-old Vetchinkin was a deputy director at the Central Aerohydrodynamics Institute (TsAGI), the leading aeronautics institution in the country. Well known nationally as a theorist of aeronautics, he also held a dual position as a professor at the Zhukovskii Academy whose students had formed the space group. Vetchinkin's most notable research work had been on the theory of propellers but, after the Revolution, like a few

15. The leading VNO student members included V. P. Kaperskii, M. G. Leiteizen, A. I. Makarevskii, M. A. Rezunov, and N. A. Sokolov-Sokolenok. ARAN, r. 4, op. 14, d. 197, ll. 32–33. The VNO was established in 1923 with seven subsections devoted to aviation and gliding, engines, aerodynamics, scientific organization of labor, tactics, photography, and popularization. In the technical vernacular of the 1920s, Russians considered the word "reactive" or "reaction" (*reaktivnyi*) to represent any mode of propulsion that depended on the force of exhausted particles. Forms of reactive motion included rockets (where the vehicle carries both its fuel and oxidizer) and jets (where the vehicle carries only the fuel and uses natural air as an oxidizer). Rockets were thus capable of working in vacuum whereas use of jets was limited to inside the atmosphere.

16. "Organizatsiia v S.S.S.R. O-va mezhplanetnykh soobshchenii," *Tekhnika i zhizn'* no. 12 (6 July 1924): p. 1.

17. Winter refers to an earlier organization, the Rocket Society of the American Academy of Sciences formed in Savannah, GA, in 1918. Little is known about the organization, which focused on rocketry rather than space exploration. *Prelude to the Space Age*, p. 27.

other prominent aeronautics specialists, he had begun to dabble in prognostications on space travel.[18] Vetchinkin's support of the Section added a critical ingredient to the Section: a patina of legitimacy.

Unlike Vetchinkin, Tsander did not enjoy a high governmental position, but there were few individuals in the Soviet Union who were more qualified to speak on the scientific and technological aspects of space travel than this 37-year-old Latvian savant. Tsander heard of the new student group on space through Vetchinkin and, although discouraged that they were not planning to immediately build rockets, quickly joined forces with them.

Having established a mandate, the Section's first order of business was to establish contact with one man who, because of Soviet media's search for a domestic counterpart to Oberth and Goddard, had assumed the wizened role of patron saint of the cause of spaceflight—Konstantin Tsiolkovskii. On 22 April, Moris Leiteizen, the de facto leader of the Section, composed a letter informing the old man of its recent formation, noting that 23 of its 25 members were Academy students. If Tsiolkovskii could not move to Moscow to take up leadership of the Section, then Leiteizen invited him to at least give a public talk on space travel, which they would organize at the Polytechnic Museum.[19] In a couple of letters, Tsiolkovskii responded that he was "joyous" to hear of the creation of the Section, having already heard of its existence via newspaper reports. In poor health and terrified of leaving Kaluga, Tsiolkovskii declined the offer to visit Moscow but suggested that the Section might read excerpts to students from his recent science-fiction novel, *Beyond the Earth*, two copies of which he sent to Leiteizen.[20] Less a work of literature than a dryly written polemic describing such fantastic conceptions as space suits, multistage rockets, and mooring ships in space, the novel was not known widely beyond Kaluga. Inspired by receiving the initial two copies, the Section requested the Academy administration to purchase copies of Tsiolkovskii's books for the school library since the "demand for them was [so] great."[21]

Through May, the Section continued to correspond with Tsiolkovskii on various topics, although the latter's responses often evinced a certain irritation when the young enthusiasts took action that the pioneer did not agree with. In one letter dated May 4, Leiteizen informed Tsiolkovskii that one of the immediate goals of the Section was to "study reactive engines, independent of their application . . . [and] in

18. For background to Vetchinkin, see A. M. Tarasenkov, "29 iunia—100 let so dnia rozhdeniia V. P. Vetchinkina (1888 g.)," *Iz istorii aviatsii i kosmonavtiki* 59 (1989): pp. 59–64. One of the initial organizers of the Section, Georgii Kramarov, notes that Vetchinkin was "one of the first" to support the group. Kramarov, *Na zare kosmonavtiki*, p. 12.

19. Leiteizen to Tsiolkovskii, 22 April 1924, ARAN, f. 555, op. 4, d. 356, ll. 1-1ob.

20. For Tsiolkovskii's four letters to the Section in 1924, all addressed to Leiteizen, see ARAN, f. 444, op. 3, d. 102a, ll. 1–2 (April 29), ll. 3–4 (May 14), l. 5 (May 31), and l. 6 (June 4).

21. Kramarov, *Na zare kosmonavtiki*, p. 14.

order not to complicate [matters] . . . use them in the most simple form of motion on the Earth's surface: in a reactive automobile."[22] Tsiolkovskii tersely replied that "[I]t's known that reactive automobiles . . . are playthings and won't provide any nothing new to you."[23] Despite the often strained communications, members of the Section clearly considered Tsiolkovskii some sort of spiritual guide whose works could convert the uninitiated, not just in the Soviet Union but everywhere, to the cause of space exploration. On behalf of the Section, Tsander wrote to Tsiolkovskii on 16 May, asking for permission to translate Tsiolkovskii's classic works from 1903, 1911, and 1914 into German. Amazingly, the old pioneer replied that although he was "grateful," he could not provide original copies of the said publications since he had no copies left.[24] Tsander, unable to find the originals, did not translate them, significantly limiting Tsiolkovskii's reach beyond the Soviet Union.

"How Modern Science and Technology Solves This Question"

The Section's first public task was to organize a lecture. In picking a first speaker, the members of the Section not only looked for someone with authority but also an individual who would dispel the notion that space travel was a fantasy. The Section students invited one Mikhail Lapirov-Skoblo, a propitious choice since he was personally acquainted with Lenin. An engineer by education, Lapirov-Skoblo had been part of the pre-Revolutionary technical intelligentsia but was also one of the first of that demographic to wholeheartedly put his faith behind Lenin's vision of a modern Russia. In the early 1920s, Lapirov-Skoblo was the deputy chairman of the Scientific-Technical Department of the Supreme Council of the People's Economy (VSNKh or *Vesenkha*), a job that put him in charge of intellectuals who were engaged in introducing modern technology to the Russian economy. When Lenin supervised the formation of the State Commission for Electrification of Russia (GEOLRO) in 1920, Lapirov-Skoblo was tapped to represent the *Vesenkha* on GOELRO. His many other ad hoc duties included service as head of the Department of Science and Technology of the newspaper *Pravda*.[25] Having read Tsiolkovskii's recently published works, Lapirov-Skoblo wrote the first comprehensive expositions on the topic in the post-Revolutionary era in

22. Leiteizen to Tsiolkovskii, 4 May 1924, ARAN, f. 555, op. 4, d. 356, ll. 2–3.

23. Tsiolkovskii to Leiteizen, 14 May 1924, ARAN, f. 555, op. 3, d. 102a, ll. 3–4.

24. Tsander to Tsiolkovskii, 16 May 1924, ARAN, f. 555, op. 4, d. 670, ll. 4–5.

25. Lapirov-Skoblo was also a member of the collegium of the Electrical Industry Association, an academic, a member of the State council, and member of the collegium of the People's Commissariat of Communications. For biographies, see ARAN, r. 14, op. 14, d. 197, ll. 30–30b; Kramarov, *Na zare kosmonavtiki*, pp. 22–23.

Pravda and other publications in 1924.[26] In the waning days of May 1924, members of the Section put up posters at various key intersections in Moscow proclaiming that Lapirov-Skoblo would give a talk on "Interplanetary Communications" ("How Modern Science and Technology Solves This Question"); the paid lecture would begin at 8:00 in the evening on Friday, 30 May at the main auditorium of the Polytechnic Museum. All monetary contributions would go to fund a laboratory for the Section.[27]

The event was successful beyond all expectations. Tickets sold out two days before the event; on the day of the lecture, the Museum's administrators were forced to call for the police to control the mass of people who had arrived to attend. All the literature on space travel that the students had on hand to sell—Russian editions of H. G. Wells's *War of the Worlds*, Russian science fiction from Aleksei Tolstoi and Aleksandr Beliaev, and books by the popular science writer Iakov Perel'man—had been sold out in minutes. In total, the group amassed 2,500 rubles—an astronomical amount for the time. Lapirov-Skoblo's lecture, quite possibly the first talk on space exploration in Russia open to the general public, consisted of a short history of rocketry and space exploration.

Lapirov-Skoblo approached the idea of space travel much like one would treat an arcane discussion on any branch of science or technology. He considered the different technical options for reaching space, methods of propulsion, and the choice of propellants. At the same time he also linked the promise of space travel with the future modernization of the Soviet Union, grounding his words into a narrative that privileged national development over universal significance. Like many others of the period, he took pains to emphasize that the Russian Tsiolkovskii had done decades earlier what the foreigners Oberth and Goddard were only doing now. Yet Lapirov-Skoblo was also generous to the Westerners. Using hand-built models, he explained to the rapt audience the characteristics of multistage rockets and spaceships proposed by Oberth and Goddard. On the latter, Lapirov-Skoblo was less confident that the American would be able to launch a rocket to the Moon, given the vast technical difficulties of such a project. He ended his lecture by calling on Soviet populace to focus on a more immediate goal, to build rocket engines in order to "transform into reality the centuries-old dream of flight into space."[28]

26. "Puteshestviia v mezhplanetnye prostranstva," *Pravda*, 15 April 1924: pp. 5–6; "O puteshestviiakh v mezhplanetnye prostranstva," *Molodaia gvardiia* no. 5 (1924); "O puteshestviiakh v mezhplanetnye prostranstva," *Khochu vse znat'* no. 3 (1924): p. 140.

27. For a reproduction of the poster, see Kramarov, *Na zare kosmonavtiki*, p. 23.

28. The complete text of Lapirov-Skoblo's lecture is stored in ARAN, r. 4, op. 14, d. 194, ll. 49–62. For recollections of attendees, see ARAN, r. 4, op. 14, d. 197, ll. 35–38; Kramarov, *Na zare kosmonavtiki*, pp. 25–28.

The Society for the Study of Interplanetary Communications

After the lecture Section members informed the audience that they were intent on expanding the Section into a larger Society for the Study of Interplanetary Communications (*Obshchestva izucheniia mezhplanetnykh soobshchenii*, or OIMS) not limited to Academy students but open to the general public, and invited audience members to sign up. As a result, 179 people from the audience put their names on a list. Of the 121 names preserved in the archives, 104 were men. The majority of the members (about 60 percent) were young, between the ages of 20 and 30; most of the rest were either teenagers or in their 30s. The signatories were also asked to list their professions. A total of 96 members (roughly 80 percent) were evenly split between being students and workers. A smaller number identified themselves as "scientific workers," "writers," and "scientists and inventors."[29]

We know some personal details of these young men and women. Of the Zhukovskii Academy students, the most active was 27-year-old Morris Leiteizen, the son of the old Bol'shevik, G. D. Lindov-Leiteizen, who was killed in 1919 on the eastern front. The Leiteizen family lived with the exiled Lenin in Finland before the Revolution.[30] After the October Revolution, Leiteizen worked briefly in the People's Commissariat of Foreign Affairs in Finland before opting to come back to Russia to fight in the Civil War. On the way back from Helsinki, Leiteizen was arrested by the White Finnish imperial forces and thrown into prison in Helsinki. Lenin took a personal interest in the matter, and only after a series of diplomatic moves did the Finnish free the near-dead prisoner. Leiteizen, who was fluent in German, French, and English, had been interested in astronomy since his youth, and, according to a contemporary, believed that space travel was "a matter of the relatively near future."[31]

The rank-and-file members included V. S. Berdichevskii, only 17 years old; V. L. Pul'ver, the son of a well-known musician-composer, Anatolii Beliaev, who managed a local astronomical observatory; and Nataliia Sysoeva, a young assistant who worked there. Then there was Valentin Chernov, a young musician apprentice so taken by reading the works of Tsiolkovskii that he had visited the old man in Kaluga. Unable to decide between a career in music or space travel, he had decided to do both. He enrolled in a university to get a degree in astronomy while working part-time as a violinist at the Bol'shoi theater.[32]

---

29. "Spisok chlenov obshchestva mezhplanetnykh soobshchenii," ARAN, r. 4, op. 14, d. 196, ll. 6–21. See also V. M. Komarov and I. N. Tarasenko, "20 iunia—50 let so vremeni sozdaniia v moskve obshchestva izucheniia mezhplanetnykh soobshchenii (1924g.)," *Iz istorii aviatsii i kosmonavtiki* 22 (1974): pp. 75–82; Kramarov, *Na zare kosmonavtiki*, p. 28.

30. Ia. A. Berzin, ed., *Vospominaniia o Lenine: tom 3* (Moscow: Gospolitizdat, 1960), p. 76.

31. Kramarov, *Na zare kosmonavtiki*, p. 11.

32. Kramarov, *Na zare kosmonavtiki*, pp. 38–56.

Some Society members, including theorist Tsander, popular science writer Perel'man, and Academy professor Vetchinkin, were part of the pre-Revolutionary technical intelligentsia, i.e., they were older than the main body of members but still young enough to have been energized by the possibilities opened up by the Revolution. Vetchinkin provided the necessary institutional backing; Perel'man served as a conduit to the media, and Tsander represented the spiritual nexus for the group's activities. The de facto leader of the Society, however, was Grigorii Kramarov, a 36-year-old Comintern worker who had had a long history of active work on behalf of revolutionary causes. In January 1905 he was arrested by the Tsarist police for his illegal activities—the first of six times he was incarcerated for a variety of revolutionary activities on behalf of the Russian Social Democrats. After his last arrest in February 1913, Kramarov fled Russia and escaped to the United States. During the four years he lived in San Francisco, he worked at a fabric factory and actively participated in the Russian section of the American Communist Party, and was also a member of the Committee to Aid Political Prisoners and Exiles in Siberia. Hearing of the February 1917 revolution, Kramarov immediately returned home to work with the Men'sheviks. Following active duty with the Fifth Army on the eastern front during 1919–1920, Kramarov worked abroad for Comintern before returning to the Soviet Union.[33]

The Soviet media were surprisingly attentive to the work of these men and women and, with a few exceptions, journalists favorably reported on their activities. *Vecherniaia izvestiia* (*Evening News*), a widely read evening newspaper, and the journal *Iskra* (*Spark*) gave prominent space to the Society's structure, goals, and members. The former underscored that the Society's activities were not divorced from the masses, whereas the latter emphasized that Tsiolkovskii's and Goddard's work gave, for the first time, hope to bridge the gap between aspiration and achievement.[34] The biweekly *Tekhnika i zhizn'* (*Technology and Life*), one of the most popular science journals of the day, published a special issue in July 1924 dedicated to interplanetary travel; its cover story on the OIMS included a picture of Kaperskii, Rezunov, and Leiteizen in Academy uniforms, huddled over their documents. The author of one article, A. A. Mikhailov, the director of the Pulkovo observatory, introduced his readers to the elementary concepts of rocketry and space exploration and Tsiolkovskii's various spaceship designs. He singled out Tsander's scientific contributions to the field. In another article, "The Rockets of Goddard and Oberth," Ia. Gol'berg provided illustrated summaries of various Oberth projects, including a "passenger

33. I. N. Tarasenko, "1 dekabria—90 let so dnia rozhdeniia G. M. Kramarova (1887 g.)," *Iz istorii aviatsii i kosmonavtiki* 33 (1978): pp. 173–176.

34. *Vecherniaia izvestiia* issues for 25 June and 7 July 1924. These articles are preserved in ARAN, r. 4, op. 14, d. 194, ll. 115–116.

rocket," a "station in space," and Goddard's alleged Moon rocket.[35] Other articles in the issue dealt with the biology of high-altitude flight, long-distance aviation, and "superguns." The issue was illustrated with detailed artists' impressions of rockets and spaceships taken without attribution from the popular American science magazine *Science and Invention*.[36]

The Society's Work

Having formed the core of the Society, the members now set out to define precisely what they wanted to do. On 20 June 1924 the OIMS held its first official organizational meeting in the auditorium of the astronomical observatory of the Moscow region's Department of National Education (MONO).[37] About 200 people filed into the relatively small hall adorned with portraits of Newton, Kepler, Herschel, Bredikhin, and other renowned scientists. The three main organizers of the event, Leiteizen, Kaperskii, and Rezunov, sat at the main podium. As the first order of business, all assembled unanimously elected two men as honorary members of the Society: Tsiolkovskii, the founding father of Soviet rocketry, and Perel'man, the most important publicizer of the cause of spaceflight.

The meeting then led into a spirited discussion over the future goals of the Society. Using the blueprint of the smaller student Section of a few months earlier, the OIMS drew up a six-point charter (*ustav*), the first point of which was to "work on the accomplishment of extra-atmospheric flight with the aid of a reactive vehicle." Knowing that such an ambitious goal was out of the immediate reach of the Society, members emphasized more immediate activities like its predecessor Section, such as establishing a network of people all over the Soviet Union interested in the science of interplanetary travel, collecting data from the West, and most important, disseminating "reliable information" to the "broad masses" by way of "lectures, reports, organizing libraries, exhibitions, publishing scientific and popular literature, both original and translated."[38] Outreach and publicity were central to the Society's mandate and crossed over international barriers: the members decreed that one of their main goals would be to contact Goddard and Oberth and translate into Russian their respective 1919 and 1923 monographs.[39] At the end of the meeting,

35. A. Mikhailov, "Mezhplanetnye puteshestviia" and Ia. Gol'berg, "Rakety goddarda i oberta" in *Tekhnika i zhizn'* no. 12 (6 July 1924): pp. 10–12 and 12–13.

36. Compare the illustrations on p. 11 and the inside cover of *Tekhnika i zhizn'* issue with those from Don Home, "Can We Visit the Planets?" and Raymond Francis Yates, "Picture of the Earth," both in *Science and Invention* 11 (February 1924): pp. 962, 977.

37. "Otchet ob organizatsionnom sobranii OIMS," ARAN, r. 4, op. 14, d. 196, l. 1.

38. "Ustav obshchestva mezhplanetnykh soobshchenii," ARAN, r. 4, d. 196, ll. 2–3.

39. V. Chernov, "Obshchestvo izucheniia mezhplanetnykh soobshchenii," *Iskra* no. 8 (1924): p. 35.

all the members voted to elect a central body of the OIMS, the "presidium," that included Tsander, Leiteizen (secretary), Kaperskii, Rezunov, Chernov, Veigelen, and Kramarov. For the time being, the Society's home base would be the MONO astronomical observatory.

Kramarov, who was chosen to chair the presidium, recalled 40 years later that no one had any illusions that they would soon be sending men into space. He remembered that "[I]n the work of the Society [they] all saw one more possibility to aid the Motherland, to aid in the building of socialism." They would do this not by actually building rockets, but by bringing science to the masses. They were "convinced that the Society's work would contribute to the preparation of cadres, who in the future would create the economic and scientific and technical base for solving the greatest problems."[40] Like many utopians of the day, they justified their actions by establishing the most tenuous of connections with the exigencies of the new Soviet state. The technological fascination of the era provided a perfect bridge to connect utopian dreaming with real problems.

The OIMS was officially divided into three thematic sections, which loosely reflected its activities: a scientific-research section to do experimental work; a literature and propaganda section to publish the work of the Society; and a scientific-popularization section to organize lectures at various institutions.[41]

Because of limited funds, the Society produced little in terms of technical achievement—the goal of their first section. During its lifetime, there was no unanimous agreement on the clear distribution between performing "real" science and doing outreach at the OIMS. Tsander, one who frequently let his enthusiasm exceed his reach, most passionately believed that the members should immediately move to scientific and engineering work. No one in the Society, however, had the slightest idea of how to cultivate practical work. In this vacuum, Tsander took control. In a report to the Society's "board of directors" (*pravleniia*) on July 15, he asked rhetorically "How can we perform our investigations, what should we undertake first . . . ?" He laid out a four-point plan that would begin with "laboratory work" and end with the "construction of a large vehicle to lift men into the higher layers of the atmosphere and interplanetary space itself; we hope we will be able to shake hands on other planets." His detailed laboratory plan included considerations on testing small rockets, researching the various dimensions of rocket design; designing and cooling fuel tanks and engines; ideas on propellant selection and metal selection for the body of the rocket; exhaust gas dynamics; designing

40. Kramarov, *Na zare kosmonavtiki*, p. 50.

41. "Protokol zasedaniia pravleniia o-va iz mezhplan. soobshch. ot 23 iunia 1924 g.," ARAN, r. 4, op. 14, d. 196, l. 38. The scientific research section included F. A. Tsander, M. G. Leiteizen, and M. A. Rezunov; the popular science section included M. G. Serebrennikov and G. M. Kramarov; and the literature and propaganda section was represented by V. P. Kaperskii and V. I. Chernov.

ground systems to test rocket elements; and developing life support and electrical systems. He seemed to believe that the Society's announced contest to develop a rocket "would advance [their] work greatly."[42] Unfortunately, despite talk of closely coordinating theoretical, design, and experimental work, there is no evidence that the OIMS actually established a laboratory for such tasks.

The second section's ambitions also outstripped its material limitations. In May, leading member Leiteizen invited Tsiolkovskii in a letter to contribute an article for a new monthly journal which they would call *Raketa* (*The Rocket*). Their plan was to publish serious works on interplanetary travel and related fields such as astronomy, physics, chemistry, and aeronautics. Since they were unable to find any publishers interested in issuing such a journal, the members had planned to use money collected from both the Lapirov-Skoblo talk and from book sales. Tsiolkovskii offered the first part of his "Life in the Cosmic Ether" for *Raketa*.[43] Others, including Tsander ("On a Flight Over Other Earthly Spheres") and Vetchinkin, as well as OIMS member Rezunov ("The Dream of Humanity") wrote additional articles. Society member Mozharovskii prepared an evocative cover of a rocket flying over a planet with a background of galaxies and stars.[44] In the end, no journal was published. Kramarov remembers that they were unable to "overcome many difficulties, especially insufficient money."[45] Members collectively sent a letter to a major newspaper, asking to have an announcement published, free of charge, explaining that the Society had books and information on space travel for sale at their home location at Bol'shaia lubianka street. The newspaper declined to publish the letter.[46]

The OIMS was more successful disseminating ideas about space travel via lectures. At a meeting on June 30, the scientific-popularization section rejected section head Serebrennikov's suggestions for public presentation topics on specialized topics, and instead voted to have talks that would appeal to a very *general*

42. F. A. Tsander, "Organizatsionnyi doklad inzhenera F. A. Tsandera on predpolagaemykh rabotakh Nauchno-issledovatel'skoi sektsii Obshchestva izucheniia mezhplanetnykh soobshchenii o reaktivnom dvitatele," in *F. A. Tsander: iz nauchnogo naslediia*, pp. 30–31.

43. Leiteizen to Tsiolkovskii, 21 May 1924, ARAN, f. 555, op. 4, d. 356, ll. 6–7; Tsiolkovskii to Leiteizen, 4 June 1924, ARAN, f. 555, op. 3, d. 102a, l. 6.

44. For the draft of Tsander's manuscript for *Raketa*, see F. A. Tsander, "O pereletakh na drugie zemnye shary," in *F. A. Tsander: iz nauchnogo naslediia*, pp. 44–47. For two other articles for the journal, see V. Chernov, "Raketa na lunu" and M. Rezunov, "Mechta chelovechestva" in ARAN, r. 4, op. 14, d. 194, ll. 1–7. The first issue would have featured 11 articles by Tsiolkovskii, Tsander, Kramarov, Vetchinkin, Mikhailov, Rezunov, Chernov, Perel'man, Sharonov, and Kaperskii. See ARAN, r. 4, op. 14, d. 196, ll. 34–35.

45. Kramarov, *Na zare kosmonavtiki*, p. 60.

46. Kramarov, *Na zare kosmonavtiki*, p. 34.

audience, with a focus on history and forecast.[47] However, few in the Society beyond Tsander knew anything about rocket engine design. In late July, Tsander gave a talk to the OIMS's scientific-popularization section, the first part of which dealt with the history of visions of space travel, and the second part on the mathematics of rocket velocities, complete with differential equations. By teaching the Society's own members something of both the past and the possibilities of space exploration, Tsander not only helped to disseminate ideas about the "correct" way to space, i.e., with liquid-propellant rockets, but also inculcated his young audience with the idea that three men, one Soviet (Tsiolkovskii), one German-Romanian (Oberth), and one American (Goddard), were the true founding fathers of the theory of spaceflight. He helped to establish a pantheon—a holy trinity—for a new scientific era that has endured to the current day.[48]

Mania Over Goddard

The Society not only maintained regular contact with Tsiolkovskii through its existence, but also communicated with the reclusive American rocket scientist Robert Goddard. Society records reveal that in response to an inquiry from Leiteizen, the American wrote to the Society on 16 August 1924, expressing pleasure at the formation of the Society and indicating that he would be interested in cooperating with them.[49] At one of the major talks in Moscow in early October, Tsander read Goddard's letter to the audience.[50] Goddard had also sent the Society a copy of one of his recent articles published in the journal *Monthly Weather Review* wherein he conjectured on the velocity required to send a rocket to the Moon; the Society quickly translated the article into Russian for its members, thus providing one of the most influential sources of information about Goddard's work in the Soviet Union for many years.[51]

Garbled press reports alleging that Goddard was planning to launch a rocket to the Moon fostered a mini-mania among space enthusiasts all over the Soviet

47. "Protokol zasdeaniia pravleniia O.I.M.S. ot 30/VI-1924 g.," ARAN, r. 4, op. 14, d. 196, l. 39. Serebrennikov had suggested talks with such titles as "The Rocket, Its Instruments, Production, and Military Application," "Comparing Steam, Internal Combustion, and Reactive Engines," "Flying to Great Altitudes," "The Atom and Its Energy," "Radioactive Matter," "Observational and Theoretical Astronomy," "Celestial Mechanics," and "Aerodynamics."

48. F. A. Tsander, "Doklad v nauchno-populiarnoi sektsii Obshchestva izucheniia mezhplanetnykh soobshchenii o reaktivnom dvigatele," in *F. A. Tsander: iz nauchnogo naslediia*, pp. 32–34.

49. Goddard to Leiteigen [Leiteizen], 16 August 1924, ARAN, r. 4, op. 14, d. 195, l. 16.

50. For Tsander's note on the reading, see F. A. Tsander, "Nekotorye materially k vystupleniiu na dispute, sostoiavshemsia v 1 MGU 1, 4, 5 oktiabria 1924 g.," in *F. A. Tsander: iz nauchnogo naslediia*, p. 50.

51. Goddard to OIMS, 16 August 1924, ARAN, r. 4, op. 14, d. 195, l. 16. For the sent version of the Goddard article and its Russian translation, see ARAN, r. 4, op. 14, d. 194, l. 44–45ob. The article was "The High-altitude Rocket," *Monthly Weather Review* 52 (February 1924): pp. 105–106.

Union. Original reports had suggested that Goddard would launch a rocket to hit the Moon on 4 July 1924. But in mid-June, the science reporter for *Izvestiia* noted that according to *The New York Herald*, Goddard had postponed his launch to August 1924.[52] As summer turned to fall and there was no news, Goddard fever reached such proportions that the city police once had to be called out in Moscow. Responding to the notoriety over the alleged Goddard shot, the Society asked a prominent Leningrad-based astronomer, V. V. Sharonov, to speak to the public on the Goddard project at the Main Hall of the Physics Institute of the Moscow State University. The Society printed up artful posters under the giant headline "Polet na drugie Miry" ("Flight to Other Worlds") which were put up at several major intersections in Moscow. Sharonov gave two separate lectures, both on the first day of October 1924: "The Truth on the Dispatching of Professor Goddard's Projectile to the Moon on 4 August 1924" and "Debates in the West in Connection with Sending a Projectile to the Moon." At 8:00 p.m. on the night of the talk, so many people showed up that the Moscow horse militia had to be called out to control the crowds who were unable to enter the auditorium. All the tickets, for 30 kopecks each, had been sold out. Due to popular demand, the Society asked Sharonov to repeat the talks—this time followed by public debates—on 4 and 5 October.[53]

Society-organized debates, such as the one on Goddard, were quite common in the fall of 1924. Some were limited to the membership, such as the one on "Cosmogony Hypotheses" by member V. S. Berdichevskii on September 24. Others were open to the public for a small fee—about 30 kopecks. Usually, these public lectures had two components, a first talk by a distinguished guest and then a second talk and open debate (*disput*) on a controversial topic. For example, the Society held highly publicized lectures at the Russian Polytechnical Museum on 31 October and 2 November, this time by Vladimir Vetchinkin, the Zhukovskii Academy professor and aeronautics expert who had helped to establish the Society. His two-part lecture, beginning each night at 8:00 p.m., covered both the popular and mathematical aspects of interplanetary travel. Later, the Society's own Vladimir Chernov gave a talk on "Construction of Rockets," followed by debate on the best way to do so. Those who paid the 75 kopecks to 3 rubles (depending on the seat) were privy to see not only Vetchinkin and Chernov but also to hear spirited discussion among representatives from the Moscow Society of Amateur Astronomers, Glavvozdukhflot (the civil air fleet), the ODVF (a voluntary society for aviation enthusiasts), Dobrolet (Soviet airline), and the journals *Vestnik vozdushnogo*

52. "Preslovutaia 'raketa'," *Izvestiia VTsIK*, 13 June 1924: p. 5.

53. For a reproduction of the poster for 4 October as well as Tsander's comment about the "horse militia," see Korneev, "Zhizn'," pp. 29–30. See also Kramarov, *Na zare kosmonavtiki*, pp. 54–55. Tsander also gave talks on all three days.

flota (*Air Fleet Bulletin*) and *Samolet* (*Aircraft*).[54] Their attendance was critical because it gave these often-bizarre proceedings the implicit stamp of official sanction; ODVF and Dobrolet were important elements in the Soviet government's drive to inculcate a young post-Revolution generation with up-to-date technology by equating aviation with modernity. If some in the press and elsewhere were tempted to derisively speak of the "youth who believe in such fables," others took pause and infused the talk on spaceflight with a certain sense of gravitas. Lectures elsewhere in 1924—at a number of aviation factories, at the Bauman Higher Technical School, and at the Shternberg Astronomical Institute—also expanded the sites of discussion. Society representatives, including Sharonov, Prianishnikov, and the indefatigable Tsander, took their message outside of Moscow in 1924–1925 to give talks in such disparate locales as Khar'kov (on 17 October), Leningrad (on 17 November), Riazan' (in November), Tula (in December), and Saratov (in January 1925).[55] The Society archives contain letters from enthusiasts in places as far away as Zlatoust and Khar'kov, expressing their wish to engage in joint work on interplanetary travel.[56]

THE SOCIETY AND THE ARTS

The Society produced no work of art or literature but, during its brief existence, its members actively discussed and helped to popularize space-themed artistic works of others.[57] For example, their interest in space travel was catalyzed by a number of famous science-fiction novels such as Aleksandr Bogdanov's *Krasnaia zvezda* (*Red Star*, 1908) and Aleksei Tolstoi's *Aelita* (1923). The former, the first major Russian work in the genre of science fiction to deal explicitly with spaceflight, was written before the Revolution but was widely read and discussed afterward.[58] Bogdanov (a pseudonym for Aleksandr Malinovskii) had had a much-publicized split with Lenin, but his popularity in the 1920s, which rested on *Red Star* and his other science-fiction novel *Inzhener Menni* (*Engineer Menni*), did not suffer. Reprinted in 1918 and 1929, *Red Star* was not, of course, about spaceflight, but about an

54. The Moscow Society of Amateur Astronomers (MOLA) was represented by its chairman A. A. Mikhailov, Dobrolet by engineer Kh. I. Slavernov, and *Samolet* by pilot S. I. Pokrovskii. See the poster for the talk in Kramarov, *Na zare kosmonavtiki*, p. 53. *Glavvozdukhoflot* stood for Main Administration of the Workers' and Peasants' Red Military Air Fleet, i.e., the Soviet "air force." ODVF stood for Society of Friends of the Air Fleet.

55. ARAN, r. 4, op. 14, d. 194, l. 76; Tsander, *F. A. Tsander: iz nauchnogo naslediia*, p. 6. For the texts of many of Sharonov's lectures, see ARAN, r. 4, op. 14, d. 194, ll. 77–91.

56. ARAN, r. 4, op. 14, d. 195, ll. 14–15.

57. For a detailed description of the artistic dimensions of the "space fad," see Siddiqi, "Imagining the Cosmos."

58. A. Bogdanov, *Krasnaia zvezda: roman-utopiia* (St. Petersburg: Tvorchestvo khudozhestvennoi pechati, 1908).

idealized communist utopia on the planet Mars; some have seen Bogdanov's tale as a warning on how socialism would take on distinctly totalitarian tones if sufficiently militarized.[59] In constructing a future administered by a "benevolent technocracy," Bogdanov articulated an explicitly technologically utopian vision that seemed to coincide much more with the prevailing feeling of the 1920s than the pre–World War I period when he wrote the novel.

After the Revolution, when *Red Star* was especially popular among the reading public, the novel resonated deeply with the space enthusiast community who shared Bogdanov's utopian notion that modern technology, especially space technology, could remake society in unimaginable (and positive) ways. At the height of the space fad in 1924, Society for the Study of Interplanetary Communications secretary Moris Leiteizen wrote to Tsiolkovskii, announcing that they were in communication with Bogdanov; among other things, they were particularly interested in Bogdanov's rather unusual proposal to use atomic power to propel his apocryphal spaceship.[60] The novel's lasting relevance for space enthusiasts is underscored by a review of the book published in 1934 by spaceflight popularizer Nikolai Rynin, who gushed (incorrectly) that Bogdanov was the first to predict the use of nuclear power for spaceflight.[61] In general, space enthusiasts were less likely to explore Bogdanov's philosophical arguments than his technological vision; both parties shared a view of technology as autonomous, positive, and liberating.

The space enthusiast community had a similar take on the most famous Soviet science-fiction novel of the 1920s, Aleksei Tolstoi's *Aelita: Zakat Marsa* (*Aelita: Sunset of Mars*), first published in serialized form in 1922–1923.[62] Much has been written about the novel, which is about an engineer's and a soldier's trip to Mars, the latter of whom incites a proletarian revolution in a bourgeois social setting. Aelita is the queen of Mars who falls in love with the Red Army solder. Superficially, *Aelita* has all the elements of post-Revolutionary utopian science fiction: a bourgeois enemy, a Marxist revolution, the most modern science and technology of space travel, adventure and romance borrowed from Edgar Rice Burroughs, and a utopian theme. Yet, as Halina Stephan has pointed out, the novel is also characterized by mysticism and by "themes borrowed from the theosophic and anthrosophic

59. Loren R. Graham, "Bogdanov's Inner Message," in *Alexander Bogdanov: Red Star: The First Bolshevik Utopia*, Loren R. Graham and Richard Stites, ed. (Bloomington, IN: Indiana University Press, 1979), pp. 241–253; Mark B. Adams, "Red Star: Another Look at Aleksandr Bogdanov," *Slavic Review* 48 (Spring 1989): pp. 3–15. Roy Medvedev was perhaps the most well-known historian to make this argument; See his *Let History Judge: The Origins and Consequences of Stalinism* (New York: Alfred Knopf, 1972), p. 374.

60. Leiteizen to Tsiolkovskii, 4 May 1924, ARAN, f. 555, op. 4, d. 356, ll. 2–3.

61. N. A. Rynin, "Tekhnika i fantaziia," *V boi za tekhniku* no. 8 (1934): p. 22.

62. The novel was originally published in three serialized parts in the journal *Krasnaia nov'*. In 1923, it was published as a standalone novel as A. Tolstoi, *Aelita (Zakat Marsa)* (Moscow: GIZ, 1923).

mythology."[63] The immediate reaction to the novel was rather negative since many of the more ideologically active Bol'sheviks viewed *Aelita* as a novel with an ambiguous pro-Soviet message—the Revolution on Mars had, after all, failed. Yet, the technologically-minded spaceflight enthusiasts found it inspiring: Tolstoi's novel was the first major Soviet novel of the period that used a rocket for interplanetary travel—a key aspect that attracted the attention of many. For example, the Society for the Study of Interplanetary Communications was so taken by Tolstoi's use of the rocket that they considered using the story to develop a film script, a project that was brought to fruition by others.[64]

In terms of film, the seminal *Aelita* (1924) and the lesser-known *Kosmicheskii reis* (*Space Voyage*, 1935)—the two movies which bookended the Soviet space fad—were the only film projects that reached fruition, but they were not the only attempts to bring cosmic voyages to the screen. In 1924, the Society informed Tsiolkovskii that they had contacted Proletkino, the official Soviet movie-making authority, with the idea to develop an original film script.[65] Tsiolkovskii insisted that the Society write a script for his novella *Beyond the Earth*, but Society representative Moris Leiteizen responded that they would rather not use that pre-Revolutionary story: "We have decided to give our film some lively character: the action should take place in our day in Soviet Russia."[66] The Society produced two scripts, both extraordinary not for their artistic merit but because they astutely mirrored the concerns of the Society.

Member M. G. Rezunov's four-part script closely followed Tolstoi's *Aelita* with a Martian character named Le (Tolstoi's hero was named Los'), but added an odd twist near the end. The action is initiated when Stepan, a factory worker, reads about Goddard's plan to launch a rocket to the Moon on 4 July. Gripped by the idea of spaceflight, Stepan devises a plan for a spaceship which, after presentation to the ODVF (the voluntary aviation society), the Society decides to build. The central intrigue of the story involves a "foreign" spy who disrupts the construction work. The obligatory voyage to Mars is embellished by scenes of life in space, the hazards of meteorites, befriending a Martian communist, and ultimately a return trip in which Stepan, unable to live without Le, decides to commit suicide by exiting his ship and becoming an Earth satellite. Stepan, it seems, would rather be dead in space than alive on Earth.[67]

---

63. Halina Stephan, "Aleksei Tolstoi's *Aelita* and the Inauguration of Soviet Science Fiction," *Canadian-American Slavic Studies* 18 nos. 1–2 (1984): pp. 63–75.

64. Leiteizen to Tsiolkovskii, 4 May 1924, ARAN, f. 555, op. 4, d. 356, ll. 2–3.

65. Ibid.

66. Tsiolkovskii to OIMS, 14 May 1924, ARAN, r. 4, op. 14, d. 195, ll. 3–5; Leiteizen to Tsiolkovskii, 21 May 1924, f. 555, op. 4, d. 356, ll. 6–7.

67. "Kinotsenarii," ARAN, r. 4, op. 14, d. 194, ll. 118–119.

Society chief Grigorii Kramarov's script was a brilliant synopsis of many of the tropes of the space fad. The action begins in the late Tsarist era when the imperial government forbids any talk of space exploration. A Russian revolutionary inventor, modeled on the great martyred "spaceship inventor" Nikolai Kibal'chich, produces a plan for a rocket ship.[68] He is arrested and executed, but not before bequeathing his plan to his young son, Viktor, who like his father, links the struggle against the autocracy with the struggle against gravity. With his plans and dreams in hand, the young revolutionary moves to the United States to fulfill his father's dream of spaceflight. With the help of likeminded others in America, Viktor begins building the rocket ship when a rich banker discovers the project. The banker decides to build his own ship and asks Viktor to join his company. Angry with Viktor's refusal, the banker seizes Viktor and his model and takes them to his villa, where he employs a team of engineers to finish the design. The banker's daughter, of course, falls in love with Viktor and helps him escape the depths of the banker's capitalist hell. Viktor's industrious American friends convince him that he must return to Russia, for there is a Revolution underway at home. The young man, disguised as an "engineer-expert," manages to free his vaulted ship and fly the rocket-machine to Russia where, upon landing, he is greeted with much fanfare to "start the victory over the cosmos!"[69] Here we find the patron saint (the Kibal'chich archetype), the equation of gravity with oppression, America as a place where technological dreams come true (especially if capitalists get out of the way), and the all-encompassing myth of Bol'shevik recognition of the value of spaceflight—all in one film. Although the Society approached several major studios to finance the project, including Proletkino and Mezhrabpom-Rus', they found no one to produce the movie.[70]

Even though, in 1924, Soviet filmgoers were deprived of the pleasures of watching Stepan and Viktor on screen, they might have seen the most famous Soviet science fiction in the prewar era, *Aelita*, the movie version of Tolstoi's book about a voyage to Mars, directed by the famed pre-Revolutionary Russian director Iakov Protazanov.[71] The silent movie, which was released officially on 25 September

68. Kibal'chich was a member of the People's Will terrorist organization which assassinated Tsar Aleksandr II. Before his execution in 1881, Kibal'chich devised a plan for a rocket-propelled vehicle that later Soviet historians absurdly claimed was an early design for a spaceship.

69. Kramarov, *Na zare kosmonavtiki*, pp. 62–63.

70. "Vospominaniia Kaperskogo," ARAN, r. 4, op. 14, d. 197, l. 41.

71. For Protazanov, see M. Aleinnikov, *Iakov Protazanov: o tvorcheskom puti rezhisera*, 2nd ed. (Moscow: Iskusstvo, 1957); M. Aleinnikov, *Iakov Protazanov* (Moscow: Iskusstvo, 1961); Ian Christie and Julian Graffy, eds., *Protazanov and the Continuity of Russian Cinema* (London: British Film Institute, 1993); Denise J. Youngblood, "The Return of the Native: Yakov Protazanov and Soviet Cinema," in *Inside the Film Factory: New Approaches to Russian and Soviet Cinema*, Richard Taylor and Ian Christie, eds. (London: Routledge: 1991), pp. 103–123. *Mezhrabpom-Rus'* combined *Mezhrabpom* (International Workers' Aid), a Berlin-based relief organization and *Rus'*, a Russian production company formed in 1918.

1924 at the peak of the space fad, contributed enormously to the popularization of spaceflight in Soviet culture in the 1920s. It is no coincidence that space enthusiasts such as the OIMS enjoyed their highest attendance ratings at talks soon after the movie was released.

Protazanov's goal to produce an "impartial" work was not rewarded in like. Despite widespread criticism of the film—many complained about the fact that the revolution on Mars had failed—it was an incredibly popular film; it did, after all, feature dramatic acting, an exotic planet, a glamorous princess, and a romantic story arc. Grigorii Kramarov, the head of the OIMS, effusively remembered years later that "[T]he book and film played a significant role in strengthening interest towards interplanetary communications and contributed to the development of activities of our Society."[72] The movie made a deep impression on many young people. Vladimir Chelomei, who was only 10 years old when the film came out, 45 years later named his dream project—a huge space complex to send the first Soviet cosmonauts to Mars—Aelita.[73]

THE SOCIETY'S END

In the end, lack of financial and governmental support proved to be the Society's undoing. Most Russian sources note vaguely that the Society existed for "approximately a year" and then disbanded.[74] In reality, the Society probably dissolved long before a year was over. Society secretary Leiteizen wrote to Tsiolkovskii on 16 December that the Society was "currently in a period of liquidation." As an explanation, he noted, "[I]t's not so terrible that we hurried, but we ran ahead [of ourselves by] several years, only to come back to the present." He added in resignation, "I'm also inclined to think that such a noisy Society was excessive: we can still work splendidly for now without a Society. And [perhaps] even work better."[75]

Why did the Society fall apart so soon after the successful lecture series in October and November? The most important factor was lack of official state recognition. Although the Society was sponsored by the Moscow Society of Amateur Astronomers (MOLA) and the Military-Scientific Society (VNO) of the Zhukovskii Academy, it was never an *officially* recognized organization. In late 1924, when the Society petitioned the administrative department of the Moscow

---

72. Kramarov, *Na zare kosmonavtiki*, pp. 19–20.

73. Asif A. Siddiqi, *Challenge to Apollo: The Soviet Union and the Space Race, 1945–1974* (Washington, DC: NASA, 2000), pp. 745–754.

74. See for example, Komarov and Tarasenko, "20 iiunia–50 let," p. 81.

75. Leiteizen to Tsiolkovskii, 16 December 1924, ARAN, f. 555, op. 4, d. 356, ll. 11–11ob. The former wrote his last letter to Tsiolkovskii on 7 January 1925. For the complete (eight) letters from Leiteizen to Tsiolkovskii, see ARAN, f. 555, op. 4, d. 356, ll. 1–15.

city council to officially register the organization, the city council rejected the application on grounds that the Society had "insufficient scientific strength among its members."[76] Probably the most fatal blow for the Society was that its patron, Academy Professor Vetchinkin, declined to formally become a member, thus depriving the group of legitimacy in the eyes of both the Academy and the Soviet security services who had to sign off on any formal social organization in the Soviet Union.[77] The following year, Tsander confirmed that the lack of "scientific workers" among members of the "board of directors," i.e., Tsander, Leiteizen, Kaperskii, Rezunov, Chernov, Serebrennikov, and Kramarov, was a source of dissension that contributed to the Society's dissolution.[78] The members unsuccessfully appealed the city council decision and were dealt a second blow when, upon hearing of the city council's decision, the Military-Scientific Society withdrew its support of the Society.[79] Soon after, the enthusiasts abandoned their efforts.

The Society's members also had deal with less-committed members who were unable to sustain interest in the face of both financial insecurity and the possibility that space exploration was decades away. Society head Kramarov remembers that the most common question from the audience after each lecture was "How quickly would flight to the planets be accomplished?"[80] Tsander's naïve and unbounded optimism that interplanetary travel was imminent raised the hopes of many who had no idea of the incredible technical difficulties. When it became clear that travel into space was years (if not decades) away, the "accidental members" dispersed quickly, leaving only a handful of the truly dedicated to pursue the cause.[81] And eventually even the faithful had to come down to the ground; most, such as Tsander, had little time to devote to activities that did not provide money for living. Chernov, for example, remembered later that his job as a musician forced him to abandon the Society.[82] Because the group was never officially recognized or registered by

76. Rezunov to Korneev, 19 February 1961, ARAN, r. 4, op. 14, d. 197, l. 19.

77. Ibid.

78. Tsander noted in an unpublished manuscript from August 1925 that "[T]he shortage of time which I had free and the comparatively small number of scientific workers in the board of directors is the reason why, after the Society had been approved by a number of others, one last person did not approve it . . ." See F. A. Tsander, "Materialy k knige 'Polety na drugie planety i na Lunu'," in *F. A. Tsander: iz nauchnogo nalslediia*, p. 58.

79. "Vospominaniia Kaperskogo," ARAN, r. 4, op. 14, d. 197, ll. 43–44.

80. Kramarov, *Na zare kosmonavtiki*, p. 56.

81. Russian historians Raushenbakh and Sokol'skii note that "the Society absorbed many accidental members since most of those [who] joined the Society in 1924 never contributed to the development of rocketry in any way." B. V. Rauschenbach and V. N. Sokolsky, "The First Soviet Space Flight Organisations," 49th International Astronautical Congress, 28 September–2 October 1998, Melbourne, Australia.

82. Kramarov, *Na zare kosmonavtiki*, pp. 51–52.

any state authority, it never received any funds from the state and maintained its activities by donations or from ticket imbursements. And once the members were unable to register the Society with official authorities, the two major institutions that sponsored the Society, the MOLA and the VNO, cut off their modest support.

Few members of the Society for the Study of Interplanetary Communications went on to participate in the practical development of rocketry funded by the Soviet government in the 1930s. The notable exception was Fridrikh Tsander, who in 1931 founded the Group for the Study of Reactive Motion (GIRD), the semi-amateur team that launched the first Soviet liquid-propellant rockets in 1933. Tsander served as a crucial link between the Society and the most important personality of the early Soviet space program, Sergei Korolev. Tsander's unexpected death in 1933 from poor health was a deep blow to Korolev. In the heady days after the launch of Sputnik in 1957, Korolev searched out Tsander's grave to set up a special memorial in his name.

Others from the 1924 Society for the Study of Interplanetary Communications fared worse, especially the ones who fell at the height of the Stalinist Great Terror in the late 1930s. On 16 April 1939, the NKVD (the Soviet security services) shot former Society secretary Morris Leiteizen—then 42 years old—because he was the son of an old Bol'shevik. Mikhail Lapirov-Skoblo, one of the earliest advocates for spaceflight in the 1920s and the man who gave the first rousing lecture to recruit members for the Society in 1924, also fell to the purges. After a very distinguished career as a vocal spokesperson for the Soviet scientific and technical intelligentsia, he was arrested in 1937, sentenced in 1941, and died in confinement in 1947 while working at a factory.[83]

CONCLUSIONS

The Society for the Study of Interplanetary Communications was the world's first organization dedicated to the study of spaceflight. Besides its claim to priority in the history of space exploration, the organization left behind two important legacies.

First, the Society—with the help of enthusiastic Soviet media—wrenched space travel from the discourse of fantasy and relocated it into the language of science and technology. Prior to the 1920s, in the public imagination space exploration was considered in the same breath as mythology, speculation, and mysticism. In the 1920s, by linking spaceflight with the sciences (mathematics, chemistry, metallurgy, etc.) and suggesting that space travel was possible by means familiar to most people (such as rockets), the spaceflight advocacy community brought such ideas into the realm of reasonable scientific prognostication.

83. "Rasstrely v Moskve-G," *http://mos.memo.ru/shot-13.htm* (accessed 22 January 2007); Medvedev, *Let History Judge,* 444; "Lapirov-Skoblo Mikhail Iakovlevich," in *Repressirovannoe ostekhbiuro*, E. N. Shoshkov, ed. (St. Petersburg: NITs Memorial, 1995), p. 137.

Second, by underscoring the scientific and technological bases of cosmic travel, the community equated spaceflight with "being modern." Space advocates fit their conception of space travel firmly within the Bol'shevik cause of remaking Russia into a modern, technologically capable nation. After the 1920s, spaceflight was, like aviation, a manifestation of the self-reflexive notion of twentieth-century modernity. The equation of spaceflight with science and modernity meant that space travel was now connected not only with the past—such as the Russian Cosmist philosophy which was committed to reanimating the dead by exploring space—but also with the future.[84] Mystical ideas like Cosmism were rare among space enthusiasts of the societies and exhibitions of the 1920s, which favored a fetishistic view of technology in general and space travel in particular. This fetishism, bordering on messianism, is profoundly evident in the works of the Society—particularly in the language that they used, the goals that they set out to accomplish, and the way in which they saw technology as a panacea for many if not all social ills. Writing of the Soviet Union in the 1920s, Anthony J. Vanchu notes that "[W]hile science and technology had the power to demystify religion and magic, they themselves came to be perceived as the locus of magical or occult powers that could transform the material world."[85] This kind of stance toward space travel is not so very different from the visions of space enthusiasts in entirely different social contexts, such as in the United States; many American activists also adopted an almost evangelical view of human migration from Earth into outer space—one that was not only inevitable but essential for the survival of the species.[86] The Society for the Study of Interplanetary Communication was perhaps the first organized attempt to articulate this singular vision of space travel which has continued to play an important role in those who believe—misguidedly or not—that through the power of science and technology, humankind will attain its rightful destiny in the deeper reaches of the cosmos.

84. For Cosmism, see Siddiqi, "Imagining the Cosmos."

85. Anthony J. Vanchu, "Technology as Esoteric Cosmology in Early Soviet Literature," in *The Occult in Russian and Soviet Culture*, ed. Bernice Glatzer Rosenthal (Ithaca, NY: Cornell University Press, 1997), pp. 205–206.

86. Roger D. Launius, "Perfect Worlds, Perfect Societies: The Persistent Goal of Utopia in Human Spaceflight," *Journal of the British Interplanetary Society* 56 (2003): pp. 338–349.

CHAPTER 28

C/SETI AS FICTION: ON JAMES GUNN'S *THE LISTENERS*

De Witt Douglas Kilgore

Astronomy compels the soul to look upwards and leads us from this world to another.
—PLATO, *The Republic*, 529[1]

Over the past four decades the search for extraterrestrial intelligence (SETI) has taken a small but significant place in American culture. Walter Sullivan, Frank Drake, and David Grinspoon, among many others, have written popular accounts of the science. Motion pictures such as *Contact* (1997) and *Species* (1995) have visualized the field with the tools available to filmmakers. The SETI Institute's radio program *Are We Alone*, hosted by astronomer Seth Shostak, makes the science a part of the heavily mediated environment of popular entertainment. There is also the small, robust field of historical and critical scholarship. It is in fiction, however, that SETI has found its most effective expression for general audiences. This medium provides literary arguments for why the science matters in contemporary life. I argue that it is fiction about SETI that helps us model the societal implications of extraterrestrial contact.

Fiction writers such as James Gunn and writer-scientists such as Carl Sagan have produced a literary subgenre that I call the CETI (communication with extraterrestrial intelligence) novel.[2] These writers take the science seriously and share an interest in

1. Plato, *The Republic* (New York: Vintage Books, 1991), p. 274.

2. For the purpose of distinguishing it from the science (SETI), I use for the fiction the original acronym for the project: CETI or communication with extraterrestrial intelligence. John Billingham recalls that the change in nomenclature was inspired by the establishment of the science as a NASA project in the early 1970s. David W. Swift, *SETI Pioneers: Scientists Talk about Their Search for Extraterrestrial Intelligence* (Tucson, AZ: University of Arizona Press, 1990), pp. 258–259.

both representing the project and its possible social and political effects.[3] They are also committed to dramatizing the early, founding goal of SETI—communication with extraterrestrial civilizations. This leads them to consider ideas that are always revisited in debate around the project. Could we communicate with an intelligent species that is not human? What effect would such an exchange have on us?

These questions create challenges of both form and content for the CETI novelist. How might these issues and the debates they inspire be presented to general readers, engaging the emotions as well as the intellect? One solution is the genre's focus on the future that a SETI contact could produce; the CETI novel always takes place in a recognizable present and that future is always just around the corner. It is the closeness of that fictional present to our world that makes CETI narratives credible, and it is a futurism that helps the writer emphasize the science's connection to contemporary life. It is through this link that the CETI novelist hopes to sustain the reader's interest in SETI as necessary to any future we might create.

The central premise of a CETI narrative is that the search for an artificial signal from another world is successful. What follows is a fruitful exchange of scientific and cultural ideas. Whatever arguments are used to sustain this presumption, it is liberal hope that inspires this hypothesis. Discovering the knowledge we need to make manageable the social and political problems of our world is the foundation of that hope. It forecasts a terrestrial future in which humanity survives its internal divisions and its Malthusian consumption of Earth's resources. In other words, communication with extraterrestrial others would prompt us to communicate with each other. We could then work together to make the world more livable and sustainable. Thus, human civilization would break out of the historical bind in which it finds itself.

Steven Dick argues that "perhaps the classic expression" of SETI in fiction is James Gunn's 1972 novel, *The Listeners*.[4] This book is my test case for the futurist concerns I have outlined—my model for the CETI novel as a distinct literary genre. Although other works in fiction and nonfiction preceded it, Gunn's achievement is to create a narrative that speaks to prevailing social concerns while introducing the reader to this new science. Here SETI is not merely a convenient backdrop for the agony of middle-class lives; it is the *raison d'être* of those lives. This emotional as well as intellectual centrality helps clarify the feeling of social mission the writer finds in the science.

---

3. Writers such as Don DeLillo and Stanislaw Lem have been interested in the science. However, they engage it both as a demonstration of the limits of science and of our ability to communicate through language. This is very different from the ideology and spirit encoded in the CETI novel as a narrative form of scientific representation. Don DeLillo, *Ratner's Star* (New York: Vintage Books, 1989 [1976]) and Stanislaw Lem, *His Master's Voice* (Evanston, IL: Northwestern University Press, 1999 [1983]).

4. Steven J. Dick, *The Biological Universe: The Twentieth-Century Extraterrestrial Life Debate and the Limits of Science* (Cambridge, U.K.: Cambridge University Press, 1996), p. 260.

In the following paragraphs I will explore Gunn's solution to the problem of introducing a new science to nonspecialist audiences. As I present it here, Gunn follows two distinct but conjoined strategies: he creates a form that provides the historical and philosophical background against which to view SETI, and he tells a story in which difficult social debates are resolved by lessons learned from science. The novelist's goal is to strike a balance between the didactic demands of his subject and familiar fictional conventions. Thus, I argue, Gunn makes a fiction in which SETI is part of our ordinary culture—the herald of an extraordinary future.

ON FORM

Gunn's *The Listeners* may be viewed as an experimental novel within the context of genre practice during its composition. It was written serially, as a series of short stories for *Galaxy Magazine* and *The Magazine of Fantasy and Science Fiction*, from 1968 to 1972. Marketed and published as general fiction, not science fiction, in 1972, the novel has been considered a "fix-up," defined by Clute and Nicholls as "a book composed of previously written stories which have been cemented together."[5] The cement, in this case, is what Gunn calls "computer runs," interchapters that provide a historical and philosophical overview of SETI and the future posited by the novel. The usual practice with fix-ups, such as Ray Bradbury's 1951 collection *The Illustrated Man*, is to provide fictive interchapters linking together stories that otherwise have no compositional relation to one another. Gunn's strategy is more like that used by John Steinbeck in his 1939 novel *The Grapes of Wrath*; it offers the reader a social and political context within which to understand the narrative portions of the work.

The Listeners presents SETI as a humanistic science, a research agenda that boldly addresses our social and philosophical concerns through the question of whether sentient life exists elsewhere. Gunn's formal strategy is to create a narrative in which science-based exposition and debate are entangled with conflicts around the politics of race, between religion and science, between a middle-class status quo and a utopian future, and around the private misunderstandings of husbands and wives, fathers and sons. In this narrative context, Gunn's computer runs present three lines of argument: history and futurist speculation, inspirational epigraphs from literature, and statements from SETI pioneers. These arguments create a dialectic that defines the science as a movement that can change the course of individual and collective lives.

Through historical reconstruction and speculation, Gunn gives us a select history of science. We learn that SETI was born when Cocconi and Morrison discussed the feasibility of sending interstellar messages during work on "gamma

5. John Clute and Peter Nicholls, *The Encyclopedia of Science Fiction* (London: Orbit, 1993), p. xxxiii.

ray production from the Crab Nebula."[6] The Big Bang and inflationary universe theories are presented, in free verse.[7] The course of future progress in cryonics, genetics, psychiatry, and weather control is indicated. Advances in technology such as a milk-making machine (heralded as a part of recycling waste and not as a way to eliminate hunger), computer-generated art, and the first Mars landing are cited. We are given snippets foretelling the future of media technology from holovision to the responsive environment ("respen") to a full-spectrum sensorium ("V-A-F-S"), a scenario that anticipates the computer-based concept of virtual reality. Lists of stars, galaxies, and great astronomers round out Gunn's sketch of knowledge acquisition and engineering prowess. The novel's Listening Project is presented as a significant development in this hymn of science and invention.

In mid-twentieth-century America, a common concern was that social and political progress did not keep pace with advances in science and technology. Gunn's computer runs offer a future in which society and politics keep pace with technoscience. Wealth is more equitably distributed, Arabs and Israelis work together, the prison–industrial complex fades into irrelevance, the 4-day week and 13-week vacations become standard, and an increase of leisure means that the amateur is reconstituted as a respected and productive social position. Gunn anchors this speculation in a bulletin on the growth of national and global wealth:

> The [United States Bureau of Economics] attributes the solution of many of the problems that troubled the world a half century ago to the dramatic growth in the Gross World Product through automation, fusion power, greater use of computers and cybernation, and new educational methods.[8]

Here the link between social and technological progress is completed by economic development. Technoscientific progress sponsors wealth creation, which in turn empowers and sustains the resolution of social and political problems. Given current conditions, this scenario may be optimistic, but is powerful nonetheless because of both the utopian desire it encodes and its resonance with familiar hopes in a culture resistant to socialist alternatives. It is the accumulation of wealth under an efficient, global capitalism guided by a social democratic liberalism that empowers and sustains Gunn's vision of social and technological progress.[9]

6. James Gunn, *The Listeners* (Dallas, TX: BenBella Books, 2004 [1972]), pp. 26–27.

7. Ibid., pp. 173–174.

8. Ibid., p. 141.

9. David Potter's landmark study, *People of Plenty*, states this position in a more formal context. Potter argues that a "culture of abundance" is the thing that explains American character and gives the nation's people a unique place in history. The United States will survive if it can avoid the pitfalls of scarcity. David M. Potter, *People of Plenty: Economic Abundance and the American Character* (Chicago: University of Chicago Press, 1958).

The bottom line here is that the world is reformed, socially and politically. However, it is important to note that Gunn's future describes liberal reform within political and economic institutions that do not change. In the near term, the world does not become unrecognizable from a mid-twentieth-century capitalist perspective, as it does in Kim Stanley Robinson's post-Marxist *Mars Trilogy*.[10] The hope is that the United States of the twentieth century will inherit any extraterrestrial knowledge, surviving its own problems and serving as a model for the rest of the world.

In Gunn's interchapters the mixture of history with futurist speculation deploys a historiographic vocabulary familiar in science fiction.[11] By the mid-1960s the gesture is arguably a cliché. Gunn's desire to teach, however, leads him to directly include the results of his research into the early history of SETI. The computer runs are also an archive of the scientific commentary that inspires the narrative. It is here that we are introduced to first generation of C/SETI thought. Extracts from the seminal paper by Giuseppe Cocconi and Philip Morrison, Cocconi's letter to Bernard Lovell, estimates of the likelihood of living planets and intelligent species by Sebastian Von Hoerner, and Carl Sagan's, Frank Drake's, and Ronald Bracewell's early statements on the kinds of searches possible and the time scales involved are imbricated with fictive explication and history. Gunn combines scientific propositions with belletristic rumination from speculative essays by Eiseley, Shapley, and Dyson. Here we have scientists functioning as writers—intellectuals who can speak authoritatively about why thinking about sentient extraterrestrials is important to our sense of self and destiny. Their words are carefully dated, allowing us to locate utterances from the late 1950s through the 1960s. Gunn's strategy here is to support and authorize his future history with direct reference to the thought of those scientists responsible for both creating SETI and promoting the science in print for nonspecialist audiences.

What Gunn has made is a model for integrating fiction and science writing. Coupled with the emotional resonances produced in the novel's fictional chapters, science (particularly new science) is made compelling or, at least, interesting for general audiences. The goal here is to gain cultural credibility for SETI, building a case that makes it look like more than science fiction.[12] In the computer runs, it is Gunn's references to literature that supply the cultural capital necessary for

10. Kim Stanley Robinson, *Red Mars* (New York: Spectra/Bantam Books, 1993), *Green Mars* (New York: Spectra/Bantam Books, 1994), *Blue Mars* (New York: Spectra/Bantam Books, 1997).

11. Isaac Asimov's Foundation Trilogy (1942–1950), A. E. van Vogt's *The Voyage of the Space Beagle* (1950), and Robert Heinlein's future history stories (1939–1941, 1945–1950) are paradigmatic in this regard.

12. For science, especially science that wrestles with pre-existing cultural tropes such as the idea of extraterrestrial life, science fiction is a problem as well as an invaluable resource. Sensitivity about the standing of science in relation to science fiction is indicated by Giuseppe Cocconi's reluctant admission to Bernard Lovell that the notion of life and intelligence on other planets might, "at first sight [look] like science fiction." He goes on to argue that it is not (Gunn, p. 28).

that credibility. British and American authors Walt Whitman, Alice Meynell, Gerard Manley Hopkins, and Edward Lear link the reader to the thread of fanciful or religious speculation that has always been a part of Western (English-language) literature. Within this tradition Gunn also takes care to include H. G. Wells, Murray Leinster, A. E. van Vogt, and Theodore Sturgeon—authors who have helped formulate the modern conception of extraterrestrial sentients and have had powerful influence on how to present narratives of human–alien contact.

For analytical purposes I have split apart these literary (and nonfictional) threads. Gunn's strategy, however, is to weave them together. He juxtaposes religious texts (the Judeo-Christian Bible, the Hindu Rig Veda) with modern literature with history of science with science writing, speculative essays, futurist scenarios, and government reports. Differences and historical context are indicated by changes in typography and dated attribution. The effect is a cultural babble that mirrors the cacophony of radio voices depicted in his fiction. In this orchestration no particular utterance is given prestige or priority above another; each saying builds on and offers support to the next. As Gunn's science heroes must decode the Message sent by the Capellans, so the reader is to decipher the cultural puzzle of the interchapters.

The book's final chapter, "The Computer," integrates fiction, nonfiction, and literary and science history. Having taught us how to read C/SETI as science and culture in the previous chapters and computer runs, Gunn hopes that we are ready to interpret the Message of the novel and to integrate it into our own sense of how the world does and will work. The effect of his stylistic formula is to indicate that there is more at stake here than either disinterested curiosity or personal salvation. Nothing less than the future of the human race is at issue. However melodramatic this might sound, Gunn's effort provides insight into the mid-twentieth-century anxieties and hopes that informed the creation and reception of SETI science.

ON CETI AND RACE

During the composition of *The Listeners* in the late 1960s and early 1970s, a great deal of criticism was leveled at the U.S. space program, for example, cultural critic Lewis Mumford's contemporaneous judgment that spaceflight dreams are "infantile fantasies."[13] But as I have noted elsewhere, "[T]he political voice that served as the iconic focus of liberal/left dissent from the space vision and its economics came from the nation's black community."[14] The Southern Christian Leadership Conference, for example, organized a protest march during the Apollo 11 launch (on 16 July 1969) that questioned the expense and the societal importance of the

13. Stewart Brand, ed., *Space Colonies* (San Francisco: Waller Press/Penguin, 1977), p. 34.

14. De Witt Douglas Kilgore, *Astrofuturism* (Philadelphia: University of Pennsylvania Press, 2004), p. 198.

venture.[15] Astrofuturists, such as science fiction writer Ben Bova, felt it necessary to directly confront African American criticism, offering the exploration of space as a solution to the racial conflicts of the time.[16] Gerard O'Neill notes in his 1978 book, *The High Frontier*, that the ability of science to address contemporary social problems was also in question on university campuses.[17] Working as a public relations officer at the University of Kansas at that time, James Gunn was well aware of the problem. Indeed, in his preface to the 2004 edition of the novel he tells us that it was his job "to explain student unrest to the various University publics."[18]

It is in *The Listeners* that Gunn makes his most compelling contribution to that social and political debate. In the chapter "Andrew White," the author offers a description of how the discovery of an extraterrestrial civilization might affect the United States. However, how race is seen (as a social and political problem) must be resolved before the future that follows unfolds. In this novel the foundation of political hope is SETI's potential for changing or eliminating race's historical standing as an ineradicable mark of human difference.

Gunn draws our attention to race by filtering the reader's view of C/SETI through the thoughts of Andrew White, the African American president of Gunn's near-future United States. This is a critical and perhaps bold move inspired by the racial tensions of the 1960s and early 1970s. Here Gunn conducts a literary thought experiment that highlights problems while setting the stage for the solution. The hope that social and political advance will move in company with progress in science is fundamental here. Political progress is signaled by the deliberations of an African American occupant of the Oval Office; scientific progress is represented by the Listening Project's success in discovering an extraterrestrial message. Gunn creates this character not to examine political racism in high office (as Irving Wallace did in his near-contemporary novel, *The Man* [1964]), but to raise at least three pertinent questions: What would it mean to speak for humanity to an extraterrestrial civilization? Who could authorize such a communication and to what end? And, is civilized communication truly possible between different races, cultures, civilizations

15. Courtney G. Brooks, James M. Grimwood, and Loyd S. Swenson, *Chariots for Apollo: A History of Manned Lunar Spacecraft*, NASA History Series, 1979, *http://history.nasa.gov/SP-4205/ch14-2.html#source2* (accessed 14 August 2006).

16. Ben Bova, *The High Road* (New York: Pocket Books, 1983).

17. Kilgore, pp. 157–59.

18. Gunn, p. xvii.

or generations? [19] What binds these issues together in *The Listeners* is the kind of social future we hope for and how the past informs it.

Gunn's President White must decide whether the novel's CETI project may share news that a message has arrived from a planet orbiting the double star Capella and send a reply. This issue is complicated by an argument with his son over the meaning and relevance of race now and in the future. What is at stake is whether the president can accept the Capellan Message as a furtherance of the political evolution that he represents.

As he sets up White's view of C/SETI, Gunn is careful to establish the president's authority as holder of the nation's highest office. We first see White in the Oval Office and our attention is drawn to its size:

> The office was big . . . . Across the broad blue, deep-piled carpet with the woven seal in the middle to the carved white door by which visitors entered was a good twenty yards—that was bigger than the entire flat in which he had been born . . . .[20]

This scene setting makes the Oval Office Andrew White's natural environment, a move underscored by the reference to humble beginnings. White's log cabin story has him escaping from urban poverty, but otherwise cleaves to this familiar aspect of American political hope: that the small, the weak, and the disenfranchised can achieve greatness in this country. And, as befits a storybook president who is entrusted with national destiny, Gunn makes his chief executive larger than life:

> He stretched his shoulders and felt the long, flat muscles rippling under the layer of fat too many chicken dinners had put there, and he knew he was a big man, an imposing president, six-feet-six from the soles of his feet to the top of his close-cropped hair, and the physical equal of any of them, of anyone who came through that door.[21]

Gunn's goal here is to write a president whose authority we do not question. His race is important to the narrative but is not represented as a problem that undermines his effectiveness or longevity in office.

President White's race also signals that he represents a future in which social change is possible. The writer's strategy of linking his decision-making to an

---

19. Andrew White, Gunn's African American president, is not a unique fictional invention; several other popular writers responded to the "race problem" of the 1960s by mounting fictional speculations on a black presidency. Irving Wallace's early 1960s best seller, *The Man*, predates Gunn's story by several years. A film based on book, starring James Earl Jones as the first African American president, was released in 1972. Gunn's short story, "Andrew White," was published that year in the January-February issue of *Galaxy Magazine*. Gunn has stated, however, that he has not read the Wallace book. He is, however, familiar with the movie (private conversation with author, 7 July 2007). The connection, therefore, has more to do with contemporary social thinking rather than a specific influence.

20. Gunn, p. 101.

21. Ibid.

historical and iconic president, Theodore Roosevelt, underscores this. Gunn has White remember Roosevelt's reputation for making up "his mind somewhere in the region of his hips."[22] Whatever we remember about Roosevelt's racial beliefs connecting White to this popular historical figure, and by association to the relatives who followed him as president and first lady, indicates how we are to view White as an author of progressive social reform.[23] In this narrative he is authorized as a man, as a father, as president, and as the representative of an historically aggrieved population who can open the way for a radical change in human relations.

The dynamic interchange Gunn stages between SETI and the personal and political investments of his cast allow him to demonstrate how science might solve race as a social problem. For Andrew White, who was "wound up in the ghetto" and who used the White House to "right the ancient wrongs," race and racism are fundamental aspects of human character. As a career politician, he understands "that there were differences between men and these differences inevitably affected how they felt about each other, about themselves." These differences, coupled with the human capacity for, as the president argues elsewhere, "malice and greed . . . the persistent search for advantage" make him pessimistic about the state of the future he is helping to create.[24]

On the other hand, the president's son, John White, is convinced that the past his father remembers "is irrelevant" to the world in which he lives. If his father argues that the battle for civil rights is not finished, John believes that it is. Impatient with his father's account of past troubles, John argues that "[B]eing human is being able to do something different." It is at this point that Gunn has his fictional president reflect, "*The boy was an alien, but somehow he had to communicate.*"[25] Here Gunn emphasizes the metaphorical link between speculation about extraterrestrials and the "aliens" we live with here on Earth. That connection illuminates SETI's potential for changing how we see contemporary social problems. In *The Listeners*, resolving this generational conflict over race is a matter for science, not politics.

In this novel there is no question that discovery of an extraterrestrial civilization will have tremendous impact on our world. Considering the prospect, President White comes to the conclusion that it is "bad news." He tells Robert MacDonald, the humanist and engineer who directs the Listening Project, that an alien message could sow "the dragon's teeth of dissension and strife which may well destroy this country." His concern is that public knowledge of the Capellan Message would disrupt the domestic tranquility necessary for reform: "We have solved many of the problems that threatened to tear this nation apart at the time the Project was started and trouble will keep us from solving the others. We need, calm, serenity . . . ," not

22. Ibid., p. 104.

23. For more on Theodore Roosevelt's complex understanding of race, see Thomas G. Dyer, *Theodore Roosevelt and the Idea of Race* (Baton Rouge, LA: Louisiana State University Press, 1980).

24. Gunn, pp. 110, 114, 126.

25. Ibid., pp. 105, 110.

change that is "unmanageable."[26] Director MacDonald's answer to these concerns is that the act of communication, particularly scientific and cultural communication, is our guarantee that social progress would be fostered and not hindered should SETI achieve its aims. Gunn has his scientific humanist visionary list at least three arguments in favor of this position: (1) if the president allows scientists to release the Message and their interpretation of it, then public understanding will follow; (2) communication is a civilized act that precludes violent intent by sender or receiver; and (3) once the public learns the facts, it will realize that the benefits of communication outweigh the risks.[27] These arguments outline a science-based hope. It is a brief that embraces the changes that would follow the discovery of an advanced extraterrestrial civilization.

The Listeners is a book written in praise of science as an activity that fosters communication between people who differ by reason of border or birth. SETI is presented as an inquiry that ennobles the human species and guarantees its future. This is a minority opinion in a popular culture in which thrilling horror-from-the-stars stories function as an entertaining commonsense. *Species* (1995), for example, is a film that directly mocks SETI (with what purports to be footage of astronomers sending the 1974 Arecibo message) as the folly of naïve scientists. The message received by the film's protagonists produces a dangerous white-black-Asian hybrid, a monstrous miscegenation that preys on "normal" humans. Communication with an extraterrestrial civilization is dangerous and sets up an interspecies conflict (rendered in visual terms as traditionally racial) that could lead to the extinction of terrestrial intelligence. This is "the boogeyman from the stars" cliché at which Gunn's project director casts an indulgent smile.[28] If popular narratives insist that communication between different races is impossible, then narratives inspired by science (SETI science, at least) argues that communication is possible and necessary for our survival. In American cultural discourse this position allows SETI to stand out as a science that seeks new discoveries; it does so in part because it runs counter to popular narratives around race and the fears they express.

This takes us to the hope that Gunn engages through the thematics of race and communication: the progress of human civilization and its long-term survival. If human beings can talk to one another, we may also engage in conversations with beings from other worlds; if human civilization can mature, it could join what Ronald Bracewell calls the "galactic club."[29] If we accept these propositions, then we find ourselves considering a "universal" future—an era in which we create an ethical reality that transcends what Edward Said calls "the easy certainties provided us

26. Ibid., pp. 115, 107, 122, 132.

27. Ibid., pp. 124, 125.

28. Ibid., p. 125.

29. Ronald N. Bracewell, *The Galactic Club: Intelligent Life in Outer Space* (Stanford, CA: Stanford Alumni Association, 1974).

by our background, language, nationality, which so often shield us from the reality of others."[30] Thus, the narrative resolution of conflict between races and generations in *The Listeners* provides a social and political justification for SETI: that through this science we may discover the material grounds for a universal language, an episteme that does not assume incommunicability of experience across differences of culture and biology.

Gunn underscores this goal by having his president perceive a truth and give up his son. The truth comes to him graphically in the reply that MacDonald would like to send to the Capellans. It is a simple pictogram of dots and spaces, a message that contains some basic biological and astronomical information. President White laughs when he sees it:

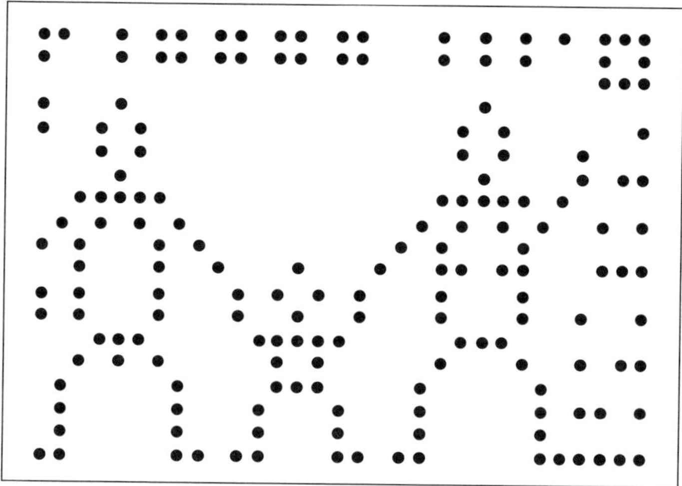

Figure 28.1—Listening Project Message (*James Gunn*, The Listeners [*1972*]).

> "I'm sorry . . . I wasn't laughing at the answer. I don't begin to understand half of what's here. But that's obviously a father and a mother and a son—a child—and the Capellans would have no way of knowing whether they were white or black."[31]

If the president's fear is that any dialogue with the Capellans will only serve to exacerbate our parochialism, then MacDonald's answer seems to set that aside. At this level of communication, human differences are trumped by the reproductive biology that sustains us. The biological insight that humanity is a single species shapes White's decision to let the project go forward. From this point, the history foretold by *The Listeners* takes a hopeful turn.

30. Edward W. Said, *Representations of the Intellectual: The 1993 Reith Lectures* (New York: Vintage, 1996), p. xiv.

31. Gunn, p. 135.

Because Andrew White is an African American president, his decision is doubly authoritative in answering 1960s questions about science's relevance to social problems. What is at stake here is not only the acceptance of SETI by a politically active and critical African American community, but also whether the project is open to participation by its members. Gunn answers this last question by having John White, the president's son, become a contributing member of the Listening Project. After John makes an important breakthrough toward deciphering the Capellan Message, the president allows him to join the project. John eventually succeeds MacDonald as Listening Project director.[32]

In the spare, economical world of Gunn's fiction, the symbolic import of this is obvious. It underscores the message that C/SETI is (or should be) the work of a scientific community that represents all of humanity. It is through this openness that Gunn makes his most direct plea for science as a social as well as an intellectual good. Gunn has MacDonald state, "The fraternity of science is closer than the fraternity of birthplace or of language."[33] By becoming part of this fraternity, John White represents the hope for the future that Gunn sees in the search for extraterrestrial intelligence.

It is possible to dismiss the position that Gunn articulates, in championing SETI as "nothing more . . . than salvation from the stars," following George Basalla's critique in *Civilized Life in the Universe* (2006).[34] What Basalla fears, perhaps, is that SETI, as an ongoing public discourse, amounts to a ceding of scientific and political ground to unreasoning popular culture. The literary experiment that Gunn conducts, however, allows us to mount a more subtle analysis. The novel's hopes for social and political progress notwithstanding, good news from the stars is not a panacea, a *deus ex machina* that resolves all the difficulties we face. Having received the Capellan Message, humanity is given neither final answers nor a perfect social order, but simple confirmation of a way forward. History continues, in a different register, with plenty for the human race to learn and accomplish:

> The Project set up nearly one hundred fifty years before to listen for messages from the stars and the Message from Capella that it had received and deciphered had given Earth and its people ninety years of peace in which to explore the other aspects of humanity besides aggression. The problems that had seemed so difficult, virtually unsolvable, one hundred and fifty, even ninety, years before had seemed to solve themselves once the world relaxed.[35]

32. Ibid., pp. 136–138, 155–157.

33. Ibid., p. 130.

34. George Basalla, *Civilized Life in the Universe: Scientists on Intelligence Extraterrestrials* (Oxford, U.K.: Oxford University Press, 2006), p. 121.

35. Gunn, pp. 177–178

The lesson drawn here presents two effects of the Capellan Message: it helps humanity take stock, improving its situation; and the answer to our reply will help spark a change, the "something different" for which John White argues.[36]

The final chapter of the book, "The Computer," clarifies what is new about this emerging difference. Throughout the novel, the Listening Project computer is used by astronomers as an instrument and archive. MacDonald tells White that "[V]irtually the entire written history and literature of mankind—in all his written languages—is stored in there." It is a "research tool" that contains the sum of human cultural achievement in art and science: the universal knowledge necessary for a universal future. The machine represents the human race in the project of decoding any message we might receive. In the novel's closing lines we are told that when a call from a second extraterrestrial civilization arrives, the computer is "at least half Capellan."[37] In other hands, this would be prelude to xenophobic machine-breaking.[38] In the context of Gunn's narrative, however, it represents change for the better. Gunn has MacDonald's grandson, William, make the point:

> We have received a legacy more valuable than the physical possession of another world, with all its natural treasures, and the world's scientists and scholars and everyone else who wishes to explore it may spend their lifetimes studying it, interpreting it, and adding bits and pieces of it to our civilization, enriching us by a whole new world and everything it has.[39]

We are presented with a new dispensation, one in which the impact of a wiser, more advanced extraterrestrial culture produces radical change here on Earth.

Thus, *The Listeners* is a novel that fuels our expectation that, given the right pressure, the human species can change for the better. Gunn's universal computer heralds the postbiological universe that Steven Dick has theorized elsewhere.[40] Hope resides in the cultural and political evolution fostered by our technology. However,

36. Ibid., pp. 110.

37. Ibid., pp. 120, 192.

38. Indeed, Hoyle teases us with this possibility in an early novel, using SETI as a jumping-off point for a narrative of uncertainty about extraterrestrial motives. Fred Hoyle and John Elliott, *A is for Andromeda* (New York: Harper, 1962).

39. Gunn, pp. 191–192.

40. Steven J. Dick, "Cultural Evolution, the Postbiological Universe and SETI," *International Journal of Astrobiology* 2/1 (2003): pp. 65–74. This is a common idea among intellectuals who have considered the possibility of extraterrestrial intelligence. Williams Sims Bainbridge notes that in the mid-1960s Roger MacGowan and Frederick I. Ordway "argued that civilizations quickly evolve out of the biological stage and become entirely mechanistic, artificial intelligence replacing the natural intelligence that originally created the machines." William Sims Bainbridge, "Extraterrestrial Intelligence: Communication," *The Encyclopedia of Language and Linguistics*, R. E. Asher, ed. (Oxford, U.K.: Pergamon, 1994), p. 1201.

discussion of artificial intelligence, Bracewell probes, and the like is beyond the brief of this narrative. More important is Gunn's presentation of a technoscientific trajectory in which the politics of race are rendered irrelevant, leading to a species-wide transformation that fulfills the promise of scientific humanist values.

The amelioration of race is prologue to the reconfiguration of women's roles as the new (post)human reality emerges. Heretofore, women are supernumerary to the creation of the future. Their principal function is to secure the succession of great men and their ideas. However, prior to the arrival of the Capellans' final message, William MacDonald tells a waiting world that while he has "no son . . . I do have a daughter who is now a member of the Project staff."[41] Thus, the father-son succession that sustains the project directorship in the novel's history is broken. The future created by the novel's C/SETI project lays the groundwork for both racial and gender equality. On the eve of the women's movement, Gunn imagines that the institutions of science represented by SETI are the vanguard of progressive social and political change.

Whatever we make of these positions, it is important to note the power of the technoscientific speculation that authorizes them. Here scientific and technological invention are the vanguard of social and political change. This sentiment is driven by the hope that progressive, evolutionary trends in science and technology, history, and society will allow our civilization to transcend human frailty, poor judgment, and bad luck.

Gunn's contribution is to connect formal invention and his reading of race with the scientific and social debates current during SETI's first decade. Thus, *The Listeners* has the complexity and depth it needs to serve as a dress rehearsal for how America might react to the news that "we are not alone." With its interchapters containing quotations from SETI's pioneering generation, *The Listeners* is also a model for how the traffic between science, society, and literature moves at particular times in our history. This presentation of SETI's social mission makes the science credible within the social investments important to a general reading public. It also, as SETI League executive director Paul Shuch says in his introduction to the 2004 edition, "inspired a generation of SETI scientists to pursue the seemingly impossible."[42] It is the desire for a future occupied by intellects other than our own that gives C/SETI narratives the intellectual, formal cohesion necessary to argue for the social relevance of its authorizing science. As presented here, this branch of astronomy is neither apolitical nor disinterested in the nature and course of human life. Its practitioners hope that by searching for and gaining more knowledge about the standing of life and intelligence in the cosmos, old-fashioned values such as tolerance, equity, peace, and freedom will be the anchoring virtues of any global destiny. It is a code of moral and intellectual conduct as well as a disciplined way of fixing our celestial knowledge.[43]

---

41. Gunn, p. 183.

42. Ibid., p. xi.

43. For more on science as a moral force against social oppression, see Neil de Grasse Tyson, *The Sky Is Not the Limit: Adventures of an Urban Astrophysicist* (New York: Doubleday, 2000).

At the end of the novel, humanity stands in relation to an extraterrestrial civilization in the way that the Renaissance was influenced by classical antiquity: intellectual and cultural beneficiaries of a long-dead civilization. What is achieved is not utopia but, perhaps, a better world. *The Listeners* certainly argues for this possibility. It has produced or reinforced a pattern of expectation that would lead some to embrace rather than fear communication with another intelligent species.

CHAPTER 29

Reclaiming the Future: Space Advocacy and the Idea of Progress

Taylor E. Dark III

The idea of progress is clearly central to American national identity, yet the popularity and credibility of the idea have undergone significant fluctuations over the course of American history. This essay assesses how the onset of space travel stimulated an attempted revitalization of ideas about progress that, by the late 1960s, were coming under increased attack. The idea of progress has typically advanced three claims: (1) there are no fundamental limits on the human capacity to grow, however growth is defined; (2) advancements in science and technology foster improvements in the moral and political character of humanity; and (3) there is an innate directionality in human society, rooted in societal, psychological, or biological mechanisms, that drives civilization toward advancement. American believers in progress quickly embraced space travel, viewing it as a vindication of the doctrine's original claims about the near-inevitability of human improvement. With space travel understood in this fashion, the fate of the space program took on a far greater meaning than developments in other areas of technological endeavor, as it became symbolic of the entire directionality of human civilization. The early and astonishing success of Apollo, followed almost immediately by signs of disarray, served to stimulate a new vision of progress and then quickly threaten it. In this context, a space advocacy literature arose that was simultaneously grandiose about the human future yet intensely fearful about missed opportunities. This confluence of ambition and anxiety continues to characterize both the pro-space movement and the larger debate about the American future in space. As the historian Charles Beard noted: "Into the mood of the American people . . . the idea of progress fit with extraordinary precision . . . . It remains, and will remain, a fundamental tenet of American society, and while vigor is left in the race it will operate with all the force of a dynamic idea rooted in purpose, will, and opportunity."[1]

1. Charles Beard, "Introduction," in J. B. Bury, *The Idea of Progress: An Inquiry Into Its Origins and Growth* (New York: Dover Publications, 1932), pp. xxxvi–xxxvii.

Americans have been more deeply wedded to the idea of progress than perhaps any people on the face of Earth. The key claim of the idea of progress—that human civilization has moved and will continue to move in a desirable direction—has been central to American culture and identity for virtually all of the nation's history.[2] Indeed, many would argue that a profound faith in progress has been one of the key features distinguishing Americans from people elsewhere, and a recurring source of the country's distinctive appeal across the globe. What happens, then, when the idea of progress starts to lose credibility? How do Americans react when they become fearful that the direction of society has become negative rather than positive? The period of the late 1960s and early 1970s provides an example of a time when the American faith in progress started to unravel. Although belief in progress has arguably recovered partially, most observers would still view the late 1960s as a turning point after which the idea came under siege in ways that had not been experienced previously.

The argument of this essay is that the rise of new forms and doctrines of space advocacy reflects exactly this crisis in the idea of progress. If the forward march of humanity (with America noticeably in the lead) had been halted, something had to be done. A movement into space was proposed as the solution. Thus was born the modern pro-space movement, and the contemporary fusion of the idea of progress with ideas about space travel, space development, and, most of all, space colonization.

Over the last 40 years, space advocates have constructed a set of doctrines that address all the key components found in the idea of progress since it first took modern form during the Enlightenment. This new pro-space ideology was a reaction to the problems that had become apparent by the time of the first Moon landing, namely, environmental crises, limits to economic growth, and fears of cultural decay. Space advocates proposed solutions to these problems and others. They concluded that an expanded space program was the essential condition to revive both the idea and reality of progress. The irony was that *they embraced this belief at the very moment that the Apollo program was coming to a close*, and the future of NASA and space travel becoming increasingly uncertain. Thus, a strong edge of anxiety and urgency was introduced into the writings of space advocates. The means to ensure progress had been found but would soon be lost forever if government policy was not properly adjusted. This combination of *certainty* about the path toward redemption alongside *anxiety* about the possibility of missing a singular opportunity energized the new pro-space literature and encouraged the growth of an accompanying space advocacy movement.

2. The definition here is derived from Bury, *Idea of Progress*, p. 2. For the importance of the idea of progress in America, see Clarke A. Chambers, "The Belief in Progress in Twentieth-Century America," *Journal of the History of Ideas* 19, no. 2 (1958); Hugh De Santis, *Beyond Progress: An Interpretive Odyssey to the Future* (Chicago: University of Chicago Press, 1992); Samuel Huntington, *American Politics: The Promise of Disharmony* (Cambridge, MA: Harvard University Press, 1981), pp. 259–262; Christopher Lasch, *The True and Only Heaven: Progress and Its Critics* (New York: W.W. Norton & Co., 1981); Seymour Martin Lipset, *American Exceptionalism: A Double-Edged Sword* (New York: W.W. Norton & Co., 1996), p. 37; and Robert Nisbet, *History of the Idea of Progress* (New York: Basic Books, 1980).

The Idea of Progress

Despite the many controversies over whether progress is actually taking place, there is little dispute about how to define the idea itself. J. B. Bury's classic formulation of 1920, with its succinct assertion that the idea "means that civilization has moved, is moving, and will move in a desirable direction," remains as useful as any proposed since.[3] But theories of progress have always tried to do far more than simply attach a normative gloss to the passage of time. By necessity, they have embraced a set of larger claims, captured in the following mutually reinforcing and interlocking premises:

1. *No Limits.* There are no fundamental limits—nor should there be—on the collective human capacity to grow, no matter how growth is defined (which may be in terms of knowledge, wealth, power, population, or morality). Progress is endless (or at least indefinite for all practical purposes).
2. *All Good Things Go Together.* Advancements in science and technology, and the resulting mastery over nature, expand our knowledge, wealth, and power. In so doing, they bring improvements in the moral, political, and spiritual character of the human race. The elements of progress are linked to one another and are mutually reinforcing.
3. *Innate Directionality.* There exist developmental tendencies, rooted in societal, psychological, or biological mechanisms, that make it far more likely that human civilization will move "upward" toward greater control and understanding of nature and ourselves, rather than "downward" toward chaos and entropy. Progress is, if not inevitable, always highly probable.

Whether one was a liberal who embraced science, markets, and technology; a Marxist who saw class conflict at work in capitalist society; or an evolutionist who focused on the winnowing effects of natural selection, these three premises were always addressed, either directly or implicitly, in the great nineteenth-century theories of progress.

Unfortunately for defenders of progress, all three of these premises became less convincing during the course of the twentieth century.[4] Major limits on economic, demographic, and even intellectual growth were identified, and their importance was

3. Bury, *Idea of Progress*, p. 2. For contemporary variations, see Nisbet, *History of the Idea of Progress*, p. 4, and Charles Van Doren, *The Idea of Progress* (New York: Praeger, 1967), p. 7.

4. For discussions of the declining faith in progress, especially among intellectuals, see Gabriel Almond, Marvin Chodorow, and Roy Harvey Pearce, eds., *Progress and Its Discontents* (Berkeley, CA: University of California, 1982); De Santis, *Beyond Progress*; John Horgan, *The End of Science: Facing the Limits of Knowledge in the Twilight of the Scientific Age* (New York: Broadway Books, 1996); Lasch, *True and Only Heaven*; Leo Marx and Bruce Mazlish, eds., *Progress: Fact or Illusion?* (Ann Arbor, MI: University of Michigan Press, 1996); and Nisbet, *History of the Idea of Progress*.

widely proclaimed in both elite and popular culture. A belief that there are innate human tendencies toward progress was undermined by the evidence that humans were inherently violent or even self-destructive due to deep-rooted psychological or biological drives. Science and technology, it was suggested, have advanced far out of proportion to the capacity of humans to control them rationally, making total self-destruction as likely as benign development. Advancements in humanity's control over nature were thus no longer necessarily viewed as automatic *improvements*, and could even be seen as detrimental, contributing to the further estrangement of humanity from its natural environment. For many, gains in science, economics, and technology seemed to have corrupted humanity, producing a culture of widespread pornography, greed, and violence, and a political sphere dominated by the trivialized discourse of mass marketing. Summarizing a vast literature, physicist and historian Gerald Colton concluded in 1980: "Future historians will probably record that from the mid-twentieth century on, it was difficult for anyone to retain faith in the idea of inevitable and continuing progress. People increasingly use the word in quotation marks or with mocking sarcasm or speak not of progress in civilization but in barbarism."[5]

Enter the Space Advocates

It is in this context that the space advocates enter the debate, with their own agenda to revitalize the idea of progress and, in so doing, to buttress a central component of American national identity. Although space advocacy ideologies and organizations have existed in various forms since the first decades of the twentieth century, the 1970s saw the formation of a pro-space movement that was determined to build upon the successes of Apollo by creating the setting for an imminent migration into space on a massive scale. The most notable of these new advocates was Princeton physicist Gerard K. O'Neill, who foresaw the creation of an expanding human civilization in space by the early 1990s. At the core of his vision was the establishment of gigantic rotating colonies, shaped liked a sphere or cylinder, in which tens of thousands of inhabitants living on the interior surface would experience the sensation of gravity produced by centrifugal force. With sunlight beamed in by massive mirrors, these colonies could provide park-like conditions for their inhabitants and a lifestyle that resembled that of a comfortable American suburb. The economic rationale for this massive endeavor would come from the construction of enormous grids for collecting solar energy, which would then be converted into microwaves and beamed down to receptors on Earth. In this manner, O'Neill anticipated that the movement into space could provide clean, limitless, and inexpensive power for a world that was facing worsening shortages of energy and raw materials. The Space Shuttle, which promised to soon provide low-cost access to space, would be one of several vehicles that would

5. Joel Colton, "Foreword," in *Progress and Its Discontents*, ed. Almond, Chodorow, and Pearce, p. xi.

ferry construction crews to space, eventually arriving at Lagrange point 5 (L-5), the location where the balancing of the gravity of Earth and the Moon would allow a colony to remain stationary.

This bold vision of a profitable and practical future in space prompted the formation of an interest group called the L-5 Society, which was set up in 1975 in order to promote the creation of O'Neill-style colonies as soon as possible. The group embraced the slogan "L-5 in 1995" and proclaimed that its goal was "to disband the Society in a mass meeting at L-5." Within a few years, the organization had attracted considerable attention, and the O'Neill vision of a new human future in space was the object of serious study in Congress and NASA. By the mid-1980s, however, the limited success of the Space Shuttle, as well as the subsidence of energy prices and the avoidance of other predicted catastrophes, encouraged the postponement of the space colonization dream to a rather more distant future. By 1987, a demoralized L-5 Society had merged with the National Space Institute, a more subdued pro-space group established in 1974 with closer ties to NASA, and together formed the National Space Society (NSS), which continues as an active pro-space group to the present day. The NSS has eschewed the vision of L-5 colonies as the only desirable means to populate space, and toned down the messianic rhetoric that made the L-5 Society somewhat notorious, but it remains committed to the goal of "people living and working in thriving communities beyond the Earth."[6]

While NSS and its supporters delayed their dreams, a second group of space advocates emerged in the 1990s with much of the same élan that had characterized the pro-space partisans of the 1970s. For these activists, the plan was to colonize Mars rather than to create artificial spheres in space. As such, their program was simpler: they advocated immediate steps to begin the settlement of the Red Planet, hoping to send the first humans in the early decades of the twenty-first century. Supporters of Mars colonization believe that the long-term project of developing a human civilization on Mars will promote social cohesion and economic growth on Earth, and create a new branch of civilization that will rival in its accomplishments anything that mankind has done previously. The most enthusiastic proponent of this plan has been the engineer and author Robert Zubrin, who wrote the 1996 book *The Case for Mars* and was instrumental in the founding of the Mars Society, an organization dedicated to making Mars settlement the main goal of America's space policy.

A third contemporary pro-space group is the Planetary Society, founded by scientists Carl Sagan, Bruce Murray, and Louis Friedman in 1980 to encourage "the exploration of the solar system and the search for extraterrestrial life." With more than 100,000 members from numerous countries, it is the largest single pro-space interest group, but has largely promoted exploration as a value in its own right, showing little

6. National Space Society, "Statement of Philosophy," available online at *http://www.nss.org/about/philosophy.html* (accessed 7 October 2006).

interest in the grandiose vision of space industrialization and colonization endorsed by NSS and the Mars Society. Nonetheless, the society has strongly supported the value of a human mission to Mars and has worked closely with other space groups to support President George W. Bush's Vision for Space Exploration initiative. And Sagan himself became a prominent advocate of the larger goal of space colonization in his writings of the 1990s, providing influential and eloquent support for a continued human role in spaceflight and the long-term goal of space migration.

SPACE DEVELOPMENT AND THE IDEA OF PROGRESS

Although the various pro-space intellectuals and space advocacy groups often endorse different practical programs and strategies, they use similar arguments in justifying their vision of the human future in space. As they have attempted to persuade a skeptical public of the desirability of their (rather expensive) programs, they have come to address all three of the classical premises of the idea of progress.

No Limits

The modern view of progress is resolutely opposed to the idea of limits, including limits in space and time on the growth of the human species. It is perfectly logical, therefore, that many authors see the rise in recent decades of the idea of "limits to growth" as signaling the decisive end of the idea (and reality) of progress in our time. "The belated discovery that the Earth's ecology will no longer sustain an indefinite expansion of productive forces deals the final blow to the belief in progress," Christopher Lasch confidently asserts.[7]

Space advocates will have none of this. Responding to the first appearance of the "limits to growth" idea, pro-space intellectuals of the 1970s were eager to assert that space could be the source of limitless new reserves of energy and natural resources. In promoting his scheme of magnificent L-5 colonies, Gerard K. O'Neill wrote:

> The human race stands now on the threshold of a new frontier, whose richness surpasses a thousand-fold that of the new western world of five hundred years ago. That frontier can be exploited for all of humanity, and its ultimate extent is a land area many thousands of times that of the entire Earth. As little as ten years ago we lacked the technical capability to exploit that frontier. Now we have that capability, and if we have the willpower to use it we cannot only benefit all humankind, but also spare our threatened planet and permit its recovery from the ravages of the industrial revolution.[8]

7. Lasch, *True and Only Heaven*, p. 529.

8. Gerard K. O'Neill, *The High Frontier: Human Colonies in Space* (New York: Bantam Books, 1977), p. 8.

Based on his calculations (which assumed a robust and cost-efficient Space Shuttle fleet), O'Neill thought it quite possible for a space colony to be "in place, with its productive capacity benefiting the Earth, before 1990."[9]

More recent space advocates have also been insistent that the development of space resources would overcome all resource constraints. Robert Zubrin, in his 1996 vision of Mars colonization, writes: "We can establish our first output on Mars within a decade, using well-demonstrated techniques of brass-tack engineering backed up by our pioneer forebears' common sense." Once settled on Mars, the colonists would find the opportunities for growth to be immense: "Virtually every element of significant interest to industry is known to exist on the Red Planet." Eventually, the Mars colonists would venture off to the nearby asteroid belt, where they would find vast mineralogical resources just waiting to be tapped. The ultimate outcome would be a "triangle trade" similar to that which existed between Britain, the North American colonies, and the West Indies during the eighteenth century. In the twenty-first-century version, Earth would supply "high-technology manufactured goods to Mars," Mars would supply "low-technology manufactured goods and food staples to the asteroid belt," and the workers in the asteroid belt would send precious metals back to Earth. The new Martian civilization would also be a "hotbed of invention," producing "wave after wave of invention in energy production, automation and robotics, biotechnology, and other areas." As a result, Mars colonization "will dramatically advance the human condition in the twenty-first century."[10]

Space advocates argue that even if such activities eventually deplete the resources of nearby planets and asteroids (an unlikely eventuality for thousands of centuries), humans can simply move on to other solar systems. Carl Sagan suggests that human communities might migrate from planet to planet, exhausting the resources of each before moving on, much as earlier nomadic human communities exhausted local agricultural and animal resources and then migrated elsewhere. "We might call it 'pioneering' or 'homesteading,'" Sagan observes, while acknowledging that "a less sympathetic observer might describe it as sucking dry the resources of little world after little world. But there are a trillion little worlds in Oort Comet Cloud."[11] Zubrin puts it simply: "The universe is vast. Its resources, if we can access them, truly are infinite."[12]

The only danger these theorists see, not surprisingly, is the possibility that mankind will fail to grasp the opportunities before it. Zubrin is emphatic that humanity needs to be constantly on the move, constantly growing, or it will stagnate and die:

9. Ibid., p. 10.

10. Robert Zubrin, *The Case for Mars: The Plan to Settle the Red Planet and Why We Must* (New York: The Free Press, 1996), pp. xix, xi, 236, 225, 301.

11. Carl Sagan, *Pale Blue Dot: A Vision of the Human Future in Space* (New York: Ballantine Books, 1994), p. 321.

12. Zubrin, *Case for Mars*, p. 305.

> The key is not to let the process stop. If it is allowed to stop for any length of time, society will crystallize into a static form that is inimical to progress. That is what defines the present age as one of crisis. Our old frontier is closed. The first signs of stagnation are clearly visible. Yet progress, while slowing, is still extant: Our people still believe in it and our ruling institutions are not yet incompatible with it.[13]

There is hope for the future, Zubrin avers, but only if the people act at this crucial movement to reverse decline by opening a new frontier. The stakes, in his view, could hardly be any higher:

> The creation of a new frontier thus presents itself as America's and humanity's greatest social need. *Nothing is more important*: Apply what palliatives you will, without a frontier to grow in, not only American society, but the entire global civilization based upon values of humanism, science, and progress will ultimately die.[14]

Zubrin specifically endorses the theories of Frederick Jackson Turner, agreeing that what makes America free, egalitarian, and innovative can all be traced to the existence of a frontier. Without a frontier, America will lose those traits. As evidence of decline, Zubrin cites the "increasing fixity of the power structure and bureaucratization of all levels of life; impotence of political institutions to carry off great projects; the proliferation of regulations affecting all aspects of public, private, and commercial life; the spread of irrationalism; the banalization of popular culture; the loss of willingness by individuals to take risks, to fend for themselves or think for themselves; economic stagnation and decline; the deceleration of the rate of technological innovation . . . . " Despite his purported faith in progress, science, technology, and the American Dream, Zubrin believes that the contemporary United States is in deep trouble: "Everywhere you look, the writing is on the wall."[15] But he is not too worried since he knows the real explanation for his long list of maladies: the frontier is gone. Once we get it back, all those measures of societal health will reverse direction and start trending upward again. Zubrin is confident of this because he embraces the assumption that all facets of civilization will either rise or decline together as a function of the presence or absence of a frontier experience.

Even the limits posed by celestial threats to Earth can be overcome, space advocates say, but only if we colonize space quickly. Since the 1970s, space advocates

13. Ibid.

14. Ibid., p. 297, emphasis added.

15. Ibid.

have become increasingly fond of justifying space or Mars colonization on the grounds that it is the only means to ensure the survival of the human race in a universe where an asteroid could slam into Earth at any moment. Space advocates emphasize that the dinosaurs became extinct because they failed to master their environment—without a space program, they were left at the mercy of untoward celestial encounters. Such encounters await Earth again in the future, space advocates note, and suitable preparation is therefore required. As Sagan puts it, we live in a solar system marked by "routine interplanetary violence." He estimates that "[T]he chance is one in a thousand that much of the human population will be killed by an impact in the next century."[16] It was this possibility that prompted Sagan, long a skeptic of human spaceflight, to finally endorse an active program of space colonization in the near future. He writes: "The asteroid hazard forces our hand. Eventually, we must establish a formidable human presence throughout the inner solar system. On an issue of this importance I do not think we will be content with purely robotic means of mitigation."[17]

Sagan notes that as humanity spreads to other planets, it will then possess a form of "planetary insurance" against the possibility that an asteroid deflection system might break down or that other catastrophes on Earth might destroy humanity.[18] When life is spread widely, he writes, it can never be killed. Thus, Earthlife will become immortal or at least will find the means to survive as long as the universe itself. In this sense, space advocates see their own agenda as ultimately more life-affirming than any conceivable program of political action on the agenda today. As former astronaut John Young puts it: "Knowing what we know now, we are being irresponsible in our failure to make the scientific and technical progress we will need for protecting our newly discovered severely threatened and probably endangered species—us. NASA is not about the 'Adventure of Human Space Exploration,' we are in the deadly serious business of saving the species."[19]

But, alas, time is short. Although all space advocates will acknowledge that the odds of a massive asteroid collision in the next century are extremely slim, they nonetheless argue for taking immediate steps to establish a permanent human presence in space. Their reasoning is that humanity may be going through a "critical period" in which it simultaneously has the capability both to destroy itself and yet

16. Sagan, *Pale Blue Dot*, p. 259.

17. Ibid., p. 264.

18. Ibid, p. 310. For a similar formulation, see William K. Hartmann, Ron Miller, and Pamela Lee, *Out of the Cradle: Exploring the Frontiers Beyond Earth* (New York: Workman Publishing, 1984), pp. 37–42.

19. John Young, "The Big Picture," online at *http://members.aol.com/ramjetuww1/private/Space/Young.html* (accessed 16 September 2006).

also to establish a beachhead in outer space.[20] There may be a narrow window in human history, space advocates argue, in which humanity has the opportunity to move into space. This window has only recently been opened and it may swing shut quite soon, due to the surprise arrival of an asteroid, nuclear war, biological war, or a general societal breakdown. The lesson: Act now, while you can, or forever be sorry. As Sagan writes: "The more of us beyond the Earth, the greater the diversity of the worlds we inhabit, the more varied the planetary engineering, the greater the range of societal standards and values—then the safer the human species will be."[21]

Thirty years after he first walked on the Moon, Neil Armstrong commented upon the Apollo program's significance: "The important achievement of Apollo was a demonstration that humanity is not forever chained to this planet. Our visions go rather further than that and our opportunities are unlimited."[22] In this statement he expressed the abiding faith in an unlimited future that has historically been central to the idea of progress, and which space advocates insist, contrary to a host of critics, remains equally valid today.

All Good Things Go Together

In a review of Enlightenment ideas about progress, political theorist Nannerl Keohane writes: "In its most robust and purest form, the belief in progress affirms that increases in human knowledge, the establishment of human control over nature, and the perfecting of the moral excellences of the species will guarantee one another, with a concomitant increase in human happiness." But it is precisely this faith, she argues, that "good things come in clusters" that is "the Enlightenment addition to the theory of progress that is most problematical today."[23] The space advocates, however, are not worried. In general, they see advances in one area—science and technology—as contributing only to advances, not problems, in other spheres of human life. Space advocates cheerfully maintain the old, untroubled conviction that "good" things and ultimate values do not collide but, rather, *reinforce* each other. The human movement into space, consequently, is not expected to bring any unforeseen problems, but will instead only contribute to massive improvements in other aspects of human existence.

"Space colonization appears to offer the promise of near-limitless opportunities for human expansion, yielding new resources and enhancing human wealth," concluded a 1977 NASA study on the possibility of space settlements. The study asserted,

20. Sagan, *Pale Blue Dot*, p. 312; see also J. Richard Gott III, *Time Travel in Einstein's Universe: The Physical Possibilities of Travel Through Time* (New York: Houghton Mifflin, 2002), p. 231.

21. Sagan, *Pale Blue Dot*, p. 310.

22. Reuters, "Moonwalkers Gather to Remember Apollo 13," *The Daily Yomiuri* [Tokyo], 18 July 1999.

23. Nannerl Keohane, "The Enlightenment Idea of Progress Revisited," in *Progress and Its Discontents*, Almond, Chodorow, and Pearce, eds., pp. 26, 37. Keohane's discussion is based on the more extensive critique developed in Isaiah Berlin, *Four Essays on Liberty* (Oxford, U.K.: Oxford University Press, 1969).

> The opening of new frontiers, as it was done in the past, brings a rise in optimism to society .... It has been argued that it may also enhance the prospects of peace and human well-being. Just as it has been said that affluence brings a reduction in the struggle for survival, many have contended that expansion into space will bring to human life a new spirit of drive and enthusiasm. [The benefits of space colonization will be experienced by all, for] successful exploitation of the extraterrestrial environment is expected to enhance the standard of living not only of the population in space but the population remaining on Earth as well .... [The opening of space is so significant that] this new vista, suddenly open, changes the entire outlook on the future, not only for those who eventually want to live in extraterrestrial communities but also for those who want to remain on Earth.[24]

The NASA authors clearly see this change in outlook as a transformative one—a shift from a negative, cramped view of the future to a vision in which limits are overcome and countless new opportunities are created. The opening of the space frontier, they suggest, will improve human civilization in all respects—not just by bringing economic and technological advancement, but also by enhancing the spiritual, political, and cultural health of all humanity.

Others have argued that the diffusion of human beings off the planetary surface will open up new opportunities for social experimentation—opportunities that were last seen, they suggest, in the original settlement of the New World and the American frontier. "On Earth it is difficult for ... people to form new nations or regions for themselves," science author T. A. Heppenheimer observed. "But in space it will become easy for ethnic or religious groups, and for many others as well, to set up their own colonies .... Those who wish to found experimental communities, to try new social forms and practices, will have the opportunity to strike out into the wilderness and establish their ideals in cities in space." In a burst of multicultural enthusiasm, Heppenheimer even suggests that "[W]e may see the return of the Cherokee or Arapaho nation—not necessarily with a revival of the culture of prairie, horse, and buffalo, but in the founding of self-governing communities which reflect the Arapaho or Cherokee customs ...."[25] Carl Sagan also sees more cultural diversity as humanity establishes new civilizations on different planets and other celestial bodies: "Each society would tend to be proud of the virtues of its world, its planetary engineering, its social conventions, its hereditary predispositions. Necessarily, cultural differences

24. Richard D. Johnson and Charles Holbrow, eds., *Space Settlements: A Design Study* (Washington, DC: NASA SP-413, 1977) pp. 175–176; also available online at *http://www.nas.nasa.gov/About/Education/SpaceSettlement/75SummerStudy* (accessed 16 September 2006).

25. T. A. Heppenheimer, *Colonies in Space* (New York: Warner, 1978), pp. 279–280.

would be cherished and exaggerated. This diversity would serve as a tool of survival."[26] Zubrin likewise claims that Mars colonization will promote cultural diversity in a world where it is increasingly threatened by proximity and overcrowding.

Space migration will also enlarge the pool of positive images of the future available to humanity—images that space advocates consider essential to motivate and guide purposeful activity. Many space advocates complain that optimistic images of the future have been displaced in recent decades by far more negative views. Sagan writes: "Where are dreams that motivate and inspire? Where are the visions of hopeful futures, of technology as a tool for human betterment and not a gun on a hair trigger pointed at our heads?" A rare exception to the spread of gloomy visions, according to Sagan, was the space program of the 1960s: "Apollo conveyed a confidence, energy, and breadth of vision that did capture the imagination of the world .... It inspired an optimism about technology, an enthusiasm for the future .... With Apollo, the United States touched greatness."[27] With a renewed commitment to space, the psychological and cultural health of America and humanity in general would surely improve.

Space advocates also foresee a new era of peace and mutual understanding arising as a result of space travel. Sagan writes that "the unexpected final gift of Apollo" was "the inescapable recognition of the unity and fragility of the Earth." Sagan continues: "I'm struck again by the irony that spaceflight—conceived in the cauldron of nationalist rivalries and hatreds—brings with it a stunning transnational vision. You spend even a little time contemplating the Earth from orbit and the most deeply ingrained nationalisms begin to erode. They seem the squabbles of mites on a plum."[28] Another space enthusiast, Frank White, argues for the existence of what he calls an "overview effect" in which humans who are launched into space achieve a veritable breakthrough in human consciousness:

> [Those living in space] will be able to see how everything is related, that what appears to be 'the world' to people on Earth is merely a small planet in space, and what appears to be 'the present' is merely a limited viewpoint to one looking from a higher level. People who live in space will take for granted philosophical insights that have taken those on Earth thousands of years to formulate. They will start at a place we have labored to attain over several millennia .... [Space dwellers will become aware that] we are one; we are all in this together; war and strife solve nothing .... [T]he multiplier effect means that sending a

26. Sagan, *Pale Blue Dot*, p. 317; for similar views, see Freeman Dyson, *Disturbing the Universe* (New York: Harper & Row, 1979), pp. 233–234.

27. Sagan, *Pale Blue Dot*, pp. 70, 171.

28. Ibid., pp. 171, 174–175.

> limited number of people into space can lead to a broad-based social transformation. The experiences of the few become new information for the many, serving as fuel for social evolution.[29]

Sagan argues that the need to detect and deflect threatening asteroids will encourage the formation of some form of world government. Since the technology needed to deflect asteroids is potentially very dangerous (not least of all because it could be used to send asteroids hurtling *into* Earth instead of away), it will be necessary to develop much stronger international institutions to develop and control this capability. "The existence of interplanetary collision hazards, when widely understood, works to bring our species together . . . . [T]he small near-Earth worlds provide a new and potent motivation to create effective transnational institutions and to unify the human species. It's hard to see any satisfactory alternative."[30]

With similar esprit, the Planetary Society once stated in its official Web site that it hoped to "reach out into the low-energy universe, investigate and understand its many splendors, travel to and perhaps settle its distant shores, seek those unfound 'others' and, in the process, *advance the cause of world citizenship here at home*."[31] World citizenship: it has that universalistic ring that has informed theories of progress from the very beginning.

The only real impediment to the realization of these noble dreams, Sagan and other space advocate suggest, is the rise of ungrounded fear of science. Writing in the 1990s, Sagan describes a "demon-haunted world" where popular understanding of science is under siege by believers in alien abductions, astrology, crop circles, crystal power, and so on. "We risk becoming a nation of suckers, a world of suckers, up for grabs by the next charlatan who saunters along," Sagan warns.[32] His own message is simple: the more science and technology, the better. Sagan writes: "There's no turning back from science. Many will have to become scientifically literate. We may have to change institutions and behavior. But our problems, whatever their origins, cannot be solved apart from science."[33] The faith in science is fundamental because the space advocates believe, as believers in progress always have, that science and technology will have predominantly beneficial effects for humanity. Accordingly, moving into space will not have any unintended, negative consequences for either the humans living there or those remaining on Earth.

29. Frank White, *The Overview Effect: Space Exploration and Human Evolution* (Boston: Houghton Mifflin, 1987) pp. 4, 50, 53.

30. Sagan, *Pale Blue Dot*, p. 263.

31. The Planetary Society, "Facts About The Planetary Society's History," *http://www.planetary.org/society/society-history.html* (accessed 24 May 2000), emphasis added.

32. Carl Sagan, *The Demon-Haunted World: Science as a Candle in the Dark* (New York: Ballantine Books, 1996), p. 39.

33. Sagan, *Pale Blue Dot*, p. 317.

Sagan sums up the prevailing view with his usual flair: "I think that, after some debugging, the settlement of the Solar System presages an open-ended era of dazzling advances in science and technology; cultural flowering; and wide-ranging experiments, up there in the sky, in government and social organization."[34] Zubrin similarly concludes that "Mars may someday provide a home for a dynamic new branch of human civilization, a new frontier, whose settlement and growth will provide an engine of progress for all of humanity for generations to come."[35] In such words we see a virtually perfect expression of the faith that all good things go together, and that profound and tragic choices between equally valued goals can be easily avoided.

Innate Directionality

Supporters of the idea of progress have usually insisted that there is some kind of mechanism or force—perhaps even a divine force—that keeps history on track, moving it forward toward betterment. In their own revitalization of the idea of progress, space advocates have also identified certain innate developmental tendencies that, in their view, are likely to drive history forward. As has historically been common with theorists of progress, the most popular mechanism is one based on the idea of evolution through natural selection. Sagan suggests that a human expansion into space is ultimately rooted in those traits that tens of thousands of years of natural selection have ingrained in mankind. Humans began as nomadic hunters and foragers, and those who were the most adventuresome, who courageously sought new sources of food and water, were the ones who survived. "Even after 400 generations in villages and cities, we haven't forgotten. The open road still softly calls, like a nearly forgotten song of childhood . . . . This appeal, I suspect, has been meticulously crafted by natural selection as an essential element in our survival." The exploratory urge is simply built-in to humanity, although not everyone possesses it in equal measure. "Your own life, or your band's, or even your species' might be owed to a restless few—drawn, by a craving they can hardly articulate or understand, to undiscovered lands and new worlds." This is the drive—the causal mechanism—that almost inevitably will eventually lead to an expanding human civilization in space. "We're the kind of species that needs a frontier—for fundamental *biological* reasons," Sagan writes.[36] And it is in space, nowhere else, that the best and final frontier can be found.

The same kind of evolutionary reasoning is employed by historian Louis Halle, who writes in *Foreign Affairs* in 1981 that there is "reason to believe that, in its progressive evolution, life has at last reached the point where it is about to expand into outer space, *as if it had been programmed in advance*. For, as evolution has a direction,

34. Ibid., p. 318.

35. Zubrin, *The Case for Mars*, p. 1.

36. Sagan, *Pale Blue Dot*, pp. xiv, 230, emphasis added.

it has an implicit destiny in that direction; although it may fail to realize that destiny as an infant may be killed by an accident before it had realized its own *destiny* of achieving adulthood."[37] Halle specifically endorses O'Neill's then-popular plan for space colonies, and suggests that the movement into space is functionally similar to the movement of life from the seas onto dry land some 350 million years ago. Wernher von Braun similarly exclaimed that the first Moon landing was "equal in importance to that moment in evolution when aquatic life came crawling up on the land."[38] Neil Armstrong himself joined in the naturalistic imagery, telling *The New York Times* that humans must move into space "just as salmon swim upstream."[39]

Such views typically reveal a classically teleological style of reasoning, so popular in theories of progress, in which all of human history is seen as motivated and guided by an ultimate purpose or final end. This doctrine imputes a certain consciousness and directionality—a telos—to a process that, on the face of it, might seem dominated by arbitrariness and accident. As one critic has noted, such thinkers claim that space migration "must be understood as an evolutionary stage, a natural development, not just comparable to but homologous with the emergence of life on Earth from the water, or the separation of a child from its mother."[40] It is not surprising, then, that the directionality identified by space advocates often shades off into a prediction of outright inevitability. "Maybe not right now, or next year, or even in 1990, but the space solution is inevitable, and, as shocking as it might at first appear, it is too late to debate the right-or-wrong of it," former NASA astronaut Bryan O'Leary informs us in his 1981 work, *The Fertile Stars*. "It is fruitless to make value judgments about whether we *should* go into space," O'Leary continues, "whether mining the Moon and asteroids is the *right* thing to do, or whether we *ought* to build space shuttles or receive solar power from space—for we are in space to stay."[41] Still, O'Leary does spend the rest of his book rehearsing the various arguments—both economic and spiritual—for why space colonization should, in fact, be pursued. Much like other theorists of progress, O'Leary is convinced of the inevitability of his dream but sees no contradiction in calling for conscious action to bring about its realization.

37. Louis Halle, "A Hopeful Future for Mankind," *Foreign Affairs* (Spring 1981); see also Louis Halle, *Out of Chaos* (Boston: Houghton Mifflin, 1977).

38. Quoted in William Sims Bainbridge, *The Spaceflight Revolution* (New York: John Wiley & Sons, 1976), p. 1.

39. Quoted in Marina Benjamin, *Rocket Dreams: How the Space Age Shaped Our Vision of a World Beyond* (New York: Free Press, 2003), p. 57.

40. David Lavery, *Late for the Sky: The Mentality of the Space Age* (Carbondale, IL: Southern Illinois University Press, 1992), p. 108.

41. Brian O'Leary, *The Fertile Stars* (New York: Everest House, 1981), p. 18, emphasis in original.

DISCUSSION

In truth, most of the claims of the new pro-space ideology of progress do not withstand critical scrutiny. A massive program of space development and colonization was not and is not a rational response to rising energy costs. There are no known resources at the present time that could be obtained more profitably in space than on Earth. Likewise, arguments that America is stagnating scientifically, culturally, or economically, and therefore needs the stimulus of a major space endeavor have no backing in hard evidence. The notion that a frontier is crucial for freedom or innovation is not a finding accepted by either historians or social scientists. Claims that space travel will broaden human sensibilities, thus reducing nationalist and chauvinist impulses, are about as convincing as the belief, once widely held, that the view from high-flying airplanes would alter consciousness in a more cosmopolitan direction. New forms of diversity and social organization in space might eventually occur, but any effort at space colonization in the near term is likely to involve a degree of regimentation and control that is hardly conducive to social experimentation. The oft-mentioned need to deflect asteroids could be handled effectively by robotic spacecraft or by small piloted missions. And if there is a compelling need for an off-planet civilizational "back-up" source, a small base on the Moon would suffice (no massive colony required). Finally, claims that humanity is programmed to explore the universe fall prey to the fallacy of composition: just because *individuals* like to explore does not mean that the larger group of which they are a part (in this case, the human race) has a need to *collectively* explore. As some space advocates themselves point out, history is full of examples of human societies that decided to stay put, choosing to remain within a viable local habitat indefinitely. There is no known innate human tendency that will require human beings to explore space, any more than there is a program forcing humans to settle Antarctica or the sea floor. To the extent that space migration does occur, it will be as a result of conscious human decisions, just as human migration has been in the past.

Without question, then, much of the pro-space literature is deeply flawed.[42] Ironically, it is also in tension with other forms of technological optimism. For those committed to the notion that humanity must move into space to survive, signs of growing success on Earth are disconcerting. If advances in biotechnology, artificial intelligence, and nanotechnology allow humanity to prosper on Earth

42. Fascinating discussions and critiques can be found in Benjamin, *Rocket Dreams*; Stewart Brand, ed., *Space Colonies* (San Francisco: Walker Press, 1977); De Witt Douglas Kilgore, *Astrofuturism: Science, Race, and Visions of Utopia in Space* (Philadelphia: University of Pennsylvania, 2003); Roger D. Launius, "Perfect Worlds, Perfect Societies: The Persistent Goal of Utopia in Human Spaceflight," *Journal of the British Interplanetary Society* 56 (2003); Roger D. Launius, "Perceptions of Apollo: Myth, Nostalgia, Memory, or All of the Above?" *Space Policy* 21 (2005); Lavery, *Late for the Sky*; and Howard McCurdy, *Space and the American Imagination* (Washington, DC: Smithsonian Institution Press, 1997).

to a greater extent than ever before, the urgency of the space endeavor is lost. In fact, if one has faith that terrestrial technology will continue to advance, the idea of spending billions of dollars on unprofitable space ventures becomes even less attractive. Why not just wait until new technologies reduce the cost of spaceflight to reasonable levels? At that point, normal market mechanisms (such as tourist demand) may allow major increases in human spaceflight without government intervention. But then, of course, no grandiose ideology of progress will be required, any more than such an ideology was required to populate the formerly arid deserts of the American southwest once water and air conditioning became widely available.

Despite all this, pro-space ideology retains the power that only a full-blown theory of progress can possess. NASA's own discourse, not to mention that of President George W. Bush and other politicians, frequently invokes the language of progress, with talk of destiny and inevitability, the overcoming or abolition of limits, and the virtuous cycles of knowledge and goodness that space development will surely bring forth. It is reasonable, therefore, to expect that the discourse surrounding space will remain one of the primary expressions of the idea of progress for a very long time to come.

Conclusion

The rise of a post-Apollo space advocacy literature, and its integral relationship to the idea of progress that has been at the core of Western civilization, surely counts as one of the more provocative societal consequences of the success of space travel. Although its concerns are currently on the margins of public debate, the potency of the modern space advocacy synthesis suggests that it will continue to draw adherents and influence the thinking of policy-makers. In comparison to earlier doctrines of progress, pro-space ideology is more grandiose, with its vision of planetary engineering and cosmic expansion, yet also more fearful, with its suggestion that the end times may be near if the space frontier is not soon conquered. This peculiar confluence of ambition and anxiety is likely to influence the pro-space movement and the larger debate about the American future in space for many years to come. Given the deep commitment of Americans to ideas about progress, such ideological concerns are as likely to affect policy as any rational assessment of scientific or economic need. As the development of the American space program itself attests, the capacity of the idea of progress to drive politics—and history—in unexpected directions should not be underestimated.

CHAPTER 30

Space Activism as an Epiphanic Belief System

Wendell Mendell

Within the halls of space agencies, common wisdom holds that a space program will inspire young people to study science and mathematics. The word "inspire" has its roots in an old belief that a supernatural spirit could infuse a person with a vision or an insight into truth. Along those lines, one modern meaning concerns fostering creativity. Again, common wisdom in the United States holds that science, mathematics, and engineering are "hard" subjects, to be avoided by anyone who is not "smart." Even "smart" people see no point in working hard to enter technical professions where the remuneration may not match such fields as business or law. On the other hand, policy-makers at the national level understand that the ultimate competitiveness of the United States is related to its ability to generate and utilize advanced technologies, which requires a workforce educated in science, mathematics, and engineering.

If it is true that a vigorous and exciting space program can motivate a subset of young people to switch to technical studies, then the program will serve a national need. But why should a space program uniquely motivate young people to pursue a technical education? The methodologies of the robotic space science missions do not differ so much from other forms of research, and the explanations of their findings are esoteric. Human space operations are intellectually more accessible to a young person, but few people actually become astronauts.

It is true that people do become passionate about space. We all know the terms "space cadet" and "Trekkie." Yet, both these terms have origins in entertainment media. Are "space cadets" only a type of groupie, a product of celebrity worship? Although people can also be passionate about hobbies and can be motivated to study to become a specialized expert, space seems to have very catholic appeal over all cultures and genders. Is there something different about space that can motivate people, particularly young people, to form a life plan directed toward involvement at some level in the space program?

RATIONALES

In 1982 I became the junior member of a Gang of Three whose aim was to initiate discussion within NASA concerning a human lunar base for the era following the low-Earth orbit space station. At the time, NASA had no long-term planning organization, and was focused on getting approval for a space station. Talk of a lunar base was met with indifference if not hostility.

Over the next several years, our advocacy group was successful in attracting interest both within and external to the Agency for discussion of human exploration of the Moon. Within our little cabal, my role usually was that of a spokesman at conferences and workshops. We developed a more or less standard presentation or "pitch" on overhead transparencies. The title chart was followed by a Rationale or Justification chart, containing four to six bullets (talking points) that were intended to convince the audience of the importance and truth of our cause. The Justification chart is required by all managers for project presentations and represents the customer requirements. In an engineer's ideal world, the Rationale comes down the chain of command, generated at some higher pay grade.

Early in our process, we developed Rationales based on our perception of strategic goals for NASA. As we consulted with a wider audience, we began to adopt Rationales that might address national (or societal) needs. For example, the resources available on the Moon could supplement decreasing inventories of raw materials on Earth. Our presentations worked pretty well with passive audiences, but one-on-one encounters with senior experts on national policy were a different kettle of fish. I quickly learned that any of my naïve generalizations could be savaged by an appropriate authority unsympathetic to the "space solution." I do not mean to imply that our ideas were bad or wrong, but our arguments needed much more sophistication and nuance.

Within a very few years, my Rationales had moved to grand concepts such as humanity moving into the solar system, or intangibles such as inspiration of youth. Many essayists and professional authors have written about the space program in terms that express the broadest human issues. While NASA was formulating the response to the 1989 speech by President George H. W. Bush in which he outlined his Space Exploration Initiative, a number of philosophical discussions took place over how the space program serves the nation; some engineers were very uncomfortable talking about unquantifiable "requirements."

As the lunar base movement became more widespread prior to that presidential speech, I became identified in some quarters of the space community as a leading advocate for settling the Moon. Consequently, I received invitations to speak at a number of events, particularly those sponsored by space activist groups. The audiences were thrilled with artist concepts of lunar facilities and loved to hear what kind of activities would take place on the Moon. The agendas of these conferences always included discussions of the "message" needed to encourage more spending on space programs. I noted that the advocate groups brainstormed to generate rationales for space investment, to be expressed in bullet format on charts.

Epiphanies

Individual conference attendees always expressed a great passion for space activities and carried a vision of a future where living in space was commonplace. Over the years I also became aware that many, if not most, of them could tell a story of some specific event in their lives that marked the beginning of their intense interest in space exploration and development. People said, for example, "I bought a science fiction book to read at Girl Scout camp, and I have been interested in space ever since." "When I was five years old, my dad told me to come downstairs and watch men walking on the Moon."

As I was absorbing this information, my wife was taking training to become more involved in evangelism for her Lutheran church. I heard portions of training videos with discussions about Christian outreach. I noted certain similarities and differences between the two advocacies. In both cases, congregations hear sermons expressing certain truths upon which they all agree. In both cases, people inside the movement cannot understand why everyone is not part of their belief system. In both cases, the group concludes that outsiders simply lack the facts.

The two communities differ dramatically in their form of outreach. The church encourages individual Christians to explain to outsiders how their lives have been changed for the better through adoption of the church's belief system. In fact, many Christians believe they have had transformational experiences—epiphanies, if you will—that have changed their lives, and they tell the unchurched that similar experiences will be accessible if they enter into the Christian belief system. On the other hand, space advocates believe that their conclusions are based on reasoning, and some of them use the overhead chart with Rationales. Their success is limited.

I wondered whether many space advocates also had had their interest awakened by some inspirational experience. If so, it would suggest an outreach strategy based on involving more people in space activities in a personal way. Presumably, opportunities would present themselves from which people would discover an interest in a space future.

Frank White has written of an "overview effect" experienced by astronauts on orbit from which they experience a profound change in personal perspective toward life. I am not convinced that the overview effect is an example of a space-derived epiphany I am thinking about. Rather, I associate it with a kind of perspective alteration associated with extreme experiences, such as a brush with death or personally witnessing for the first time grinding poverty or deep suffering.[1]

I had an opportunity to take a step toward exploring this hypothesis at the 2003 meeting of the International Lunar Exploration Working Group (ILEWG) in Hawaii.

1. Frank White, *The Overview Effect: Space Exploration and Human Evolution*, 2nd ed. (Reston, VA: American Institute for Aeronautics and Astronautics, 1998).

ILEWG

In 1989 President George H.W. Bush launched his Space Exploration Initiative (SEI), which charged NASA with returning to the Moon to stay and journey onward to Mars. NASA composed a reply with scenarios for future human exploration missions and hired a new Associate Administrator for Exploration by the name of Dr. Michael Griffin. By 1992, the SEI was politically dead in the United States and NASA had a new Administrator by the name of Dan Goldin. For space pundits in the United States, that was that.[2]

However, other nations had reacted to the U.S. initiative. The European Space Agency (ESA) produced a collection of studies of potential scientific investigations on the Moon, which led to a technical session at the World Space Congress in 1992. In 1994, ESA convened a workshop on a future lunar exploration and development strategy in Beatenberg, Switzerland. At that workshop, the Japanese delegation offered to host a subsequent topical workshop in 1996 in Kyoto, Japan.

After the Beatenberg event, the NASA Associate Administrator for Science proposed an International Lunar Exploration Working Group (ILEWG) to the Interagency Coordinating Group (IACG). The 1996 Kyoto workshop was convened as an ILEWG meeting. Much to the surprise of many of the attendees, the Japanese presented a large number of papers on future lunar exploration plans, including the first official announcement of the SELENE mission. (The lunar orbital scientific spacecraft, recently renamed Kaguya, will finally be launched in 2008.) However, a scientific finding suggesting evidence for past life in a Martian meteorite had been announced only two months before. NASA Administrator Goldin was looking past the Moon toward Mars. One of the official NASA representatives to ILEWG in Kyoto gave a rather lengthy commentary on reasons to forget about the Moon and focus on Mars.

After Kyoto, official NASA representation disappeared from ILEWG. Although meetings were held, the group became more of a gathering of lunar exploration advocates than a true interagency coordinating body. However, the Japanese were still treating ILEWG as a recognized body and ESA continued to fund travel for a few representatives. Dr. Bernard Foing of ESTEC (the European Space Research and Technology Centre, which is the design hub for most ESA spacecraft and technology development, and is situated in Noordwijk, The Netherlands) became a major advocate for ILEWG and eventually was named Project Scientist of the ESA Smart-1 lunar technology mission.

2. On the Space Exploration Initiative see, George H. W. Bush, "Remarks by the President at 20th Anniversary of Apollo Moon Landing, National Air and Space Museum, July 20, 1989"; Dwayne A. Day, "Doomed to Fail: The Birth and Death of the Space Exploration Initiative," *Spaceflight* 37 (March 1995): pp. 79–83; Thor N. Hogan, "Mars Wars: A Case History of Agenda Setting and Alternative Generation in the American Space Program," Ph.D. dissertation, George Washington University, 2003.

Steve Durst, head of Space Age Publishing and a long-time advocate of human settlement of the Moon, offered to host the 2003 ILEWG gathering at his home base in Hawaii. I attended the meeting, but NASA sent no official representative. Dr. Michael Duke sat at the table as the senior U.S. presence, but he had retired from NASA. On the other hand, Indian representatives attended, and the Chinese Space Agency tried to send a large delegation. No Chinese actually attended because they could not obtain U.S. visas. The usual small number of Japanese and European delegates were there. The remainder of the attendees included various lunar scientists, advocates, students, and a few space celebrities, such as astronaut John Young, invited by Durst. The good attendance possibly had something to do with the attractive venue in Hawaii. I have known Steve Durst for more than 20 years through his interest in lunar development, and I also knew almost all the senior people at the meeting because the list of usual suspects in the lunar advocacy community is not large.

One of the luncheons at the meeting was sponsored by the aerospace company SpaceDev and its founder, Jim Benson, was the featured speaker. I have heard Jim speak several times, and he often begins by pulling out his 1955 membership card from the Science Fiction Book Club, where he encountered the writings of Isaac Asimov and was "turned on" to space. Jim was trained as a geologist and by chance entered the computer software industry in its early days. He later sold a successful software business and used the money to start SpaceDev as a company to build small spacecraft for planetary science missions. Here was another of my space stories.

At the next luncheon, two young people spoke about their interest in the Moon. One was Yuki Takahashi, who had contacted me a year or two earlier with a great passion to build a telescope on the Moon. He went from an undergraduate degree at Caltech to a doctorate program in astrophysics at the University of California at Berkeley. At one point he and I exchanged e-mails about his desire to abort his studies in astrophysics to work somewhere on a lunar telescope project. (I urged him to complete his doctorate!) At the luncheon, he spoke of a trip to a library with his parents in Japan at age 9. He discovered there an entire encyclopedia on astronomy. He devoured it and became absorbed with the idea of space exploration. Here was another of my space stories.

At an afternoon break that day, I rudely grabbed the microphone from Steve Durst and told the people in the room that I wanted them to help me with an experiment. I briefly explained my idea about transforming experiences triggering an interest in space. At the hotel meeting room, every seat had a small notepad and a pen. I asked them to use the notepad to first answer the question, "Did you have a specific event in your life that you remember as catalyzing your passion for space?" I asked them to write Yes or No on the pad. If the answer was Yes, I asked for a brief description of the event. I also asked for the person's age at the time of the event and the person's current age (at the time of the meeting in 2003).

DATA

I eventually received 50 submissions. Adding Benson and Takahashi and myself, I have 53. I forgot to ask for gender, but six submissions are clearly from women. Current ages (in 2003) start at the early 20s. I have four from men older than 75. In the text below, I include three figures that are scans of actual submissions to me in order to illustrate the style and informality of typical responses.

Three submissions (including my own) simply answered in the negative. Everyone else wrote a few words of explanation. Seven of them declared no real moment of "awakening" but elaborated on the beginnings of an evolution of interest. Two men grew up in families involved in the space program or space science and just followed their parents into the field. Another man has a stepfather who flew helicopters that picked up astronauts after landing in the water. That memory dates from when he was 8 or 9 years old.

A man (I presume) states that he had no epiphany but remembers lying on his back, looking at the summer sky, and talking about the stars with his friends at age 10. He joined the high school astronomy club. After a B.Sc. in physics, he eventually acquired a doctorate in economics while maintaining a lifelong interest in astronomy. Although he enters himself in the "negative" column, his commentary is not qualitatively different from that of another man, who writes simply, "Astronomy merit badge Age 14." The latter submitter presumably intended that brief statement to indicate an event that brought him into the space community.

A woman who denies having an "epiphany" nevertheless mentions that "[T]he movie *Contact* did inspire me to want to continue our space exploration efforts for the pure scientific joy if nothing else." A 22-year-old woman who responded positively to the question recalls a transient interest after watching the movie *Space Camp* when she was about 8 years old.[3] However, she attributes her current involvement to learning at age 16 that "a family member was involved in Apollo 11."

A native Hawaiian woman writes: "When I was a small girl, maybe 3 or 4 years old, I noticed the moon would follow me wherever I went. Obviously, the moon liked me. It even would follow me home and wait near my house for me to come outside . . . . So I decided she was my friend and always looked forward to seeing her." How this cultural view of the Moon evolved into attendance at a space conference is not explained. Another woman who describes herself as "a non-american female" writes: "I used to watch the Moon & stars with my father and he & my brothers used to tell me about astronauts, etc. when I had to go to sleep. So I always wanted to dedicate my life to space exploration. I am now 30 years old. And as far as I can remember I always looked up to the Moon & the stars."

---

3. *Space Camp* [film], Metro-Goldwyn-Mayer, 1986, directed by Harry Winer, starring Kate Capshaw, Lea Thompson, Kelly Preston.

A 42-year-old man denies any single event but adds, "I think that generally Rocket Launches are the thing that drives [*sic*] my interest the most." Another man, born in India but now a U.S. citizen, describes his triggering event as: "1963—six years old(.) Nike-Apache sounding rockets blasting off from Thumba Equatorial Launching Station (TERLS) in Trivandrum, India." For both, rocket launches are a common focus, but one writer sees them as triggering his interest in space whereas the other doesn't.

One interesting catalyzing event was mentioned by a 29-year-old from the Netherlands: "I never had a specific moment to determine my interest. [It] has always been there. However, I never believed that I could do something with it. Untill [*sic*] I helped organize a conference (1998), then my life changed and got more direction to where I am Now [*sic*]." Here a person believes he went from a passive to an active role through volunteering to organize an activity.

Among those who responded positively to the basic question, certain themes are repeated. Three (older) people mention Sputnik. Many inputs reference early human missions, particularly Gemini and (of course) Apollo. A tale of Gemini 8 is reproduced as figure 30.1.

The Apollo 11 first step on the Moon is listed by several. One woman's memory of Apollo 11 is reproduced here as figure 30.2.

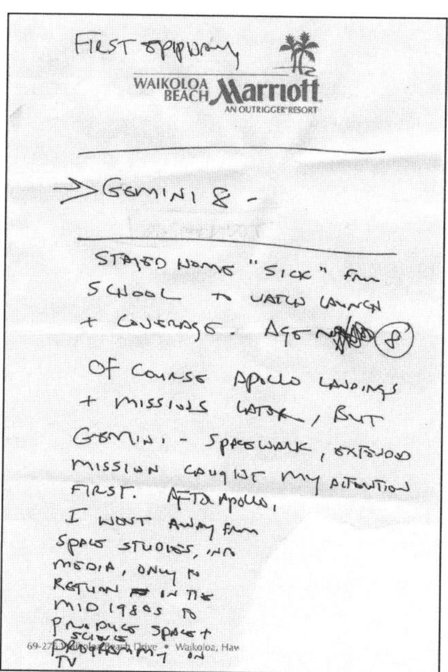

FIRST EPIPHANY

WAIKOLOA BEACH Marriott
AN OUTRIGGER RESORT

→ GEMINI 8 -

STAYED HOME "SICK" FROM SCHOOL TO WATCH LAUNCH + COVERAGE. AGE 8

OF COURSE APOLLO LANDINGS + MISSIONS LATER, BUT GEMINI - SPACEWALK, EXTENDED MISSION CAUGHT MY ATTENTION FIRST. AFTER APOLLO, I WENT AWAY FROM SPACE STUDIES, INTO MEDIA, ONLY TO RETURN IN THE MID 1980s TO PRODUCE SPACE + SCIENCE PROGRAMMING ON TV

69-275 Waikoloa Beach Drive • Waikoloa, Haw

Figure 30.1—Here, a person was energized by a pre-Apollo human mission. Note the comment about leaving space studies after Apollo. Many people were disillusioned by the cancellation of the Apollo program by President Nixon.

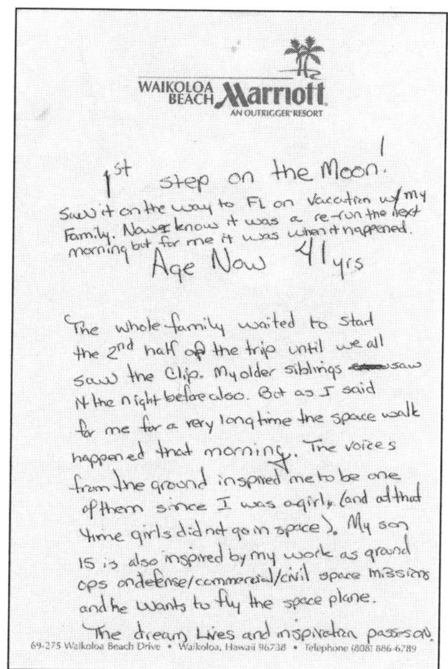

WAIKOLOA BEACH Marriott
AN OUTRIGGER RESORT

1st step on the Moon!

Saw it on the way to FL on vacation w/my Family. Now I know it was a re-run the next morning but for me it was when it happened.

Age Now 41 yrs

The whole family waited to start the 2nd half of the trip until we all saw the clip. My older siblings saw it the night before also. But as I said for me for a very long time the space walk happened that morning. The voices from the ground inspired me to be one of them since I was a girl. (and at that time girls did not go in space). My son 15 is also inspired by my work as ground ops on defense/commercial/civil space missions and he wants to fly the space plane.

The dream lives and inspiration passes on.

69-275 Waikoloa Beach Drive • Waikoloa, Hawaii 96738 • Telephone (808) 886-6789

Figure 30.2—The first human landing on the Moon is widely considered to be an iconic event in the history of civilization. Not surprisingly, many people recall it as memorable, often mentioning watching the first lunar step with family.

For a man, Apollo 17 was the most important. His memoir:

Launch of Apollo XVII, December 7, 1972

- I was 6 years old in the first grade.
- I had never stayed up so late to watch TV.
- I remember NBC's program theme music for its coverage of the mission.
- I met Gene Cernan when I was 12.
- I will meet Jack Schmitt this Thursday.

Steve Durst's daughter, who was a graduate student at the time of the ILEWG meeting, writes that she was raised in a family that talked about space a great deal. However, her personal interest was catalyzed by the LDEF mission, a technology experiment that returned various material samples to Earth after exposure to the low-Earth orbit environment for five years. For Ms. Durst, the most interesting returned cargo was tomato seeds that were distributed to schools for planting.

A 38-year-old man cites a three-hour chat across the aisle on an airplane flight with Tom Stafford at age 28.

A 48-year-old man by chance heard Jim Lovell speak in Hawaii at a convention two years before this survey. The writer is president of a construction company who came to the ILEWG meeting to learn more about construction on the Moon.

Books, particularly science fiction books, are important, especially around ages 8 to 15. Verne, Heinlein, Asimov, Wells, and Clarke are all mentioned. Figure 30.3 is a memorable submission. However, not only science fiction was important; Yuki Takahashi mentioned the encyclopedia of astronomy. Jack Green, a prominent pre-Apollo proponent of lunar craters as calderas and 78 years old at the time of this survey, writes, "1945 publications of J. E. Spurr on lunar volcanism." A Jet Propulsion Laboratory (JPL) technologist recalls the time "My Mom gave me a section of the *Chicago Tribune* with an article by Dr. Dan Q. Posen on space." A woman writes: "When I was in 5th grade (10 or 11 years old), I had a class in the school library where we learned about the Dewey decimal system. It was *so boring* that as I sat on the floor, I read the books I was seated next to. That happened to be the 523 section that contained the Astronomy books. I knew I was hooked when I read them all!"

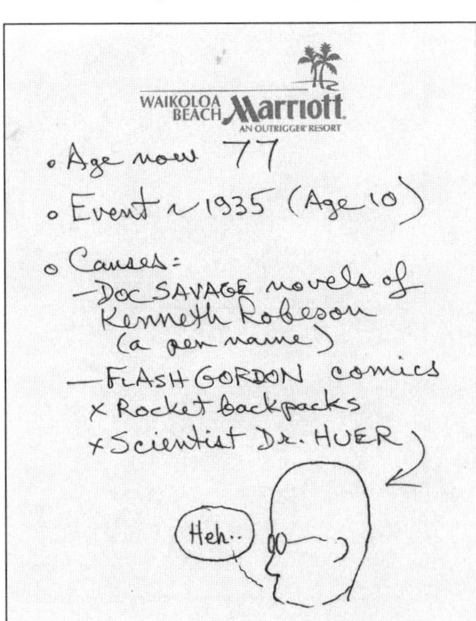

Figure 30.3—Early science fiction featured utilization of exotic technology to conquer frontiers laden with adventure. While the modern genre is more complex and sophisticated in its themes, fiction set in the future continues to inspire careers in space exploration.

A (now retired) internationally recognized scientist, who was one of the top leaders in the Apollo scientific studies, writes: "My parent bought me a book which still sits in my bookcase, 'The Omnibus of Jules Verne', which had both books 'From the earth to the moon' and 'Around the moon.' I was 11 years old. I was not aware at the time of a transformation. However, I realized during Apollo 16 that 'I had been there before'—and quickly found the page in the book which contained the specific paragraph which gave me that certainty."

Visual media also play a role. Carl Sagan's *Cosmos* series is mentioned three times. Interestingly, no one mentioned the classic Disney series from the 1950s that featured Wernher von Braun, although I noted above two women who identified specific movies. A man writes, "My interest in space came when I was about 5 years old; I saw the movie 'The Conquest of the Moon' in 1955."

The last recurring category is astronomy. Inputs already cited mention looking at the sky or reading books or achieving a merit badge. There are those who received telescopes from their parents or visited observatories. It has been known for a long time that widespread public interest in astronomy makes that subject a good way to introduce scientific concepts.

Discussion

My question to think about an epiphany may have seemed strange, but I was pleased by the enthusiastic reaction. Approximately 80 percent of the responders believed they had something to say. Even those who claimed to have nothing to say often added commentary. Several thanked me for the opportunity to write down their experiences.

On the surface, nothing in these notes is new to those familiar with the space program. There are many different paths to an interest in space, yet no one describes his or her involvement in the form of a reasoned argument. Two prominent lunar geologists come closest. One was a graduate student in geology, napping one day when a fellow student pounded on his door to tell him the Russians had "launched a 180 pound satellite." The other geologist was a young Ph.D. who "watched Neil and Buzz step onto the Moon" and "shortly thereafter . . . saw my 1st Apollo 11 rock." The fact that so many people found my question interesting rather than bizarre and that such a large percentage remember their first awareness of space suggests an emotional attachment, which is part of what I was seeking. A next step would be to ask these responders to elaborate on their vision of humanity's future in space. Launch technology evolves so slowly that they might not actually see themselves in space.

I chose the word epiphany deliberately because I find the passion behind many peoples' interest in space to be almost religious in nature. (I am not the only one to notice this.) I have encountered many individuals or groups who have some transcendent vision of future activities or assets in space. This basis for these beliefs becomes a particularly important issue today because the vision of an economic sphere in cislunar space and on the Moon has become an element of national policy and a little-understood element of the President George W. Bush's Vision for Space Exploration initiative.[4] Traditionally, people who believe in a private economy in space have been belittled by the space establishment such as NASA employees or major aerospace contractors. The latter argue that there is no business for major investors beyond government contracting.

Today, of course, we see a new phenomenon whereby so-called high net worth individuals (HWNIs) are investing significant amounts in ventures to build new launch systems, space vehicles for tourism, and even a hotel. Robert Bigelow has committed $500M million of his personal fortune to build an orbiting facility and has already launched the first test module. If the business case is so weak, why are these people in the game? I suggest that they are "believers" who happen to have the money to pursue their dreams.[5]

CONCLUSIONS

This simple, serendipitous sampling suggests that further studies by professionals in social science might prove interesting. However, one must remember that this particular meeting was convened by a visionary with the stated purpose of discussing future activities on the Moon, and the attendees may well be self-selecting for this kind of survey. We may be looking at a subculture rather than the space movement as a whole.

In 1995 in Stockholm, Sven Grahn of the Swedish Space Corporation and I led a team project titled "Vision 2020" at the summer session of the International Space University (ISU). About 40 young space professionals from a variety of countries and a variety of disciplines volunteered to be part of an attempt construct a vision of what space activities would be important 25 years into the future. As leader of the exercise, I went to great lengths to avoid providing structure that might suggest the "right answer" to the students. We worked with an academic in Future Studies to employ visioning, a technique for creating desirable future scenarios. The lack of structure given to the students had the unfortunate effect of producing a lack of structure in their deliberations. When time came to present an outline of the final product to the ISU faculty, the presentations seemed unimaginative.

4. On the Vision for Space Exploration, see Alain Dupas and John M. Logsdon, "Creating a Productive International Partnership in the Vision for Space Exploration," *Space Policy* 23 (February 2007): pp. 24–28; Craig Cornelius, "Science in the National Vision for Space Exploration: Objectives and Constituencies of the 'Discovery-Driven' Paradigm," *Space Policy* 21 (February 2005): pp. 41–48; Frank Sietzen, Jr., and Keith L. Cowing, *New Moon Rising: The Making of the Bush Space Vision* (Burlington, Ontario: Apogee Books, 2004); Craig Cornelius, "Science in the National Vision"; Wendell Mendell, "The Vision for Human Spaceflight," *Space Policy* 21 (February 2005): pp. 7–10.

5. The alt.space community has enjoyed considerable positive publicity. An example of this is Elizabeth Weil, *They All Laughed at Christopher Columbus: An Incurable Dreamer Builds the First Civilian Spaceship* (New York: Bantam Books, 2002).

I chaired a faculty meeting to provide feedback to the students. Sitting in on the meeting were about half a dozen ISU alumni who were in town for the annual Alumni Weekend. The faculty criticism was savage. I took notes and struggled to understand how I might communicate the feedback in a constructive way. However, one by one, the alumni began to assail the faculty, accusing them of being part of a bygone and obsolete "Apollo generation." The young people claimed that space programs of today were not built around grand visions but should be directed to benefit the quality of life on Earth—a theme that was being carried within the project. Clearly, there was a generation gap in expectations in this exercise.

When one looks around the world, one can see space projects aimed at disaster management or provision of services to the population or scientific investigation. The grand structure of the Vision for Space Exploration initiative stands out for its bold assertion of a potential future society in space. Is this just a U.S. cultural product, or is it really a more generally held vision? I personally find a response to the vision around the world, but I believe more study of societal response to visions of a future space civilization is important to the promoters of these policy goals. I must leave such studies to professionals in the social sciences.

CHAPTER 31

Flights of Fancy: Outer Space and the European Imagination, 1923–1969

Alexander C. T. Geppert

The only interplanetary voyages made by our species to date have been flights of fancy.

A.V. Cleaver
British Interplanetary Society
11 October 1947[1]

Historicizing the European Space Effort

"Historians need to explore . . . the cultural meaning of space exploration, and the space programs of other countries" besides those of the United States, American scholar Pamela E. Mack has suggested.[2] Since she made this statement in 1989, i.e., almost 20 years ago, the state of research in Europe has gradually, yet not absolutely, improved. The publication of the official history of the European Space Agency (ESA), on the occasion of its 25th anniversary in 2000, certainly marked a milestone in this process. In much detail and on more than 1,100 pages,

1. A.V. Cleaver, "The Interplanetary Project," *Journal of the British Interplanetary Society* 7, no. 1 (January 1948): pp. 21–39, here p. 21. This paper presents the first findings of a comprehensive research project on the cultural history of outer space, astrofuturism and extraterrestrial life in the European imagination of the twentieth century, approximately between the publication of Hermann Oberth's pioneering *Die Rakete zu den Planetenräumen* (1923) and the establishment of ESA a half-century later (1975). I would like to express my sincere gratitude to the editors of this volume for their kind invitation to me to discuss parts of my work at such an early stage, and to the Fritz Thyssen Stiftung for generously supporting this *Habilitationsprojekt*. Heartfelt thanks are also due to Dorothee Dehnicke, Rita Hortmann, Heinz Hermann Koelle, Tessa Mittelstaedt, and Paul Nolte.

2. Pamela E. Mack, "Space History," *Technology and Culture* 30, no. 3 (July 1989): pp. 657–665, here p. 658.

this multinational institution's complex beginnings and the development of its organizational structure were meticulously traced back to the formation of its two precursors, the European Launcher Development Organisation (ELDO) in 1962 and the European Space Research Organisation (ESRO) in 1964, while deliberately leaving aside broader social, cultural, and imaginative issues.[3] What is true for the history of Europe at large, however, is applicable here as well: European history and the history of Europe are not identical, but overlap at best. The in-depth analysis of a single European institution—thoroughly researched and, without the slightest doubt, enormously valuable as it now stands—should not be taken for a social and cultural history of European practices and representations of outer space per se. In other words: Because it lacks a direct equivalent to the NASA History Office and the Space History Division of the Smithsonian National Air and Space Museum—with their concerted activities and unparalleled research programs which effectively created and later shaped the entire field in the United States—in Europe, space history is a much smaller, more fragmentary, and by far less established area of historiographical research.[4]

In direct consequence, much research on the cultural history of the European space effort, its historical significance, and societal impact still remains to be done, especially with regard to the 12 years after Peenemünde and prior to Sputnik—a period which many consider the Golden Age of Space Travel *avant la lettre*. Therefore, the following tentative reflections are limited to presenting and discussing three broader issues and provisional hypotheses. They are intended to ask the right questions rather than to provide possibly premature answers. First, where do historians have to look in order to locate what kind of possible repercussions of spaceflight? Why is it so complex a task to identify and measure any kind of impact spaceflight might have had on culture and society in general? Second, how can Europe's role and position in this entire scenario be described and assessed? Has a distinctly European version of outer space evolved gradually, especially since 1945? Or were, for instance, the Apollo flights already subject to

3. John Krige, Arturo Russo, and Lorenza Sebesta, *A History of the European Space Agency*, 2 vols. (Noordwijk, The Netherlands: European Space Agency Publications, 2000). "Little has, in fact, been written on the European space effort," the authors conceded in their preface, and went on to ask for "more reflective and comparative studies" to contextualize their results and to relate them to other research in the field (vol. 1, p. xvi). Walter A. McDougall, "Space-Age Europe: Gaullism, Euro-Gaullism, and the American Dilemma," *Technology and Culture* 26, no. 1 (January 1985): pp. 179–203. For a recent fact-oriented summary of European activities in outer space, see Brian Harvey, *Europe's Space Programme: To* Ariane *and Beyond* (Chichester, U.K.: Springer-Praxis Publishing, 2003); see also ESA's own historical series, called History Studies Reports. The most recent of these more than 30 working paper-type publications are to be found at *http://www.esa.int/esapub/pi/hsrPI.htm* (accessed 17 January 2007).

4. W. D. Kay, "NASA and Space History," *Technology and Culture* 40, no. 1 (1999): pp. 120–127; Roger D. Launius, "NASA History and the Challenge of Keeping the Contemporary Past," *The Public Historian* 21, no. 3 (Summer 1999): pp. 63–81.

Figure 31.1—"Berlin, We Have a Problem." National advertising campaign, German space industry (2005).

such globalized, carefully orchestrated, and extensive media coverage that they subsequently gave rise to the same process of myth-making both in the United States *and* in Europe? And, third, while the persistence of the so-called American spacefaring vision has, time and again, been attributed to the deeply rooted frontier myth in popular culture and in the public imagination, is the European vision of outer space best characterized by the complete lack of such a key trope and the absence of a commonly shared belief system—or is there any equivalent?

"Berlin, wir haben ein Problem," ("Berlin, we have a problem") read one of the most popular slogans in a recent national advertising campaign, entitled "Deutschland braucht Raumfahrt" ("Germany needs spaceflight"). Financed by the German space industry, this campaign lamented the lack of government support and endeavored to raise public awareness of the industry's pecuniary needs and oft-formulated appeal for increased government funding (figure 31.1). It is more than remarkable that such a tribute to the famous Apollo 13 phrase "Houston, we've had a problem"—originally uttered by Commander James Lovell in April 1970 but popularized by Tom Hanks in the Oscar-winning Hollywood blockbuster 25 years later—did not require an explanation, even in a German context, proving to be sufficiently effective in an entirely different public sphere. Apparently, the responsible publicity agency had every reason to take for granted the existence of the necessary historical background information. The well-informed spectator would be rewarded with a smile for his successful knowledge transfer from one national context to another and applying it to quite a different problem. In the *present* academic context, however, this witty motif draws our attention to the potency of American popular culture myths far beyond national borders, where, in fact, they may be interpreted quite differently. In the end, "Berlin, wir haben ein

Problem" points to the far-reaching importance of international and crosscultural perspectives in analyzing both science fact and science fiction. It reminds us, in short, that the societal impact of spaceflight and its imaginative dimension cannot be properly historicized without considering transatlantic references, perceptual interdependencies, and transnational interrelations.

SPACEFLIGHT, THE IMPACT QUESTION, AND THE SPATIAL TURN: THREE HEURISTIC REMARKS

Without going into too much detail, the first heuristic remark consists of a fundamental conceptual question: How can we define—and then diagnose—the societal impact of spaceflight in an historical and international perspective? How do we describe, assess, and characterize it once such impact has been recognized? And what set of criteria do we employ in order to compare views and attitudes toward space and spaceflight across different cultures worldwide?

The challenging aspect of this question lies, of course, in the paradigmatic inversion and the complete change in historiographical perspectives it proposes. For a long time one of the most fundamental question in the field of space history was how spaceflight, in a comparatively short time, developed from science fiction into science fact. Who invented the necessary technologies, where, and in what circumstances? What kind of institutions were established, who financed them, and how did they produce, structure, and organize knowledge? Which scientific fictions became, at what point, predominant and were transformed into actual science—whereas others remained merely fictitious? Today, historians seem far less interested in compiling yet another, sometimes semi-hagiographical, often oversimplistic, yet almost always teleological genealogy of space travel thought, its pioneers, and its practitioners. The new perspective is diametrically reversed. It is also less prone to endorsing undercomplex master narratives. What was formerly the dependent variable now becomes the independent one. We take spacefaring as given, and then aim to understand and explain the impact and possible effects that outer space, space travel and space exploration have had on culture and society at large.

From the viewpoint of a cultural historian and recent newcomer to this field of historical research, such a general change of perspective seems more than overdue. Almost inevitably it should lead to less specialist and fewer internalist analyses, to the advantage of broader, more integrative perspectives, by paying greater attention to historical context and culturally ascribed meaning. Multifaceted links and manifold connections are conceivable if space history breaks with its self-chosen splendid isolation and decides to open itself up to different historiographical branches such as social, cultural and intellectual history, but also to less obviously related disciplines such as science studies, social and cultural geography, sociology of knowledge, literary criticism, art history, and contemporary archaeology. Most probably, such an extension will eventually lead to a complete transformation of the entire field and enable space history to harmonize far better with presently "hot" historiographical debates on transnationality, globalization, and "entangled history" as the key features of a state-of-the-art *histoire du temps présent*.[5]

At the same time, space historians must remain cautious. "Impact" is a very broad and imprecise term that can denote many different aspects simultaneously. Space-related cultural artifacts, such as Hermann Oberth's seminal 1923 book *Die Rakete zu den Planetenräumen* and Frith Lang's 1929 film *Frau im Mond*, certainly had an immense public impact; so had Willy Ley's numerous public lectures and Wernher von Braun's countless media appearances, not even to mention the actual events such as the Sputnik launch in 1957, the first weather satellite launch three years later, the Mercury and Gemini flights, or the Apollo landings which, in many ways, marked the end of an epoch rather than the beginning. After defining and refining the original "impact question," space historians thus need to sharpen their analytical tools by differentiating carefully between different kinds of impact, and by endeavoring to categorize, classify, and contextualize. Whom or what was affected, when and where, immediately or on a long-term basis? And did this impact have a lasting effect or did it fizzle out soon after? Different notions such as "repercussions," "influence," "consequences," and "effects" on the one hand, and "perception," "reception," "consumption," and "appropriation" on the other, come to mind and could be considered possible alternatives.

Conceptual problems of an almost identical kind are by no means specific to this field of historical research. They have been discussed by neighboring disciplines and branches of research—such as reception aesthetics—at least since the 1970s, if not even earlier. Quite clearly, this constitutes another good reason why space history can only profit from widening its thematic focus and sharpening its analytical tools, including the incorporation of innovative methodological achievements made elsewhere.[6]

Moreover, in order to locate and specify societal and cultural impacts of spaceflight and space exploration, it might prove necessary to search elsewhere, in areas where effects, consequences, and repercussions are not readily expected and hence are

5. John M. Staudenmaier, "Recent Trends in the History of Technology," *American Historical Review* 95, no. 3 (June 1990): pp. 715–725; Michael Werner and Bénédicte Zimmermann, "Vergleich, Transfer, Verflechtung: Der Ansatz der *Histoire croisée* und die Herausforderung des Transnationalen," *Geschichte und Gesellschaft* 28, no. 4 (2002): pp. 607–636; Michael Werner and Bénédicte Zimmermann, "Penser l'histoire croisée: Entre empire et réflexivité," *Annales HSS* 58, no. 1 (2003): pp. 7–36; Michael Werner and Bénédicte Zimmermann, "Beyond Comparison: *Histoire croisée* and the Challenge of Reflexivity," *History and Theory* 45, no. 1 (February 2006): pp. 30–50. The numerous contributions to Jürgen Kocka's recent *Festschrift* (Gunilla Budde, Sebastian Conrad and Oliver Janz, eds., *Transnationale Geschichte: Themen, Tendenzen und Theorien* [Göttingen: Vandenhoeck & Ruprecht, 2006]) aim to make an interim stock-taking of these debates on the globalization of historiography itself through global and transnational history.

6. See, for example, Martyn P. Thompson, "Reception Theory and the Interpretation of Historical Meaning," *History and Theory* 32, no. 3 (1993): pp. 248–272; Marian Füssel, "Die Kunst der Schwachen: Zum Begriff der 'Aneignung' in der Geschichtswissenschaft," *Sozial. Geschichte* 21, no. 3 (2006): pp. 7–28; Jennifer Wallace, *Digging the Dirt: The Archaeological Imagination* (London: Gerald Duckworth, 2004).

less obvious than, say, in utopian literature, pop culture, or science fiction TV shows and movies. If the implicit assumption of this volume is correct—that spaceflight was, at least for several decades during the twentieth century, an emblematic and absolutely central element to the project of Western modernity—historians should be able to locate evidence for its lasting effectiveness far *beyond* the established circles of the international spaceflight community and their widespread activities, and as independently of its historical and contemporaneous advocates as possible.

Last but not least, it is somewhat surprising to observe that the notion of "space" itself has hardly been problematized and how little theoretical attention it seems to have received, despite the classic works by geographers, philosophers, sociologists, and ethnographers such as Gaston Bachelard, Henri Lefebvre, Yi-Fu Tuan, Pierre Nora, Marc Augé, and numerous others.[7] How can we explain that the current upsurge of interest in space as a specific category of historical analysis—which is frequently labeled *topographical* or *spatial turn*—and the historiographical field of space history have not at all been seen in relation to each other so far? "What are you buying when you go to outer space?" someone asked rhetorically during the conference which inspired the present volume, and then replied to his own question, seemingly surprised, by exclaiming in despair: "There is simply nothing there!" Yet, outer space is by no means a spatial vacuum; it changes constantly and is always formed and refashioned. It is precisely this static and ahistorical notion of space that must be problematized by asking questions such as: "Where" and "what" was outer space at which point in time? How was it represented and imagined? Has the perception of space changed over time? And in what way were changing conceptions in turn affected by the continuous and ongoing exploration of outer space? Even if it may sound paradoxical, it would seem high time to apply the so-called spatial turn to space history, and to write a social and cultural history of outer space with special emphasis on: space.[8]

EUROPE'S OUTER SPACE IN THE POSTWAR PERIOD: 1945 AND 1968/69 AS GLOBAL TURNING POINTS

Is there a genuinely European variant of outer space? In fact, is there a specific perspective on outer space which could be called European? Is there a set of commonly shared assumptions about space and the specific role of Europe in the history

7. Gaston Bachelard, *La Poétique de l'espace* (Paris: Presses Universitaires de France, 1957); Henri Lefebvre, *La Production de l'espace* (Paris: Éditions anthropos, 1974); Yi-Fu Tuan, *Space and Place: The Perspective of Experience* (London: Edward Arnold, 1977); Pierre Nora, ed., *Les Lieux de mémoire*, 3 vols. (Paris: Gallimard, 1984–1992); Marc Augé, *Non-Lieux: Introduction à une anthropologie de la surmodernité* (Paris: Éditions du Seuil, 1992).

8. Alexander C. T. Geppert, Uffa Jensen, and Jörn Weinhold, "Verräumlichung: Kommunikative Praktiken in historischer Perspektive, 1840–1930," in *Ortsgespräche: Raum und Kommunikation im 19. und 20. Jahrhundert*, Alexander C. T. Geppert, Uffa Jensen, and Jörn Weinhold, eds. (Bielefeld: Transcript, 2005), pp. 9–49, 361–371, with numerous suggestions for further reading.

of spaceflight? For the period prior to 1945 at least, these questions are not highly problematic. Historians agree as to the fundamental significance of the various European space fads of the 1920s, notably in Germany and Russia, and, in consequence, the far-reaching effects of the slow yet steady formation of various interconnected networks of international space experts. Also in other European nations, such as Great Britain or France, transnational personal contacts among rocket scientists and space experts frequently preceded the establishment of the various amateur societies.[9]

Yet, by the time the Golden Space Age was in full swing in the 1950s—with actual space travel remaining far from a reality—Europe's position was secondary at best. Hence, the crucial question in this context is precisely when and through which unilateral transfer processes such cultural hegemony was lost in the aftermath of World War II, not least as a concomitant of the rapidly emerging U.S.–USSR polarization. Analytically, the problem is further complicated by the fact that many of the new key players then residing in the United States, such as the engineer and rocket scientist Wernher von Braun (1912–1977), the writer and space science popularizer Willy Ley (1906–1969), and the technician Krafft Arnold Ehricke (1917–1984), were clearly of European origin, even if they may have gradually developed a somewhat different attitude concerning their own nationality. Von Braun, for instance, did not participate in any of the early international congresses on astronautics which began in Europe annually after 1950, and did not (even temporarily) return to Germany prior to the autumn of 1957.[10] Be this as it may, with regard to Europe's (self-)positioning vis-à-vis outer space, the end of World War II was definitely the first decisive turning point whose long-lasting impact can still be felt today.

Indications of this fundamental displacement of cultural hegemony can, for instance, be observed in the realm of popular culture, especially with regard to imagery and iconography. More than 10 years after its 1929 premiere, film stills originally produced for Fritz Lang's epochal *Frau im Mond* were used time and again to illustrate numerous articles on rocketry and spaceflight appearing in a wide array of international journals and magazines, simply due to the lack of visual material otherwise available.[11] By the early 1950s, however, the situation had drastically changed.

9. Michael J. Neufeld, "Weimar Culture and Futuristic Technology: The Rocketry and Spaceflight Fad in Germany, 1923–1933," *Technology and Culture* 31, no. 4 (1990): pp. 725–752. See my forthcoming article "Space *Personae*: Scientific Networks of Remote Knowledge, 1923–1969," *Journal of Modern European History* 6, no. 1 (2008).

10. See, for instance, von Braun's letter to Frederick C. Durant III, 23 December 1953, Wernher von Braun Papers, Manuscript Division, Library of Congress, Washington, DC, 1/3.

11. Willy Ley Collection, National Air and Space Museum Archives, Smithsonian Institution, Washington, DC, 33/9. Milton Fairman, "The Race to Explore," *Popular Mechanics* 53 (March 1930): pp. 286–289; Kenneth W. Gatland, "Development of Rocket Flight: A Review of the Film Shown to the British Interplanetary Society in London on February 14th, 1948," *Journal of the British Interplanetary Society* 7, no. 2 (March 1948): pp. 112–119; Kenneth W. Gatland, "History of Rocket Development: A Review of the Film Shown to Fellows of the British Interplanetary Society at the War Office, Whitehall, on July 16, 1949," *Journal of the British Interplanetary Society* 9, no. 2 (1950): pp. 64–70; Heinz Gartmann, "Rakete und Raumflug im Film," *Weltraumfahrt* 1, no. 3 (Juni 1950): pp. 86–91.

Figure 31.2—British Interplanetary Society Christmas Card (*Annual Report of the British Interplanetary Society/Journal of the British Interplanetary Society* 11 [1952]: p. 364).

At that time, a direct equivalent to the enormously successful *Collier's*/ Disney complex simply did not exist in any European context. Popular iconography widely available in the European public sphere was far less elaborate, less futuristic, and by no means as imaginative as its American counterpart—just compare, for instance, the visual expressiveness of Chesley Bonestell's (1888–1986) famous illustrations for *Collier's* magazine with a number of European samples from the same period (figures 31.2 and 31.3). The first example, showing a spaceship on the Moon with its crew about to begin their exploration, was distributed by the British Interplanetary Society to its approximately 2,300 members as their official Christmas Card for 1952; the second stems from one of the numerous publications of Heinz Gartmann (1917–1960), a German engineer, author, and well-connected editor of *Weltraumfahrt: Beiträge zur Raketenentwicklung*, and was based on a painting by graphic artist Klaus Büergle (1926–). It is hardly surprising that these illustrations proved far less effective and by no means as trendsetting as their American counterparts.

On the other hand, the lavishly illustrated, large-format books into which the American *Collier's* series had swiftly

Figure 31.3—Frontispiece to a 1950 German magazine article. (Heinz Gartmann and Klaus Büergle, "Zwischenfall mit Heliopolis/Weltraumstation im Bau," *Das Neue Universum* 67 [1950]: pp. 173–191.)

been transformed were soon translated into numerous foreign languages, including Japanese, Italian, Dutch, Swedish, Finnish, German, French, and Spanish, and sold especially well in Europe.[12] In Germany, for instance, S. Fischer, one of the best-reputed publishing houses, soon acquired the copyrights and, before long, sold a considerable number of copies, running into several thousands. Thus, geopolitical and iconographical shifts went hand in hand. From a European perspective, the 1950s space race was accompanied by a new sociocultural space fad in the public imagination, although this time of transatlantic rather than domestic origin.

Ever since then, autonomy—or, rather, the quest for it—has been the predominant theme of the European space effort. Since the first "philosophical" reflections, undertaken by the Italian physicist Edoardo Amaldi (1908–1989) in April 1959, on the urgent need to establish an official and independent European space organization, "encouraging" or even "increasing" European unity "from without" has been the explicit political aim in all common activities in space. In the late 1980s five European research institutes came to the conclusion that a European "space spirit" had gradually developed, largely as a consequence of ESA's activities. They argued that to undertake such activities would give the European Union (EU) nothing less than a genuine chance to "consolidate a common identity." The EU, they concluded in their large-scale study, should hence have its "eyes and ears in space."[13] Steadily pursuing this goal has, at least at times, posed a considerable challenge to all plans of cooperation with third parties and possible partners such as the United States.

---

12. Chesley Bonestell and Willy Ley, *The Conquest of Space* (New York: Viking Press, 1949); Wernher von Braun, Joseph Kaplan, Heinz Haber, Oscar Schachter, Fred L. Whipple, and Cornelius Ryan, *Across the Space Frontier* (New York: Viking Press 1952); Cornelius Ryan, Wernher von Braun, Fred L. Whipple, and Willy Ley, *Conquest of the Moon* (New York: Viking Press, 1953); Willy Ley, Wernher von Braun, and Chesley Bonestell, *The Exploration of Mars* (New York: Viking Press, 1956). The German editions were Wernher von Braun, Joseph Kaplan, Heinz Haber, Oscar Schachter, Fred L. Whipple, and Cornelius Ryan, *Station im Weltraum: Die technischen, medizinischen und politischen Grundlagen des Raketenflugs in den Weltraum* (Frankfurt am Main: S. Fischer, 1953); Cornelius Ryan, Wernher von Braun, Fred L. Whipple, and Willy Ley, *Die Eroberung des Mondes* (Frankfurt am Main: S. Fischer, 1954); Willy Ley, Wernher von Braun, and Chesley Bonestell, *Die Erforschung des Planeten Mars* (Frankfurt am Main: S. Fisher, 1956); Wernher von Braun and Willy Ley, *Start in den Weltraum: Ein Buch über Raketen, Satelliten und Raumfahrzeuge* (Frankfurt am Main: S. Fischer, 1958).

13. Edoardo Amaldi: Introduction to the Discussion on Space Research in Europe, 30 April 1959, Historical Archives of the European Union (HAEU), Florence, Italy, COPERS 0001. Reimar Lüst, "Cooperation between Europe and the United States in Space," *ESA Bulletin* 50 (May 1987): pp. 98–104, here pp. 99, 104; Forschungsinstitut der Deutschen Gesellschaft für Auswärtige Politik (Bonn), Institut Français des Relations Internationales (Paris), Istituto Affari Internazionali (Rome), Nederlands Instituut voor Internationale Betrekkingen "Clingendael" (Den Haag), and Royal Institute of International Affairs (London), *Europas Zukunft im Weltraum: Ein gemeinsamer Bericht europäischer Institute* (Bonn: Europa Union Verlag, 1988), here pp. 1, 168: "Der Weltraum [ist, ACTG] ein wichtiger Bereich, in dem Europa eine gemeinsame Identität konsolidieren und seine Einheit verwirklichen kann . . . Europa muß seine eigenen Augen und Ohren im Weltraum haben."

Christmas 1968 marks the second, well-known turning point, with the epoch-making spaceflight of Apollo 8 and the first photographs of the entire Earth taken on this occasion. American writer Archibald MacLeish (1892–1982) and Austrian philosopher Günter Anders (1902–1992) were among the first to realize that the most profound consequence of the Apollo program was not at all the continued exploration of outer space, its scientific results, or the proof of the actual technical possibility for so doing. It was, rather, a radical change in self-perception on a genuinely global level, literally resulting in a new *Weltanschauung*, i.e., ways of viewing the world. For the first time ever, it was felt that the entire Earth could be seen—and see itself—from without and as a whole.[14]

Even if this argument, originally coined by MacLeish and Anders simultaneously but independently of each other, has been repeated and modified so many times since that it now might be considered trite, it is beyond doubt that the images originating from this particular mission (especially the famous photograph "Earthrise") produced enormous, hitherto entirely unforeseen effects. It is no exaggeration to state that these images fundamentally altered our contemporary geographical imagination. Thus, the major television event of the 1960s, the 1969 Apollo Moon landing, must, first and foremost, be seen as a carefully orchestrated, global media event. The fact, that it could be witnessed by approximately 600 million people, i.e., approximately a fifth of the world population, live on TV in 49 countries worldwide, symbolizes the central role of space exploration in the process of globalization. Taking place only seven months after Apollo 8, it was cause and effect at the same time: Technically only possible because of the recently set up Intelsat system, the Moon landing proved an event which, in itself, had considerable globalizing consequences difficult to overestimate in retrospect. Not only does this second turning point afford a perfect example of how the exploration of outer space substantially affects and alters conceptions of "inner," (i.e., Earthly) space, it also makes the question of Europeanness even more complex.[15]

---

14. Archibald MacLeish, "A Reflection: Riders on Earth Together, Brothers in Eternal Cold," *The New York Times,* 25 December 1968: p. 1; Günther Anders, *Blick vom Mond: Reflexionen über Weltraumflüge,* 2nd ed. (München: C. H. Beck, 1994), p. 12: "Das entscheidende Ereignis der Raumflüge besteht nicht in der Erreichung der fernen Regionen des Weltalls oder des fernen Mondgeländes, sondern darin, daß die Erde zum ersten Mal die Chance hat, sich selbst so zu sehen, sich selbst so zu begegnen, wie sich bisher nur der im Spiegel sich reflektierende Mensch hatte begegnen können."

15. See, for instance, Frank White, *The Overview Effect: Space Exploration and Human Evolution* (Boston: Houghton Mifflin, 1987); Wolfgang Sachs, "Satellitenblick: Die Ikone vom blauen Planeten und ihre Folgen für die Wissenschaft," in *Technik ohne Grenzen,* Ingo Braun and Bernward Joerges, eds. (Frankfurt am Main: Suhrkamp, 1994), pp. 305–346; Denis Cosgrove, "Contested Global Visions: 'One-World,' 'Whole-Earth,' and the Apollo Space Photographs," *Annals of the Association of American Geographers* 84, no. 2 (June 1994): pp. 270–294; Denis Cosgrove, *Apollo's Eye: A Cartographic Genealogy of the Earth in the Western Imagination* (Baltimore, MD: The Johns Hopkins University Press, 2001); Andreas Rosenfelder, "Medien auf dem Mond: Zur Reichweite des Weltraumfernsehens," in *Medienkultur der 60er Jahre: Diskursgeschichte der Medien nach 1945,* vol. 2, Irmela Schneider, Torsten Hahn, and Christina Bartz, eds. (Opladen: Westdeutscher Verlag, 2003), pp. 17–33.

The same tension—a certain form of nationalism or Europeanness on one hand, versus a specific variant of internationalism or globalism *avant la lettre* on the other—was already a central element in many of the programmatic self-descriptions of the early space advocates, often proving a highly problematic issue. Directly after its inception in 1934, members of the British Interplanetary Society, for instance, discussed their self-image as expressed in the Society's name on several occasions. Although it was at first disputed whether British Rocket Society would not be a more appropriate name than British Interplanetary Society if it was to harmonize better with its Continental and North American sister organizations, 17 years later, in 1951, the conflict had become whether the word "British" should not be omitted completely from the its title. Some of the Society's members strongly feared that it could be interpreted as an expression of just that kind of latent nationalism which they considered as contradictory to the Society's international character and detrimental to their common "vision of the coming of interplanetary communication." From their perspective, Europe was a state in between two different stages of development. Europe was to be preferred to any national context and something of a self-evident frame of reference, yet not the global society either, which they hoped eventually to create through their cosmic activities.[16]

Infinite Frontiers, Cosmic Visions, Futures Past

"The desire to explore and understand is part of our character," President George W. Bush declared on 14 January 2004, announcing his new "Vision for Space Exploration" initiative at NASA Headquarters in Washington, DC. "Mankind is drawn to the heavens for the same reason we were once drawn into unknown lands and across the open sea," he continued, in order to justify space travel by making bold psychological statements and indicating broad historical parallels: "We choose to explore space because doing so improves our lives, and lifts our national spirit. So let us continue the journey."[17] Thus Bush, in a few simple words, connected

16. P. E. Cleator, "What's in a Name?" *Journal of the British Interplanetary Society* 4, no. 1 (October 1934): p. 34: "The *raison d'être* of the Society—however remote it may seem at present—is to achieve the conquest of space, and thence interplanetary travel. There can be no question, therefore, but that the term 'interplanetary' is a fitting designation." H. E. Ross, "Gone with the Efflux," *Journal of the British Interplanetary Society* 9, no. 3 (May 1950): pp. 93–101, here p. 95; D. F. Martyn, "Correspondence: Change of the Society's Name?" *Journal of the British Interplanetary Society* 10, no. 2 (March 1951): p. 93f.

17. "President Bush Announces New Vision for Space Exploration Program: Remarks by the President on U.S. Space Policy, NASA Headquarters, Washington, DC," 14 January 2004, *http://www.whitehouse.gov/news/releases/2004/01/20040114-3.html* (accessed 3 January 2007).

individual with collective psychology, history with the future, and questions of the quality of life with the welfare of the entire nation. He also reintroduced one of the oldest and most frequently advanced arguments for investing extensive resources in human spaceflight: outer space as humankind's final frontier, as an enormous, quite natural task only to be undertaken by all concerned collectively, yet obviously and exclusively under his direction and leadership.

By no means was George W. Bush the first to envisage the exploration of outer space as a direct consequence of America's alleged predisposition to continuously open new frontiers and steadily seek new discoveries. Originally, this famous argument goes back to historian Frederick Jackson Turner (1861–1932) who, on the occasion of the epochal World's Columbian Exposition 1893 in Chicago, had tried to explain that many characteristic features of American society developed during the exploration of the American West and later persisted, even after the actual process of conquest and settlement was long completed. Turner's so-called frontier thesis has, ever since, been attacked by legions of professional historians who have severely and successfully criticized his argument from various angles, eventually debunking the thesis as a myth. It was, nevertheless, successfully politically exploited in the 1960s in order to strengthen unity and the nation's self-image by conjuring one great common task.[18] Thus, Bush did not add a new element to the debate but, rather, took over and gave fresh justification to exactly the same kind of frontier rhetoric that numerous other space advocates, including John F. Kennedy and John Glenn, had already used long before him, and in which current NASA Administrator Michael Griffin also seems willing to invest.[19]

Whether true or not, from the perspective of a professional historian it is quite remarkable how present a simplified and vulgarized version of Turner's original thesis still is in everyday public discourse, and what considerable influence it continues to have in the American public sphere. It might be best described as an historical argument of the second order, i.e., an argument that is historically far from accurate but is retrospectively *believed* to be true and, despite its original falseness, thus proves enormously effective (in this case, up to the present day). When applied within a European context, its shortcomings become even more obvious. Although Europe

18. Frederick Jackson Turner, "The Significance of the Frontier in American History," in Frederick Jackson Turner, *The Frontier in American History* (New York: Henry Holt, 1921), pp. 1–38; Janice Hocker Rushing, "Mythic Evolution of 'The New Frontier' in Mass Mediated Rhetoric," *Critical Studies in Mass Communication* 3, no. 3 (September 1986): pp. 265–296; Matthias Waechter, *Die Erfindung des amerikanischen Westens: Die Geschichte der Frontier-Debatte* (Freiburg im Breisgau: Rombach, 1996).

19. For instance, in a recent interview, Christian Schwägerl, "'Wir könnten längst unterwegs sein': Ein Gespräch mit dem Nasa-Chef Michael Griffin über unsere Zukunft im Weltraum," *Frankfurter Allgemeine Zeitung* 22 (26 January 2007): p. 44; Roger D. Launius, "Perfect Worlds, Perfect Societies: The Persistent Goal of Utopia in Human Spaceflight," *Journal of the British Interplanetary Society* 56 (September/October 2003): pp. 338–349.

has always had a variety of different and oft-changing frontiers and open spaces, both real and imagined (most of which were located overseas), a direct equivalent to the American frontier has never existed, neither in *Realgeschichte* nor in myth. Even more than in the United States, some time in the early 1970s the theme of "space" lost much of its popular and futuristic appeal, as did nuclear power and atomic energy.[20]

From the aftermath of World War II through the early 1970s, however, outer space clearly had constituted one of the "major sites of utopian thinking," not only in the United States but also in Europe.[21] During these three decades it was a widely shared assumption that the future would—literally—take place in outer space. While also fostering actual technological developments, scientists, technicians, and engineers drafted grand schemes of how the world would soon appear. They attempted to anticipate a future in space and at the same time tried to interpret and make this understandable from a philosophical perspective. Some of the main protagonists, such as the English science-fiction author Arthur C. Clarke (1917–), philosopher Olaf Stapledon (1886–1950), aeronautical engineer A. V. Cleaver (1917–1977), and the already mentioned technician Krafft A. Ehricke (1917–1984), considered themselves explicitly as agents of an ongoing spatial revolution—the "interplanetary project"—which would change humankind forever. Long before Sputnik they already declared the advent of the "Space Age" or the "orbital age," featuring an "interstellar" or "multi-planetary society" peopled by "*Homo extraterrestris*" or "interplanetary man," and diagnosed the dawn of a new "3-dimensional civilization." They discussed potential repercussions of spacefaring on the human psyche, demanded the establishment of a United Nations organization control, and debated with great verve not only whether "on Mars life (such as we know) has evolved to the human level" but also how communicative exchange with extraterrestrials might be made possible. Reflecting upon the necessity for a "common field of semantic reference (C.F.S.R.)" as a condition for any exchange between Earthlings and their "extra-terrestrial neighbours (E.T.N.)," under the heading of "Astraglossa" some enthusiasts even set out to take "first steps in celestial syntax," just in case genuine long-distance communication "with beings with whom we presumptively share no idiom of discourse" had to be initiated (figure 31.4).[22]

20. "European Interest in Apollo Dwindles," *The New York Times*, 10 February 1971: p. 24.

21. Constance Penley, *NASA/Trek: Popular Science and Sex in America* (New York: Verso, 1997), p. 22.

22. Arthur C. Clarke, "The Challenge of the Spaceship (Astronautics and Its Impact upon Human Society)," *Journal of the British Interplanetary Society* 6, no. 3 (December 1946): pp. 66–81; A.V. Cleaver, "The Interplanetary Project," *Journal of the British Interplanetary Society* 7, no. 1 (January 1948): pp. 21–39; Olaf Stapledon, "Interplanetary Man?" *Journal of the British Interplanetary Society* 7, no. 6 (November 1948): pp. 213–233; Krafft A. Ehricke, "The Anthropology of Astronautics," *Astronautics* 2, no. 4 (November 1957): pp. 26–29, 65–68; Siegfried Johannes Gerathewohl, "Zur Psychologie der Raumfahrt: Eine kleine Studie über unsere Vorstellung von Raum und Zeit," *Weltraumfahrt* 5, no. 3 (Juli 1954): pp. 65–68; A. W. B. Hester, "Some Political Implications of Space-Flight," *Journal of the British Interplanetary Society* 14, no. 6 (November/December 1955): pp. 314–319; Lancelot Hogben, "Astraglossa or First Steps in Celestial Syntax," *Journal of the British Interplanetary Society* 11, no. 6 (November 1952): pp. 258–274, here p. 259.

Verständigung mit den Marsbewohnern.

Figure 31.4—In the 1920s, it was suggested to communicate with the Martians by means of a newly-invented "light-telephone" which is here shown in action. The question of whether the inhabitants of planet Mars (if they existed) would understand the constitution of such a signal, set up a corresponding station of their own, and then respond in the same language, remained controversial. ("Communication with Martians," *Die Umschau: Illustrierte Wochenschrift über die Fortschritte in Wissenschaft und Technik* 28, no. 16 [19 April 1924]: p. 284).

Controversies on the existence of extraterrestrial life are by no means an epiphenomenon of postmodernity; they go back at least to the Copernican revolution.[23] Even if early experts such as Willy Ley, Wernher von Braun or the numerous members of the German *Gesellschaft für Weltraumforschung* (GfW), under the direction of engineer Heinz Hermann Koelle (1925–), often did their very best to steer clear of such muddy waters and preferred, mainly for tactical and strategic reasons, *not* to take an openly positive stand, the quest for extraterrestrial life has always been one of the most central, yet often hidden, motives behind many scientific debates and space-related enterprises; not surprisingly, it still is today. It is not a coincidence that so many contemporary space exploration projects are motivated by the search for water, believed to be a necessary precondition for extraterrestrial life-forms.

23. Karl S. Guthke, *Der Mythos der Neuzeit: Das Thema der Mehrheit der Welten in der Literatur- und Geistesgeschichte von der kopernikanischen Wende bis zur Science Fiction* (Bern/München: Francke, 1983); Lucian Boia, *L'Exploration imaginaire de l'espace* (Paris: La Découverte, 1987); Steven J. Dick, *The Biological Universe: The Twentieth-Century Extraterrestrial Life Debate and the Limits of Science* (Cambridge, U.K.: Cambridge University Press, 1996); Steven J. Dick, *Life on Other Worlds: The 20th-Century Extraterrestrial Life Debate* (Cambridge, U.K.: Cambridge University Press, 1998)

Ultimately, it can be argued that a vast number of the debates fought out during the twentieth century on the social and cultural rationale behind the space effort oscillated between two distinct poles—or, as it were, "discursive vanishing points"—other than the myth of the frontier. First, there is that entire argumentative strand for which literary scholar De Witt Douglas Kilgore has found the neat label "astrofuturism," i.e., the widely shared belief system that the future would necessarily take place in outer space and that it would entail an encounter with alien life-forms as the twentieth century's most radical version of imagined alterity but yet an "other" unlike any other before.[24] A German journalist summarized expectations for the future in 1951: "The steady progress of science and technology obscure the boundaries. More than ever the world of tomorrow is a concern of the present, and even to visit a secret laboratory can today become the beginning of an adventurous voyage of discovery into the future."[25]

Second, in addition to this utopian component there is a strong religious, spiritual, and/or mythical strand to be detected behind these numerous debates and activities. How did changing images and conceptions of outer space, "other worlds," and the entire cosmos impinge not only on competing versions of the future but also on religion and transcendental beliefs? As early as the late 1920s, contemporary critics spoke of the "religion of spacecraft," calling it a "technological ersatz religion." Although such an aspect continues into the present, it has hitherto not been adequately examined, sometimes quite dramatically so. Both factors—the utopian/futuristic strand and the spiritual/religious component—must be made central categories when historicizing outer space.[26] Last but not least, it is Archibald MacLeish and Günter Anders who deserve credit for drawing our attention to the anthropocentrism so deeply inherent in all variants of space exploration. They made us realize that the Copernican revolution never took place in the realm of the imaginary. The more elaborate and far-fetched our conceptions and images of outer space, of other worlds "out there," and extraterrestrial life-forms have become, the more anthropocentric they remain. Indeed, distinguishing between space exploration and its impact can only constitute a temporary analytical separation. In fact, spaceflight already affected European culture and society long before it ceased to be a mere "flight of fancy."

24. De Witt Douglas Kilgore, *Astrofuturism: Science, Race, and Visions of Utopia in Space* (Philadelphia: University of Pennsylvania Press, 2003); cf. in this context especially C. G. Jung, "Ein moderner Mythus: Von Dingen, die am Himmel gesehen werden [1958]," C. G. Jung, *Gesammelte Werke*, vol. 10: *Zivilisation im Übergang* (Düsseldorf: Walter, 1995), pp. 337–474.

25. Rosemarie Knittel, "Gesellschaftsreisen in die Zukunft," *Alles aus der Welt* (Mai 1951): pp. 33–38, here p. 38: "Die schnellen Fortschritte von Wissenschaft und Technik verwischen die Grenzen. Mehr denn je ist die Welt von morgen eine Angelegenheit der Gegenwart, und schon der Besuch eines geheimen Laboratoriums kann heute zu einer abenteuerlichen Entdeckungsreise in die Zukunft werden."

26. Walter Deubel, "Die Religion der Rakete," *Deutsche Rundschau* 55 (Oktober 1928): pp. 63–70.

SECTION VII

AFTERWORD

CHAPTER 32

Are We a Spacefaring Species? Acknowledging Our Physical Fragility as a First Step to Transcending It

M. G. Lord

In a preface to *The Man Who Sold the Moon*, a 1950 collection of short stories, Robert Heinlein emphasized that science-fiction authors do not generate "prophecy."[1] They write speculative fiction, a riff on the notion of "What if?" Often they invent alternative realities to critique aspects of the world in which they live. Yet our aspirations to become a spacefaring species—and our assumptions about our abilities to do so—were defined by science fiction, from the cannon-ball-shaped spaceship that blasted off in Jules Verne's 1865 fantasy *From the Earth to the Moon* to the canonical stories and screenplays produced by various authors—Ray Bradbury, Isaac Asimov, Frank Herbert—during the Cold War.

What interests me is the disparity between who we are in our imaginations and who we are in real life. In fiction, we have mastered extraterrestrial flight; our technology enables us, to quote a classic split infinitive, "to boldly go" anywhere. In reality, however, we are fragile creatures that do not thrive outside Mother Earth's atmosphere, gravity, and magnetic field. In the last 40 years, robotic scouts—many designed and flown by NASA's Jet Propulsion Laboratory (JPL)—have sent home tantalizing glimpses of new worlds. In the next 40 years, we ourselves will strike out for these worlds—planets many times more distant than our Moon, worlds truly apart.

Engineers have demonstrated the know-how to build a craft that will transport a crew to, say, Mars. But there is a catch. In recent months, biomedical researchers working in relative obscurity have begun to raise a big unanswered question: Can the human body withstand a prolonged journey into deep space? The constraints imposed by our biology could have profound societal implications. What does it mean to our collective dreams if a major obstacle exists to realizing them? And how much risk—as well as financial expenditure—will society tolerate in their pursuit?

1. Heinlein, Robert, *The Man Who Sold the Moon* (New York: Signet, 1950), pp. v–vi.

Galactic cosmic rays are the "dragons" that lurk beyond Earth's magnetic field, the biggest threat to human spaceflight. And they are not just of concern to a few Chicken Littles. They trouble, for example, Shannon Lucid, the plucky astronaut who spent 223 days, 2 hours, and 50 minutes in space, the longest stint of any American woman. "Radiation could be a showstopper," she said.[2]

This is a far cry from the way radiation appeared in literature at the dawn of the space program. In *Americans into Orbit*, a 1962 book aimed at young readers, radiation was simply one more hurdle that a can-do astronaut could easily surmount.[3] No air in space? We'll take it with us. No gravity? We'll spin the spacecraft and make some. And to reverse the decline of muscle and bone caused by weightlessness, we'll lash ourselves to a treadmill and break a sweat.

In 1961, Marvel Comics writer Stan Lee and artist Jack Kirby actually cast radiation as benign, suggesting it could produce positive mutations. After exposure to a solar flare during the flight of an experimental spacecraft, the Fantastic Four—scientist Reed Richards and his fellow astronauts—morphed into crime-fighting superheroes: Mr. Fantastic, the Thing, the Human Torch, and Invisible Woman (who back then was called Invisible Girl).

Perhaps the fiction most responsible for the dream of safe, easy travel was Robert Heinlein's *Rocket Ship Galileo*, a 1947 tale of two boys who, with the help of one boy's scientist uncle, cobble together a rocket in their back yard and blast off for the Moon. Nearly every engineer I interviewed for *Astro Turf*, my recent book on JPL, recalled Heinlein's impact, as did many of today's space entrepreneurs: X-Prize founder Peter Diamandis, XCOR chief Jeff Greason, and even *SpaceShipOne* financier Paul Allen.

Heinlein was not unaware of radiation, and he knew it could cause mutations. But he, too, cast them as benign. His character Joe-Jim, for example, a two-headed beefcake who leads a mutant army in *Universe* (1951), is none the worse for his genetic anomaly. If anything, because Joe-Jim has the thinking power of two brains, he is better.

What exactly is a galactic cosmic ray and why does it pose a threat? It is a high-energy heavy ion that originates from an exploding supernova outside our galaxy. Such atoms travel very fast, about the speed of light, and they are said to be "ionized," meaning that they have gained or lost electrons. As a consequence, they carry a positive or negative charge. They can be any element up to the atomic weight of iron. We don't encounter them on Earth because our atmosphere protects us. Likewise, the Earth's magnetic field, which extends beyond the atmosphere, functions as a shield—except for a small, quirky patch of orbit over Brazil known as the "South Atlantic anomaly," where it does not.

2. Interview with Shannon Lucid, Johnson Space Center, 5 June 2005.

3. Gurney, Gene, *Americans Into Orbit: The Story of Project Mercury* (New York: Random House, 1962), p. 28.

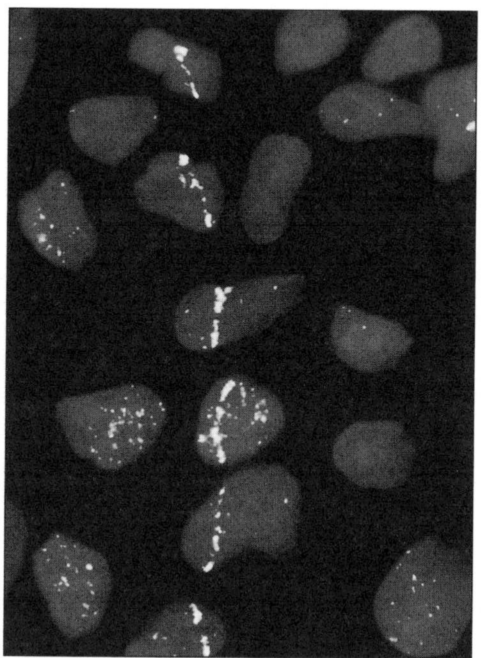

Figure 32.1—This image shows a "track" made by a heavy ion passing through a brain cell in vitro. (*Image courtesy of Brookhaven National Laboratory.*)

One might expect lead—that familiar prophylactic against dental x-rays—to provide shielding. But a lead "shield" actually compounds the problem. When a heavy ion strikes an atom of lead, it dislodges a cascade of charged particles that can cause destruction similar to that of the heavy ion.

Think of a cosmic ray as "a bullet flying around with speed and mass," Marcelo Vazquez explained.[4] He is a researcher at Brookhaven National Laboratory on Long Island who studies the effect of radiation on brain cells and other tissue. "When the rays go through matter—it can be a rock, a body, or your brain—they produce so much energy and charge that they produce a kind of hole. But they also produce secondary particles, like a shower." These secondary particles are called delta rays.

Hydrogen compounds offer the best protection against secondary particles because hydrogen atoms contain one electron for one proton. The current best solution for a Mars-bound ship is an aluminum or carbon-composite shell, with 2 to 4 inches (5 to 10 cm) of polyethylene (a hydrogen compound) around the crew compartment. Water is another efficient shield. But nothing—not even an artificial electromagnetic shield around the spacecraft—will block the rays entirely.

Shielding matters because cosmic rays kill brain cells (or, in any case, brain cells in vitro). In Vazquez's laboratory experiments, cosmic rays induced what is called programmed cell death in neurons. The bombarded brain cell assesses its degree of damage and decides to die rather than try to fix itself. Moreover, as a heavy ion passes through a column of cells, it forms a "track," which you can see in the microscopic image in figure 32.1. Dislodged particles career out from this track, inflicting yet more damage on adjacent tissues.

4. Interview with Marcelo Vazquez, Brookhaven National Laboratory, 10 November 2005. All Vazquez quotations are from this interview; some of Vazquez's observations are cited in M. G. Lord, "Impossible Journey," *Discover* (June 2006): 38–45.

Vazquez has written papers calculating the amount of brain cells that will be struck by cosmic rays on a two-and-a-half year voyage beyond the Earth's magnetic field—in other words, on a Mars trip as it is currently conceived by NASA planners.

> Depending upon the size of the cell, we estimate that between 13 percent and 40 percent of brain cells will be hit once by cosmic rays .... We have billions and billions of cells in our brain. But 40 percent is a lot. Some areas are very tiny, but they play an important role in functioning. If you wipe out those cells, you don't need to worry about the hundreds of billions of cells. A few million—you're gone. [Alzheimer's patients lose about 5 percent of their brains annually.][5]

Radiation, we know from studying atomic bomb survivors, put people at risk for long-latency soft-tissue tumors. But, Vazquez pointed out, leukemia can develop more quickly, in as few as two years, raising the possibility of a severely brain-damaged, fatally sick astronaut barely limping back to Earth.

Of course, what happens to cells in vitro does not always reflect what happens in the body, where optimists hope repair will be possible. "To go from Petri dishes to humans is a big jump," Vazquez said. But his experiments on animals do not bode well for astronauts. Rats exposed to cosmic rays, as well as rats exposed to conventional radiation, showed major impairment of motor skills, measured by their ability to move around in a box. Over a period of 11 months, the rats exposed to x-rays and gamma rays recovered some of their lost coordination, but the rats exposed to cosmic rays appeared never to recover.

On the bright side, NASA scientists believe they have solved one major radiation problem: solar particle events. These are sudden bursts of hot plasma that periodically spew out at high speeds from our Sun. When astronauts return to the Moon, they will most likely tour its surface in a vehicle equipped with a solar-particle shield. Or they will carry equipment for building a temporary storm shelter—a concept that will also work on Mars where, because the planet is farther away from the Sun, astronauts will have more time to secure refuge. The most acute problem remains chronic exposure to cosmic rays.

Not only does chronic radiation exposure hurt people, it also degrades the drugs used to treat them. The antibiotics NASA currently uses on Shuttle missions (Bactrim, Cipro, and Augmentin) lose their potency after about two weeks in space, according to Lakshmi Putcha, a pharmacotherapeutics researcher at the Johnson Space Center.[6] If Mars-bound astronauts equipped with these drugs in their current formulations became ill, she said, "We would have no way of treating them."

5. Lord, "Impossible Journey."

6. Interview with Lakshmi Putcha, Johnson Space Center, 7 June 2005.

The problem of drug degradation, however, is not insoluble, and its resolution could come through chemical engineering—coating the active ingredients of a drug with a substance or material that will preserve its effectiveness. For example, lidocaine (an anesthetic used on the Space Shuttle) is suspended in water—a radio-resistant substance—so it tends not to degrade in flight. Dosage for astronauts is also a hit-and-miss affair, particularly for substances ingested orally. Gastrointestinal motility slows down in space and it takes longer for medications to be absorbed.

Space medicine is a long way from the ideal projected in *Star Trek: Voyager* where a holographic doctor, the Mark I Emergency Medical Hologram, performs nifty repairs without ever making a cut. Indeed, laparoscopic surgery, because it does not release buckets of blood, is well-suited to microgravity. Astronaut and medical doctor Dafydd Williams and Mehran Anvari, a Canadian scientist specializing in telerobotic surgery, have been experimenting with telerobotic medical techniques that might be used at, say, a Moon base.[7] But because of the 20-minute lag for radio signals between Earth and Mars, telerobotics will not be of much use to a crew on Mars.

Space medicine must advance before a Mars flight because the human body—even one belonging to a robust astronaut—is extremely vulnerable in space. In a test on an early Shuttle mission, astronauts were pricked on the arm with various antigens, to which they showed diminished immune response. Latent viruses also express themselves in space. NASA microbiologist Duane Pierson has published several papers documenting the presence in astronaut saliva of various viruses, including Epstein-Barr, which has been linked to human mononucleosis.[8] Other common latent viruses could be yet more uncomfortable if they reactivated on a long spaceflight—chicken pox, for example, which usually returns as shingles. And no data exist on whether antiviral drugs, such as acyclovir, work in flight.

The main issue, which NASA scientists have been unable from their current data to determine, is: Does the immune system remain severely depressed, or, in the words of NASA immunologist Clarence Sams, does it "adjust to a new normal?"[9] Some researchers are optimistic that if they could collect data throughout the mission, instead of merely at takeoff and landing, their results might be different. But stress of the sort experienced by, say, researchers in Antarctica also degrades the immune system. And radiation is a big wild card, which may intensify the weakening. Vazquez was blunt in his assessment of a Mars mission: "No matter what you do, you put a guy in a can for six months and it's a big stress."

Skeptics often ask, Why bother to leave Earth? Science fiction answers that question with a bigger question, a question that has haunted human society since its first members looked skyward: Are we alone in the universe?

7. Interview with Dafydd Williams, Johnson Space Center, 6 June 2005.

8. Interview with Duane Pierson, Johnson Space Center, 7 June 2005.

9. Interview with Clarence Sams, Johnson Space Center, 7 June 2005.

In the 1980s, NASA began a ground-based program, the Microwave Observing Project, to address this question. It involved a systematic sweep of the heavens with radio telescopes. The project was of immense concern to scientists, artists, theologians—just about everyone but former Senator William Proxmire, who tarred it with his "Golden Fleece" award for wasting public money. In 1982, Congress cut the project's funds, but the public's yearning for an answer was more powerful than Proxmire's scorn. Privately financed, the sweep continues.

Robots also seek evidence of life. At JPL, the mission architect's mantra—"Follow the water"—is code for "Find clues linked to life." In 1976, scientist Carl Sagan was involved with Project Viking, the Martian probe that some scientists hoped would detect traces of life. Such a wish, however, was never made explicit. Nor was it gratified, which may have led Sagan to fulfill it through science fiction. Nine years later he published *Contact*, a novel describing a fictive alien encounter.

Scientists approach the hunt for life with a mixture of hope and dread, their expectations molded by science fiction. Hope is inspired by the aliens who came in peace; dread, by those who waged war. The film *The Day the Earth Stood Still* (1951), directed by Robert Wise, is a hopeful classic; benevolent aliens try to stop us from destroying ourselves. They stand in sharp contrast to the weapon-toters who touched down in H. G. Wells's *War of the Worlds* (1898). And I find it hard not to view Robert Heinlein's *The Puppet Masters* (1951), about an invasion of parasitic extraterrestrial slugs, as a prescient book. In one scene, scientists view a map of this country. Its central states are designated "red"—indicating the residents' brains have been taken over by alien slugs.[10]

The most interesting fantasy, however, involves the commingling of human DNA with that of a benign alien species—the subject of Octavia Butler's Xenogenesis Trilogy, whose first volume, *Dawn,* appeared in 1987. Butler, an idiosyncratic loner, may not consciously have sought to comment on the zeitgeist, but her theme reflected the times. With the Cold War nearing an end, Americans were less paranoid about the alien, or—to strip away a metaphor—the communist "Other."

In contrast, at the height of the Cold War, alien motives were highly suspect. Arthur C. Clarke's *Childhood's End* (1952) is a chilling fantasy of alien "Overlords" who dupe us into relinquishing our autonomous future. Perhaps the funniest—and the cruelest—comment on our eagerness to trust was a 1962 television episode of the *Twilight Zone* adapted by Rod Serling from a short story by Damon Knight. Titled "To Serve Man," the show introduced an extraterrestrial race devoted to advancing human health and agriculture. All that human cryptographers know about this race is the title of what appears to be their bible: *To Serve Man*—which, as it turns out, is a cookbook. [11]

---

10. Robert Heinlein, *The Puppet Masters* (New York: Ballantine, 1990), p. 151.

11. Marc Scott Zicree, *The Twilight Zone Companion* (Los Angeles: Silman-James Press, 1982), pp. 235–236.

Since the beginning of the Cold War, gender roles have also been rethought. No longer, for instance, is the idea of women astronauts ridiculed as it was in 1962, when scientist Wernher von Braun responded dismissively to a question about women in space. His boss at the Agency, he quipped, might allow room on a future mission for "110 pounds of recreational equipment."[12]

Curiously, during the Cold War, when the private lives of American engineers were often rigidly sexist, the science fiction they consumed explored unconventional gender identities.[13] Heinlein, for instance, was almost subversive in the way he proselytized for women as equals, anticipating the careers of astronauts such as Shannon Lucid. "Delilah and the Space Rigger," his short story from 1949, featured a woman engineer who coeducates an all-male space station, transforming her misogynistic boss into a cautious feminist because her presence increases productivity.

But at odds with fiction is this troubling fact: Women are in certain ways less suited than men to long-duration space travel. Breasts and ovaries are vulnerable to radiation-linked cancers in ways the prostate, for example, isn't.[14] And women's metabolisms are different. Promethazine, NASA's motion sickness drug, takes longer to reach a level of effectiveness in women. Women also tend to show more orthostatic intolerance—light-headedness after getting up from bed or being in space.

Walter Sipes, a Johnson Space Center psychological researcher, believes a mixed-gender crew may have psychosocial advantages on a Mars trip.[15] But women will have to ask themselves: Is this worth the risk?

In an e-mail sent in September 2006, Vazquez reminded me that solutions to the radiation problem will mostly likely not come exclusively through engineering. They will also be operational and medical, and they will always involve *minimizing* rather than eliminating risk.

Francis Cucinotta, chief scientist of NASA's radiation program, explains his Agency's goal: Limit an astronaut's radiation exposure so that his or her likelihood of developing cancer will be no more than 3 percent greater than that of an average person.[16] Astronauts, as a result, wear dosimeters to monitor the radiation they receive during flight. But the 3 percent is a murky number because data on soft-tissue cancer formation, gleaned largely from atom bomb survivors, are imprecise. Likewise, the dose rate, the type of radiation, and the differences between the populations exposed play a role in computing probability, as do individual dietary, environmental, and genetic factors.

12. Wernher von Braun quoted in Howard E. McCurdy, *Space and the American Imagination* (Washington, DC: The Smithsonian Institution, 1997), p. 224.

13. M. G. Lord, *Astro Turf: The Private Life of Rocket Science* (New York: Walker, 2004), pp. 185–187.

14. Vazquez talked about women's cancer risks in the interview cited earlier.

15. Interview with Walter Sipes, Johnson Space Center, 7 June 2005.

16. Interview with Francis Cucinotta, Johnson Space Center, 9 June 2005.

These are some solutions that might make for a safer Mars trip:

1. Travel faster to reduce exposure. Power the spacecraft with a nuclear reactor placed far away from a well-shielded crew compartment.
2. Surround the spacecraft with an artificial magnetic field—an engineering solution that has not yet proven feasible.
3. Surround the crew quarters with a shield of water—currently impractical because of the expense of getting an immense weight, or mass, of water off the Earth.
4. Ingest foods or drugs that shield against radiation. The flavonoids in blueberries and strawberries have antioxidant properties. They don't yet protect the brain from radiation, but in 30 years there may be a breakthrough.[17]
5. Finally, send a bunch of aging space cowboys. If, after 10 years, astronauts in their seventies developed cancer, one could hardly say they were struck down in the prime of life. "My kids said that NASA should send me to Mars," Shannon Lucid joked. "They said, if NASA would send you when you're 80, Mom, then you could live up there, do something, and they wouldn't even have to worry about bringing you back." Of course, with the nearest emergency room 30 million miles away, NASA had better make progress on that holographic doctor.[18]

The most radical solution, and the one with the greatest societal impact, however, may involve genetic engineering. In defiance of the odds, some atom bomb survivors have not developed soft-tissue cancers. Scientists need to examine their genetic material to determine why. Resistance to radiation would be a strong asset to an astronaut on a long-duration mission. Because of concerns over the medical privacy of astronauts, it's currently illegal to screen a crew based on genetics, but this might change for a Mars mission. Or, in the more distant future, a private agency could grow radio-resistant astronauts through the miracles of genetic engineering—an idea with a science-fiction precedent.

Gattaca, a 1997 movie written and directed by Andrew Niccol, describes a space program in which astronauts are chosen based on genetic superiority. In the movie, this is a bad thing. The hero is a love child, not the product of a eugenics exercise. Yet the hero's short-term triumph—securing a spot on a desirable space mission—may seem less admirable if his weak genes expose him to a fatal illness. In the 1970s, in vitro fertilization procedures were uncommon as well as ethically suspect. Today they are performed frequently and not considered an ethical problem. Likewise, other eugenics procedures that are not approved today may become commonplace in the future.

17. Vazquez discussed flavonoids during his 2005 interview cited earlier.

18. A variety of solutions are discussed in Eugene N. Parker, "Shielding Space Travelers," *Scientific American* (March 2006): 41–47; also Lord, "Impossible Journey."

At an International Space Development Conference in Los Angeles in May 2006, I chatted informally with some people from the Space Frontier Foundation. We discussed space tourism—the fact that engineer Burt Rutan's suborbital spaceship and Los Vegas entrepreneur Robert Bigelow's orbiting hotel, things that science fiction writers had forecast, were hurtling toward reality.

When Heinlein's 1955 novel *Tunnel in the Sky* came up, however, some space fans said things that disturbed me. In expressing their desire to emulate the novel's characters, they failed to make a distinction between plausible reality and fantasy. Set in the future, the novel deals with a group of high school students taking a survival test who end up stranded on a hostile planet. One of the kids who weathers the ordeal returns to Earth but later goes back to the planet with other settlers to found a colony.

I understand the allure of space settlement and share a passion to achieve it. But there is no "tunnel in the sky"—a safe passage without radiation to distant worlds. On Mars, the preferred destination of today's settlers, there is no magnetic field to shield people from cosmic rays. The most disturbing comment I heard in a space settlement workshop was "When are we going to see babies born in space?" Not soon, I hope, given the damage cosmic rays are known to cause to genetic material.

Robert Zubrin, the engineer-founder of the Mars Society who favors a trip to Mars with Apollo-era technology, has pooh-poohed the threat of radiation. In a 2003 letter responding to a *New York Times* article on space radiation, he pointed out that astronauts who spent months in Earth orbit have not been debilitated.[19] Yet oddly, he did not adequately consider the fact that low-Earth orbit is within the Earth's magnetic field, and Apollo astronauts never spent more than a couple of weeks outside it.

John Charles of the NASA Space Life Sciences Division seems to have a firmer grip on reality. "Lots of internal discussion is going on now about what level of risk is acceptable for trips like the Mars flight," he told me, and added,

> What is it going to mean in terms of real-world manifestations, including: What is the likelihood of losing a person—having somebody die on a trip to Mars? We may have a case where they only have so much morphine and so many antibiotics so if somebody's really sick, do you just keep pumping them full of morphine that somebody else might need tomorrow? Or does something else have to happen? And what that something else is, we all dance around because nobody wants to talk about it . . . . We constantly remind each other that our examples

19. Robert Zubrin, "New York Times Misrepresents Mars Radiation Danger," online at *http://www.marssociety.org/news/2003/1210.asp*, accessed September 2006.

> would be people like those who settled the North American continent. [When they struck out for the West], they weren't planning on coming home. If they got sick along the way, somebody buried them.[20]

Astronauts have a different relationship to risk than normal people. I got my first sense of this reading the Astronaut Ten Commandments on a Web site for Astronaut Hopefuls (or As Hos, as they call themselves). "Keep your weaknesses to yourself," says Commandment Three. "If you don't point them out to others they will never see them."[21]

This sense was confirmed when I raised cosmic-ray concerns at a recent dinner in Hollywood. An astronaut visiting the host rolled her eyes dismissively. "Doctors have been wrong in the past," she said. "They used to think you couldn't swallow in space." And here sat one astronaut, she assured me, who would leap for a spot on a one-way Mars mission.

Public perception, however, is another story. This has profound implications for society. Society must determine, through public discussion, how much risk it is willing to tolerate or, in any event, how much it is willing, through tax dollars, to underwrite. I'm not sure a federal agency should use public money to place civilian astronauts at high risk. Soldiers maybe, but not civilians. The burgeoning private space industry may best accommodate those who choose to place themselves in extreme jeopardy. According to Federal Aviation Administration guidelines, a tourist on a suborbital flight, for example, does not have the same assurance of safety as a passenger on a commercial airline. He or she understands and chooses the hazard. Likewise, explorers on a privately funded space mission could imperil themselves in any way they want, irrespective of society's disapproval.

This evaluation of risk reminds me of the way scientists discussed cosmic rays in the 1970s and of the unusual means by which a controversy about high-energy heavy ions was resolved. When a heavy ion passes through an astronaut's head, the astronaut sees a burst of light. This is called a retinal flash. Astronaut Buzz Aldrin first reported the phenomenon in 1969, when he returned from the Moon, but scientists studying cosmic rays had anticipated it. In the early 1950s, University of California at Berkeley biophysicist Cornelius Tobias posited a link between cosmic rays and such visual fireworks. In 1970 (a time when guidelines for experimenting on oneself were more lax than they are today), Tobias placed his own head in the path of a high-energy heavy ion. He wore a black hood when the beam sliced through him; ambient light, he feared, would distort his perception. Tobias believed

20. Interview with John Charles, Johnson Space Center, 6 June 2005. All Charles quotes are from this interview.

21. "Ascan Ten Commandments," online at *http://www.ashos.org/astro_ten_commandments.html*, accessed September 2006.

Figure 32.2—A high-energy heavy ion passes through the head of Cornelius Tobias (wearing a hood) at the Lawrence Berkeley Laboratory in 1970.

strongly that he would see a pyrotechnic streak, and what he saw would confirm his thesis. The experiment accomplished this and more: "It was as if I were looking into the universe itself," he said.[22]

I hope the optimists are right: that the human body can survive chronic exposure to cosmic rays. I hope our flesh does not forever curtail our dreams. But to prove this, the optimists might, in the manner of Cornelius Tobias, have to place themselves in the line of fire. And unless society becomes dramatically less risk-averse, they might have to do it on their own dime, without popular support.

22. Tobias's experiment is discussed in "Cosmic Ray Questions: Studies at Lawrence Berkeley Laboratory's Bevalac Will Help Resolve Uncertainties about Radiation Risks to Astronauts," online at *http://imglib.lbl.gov/LBNL_Revs/RR_online/91fw.html,* accessed September 2006.

CHAPTER 33

Production and Culture Together: Or, Space History and the Problem of Periodization in the Postwar Era

Martin J. Collins

In 1956, philosopher Hannah Arendt delivered at the University of Chicago a series of lectures entitled "Vita Activa." She reworked these reflections, yielding the classic *The Human Condition*, published in 1958.[1] In these works, Arendt's organizing question centered on whether the Western experience in the previous two centuries might confound rather than aid in working through the challenges of a "modern world . . . born with the first atomic explosion."[2] She argued that historically ingrained modes of economic and cultural life had created "a theoretical glorification of labor" and "a factual transformation of the whole of society into a laboring society."[3] As labor ascended as a social organizing principle, ends rather than means guided value judgments, the practical overshadowed the theoretical, and (enlisting a pairing from classical philosophy) action (*vita activa*) trumped contemplation (*vita contemplativa*) as a principle of self- and community-attainment.

In "laboring society," Arendt claimed, the very intellectual resources we needed to judge, weigh, and direct our technical creations had withered. Events such as the "first atomic explosion" signaled that thought and "know-how" had "parted company for good," and that we risked becoming "helpless slaves, not so much of our machines as [that] know-how, thoughtless creatures at the mercy of every gadget that is technically possible."[4] In this concern, she drew on more than a century of social and political critique of the fusion of science, technology, industrialization, and capitalism—the influences of Marx and Weber are prominent, and Heidegger and Mumford loom just offstage. To this quartet of historical vectors, she conjoined

1. Hannah Arendt, *The Human Condition* (Chicago: University of Chicago Press, 1998).

2. Ibid, p. 6.

3. Ibid, p. 5.

4. Ibid, p. 4.

stepped-up postwar levels of advertising and mass consumption to reinforce her claims about the scope and entrenchment of a "laboring society."[5] Collectively, these deep cultural transformations held "great political significance"—not only in terms of the decision making of nation-state elites but also as a rupture among citizens, increasingly disinclined and unable to assess, weigh, and manage their circumstances. Her narrative, thus, fit into a well-developed genre of alienation in modernity; of new modes of social being conditioning human consciousness in ways that threatened the quintessentially human; and of technological supplanting humanistic values.[6]

As historians interested in spaceflight, what might catch our eye about Arendt's analysis and this literature of sociological and philosophic critique? One fact I have withheld. In that short interval between her Chicago lectures and their revision as *The Human Condition*, Sputnik occurred, and in her prologue to the book it becomes the dominant symbol of this condition of alienation. Sputnik, viewed through the Arendtian lens, was neither sui generis nor an exclamatory act in the superpower-centered geopolitical drama of the cold war. It was, rather, a manifestation and symbol of deeper structures of economic and cultural order, an analysis deriving broadly from Marxian sociology (in distinction from Marxian politics).[7]

This blend of theory and historiography provides the vantage for this essay. Marx's sociology often is seen with a hard, deterministic slant ("[T]he hand mill gives you society with the feudal lord; the steam mill society with the industrial capitalist"), but is used here less rigidly and as a methodological injunction: In assessing the "social" in spaceflight, look for the relations between modes of production (often with emphasis on the economic) and the cultural.[8] This mantra

5. Although Arendt does not cite either Theodor Adorno or Marshall McLuhan, one suspects she probably was familiar with their claims on the enhanced integration of mass production and mass consumption in Western societies. Adorno's well-known essay, "The Culture Industry," was published in 1947 and McLuhan also had begun a long run of media and advertising critiques. See Theodor W. Adorno, *The Culture Industry: Selected Essays on Mass Culture* (New York: Routledge, 2001) and Marshall McLuhan, *The Mechanical Bride: Folklore of Industrial Man* (New York: Vanguard Press, 1951).

6. This critique took a sharper turn in the 1960s. See, as an important example, Herbert Marcuse, *One-dimensional Man: Studies in the Ideology of Advanced Industrial Society* (Boston: Beacon Press, 1964). For a useful assessment of this genre and its meaning for the 1960s and beyond, see Marianne DeKoven, *Utopia Limited: The Sixties and the Emergence of the Postmodern* (Durham, NC: Duke University Press, 2004).

7. For an assessment of Marx from a contemporary stance, especially as regards the distinction between sociology and politics in Marx's thought, see Göran Therborn, "After Dialectics: Radical Social Theory in a Post-Communist World," *New Left Review* 43 (2007): pp. 63–114. See also Anthony Giddens, *Capitalism and Modern Social Theory: An Analysis of the Writings of Marx, Durkheim and Max Weber* (Cambridge, U.K.: Cambridge University Press, 1999).

8. Marx's notion of culture, of course, was not that of the post-World War II academy in the West. For a classic, brief etymological essay on "culture," see Raymond Williams, *Keywords: A Vocabulary of Culture and Society* (New York: Oxford University Press, 1976). On Marx's notion of culture and its relation to the broader genealogy of the concept, see Michael Denning, *Culture in the Age of Three Worlds* (London: Verso, 2004).

is intended only as a point of orientation. As analytical terms, "production" and "culture" run at the edges, embracing different elements, with different meanings, in different contexts. They only get life and clarity through historical specificity. For the historian, an Arendt-type or Marxian sociological analysis carries a burden: Details of difference and situatedness yield to the bright line-making of theory. Despite such semantic and methodological liabilities, I will use this high-level talk to highlight what I regard as a central historiographic problem in the field: the place of culture in our understanding of the development of spaceflight, a problem bound to its theoretical sibling, production. I will approach this claim not through the space history literature but through the broad, postwar intellectual reorientation of the humanities, in which the concepts of production and culture became key sites of theoretical engagement. The underlying question for space history is whether this theorizing and its implications, only loosely grounded in the empirical, provides a means for looking afresh at the field.[9]

In distilling the postwar theoretical landscape I will necessarily (dangerously) set the "Google map" to a wide view, emphasizing selected features. Soon after Arendt's *The Human Condition*, notions of production and culture took on new connotations. Production as a form of power, control, and as a technique for replicating and perpetuating particular forms of social life (through capitalist configurations of economic ownership, technology, and labor) came to include also (via the influence of Foucault's writings) the power attached to scientific disciplines, professions, and the academy, especially when seen historically as coincident with the formation and consolidation of nation-states.[10] In concert, culture as an analytic construct enjoyed luxuriant growth, in narrow application (subcultures of every stripe, corporate culture, institutional culture, and the like—think of how unquestionably natural it now seems to speak of "NASA culture"), and in the expansion of its ambit, carried along by the ever-widening reach of commodities and consumer values. Both trends helped to erase the older distinctions of high and low culture and to establish culture as a pervasive conditioner of human affairs. Culture, thus, with protean alacrity, adhered to production in the small (say, within a community, an institution) and in the transnational-spanning large. It served, at once, as energetic contributor to capitalism and as a potential source of resistance. The dominant thread in the mix, at

9. This essay may be read as a companion to Martin Collins, "Community and Explanation in Space History (?)," in *Critical Issues in the History of Spaceflight*, Stephen J. Dick and Roger Launius, eds. (Washington, DC: NASA, 2006): pp. 603–613.

10. The works generally cited as most influential in the United States are Michel Foucault, *The Birth of the Clinic: An Archaeology of Medical Perception* (New York: Pantheon Books, 1973), and Michel Foucault, *Discipline and Punish: The Birth of the Prison* (New York: Pantheon Books, 1977). For a useful instance of Foucault-style analysis regarding states and production, see James C. Scott, *Seeing Like a State: How Certain Schemes to Improve the Human Condition Have Failed* (New Haven, CT: Yale University Press, 1998).

least as regards theory, focused on big capital, commodities, and consumption.[11] Culture, in this sense, not only was bound to a particular era of capitalist production, but also participated in an important epistemological shift: In a world of circulating commodities, representations and meanings readily detach from the circumstances of their creation (a basic tenet of poststructuralism, deconstructionism, and postmodernism), become malleable, locally or individually defined. In Western and global capitalism, culture, then, is expressive of production and constitutes a phenomenon that envelops and conditions it. Or, simply, culture becomes self-referential.[12]

This theoretical neighborhood informed a range of cultural and social fields that emerged in the 1960s and has dominated academic critique since, in part,

11. The point here is to emphasize the emergence of culture as a widely used conceptual category. Different definitions of the cultural coexisted under this umbrella. Two of the most influential came from Raymond Williams (culture as a creative intersection between tradition and new experience) and Clifford Geertz (culture as a semiotic system). Particularly relevant here, Williams's work served as an exemplar in connecting culture and modes of production; see Raymond Williams, *Culture and Society, 1780–1950* (New York: Columbia University Press, 1983) [originally published 1958]. For Geertz, see Clifford Geertz, *The Interpretation of Cultures: Selected Essays* (New York: Basic Books, 1973). Not discussed here is the relation between the "social" and the "cultural" in the academy in this time period; see Victoria Bonnell and Lynn Hunt, eds., *Beyond the Cultural Turn: New Directions in the Study of Society and Culture* (Berkeley, CA: University of California Press, 1999). For a broad review of notions of culture, see Peter Burke, *Varieties of Cultural History* (Ithaca, NY: Cornell University Press, 1997). For the important argument on the historically specific character of the postwar use of culture, see Denning, *Culture in the Age of Three Worlds*; making a similar case is Geoff Eley, *A Crooked Line: From Cultural History to the History of Society* (Ann Arbor, MI: University of Michigan Press, 2005). For a view that emphasizes the connection between protest politics and the rise of culture, see Terry Eagleton, *The Idea of Culture* (Oxford, U.K.: Blackwell, 2000). On cold war culture, see Stephen J. Whitfield, *The Culture of the Cold War* (Baltimore, MD: Johns Hopkins University Press, 1996); and Peter J. Kuznick and James Gilbert, eds., *Rethinking Cold War Culture* (Washington, DC: Smithsonian Institution Press, 2001). On consumption, see Richard Wightman Fox and T. J. Jackson Lears, *The Culture of Consumption: Critical Essays in American History, 1880–1980* (New York: Pantheon Books, 1983); Sande Cohen and R. L. Rutsky, eds., *Consumption in an Age of Information* (Oxford: Berg, 2005); David B. Clarke, *The Consumer Society and the Postmodern City* (London: Routledge, 2003); John N. Duvall. ed., *Productive Postmodernism: Consuming Histories and Cultural Studies* (Albany, NY: State University of New York Press, 2002); and Jean Baudrillard, *The Consumer Society: Myths and Structures* (London: Sage Publications, 1998). On the important question of the international reach of American culture and its effects, see Rob Kroes, *If You've Seen One, You've Seen the Mall: Europeans and American Mass Culture* (Urbana, IL: University of Illinois Press, 1996), and Rob Kroes, ed., *High Brow Meets Low Brow: American Culture as an Intellectual Concern* (Amsterdam: Free University Press, 1988).

12. It is important to distinguish two threads of this claim. One is a general epistemological claim on the foundations of knowledge: the view that we have no privileged a priori or empirical means to assess the validity of propositions about the world. The other is historically grounded: that an emergent postwar commercial and communications culture created a new relation between signs and the real and the cultural status and use of such signs. This work built on Marx's early arguments on the commodity as "fetish." This view is most thoroughly developed in Jean Baudrillard, *For a Critique of the Political Economy of the Sign* (St. Louis, MO: Telos Press, 1981).

because it offered a first-order cut at defining and investigating basic questions.[13] It pushed to the fore questions of causation: How and why modes of production change and, in turn, are related to and interact with reshaped social or cultural orders. In highlighting causation, it made questions of periodization, of alertness to patterns of change, an explicit, necessary project.

Consider the colloquialism that provided the organizing idea for this conference (and as an assumption only is occasionally engaged by the papers)—that spaceflight "impacts" society and by implication that such effect is historically descriptive and serves as an important basis for conceptualizing research. The first part of this claim is trivially true in a limited way but obscures a more cogent historical question: How do we theorize the relation between spaceflight and the social? Or, more aptly and fundamentally, to build on the interpretive sketch advanced here, how do we theorize spaceflight in and as part of a causally informed history?

In space history, "impact" seems a natural category because of the privileged role ascribed to state-centered action in cold war historical accounts. The invigoration of government-directed power and action postwar (especially in the United States) gives credence to a rough causal formula of "the state acts, society receives." Or, to venture a touch more complexity, each may affect the other, but each stands as a relatively distinct sphere of activity. Few, if any, historians would subscribe overtly to this cartoonish causal sketch, yet the field has been noticeably slow to articulate more robust frames of analysis. In this conference, authors joined social and cultural with politically centered narratives but leaned away from considering underlying causal issues or broader frames of theoretical interpretation. In terms of engagement

13. This thumbnail is not meant to supplant or diminish other key analytic concepts such as class, gender, or race, or the related pairing of identity and difference. Culture and production may complement or serve as tools for understanding how these other categories may be constituted. Too, the emphasis on culture highlights the way in which, over this time period, anthropology (in particular) and sociology have been the crucial intellectual sites for working through the implications of theoretical claims generated in philosophy, literary criticism, and intellectual history. A focal point, relevant to this essay, for examining this process is the material culture literature, which largely originated in the 1960s. See especially the work of Daniel Miller, founding editor of the journal *Material Culture*; books include Daniel Miller, *Capitalism: An Ethnographic Approach* (Oxford: Berg, 1997); Daniel Miller, ed., *Material Cultures: Why Some Things Matter* (Chicago: University of Chicago Press, 1998); Daniel Miller, ed., *Materiality* (Durham, NC: Duke University Press, 2005). In this vein, see also Arjun Appadurai, ed., *The Social Life of Things: Commodities in Cultural Perspective* (Cambridge, U.K.: Cambridge University Press, 1986); Bill Brown, ed., *Things* (Chicago: University of Chicago Press, 2004); Nicholas Thomas, *Entangled Objects: Exchange, Material Culture, and Colonialism in the Pacific* (Cambridge, MA: Harvard University Press, 1991); and Fred R. Myers, ed., *The Empire of Things: Regimes of Value and Material Culture* (Santa Fe, NM: School of American Research Press, 2001). On theory in relation to material culture, see especially Pierre Bourdieu, *The Logic of Practice* (Stanford, CA: Stanford University Press, 1990), and Jean Baudrillard, *Selected Writings* (Stanford, CA: Stanford University Press, 2001.

with other historical subfields, this represents a measure of progress but leaves unaddressed the deeper challenges of using available theoretical tools to clarify the field's explanatory aims—of placing spaceflight *in* history.[14]

And that brings us back to Arendt, the tradition of Marxian sociological critique of which her work is part, and the analysis presented here.[15] This analytic frame has deep relevance to post–World War II historiography, including and especially for spaceflight. It cautions us to de-center the importance given to state political action and actors in trying to understand historical change; that is, it suggests that the crucial unit of analysis is not the state in isolation, or even the "contract state" or "military-industrial-university complex" (as widely used constructs attentive to government–market interdependencies).[16] Rather, it directs our attention to the largest configurations of production—of the state in conjunction with capitalism writ large (to include the academy as it constitutes disciplines and professions).[17] Arendt's fleeting attention to Sputnik was to make this theoretical point and to situate the event historically—to ask how the first satellite and the emerging effort of spaceflight fit into a *particular* configuration of production (state and capital) and culture. That question should be as paramount for space history as it was for Arendt.

In the 1960s and after, a torrent of literature concerned with production and culture yielded theoretically infused, historically specific, profound claims regarding the postwar years: that the basis of production *and* the basis of culture had been transformed, reconfiguring each and, to use Arendt's labeling, the human condition.[18] Business, markets, and nation-state prerogatives and power changed; postcolonial and new forms of global politics and geographies emerged; and, above all (especially

14. For a related example of this type of exercise, reflecting a different historiographic moment, see Eric Hobsbawm, "From Social History to the History of Society," *Daedalus* 100 (1971): pp. 20–45. For a review of the space history literature that samples a range of current methodologies but is relatively silent on the issue of relating the field to broader theoretical accounts, see Roger D. Launius, "Interpreting the Moon Landings: Project Apollo and the Historians," *History and Technology* 22 (2006): pp. 225–255.

15. I loosely confederate under the banner of Marxian sociology a broad range of authors and perspectives, including those that fall on a spectrum of sympathy or antipathy to Marxian *politics* (consider, say, the work of Frederic Jameson and Daniel Bell). The circumstance that this essay points to is the deep sway of Marxian modes of explanation in the historical and cultural studies fields and the still-powerful utility of a structural analysis in probing issues of causation. Even the poststructuralist and postmodern literatures that see history as composed of texts and representations loosened from any notion of a substantial historical reality don't quite succeed in diminishing history to mere narrative.

16. On the notion of the United States as a contract state, see H. L. Nieburg, *In the Name of Science* (Chicago: Quadrangle Books, 1966).

17. For a useful instance of engagement with issues of causation and periodization relevant to this essay, see Ann Douglas, "Periodizing the American Century: Modernism, Postmodernism, and Postcolonialism in the Cold War Context," *Modernism/Modernity* 5 (1998): pp. 71–98. Douglas is skeptical of the various "isms" of critique but relies fundamentally on the notion of Marxian sociology sketched here.

18. The clearest exposition of these claims is David Harvey, *The Condition of Postmodernity: An Enquiry into the Origins of Cultural Change* (Oxford, U.K.: Blackwell Publishers, 1989).

given the topic of this conference), that culture (as embodied in commodities, transnational business practices, and ever more ubiquitous communications and media technologies) was not merely a glossy coat of paint on capitalism's formidable machine but integral to it and constituted a potent, complementary reality. The insights relating to causation and periodization in this swath of research and theory have remained at the margins of space history, despite the obvious: The changes mapped and claimed are coeval *and* intimately bound to the development of spaceflight—in its many dimensions as technology, site of knowledge creation, state activity, business undertaking, military venture, global utility, and national and international cultural trope extraordinaire. Can space history be history without more fundamentally assessing and testing itself against this theoretical frame, seemingly, at first pass, deeply relevant to the field's research challenges (allowing for many the points of disputation on methodology, cause, and consequence contained in this unwieldy set of literatures)? Not just to import methods into the field, but to wrestle with problems of cause and change alongside other humanistic disciplines? The state-centered narratives that seemed passably suitable in the early years of the field now seem increasingly inadequate for reinterpreting the early cold war and, especially, to organize the history of recent decades. The rest of this essay will look briefly at the two nodes of this analysis—production and culture—and their implications for situating spaceflight in this broader landscape of the postwar historical experience.

After World War II, the United States, in a series of steps in collaboration with business and the academy, created, to use Daniel Yergin's coinage, a national security state.[19] Assessments of its relation to modes of production and shifts in the cultural landscape surfaced in short order.[20] But broad views that an assertive policy of massive government spending on research, development, discipline-centered knowledge, and technology was providing an opportunity to refashion the very basis of capitalism took longer to arise.[21] Emblematic in this regard was the work of the politically liberal economist John Kenneth Galbraith and the conservative

19. On the national security state, see Daniel Yergin, *Shattered Peace: The Origins of the Cold War and the National Security State* (Boston: Houghton Mifflin, 1977). My analytic focus here is the United States. The goal is not a comprehensive historiographic review and argument; rather, it is only to suggest underlying research problems for the field. It should be noted, though, that the literature on the transformation of capitalism in the 1960s and 1970s is U.S.-centric, given the United States' dominant economic position.

20. Critiques and targeted assessments were prevalent throughout the late 1940s and 1950s. See, as but one example, C. Wright Mills, *The Power Elite* (New York: Oxford University Press, 1959).

21. In my reading, most of the early literature envisions the military-industrial-academic complex as an adjunct to the broader economic system of capitalism. The focus tends to be on distortion rather than on a reconstitution of economic structures. The distortion view is the organizing idea of the classic study by RAND economist Charles Hitch. See Charles Johnston Hitch, *The Economics of Defense in the Nuclear Age* (Cambridge, MA: Harvard University Press, 1960).

sociologist Daniel Bell. By 1960, both had examined capitalism's condition sans consideration of the cold war elephant in the room. Galbraith's 1958 *The Affluent Society* explored the cultural changes accompanying widespread consumerism (a theme already much in the air, as indicated by Arendt's exposition); Bell's 1960 *The End of Ideology: On the Exhaustion of Political Ideas in the Fifties* claimed that the long-standing tussle between capital and labor had petered out.[22] A triangulation of accommodation among labor, business, and the state, he argued, dampened the conflicting ideological positions of Marxian and capital-oriented politics. In short, both authors saw the defining issues of the 1950s in conceptual terms consistent with the political economy of the early twentieth century. But each, as a consequence of these researches, immediately began a reappraisal of the postwar landscape.

By the late 1960s, Galbraith and Bell made their central research concern the question of whether capitalism and the state, through cold war scientific and technological activism, had revamped existing modes of production. Galbraith's 1967 *The New Industrial State* and, especially, Bell's 1973 *The Coming of Post-Industrial Society: A Venture in Social Forecasting* began to make the case that (among other factors) the federal research and development engine had facilitated a fundamental change in business practice, elevating the role of knowledge, and thereby knowledge professions, serving to recast the basic relationship between labor and capital.[23] This reconfiguration had many other elements, including a policy shift that loosened government controls over markets and corporations (nationally and internationally), the burgeoning possibilities of information technologies, and the enhanced role

22. New additions with updated introductions by the authors help in approaching these texts; see John Kenneth Galbraith, *The Affluent Society* (Boston: Houghton Mifflin, 1984) and Daniel Bell, *The End of Ideology: On the Exhaustion of Political Ideas in the Fifties: With "The Resumption of History in the New Century"* (Cambridge, MA: Harvard University Press, 2000).

23. John Kenneth Galbraith, *The New Industrial State* (Boston: Houghton Mifflin, 1967); Daniel Bell, *The Coming of Post-industrial Society: A Venture in Social Forecasting* (New York: Basic Books, 1999).

of consumption in economic and cultural life.[24] A range of authors came to see this complex of changes as the seedbed of globalism (conceived as the relative enhancement of the power of markets in relation to states across the transnational landscape) and postmodernism (conceived as a new cultural condition associated with this mode of production).[25] This spare delineation of this significant shift, played out over two-plus decades, is intended only for a limited point: that spaceflight—in its civil, military, commercial dimensions—developed and participated in fundamental changes in state-market regimes. Thus, if explanation (whether of "impact," of causes, or patterns of change) is a goal, then such explanation needs to encompass, as a first-order concern, how nation-state-centered accounts relate to the different frames of analysis suggested by globalism and postmodernism.[26] But these points bring us back to the other, crucial question of this review: What is the relation between modes of production and culture in the age of spaceflight?

---

24. The best delineation of the shift from "organized" to "disorganized" capitalism (and, in the Marxian tradition, correlating this change to distinctive cultural orders) is Scott Lash, *The End of Organized Capitalism* (Cambridge, U.K.: Polity, 1987); also see Nick Heffernan, *Capital, Class and Technology in Contemporary American Culture: Projecting Post-Fordism* (London; Sterling, VA: Pluto Press, 2000). For a dense, contemporaneous account, less attuned to the enhanced status of knowledge seen by Galbraith and Bell, see Ernest Mandel, *Late Capitalism* (London: NLB, 1975) [first published as Der Spätkapitalismus, Suhrkamp Verlag, 1972]. For an overview of changes from the 1960s in corporate structure and strategy as firms moved from primarily national to broadly transnational modes of operation, see Naomi R. Lamoreaux, Daniel M. G. Raff, and Peter Temin, "Beyond Markets and Hierarchies: Towards a New Synthesis of American Business History," *American Historical Review*, 108 (April 2003): pp. 404–433. On the rise of private market ideology and accompanying policy reorientation in this period, see Daniel Yergin and Joseph Stanislaw, *The Commanding Heights: The Battle between Government and the Marketplace That Is Remaking the Modern World* (New York: Simon & Schuster: 1998). On changes in modes of production from the perspective of history of science and technology, see Paul Forman, "The Primacy of Science in Modernity, of Technology in Postmodernity, and of Ideology in the History of Technology," *History and Technology* 23 (2007): pp. 1–152; and Philip Mirowski, *The Effortless Economy of Science?* (Durham, NC: Duke University Press, 2004).

25. See, as a range of examples, Arjun Appadurai, ed., *Globalization* (Durham, NC: Duke University Press, 2001); Francis Fukuyama, *The End of History and the Last Man* (New York: Perennial, 2002); Anthony Giddens, *The Consequences of Modernity* (Stanford, CA: Stanford University Press, 1990); Harvey, *The Condition of Postmodernity;* Fredric Jameson, *Postmodernism, Or, the Cultural Logic of Late Capitalism* (Durham, NC: Duke University Press, 1991); Jonathan Xavier Inda and Renato Rosaldo, eds., *The Anthropology of Globalization: A Reader* (Malden, MA: Blackwell Publishers, 2002); and Frank Webster, *Theories of the Information Society* (London: Routledge, 2002). Several case studies in space history engage these changes. See Martin Collins, "One World One Telephone: Iridium, One Look at the Making of a Global Age," *History and Technology* 21 (2005): pp. 301–324; Lisa Parks, *Cultures in Orbit: Satellites and the Televisual* (Durham, NC: Duke University Press, 2005); and Peter Redfield, "The Half Life of Empire in Outer Space," *Social Studies of Science* 32 (2002): pp. 791–825. More broadly, in the historical profession these changes have given new life to the transnational (in contrast to the long-standing preference for the "national") as a key unit of analysis. See Thomas Bender, ed., *Rethinking American History in a Global Age* (Berkeley, CA: University of California Press, 2002).

26. Much of the globalism literature makes the case for the diminishment of nation-state power in face of the market. A cogent analysis of the relation among markets, states, and ideology is Therborn, "After Dialectics."

This question vexes space history—all the more for lack of explicit *theoretical* attention.[27] For it is in the very period in which spaceflight develops that culture becomes an historically specific, key concept—as an analytical instrument within the academy and as a descriptor of the broad phenomenological effect of commerce, commodities, and communications on our day-to-day perceptions and constitution of the world.[28] In this latter regard, it embodies deep claims, ontological and psychological: that time, distance, what counts as global or local, and how identities are constituted (ours and those of distant others) all are conceptualized and experienced in new ways. Such propositions became commonplace in the globalism and postmodernism literatures.[29] Philosopher Langdon Winner, as one example, captured this melding of productive regimes, technology, global-local perceptions, and individual identity and their close association with the construct of culture:

> The map of the world shows no country called Technopolis, yet in many ways we are already its citizens. If one observes how thoroughly our lives are shaped by interconnected systems of modern technology, how strongly we feel their influence, respect their authority and participate in their workings, one begins to understand that, like it or not, we have become members of a new order in history. To an ever-increasing extent, this order of things transcends national boundaries to create roles and relationships grounded in vast, complex instrumentalities of industrial production, electronic communications, transportation, agribusiness, medicine, and warfare. Observing the structures and processes of these vast systems, one begins to comprehend a distinctively modern form of power, the foundations of technopolitan culture.[30]

27. Of course, the cultural has long been on space history's radar. In Howard McCurdy's oft-cited work, the claim is advanced that popular culture helped political elites accept the idea of spaceflight. Missing is theoretical understanding of why that might have been so. See Howard McCurdy, *Space and the American Imagination* (Washington, DC: Smithsonian Institution Press, 1997). Closer to the mark methodologically is Michael L. Smith, "Selling the Moon: The U.S. Manned Space Program and the Triumph of Commodity Scientism," in *The Culture of Consumption: Critical Essays in American History, 1880–1980,* Richard Wightman Fox and T. J. Jackson Lears, eds. (New York: Pantheon, 1982).

28. Again, Denning, *Culture in the Age of Three Worlds,* is the best historical account of this change. The touchstone theoretically is Jameson, *Postmodernism.* See also Zygmunt Bauman, *The Individualized Society* (Cambridge, U.K.: Polity, 2001).

29. A useful assessment and critique of these claims is Anna Tsing, "The Global Situation," in *The Anthropology of Globalization,* ed. Jonathan Xavier Inda and Renato Rosaldo (New York, Blackwell Publishers, 2001), pp. 453–485.

30. Langdon Winner, *The Whale and the Reactor: A Search for Limits in an Age of High Technology* (Chicago: The University of Chicago Press, 1986), p. ix. Winner presented this notion as a preamble to his study and not as the focus of his research. A kindred outlook, richer in sociological claims, is Ulrich Beck, *Risk Society: Towards a New Modernity* (London: Sage Publications, 1992).

In its emphasis on the constitutive role of technology, one easily can read into this description the multiple ways in which the various aspects of spaceflight intersect and participate in this reinvention of the human experience on a global scale. Less evident in Winner's characterization are the myriad sites of contestation to these changes, the ways in which the local and global get interpreted, defined, and formed by individuals, communities, and nations.[31] Culture, in this argument, is as pervasive as capital and the technological systems with which it is one—not resolutely dominant, but "there" as an element joined, adapted, or countered by many local cultures.[32] It becomes a form of political power and a framework of meanings that may reinforce modernist, nation-state-centered notions of progress, or, or alternatively, in postmodern fashion refer back on itself.[33]

The linkage between culture and capitalism has been strongest in the postmodernist critique, reflecting the seminal influence of Frederic Jameson's 1984 essay, "Postmodernism, Or, the Cultural Logic of Late Capitalism."[34] Writing in 1991, Jameson gave the condition described by Winner a different inflection, shifting emphasis from technological systems vast in extent to commodities as key in understanding the status of culture and its relation to changes in the human experience:

> Postmodernism is what you have when the modernization process is complete and nature is gone for good. It is a more fully human world than the older one, but one in which "culture" has become a veritable "second nature." Indeed, what happened to culture may well be of the more important clues for tracking the postmodern: an immense dilation of its sphere (the sphere of commodities), an immense and historically original acculturation of the real "[C]ulture" has become a product in

31. The anthropological and postcolonial literatures have covered this issue in depth. See, as examples, Arjun Appadurai, *Modernity at Large: Cultural Dimensions of Globalization* (Minneapolis, MN: University of Minnesota Press, 1996); Edward W. Said, *Culture and Imperialism* (New York: Vintage Books, 1994); James Clifford, *Routes: Travel and Translation in the Late Twentieth Century* (Cambridge, MA: Harvard University Press, 1997); Frederick Cooper, *Colonialism in Question: Theory, Knowledge, History* (Berkeley, CA: University of California Press, 2005); *The Anthropology of Globalization: A Reader* (Malden, MA: Blackwell Publishers, 2002); and Bryan S. Turner, *Orientalism, Postmodernism, and Globalism* (London: Routledge, 1994).

32. Valuable in sorting through these issues from a sociological perspective is Ulrich Beck, *Power in the Global Age: A New Global Political Economy* (Cambridge, U.K.: Polity, 2005); and Ulrich Beck, *The Cosmopolitan Vision* (Cambridge, U.K.: Polity, 2006).

33. The modernist ideals of progress, especially as adopted by less-developed countries, are a crucial part of geopolitics in this dynamic. They are a resource in defining these nation-states contra to the West. See, for example, Gyan Prakash, *Another Reason: Science and the Imagination of Modern India* (Princeton, NJ: Princeton University Press, 1999).

34. Jameson's 1984 essay provided the title for his 1991compilation of essays and was republished therein: Fredric Jameson, *Postmodernism, Or, the Cultural Logic of Late Capitalism* (Durham, NC: Duke University Press, 1991). On Jameson's singular standing in postmodern and critical thought after 1980, see Perry Anderson, *The Origins of Postmodernity* (London: Verso, 1998).

> its own right; the market has become a substitute for itself and fully as much a commodity as an any of the items it includes within itself.[35]

The "dilation" of the congruent spheres of commodities and culture implied not only that more stuff circulated ever more widely through capitalism's arteries, but also that ever more realms of life—ideas, images, emotions, spirituality, politics—had become commoditized, packaged, and divorced from the circumstances of their creation, infused with the values of the market, including culture itself.

Jameson's analysis, Marxist and pointedly historical, was not wholly original. In 1961, the more politically and theoretically measured Daniel Boorstin emphasized, too, culture's historically new and ascendant status, especially as connected to the use of images.

> In . . . nineteenth-century America the most extreme modernism held that man was made by his environment. In twentieth-century America, without abandoning the belief that we are made by our environment, we also believe our environment be made almost wholly by us. This is the appealing contradiction at the heart of our passion for pseudo events: for made news, synthetic heroes, prefabricated tourist attractions, homogenized forms of art and literature (there are no "originals," but only the shadows we make of other shadows). We believe we can fill our experience with new-fangled content. Everything we see and hear and do persuades us that this power is ours.[36]

The primary distinction between the historical-empirical Boorstin and critically inclined Jameson was a profound shift in attitude. Boorstin lamented this undermining of the real; Jameson accepted it as a constitutional feature of capitalism in the latter half of the twentieth century.[37]

What might we make of these claims binding together epistemology, commodities, and culture conceived as a "second nature"? First is to reemphasize that these are claims about periodization—the making of an era which included spaceflight. As a start, they might encourage us to overlay this analytic on space history's chronology and look for reorientations that include state-centered politics

35. Jameson, *Postmodernism,* pp. ix–x.

36. Daniel J. Boorstin, *The Image: A Guide to Pseudo-events in America* (New York: Atheneum, 1971), pp. 182–183. Boorstin's analysis was roughly contemporaneous with Marshall McLuhan's first articulations of the notions of the global village and the medium as message in the late 1950s, early 1960s.

37. To Daniel Bell, writing in 1976, Boorstin's concern loomed large. The abandonment of the real meant abandoning the Protestant ideals of restraint and soberness (borrowing from Max Weber's classic argument), and thus hollowing out a key pillar of capitalism. Jameson's account, on this point, seems to have weathered best. See Daniel Bell, *The Cultural Contradictions of Capitalism* (New York: Basic Books, 1996 [a 20th anniversary edition]).

and happenings in production and culture broadly conceived. A reconceptualization in this vein of the 1960s, for example, might offer interpretations that correlate culture and politics within an integrated frame, rather than as partially overlapping but separate orders of experience. This integration already can be found in several examples of literary criticism that have plumbed space- and business-themed literature. These works employ the same conceptual structure presented here (not surprising) and stand as examples of the explanatory possibilities of this type of analysis.[38] And they raise a deeper point: that in an era of culture, literary tools and modes of analysis may be a necessary complement to history in understanding the human condition. In modernism, the view prevailed that artists stood as critics positioned athwart the culture of their time. In postmodernism and an era of culture as "second nature," the implied claim is that all historical actors are so suffused with the cultural that no one stands beyond it—we all, in different ways, are users, interpreters, and refashioners in the tide of symbols.

This deep, prosaic sense of culture in the postwar period is the essence of Jameson's observation. This view marries well with my own research on multinational business and the development of a global satellite telephony system by the Fortune 500 company, Motorola—a subject distant from literary theory but richly indicative of these claims about the status of culture. Culture as preoccupation, conceptual category, and tangible reality suffused Motorola and its development of this satellite project in the later 1980s and 1990s. Such an outlook came to seem essential in running a business with tens of sites around the globe, managing a complex organization, creating new modes of engineering project management, and operating in a media and symbolic environment over which the corporation only had partial control. This culture fixation led the firm to create Motorola University in 1989, with the charge to research and manage the many intersections of culture and corporate practice that now seemed central to the firm's success. Within the university sat the Center for Technology and Culture, led by an anthropologist who established a range of links to academic anthropology. As the work of these units became integrated into the Motorola business enterprise, they reinforced the view that a necessary relationship existed among culture, the corporation, and an ability to act across the world stage.[39] Motorola University was one notable instance

38. See William D. Atwill, *Fire and Power: The American Space Program as Postmodern Narrative* (Athens, GA: University of Georgia Press, 1994); Thomas Peyser, *Utopia & Cosmopolis: Globalization in the Era of American Literary Realism* (Durham, NC: Duke University Press, 1998); Joseph Tabbi, *Postmodern Sublime: Technology and American Writing from Mailer to Cyberpunk* (Ithaca, NY: Cornell University Press, 1995); and Graham Thompson, *The Business of America: The Cultural Production of a Post-war Nation* (London: Pluto Press, 2004).

39. For a comic, often polemical, and sometimes accurate account of the relation between academic cultural studies and business uses of culture, see Thomas Frank, *One Market Under God: Extreme Capitalism, Market Populism, and the End of Democracy* (New York: Doubleday, 2000), chapter 8.

of a larger trend: over a decade, from the mid 1980s to mid 1990s, more than a thousand corporate universities were created in the United States—all of which were a response, in one fashion or another, to the perceived culture problem and its relation to competition in transnational markets.[40]

This example suggests the ways in which culture—and the pursuit of space history—may be intimately bound to configurations of production and politics.[41] More broadly, it suggests the possibilities for space history when we widen the frame of analysis. In emphasizing theory, the argument in this essay is not a call to reductionism, to suggest that richness of historical experience get shoehorned into an empirically contestable structure of critique. Nor is it to minimize the limitations of this theoretical literature. It, by and large, reflects its origination in the Western academy and life experience. Rather, this essay presents a gentle imperative: Be attentive to the possibilities of thought in the humanities that has seen production and culture as a historical problem of the first rank. The notions limned here, as propositions and analytic constructs, provide a basis for reassessing space history's domain of problems and questions, such as the meaning and status of spaceflight since World War II; its political possibilities; its place in rhetoric and thought in the period; its relation to capital, culture, and commodities; and the combinations of the real, the symbolic, and the imaginary that have gained cultural preference.

In making a profound *historical* claim that a new configuration of epistemology, production, and culture marks the very period coincident with the development of spaceflight, this body of thought poses a deep challenge to space history. To understand the functioning of symbols and images surfaces as a crucial problem. In what ways, for example, do spaceflight and space fiction—in what balance, with what blurring of genres, of the real and the semiotic—speak to the preoccupations of particular regimes of production and cultural experience? What satisfies and why? What period themes—whether utopia, Earthly escape, the human body, identity, difference—find resonance in the narratives of spaceflight?[42] These are substantial questions that deserve theoretical articulation. Otherwise, our understanding of

40. For an overview of this trend from a policy perspective, see Stuart Cunningham et al., "The Business of Borderless Education," a report commissioned by the Department of Education, Youth, and Training, Canberra, Commonwealth of Australia, 2000.

41. This perspective is partially realized in Constance Penley, *NASA/Trek: Popular Science and Sex in America* (London: Verso, 1997).

42. Jameson, the doyen of the capitalist-articulated postmodern, has made science fiction a major theme of his research, reflecting a judgment that the genre is deeply expressive of the period. See Frederic Jameson, *Archaeologies of the Future: The Desire Called Utopia and Other Science Fictions* (London: Verso, 2005). For an overview of culture-inflected literature in space historiography, see Asif Siddiqi, "American Space History: Legacies, Questions, and Opportunities for Future Research," in *Critical Issues in the History of Spaceflight*, Stephen J. Dick and Roger Launius, eds. (Washington, DC: NASA, 2006), pp. 433–480. But, again, the question of how culture might relate to broader problems of explanation in the postwar period are not explicitly addressed.

spaceflight (whether as a state or market undertaking) may remain disconnected from analyses and insights that view production more comprehensively. Without an invigorated conceptual toolbox, the intersection of the cultural and spaceflight may seem primarily as kitsch (Star Trek *Enterprise*/Space Shuttle *Enterprise*) or as haphazard and of elusive meaning—rather than as a critical site for investigating the postwar experience.

This analysis offers one path toward a more considered balance between theory and the empirical, toward thinking about spaceflight in history, as a telling angle on our understanding of the late twentieth and early twenty-first centuries.

About the Authors

James T. Andrews is an associate professor of modern Russian and comparative European history in the Department of History at Iowa State University (ISU), where he is co-director of ISU's Ph.D. program and Center for the Historical Studies of Technology and Science. At ISU, he has also been director of Russian, East european, and Central Asian Studies and director of graduate studies in history. He holds a Ph.D. in modern Russian/Soviet history from the University of Chicago and has taught as a visiting professor at several research institutions, including the University of Texas at Austin. Since the summer of 1995, he has been affiliated as a senior research associate with the Russian Academy of Sciences Institute for the History of the Natural Sciences and Technology in Moscow and St. Petersburg. Professor Andrews's books and numerous articles have analyzed the intersection of science and technology, society, and public culture in modern Russia and in a comparative Eurasian framework. He is the author of *Science for the Masses: The Bolshevik State, Public Science, and the Popular Imagination, 1917–34* (Texas A&M University Press, 2003) and editor of *Maksim Gor'kii Revisited: Science, Academics and Revolution* (1995). His newest book, to be published in the Texas University Press Centennial of Air-Flight Series, is entitled *Visions of Space Flight: K. E. Tsiolkovskii, Russian Popular Culture, and the Birth of Soviet Cosmonautics, 1857–1957* (forthcoming, 2008). He is also currently completing a monograph (overview synthesis) entitled *Science and the Public Sphere: Technology, Science, and European Public Culture, 1543–Present.*

Glen Asner joined the NASA Headquarters History Division in December 2004. He recently completed his doctoral dissertation, titled "The Cold and American Industrial Research," at Carnegie Mellon University. Among his publications is an article, "The Linear Model, the U.S. Department of Defense, and the Golden Age of Industrial Research," in *The Science-Industry Nexus*, proceedings of the 123rd Nobel Symposium, edited by Karl Grandin, Nina Wormbs, and Sven Widmalm. In collaboration with Stephen Garber, Glen is currently writing a history of the development of the Vision for Space Exploration. Glen served as a contract historian prior to joining NASA on projects for the U.S. Department of Defense, the U.S. Department of Energy, and the Hagley Museum and Library.

Linda Billings is a Washington-based research associate with the SETI (Search for ExtraTerrestrial Intelligence) Institute of Mountain View, California. She has been conducting science and risk communication research for NASA's Planetary Protection Office since September 2002. From September 1999 through August

2002, she was director of communications for SPACEHAB, Inc., a builder of space habitats. Dr. Billings earned her Ph.D. in mass communication from Indiana University's School of Journalism in September 2005. Her expertise is in mass communication, science communication, risk communication, rhetorical analysis, and social studies of science. Her research has focused on the role that journalists play in constructing the cultural authority of scientists and the rhetorical strategies that scientists and journalists employ in communicating about science. She earned her B.A. in social sciences from the State University of New York at Binghamton and her M.A. in international transactions from George Mason University. Dr. Billings has worked for 25 years in Washington, DC as a researcher, journalist, freelance writer, communication specialist, and consultant to the government. As a researcher, she has worked on communication strategy, risk communication, and audience studies for NASA's Planetary Protection program. As a corporate communications director, she was responsible for communications with the media, the public, and the financial community. As a journalist, she covered energy, the environment, labor relations, and aerospace, primarily for the trade press. She was the founding editor of *Space Business News* (1983–1985) and the first senior editor for space at *Air & Space/ Smithsonian* magazine (1985–1988). She also was a contributing author for *First Contact: The Search for Extraterrestrial Intelligence* (New American Library, 1990). Dr. Billings was a member of the staff for the National Commission on Space (1985–1986). For the National Science Foundation and NASA, she has worked as a policy analyst, communications specialist, education and outreach planner, and writer and editor. Her freelance articles have been published in outlets such as the *Chicago Tribune* ("Space Station Is Good for More Than Star-Gazing," 8 October 1998), the *Washington Post Magazine* ("Realtime: Pre-Life Sciences," 11 August 1996) and *Space News* ("Aim for Exploration, Not Exploitation," 14–20 October 1996).

Kevin M. Brady is a doctoral candidate in history at Texas Christian University in Fort Worth, Texas, and is studying with Dr. Gregg Cantrell. Brady's classwork and research have focused on Texas history, science and technology, history of NASA, and military history. Currently, Brady is working on a dissertation titled "NASA Launches Houston into Orbit: The Economic, Political, and Social Impact of the Aerospace Program on Southeast Texas, 1961–1969." In May 2000, Brady earned a B.S. degree in secondary education with a specialization in American history from Baylor University in Waco, Texas. After graduating from Baylor, he remained at the university to pursue graduate work in history. While at Baylor, Brady was the recipient of numerous awards for his academic excellence and dedication to the university. For example, Baylor's Student Life Division awarded him with the James Huckins Award for Outstanding Graduate Assistant. In August 2002, Brady received an M.A. in history from Baylor University. In the fall of 2002, Brady enrolled at Texas Christian University to pursue doctoral studies in history. From 2003 until 2005, he served as a teaching assistant at Texas Christian University, where he taught U.S. survey courses.

During this time, Brady also worked as an adjunct instructor at the Tarrant County Community College. Brady is the author of several articles and has also contributed to a number of reference works, including *Space Exploration and Humanity: A Historical Encyclopedia*. Currently, Brady serves as a research intern with the Universities Space Research Association (USRA) at the NASA Lyndon B. Johnson Space Center (JSC) in Houston, Texas, where he assists the staff of the JSC History Office.

Andrew Chaikin has authored books and articles about space exploration and astronomy for more than two decades. He is also active as a lecturer at museums, schools, and corporate events, and in radio and television appearances. Chaikin is best known as the author of *A Man on the Moon: The Triumphant Story of the Apollo Space Program,* first published in 1994. This acclaimed work was the main basis for Tom Hanks's HBO miniseries, *From the Earth to the Moon*, which won the Emmy for best miniseries in 1998. Chaikin spent eight years writing and researching *A Man on the Moon*, including hundreds of hours of personal interviews with each of the 23 surviving lunar astronauts. Apollo Moonwalker Gene Cernan said of the book, "I've been there. Chaikin took me back." A three-volume, fully illustrated edition of *A Man on the Moon* was published by Time-Life Books in 1999. Chaikin co-edited *The New Solar System*, a compendium of writings by planetary scientists, now in its fourth edition. He is also the author of *Air and Space: The National Air and Space Museum Story of Flight*, published in 1997 by Bulfinch Press. He collaborated with Moonwalker-turned-artist Alan Bean to write *Apollo: An Eyewitness Account*, published in 1998 by the Greenwich Workshop Press. Chaikin co-authored the text for the highly successful collection of Apollo photography, *Full Moon*, which was published by Knopf in 1999. His book, *SPACE: A History of Space Exploration in Photographs*, was published in 2002 by Carlton Books. In 2004 he authored the chapter on human spaceflight in *The National Geographic Encyclopedia of Space.* From 1999 to 2001 Chaikin served as Executive Editor for Space and Science at SPACE.com, the definitive Web site for all things space. He was also the editor of SPACE.com's print magazine, *Space Illustrated*. Chaikin is a commentator for National Public Radio's *Morning Edition*, and has appeared on *Good Morning America*, *Nightline*, and the NPR programs *Fresh Air* and *Talk of the Nation*. He has been an advisor to NASA on space policy and public communications. A former editor of *Sky & Telescope* magazine, Chaikin has also been a contributing editor of *Popular Science* and has written for *Newsweek*, *Air&Space/Smithsonian*, *World Book Encyclopedia*, *Scientific American*, and other publications. A graduate of Brown University, Chaikin served on the Viking missions to Mars at NASA's Jet Propulsion Laboratory, and was a researcher at the Smithsonian's Center for Earth and Planetary Studies before becoming a science journalist in 1980.

Martin J. Collins is a curator in the Smithsonian National Air and Space Museum. He received his Ph.D. from the University of Maryland in history of science and technology. Book publications include *Cold War Laboratory: RAND, the Air Force, and the American State* (2002) and *Showcasing Space*, edited with Douglas Millard, volume 6 of the Artefacts Series: *Studies in the History of Science and Technology* (2005). He is currently working on a history of the Iridium satellite telephone venture.

Erik Conway is the Historian at the Jet Propulsion Laboratory (JPL) of the California Institute of Technology. He is completing a history of atmospheric research at NASA that he started as a contract historian at Langley Research Center in Virginia, and is beginning a history of robotic Mars exploration in the 1980s and 1990s. He has been at JPL since 2004.

Taylor E. Dark, III received his Ph.D. in political science from the University of California, Berkeley and currently teaches in the Department of Political Science at California State University, Los Angeles. He was formerly Associate Dean of the Graduate School of American Studies at Doshisha University in Kyoto, Japan, where he taught for eight years. Professor Dark is the author of *The Unions and the Democrats: An Enduring Alliance* (1999) and has previously published in *The National Interest*, *Journal of Labor Research*, *Labor History*, *Political Science Quarterly*, *Political Parties*, *PS: Political Science and Politics*, and *Presidential Studies Quarterly*. He is currently completing a book manuscript on philosophies of space exploration and development in American culture titled "Reclaiming the Future: The Space Program and the Idea of Progress."

Steven J. Dick is the Chief Historian for NASA. He obtained his B.S. in astrophysics (1971) and his M.A. and Ph.D. (1977) in history and philosophy of science from Indiana University. He worked as an astronomer and historian of science at the U.S. Naval Observatory in Washington, DC for 24 years before coming to NASA Headquarters in 2003. Among his books are *Plurality of Worlds: The Origins of the Extraterrestrial Life Debate from Democritus to Kant* (1982), *The Biological Universe: The Twentieth Century Extraterrestrial Life Debate and the Limits of Science* (1996), and *Life on Other Worlds* (1998). The latter has been translated into Chinese, Italian, Czech, Greek, and Polish. His most recent books are *The Living Universe: NASA and the Development of Astrobiology* (2004) and a comprehensive history of the U.S. Naval Observatory, *Sky and Ocean Joined: The U.S. Naval Observatory, 1830–2000* (2003). The latter received the Pendleton Prize of the Society for History in the Federal Government. He also is editor of *Many Worlds: The New Universe, Extraterrestrial Life and the Theological Implications* (2000) and (with Keith Cowing) of the proceedings of the NASA Administrator's symposium *Risk and Exploration: Earth, Sea and the Stars* (2005). He is the recipient of the Navy Meritorious Civilian Service Medal. He received the NASA Group Achievement Award for his role in NASA's multidisciplinary program in astrobiology. He has served as chairman of the Historical Astronomy Division of the American Astronomical Society and as president of the History of Astronomy Commission of the International Astronomical Union, and he is the immediate past president of the Philosophical Society of Washington. He is a member of the International Academy of Astronautics.

Andrew Fraknoi is the chair of the Astronomy Program at Foothill College near San Francisco. For 14 years, he served as the executive director of the Astronomical Society of the Pacific, the largest and oldest organization devoted to astronomy education. He founded and directed Project ASTRO, a national program that partners volunteer astronomers with fourth- through ninth-grade teachers. A branch of the project has developed family astronomy games and kits and trained educators and amateur astronomers to use them in regional sites from Boston to Hawaii. Fraknoi organized and moderated more than 20 workshops on teaching astronomy in grades 3–12 and four national symposia on teaching introductory astronomy to college nonscience majors. Educated at Harvard and the University of California, Berkeley, Fraknoi has also taught astronomy and physics at San Francisco State University. With Sidney Wolff, he is co-editor of *Astronomy Education Review,* a Web-based refereed journal on education and outreach. Among the books he has written and edited are *Voyages through the Universe* (a college astronomy text), *The Universe at Your Fingertips* (a collection of K–12 activities and resources), and *The Planets* and *The Universe*, two collections of science and scientifically accurate science fiction. Fraknoi serves on the board of trustees of the SETI Institute and is a fellow of the Committee for the Scientific Investigation of Claims of the Paranormal, specializing in debunking astrology. Awards he has won include the Annenberg Prize of the American Astronomical Society for his contributions to astronomy education. The International Astronomical Union has named Asteroid 4859 Asteroid Fraknoi to recognize his work in astronomy outreach.

Alexander Geppert teaches modern European history at Freie Universität Berlin. He studied history, philosophy, and psychology at Universität Bielefeld, Johns Hopkins University (M.A., 1995) in Baltimore, Georg-August-Universität Göttingen (M.A., 1997), and the University of California, Berkeley. From 1997 to 2004, he was a Ph.D. candidate and research associate at the European University Institute in Florence, where he wrote and defended his dissertation, "London vs. Paris: Imperial Exhibitions, Transitory Spaces and Metropolitan Networks, 1880–1930," under the supervision of Professors John Brewer, Luisa Passerini, and Bernd Weisbrod. He has held various long-term fellowships at the École des Hautes Études en Sciences Sociales in Paris (1999), the German Historical Institute in London (2000), the Internationales Forschungszentrum Kulturwissenschaften in Vienna (2001–2002), and the Kulturwissenschaftliches Institut in Essen (2002–2005). His work on oral, visual, and urban history (particularly on the history of international expositions); on the history of sexuality; and on the occult has been published in German, English, Italian, and Chinese. Currently, he is developing a new research project tentatively titled "Outer Space and the European Imagination, 1923–1969." Publications include five edited or co-edited volumes: *European Ego-Histoires. Historiography and the Self, 1970–2000* (Athens, 2001); *Orte des Okkulten* (Vienna, 2003); *Esposizioni in Europa tra Otto e Novecento. Spazi, organizzazione, rappresentazioni* (Milan, 2004); *Ortsgespräche. Raum und Kommunikation im 19. und 20. Jahrhundert* (Bielefeld, 2005); *New Dangerous Liaisons. Discourses on Europe and Love in the Twentieth Century* (Oxford/New York, 2006). Geppert is also the author of *Brief Cities: Imperial Expositions in Fin-de-siècle Europe* (London, 2006).

Roger Handberg is professor of political science and chair at the University of Central Florida. His published work in space policy and history includes works on NASA, international space commerce, and military space activities. He has published 154 articles and book chapters, 124 professional papers, and 8 books, the most recent being *International Space Commerce: Building from Scratch*, published in June, 2006 by the University Presses of Florida. Current work involves a study of the Chinese space program and work on nanotechnology policy. Dr. Handberg received his B.A. from Florida State University and Ph.D. from the University of North Carolina.

James R. Hansen is professor of history in the Department of History at Auburn University in Alabama, where he teaches courses on the history of flight, history of science and technology, space history, and the history of technological failure. He has published nine books and three dozen articles on a wide variety of technological topics, ranging from the early days of aviation to the first nuclear fusion reactors, to the Moon landings, to the environmental history of golf course development. His books include *First Man: The Life of Neil A. Armstrong* (2005); *The Bird Is on the Wing: Aerodynamics and the Progress of the Airplane in America* (2003); *The Wind and Beyond: A Documentary Journey through the History of Aerodynamics in America* (Vol. 1, 2002), *Spaceflight Revolution* (1995), *From the Ground Up* (1988), and *Engineer in Charge* (1987). Hansen earned a B.A. degree with high honors from Indiana University (1974) and an M.A. (1976) and Ph.D. (1981) from The Ohio State University. He served as historian for NASA Langley Research in Hampton, Virginia, from 1981 to 1984, and as a professor at the University of Maine in 1984–1985. Professor Hansen has received a number of citations for his scholarship, including the National Space Club's Robert H. Goddard Award, the Air Force Historical Foundation's Distinction of Excellence, the American Astronautical Society's Eugene Emme Prize in Astronautical Literature (twice), the American Institute of Aeronautics and Astronautics's History Book Award; and AIAA Distinguished Lecturer. He has served on a number of important advisory boards and panels, including the Research Advisory Board of the National Air and Space Museum, the Editorial Advisory Board of the Smithsonian Institution Press, the Advisory Board for the Archives of Aerospace Exploration at Virginia Polytechnic Institute and State University, the Museum Advisory Board of the U.S. Space and Rocket Center in Huntsville, Alabama, and the board of directors of the Space Restoration Society. He is a past vice president of the board of directors of the Virginia Air and Space Museum and Hampton Roads History Center in Hampton, Virginia.

Glenn Hastedt holds a Ph.D. in political science from Indiana University. Formerly the chair of the Political Science Department at James Madison University, he is now the director of the Center for Liberal and Applied Social Sciences. He is the author of *American Foreign Policy: Past, Present, Future*, 6th edition (2005) and has recently authored two articles on intelligence policy, "Public Intelligence: Leaks as Policy Instruments—The Case of the Iraq War," *Intelligence and National Security* (2005), and "Estimating Intentions in an Age of Terrorism," *Defense Intelligence Journal* (2005). Along with Kay Knickrehm, he is co-author of *International Politics in a Changing World* (2003). With Tony Eksterowicz, he is co-editor of *White House Studies*. He contributed a chapter, "Sputnik and Technological Surprise," to Roger Launius et al., *Reconsidering Sputnik* (2000).

Henry R. Hertzfeld is a research professor of Space Policy and International Affairs at the Space Policy Institute, Center for International Science and Technology Policy, Elliott School of International Affairs, George Washington University. He is an expert in the economic, legal, and policy issues of space and advanced technological development and teaches a course in space law as well as one in microeconomics. Dr. Hertzfeld has served as a senior economist and policy analyst at NASA and the National Science Foundation. He has been a consultant to both U.S. and international agencies and organizations, including the Organization for Economic Cooperation and Development, NASA, the Aerospace Corporation, and private companies. He was the co-editor of *Space Economics* (AIAA, 1992) and co-author of *A Study Guide to Managerial Economics* (2005) as well as many articles on the economic and legal issues concerning space and advanced technology. Dr. Hertzfeld has a B.A. from the University of Pennsylvania, an M.A. from Washington University, and a Ph.D. in economics from Temple University. He holds a J.D. degree from the George Washington University and is a member of the bars in Pennsylvania and the District of Columbia.

Stephen B. Johnson is an associate research professor with the Institute for Science and Space Studies at the University of Colorado at Colorado Springs; he is also a health management systems engineer for the Advanced Sensors and System Health Management Branch, NASA Marshall Space Flight Center. He was a faculty member in the University of North Dakota's Department of Space Studies from 1997 to 2005, teaching military space, space history, and management and economics of space endeavors. He is the author of *The United States Air Force and the Culture of Innovation, 1945–1965* and *The Secret of Apollo: Systems Management in American and*

European Space Programs, both published in 2002. He was also the editor of *Quest: The History of Spaceflight Quarterly* from 1998 to 2005 and is currently the general editor for a two-volume encyclopedia of space history to be published in 2007 by ABC-CLIO, *Space Exploration and Humanity: A Historical Encyclopedia.* His current research involves dependable space system design and operations, space industry management and economics, the history of space science and technology, and the history of cognitive psychology and artificial intelligence. He received his bachelor degree in physics from Whitman College in 1981 and his doctorate in 1997 in the history of science and technology from the University of Minnesota, where he was also the associate director of the Babbage Institute for the History of Computing. Prior to 1997, he worked for Northrop and Martin Marietta and was co-owner of his own small business managing computer simulation laboratories, designing space probes, and developing engineering processes.

De Witt Douglas Kilgore is associate professor of English and American Studies at Indiana University. He is the author of *Astrofuturism: Science, Race and Visions of Utopia in Space* (2003). In 2001, he received the Science Fiction Research Association's Pioneer Award for Excellence in Scholarship. Within the field of twentieth-century American literature and culture, he is particularly concerned with exploring the political (utopian) hopes expressed by our society through its projects in science and technology. He is interested in race as both a social and an analytic category. His first book, *Astrofuturism,* is an incisive engagement with the science writing and science fiction produced by the modern spaceflight movement. As a history, it takes seriously the (sometimes progressive) hopes of those scientists and engineers who wrote the Space Age into being as a great cultural project. As a critique, it turns a cold eye on those narratives of disciplined futurism. Kilgore's current work engages the fiction and science writing that has emerged from SETI (Search for ExtraTerrestrial Intelligence).

John Krige is the Kranzberg Professor in the School of History, Technology and Society at the Georgia Institute of Technology in Atlanta. Krige is a long-time member of the History of Science Society (HSS) and the Society for the History of Technology (SHOT) and has published in the Societies' journals, *Isis* and *Technology and Culture.* He is also the editor of the journal *History and Technology,* published by Routledge (UK). Krige's main focus of research is the relationship between foreign policy and science and technology. He played a major role in two international projects that resulted in a three-volume history of CERN (the European Laboratory for Particle Physics) and a two-volume history of ESA (the European Space Agency). Since moving to the United States in the summer of 2000, he has placed increasing emphasis on U.S.–European relationships in science and technology. He is co-editor (with Kai-Henrik Barth of Georgetown University, Washington, DC) of *Global Power Knowledge. Science and Technology in International Affairs, Osiris* 21 (2006), and his most recent book is *American Hegemony and the Postwar Reconstruction* *of Science in Europe* (2006). During academic year 2004–2005, Krige was the Charles A. Lindbergh Professor in Aerospace History at the Smithsonian National Air and Space Museum, Washington, DC. He took this opportunity to initiate research on his next project, which will deal with the role of space technology as an instrument of U.S.–European relations in the 1960s and early 1970s. His preliminary findings were presented as "Technology, Foreign Policy, and International Cooperation in Space," in *Critical Issues in the History of Spaceflight*, ed. Steven J. Dick and Roger D. Launius (2006). This project was at the core of Krige's successful application for a fellowship at the Davis Center, Department of History, Princeton University. Krige was awarded the biennial Dickinson Medal by the (UK) Newcomen Society for the Study of the History of Engineering and Technology in London in May, 2005.

W. Henry "Harry" Lambright is a professor of public administration and political science, and director of the Science and Technology Policy Program at the Maxwell School of Citizenship and Public Affairs at Syracuse University. He teaches courses at the Maxwell School on the intersections of technology, politics, energy, environment, and resources policy. He is a fellow of the American Association for the Advancement of Science. Dr. Lambright has served as a guest scholar at the Brookings Institution, director of the Science and Technology Policy Center of the Syracuse Research Corporation, and director of the Center for Environmental Policy and Administration at the Maxwell School of Syracuse University. He has served as an adjunct professor in the Graduate Program of Environmental Science and Forestry at the State University of New York. He has testified before Congress and consulted for various governmental and private-sector organizations, including the Columbia Accident Investigation Board. He has performed research under support from NASA, the National Science Foundation (NSF), the State Department, the Department of Energy (DOE), the Environmental Protection Agency (EPA), the Department of Defense (DOD), IBM, and other sponsors. A longtime student of large-scale technical projects, and particularly space policy, Dr. Lambright worked for NASA early in his career as a special assistant in its Office of University Affairs. He has been a member of the NASA History Advisory Committee and has written extensively on NASA over many years. He is author of *Powering Apollo: James E. Webb of NASA* (1995) and editor of *Space Policy in the 21st Century* (2003). He has written or edited 5 other books and more than 275 articles, reports, and papers. His doctorate is from Columbia University, where he also received his master's degree. Dr. Lambright received his undergraduate degree from Johns Hopkins University.

Roger D. Launius is a member of the Division of Space History at the Smithsonian Institution's National Air and Space Museum in Washington, DC. Between 1990 and 2002, he served as Chief Historian of NASA. A graduate of Graceland College in Lamoni, Iowa, he received his Ph.D. from Louisiana State University, Baton Rouge, in 1982. He has written or edited more than 20 books on aerospace history, including *Space: A Journey to Our Future* (2004); *Space Stations: Base Camps to the Stars* (2003), which received the AIAA's History Manuscript Prize; *Flight: A Celebration of 100 Years in Art and Literature* (2003); *Reconsidering a Century of Flight* (2003); *To Reach the High Frontier: A History of U.S. Launch Vehicles* (2002); *Imagining Space: Achievements, Possibilities, Projections, 1950–2050* (2001); *Reconsidering Sputnik: Forty Years Since the Soviet Satellite* (2000); *Innovation and the Development of Flight* (1999); *Frontiers of Space Exploration* (1998; rev. ed., 2004); *Spaceflight and the Myth of Presidential Leadership* (1997); and *NASA: A History of the U.S. Civil Space Program* (1994; rev. ed. 2001). He is frequently consulted by the electronic and print media for his views on space issues. His research interests encompass all areas of U.S. and space history and policy history, especially cultural aspects of the subject and the role of executive decision makers and their efforts to define space exploration.

John M. Logsdon is the director of the Space Policy Institute at George Washington University's Elliott School of International Affairs, where he is also professor of international affairs. He holds a B.S. in physics from Xavier University (1960) and a Ph.D. in political science from New York University (1970). Dr. Logsdon's research interests focus on the policy and historical aspects of U.S. and international space activities. Dr. Logsdon is the author of *The Decision to Go to the Moon: Project Apollo and the National Interest* (1970) and is the general editor of the eight-volume NASA History series *Exploring the Unknown: Selected Documents in the History of the U.S. Civil Space Program* (1995–). He has written numerous articles and reports on space policy and history. He is frequently consulted by the electronic and print media for his views on space issues. In 2003, Dr. Logsdon served as a member of the Columbia Accident Investigation Board. He is a member of the NASA Advisory Council and of the Commercial Space Transportation Advisory Committee of the Department of Transportation. He is a recipient of the NASA Public Service and Distinguished Public Service Medals, the 2005 John F. Kennedy Award from the American Astronautical Society (AAS), and the 2006 Barry Goldwater Space Educator Award from the American Institute of Aeronautics and Astronautics (AIAA). He is a fellow of the American Institute of Aeronautics and Astronautics and of the American Association for the Advancement of Science (AAAS), and he is a member of the International Academy of Astronautics.

M. G. Lord is a cultural historian and investigative journalist. She is the author of *Astro Turf: The Private Life of Rocket Science* (2005) and *Forever Barbie: The Unauthorized Biography of a Real Doll* (1994, 2004). Since 1995 she has been a frequent contributor to *The New York Times Book Review* and *The New York Times* Arts & Leisure Section. Her work has appeared in numerous publications, including *The Wall Street Journal, The Los Angeles Times Book Review, The New Yorker, ArtForum*, and *Vogue*. For 14 years she was a syndicated political cartoonist based at *Newsday*. She currently lives in Los Angeles, where she teaches a nonfiction workshop in the Master of Professional Writing Program at the University of Southern California.

Howard E. McCurdy is a professor of public affairs at American University in Washington, DC. Author or co-author of six books on the U.S. space program, he is best known for *Space and the American Imagination* (1997). He authored *Faster, Better, Cheaper* (2001), a critical analysis of cost-cutting initiatives in the U.S. space program, and an earlier study of NASA's organizational culture, *Inside NASA*, won the 1994 Henry Adams Prize for that year's best history on the federal government. He is often consulted by the media on public policy issues and has appeared on national news outlets such as *NewsHour with Jim Lehrer*, National Public Radio, and *NBC Nightly News*. Professor McCurdy received his bachelor's and master's degrees from the University of Washington and his doctorate from Cornell University.

Kim McQuaid is a professor of history at Lake Erie College in Painesville, Ohio. He received his undergraduate degree in history at Antioch College and his master's degree and Ph.D. in U.S. history at Northwestern University. He was a Woodrow Wilson Fellow in 1970 and has held several Fulbright overseas teaching positions: first as Mary Bell Washington Visiting Professor of U.S. History at the University College Dublin in 1985–1986, and later as a visiting lecturer in U.S. history at the University of Science in Penang, Malaysia's second-oldest national university campus, in 1995–1996. McQuaid's books are *Creating the Welfare State* (co-authored with Edward Berkowitz); *Big Business and Presidential Power* (1982); *The Anxious Years: America in the Vietnam-Watergate Era* (1989); and *Uneasy Partners: Big Business in American Politics, 1945–1990* (1993): and *A Response to Industrialism: Liberal Businessmen and the Evolving Spectrum of Capitalist Reform, 1886–1960* (2003). He is a contributor to academic and nonacademic journals. His most recently published academic piece is "Selling the Space Age: NASA and Earth's Environment, 1958–1990," which appeared in *Environment and History*, a journal published in the United Kingdom. He is a member of three space advocacy groups and as many environmental groups. He is also a regular visitor to the high Arctic of Canada and, most recently, Norway; most of these trips have been wilderness travel. He began to try and grow old gracefully on a Russian scientific survey vessel with an ice-hardened hull. He is currently at work on a social history of the Space Age. A native of rural Maine, he has worked as an antiquarian bookseller and a psychiatric social worker in addition to being an academic.

Wendell Mendell is a planetary scientist serving as the manager of the Office for Human Exploration Science of the NASA Johnson Space Center, where he has been employed since 1963. He is married and has four children. Dr. Mendell has a B.S. in physics from the California Institute of Technology, an M.S. in physics from the University of California, Los Angeles (UCLA), and an M.S. in space science and a Ph.D. in space physics and astronomy from Rice University. His scientific research focus is remote sensing of planetary surfaces, particularly specializing in thermal emission radiometry and spectroscopy of the Moon. Since 1982, he has worked at NASA on planning and advocacy of human exploration of the solar system, especially on the establishment of a permanent human base on the Moon.

His interests in this regard lie as much with policy issues as with technical solutions. He is most well known as the editor of the volume *Lunar Bases and Space Activities of the 21st Century*, and he received the 1988 Space Pioneer Award for Science and Engineering from the National Space Society for this work. Currently, Dr. Mendell splits his time between communicating the principles of the human exploration of the solar system to both lay and technical audiences and working on lunar research. He is a member of the College of Teachers of the International Space University (ISU). At the ISU, he has led "Design Projects for an International Lunar Base" (1988); "International Mars Mission" (1991); "International Lunar Farside Observatory and Science Station" (1993); "Vision 20/20" (1995), a sampling of the future as seen by young space professionals; and "Space Tourism: From Dream to Reality" (2000). He belongs to several professional scientific and engineering societies. He is most active in the International Academy of Astronautics, where he has served as secretary of the Cosmic Study on International Human Exploration of Mars and is currently serving on Academic Commission III; and in the AIAA, where he has chaired the Space Science and Astronomy Technical Committee and sits on the International Activities Committee. He served on (and chaired) the Executive Committee of the Aerospace Division of the American Society of Civil Engineers. He has served as editor for nine technical volumes and has published more than 40 articles in professional journals and conference proceedings. He is also the author of numerous abstracts and short papers presented at technical conferences.

Ron Miller is the author/illustrator of some 40 books, most of them dealing with space exploration, astronomy, and other sciences. These include the award-winning Worlds Beyond series of astronomy books for young adults and *The Art of Chesley Bonestell* (2000), a biography of the grand master of astronomical art. He has also collaborated on five books with noted astronomer William K. Hartmann. These include *The Grand Tour, Cycles of Fire, In the Stream of Stars*, and *The History of Earth* (all published by Workman Publishing Co.). Considered an authority on Jules Verne, Miller has translated and illustrated new, definitive editions of Verne's *20,000 Leagues Under the Sea* and *Journey to the Center of the Earth*. A book published in July, 1993, *The Dream Machines* (a comprehensive, quarter-million-word, 744-page history of manned spacecraft) was nominated for the prestigious IAF Manuscript Award and won the Booklist Editor's Choice Award for 1994. He designed a set of 10 commemorative space exploration stamps for the U.S. Postal Service, one of which is on-board the New Horizons spacecraft bound for Pluto. A current project is the reprinting of classic and little-known space travel novels from the past 200 years.

Valerie Neal has been a space history curator at the Smithsonian National Air and Space Museum since 1989, where she engages in scholarly research, exhibit development, artifact collection, and public service. She has edited two books on space exploration and curated two major exhibitions on the space race and the challenges of future exploration. Most recently, she led the effort to restore the Space Shuttle *Enterprise* for permanent display and to acquire *SpaceShipOne* for the national collection. Her recent articles on Shuttle history have appeared in *History and Technology* and *Space Policy*. Her current projects are a book and exhibition on the Space Shuttle era. Before joining the Smithsonian, Neal spent a decade as a writer, editor, and manager for some 50 NASA publications on Shuttle and Spacelab missions, the Great Observatories, the space sciences, and NASA history. She also participated in astronaut training activities and worked on the mission management team for four Shuttle missions. Neal earned graduate degrees in american studies from the University of Southern California (M.A.) and the University of Minnesota (Ph.D.). She has taught at the University of Minnesota, the University of Alabama in Huntsville, and Vanderbilt University.

Jennifer Ross-Nazzal currently serves as the historian for the NASA Johnson Space Center (JSC) in Houston, Texas and is a member of the editorial board for *Quest: The History of Spaceflight Quarterly*. Ross-Nazzal began working for JSC in the summer of 2000 when she received an internship with the Johnson Space Center Oral History Project. In her position as historian, she has conducted more than 86 interviews for the Johnson Space Center Oral History Project and the NASA Headquarters History Office. From 2002 to 2004, she served as the oral history editor for *Quest*. Ross-Nazzal also participated in the *Columbia* Recovery Oral History project and later took the lead in a series of interviews to explore NASA's role in the development of microelectromechanical systems. In addition to her work for the JSC History Office, she works as an adjunct instructor for the University of Maryland University College, where she teaches U.S. and women's history. Ross-Nazzal received her M.A. in history from New Mexico State University in 1996 and her Ph.D. in history from Washington State University in May, 2004. Her dissertation, which was nominated for the David H. Stratton Award for the department's best dissertation, the Lerner-Scott Dissertation Prize, and the Phi Alpha Theta/Westerners International Dissertation Prize, focused on the life and times of suffragist Emma Smith DeVoe. Ross-Nazzal is currently revising her dissertation for publication. Two articles spun off from her dissertation have been published by the *Pacific Northwest Quarterly* and *South Dakota History*. Among her many honors and awards, she received a practicum grant from the Woodrow Wilson National Fellowship Foundation, and she is a former graduate fellow at the Thomas S. Foley Institute for Public Policy and Public Service. Most recently, she received the HRA New Professional Award from the National Council on Public History.

Philip Scranton is University Board of Governors Professor, History of Industry and Technology, at Rutgers University, where he directs the M.A. history program (Camden) and assisted in developing a Ph.D. field in the History of Technology, Environment and Health (New Brunswick). He served the Georgia Institute of Technology as Kranzberg Professor of the History of Technology and Science from

1997 to 1999. Professor Scranton also directs the Hagley Museum and Library's research arm, the Center for the History of Business, Technology and Society, with responsibility for a seminar series, twice-yearly conferences, short-term fellowships, and consultation on collections, programs, and planning. During 2003–2004, he held the Lindbergh Chair in Aeronautic and Aerospace History at the Smithsonian's National Air and Space Museum. His publications include 5 books and 35 scholarly articles; multiple contributions to museum catalogs; and numerous reviews of books, conferences, and exhibits. Since 1985 he has presented research papers at 38 international conferences in Europe, Canada, and Japan. Most recently, Princeton University Press released his *Endless Novelty: Specialty Production and American Industrialization, 1865–1925* (1997, paperback 2000, Japanese translation, 2004). Earlier monographs include *Proprietary Capitalism* (1983) and *Figured Tapestry* (1989), which received the SHEAR and Taft prizes, respectively. At present, Scranton is editor or co-editor of two book series: Studies in Industry and Society (Johns Hopkins University Press) and Hagley Perspectives on Business and Society (formerly Routledge, presently Macmillan, with Roger Horowitz). He has edited or co-edited three volumes in the Hagley Series. In addition, Scranton served the Business History Conference as president (2002–2003) and its journal, *Enterprise and Society* (Oxford University Press), for four years as its initial associate editor for reviews. He is a member of the editorial boards of *Technology and Culture*, *Business History Review*, and *Pennsylvania History*. Born in 1946 in western Pennsylvania, Scranton received undergraduate and graduate degrees in history from the University of Pennsylvania (Ph.D., 1975). He taught at the Philadelphia College of Textiles and Science (1974–1984) before joining the faculty at Rutgers-Camden (1984–1997), moving to Georgia Tech, then returning to Rutgers in fall, 1999. His current research project examines the course of specialty manufacturing in the United States during the Cold War, with a focus on jet propulsion and NASA space capsules.

Asif A. Siddiqi is an assistant professor of history at Fordham University in New York. He specializes in the social and cultural history of technology and modern Russian history. His forthcoming book, *The Rockets' Red Glare: Spaceflight and the Russian Imagination, 1857-1957*, will be published in 2008. He is also working on a project on technology, authenticity, and the evolution of rock'n'roll.

Rick W. Sturdevant is the deputy director of history at Headquarters Air Force Space Command, Peterson Air Force Base, in Colorado Springs, Colorado. He holds B.A. (1969) and M.A. (1973) degrees in history from the University of Northern Iowa and a Ph.D. (1982) from the University of California, Santa Barbara. Dr. Sturdevant joined the United States Air Force history program in April, 1984 as the chief historian for Air Force Communication Command's Airlift Communications Division at Scott Air Force Base, Illinois. In April, 1985 he moved to Peterson Air Force Base, Colorado to become the chief historian for Space Communications

Division. He held that position until the division was inactivated in 1991, at which time he moved to the Air Force Space Command History Office. Dr. Sturdevant has published extensively on the subject of military aerospace history in such periodicals as *Space Times*, *Journal of the British Interplanetary Society*, *Air & Space/Smithsonian*, *Quest: The History of Spaceflight Quarterly*, *Air Power History*, *Air Force Magazine*, *High Frontier: The Journal for Space & Missile Professionals*, and *Journal of the West*. He has authored or co-authored chapters or essays in *Beyond the Ionosphere: Fifty Years of Satellite Communication* (1997); *Organizing for the Use of Space: Historical Perspectives on a Persistent Issue* (1995); *Golden Legacy, Boundless Future: Essays on the United States Air Force and the Rise of Aerospace Power* (2000); *Air Warfare: An International Encyclopedia* (2002); *To Reach the High Frontier: A History of U.S. Launch Vehicles* (2002); *The Limitless Sky: Air Force Science and Technology Contributions to the Nation* (2004); and *Encyclopedia of 20th-Century Technology* (2005). A frequent lecturer on space history topics, he has taught Elderhostel and Pillar courses under the sponsorship of Pikes Peak Community College in Colorado Springs and was a guest speaker at the 2005 High Plains Chautauqua in Greeley, Colorado. Dr. Sturdevant is an active member of the American Institute of Aeronautics and Astronautics (AIAA), British Interplanetary Society (BIS), Air Force Space Operations Association (AFSOA), Society for the History of Technology (SHOT), and American Astronautical Society (AAS). Serving on the AAS History Committee, he has participated for nearly a decade in selecting the annual recipient of the Eugene M. Emme Astronautical Literature Award. He is a recipient of the Air Force Exemplary Civilian Service Award and the AAS President's Recognition Award.

James A. Vedda joined The Aerospace Corporation in March 2004 to perform research and analysis on national security, civil, and commercial space issues. Previously, he spent six and a half years at ANSER Inc. in Arlington, Virginia, assigned full-time to the Office of the Secretary of Defense. This included two years with the assistant secretary for Homeland Defense and four and a half years in the Space Policy Directorate. While at ANSER, Dr. Vedda received the company's highest employee award, the Alan S. Boyd Award for Professional Development, in 2002; an annual Trustee's Award and a quarterly Team Excellence Award in 2003; and several awards for publications throughout his tenure. Jim received his Ph.D. in political science from the University of Florida. His dissertation analyzed the evolution of post-Apollo space policy making in the executive and legislative branches. He also has an M.A. in science, technology, and public policy from George Washington University in Washington, DC and a B.S in business administration from John Carroll University in Cleveland, Ohio. He has been a member of the American Astronautical Society since 1997, serving as its vice president for public policy from July 2002 to November 2004, and as a member of its board of directors since then. From 1987 to 1993, Jim was a professor in the Department of Space Studies at the University of North Dakota, where he taught courses on civil, commercial, and military space policy to undergraduate and graduate students. He was one of the founding members of the faculty, helping to create the curriculum for the M.S. in Space Studies degree. He was associate director of North Dakota's participation in the NASA Space Grant program, served for a period as department chairman, and pioneered the department's use of multimedia teaching techniques. Jim's writing has appeared in publications such as *Space Policy*, *Space News*, *Space Times*, *Ad Astra*, *Space Energy and Transportation*, and *Space Business News*. He has presented conference papers and commentary for the International Astronautical Federation, the Midwest Political Science Association, the Public Members Association of the Foreign Service, and CNN.

Peter J. Westwick is visiting researcher in history, University of California, Santa Barbara. He was previously an Olin Fellow in international security studies at Yale University and, from 2000 to 2004, senior research fellow in humanities at the California Institute of Technology. He is the author of *The National Labs: Science in an American System, 1947–1974* (2003), which won the 2004 Book Prize of the Forum for the History of Science in America (for best first book), and *Into the Black: JPL and the American Space Program, 1976–2004* (2006). He is currently working on a history of the Strategic Defense Initiative. He received his B.A. in physics and Ph.D. in history from the University of California, Berkeley. He is a moderator of the working group on science and technology at the Institute on California and the West, University of Southern California (USC)/Huntington Library, where he is helping to start an initiative to document the history of the aerospace industry in southern California.

David J. Whalen is vice president, satellite systems consulting at IOT Systems LLC. He has been an engineer and engineering manager in the communications satellite industry for more than 30 years. He has also worked on weather satellites (INSAT, GOES-NEXT), Earth observing satellites (Landsat), and science satellites (GRO, Hubble). He is an associate fellow of the American Institute of Aeronautics and Astronautics (AIAA). He has been a member of the AIAA History Technical Committee and is currently a member of the AIAA Communications Systems Technical Committee. Over the last 20 years he has written about space history and space policy in addition to his engineering work. He holds a B.A. in astronomy from Boston University, an M.S. in astronomy from the University of Massachusetts, an M.B.A. from the College of William and Mary, and a Ph.D. in science, technology, and public policy from George Washington University. He has taught university and industrial courses in orbit determination and maneuver planning, satellite communications, space policy, and the history of technology. His book, *Origins of Satellite Communications 1945-1965*, was published by the Smithsonian Institution Press in 2002. He is currently at work on a book about COMSAT Corporation. He has made many presentations at AIAA, PTC, CASBAA, AHA, SHFG, and NASA conferences. He has published articles and book reviews on space history, space policy, and space technology in a variety of publications, including *IEEE Technology and Society* and *Technology and Culture*.

Ray A. Williamson is research professor of space policy and international affairs in the Space Policy Institute, focusing on the applications of geospatial information for the management of natural and cultural resources. He is principal investigator for a NASA/NOAA-funded study of the socioeconomic benefits of earth science. He was also co-investigator of *Bridging the Gap: European C4ISR Capabilities and Transatlantic Interoperability* (2004), and principal investigator of a two-year study of dual-purpose space technologies (satellite communications; remote sensing; and position, navigation, and timing) for a private foundation. He serves on the National Academy of Sciences Committee on Space-based Global Precipitation Measurements and the Independent Committee to Assess the National Space Weather Program. Dr. Williamson is also a member of the NOAA Advisory Committee on Commercial Remote Sensing (ACCRES). From 1979 to 1995, he was a senior associate and project director in the Office of Technology Assessment (OTA) of the U.S. Congress. While at OTA, Dr. Williamson was project director for space policy reports that focused on satellite remote sensing. Dr. Williamson is an external faculty member of the International Space University (ISU), Illkirch, France, teaching general space policy and remote sensing for the ISU Masters of Space Studies and summer session programs. He has lectured on remote sensing policies and markets and the applications of geospatial technologies in regional, national, and international forums. Dr. Williamson received his B.A. in physics from the Johns Hopkins University and his Ph.D. in astronomy from the University of Maryland, and spent two years on the faculty of the University of Hawaii studying diffuse emission nebulae. He taught philosophy, literature, mathematics, physics, and astronomy at St. John's College, Annapolis for 10 years, the last 5 of which he also served as assistant dean of the college. Dr. Williamson is editor of *Imaging Notes* and a contributing editor to the journal *Space Policy*. From 1998–2001 he was a member of the Aeronautics and Space Engineering Board of the National Academy of Engineering. He is also a member of the International Academy of Astronautics. His published books include *Bridging the Gap: European C4ISR Capabilities and Transatlantic Interoperability*, ed. with Gordon Adams, Guy Ben-Ari, and John M. Logsdon (2004) and *Commercial Observation Satellites: At the Leading Edge of Global Transparency*, ed. with John C. Baker and Kevin O'Connell (2001).

The NASA History Series

Reference Works, NASA SP-4000:

Grimwood, James M. *Project Mercury: A Chronology*. NASA SP-4001, 1963.

Grimwood, James M., and C. Barton Hacker, with Peter J. Vorzimmer. *Project Gemini Technology and Operations: A Chronology*. NASA SP-4002, 1969.

Link, Mae Mills. *Space Medicine in Project Mercury*. NASA SP-4003, 1965.

Astronautics and Aeronautics, 1963: Chronology of Science, Technology, and Policy. NASA SP-4004, 1964.

Astronautics and Aeronautics, 1964: Chronology of Science, Technology, and Policy. NASA SP-4005, 1965.

Astronautics and Aeronautics, 1965: Chronology of Science, Technology, and Policy. NASA SP-4006, 1966.

Astronautics and Aeronautics, 1966: Chronology of Science, Technology, and Policy. NASA SP-4007, 1967.

Astronautics and Aeronautics, 1967: Chronology of Science, Technology, and Policy. NASA SP-4008, 1968.

Ertel, Ivan D., and Mary Louise Morse. *The Apollo Spacecraft: A Chronology, Volume I, Through November 7, 1962*. NASA SP-4009, 1969.

Morse, Mary Louise, and Jean Kernahan Bays. *The Apollo Spacecraft: A Chronology, Volume II, November 8, 1962–September 30, 1964*. NASA SP-4009, 1973.

Brooks, Courtney G., and Ivan D. Ertel. *The Apollo Spacecraft: A Chronology, Volume III, October 1, 1964–January 20, 1966*. NASA SP-4009, 1973.

Ertel, Ivan D., and Roland W. Newkirk, with Courtney G. Brooks. *The Apollo Spacecraft: A Chronology, Volume IV, January 21, 1966–July 13, 1974*. NASA SP-4009, 1978.

Astronautics and Aeronautics, 1968: Chronology of Science, Technology, and Policy. NASA SP-4010, 1969.

Newkirk, Roland W., and Ivan D. Ertel, with Courtney G. Brooks. *Skylab: A Chronology*. NASA SP-4011, 1977.

Van Nimmen, Jane, and Leonard C. Bruno, with Robert L. Rosholt. *NASA Historical Data Book, Volume I: NASA Resources, 1958–1968*. NASA SP-4012, 1976, rep. ed. 1988.

Ezell, Linda Neuman. *NASA Historical Data Book, Volume II: Programs and Projects, 1958–1968.* NASA SP-4012, 1988.

Ezell, Linda Neuman. *NASA Historical Data Book, Volume III: Programs and Projects, 1969–1978.* NASA SP-4012, 1988.

Gawdiak, Ihor Y., with Helen Fedor, compilers. *NASA Historical Data Book, Volume IV: NASA Resources, 1969–1978.* NASA SP-4012, 1994.

Rumerman, Judy A., compiler. *NASA Historical Data Book, 1979–1988: Volume V, NASA Launch Systems, Space Transportation, Human Spaceflight, and Space Science.* NASA SP-4012, 1999.

Rumerman, Judy A., compiler. *NASA Historical Data Book, Volume VI: NASA Space Applications, Aeronautics and Space Research and Technology, Tracking and Data Acquisition/Space Operations, Commercial Programs, and Resources, 1979–1988.* NASA SP-2000-4012, 2000.

Astronautics and Aeronautics, 1969: Chronology of Science, Technology, and Policy. NASA SP-4014, 1970.

Astronautics and Aeronautics, 1970: Chronology of Science, Technology, and Policy. NASA SP-4015, 1972.

Astronautics and Aeronautics, 1971: Chronology of Science, Technology, and Policy. NASA SP-4016, 1972.

Astronautics and Aeronautics, 1972: Chronology of Science, Technology, and Policy. NASA SP-4017, 1974.

Astronautics and Aeronautics, 1973: Chronology of Science, Technology, and Policy. NASA SP-4018, 1975.

Astronautics and Aeronautics, 1974: Chronology of Science, Technology, and Policy. NASA SP-4019, 1977.

Astronautics and Aeronautics, 1975: Chronology of Science, Technology, and Policy. NASA SP-4020, 1979.

Astronautics and Aeronautics, 1976: Chronology of Science, Technology, and Policy. NASA SP-4021, 1984.

Astronautics and Aeronautics, 1977: Chronology of Science, Technology, and Policy. NASA SP-4022, 1986.

Astronautics and Aeronautics, 1978: Chronology of Science, Technology, and Policy. NASA SP-4023, 1986.

Astronautics and Aeronautics, 1979–1984: Chronology of Science, Technology, and Policy. NASA SP-4024, 1988.

Astronautics and Aeronautics, 1985: Chronology of Science, Technology, and Policy. NASA SP-4025, 1990.

Noordung, Hermann. *The Problem of Space Travel: The Rocket Motor. Edited by Ernst Stuhlinger and J. D. Hunley, with Jennifer Garland*. NASA SP-4026, 1995.

Astronautics and Aeronautics, 1986–1990: A Chronology. NASA SP-4027, 1997.

Astronautics and Aeronautics, 1990–1995: A Chronology. NASA SP-2000-4028, 2000.

MANAGEMENT HISTORIES, NASA SP-4100:

Rosholt, Robert L. *An Administrative History of NASA, 1958–1963*. NASA SP-4101, 1966.

Levine, Arnold S. *Managing NASA in the Apollo Era*. NASA SP-4102, 1982.

Roland, Alex. *Model Research: The National Advisory Committee for Aeronautics, 1915–1958*. NASA SP-4103, 1985.

Fries, Sylvia D. *NASA Engineers and the Age of Apollo*. NASA SP-4104, 1992.

Glennan, T. Keith. *The Birth of NASA: The Diary of T. Keith Glennan. J. D. Hunley, editor*. NASA SP-4105, 1993.

Seamans, Robert C., Jr. *Aiming at Targets: The Autobiography of Robert C. Seamans, Jr.* NASA SP-4106, 1996.

Garber, Stephen J., editor. *Looking Backward, Looking Forward: Forty Years of U.S. Human Spaceflight Symposium*. NASA SP-2002-4107, 2002.

Mallick, Donald L. with Peter W. Merlin. *The Smell of Kerosene: A Test Pilot's Odyssey*. NASA SP-4108, 2003.

Iliff, Kenneth W. and Curtis L. Peebles. *From Runway to Orbit: Reflections of a NASA Engineer.* NASA SP-2004-4109, 2004.

Chertok, Boris. *Rockets and People, Volume 1*. NASA SP-2005-4110, 2005.

Laufer, Alexander, Todd Post, and Edward Hoffman. *Shared Voyage: Learning and Unlearning from Remarkable Projects*. NASA SP-2005-4111, 2005.

Dawson, Virginia P. and Mark D. Bowles. *Realizing the Dream of Flight: Biographical Essays in Honor of the Centennial of Flight, 1903-2003*. NASA SP-2005-4112, 2005.

Mudgway, Douglas J. *William H. Pickering: America's Deep Space Pioneer*, NASA SP-2007-4113, 2007.

PROJECT HISTORIES, NASA SP-4200:

Swenson, Loyd S., Jr., James M. Grimwood, and Charles C. Alexander. *This New Ocean: A History of Project Mercury*. NASA SP-4201, 1966; rep. ed. 1998.

Green, Constance McLaughlin, and Milton Lomask. *Vanguard: A History*. NASA SP-4202, 1970; rep. ed. Smithsonian Institution Press, 1971.

Hacker, Barton C., and James M. Grimwood. *On the Shoulders of Titans: A History of Project Gemini*. NASA SP-4203, 1977.

Benson, Charles D., and William Barnaby Faherty. *Moonport: A History of Apollo Launch Facilities and Operations*. NASA SP-4204, 1978.

Brooks, Courtney G., James M. Grimwood, and Loyd S. Swenson, Jr. *Chariots for Apollo: A History of Manned Lunar Spacecraft*. NASA SP-4205, 1979.

Bilstein, Roger E. *Stages to Saturn: A Technological History of the Apollo/Saturn Launch Vehicles*. NASA SP-4206, 1980, rep. ed. 1997.

SP-4207 not published.

Compton, W. David, and Charles D. Benson. *Living and Working in Space: A History of Skylab*. NASA SP-4208, 1983.

Ezell, Edward Clinton, and Linda Neuman Ezell. *The Partnership: A History of the Apollo-Soyuz Test Project*. NASA SP-4209, 1978.

Hall, R. Cargill. *Lunar Impact: A History of Project Ranger*. NASA SP-4210, 1977.

Newell, Homer E. *Beyond the Atmosphere: Early Years of Space Science*. NASA SP-4211, 1980.

Ezell, Edward Clinton, and Linda Neuman Ezell. *On Mars: Exploration of the Red Planet, 1958–1978*. NASA SP-4212, 1984.

Pitts, John A. *The Human Factor: Biomedicine in the Manned Space Program to 1980*. NASA SP-4213, 1985.

Compton, W. David. *Where No Man Has Gone Before: A History of Apollo Lunar Exploration Missions*. NASA SP-4214, 1989.

Naugle, John E. *First Among Equals: The Selection of NASA Space Science Experiments*. NASA SP-4215, 1991.

Wallace, Lane E. *Airborne Trailblazer: Two Decades with NASA Langley's Boeing 737 Flying Laboratory*. NASA SP-4216, 1994.

Butrica, Andrew J., editor. *Beyond the Ionosphere: Fifty Years of Satellite Communication*. NASA SP-4217, 1997.

Butrica, Andrew J. *To See the Unseen: A History of Planetary Radar Astronomy*. NASA SP-4218, 1996.

Mack, Pamela E., editor. *From Engineering Science to Big Science: The NACA and NASA Collier Trophy Research Project Winners*. NASA SP-4219, 1998.

Reed, R. Dale, with Darlene Lister. *Wingless Flight: The Lifting Body Story*. NASA SP-4220, 1997.

Heppenheimer, T. A. *The Space Shuttle Decision: NASA's Search for a Reusable Space Vehicle*. NASA SP-4221, 1999.

Hunley, J. D., editor. *Toward Mach 2: The Douglas D-558 Program*. NASA SP-4222, 1999.

Swanson, Glen E., editor. *"Before this Decade Is Out . . .": Personal Reflections on the Apollo Program*. NASA SP-4223, 1999.

Tomayko, James E. *Computers Take Flight: A History of NASA's Pioneering Digital Fly-by-Wire Project*. NASA SP-2000-4224, 2000.

Morgan, Clay. *Shuttle-Mir: The U.S. and Russia Share History's Highest Stage*. NASA SP-2001-4225, 2001.

Leary, William M. *"We Freeze to Please": A History of NASA's Icing Research Tunnel and the Quest for Flight Safety*. NASA SP-2002-4226, 2002.

Mudgway, Douglas J. *Uplink-Downlink: A History of the Deep Space Network 1957–1997*. NASA SP-2001-4227, 2001.

Dawson, Virginia P. and Mark D. Bowles. *Taming Liquid Hydrogen: The Centaur Upper Stage Rocket, 1958-2002*. NASA SP-2004-4230, 2004.

Meltzer, Michael. *Mission to Jupiter: A History of the Galileo Project*. NASA SP-2007-4231.

Heppenheimer, T. A. *Facing the Heat Barrier: A History of Hypersonics*. NASA SP-2007-4232, 2007.

CENTER HISTORIES, NASA SP-4300:

Rosenthal, Alfred. *Venture into Space: Early Years of Goddard Space Flight Center*. NASA SP-4301, 1985.

Hartman, Edwin P. *Adventures in Research: A History of Ames Research Center, 1940–1965*. NASA SP-4302, 1970.

Hallion, Richard P. *On the Frontier: Flight Research at Dryden, 1946–1981*. NASA SP-4303, 1984.

Muenger, Elizabeth A. *Searching the Horizon: A History of Ames Research Center, 1940–1976*. NASA SP-4304, 1985.

Hansen, James R. *Engineer in Charge: A History of the Langley Aeronautical Laboratory, 1917–1958*. NASA SP-4305, 1987.

Dawson, Virginia P. *Engines and Innovation: Lewis Laboratory and American Propulsion Technology*. NASA SP-4306, 1991.

Dethloff, Henry C. *"Suddenly Tomorrow Came . . .": A History of the Johnson Space Center.* NASA SP-4307, 1993.

Hansen, James R. *Spaceflight Revolution: NASA Langley Research Center from Sputnik to Apollo*. NASA SP-4308, 1995.

Wallace, Lane E. *Flights of Discovery: 50 Years at the NASA Dryden Flight Research Center*. NASA SP-4309, 1996.

Herring, Mack R. *Way Station to Space: A History of the John C. Stennis Space Center.* NASA SP-4310, 1997.

Wallace, Harold D., Jr. *Wallops Station and the Creation of the American Space Program.* NASA SP-4311, 1997.

Wallace, Lane E. *Dreams, Hopes, Realities: NASA's Goddard Space Flight Center, The First Forty Years*. NASA SP-4312, 1999.

Dunar, Andrew J., and Stephen P. Waring. *Power to Explore: A History of the Marshall Space Flight Center*. NASA SP-4313, 1999.

Bugos, Glenn E. *Atmosphere of Freedom: Sixty Years at the NASA Ames Research Center.* NASA SP-2000-4314, 2000.

Schultz, James. *Crafting Flight: Aircraft Pioneers and the Contributions of the Men and Women of NASA Langley Research Center*. NASA SP-2003-4316, 2003.

GENERAL HISTORIES, NASA SP-4400:

Corliss, William R. *NASA Sounding Rockets, 1958–1968: A Historical Summary.* NASA SP-4401, 1971.

Wells, Helen T., Susan H. Whiteley, and Carrie Karegeannes. *Origins of NASA Names.* NASA SP-4402, 1976.

Anderson, Frank W., Jr. *Orders of Magnitude: A History of NACA and NASA, 1915–1980*. NASA SP-4403, 1981.

Sloop, John L. *Liquid Hydrogen as a Propulsion Fuel, 1945–1959*. NASA SP-4404, 1978.

Roland, Alex. *A Spacefaring People: Perspectives on Early Spaceflight*. NASA SP-4405, 1985.

Bilstein, Roger E. *Orders of Magnitude: A History of the NACA and NASA, 1915–1990.* NASA SP-4406, 1989.

Logsdon, John M., editor, with Linda J. Lear, Jannelle Warren-Findley, Ray A. Williamson, and Dwayne A. Day. *Exploring the Unknown: Selected Documents in the History of the U.S. Civil Space Program, Volume I, Organizing for Exploration.* NASA SP-4407, 1995.

Logsdon, John M., editor, with Dwayne A. Day and Roger D. Launius. *Exploring the Unknown: Selected Documents in the History of the U.S. Civil Space Program, Volume II, Relations with Other Organizations.* NASA SP-4407, 1996.

Logsdon, John M., editor, with Roger D. Launius, David H. Onkst, and Stephen J. Garber. *Exploring the Unknown: Selected Documents in the History of the U.S. Civil Space Program, Volume III, Using Space.* NASA SP-4407, 1998.

Logsdon, John M., general editor, with Ray A. Williamson, Roger D. Launius, Russell J. Acker, Stephen J. Garber, and Jonathan L. Friedman. *Exploring the Unknown: Selected Documents in the History of the U.S. Civil Space Program, Volume IV, Accessing Space.* NASA SP-4407, 1999.

Logsdon, John M., general editor, with Amy Paige Snyder, Roger D. Launius, Stephen J. Garber, and Regan Anne Newport. *Exploring the Unknown: Selected Documents in the History of the U.S. Civil Space Program, Volume V, Exploring the Cosmos.* NASA SP-2001-4407, 2001.

Siddiqi, Asif A. *Challenge to Apollo: The Soviet Union and the Space Race, 1945–1974.* NASA SP-2000-4408, 2000.

Hansen, James R., editor. *The Wind and Beyond: Journey into the History of Aerodynamics in America, Volume I, The Ascent of the Airplane.* NASA SP-2003-4409, 2003.

Hogan, Thor. *Mars Wars: The Rise and Fall of the Space Exploration Initiative.* NASA SP-2007-4410, 2007.

Hansen, James R., editor. *The Wind and Beyond: Journey into the History of Aerodynamics in America, Volume II, Reinventing the Airplane.* NASA SP-2007-4409, 2007.

Monographs in Aerospace History, NASA SP-4500:

Launius, Roger D. and Aaron K. Gillette, compilers, *Toward a History of the Space Shuttle: An Annotated Bibliography.* Monograph in Aerospace History, No. 1, 1992.

Launius, Roger D., and J. D. Hunley, compilers, *An Annotated Bibliography of the Apollo Program.* Monograph in Aerospace History, No. 2, 1994.

Launius, Roger D. Apollo: A Retrospective Analysis. Monograph in Aerospace History, No. 3, 1994.

Hansen, James R. *Enchanted Rendezvous: John C. Houbolt and the Genesis of the Lunar-Orbit Rendezvous Concept*. Monograph in Aerospace History, No. 4, 1995.

Gorn, Michael H. Hugh L. *Dryden's Career in Aviation and Space*. Monograph in Aerospace History, No. 5, 1996.

Powers, Sheryll Goecke. *Women in Flight Research at NASA Dryden Flight Research Center, from 1946 to 1995*. Monograph in Aerospace History, No. 6, 1997.

Portree, David S. F. and Robert C. Trevino. *Walking to Olympus: An EVA Chronology*. Monograph in Aerospace History, No. 7, 1997.

Logsdon, John M., moderator. *Legislative Origins of the National Aeronautics and Space Act of 1958: Proceedings of an Oral History Workshop*. Monograph in Aerospace History, No. 8, 1998.

Rumerman, Judy A., compiler, *U.S. Human Spaceflight, A Record of Achievement 1961–1998*. Monograph in Aerospace History, No. 9, 1998.

Portree, David S. F. *NASA's Origins and the Dawn of the Space Age*. Monograph in Aerospace History, No. 10, 1998.

Logsdon, John M. *Together in Orbit: The Origins of International Cooperation in the Space Station*. Monograph in Aerospace History, No. 11, 1998.

Phillips, W. Hewitt. *Journey in Aeronautical Research: A Career at NASA Langley Research Center*. Monograph in Aerospace History, No. 12, 1998.

Braslow, Albert L. *A History of Suction-Type Laminar-Flow Control with Emphasis on Flight Research*. Monograph in Aerospace History, No. 13, 1999.

Logsdon, John M., moderator. *Managing the Moon Program: Lessons Learned From Apollo*. Monograph in Aerospace History, No. 14, 1999.

Perminov, V. G. *The Difficult Road to Mars: A Brief History of Mars Exploration in the Soviet Union*. Monograph in Aerospace History, No. 15, 1999.

Tucker, Tom. *Touchdown: The Development of Propulsion Controlled Aircraft at NASA Dryden*. Monograph in Aerospace History, No. 16, 1999.

Maisel, Martin D., Demo J. Giulianetti, and Daniel C. Dugan. *The History of the XV-15 Tilt Rotor Research Aircraft: From Concept to Flight*. NASA SP-2000-4517, 2000.

Jenkins, Dennis R. *Hypersonics Before the Shuttle: A Concise History of the X-15 Research Airplane*. NASA SP-2000-4518, 2000.

Chambers, Joseph R. *Partners in Freedom: Contributions of the Langley Research Center to U.S. Military Aircraft in the 1990s*. NASA SP-2000-4519, 2000.

Waltman, Gene L. *Black Magic and Gremlins: Analog Flight Simulations at NASA's Flight Research Center*. NASA SP-2000-4520, 2000.

Portree, David S. F. *Humans to Mars: Fifty Years of Mission Planning, 1950–2000.* NASA SP-2001-4521, 2001.

Thompson, Milton O., with J. D. Hunley. *Flight Research: Problems Encountered and What They Should Teach Us.* NASA SP-2000-4522, 2000.

Tucker, Tom. *The Eclipse Project.* NASA SP-2000-4523, 2000.

Siddiqi, Asif A. *Deep Space Chronicle: A Chronology of Deep Space and Planetary Probes, 1958–2000.* NASA SP-2002-4524, 2002.

Merlin, Peter W. *Mach 3+: NASA/USAFYF-12 Flight Research, 1969–1979.* NASA SP-2001-4525, 2001.

Anderson, Seth B. *Memoirs of an Aeronautical Engineer—Flight Tests at Ames Research Center: 1940–1970.* NASA SP-2002-4526, 2002.

Renstrom, Arthur G. *Wilbur and Orville Wright: A Bibliography Commemorating the One-Hundredth Anniversary of the First Powered Flight on December 17, 1903.* NASA SP-2002-4527, 2002.

No monograph 28.

Chambers, Joseph R. *Concept to Reality: Contributions of the NASA Langley Research Center to U.S. Civil Aircraft of the 1990s.* SP-2003-4529, 2003.

Peebles, Curtis, editor. *The Spoken Word: Recollections of Dryden History, The Early Years.* SP-2003-4530, 2003.

Jenkins, Dennis R., Tony Landis, and Jay Miller. *American X-Vehicles: An Inventory-X-1 to X-50.* SP-2003-4531, 2003.

Renstrom, Arthur G. *Wilbur and Orville Wright: A Chronology Commemorating the One-Hundredth Anniversary of the First Powered Flight on December 17, 1903.* NASA SP-2003-4532, 2002.

Bowles, Mark D. and Robert S. Arrighi. *NASA's Nuclear Frontier: The Plum Brook Research Reactor.* SP-2004-4533, 2003.

Matranga, Gene J. and C. Wayne Ottinger, Calvin R. Jarvis with D. Christian Gelzer. *Unconventional, Contrary, and Ugly: The Lunar Landing Research Vehicle.* NASA SP-2006-4535.

McCurdy, Howard E. *Low Cost Innovation in Spaceflight: The History of the Near Earth Asteroid Rendezvous (NEAR) Mission.* NASA SP-2005-4536, 2005.

Seamans, Robert C. Jr. *Project Apollo: The Tough Decisions.* NASA SP-2005-4537, 2005.

Lambright, W. Henry. *NASA and the Environment: The Case of Ozone Depletion.* NASA SP-2005-4538, 2005.

Chambers, Joseph R. *Innovation in Flight: Research of the NASA Langley Research Center on Revolutionary Advanced Concepts for Aeronautics*. NASA SP-2005-4539, 2005.

Phillips, W. Hewitt. *Journey Into Space Research: Continuation of a Career at NASA Langley Research Center*. NASA SP-2005-4540, 2005.

Rumerman, Judith A., compiler, with Chris Gamble and Gabriel Okolski. *U.S. Human Spaceflight: A Record of Achievement, 1961-2006. NASA SP-4541, 2007.*

ELECTRONIC MEDIA, NASA SP-4600:

Remembering Apollo 11: The 30th Anniversary Data Archive CD-ROM. NASA SP-4601, 1999.

The Mission Transcript Collection: U.S. Human Spaceflight Missions from Mercury Redstone 3 to Apollo 17. NASA SP-2000-4602, 2001.

Shuttle-Mir: The United States and Russia Share History's Highest Stage. NASA SP-2001-4603, 2002.

U.S. Centennial of Flight Commission Presents Born of Dreams—Inspired by Freedom. NASA SP-2004-4604, 2004.

Of Ashes and Atoms: A Documentary on the NASA Plum Brook Reactor Facility. NASA SP-2005-4605, 2005.

Taming Liquid Hydrogen: The Centaur Upper Stage Rocket Interactive CD-ROM. NASA SP-2004-4606, 2004.

Fueling Space Exploration: The History of NASA's Rocket Engine Test Facility DVD. NASA SP-2005-4607, 2005.

CONFERENCE PROCEEDINGS, NASA SP-4700:

Dick, Steven J., and Keith L. Cowing, editors. *Risk and Exploration: Earth, Sea and the Stars*. NASA SP-2005-4701, 2005.

Dick, Steven J., and Roger D. Launius, editors. *Critical Issues in the History of Spaceflight*. NASA SP-2006-4702, 2006.

Index

A

B

C

E

F

G

H

J

K

M

N

O

P

Q

R

S

T

U

V

W

X

Y

Z